HANDBOOK OF SNOW

PRINCIPLES, PROCESSES,
MANAGEMENT & USE

HANDBOOK OF SNOW

PRINCIPLES, PROCESSES, MANAGEMENT & USE

Edited by

D. M. Gray
D. H. Male

Division of Hydrology,
University of Saskatchewan,
Saskatoon, Canada.

THE BLACKBURN PRESS

Reprint of the 1981 Edition by Pergamon Press Canada Ltd.

Handbook of Snow
Principles, Processes, Management & Use

ISBN-10: 1-932846-06-9
ISBN-13: 978-1-932846-06-5

Library of Congress Control Number: 2004110241

THE BLACKBURN PRESS
P. O. Box 287
Caldwell, New Jersey 07006 U.S.A.
973-228-7077
www.BlackburnPress.com

CONTENTS

PART II

SNOWFALL AND SNOWCOVER   127

PART III

SNOW AND ENGINEERING 519

PART IV

SNOW AND RECREATION   707

LIST OF CONTRIBUTORS

K.M. ADAM Interdisciplinary Engineering Company, Winnipeg, Manitoba, Canada, R3T 4M5.

W.P. ADAMS Graduate Studies, Trent University, Peterborough, Ontario, Canada, K9J 7B8.

M.O. BERRY Applications and Impact Division, Canadian Climate Centre, Atmospheric Environment Service, Downsview, Ontario, Canada, M3H 5T4.

D.W. BOYD (retired), Atmospheric Environment Service, Environment Canada, Ottawa, Ontario, Canada, K1A OR6.

D.R. BROHM Maintenance Operations Office, Maintenance Branch, Ontario Ministry of Transportation and Communications, Downsview, Ontario, Canada, M3M 1J8.

S. COHEN Maintenance Operations Office, Maintenance Branch, Ontario Ministry of Transportation and Communications, Downsview, Ontario, Canada, M3M 1J8.

H.L. FERGUSON Air Quality and Inter-Environmental Research Branch, Atmospheric Environment Service, Downsview, Ontario, Canada, M3H 5T4.

B. GLENNE Department of Civil Engineering, University of Utah, Salt Lake City, Utah, U.S.A., 84112.

B.E. GOODISON Hydrometeorological Division, Canadian Climate Centre, Atmospheric Environment Service, Downsview, Ontario, Canada, M3H 5T4.

D.M. GRAY Division of Hydrology, University of Saskatchewan, Saskatoon, Saskatchewan, Canada, S7N OWO.

L.M.E. HAWKINS Airport Facilities Branch, Transport Canada, Ottawa, Ontario, Canada, K1A ON8.

J. HODE KEYSER Section Génie Urbain et Transport, Faculté d'Amenagement et Ecole Polytechnique, Université de Montréal, Montréal, Quebec, Canada, H3C 3A7.

R.J. KIND Department of Mechanical and Aeronautical Engineering, Carleton University, Ottawa, Ontario, Canada, K1S 5B6.

E.J. LANGHAM Snow and Ice Division, National Hydrology Research Institute, Environment Canada, Ottawa, Ontario, Canada, K1A OE7.

D.H. MALE Division of Hydrology, University of Saskatchewan, Saskatoon, Saskatchewan, Canada, S7N OWO.

J.B. MAXWELL Arctic Meteorology Section, Canadian Climate Centre, Atmospheric Environment Service, Downsview, Ontario, Canada, M3H 5T4.

G.A. McKAY Canadian Climate Centre, Atmospheric Environment Service, Downsview, Ontario, Canada, M3H 5T4.

Contributors

L.D. MINSK   Applied Research Branch, U.S. Army Cold Regions Research and Engineering Laboratory, Hanover, New Hampshire, U.S.A., 03755.

R. PERLA   Snow and Ice Division, National Hydrology Research Institute, Environment Canada, Canmore, Alberta, Canada, TOL OMO.

P. A. SCHAERER   Division of Building Research, National Research Council of Canada, Vancouver, British Columbia, Canada, V6R 1P5.

R. S. SCHEMENAUER   Cloud Physics Research Division, Atmospheric Research Directorate, Atmospheric Environment Service, Downsview, Ontario, Canada, M3H 5T4.

W. R. SCHRIEVER   Division of Building Research, National Research Council of Canada, Ottawa, Ontario, Canada, K1A OR6.

H. STEPPUHN   Division of Hydrology, University of Saskatchewan, Saskatoon, Saskatchewan, Canada, S7N OWO.

D. A. TAYLOR   Division of Building Research, National Research Council of Canada, Ottawa, Ontario, Canada, K1A OR6.

R.W. VERGE   (retired), Meterological Applications Branch, Central Services Directorate, Atmospheric Environment Service, Downsview, Ontario, Canada M3H 5T4.

G.P. WILLIAMS   Division of Building Research, National Research Council of Canada, Ottawa, Ontario, Canada, K1A OR6.

PREFACE

Snow is central to activities in temperate and polar latitudes over a very significant part of each year. With the arrival of snow, modes of travel, working and living are transformed. The snow environment makes obsolete the technologies, tactics and most outdoor activities of the snow-free season and imposes new challenges and opportunities. This book is dedicated to the enhancement of life in a snow environment. It was planned in recognition of the need for an introductory text on snow to serve those who must work with the practical aspects of snow management and to stimulate further interest, careers, research and the development of educational programs concerning this precious, though frequently demeaned, resource.

The initial concept and planning of the book was undertaken by a Working Group on Snow Engineering of the Snow and Ice Subcommittee of the Associate Committee on Geotechnical Research, National Research Council of Canada. The objective of the project was to provide to practicing engineers and educators information concerning snow and snow problems. During preparation its scope was enlarged to make its contents of interest to disciplines other than engineering, e.g., agriculture, geography, life sciences, meteorology and others. Scientists and engineers working on snow problems in Canada and the United States have contributed to its writing.

The material has been organized into four parts - Snow and the Environment, Snowfall and Snowcover, Snow and Engineering and Snow and Recreation. In Part I, on the environment, the material stresses the impact and interaction of snow with living things, climate and agriculture. In Part II, on snowfall and snowcover, a comprehensive review of the phenomenological aspects of snow during its formation, drifting and ablation is given, as well as a compilation of physics and physical properties. This Part, although probably of most interest to academics in the natural sciences, provides the background material on which snow management principles are founded. In addition, Part II presents information on snow measurement and the special cases of Avalanches and Snow and Ice on Lakes. Part III on engineering emphasizes practical applications in which design criteria and procedures relevant to the calculation of loads, the construction of snow roads and methods of snow and ice control are presented. In Part IV, the recreational aspects of snow are discussed, in particular, skiing and the mechanics of snow skis.

The scope of the book does not allow detailed discussions of all the aspects of snow that are presented. Certain chapters, for example; Physics and Properties of Snowcover, Snowcover Ablation and Runoff, Avalanches and others, could easily be enlarged to form separate books. An effort has been made to compensate in part for this deficiency by including a reasonably

complete list of references with each chapter. In review of these references it is worthy to note the wide range in sources of books, journals, technical and research reports, and other papers cited, indicating the interdisciplinary nature of snow research.

ACKNOWLEDGEMENTS: As mentioned above, the impetus for this manuscript originated in 1974 when a small group of scientists and engineers, comprising the Working Group on Snow Engineering, proposed a general reference text on snow covering its formation, properties, problems and control. Their continued encouragement, assistance and support of the manuscript during its preparation has been unfailing.

During preparation the style, philosophy, scope and content of the book has changed considerably. These changes often made the coordination of contributions difficult and it has only been through the enthusiastic participation of the contributors, whose cooperation was voluntary, that this book is possible.

We are particularly indebted to the financial and physical resources made available by the Atmospheric Environment Service (AES) and the University of Saskatchewan (Univ. Sask.). Special acknowledgements are due: Mr. G.A. McKay, Canadian Climate Centre, AES, for his encouragement, dedication and unfailing support to the project; Mr. M. Berry, Canadian Climate Centre, AES, for his assistance in coordinating activities and material during the initial stages of preparation; Mr. E. Truhlar, Training Branch, AES, for the thorough, comprehensive review of the final copy and Mr. G.W. Young and other personnel, Administration Branch Staff, AES, for drafting and production of the figures; Mrs. Elaine Wigham, Division of Hydrology, Univ. Sask., for the meticulous typing and retyping of drafts, her direct assistance with many other aspects of the book and her committment to its successful completion; Mrs. Edna Wilson, Librarian, College of Engineering, Univ. Sask., for the many hours spent in searching and verifying references and to Dr. L.F. Kristjanson, President, and Dr. P.N. Nikiforuk, Dean, College of Engineering, Univ. Sask. for their encouragement, financial assistance, and making available many resources of the University to the project.

Special attempt has been made throughout the text to make specific acknowledgements regarding the source of material used and any failure to do so is unintentional.

D.M. Gray
D.H. Male

PART I

SNOW AND THE ENVIRONMENT

1

SNOW AND LIVING THINGS

Section 1: SNOW AND MAN

G.A. McKAY

*Canadian Climate Centre, Atmospheric Environment Service,
Downsview, Ontario.*

Section 2: SNOW: PLANTS AND ANIMALS

W.P. ADAMS

Graduate Studies, Trent University, Peterborough, Ontario.

SNOW AND MAN

Snow is a pervasive element that may, at times, paralyze communities and stagger economies throughout the world. Appreciated for its beauty and for its usefulness to winter sports enthusiasts, snow more often than not is considered an undesirable and costly nuisance. Surprisingly, the adverse aspects of snow are accepted with relative complacency as a fact of the human environment, and there is little appreciation of either the magnitude of snow's impact on modern life or its immense value as a natural resource.

Snow and cold weather thwarted the military might of Hannibal, Napoleon and Hitler; and deterred exploitation and colonization in both Siberia and North America. Voltaire's description of Canada as "quelques arpents de neige" suggests a contempt for an inhospitable snowy North. Centuries later the situation hasn't really changed. Snow continues to thwart plans and activities. In 1976-77 the estimated reduction of the United States GNP by snow and cold weather was $20 billion (Bardin, 1977).

Technological advances have enabled complex societies to evolve and prosper in snowy regions; at the same time they have increased their vulnerabilities. For example, the problems of a snowbound megalopolis far outshadow those of a snowbound town and the massive transport of material and resources over long distances, as required by our expanding society, is increasingly vulnerable to the whims of winter weather. We need warm, safe accommodation and unconstrained, warm transportation to live comfortably in cold winter climates. The hazards created by snow can be both dramatic and traumatic and understandably have received much attention. Throughout the winter the difficulties created by heavy snows, blizzards, snow slides and drift-blocked roads, make frequent newspaper headlines. These occurrences are increasing in number and seriousness as society becomes more populous, urbanized, interdependent, and reliant on transportation. Much technology exists to counter the adverse effects of these hazards, however, all too frequently the problems created by snow are due to a lack of preparation, the snowfall being unseasonably early or late, or of unexpected intensity. In mid-February 1978 a snowstorm in New England caused about 60 deaths and estimated losses of over $1 billion (the losses sustained by the Massachusetts manufacturing, retail and service sectors alone were estimated at $441 million). More severe storms had occurred in the past but they caused less damage because the population size, the mode and need for transportation and the type of commerce were far different in earlier years.

Despite the problems it creates, snow is an essential resource for many of man's activities, e.g. as a supply of water for homes, livestock, wildlife and for agricultural production (see Ch. 3) and hydro power; as a foundation material for snow roads (see Ch. 12) over which heavy loads may be hauled with little damage to the environment, and for skiing (see Ch. 19), snowmobiling and other forms of winter recreation. A description of these applications and of the physical nature of snow and interactions of snow and man form the substance of this book. Improved knowledge and management of this valuable resource is now recognized as essential in the support of food production, in the design of rural and urban facilities (buildings, in the transportation systems, etc.), in water resource development programs (flood control and protection, water supply) and in the maintenance of environmental standards.

Adapting to Snow

Devising suitable shelters in a snowy environment and improving methods of travel over snow were two major preoccupations of early cultures. Deffontaines (1957) vividly describes the problems encountered by settlers from Europe in adapting to the winters of North America. The lower temperatures, blizzards and heavy snowfalls characteristic of the northern regions of this continent, were not expected. Only 11 of 30 Europeans survived the winter at Tadoussac (where the Saguenay River enters the St. Lawrence) in 1600. One of the factors contributing to the high mortality was the inadequate shelter from snow and cold given by the tepee or the traditional architecture of France. To withstand the environment, buildings evolved having roofs designed to support heavy snow loads (see Ch. 13), and verandahs and elevated floors to offset the frost, deep snow and the effects of spring melting. Snow was piled against the buildings to supplement the insulation provided by wood which began to replace stone as a construction material. Barns were placed and oriented so as to minimize drifts near their doors. Slowly the colonist evolved a safe, comfortable shelter that resisted the snow and cold.

The mode of travel alternated from open river in summer to snow and ice surfaces in winter. Heavy haulage was often delayed until the arrival of snow so that loads could be moved more easily across fields, or on winter trails or snow roads capable of withstanding the traffic (see Ch. 12). The difficulty of travel over snow shaped settlement patterns and society. Snowstorms and the long cold winters induced people to avoid travel over long distances and to remain indoors where they developed home crafts such as weaving and carving.

Snowshoes aided walking in soft snow and various designs were developed to accommodate different land conditions such as open lakes, trails, dense woods and flat or undulating ground. For sledding, dog teams dominated. The comitik of the Inuit performed well on the hard snow of the Arctic tundra, but a variety of sled styles evolved in the more settled areas. The toboggan was best adapted to dry, shallow snow, whereas sleds equipped with runners made of tree trunks and drawn by horses were often used for heavy haulage; the sleigh and bobsleigh were preferred for hard frozen surfaces.

Travel between settlements was hazardous because of blizzards and deep, soft snow. Since breaking and maintaining a trail was a difficult task, attempts were made to obtain a hard level surface that would support the runners of sleighs and horses and also improve the ride. Levelling and clearing of routes often created huge snowbanks on the edges of the trail which obstructed visibility. The state of the snow on the trails was also a concern since wet snow stuck to the runners and to horses' hooves, while icy surfaces were difficult for horses to grip; also trails became soft during warm weather.

The post-colonization period brought the automobile, urbanization, leisure time, new technology and new needs. More sophisticated technologies liberated man from the confining winter snows and provided freer access to remote country-sides and, in particular, to high altitude lands; but, they have not removed his vulnerability. In early times a blizzard could isolate a few settlers for days, but today it can paralyze agglomerations of industry, commerce and communication within large, highly-urbanized areas.

Not many years ago cars were placed in storage during winter. Today modern society demands that all forms of land and air transportation have virtually unrestricted and safe movement throughout the winter whatever the region - the mountains, sea coast, urban centres, or sparsely-settled prairies. A society that formerly was almost immobile during the winter now follows rigid daily schedules in travelling to and from work and in supplying markets. It also spends vast amounts of money on winter recreation facilities and on winter travel. The increasing demand for resources has made necessary access to lands that were formerly isolated by snow, ice and cold. Appropriate snow technology is essential for these activities to be undertaken safely and economically.

Vulnerability

The effects of snow are many and involve complex physical, social economic and psychological factors. The depth, density, wetness and hardness of the snowcover (see Ch. 7) are major physical factors affecting snow clearing and use. As well, the depth of snowcover and the length of time the snow remains on the ground have social, economic and environmental importance. The community-at-large is highly vulnerable whenever strong winds, low temperatures and freezing rain occur with or follow a heavy snowfall. Also, the timing of snowstorms, (e.g., at rush hours or before harvesting is completed) can be critical in determining its effects.

Blizzards are among the most dangerous winter storms. The combination of strong winds, low temperatures and poor visibility creates a major hazard. The blizzard of 1888 left 112 dead and destroyed many herds of cattle in the Northern Plain States of the United States and in Canada. In more recent times the 1941 blizzard in the United States left 39 dead in eastern North Dakota, and between 76 and 90 dead in the United States. Blizzards have immobilized rail traffic forcing the diversion of trains to alternative routes across the Canadian prairies. In 1966, a storm with high winds of 130 to 160 km/h lasted up to four days over many regions of the plains of the northern United States and central Canada. In many towns schools were closed, business suspended and traffic completely halted; some roads were closed for two weeks. In the United States the number of deaths resulting from the storm were relatively few (18); this was attributed to the relatively high temperatures, which were generally above 0° C. Livestock losses in the United States were estimated at $12 million and included 74,500 head of cattle, and 54,000 sheep. Trains were buried and buildings collapsed (Stommel, 1966). Thirty-six centimetres of snow fell in Winnipeg, Manitoba where the street cleaning bill was nearly $1,000,000.

Blizzards are confined to unforested areas without any trees to brake the wind. Certain regions, such as the lee shores of the Great Lakes, are subject to severe "snow bursts" that can cripple transportation, necessitate emergency planning and require allocation of extra funds for highway maintenance. Two hundred and fifty-seven centimetres of snow was estimated to have fallen at Oswego in New York State from January 27-31, 1966, in a single storm (Sykes, 1966).

The aggregate effect of seasonal snowfall on a country can be enormous. This is illustrated by Canadian experience in 1972. In January, a violent storm, with 115-km/h winds, severely disrupted traffic and caused deaths and property damage in Montreal. In British Columbia heavy snow, strong winds and some freezing rain damaged transmission towers and triggered snow slides that isolated highways and cut rail lines. Rogers Pass in British Columbia, a main transportation route through the Rocky Mountains, had a seasonal snowfall of 1481 cm and was closed for a total of 583 h (over 24 days) because of snow and avalanches, compared to an average annual closure time of 194 h. The cost of operating an Army artillery unit to release avalanches that season was approximately $200,000 (Nelson, 1975). Blizzards in Newfoundland disrupted the ferry service in the Cabot Strait. In February, a storm with 110-km/h winds that deposited 10 cm of snow blocked transportation routes in Alberta and Saskatchewan. Later in the year a 13-cm snowfall with winds of 80 km/h produced zero visibilities and severe drifting on sections of Ontario and Quebec highways, making them impassable; most acitivies in Montreal were reduced to a standstill in the absence of snow clearing. In March, sections of the St. Lawrence Valley were without electricity for a week following a snowstorm that lasted 72 h and was accompanied by freezing rain. The same storm deposited 300 to 600 cm of snow in New Brunswick and Prince Edward Island, forcing closure of roads and schools and leading to many accidents. In April hundreds of motorists were stranded and power transmission line breaks occurred as a result of a storm in the Red Deer/Edmonton area of Alberta.

The cities

Urban centres are particularly vulnerable to heavy snowfalls and the trend toward increased urbanization underlines the importance of considering the impact of snow in the urban planning process. Storms which leave a million dollar snow removal bill in their wake are no longer unusual. During 1967 a snowstorm in Chicago deposited 58 cm of snow on the city in a two-day period, and all but paralyzed the metropolitan area with its seven million inhabitants. Forty-five deaths resulted and business losses were estimated at $150 million for the retail and service sectors alone. The city was defenseless against outbreaks of fire, crime or disease; the snow removal bill for this storm was over $5 million, compared with an annual budget for snow removal of $0.5 million. Food ran out and looting and rioting were rampant in some sectors. Hospitals were inaccessible and helicopters were pressed into action to supply insulin, oxygen, medical supplies and aid (Smith, 1967). During January, 1979, a storm of similar magnitude deposited ~53 cm of snow on Chicago in a 30-h period.

Buffalo, N.Y. is accustomed to heavy snowfalls because of its exposure to Lake Erie. Lake storms, which deposit 120 to 180 cm of snow, can occur from October through February. A late January storm in 1977 with winds gusting to 137 km/h created snowdrifts >7.6-m deep. Thousands of motorists were stranded as 5000 cars were abandoned on roadways; buildings were buried, 29 lives were lost and the estimated economic loss from snow removal costs, lost wages and production was $250 million (Dewey, 1977).

A record storm, which paralyzed the area around Boston, Mass. in February, 1978 was estimated to have cost regional commerce $1 billion. As cities and regions grow their vulnerability to snowstorms is steadily increasing.

Snow, particularly drifting snow, has crippled most northern cities in the past decade. Cities in snowy environments are usually prepared for snow, but they are in difficulty when the snowfall characteristics exceed the conditions used in the design of programs and works; e.g. snow control and removal programs (see Chs. 14, 15 and 16) and buildings (see Ch. 13). Twenty-centimetre snow storms crippled parts of Chicago, Winnipeg and Toronto in April 1975. In Toronto the too-early removal of suburban snow fencing, conversion of equipment and termination of contracts led to the formation of drifts on exposed sections of roadways that seriously impeded traffic flow. A 70-cm storm in 1969 buried cars and paralyzed air travel in Montreal and necessitated the use of the National Guard to clear the streets in neighbouring Vermont.

Storms which release their fury during peak traffic hours compounded removal problems, but those that involve freezing rain are most damaging. This latter type of storm caused a $2.2 million loss to hydro facilities, and deprived 250,000 Quebec citizens of electricity, heat, drinking water and fire protection for an extended period in December 1973. Snow removal costs have continued to mount as haulage distances and wages increase; e.g., in Montreal the costs increased from $11.8 million in 1970 to about $30 million in 1979-80 (Fig. 1.1).

Burton (personal communication), found that the loss of income to the Toronto labour force resulting from snowstorms was as great as the dollar value spent in snow removal; in 1967-68 lost work days resulted in a loss of income of $2.9 million, compared with a removal budget of $2.5 million. Farmer (1973) reported that property damage accidents in the months of

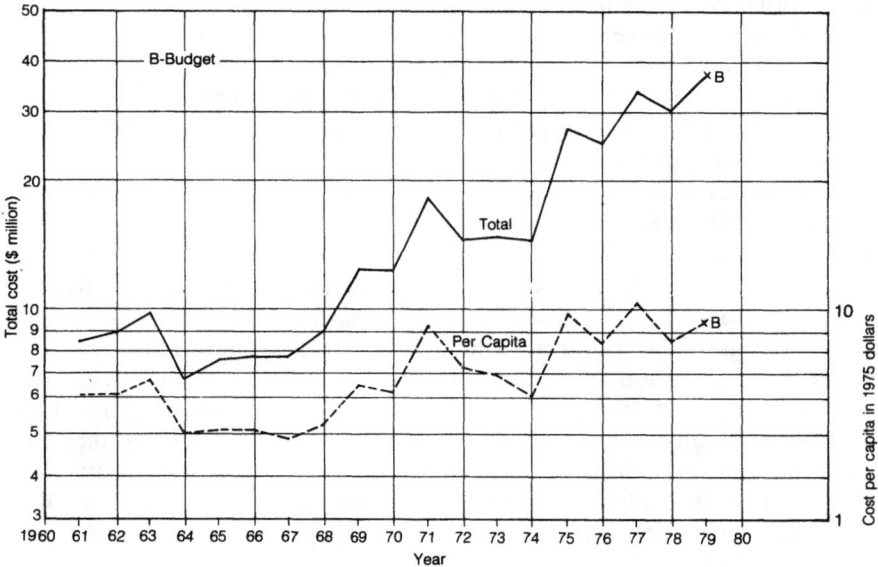

Fig. 1.1 Annual total and per capita costs of snowfall removal, Montreal, Canada.

December, January and February inclusive averaged 58% higher than the monthly average of the remainder of the year, and 48% higher than in midsummer when traffic is at its peak. The costs associated with cancelled plans, heart attacks, domestic clearing, additional energy consumption and other factors are difficult to estimate, but it is evident that the total cost of a storm may be several times larger than the costs for clearance alone.

Rural areas

An unusually cold winter usually means a long winter. Long winters may create a sensation of ennui and confinement. The threat of, as well as the actual occurrence of, snowstorms is a deterrent to travel and shipping and a continual concern to the community-at-large. Excessive snow may mean more soil moisture, runoff and ponding of water in the spring. However, the advantage of enhanced water supplies may be offset by delays and damage. Ditches and waterways may have to be cleared of snow to allow drainage and minimize flooding. Only after the snow has melted and the soil dried can the land be worked. In areas where the growing season is short, delays in seeding may critically limit the cropping options open to the farmer. When the effects of prolonged snowcover are regional in scale, such as occurred over the northern Great Plains in 1974, and in the Siberian wheat lands in 1972, the consequences can have wide-spread implications.

On the other hand the absence of snowcover can be even more serious (see Ch. 3). In both 1972 and 1974 the absence of winter snowcover in southeastern Europe left winter wheat crops highly exposed to severe cold outbreaks. The resulting loss in production due to the cold temperatures contributed to major price increases in cereals and foods throughout the world. The absence of snowcover enhances the over-winter dessication of crops and leaves them prone to wind erosion.

The devastation effected by snowstorms in rural areas, particularly by blizzards, is only partly revealed in the previous discussions of losses caused by storms. Both farm animals and wildlife are frequently killed through suffocation, starvation or exposure. The storms of 1967 which produced 150 to 200 cm of snow in southern Alberta caused a month's delay in spring seeding, increasing the risks to scheduled specialty crops. Hay had to be air-dropped to cattle and wildlife in isolated areas to prevent them from starving (Janz and Treffry, 1968). In the wake of severe storms cattle are left in a weakened state and cows are often barren. The meltwaters of heavy or delayed snows frequently lead to severe flooding and soil erosion for which the total cost of resultant damage can only be speculated about. An April 1976 storm in the Black Hills of South Dakota heavily damaged conifers over 250,000 hectares. The wet snow reached depths of 75 cm. It stuck to crowns breaking many trees in the 10 to 15 cm diameter range (Miller and Halligan, 1979).

Rural electric power outages caused by storms are common and as a rule, difficult to service. Sixty centimetres of wet snow fell in 12 h over southwestern Saskatchewan in May 1974, damaged 113 km of transmission lines as a result of the accumulations on the lines (up to 15 cm) and wind induced vibrations. Snowcover made it virtually impossible for repair crews to reach the downed lines, and the restoration of power took 5 to 6 days at a cost of $205,000 (Cooper, 1974). The cost to the consumers probably exceeded that for repairs.

Construction

Snow loads on buildings and the control of snow so as to maintain optimal access are still major problems in building design and location (see Ch. 13). Roof failures have been a major concern. The roof of a cinema in Washington, D.C. collapsed following a 71-cm snowstorm in 1922, resulting in the death of 100 viewers (Ludlum, 1962). The roofs of many buildings located in the Fraser Valley of British Columbia collapsed during a storm in January 1935 when rain followed a 17.5-cm snowfall. In 1977, roof failures were numerous in New

England following a 60-cm snowstorm. Building codes are being revised continually to obviate this hazard through the use of new data, a better understanding of snow loading and improved construction practices (see Ch. 13). The design of buildings for comfort, economy or maintenance and improved urban architecture to minimize the inconvenience, hazards and costs associated with snow and its removal, remain other prime areas of concern. The use of data about drift patterns, both naturally occurring and as modified by structures, can also be exploited to reduce the inconvenience of snow accumulation.

In many instances snowcover, because it is highly insulating, is a major factor in controlling frost penetration in soils (see Chs. 2, 3). Such control is desirable wherever the uneven movement of soil must be prevented, pipelines must be kept below the frost level, or excavation work is to be undertaken.

Transportation

Most industrial activities can benefit from the use of snow technology in design and planning, as well as in operations and maintenance. This is particularly true of transportation. The day-to-day activities of industry, commerce and schools; the day-to-day affairs of a community; the need to maintain basic supplies, services and the exploitation of resources make necessary the development and maintenance of a viable transportation system that can operate on a year-round basis. Snow technology is used extensively in the planning, design and maintenance of many elements related to transportation, such as:

1) Snow and control measures for the clearing of walkways, railways, highways and runways (see Ch. 18),
2) The construction of snow roads and runways (see Ch. 12),
3) The selection of routes and sites for facilities,
4) The development and evaluation of powered vehicles, sleighs, skis and other means of transportation designed to travel over snow (see Ch. 12),
5) The protection against and the control of avalanches (see Ch. 11),
6) Military strategies and operations on snow-covered terrain, and
7) The design and maintenance of flood control works and drainage systems (see Ch. 9).

Movement and travel during or following a major snowstorm is a serious problem. The handling of emergencies and the clearing of snow to maintain communications and essential services receive top priority in most major storms. But the fallen snow must not unduly inconvenience the community. Hence, the task of clearing may begin before the snowfall occurs through stationing equipment at strategic points and salting the streets (see Chs. 14, 15,

17 and 18). With heavy snow, parking areas must be controlled to enable efficient clearing. The costs of these operations are high and escalate rapidly as haulage distances, restrictions on disposal areas and energy costs increase (McKay and Allsopp, 1980).

Keeping a northern city functional throughout the winter is a major undertaking which requires extensive decision-making, planning and other functions (see also Chs. 14 and 18). Bremner (1977) has described how Toronto reponds. Preparatory work includes cleaning and restoring dumps and equipment, arranging for and stockpiling salt, and arranging for alternative routes, weather forecasts, advisories, and for manpower before the onset of winter. Toronto combats ice and lighter snowfall (<10 cm) by means of salt spreaders equipped with underbody plow blades (see Ch. 17). Given advance warning these are loaded and dispatched to strategic locations. Spreaders must operate at speeds of at least 7 m/s (efficient and effective removal is not possible in peak traffic hours); plows remove much of the snow and allow economical use of salt. In most cities, advantage is taken of forecasts of milder weather that might remove snow naturally.

Snowfalls in excess of 10 cm require plowing and windrowing. The windrows are removed first from emergency and primary routes and trucked to dumps where the snow is ramped by bulldozers. The clean-up after heavy snowfalls can last for days and is costly, e.g., in Toronto during the 1975-76 snow season, five major snowfalls that required plowing consumed 34% of the annual snow budget. Of the annual budget 18% was spent on clearing sidewalks for senior citizens and the handicapped; 43% was spent on the other removal and control practices, i.e. clearing, salting, snowfencing, etc.

In 1976-77 maintenance of Toronto's hundreds of kilometres of roadways consumed over 33,600 t of rock salt. Application rates are adapted to typical winter conditions rather than the temperature during a particular interval, and they are usually just sufficient to loosen the bond between the ice/snow and the pavement.

The mass of snow removed varies from year to year, e.g., for Toronto, the amounts range from 9,329 t in 1974-75 to 473,956 t in 1968-69. The city has options for disposing of the snow: by dumping it in Lake Ontario, on land sites and in sewers. These and numerous other options must be evaluated by technical, economic and environmental criteria before suitable plans are made. The use of land areas as dump sites is promoted for environmental reasons; the main pollutants within the city's snowcover are salt and lead, but analyses of soils at snow dump sites have also disclosed the presence of other heavy metals, e.g., iron and zinc (Bremner, 1977). Model studies of the last 60 years' climate have been made to evaluate interannual differences in removal requirements.

Resource Use

A snowcover influences the energy and water balances at the earth's surface (see Ch. 2) so that its sound management is important for agricultural (see Ch. 3), economic and ecological reasons. A land surface covered by snow acquires a different relief, texture, erodability and, most importantly, albedo. Clean, dry snow reflects 80% of incident sunlight, compared with 15 to 30% by meadows, and 5 to 18% by forests. As a consequence, the change from a vegetative surface to a snowcover has a dramatic effect on the climate, and thereby on animals, plant life, and the energy exchange between the ground and the atmosphere (see Ch. 2).

Snowcover is a major factor in water supply and conservation programs. Water from melting snow provides the spring freshet and most mountain runoff. It replenishes stock ponds, sloughs, infiltrates the soil, and recharges ground water supplies (see Ch. 9). It is manageable on a local scale, and in the more arid regions is used accordingly to augment soil moisture and farm water supplies. Management techniques (see Ch. 3) exploit the power of the wind to transport snow; snow is eroded from exposed areas and redeposited in relatively sheltered locations (Chs. 5, 8 and 16). A good snowcover usually means a good crop.

The desire and need of modern society for increased leisure time has catapulted the winter recreation industry into a position of major social and economic importance (see Chs. 19 and 20). In 1970-71 Canadians made 7.4 million visits to ski areas, and spent an estimated $287 million on ski equipment and transportation to ski areas. Capital investments for ski facilities amounted to $12 million and operating costs, about $60 million (Canada Dept. of Industry, Trade and Commerce, 1972). Another burgeoning activity is the manufacture and sale of oversnow vehicles (see Ch. 12). Snow is the basis for most outdoor winter recreation. Its existence, quality, manufacture and management are vital to the success of many winter sports. An understanding of these factors is basic to planning, site selection, placement of facilities and economic operation. Snow-making operations have evolved since 1950 in response to a need to stabilize the ski industry and are becoming more important in maintaining a snowcover for other uses.

SNOW: PLANTS AND ANIMALS

It is impossible to isolate a single phenomenon or group of phenomena or a single process or group of processes from the complex interrelationships of the global ecosystem. Thus, although it is often argued that snow is the dominant feature of the winter environment of large parts of the earth's surface, it should always be borne in mind that this is only an argument, a point of view. Snow *is* an important part of the environment of many living things but its effects on those living things and the effects of the living things on it cannot in any real way be considered apart from other ramifications of the environmental system of which both the snow and the things are a part.

Further, although extreme situations provide clear illustrations of interactions between snow and plants and animals the most far reaching effects of a snowcover, may well have to do with the simple fact that it is there, that it was there last year and that it will (or should) be there again next year. Plants and animals inhabiting an area *must* interact and respond to it. A year of exceptionally heavy snowfalls may highlight the role of snow as a vegetation-damaging agent; a year of little or no snow in a normally nivean environment may clearly illustrate the role of a snowcover as an insulator of the soil in which plants grow, and as a control of air temperature in the atmospheric layer in which animals live. But extreme events, important as they may be, should be seen for what they are—limiting conditions to a field of nivean situations in which plants, animals and snow mutually interact.

Consideration of interrelationships between living things and snow is made more complex by the fact that both are highly dynamic. Snow is a form of ice which is very susceptible to change at rates which are fast in terms of the life cycle of many living things. Its properties are greatly affected by atmospheric weather conditions, especially by changes of temperature. Thus interactions between living things and snow are best viewed as interactions between variables that constantly change. The burrowing of a vole in a snowpack affects the evolution of snow that allows the burrow to be made and life within it to exist. While life carries on inside the burrow, the snowcover properties that make it a useful medium for making burrows are constantly changing. Activities in the burrow also affects these changes.

However, in those areas where snowcover is a normal part of the environment it is clear that no animal can operate in isolation from something which may, among other things, cover its food, expose or hide it from predators, control the ambient temperature, light regime and gaseous mixture within which it lives. Similar points could be made for plants.

Falling, Blowing and Intercepted Snow

Falling snow does, of course, affect plants and animals in a variety of ways. Examples of this are the damage incurred by trees as a result of loading from heavy, dense snowfalls or animals dying as the insulating properties of their fur are reduced during falls of wet snow.

Tree shape and size may be considerably affected by snow loads (it has been suggested the "Christmas Tree" shape of many boreal conifers is an adaptation to a nivean environment) and by abrasion from wind-blown snow that stunts the growth of a tree. The abrasive action may produce a "wind gap", i.e., a section of tree trunk located just above the prevailing snow depth which is relatively bare of branches. In central Labrador-Ungava, observations of this gap were used as a means of predicting average snow depths before measurements were available (Hustich, 1954). Mechanical abrasion by drifting snow is also an important control of the size, shape and spacing of vegetation in tundra areas where local topographical features that minimize this process are favoured by plants. The dwarfing of some tundra plant species and the mat, rosette and tussock forms of others are examples of adaptations which assist survival in situations where snow abrasion is severe.

Snow intercepted by trees (sometimes subsumed by the Inuit term: qali) can take many forms, including substantial slabs of snow with densities of ~300 kg/m^3. Where low temperatures persist the accumulations may build up on trees in sheltered areas burdening them with heavy loads. Under different weather and exposure conditions the intercepted snow may totally evaporate or sublimate or be blown from the trees thereby affecting the snowcover distribution pattern of a forest and the local environments of plants and animals.

One of the biological roles of intercepted snow has been suggested by Pruitt (1970): it serves as "one of the agents initiating forest succession". Spruce trees are particularly susceptible to accumulation so that their branches or trunks may be broken. Consequently, a small opening in the forest canopy forms, increasing the exposure of the trees adjacent to the fallen spruce to solar radiation. The increased growth of those branches receiving direct sunlight causes the trees to grow away from the vertical, thus encouraging interception and breakage. In this manner a glade grows until it reaches such a size as to be open to the wind and solar radiation, discouraging accumulation of intercepted snow. Deciduous trees then invade the glade, and eventually mature and die. Subsequently, conifers germinate once again and the cycle is repeated.

Snow loads on deciduous trees cause them to bend so that their tips are brought near the ground thereby providing a source of food for animals, e.g., the snowshoe hare. In winters with little snow or frequent thaws (little or no interception), hare populations must seek out other food sources—often unsuccessfully.

Intercepted snow also affects the feeding habits of birds, since deep, dense accumulations restrict foraging. A behavioural adaptation to this restriction is witnessed when birds begin to either eat from the underside of branches or move to an area where the vegetation is bare (e.g., wind swept hills), even though the vegetation species, exposure conditions and other aspects may be less favourable to them (Pruitt, 1960).

Despite its biological and hydrological significance, the evolution of the intercepted snow on trees has not been studied rigorously in terms of its basic properties and processes.

Snowcover

A snowcover may be a boon or a curse for plants and animals. In some situations it provides an insulating blanket which protects life beneath it, in others an insulating blanket which restricts life beneath it. A deep snowcover which protects subnivean life may cut supranivean animals off from their food supply. Snowcover which facilitates travel one day may lethally restrict travel the next, highlighting the fact that snow is a highly dynamic material. The nature and effects of a snowcover vary greatly from place to place and over time.

Variability of snowcover

Spatial

Spatial variability in virtually all the properties of snow (e.g., depth, density, temperature, hardness) is a characteristic feature of a snowcover, one which has profound implications for life in nivean regions. Macro-scale variations result from major storm systems and large geographical features; small-scale and short-time spatial variations in snowcovers are the result of a combination of features including properties of snowfalls, changes arising from the redistribution of snow, atmospheric conditions during and following snowfall and ground surface conditions—notably topography and vegetative cover (see Ch. 5). Within the same climatic region, there is some tendency for a specific landscape type to accumulate snow in characteristic, recurring, patterns. The local vegetation plays a role in this process, partly controlling the spatial patterns, and partly responding to them. One example is the snow that falls onto an irregular, snowfree surface with low, sparse vegetation. This snow is redistributed to fill hollows and to accumulate in lee areas leaving exposed sites bare. Additional snowfall is redistributed in response to a "new", smoother surface and itself causes further smoothing of the landscape until a situation is reached in which snow is transported great distances before accumulating. Thus a characteristic initial pattern and a characteristic modification of that pattern over time, can be envisaged. Under similar snow, terrain and weather conditions different vegetative covers produce distinctive distribution patterns; the effects of various types of forests on snowcover accumulation are well-documented (see Ch. 5).

Frequently bowl-shaped depressions form in the snowcover around the trunks of trees, the results of interception and exposure to wind and radiation.

Snow depth increases outwards from the trunk, where the soil may remain exposed. Outside the radius of the crown the depth increases abruptly exhibiting an upwind-downwind pattern around the crown perimeter. Evolution of the drifts around the tree may be greatly affected by the transport and deposition of intercepted snow. Thus the role of vegetation as a "topographic" control of snowcover is particularly clear. The snow distribution patterns which evolve affect animal life in wooded areas.

Temporal

It is the temporal evolution of a snowcover rather than the falling precipitate which determines the properties of the snowcover which form such a distinctive part of the environment of cold regions. Mechanical changes are often great during deposition; thereafter, metamorphic changes are greatly influenced by the temperature regime of the snowcover (see Ch. 7). In addition to controlling snowpack evolution it is a vital facet of the environment of living things.

It is useful, when trying to convey the interrelationships of snow, plants and animals, to view the snowcover as a type of sedimentary rock composed of consolidated layers of deposited minerals in their solid state (ice) - thus, a snowcover is analogous to a sedimentary rock formed of ice, the "mineral whose principal property is its temperature". The main characteristics of a sedimentary rock are its layering (the size, shape and distribution of the grains which make up its layers) and the resulting strengths, porosities, permeabilities, etc. of the layers and of the rock as a whole. The rock is formed over time through a process of deposition and consolidation which often involves considerable modification of the original constituents. It is ultimately destroyed by the processes of weathering, erosion and transport.

As the rates of snowcover metamorphism are very rapid, the "rock" analogy (if it is an analogy) is not useful for conveying the dynamic nature of a snowcover. For this purpose snow might be considered as being more like soil, which is also characterized by layering of particles of different grain sizes, shapes, etc., but which, more obviously than rock, contains appreciable quantities of air and water and which is highly dynamic. The spatial and temporal variability of soil also provides a useful analogy with snow. Soil conditions may be dry or wet, loose (porous) or packed (dense), hard or soft, highly layered or relatively uniform, thin or thick. Some of these are conducive to plant life, some are not; some make travel by large animals easy, some do not, and so on. Similar points could be made for snowcovers although changes in snow conditions over time tend to be much faster and snow effectively lacks the organic component which is an integral part of soil. Also, temperature, although an important property of soil and an important control of soil forming processes, is a much more dominant factor in snow metamorphism.

Keeping to the soil analogy, the snowcover is a network or skeleton of solid particles (ice) enclosing voids or pores of varying size. The size, shape and distribution of the particles affect the size and shape of the pores. The pores may be filled with moist air or liquid water. Texture is an important property of soil which is determined by the shape and size distribution of particles: a soil having a large proportion of small particles may be classified as a clay. This classification indexes how much air the soil will contain, its permeability, its thermal properties, its compactness, its strengths, etc. Snow can be viewed in a very similar fashion. The horizons which make up a soil profile are in some ways analogous to strata encountered in a snowcover; that is, a snowcover is not simply ice, in the same way that soil is not simply composed of solids. The snowcover is a portion of the atmosphere which for a period of time contains a particular combination of the three states of water, and air. Thus, from some points of view, snow, like soil, is a medium for life, a sheltered extension of the atmosphere. From other points of view, also like soil, it is a substrate for life. It is important to keep in mind a fairly clear conception of the complex, dynamic (temporally-varying) nature of snow stratigraphy when considering the implications of snowcover on plant and animal life.

Effects on plants and animals.

The spatial and temporal interaction between vegetation and snowcover are illustrated by "snowbanks" in tundra areas. Vegetation not only tends to occupy the least exposed sites, e.g., depressions, but also assists in the early accumulation of snow in them, thus improving its own niche in the landscape. During mid-winter the drift provides an insulated environment for the plants, restricting energy losses from both vegetation and soil and protecting the plants from dessication because the pores within a snowcover contain saturated air. However, persistent snowcover may slow the warming of the subnivean zone in the spring. On melting, a snowbank provides moisture, with its constituent nutrients, to the soil; such additions are important since water is often a limiting factor in plant growth in the tundra. However, a rapid snowmelt and rapid runoff rate can erode plant communities.

The combination of "good" and "bad" effects of snow drifts vary from site to site and from year to year but the depth pattern, i.e., shallow at the outer edge to deep at the centre, brings out some of the possible ramifications. Vegetation at the outer edge is least protected in winter but is exposed first in the spring. It then has a relatively long growing season with a reasonably secure water supply from snowmelt runoff from the inner core. Toward the centre of the drift, improved winter protection is more than offset by a shorter growing season and a decreased water supply during the growing season. At sites which are frequently occupied by snowbanks, a distinctive pattern of vegetation develops, (Billings and Bliss, 1959) accentuated over the years at recurring drift sites, by development of less-fertile soil towards the centre. In general, the number of species, the amount of plant cover and plant productivity decrease toward the centre of a snowdrift. The lower productivity toward the centre may be attributed to the lower amounts of radiation received there as well as to a shorter growing season and a smaller supply of soil moisture and nutrients. At some recurring drift sites, plants near the centre may become selected or adapted so as to survive more than one year of continuous snowcover.

Many plants begin to grow while still covered with as much as a half metre of snow. Some metamorphic processes tend to increase grain size (see Ch. 7) so that greater penetration of light can occur allowing photosynthesis at greater depth.

Thus, a snowcover profoundly affects the living space of plants by altering the temperature regime of the underlying soil and the ambient temperature of the entire environment of smaller plants; by providing moisture and nutrients; and by lessening the effects of fluctuations in the surface climatic conditions. Such effects vary with the changing physical properties of the snowcover, e.g., the effects of a snowcover on energy transfer mechanisms change with its evolution. The snowdrift is a specific example of the effect of spatial *and* temporal variations that are normal features of snowcover.

Depth is frequently used as an indicator of the properties of a snowcover, but it should be borne in mind that conditions within and under the snow are greatly affected by thermal, optical and other properties of the snowcover (see Ch. 7). In the example of a snowdrift it was implicitly assumed there was a systematic relationship between depth and the various properties of the cover affecting plants and that deep snow provided a better insulation. In general, mid-winter temperature gradients are steeper and temperature-gradient metamorphism more pronounced in a shallow cover. However, snowcovers of given depth, but in different metamorphic states may have different temperature regimes even though subjected to the same climatic regime.

Even freshly fallen snows may vary widely in their properties so that equal depths of different snows may have quite different environmental implications. For example, twenty centimetres of newly-fallen snow of single-grain ice needles forms a dense, compact layer that can easily support an animal but is a very poor insulator; however, twenty centimetres of newly-fallen snow of fluffy flakes of spatial dendrites does not support a walker but is an effective insulator, because of its large air porosity. With time, the depth of a snowcover decreases since the overall tendency is for metamorphism to reduce the complexity of the initial precipitate, thereby compacting it. In this general scheme, shallower depth *does* mean higher density, higher thermal conductivity, greater "strength", etc. This fact re-emphasizes the importance of realizing that a snowcover is a dynamic system—not simply an inert layer on the ground. The effects of snowcover on vegetation and soil can only be properly determined by taking into account the nature and state of the snowcover during the period between its formation and disappearance. For example, the *duration* of snowcover was an important feature of the snowbank.

Another illustrative example of the interaction of snow and plants, is that of plants located on routes used by oversnow vehicles (see also Ch. 12). The environmental effects of snowmobile trails and of winter roads have attracted a good deal of attention in Canada in recent years. In winter road construction snow is artificially compacted to produce a hard, dense layer that will support oversnow vehicles. Compaction increases snow density and thermal conductivity and reduces air permeability - the air spaces in a snowcover begin to cease to be interconnecting when the density exceeds ~600 kg/m³. The snow density of a snowmobile trail rarely exceeds 500 kg/m³, its maximum value being reached after the first few passes of the vehicle; winter snow roads may be constructed to densities of ~600 kg/m³ by compacting processed snow. Masyk (1973) found that changes in densification and related processes, including sintering, caused by snowmobile traffic increased the thermal conductivity of the snow by almost twelve times. Also, because of the increase in thermal conductivity the soil temperatures are lowered, freezing the roots of perennial plants and causing their destruction by desiccation or front heave (see Ch. 3). In the study cited, the temperature of a soil beneath a snowmobile trail fell to the freezing point more than six weeks earlier than that under undisturbed snowcover. Decreases in the air porosity, the air permeability and the exchange of gases with the underlying soil have a detrimental effect on bacterial activity which is important in nutrient cycles and humus formation of the plant food cycle. Masyk reported that soil bacteria are reduced one hundredfold on compaction by snowmobile traffic.

The soil under a trail takes longer to warm in the spring because of its larger heat deficit and the higher albedo of the icy trail. Also, since snow roads and snowmobile trails are often located in areas unexposed to direct solar radiation so as to minimize winter thaws, they melt later than the surrounding snowcover so that the growing season of the underlying vegetation is shortened. The effects of oversnow traffic on plant growth vary widely. Adam and Hernandez (1977) and Adam (1978) found that after ice and snow roads disappeared the percentage of live-plant cover significantly decreased, tamarack and spruce seedlings became non-existent and numerous other saplings and shrubs were either damaged or unhealthy (retarded growth).

The nature of snow is such that a single snowmobile pass is sufficient to alter the physical environment of the soil and snow layers. The immediate effects are most obvious in the snowcover itself which is the winter habitat for many small animals.

The subnivean environment

Within an undisturbed, non-melting snowcover, the subnivean environment maintains relatively warm and stable temperatures and the air is relatively moist. However, once a snowcover is compacted subnivean animals may be subjected to certain stress conditions brought about by low temperatures, rapid changes in temperature gradients and restricted air movement. Densities under both light- and heavy-use snowmobile trails may range from 390 to 490 kg/m^3, high enough to prohibit animal movement within the snowcover. Schmid (1971) found that deer mice, shrews and meadow voles frequently must cross the surface of the snowmobile track "to get to the other side". If they leave their subnivean environment, these animals are easy victims for predators. On the other hand, the increased surface hardness of snow trails and roads allows fox, deer and other large mammals to move easily in search of food.

The travel of an oversnow vehicle provides a particularly striking example of effects of a decrease in snow depth, in this case induced by external stress. Snow is also compacted in nature (with time snowcovers tend to become denser). Catastrophic rates of change in snow depth such as those associated with a snowmobile pass do occur in nature, e.g., as a result of avalanches, snow falling from trees or snow blown from an area by a storm.

Many of the general observations made on the effects of snowcover on plants apply also to animals which live in or under the snow. They live in the same environment as the plants, experience the same temperature and light regimes, breathe the same saturated inter-granular air and are affected by snow compaction. Like plants, animals are affected by the duration of the snowcover, by its stratigraphic properties (depth, density, grain size and shape, stratification, temperature), by its physical properties (thermal conductivity, hardness, bearing strength, optical transmission, air and water permeability during the winter), and by the manner it finally melts and runs off.

Both the snow and subnivean mammals must be viewed as being dynamic in a more obvious way than plants. Ideally, it would be possible to consider the relationship between snow and small mammals at the scales of the individual animal, his community and his species. This would involve considering time scales ranging from hours (e.g., the diurnal cycle) to hundreds of thousands or millions of years necessary for physical evolutionary adaptive changes. Ideally, also, it would be possible to consider interrelationships within the entire ecosystem of which the mammal is a part—rather than simply dealing with animals or snow separately. The approach used below for discussion is similar to that used previously, employing selected illustrative examples.

The role of snow in the life of small mammals provides a good example of the difficulty of generalizing concisely about its environmental roles.

Snowcover is often described as a "blanket", keeping life under it warm. However, its effectiveness in this and other purposes can only be fully assessed by measuring its vertical (stratigraphic), horizontal and temporal variations. Generally a snowcover about 15-cm deep is sufficient for small animals to construct burrows. Marked temperature differences between the relatively warm ground surface and the air and the associated vapor pressure gradients in snowcovers promote the development of depth hoar (see Chs. 7 and 19) at the snow base, facilitating tunnelling and air circulation. For a given temperature difference the gradient will be steepest at the shallowest locations which will therefore be least conducive to life (generally because of the lower temperatures). Long grass which tends to accumulate snow during the early winter, also promotes depth hoar by producing a loose, aerated layer at the snowcover base. Once a snowcover is established, many scenarios of stratigraphic evolution can .be envisaged. If ice layers or crusts or both develop, small mammals can use them as runways within the cover and on the surface, reaching them via vertical shafts near the stems of tall plants. Air movement through these shafts ventilates the burrows thereby preventing a build-up of toxic levels of CO_2.

Within the snowcover, although the air temperatures may be far below the freezing point, small animals live in a relatively-warm (near freezing), sheltered, environment breathing pure, albeit saturated, air. They live on food stored from the previous fall or they forage within the pack. In this sort of environment mice, voles, lemming and shrew live a very active life during the winter and may reproduce freely, a very different pattern from the hibernation pattern of other animals. It has been pointed out that the fur of some of these animals, including the weasel, is a less effective insulator than that of some tropical animals (Folk, 1974). To some extent, the snowcover protects such animals from their enemies but a number of their predators, such as weasels, live equally active subnivean lives, possibly having to do less work to survive in winter than in summer!

As the depth, density and associated properties of a snowcover vary considerably from place to place, even over very short distances, the effectiveness of the cover as a medium for stable, well-ventilated, warm tunnels and nests varies. A forest snowcover provides a good example of spatial variability at scales which greatly influence the lifestyles of small mammals. At the macro-scale, a forest snowcover (in comparison to, say, a nearby meadow) is perceived as being relatively deep, soft and loose. At a micro scale, however, there is great variation in its physical properties, e.g., very shallow depths occur in the bowl-shaped areas around the trunks of trees, and deep drifts in open areas. Since small mammals depend on the insulation provided by the snowcover for survival, they avoid the cold, shallow or bare zones and congregate in the drifts. In contrast, larger animals that live, at least in part, above the snow, find refuge in the snowcaves under the trees.

The temporal evolution of a snowcover also greatly influences the animals life within it. The periods during early winter when the permanent snowcover is formed and during late winter when the snow melts, are especially important, but any situations with abrupt changes in the cover, are critically important to animals living in it. A long snowfree period between the onset of cold temperatures and the formation of a permanent snowcover can seriously deplete small mammal populations and also, result in lower subnivean ground temperatures for the entire winter. Delayed snowmelt in the spring slows the warming of the environment in which small mammals live, just when the insulating effectiveness of the snowcover and the animals' fur is greatly reduced. Delayed melt also limits early plant growth that may be a critical supply of food at winter's end. Runoff, especially where the ground is frozen, may destroy the habitat and change the environment so that creatures that live within and on the snowcover and the underlying ground cannot survive; e.g., although field mice can swim, they cannot survive damp and cold conditions for long.

An abrupt deflation of snowcover in mid-winter by changing weather conditions (e.g., a chinook), may expose well-protected plants, animals and soil to the full impact of low air temperatures.

The supranivean environment

For those animals that do not migrate out of the nivean regions the snowcover presents problems for travel, forage and protection from predators. The classic work on these and other aspects of snow biology is that of Formozov (1946). He classified animals of nivean regions into Chionophiles (snow lovers, well adapted to snow), Chioneuphores (animals semi-adapted to snow), and Chionophobes (snow haters, poorly adapted to snow). Application of this scheme to above-snow animals provides an interesting framework for discussing their interactions with snow.

In order to travel in a snowcovered region, an animal either must be able to wade through or travel on the snowcover; therefore, the depth and bearing strength of the snow are critical factors. Kelsall (1969) found that the ranges of moose and white-tailed deer in New Brunswick correlated well with their chest heights, the taller moose occupying the deeper areas. Travel in snow deeper than chest height greatly increases an animal's energy consumption. Within nivean areas, the migration of wolverine, leopard, elk, wolf, muskox and chamois, from deep to shallow snow areas is well-documented (Formozov, 1946). The extent to which movement is affected by depth is modified by the bearing strength of the snowcover; an animal sinks less in denser, harder snow and its weight may be supported by a snowcover having an ice layer or crust that is strong enough. Formozov's above-snow Chionophiles tend to be animals with large feet relative to body mass so that the pressure they exert on the snow is much less than a Chionophobe of equal mass. The ratio (contact foot area:mass) is also relatively high in many Chioneuphores.

A case much cited in the literature and mentioned in connection with a behavioural adaptation to snow is the snowshoe hare. This animal has large feet (for its body size) which are surrounded in winter by stiff bristles to assist in supporting its weight. It travels by jumping, with all four feet at a time, a mode that makes more efficient use of its large foot size:body mass ratio than other gaits. Formozov (1946) reports that a 3.4-kg hare exerts a pressure of only 1.15 kPa which can be supported by a snowcover with a density of 180 kg/m^3. Even a small surface crust is sufficient to support a pressure of this magnitude. A bird that can travel on most snowcovers is the ptarmigan (also a Chionophile) whose snowshoe-like feet, exert a pressure of 1.4 kPa, much less than the 4.02 kPa exerted by the grey partridge (a Chionophobe).

Animals with low ratios of foot area:body mass are poorly adapted to travel in snow. Some small cats, deer, and antelope, and several species of birds fall into the category of Chionophobes. If the snowcover is fairly deep, these animals can only travel on it if there is a crust or ice layer to support their weight or if the snow is dense enough to prevent them from sinking too deep. The moose (a Chioneuphore) and the caribou (a Chionophile) are partially adapted for travel in nivean regions, since they have relatively broad feet which they can spread somewhat, producing pressures of 60 and 35 kPa, respectively (Kelsall and Prescott, 1971), even up to ~100 kPa—which is less than the bearing strengths of the "hardest" crusts. Long legs on an animal can offset the disadvantage of high bearing pressures (small hooves); for example a moose can sink farther before being immobilized because it can force its body through the snow.

Clearly density and "hardness", the latter referring to the presence of crusts or ice layers, greatly modify snow depth as a control of travel in nivean regions. Deer or moose that sink 35 cm into a 40-cm cover with densities of 100-190 kg/m^3 only sink 21 cm into a 44-cm cover with a density of 490 kg/m^3 (Kelsall and Prescott, 1971). A hard crust can support even large animals having a low foot area:body mass ratio. However, if animals break through crusts or ice layers they may have difficulty removing their feet from track holes or may damage their legs. The restricted movement exposes them to predation because these same snow conditions are very favourable for travel by predators, both Chionophiles and Chioneuphores, such as the wolf.

An important purpose of animal travel is the search for food. The presence of a snowcover affects this in ways other than hindering or aiding movement. Animals supported by a snowcover can browse higher on trees (as in nivean regions where the load of intercepted snow also bends tree branches). However, a deep snow (which restricts travel) or a dense, hard snow (which assists travel) makes foraging of food beneath the cover difficult for above-snow animals. These animals adapt by migrating, by switching their food choice from a buried item to an exposed item (e.g., caribou switch from ground moss and lichens to tree branches and lichens) or by digging the snow. The musk-ox is unable to feed in deep snow and therefore is limited in winter to foraging on windswept areas where only inferior, "second choice" food is available. Members of the cat family, are particularly ill-equipped for foraging beneath snow since their retractable claws are unsuitable for digging out prey in deep or hard snow.

Pruitt (1959), emphasized hardness of the densest snowcover layer as being the key control of caribou movement in winter; they are markedly affected by structural variations in snow, particularly in the basal layer. He noted that deer avoid areas with deep, layered snowcovers in favour of those with a shallow relatively uniform snowcover. Musk-oxen also suffer seriously under conditions where a hard crust or ice layer covers their food supply.

Because snowcover density tends to increase with time, older snow tends to support a traveller better than new snow. However, abrupt changes in travel conditions can be caused by new snowfalls, changes in snowcover temperatures and the changes in bearing strength accompanying metamorphism. Thus, animals of a particular region must adapt to the temporal variations of a snowcover as well as to its spatial variations. One example of the temporal aspect, analogous to the effect of prolonged snowcover in spring on plants, is the effect of spring snow on bird populations. Migrating birds, arriving in nivean areas for the nesting season are greatly affected by persistent snowcover; e.g., greater snow geese, on Bylot Island, NWT, may not mate if snow lasts two weeks or more after they arrive (Heyland, 1975; Kerbes, 1975).

The density of snow and its strength increase when it is moved or subjected to stress (see Chs. 7 and 12). Animals in nivean regions that travel in single file or follow the tracks of preceding animals take advantage of the improved travel conditions on compacted snow.

Animals that survive in nivean regions are those that consume less energy in seeking and obtaining food, keeping warm and other purposes than the amount gained from eating the food and maintain this positive balance over long periods. Combinations of snowcover conditions that balance out disadvantages and advantages of travel and food-gathering, over space and time, determine over-winter survival rates and spring breeding potential. A number of attempts at developing survival indices are reported in the literature, such as that of Verme and Ozoga (1971) which combines measures of "snow hazard" with an air chill index.

In Formozov's classification of animals (Formozov, 1946), species in all three classes are normal inhabitants of nivean areas. Also the literature is replete with examples of heavy mortality among Chionophiles. The major adaptations of animals seem to be behavioural rather than physiological or morphological so that if an animal is unable for some reason to respond to changes in it nivean environment, it becomes vulnerable. This reflects McKay's (1970) point that the flora and fauna of the north can be viewed as being only marginally adapted to their habitat.

Snow as a substrate for life

Much of the biological activity described as "subnivean" might better be described as "intranivean" since it occurs in the snow rather than beneath it. Although burrows in the soil are by no means unknown in nivean regions, the widespread frozen ground, problems of spring flooding and the insulating benefits of habitations *within* the snowcover, make life above the ground, but below the snow surface, more attractive. Many of the nests and runways that are revealed after a snowcover has melted were not constructed on the ground surface but rather some distance above it.

The "subnivean" mammals discussed above use snowcover as a medium for dwelling in a way similar to, although perhaps more elaborate than, the way that a resting caribou or ptarmigan uses it for temporary shelter or a polar bears uses it for estivation and denning or certain squirrels use it for hibernation. However, the cover is more truly *the* substrate for life for other animals and plants including algae, fungii, bacteria (see also Ch. 7) and insects. Many of these forms can live and reproduce entirely within the snowcover; some can only exist in the snow environment. Some are even present on glaciers which, in one sense at least, provide the most nivean environments on earth. Although it is not usual to think of frozen water as a substrate for life, freshwater ice, and notably sea ice, do provide important ecological niches in their particular environments (Mohr and Tibbs, 1963; Dunbar, 1977). Downes (1962, 1965) provides some interesting examples of insects living in snow, ranging from situations in which snow is used in a way essentially analogous to that of mammals to those in which the snowcover is quite literally the substrate for life.

Acknowledgements

Since this section on living things has the format of a literature review, I gratefully acknowledge the contributions of published research studies on all aspects of snow and ice biology. In Canada, I am particularly conscious of the contributions of such people as A.W.F. Banfield, L.C. Bliss, J.P. Kelsall, W.O. Pruitt and D.B.O. Savile. The pioneering work of W.O. Pruitt deserves special mention. In the production of this article, I am grateful to Mrs. M. Waiser, formerly of the University of Saskatchewan, and to some of my own students especially Sandy Johnson and Rosemary King. I am particularly grateful to Kathy Outerbridge for her patient assistance.

REFERENCES

Adam, K.M. and H. Hernandez. 1977. *Snow and ice roads: ability to support traffic and effects on vegetation.* Arctic, Vol. 30, pp. 13-27.

Adam, K.M. (ed.). 1978. *Building and operating winter roads in Canada and Alaska, north of 60.* Environ. Studies No. 4, Dept. Indian North. Affairs, Ottawa, Ont.

Adams, W.P. 1976a. *Diversity of lake cover and its implications.* Musk-Ox, Vol. 18, pp. 86-98.

Adams, W.P. 1976b. *Areal differentiation of snowcover in east central Ontario.* Water Resour. Res., Vol. 12, pp. 1226-1234.

Babin, G. 1975. *Blizzard of 1975 in western Canada.* Weatherwise, Vol. 28, pp. 70-75.

Banfield, A.W.R. 1952. *The barren-ground caribou.* Minist. Nat. Resour. and Can. Wildlife Serv., Ottawa, Ont.

Banfield, A.W.F. 1974. *The Mammals of Canada.* Univ. of Toronto Press, Toronto, Ont.

Bardin, D.J. 1977. Statement of "Hearings before the Subcommittee on Science, Technology and Space of the Committee on Commerce, Sci. and Transp., United States Senate Ninety-fifth Congress", Serial No. 95-33, Washington., D.C.

Billings, W.E. and L.C. Bliss. 1959. *An alpine snowbank environment and its effect on vegetation, plant development and productivity.* Ecology, Vol. 40, pp. 388-397.

Bird, J.B. 1974. *Geomorphic processes in the Arctic.* In Arctic and Alpine Environments (J.D. Ives and R.G. Barry, eds.), Methuen, London, pp. 703-720.

Bliss, L.C. 1962. *Adaptations of arctic and alpine plants to environmental conditions.* Arctic, Vol. 15, pp. 117-144.

Bliss, L.C. et al. 1973. *Arctic tundra ecosystems.* Annu. Rev. Ecol. Systematics, Vol. 4, pp. 359-399.

Bliss, L.C. 1975a. *Devon Island, Canada.* Structure and Function of Tundra Ecosystems, (T. Rosswell and O.W. Heal, eds.), Ecol. Bull. (Stockholm), Vol. 20, pp. 17-60.

Bliss, L.C. (ed.). 1975b. *Plant and surface response to environmental conditions in the western high arctic.* ALUR 74-75-73, Arct. Land Use Res. Prog., Dept. Indian Affairs North. Dev., Ottawa, Ont.

Bliss, L.C. 1975c. *Tundra grasslands, herblands, shrublands and the role of herbivores.* Geosci. and Man, Vol. X, pp. 51-79.

Bliss, L.C. (ed.). 1977. *Truelove Lowland, Devon Island, Canada: A High Arctic Ecosystem.* Univ. of Alberta Press, Edmonton, Alta.

Bremner, R.M. 1977. *Report of City of Toronto winter services.* Dept. Public Works, Toronto, Ont.

Connor, W.C., K.C. Crawford and E.L. Hill. 1973. *Snow forecasting in the South: occasional blizzards over the southern plains to rare Gulf Coast snows.* Weatherwise, Vol. 26, pp. 244-249.

Cooke, W.B. 1955. *Subalpine fungi and snowbanks.* Ecology, Vol. 36, pp. 24-130.

Cooper, R.E. 1974. *Report on wet snowstorms of May 1974 in southwest Saskatchewan.* Unpubl. Rep., Sask. Power Corp., Regina, Sask.

Corbet, P.S. 1972. *The microclimate of arctic plants and animals on land and in freshwater.* Acta Arctica, Vol. 18.

Danks, H.V. 1971. *Overwintering of some north temperate and arctic Chironamidae I. Can. Entomol., Vol. 96, pp. 279-307.*

Deffontaines, P. 1957. *L'homme et l'hiver au Canada.* Geogr. Humaine 27, Presses Univ. Laval, Ste-Foy, P.Q.

Department of Industry, Trade and Commerce. 1972. *The Canada Tourism Facts Book, 1972.* Dept. Ind. Trade Commer., Ottawa, Ont.

Dewey, K.F. 1977. *Lake-effect snowstorms and record breaking 1976-77 snowfall to the lee of Lakes Erie and Ontario.* Weatherwise, Vol. 30, pp. 228-231.

Downes, J.A. 1962. *What is an arctic insect?* Can. Entomol., Vol. 94, pp. 143-162.

Downes, J.A. 1965. *Adaptations of insects in the arctic.* Annu. Rev. Entomol., Vol. 10, pp. 257-274.

Dunbar, M.J. (ed.) 1977. *Polar oceans.* Proc. Polar Oceans Conf., McGill 1974. Arct. Inst. North Am., Calgary, Alta.

Everndun, L.N. and W.A. Fuller. 1972. *Light alteration caused by snow and its importance to subnivean rodents.* Can. J. Zool., Vol. 50, pp. 1023-1032.

Folk, G.E. Jr. 1944. *Textbook of Experimental Physiology.* Len and Febiger, Philadelphia, Penn.

Farmer, P.J. 1973. *The winter driver.* Proc. First Nat. Conf. Snow and Ice Control, Roads Transp. Assoc. Can., Ottawa, Ont., pp. 8-15.

Formozov, A.N. 1946. *Snow cover as an integral factor of the environment and its importance in the ecology of mammals and birds.* Moscow Soc. Naturalists, Materials for Fauna and Flora U.S.S.R. Zool. Section, New Ser., Vol. 5, pp. 1-152. [English Transl. by W. Prychodko and W.O. Pruitt, Jr., Publ. as Occas. Pap. No. 1, 1963, Boreal Inst., Univ. Alta., Edmonton].

French, H.M. 1976. *The Periglacial Environment.* Longman, Canada, Don Mills, Ont.

Fritsch, F.E. 1977. *The Structure and Reproduction of Algae, Vol. 1.* Cambridge Press, Cambridge, Mass.

Fuller, W.A., A.M. Martell, R.F.C. Smith and S.W. Speller. 1975. *High arctic lemmings (Dicrostonyx groenlandicus), I: Natural history observations.* The Can. Field-Naturalist, Vol. 89, pp. 223-233.

Fuller, W.A., A.M. Martell, R.F.C. Smith and S.W. Speller. 1975. *High arctic lemmings (Dicrostonyx groenlandicus), II: Demography.* Can. J. Zool., Vol. 53, pp. 867-878.

Geiger, R. 1961. *Das Klima der Bodennahen Luftschicht (The Climate Near the Ground).* [English Transl. by Scripta Technica Inc., Harvard Univ. Press, Cambridge, Mass., 1966].

Gill, D., J. Root and L.D. Cordes. 1973. *Destruction of Boreal forest stands by snow loading; its implication to plant succession and the creating of wildlife habitat.* In Kootenay Collection of Res. Studies in Geography, British Columbia Geogr. Ser. No. 18, Occas., Pap. in Geogr. 1, pp. 55-70.

Granberg, H.B. 1979. *Snow accumulation and roughness changes through winter at a forest-tundra site near Schefferville, Quebec.* Proc. Modeling Snow Cover Runoff (Colbeck, S.C. and M. Ray, eds.), U.S. Army Cold Reg. Res. Eng. Lab., Hanover, N.H., pp. 83-92.

Gray, D.R. 1972. *Winter research on the musk-ox (Ovibos moschatus wardi) on Bathurst Island, 1970-71.* Arct. Circ. No. 21, pp. 158-163.

Haugen, A.O. (ed.). 1971. *Proc. of the Symp. on Snow and Ice in Relation to Wildlife and Recreation.* Iowa State Univ. Press, Ames.

Hawkins, L.M.E. 1973. *Snow removal and ice control at Canadian international airports.* Proc. First Nat. Conf. on Snow and Ice Control, Roads Transp. Assoc. Can., Ottawa, Ont., pp. 166-173.

Henshaw, J. 1968. *The activities of the winter caribou in northwestern Alaska in relation to weather and snow conditions.* Int. J. Biometeorol., Vol. 12, pp. 21-27.

Heyland, J.D. 1975. *Monitoring nesting success of greater snow geese by means of satellite imagery.* Third Can. Symp. on Remote Sensing (G.E. Thompson, ed.), Can. Aeronaut. Space Inst., Ottawa, Ont.

Hoham, R.W. 1975. *The life and history and ecology of the snow alga Chloromonas pichinchae (Chlorophyta, Volvocales).* Phycologia, Vol. 14, pp. 213-226.

Hoham, R.W. and J.E. Mullet. 1977. *The life history and ecology of the snow alga Chloromonas cryophilia sp. nov. (Chlorophyta, Volvocales).* Phycologia, Vol. 16, pp. 53-68.

Hustich, I. 1954. *On forests and tree-growth in the Knob Lake area, Quebec-Labrador penninsula.* Acta. Geographica, Vol. XIII, pp. 1-60.

Ives, J.D. and R.G. Barry (eds.). 1974. *Arctic and Alpine Environments.* Methuen, London.

Janz, B. and E.L. Treffry, 1968. *Southern Alberta's paralyzing snowstorms in April 1967.* Weatherwise, Vol. 21, pp. 70-76.

Kalff, J. and H.E. Welch. 1974. *Phytoplankton production in Char Lake, a natural polar lake, and in Meretta Lake, a polluted polar lake, Cornwallis Island, Northwest Territories.* J. Fish. Res. Board Can., Vol. 31, pp. 621-636.

Kelsall, J.P. 1968. *The migratory barren-ground caribou of Canada.* Can. Wild. Serv., Queens Printer, Ottawa, Ont.

Kelsall, J.P. 1969. *Structural adaptations of moose and deer for snow.* J. Mammal., Vol. 5, pp. 302-310.

Kelsall, J.P. and W. Prescott, 1971. *Moose and deer behaviour in snow in Fundy National Park, New Brunswick.* Rep. Ser. No. 15, Can. Wild. Serv., Ottawa, Ont.

Kerbes, R.H. 1975. *The nesting population of lesser snow geese in the eastern Canadian arctic: a photographic inventory of June 1973.* Rep. Ser. No. 35, Can. Wild. Serv., Ottawa, Ont.

Ludlum, D.M. 1962. *Extremes of snowfall in the United States.* Weatherwise, Vol. 15, pp. 246-262, 278.

MacLean, S.F., Jr. 1975. *Ecological adaptations of tundra invertebrates.* In Physiological Adaptation to the Environment (F.J. Vernberg, ed.), Proc. Symp. held at the 1973 Meet. Am. Inst. Biol. Sci., Intext Educational Publishers, New York, N.Y., pp. 269-300.

MacLean, S.F., B.M. Fitzgerald and F.A. Pitkelka. 1974 *Population cycles in arctic lemmings: winter reproduction and predation by weasels.* Arct. Alp. Res., Vol. 6, pp. 1-12.

Mansfield, A.W. 1975. *Marine ecology in arctic Canada.* Proc. Circumpolar Conf. North. Ecol., Nat. Res. Counc. Can., Ottawa, Ont., pp. II.27-47.

Masyk, W.J. 1973. *The snowmobile: a recreational technology in Banff National Park.* Environ. Impact and Decision Making, Univ. Calgary, Calgary, Alta.

McKay, G.A. 1970. *Climate: a critical factor in the tundra*. Trans. R. Soc. Can., Ser. IV, Vol. 8, pp. 405-412.

McKay, G.A. and T. Allsopp. 1980. *The role of climate in affecting energy demand/ supply*. Proc. Int. Workshop on Energy/Climate Interactions (W. Beck, J. Pankrath and J. Williams, eds.), Münster, Germany, 3-7 March, 1980, D. Reidl Publishing Co., Boston. pp. 53-72.

Miller, J.R. and D.K. Halligan. 1979. *Record snowfall, April, 1976*. Weatherwise, Vol. 32, pp. 123-125.

Mohr, J.L. and J.L. Tibbs. 1963. *Ecology of ice substrates, with discussion by N.J. Wilimovsky*. Proc. Arct. Basin Symp. AINA, pp. 245-252.

Nelson, J.W. 1975. *Avalanche*. Proc. First Nat. Conf. on Snow and Ice Control, Roads Transp. Assoc. Can., Ottawa, Ont., pp. 45-56.

Peek, J.M. 1971. *Moose-snow relationships in northwestern Minnesota*. In Proc. Symp. on Snow and Ice in Relation to Wildlife and Recreation (A.O. Haugen, ed.), Iowa State Univ. Press, Ames, pp. 39-45.

Peterson, N.M. 1978. *Ecology of the Canadian Arctic Archipelago: Selected References, Volume 5 (with cumulative index for volumes 1 to 5), North of 60*. Dept. Indian North. Affairs, Ottawa, Ont.

Pruitt, W.O. Jr. 1957 *Observations on the bioclimate of some taiga mammals*. Arctic, Vol. 10, pp. 131-138.

Pruitt, W.O., Jr. 1958. *Qali: a taiga snow formation of ecological importance*. Ecology, Vol. 39, pp. 169-172.

Pruitt, W.O., Jr. 1959. *Snow as a factor in the winter ecology of the barren-ground caribou (Rangifer arcticus)*. Arctic, Vol. 12, pp. 159-180.

Pruitt, W.O., Jr. 1960. *Animals in the snow*. Sci. Am., Vol. 202, pp. 60-68.

Pruitt, W.O., Jr. 1970. *Some ecological aspects of snow*. Proc. 1966 Helsinki Symp. on Ecol. of the Subarct. Reg., Unesco Ser. Ecol. and Conserv., No. 1, Unesco, Paris, pp. 83-97.

Pruitt, W.O., Jr. 1975. *Life in the snow*. Nature Can., Oct./Dec., pp. 40-44.

Richardson, G. and F.B. Salisbury. 1977. *Plant responses to the light penetrating snow*. Ecology, Vol. 58, pp. 1152-1158.

Round, F.E. 1953. *The Biology of Algae*. St. Martin's Press, New York.

Savile, D.B.O. 1972. *Arctic adaptations in plants*. Monogr. No. 6., Can. Dept. Agric., Ottawa, Ont.

Schmid, W.D. 1971. *Modification of the subnivean microclimate by snowmobiles*. In Proc. Symp. on Snow and Ice in Relation to Wildlife and Recreation (A.O. Haugen, ed.), Iowa State Univ. Press, Ames, pp. 251-257.

Smith, J.S. 1967. *The great Chicago snowstorm of '67*. Weatherwise, Vol. 20, pp. 248-253.

Staple, W.J. 1964. *Dry land agriculture and water conservation*. Spec. Publ. No. 4, Water, Am. Soc. Agron., Madison, Wisc., pp. 15-30.

Stein, J.R. and C.C. Amundsen, 1967. *Studies on snow algae and fungi from the front range of Colorado*. Can. J. Bot., Vol. 45, pp. 2033-2045.

Steinhoff, H.W. and J.D. Ives (eds.). 1976. *Ecological impacts of snowpack augmentation in the San Juan Mountains, Colorado*. Final Rep. to Bureau Reclam., Div. Atmos. Water Resour., U.S. Dept. Interior, Washington, D.C.

Stommel, H.G. 1966. *The great blizzard of '66 on the northern Great Plains.* Weatherwise, Vol. 19, pp. 188-193, 207.

Sykes, R.B. 1966. *The blizzard of '66 in central New York State - legend in its time.* Weatherwise, Vol. 19, pp. 240-247.

Thomas, M.K. 1971. *Canadian weather features of the past century.* Weatherwise, Vol. 23, pp. 270-275.

Trueman, M.E.H. 1970. *Montreal's great snowstorm of '69.* Weatherwise, Vol. 23, pp. 270-273.

Verme, L.J. 1968. *An index of winter weather severity for northern deer.* J. Wildlife Manage., Vol. 32, pp. 566-574.

Verme, L.J. and J.J. Ozoga. 1971. *Influence of winter weather on white-tailed deer in upper Michigan.* In Proc. Symp. on Snow and Ice in Relation to Wildlife and Recreation (A.O. Haugen, ed.), Iowa State Univ. Press, Ames, pp. 16-28.

Visser, S.A. 1973. *The microflora of a snow depository in the city of Quebec.* Environ. Lett., Vol. 4, pp. 267-272.

White, R.G. 1975. *Some aspects of nutritional adaptations of arctic herbivorus animals.* In Physiological Adaptation to the Environment (F.J. Vernberg, ed.), Intext Educational Publishers. New York, N.Y., , pp. 239-268.

Williams, K. 1972. *Avalanches in our western mountains: What are we doing about them?* Weatherwise, Vol. 72, pp. 220-227.

2

SNOW AND CLIMATE

M. O. BERRY

*Canadian Climate Centre, Atmospheric Environment Service,
Downsview, Ontario.*

Large areas of the earth are snowcovered for at least a portion of the year,
producing a substantial change in surface characteristics from those
encountered when snow is absent. Since the atmosphere is sensitive to
physical changes at its lower boundary, snowcover is recognized as an
important factor influencing climate, both on the local and global scales.

EFFECTS OF SNOWCOVER ON THE ATMOSPHERE

Energy Budget

The effects of snowcover on climate are manifested by its interaction with
the energy regime of the atmosphere. Consider the net energy flux to a volume
of air with unit cross-section extending from the earth's surface to the top of
the atmosphere. The primary processes affecting the internal energy of this
volume are: the absorption of short-wave radiation by the atmosphere and the
earth's surface, the emission of terrestrial long-wave radiation to space and the
energy exchange with the surrounding atmosphere and layers beneath the
earth's surface. A simple bulk energy equation for the volume may therefore
be written:

$$dU_a/dt = Q_a + (1 - A)Q_{si} + Q_v + Q_g + Q_{le}. \qquad 2.1$$

where: dU_a/dt = rate of change of internal energy (stored in the volume),

$\quad Q_a$ = short-wave radiation flux absorbed by the atmosphere,

$\quad A$ = albedo of the earth's surface,

$\quad Q_{si}$ = short-wave radiation flux incident on the earth's surface,

$\quad Q_v$ = net energy flux (to or from the column) resulting from
exchange of latent and sensible heat with the surrounding
air,

Q_g = net energy flux between the column and layers below the earth's surface, and

Q_{le} = long-wave radiation flux to space.

Equation 2.1 lacks several components, such as the amount of kinetic energy of the wind converted to heat friction and the energy required to melt snow. Nevertheless, when used to study systems having gross time and spatial scales, where the omitted components can be assumed small, it can be applied with reasonable confidence.

Any change in the internal energy dU_a will be reflected by a change in some climatic parameter. For example, if dU_a/dt is positive there may be an increase in the ambient air temperature. When the energy budget calculation is applied on an annual basis dU_a may be considered negligible for most purposes since the energy gained during the spring and summer is approximately balanced by losses in the fall and winter.

The relative magnitudes of the terms of Eq. 2.1 may vary considerably with latitude, season, prevailing weather conditions and the nature of the earth's surface, thus making accurate evaluation of each term difficult. Table 2.1 lists the annual values of the components (expressed as the mean annual daily flux per unit area) for different latitudes. At 20° N more energy is gained by the atmosphere-earth system from absorbed short-wave radiation (the sum of Q_a and $(1 - A) Q_{si}$, than is lost by long-wave radiation Q_{le}: 285 W/m^2 compared with 242 W/m^2. Most of this energy gain is dissipated throughout the atmosphere by advection. With increasing latitude the losses through long-wave emission decreases, while the gain by short-wave absorption decreases more rapidly, resulting in a deficit in the atmospheric energy budget. This deficit is compensated mostly by a net poleward advection of heat in the atmosphere. For example, at 80° N advection is the largest source of energy on an annual basis, contributing a mean annual daily flux of 107.6 W/m^2 compared with 62.5 W/m^2 from short-wave absorption by the atmosphere and the earth's surface. A similar situation exists in the Southern Hemisphere. The net poleward fluxes of energy in the atmosphere are fundamental in determining the global circulation and hence global climate.

The mean annual amount of solar energy reaching the earth's surface Q_{si} is approximately three times the amount absorbed by the atmosphere Q_a (see Table 2.1). That portion of Q_{si} absorbed by the surface is strongly affected by the surface albedo. The albedo of a snowcover, while varying with the age of the snowcover and other factors (see Chs. 7 and 9), is considerably higher than those of most other natural surfaces (see Table 2.2). Because of the seasonal changes in areal extent of snowcover the albedo of the earth's surface exhibits seasonal trends. Continental surface albedos for North America in mid-winter and summer are compared in Table 2.3. Between 65 and 70° N, an area primarily covered by tundra, the mid-winter albedo is very high, exceeding the

Table 2.1

MEAN ANNUAL DAILY LATITUDINAL VALUES OF COMPONENTS OF THE ENERGY BALANCE OF THE ATMOSPHERE-EARTH SYSTEM, (Units of W/m²). Reprinted from PHYSICAL CLIMATOLOGY by W.D. Sellers by permission of The University of Chicago Press. © 1965 by The University of Chicago. All rights reserved. Published 1965. Printed in the United States of America.

Lat.	Q_a	Q_{si}	$(1-A)Q_{si}$	Q_v	Q_g [a]	Q_{le}
80° N	25.5	83.3	37.0	107.6	4.6	-174.7
60° N	35.9	114.5	92.6	55.5	15.0	-199.0
40° N	52.1	187.4	163.1	-9.3	13.9	-219.8
20° N	74.0	231.4	210.6	-39.3	-3.9	-241.8

[a] It is assumed the contribution of land surfaces to Q_g is negligible and the values listed represent the contribution of ocean areas.

Table 2.2

PERCENTAGE OF INCIDENT SHORTWAVE RADIATION REFLECTED BY SOME NATURAL SURFACES.

Surface	Albedo (%)	Surface	Albedo (%)
Fresh snow	75-95	Water	5-30
Old snow	40-70	Bare fields	12-25
Sea ice	30-40	Field crops	3-25
Desert	24-30	Tundra	15-20
		Forests	3-20

summer value by about 67%. This seasonal variation diminishes at lower latitudes. However, even in the belt from 40 to 45° N the mid-winter values are 20% greater than the summer values.

Table 2.4 shows the monthly variations of albedo at three locations in Canada: Vancouver (49° N), Regina (50° N), and Resolute (75° N). At Vancouver, snowcover is normally present for only a few days of any month during winter so that the mean monthly albedos (e.g., in December and January) are only slightly higher than those for the other months. In contrast, at Regina and Resolute, where snowcover is more persistent, the albedo is significantly higher in months with snow. Although seasonal differences in albedo may be expected at any location because of changes in vegetation, sun angle, cloud cover and other factors, the major variations are due to the persistence and continuity of the snowcover.

The large albedo associated with the more persistent snowcover causes less solar energy to be absorbed by the surface at higher latitudes. For example the terms, $(1 - A)Q_{si}$ and Q_{si} (see Table 2.1), indicate that 87% of the energy at 40°N is absorbed, but only 44% at 80°N. Consequently, variations in the extent of snowcovered surfaces at higher latitudes affect the magnitude of the net radiative deficit and similarly, the resultant poleward flux of energy.

Table 2.3
AVERAGE ALBEDOS FOR NORTH AMERICA
(Kung, Bryson and Lenschow, 1964).

Lat. Zone	Albedo (%)	
(°N)	mid-winter	summer
65-70	82.8	16.1
60-65	67.3	15.6
55-60	59.1	16.5
50-55	50.3	14.6
45-50	46.4	14.8
40-45	37.9	15.8
35-40	28.5	16.5
30-35	19.1	17.2
25-30	17.8	17.9
20-25	15.8	15.8
Continental mean	43.0	16.0

Table 2.4
PERCENTAGE OF INCIDENT SOLAR RADIATION REFLECTED BY THE
GROUND. MEAN VALUES FOR 1957-64.[a]

Site	Snowcover Season	J	F	M	A	M	J	J	A	S	O	N	D	YEAR
Vancouver	-	17	16	17	14	14	14	14	14	14	14	14	19	15
Regina	Nov-Apr	54	58	46	38	15	20	22	25	24	24	34	47	28
Resolute	Sept-June	--	78	80	76	69	45	27	21	45	58	--	--	51

[a] Percentages are computed values considered representative of the areas in which stations are located. Bryson and Hare (1974). Snowcover dates were taken from Potter (1965).

The presence of snow, because of its low thermal conductivity, sharply reduces the net heat exchange between the ground and atmosphere. This reduction can significantly influence the soil temperature regime. However, since the heat flux Q_g from snow-free land surfaces is usually small relative to the other terms in Eq. 2.1 (Geiger, 1961), the reflective rather than insulative properties of the snow are of principal concern in the energy balance of the atmosphere. An exception to this general rule occurs at high latitudes over ice-covered waters where the upward heat flux through the snow-covered ice can represent a major energy source in winter months when the insolation is very low or non-existent. Vowinckel and Orvig (1961) estimate that over the Arctic Ocean during January, about one quarter of the energy available to the earth's atmosphere is obtained from this source.

Air Mass Formation and Weather Systems

In the absence of phase changes it can be assumed that any change in the internal energy of the atmosphere is shown as a change in temperature. In this case the energy-conservation equation for a column of the atmosphere with unit cross section may be approximated by:

$$\Delta U_a / \Delta t = (C_p / g) \, (\Delta \bar{T} / \Delta t) \, \Delta p \sim 10^4 \, \Delta \bar{T} / \Delta t \qquad\qquad 2.2$$

where ΔU_a = change in internal energy (kJ/m^2) during the time interval Δt (s),

   C_p = specific heat capacity of the air at constant pressure $(J/kg^\circ C)$,

   g = acceleration due to gravity (m/s^2),

   $\bar{T}$ = mean temperature of the volume of air $(^\circ C)$, and

   Δp = thickness of the column of the atmosphere expressed as an equivalent pressure thickness (kPa).

According to Eq. 2.2 a change in internal energy of the column by 10,000 kJ would change its mean temperature $\sim 1^\circ C$.

While the earth gains its energy from solar radiation with wavelengths 0.3 - 2.2 μm, it loses energy to space by long-wave radiation Q_{le} with wavelengths 6.9 - 100 μm. The characteristically high reflectivity of the snowcover to solar radiation changes so rapidly with wavelength that at the longer wavelengths snow is a poor reflector but a good emitter. Although much of the long-wave radiation emitted by the earth's surface is returned to it because of absorption and emission by the atmosphere, a significant proportion (a global average of about 20 percent is eventually lost to space. If this loss is not compensated for by energy gained from other sources the net effect is a reduction in air temperature, particularly in the lower layers of the troposphere. This situation often prevails during winter over large land areas at high latitudes, where the reflective and insulative characteristics of the snowcover result in small values of the terms Q_a, $(1 - A)Q_{si}$ and Q_g. Typically, where the net daily energy loss is of the order of 10,000 - 15,000 kJ/m^2 rapid cooling of the air takes place, but the rate of cooling varies with height. The temperature profile of air subjected to radiational cooling for long periods shows a very low

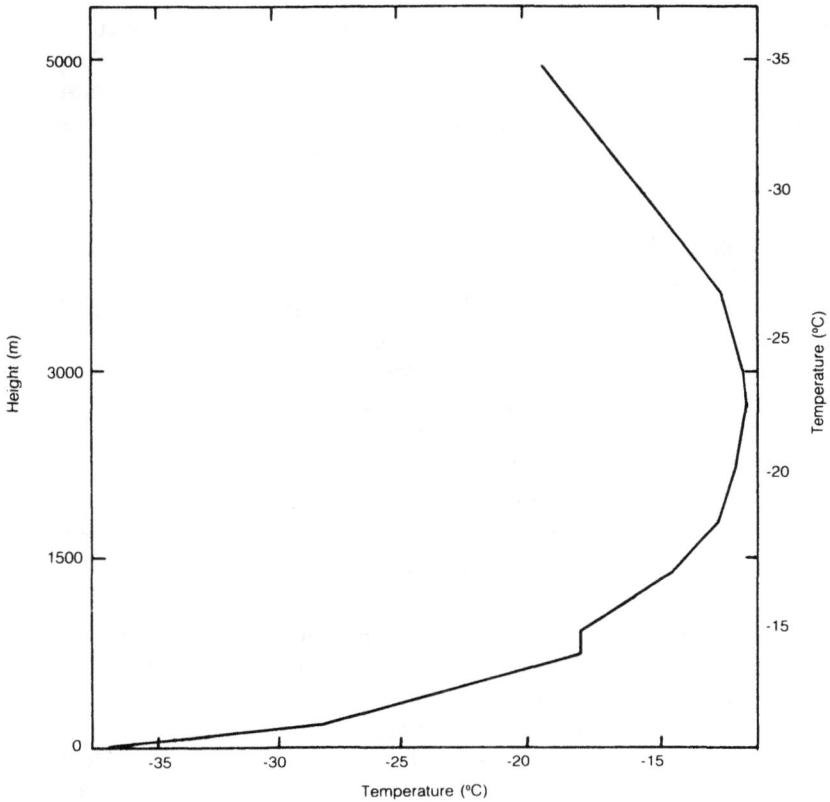

Fig. 2.1 Typical inversion resulting from intense radiational cooling (Burns, 1973).

temperature at the surface with an increase with height (see Fig. 2.1).

Siberia, northwestern North America and Antarctica are among the regions of the world where intense radiative cooling occurs, resulting in the formation of air masses characterized by very low surface temperatures, increasing temperature with height (inversion) up to 1 or 2 km, light winds and clear skies. Apart from the extreme cold, one of the important characteristics of these air masses, stemming from the combination of light winds and a strong inversion, is a limited ability to disperse pollutants and fog. These air masses, frequently termed continental polar or continental arctic, migrate because of changes in the upper, large-scale circulation. In the Northern Hemisphere, this migration is frequently in a southeastward direction, so that the snowcovered northern regions act as source regions for the cold air masses which move to mid-latitudes. In addition to the effect on mid-latitude winter temperatures, the southward migration of these cold air masses frequently cause major storms particularly in coastal areas where sharp temperature differences between the cold continental air and warmer maritime air may occur. The development and movement of major storms along these baroclinic zones can cause heavy precipitation on the eastern coast of North America.

The cooling effects of snowcover are not limited to those regions which favour the development of large-scale, cold air masses. In mid-latitudes it is difficult to distinguish the effect of snowcover from the effects of other factors which determine atmospheric temperature. Namias (1963) described a period in which the snowcover over eastern North America extended abnormally far south, reaching a position paralleling the states of Oklahoma and Virginia. He estimated that for a one-month period the average temperature at the southern border of the snow was about 5° C lower than it would have been if the anomaly had not been present.

Global Climate

Since snowcover has a major influence on the amount of solar energy which is retained as heat by the atmosphere, as opposed to being returned to space, its extent and distribution are important variables in the global energy budget and hence the world's climate. The mechanics of the earth's atmospheric circulation is highly complex and only partially understood, which makes numerical simulation difficult. Hence, it is difficult to describe rigorously the role of snow as it affects global climate. For example, as an extreme case, during the last full glaciation, when snow covered about twice the surface of the Northern Hemisphere that it covers at the present time, the world climate was significantly different: the global mean temperature was substantially lower, and the climatic zones had greatly different locations and sizes. The higher albedo of the extensive snowcover contributed greatly to these differences. However, other factors were important, e.g., topographic changes of the earth's surface caused by the ice sheets, and the covering of portions of the ocean by ice, which subsequently melted. The precise effects of any of these factors on climate is difficult to assess.

Considerable evidence suggests that snowcover plays an important role in the formation and the growth of glaciers which can lead to major changes in world climate (Flohn, 1974). Because of the high albedo of snow, tropospheric cooling may occur over large areas that are snowcovered for long periods. If other meteorological factors are favourable, this cooling causes a trough to form in the circulation aloft. This upper flow pattern enhances the tendency for lower temperatures to prevail over much of the snowcovered area, discouraging melting, but increasing snowfall near its southeastern and eastern boundaries (Lamb, 1955). A similar sequence has been used to explain the development, during the last glaciation, of the Laurentide ice sheet (which originated in the Keewatin-Quebec-Labrador area). Flohn (1974) accounts for the development in terms of an initial perturbation (several causes of which are possible) causing the snowcover to remain throughout summer. Once this occurs a positive feedback process is established: snow surface with high albedo → tropospheric cooling → cold trough aloft → increased snowfall → growth of glaciers. Using a numerical model of the atmospheric circulation, Williams (1975) studied a situation in which it was assumed that the ice-covered areas of the globe during the last glaciation, were also snowcovered in July. She found that the simulated circulation patterns largely resembled those associated with the feedback process.

The importance of the albedo-temperature relationship has been emphasized by the application of energy balance models to study the stability of world climate. Using a semi-empirical climate model, Schneider and Gal-Chen (1973) estimated the response of global temperature T to changes in the amount of solar radiation reaching the earth to be as follows:

Solar Input	T (°C)
no change	14.3
+1.0%	17.1
-1.0%	8.9
-1.6%	-97.6.

The catastrophic decline in temperature (associated with global glaciation and an increase in planetary albedo from about 0.36 to 0.85) resulting from a small decrease in solar energy input is closely linked to the tendency for decreasing temperature and enlarging snowcovered areas to be mutually sustaining, i.e., a strong positive feedback links changes in albedo with surface temperature. Because of the difficulty of modelling global climate, the values given by Schneider and Gal-Chen (1973) must be considered as speculative. However, studies of this type strongly suggest that the earth's climate, in its present state, is highly sensitive to changes in the amount of solar energy absorbed by the atmosphere or the earth's surface, and to perturbations such as an extended anomaly in climate in a particular region.

Analyses of satellite data indicate that the mean global snow and ice cover varies by as much as 12 percent. There is evidence to suggest that these variations may be related to changes in the mean tropospheric temperatures, and to the occurrence of anomalous weather patterns (Kukla and Kukla, 1974). However, on account of the small amount of data available, and the present gaps in understanding atmospheric energy exchange processes, it is difficult to determine whether such relationships are real or coincidental.

Effects of Snowcover on other Weather Elements

In addition to affecting the global energy budget and large-scale weather patterns, snowcover influences other weather elements including wind, fog and precipitation.

Wind dispersion potential

Usually, at the gradient level (the level above which the wind is not influenced by local terrain features, usually 300 to 600 m above the ground) the horizontal air flow can be closely approximated by the geostrophic wind speed U_g, where:

$$U_g = \alpha \Delta p \bar{k} / f, \qquad\qquad 2.3$$

and α = specific volume,
    Δp = horizontal pressure gradient,
    $\bar{k}$ = unit vector along the vertical, and
    f = Coriolis parameter, a constant for a given latitude.

The geostrophic wind blows perpendicular to the horizontal pressure gradient with a speed proportional to the gradient, with the low pressure to the left of an observer facing downstream in the Northern Hemisphere.

The relationship between the wind velocity at gradient level and that at a lower level, down to the lowest few metres above the ground, is usually fairly well represented by the power law,

$$U_z / U_g = (z / z_g)^n, \qquad\qquad 2.4$$

where U_z is the wind velocity at height z, z_g is the height of the geostrophic wind and n is an exponent whose magnitude depends on the roughness of the underlying surface and the stability (indicated by the vertical temperature gradient; neutral conditions $\sim$ -10°C/km) of the air. Using measurements taken over rolling, partially tree-covered terrain, DesMarrais (1959) calculated average values of n ranging from 0.19 for superadiabatic conditions ($\partial T / \partial z > -11$°C/km) to 0.49 for inversion conditions ($\partial T / \partial z > 0$). Frost (1948), using data measured at an airfield at Cardington, England under strong inversions ($\partial T / \partial z > 16$°C/km) reported an average value for n equal to 0.77. Although these values of n are somewhat site-specific, they illustrate the effects of stability on the wind speed at a given height to that at the surface; e.g., the more unstable the air mass, the larger the ratio U_z / U_g.

Because of its insulative and radiative properties snowcover tends to increase the stability of the adjacent atmosphere. Many factors influence the wind regime, so that caution must be exercised in interpreting variations in wind speed purely in terms of stability variations induced by the presence or absence of snowcover. Figure 2.2 shows the average monthly wind speeds and

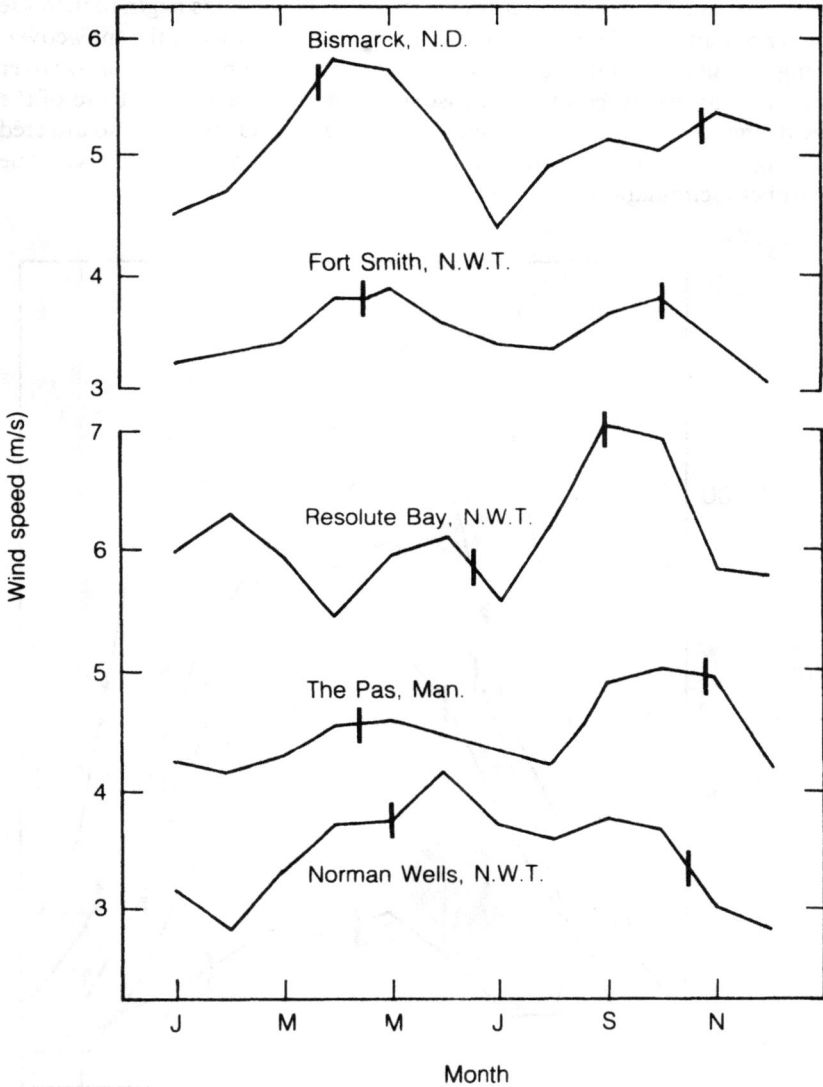

Fig. 2.2 Average monthly wind speeds. Vertical bars indicate average first and last
    dates of snowcover. (Potter, 1965; Bryson and Hare, 1974; Environment
    Canada, 1975).

the dates of occurrence and disappearance of snowcover for five stations
located from North Dakota to the High Arctic. It can be observed that in the
fall winds begin to decrease approximately on the average date of occurrence

of first permanent snowcover, however, in the spring, winds begin to increase about two months before the average date of disappearance of the snowcover. During spring the stability of the air begins to decrease before the snowcover disappears, primarily because the insolation increases and the albedo of the snow decreases. The seasonal variation in surface wind speed is also affected by changes in the upper winds resulting from variations in the large-scale atmospheric circulation patterns.

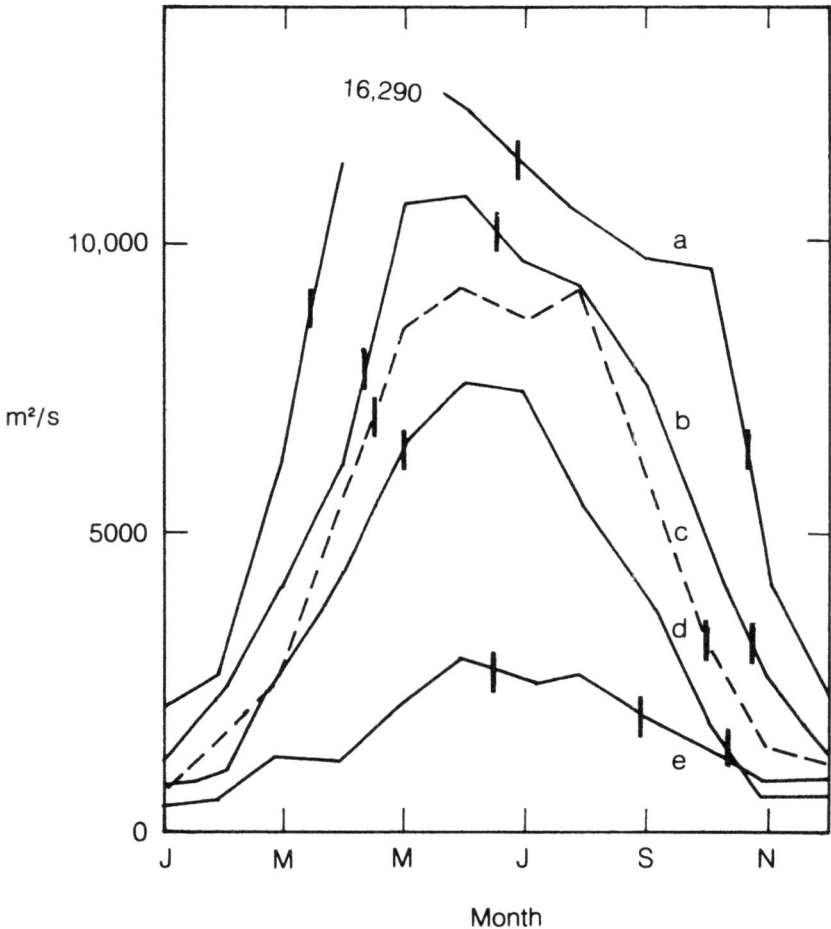

Fig. 2.3 Ventilation coefficients for (a) Bismarck, N.D., (b) The Pas, Man., (c) Fort Smith, N.W.T., (d) Norman Wells, N.W.T., (e) Resolute, N.W.T. Vertical bars indicate average first and last dates of snowcover (Potter, 1965; Portelli, 1978).

Wind and stability near the earth's surface determine the atmosphere's ability to disperse pollutants. An index of this ability is the ventilation coefficient, defined as the product of the thickness of the layer through which emissions are dispersed (determined by the vertical extent of an unstable temperature profile) and the layer's mean wind speed (Holzworth, 1974). The larger the value of the ventilation coefficient, the greater the dispersion potential of the air. Figure 2.3 shows the annual variations of this coefficient for the same stations whose seasonal wind-speed distributions are given in Fig. 2.2, illustrating typical patterns for inland stations; the lowest values of dispersion occur during the period when snowcover is present. Holzworth (1974) presented data indicating that a mid-winter minimum also occurs at inland stations where snowcover is infrequent. Portelli (1978) suggests that the variation in ventilation coefficient is strongly influenced by the amount of surface heating and hence by the seasonal variation in isolation. Snowcover, when present, serves to reduce the proportion of heat available for surface heating, and therefore to intensify the winter minimum. However, at sites adjacent to large open bodies of water, other factors including the temperature difference between snowcovered land and water may result in quite different seasonal patterns, having a maximum dispersion potential in the winter months (Portelli, 1978).

Winds near the ground are also affected by the roughness of the underlying surface, that is, the value of n in Eq. 2.4 depends on the frictional drag exerted by the surface. For example, when grass acquires a smooth cover of snow the frictional drag is reduced so that other factors being equal, the wind speed near the surface increases. Surface roughness is often quantified in terms of a roughness parameter or height z_0 whose magnitude varies with the average height of obstacles on an otherwise smooth surface.

The dependency of wind speed on surface roughness is difficult to determine mainly because it has a complex relationship with stability. Under neutral conditions ($\partial T/\partial z \sim -10°C/km$) the wind speed profile over relatively smooth surfaces, up to a height of about 30 m, can be represented by the logarithmic profile:

$$U_z = U^* \ln (z/z_0)/k \qquad\qquad 2.5$$

where: U_z = wind velocity at a height, z, above the surface,
U^* = friction velocity (= $\sqrt{\tau/\rho}$ where τ is the shear stress and ρ is the air density),
z_0 = roughness height, and
k = von Karman's constant (usually taken as 0.4).

Snowcover can alter the wind speed at levels above the ground surface by changing the roughness height. It may also alter the wind profile if drifting (saltation) occurs. Some representative values of the roughness height of different surfaces given by Deacon (1953), and Munn (1966) are:

Surface	z_0 (cm)
smooth mud flats	0.001
snow	0.001 to 0.07
smooth snow on short grass	0.005
short grass (1.5 to 3 cm)	0.2 to 0.7
long grass (60 to 70 cm)	3.7 to 9.0.

The effect of snowcover on surface roughness and the wind-speed profile can be demonstrated by means of Eq. 2.5. For example, if short grass ($z_0 = 0.4$) becomes snowcovered ($z_0 = 0.005$) the ratio of the wind speeds at 10, 5, 1 and 0.1 m above the surface to that at 20 m, changes from 0.92, 0.84, 0.65 and 0.38 to 0.95, 0.89, 0.76 and 0.59, respectively. Thus the wind speeds over the relatively smooth snowcovered surface are higher than over the grass surface with the differences becoming greater with decreasing height. However, this effect is difficult to generalize since moderate or strong winds during or after a snowfall cause drifting, which may change both the roughness and the velocity profile.

Fog

Fog usually forms when the lowest layer of the atmosphere is cooled to its saturation point, for example, by radiative cooling, or by the advection of warm air over a relatively cool surface. However, in spite of the intense cooling which can occur over a snow surface, fog is usually not observed.

The relationship between fog and snowcover is discussed in detail by Petterssen (1956). Figure 2.4 shows that at temperatures below 0°C the saturated vapor pressure of air over water is greater than that over ice (Curve Δe). Air cooled over a snowcover becomes saturated with respect to ice (for example; ~91% relative humidity at -10°C; ~82% at -20°C; see Curve RH), and because of the difference in the saturation pressure, water vapor sublimates on the snow surface. Generally, these cooling and moisture transfer processes do not promote the formation of fog except under conditions where a large inflow of moisture more than compensates the sublimation loss. At temperatures below -35°C fog usually consists of ice particles rather than water vapor droplets. Figure 2.5 shows the relative frequency of occurrence of fog in January reported at Frobisher Bay, NWT: fog is most frequent when the air temperature is within the range -33 to -58°C, where atmospheric water vapor changes to ice.

Ice fog can be extremely persistent if moisture is available and the atmospheric dispersion is low, and frequently forms in and around urban areas where moisture is supplied to the atmosphere by evaporation, combustion and other processes. It may also develop independently of moisture sources associated with human activity, e.g., near areas of open water (Berry and Lawford, 1977).

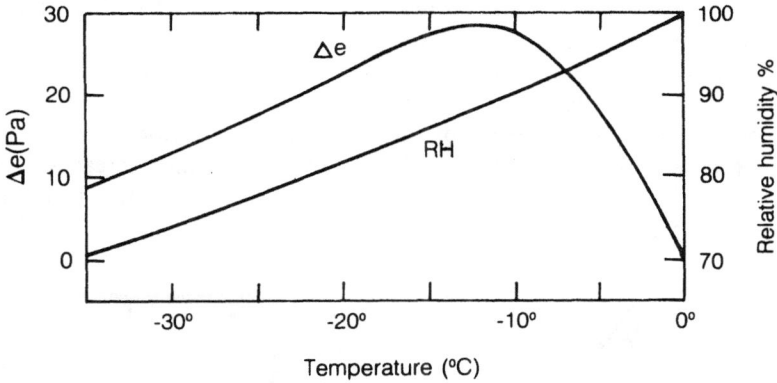

Fig. 2.4 Difference in saturation pressure over water and over ice (curve Δe); and relative humidity (curve RH) of air that is saturated over ice. From WEATHER ANALYSIS AND FORECASTING by S. Petterssen. Copyright © 1956, McGraw-Hill Book Company, Inc. Used with the permission of McGraw-Hill Book Company.

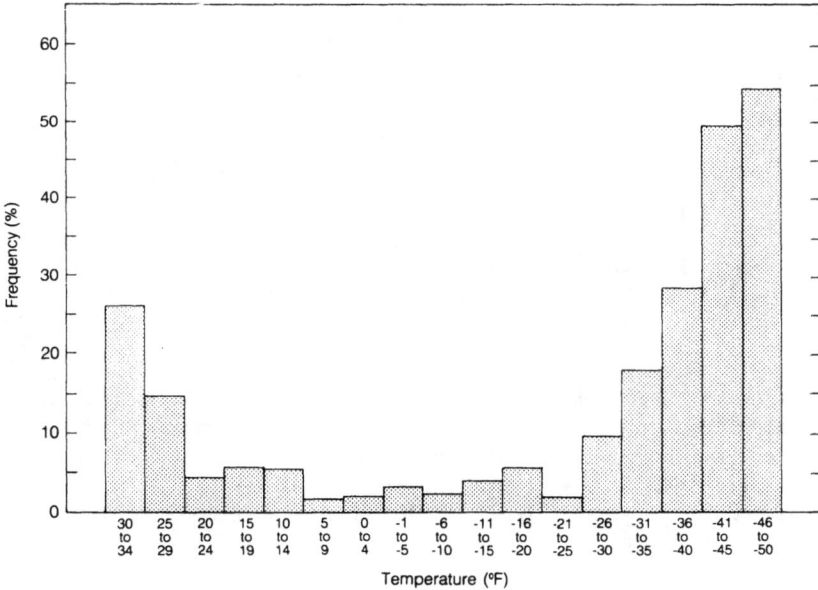

Fig. 2.5 Percentage frequency of fog occurrence in January at Frobisher Bay, N.W.T., according to temperature range.

During spring, or whenever air temperatures are above freezing while the snow is melting, a snowcover can act as either a source or sink for atmospheric moisture, depending on the vertical gradient of moisture.   Figure 2.6 (Petterssen, 1956) illustrates the interrelation between temperature and relative humidity and condensation and evaporation processes. Since fog is a product of saturated air, the possibility of it forming decreases as the temperature increases above freezing. Near the freezing point, advection of moisture into the area may sustain fog.

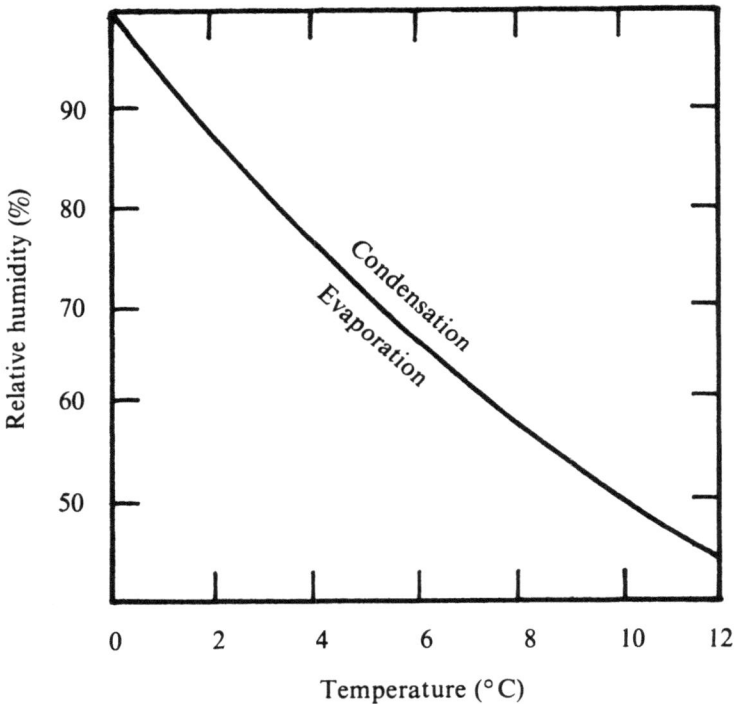

Fig. 2.6 Limiting value of relative humidity. If the relative humidity at some distance above the snow is higher than that indicated by the curve, water vapor condenses on the snow (Petterssen, 1956). From WEATHER ANALYSIS AND FORECASTING by S. Petterssen. Copyright © 1956, McGraw-Hill Book Company, Inc.   Used with the permission of McGraw-Hill Book Company.

Precipitation

The amount of moisture contained in the atmosphere, and the frequency and strength of weather disturbances, which trigger the conversion of water vapor into rain, snow or other types of precipitation, are major determinants of the amount of precipitation. Snowcover mainly affects precipitation and its formation by enhancing the cooling of the overlying atmosphere, thereby reducing its moisture holding ability. On a large scale this cooling is associated with the southward displacement of precipitation-producing weather systems - in North America, to southern Canada and the United States. Either a deficiency of water vapor or the nonoccurrence of major weather disturbances, or both, are mainly responsible for the precipitation minimum observed during the snowcover period at most locations in the Canadian prairies and the Northwest Territories (see Table 2.5). In southeastern Canada and the northeastern United States, the winter minimum is less pronounced or nonexistent. At some stations, (e.g., St. John's, Nfld.), the maximum monthly precipitation amounts occur in winter, and are much greater than those recorded in the Prairies or the North. Along the eastern seaboard cold air masses, which have developed over snowcovered continental areas, come in contact with relatively warm maritime air and precipitation-bearing weather disturbances frequently result.

Table 2.5
MEAN MONTHLY PRECIPITATION (mm) 1931-60 (Bryson and Hare, 1974).

	J	F	M	A	M	J	J	A	S	O	N	D
Edmonton, Alta. 54°N 114°W	24	20	21	28	46	80	85	65	34	23	22	25
Norman Wells, NWT 65°N 127°W	19	16	12	13	15	34	49	64	38	24	24	20
Baker Lake, NWT 64°N 96°W	5	4	6	9	8	21	40	45	34	20	9	7
Winnipeg, Man. 50°N 97°W	26	21	27	30	50	81	69	70	55	37	29	22
Montreal, Que. 46°N 74°W	87	76	86	83	81	91	102	87	95	83	88	89
Portland, Me. 44°N 70°W	111	97	110	95	87	81	73	61	89	81	106	98
St. John's, Nfld. 48°N 53oW	153	163	135	121	99	94	89	102	120	138	163	174

SNOWCOVER AND THE LAYER BELOW

Snowcover serves as the upper boundary of the earth's surface thereby affecting the "climate" of the ground. Changes in the climate of both the upper layers of the ground and the atmosphere are intimately related. However, because of differences between the properties of air and ground the influence of snowcover on the two regimes is markedly different.

Thermal Regime of the Ground

Consider the ground to consist of two layers, an upper layer a few metres deep and a layer underneath. In the lower layer, the vertical temperature (geothermal) gradient is determined primarily by the thermal conductivity of the material and the heat flow from the earth's interior. It is usually assumed that the lower layer is unaffected by annual or shorter-term variations in weather and snowcover, although long-term changes in climate may affect the temperature of the layer. Conversely, the layer adjacent to the ground surface is greatly influenced by the weather, or more accurately, by the surface energy exchange processes.

The flux of heat by conduction per unit cross-section Q in the vertical direction through an isotropic homogeneous medium under steady-state conditions can be described by

$$Q = -k \ (\partial T / \partial z), \qquad\qquad 2.6$$

where: k = thermal conductivity,
         T = temperature, and
         z = vertical distance.

The thermal conductivity of unfrozen soil is mainly a function of its moisture content (see Table 2.6).

The energy equation defining the heat flow by conduction in one dimension through an isotropic, homogeneous medium is

$$\partial T / \partial t = \kappa (\partial^2 T / \partial z^2), \qquad\qquad 2.7$$

where: t = time, and
         κ = $k / \rho C$, the soil thermal diffusivity, in which ρ is the density of the soil and C is the specific heat capacity.

Some typical values of κ and the volumetric heat capacity C_v (= ρC) are also given in Table 2.6.

Since the annual variation in the air temperature tends to be cyclic, the annual temperature at the surface of the ground can be approximated by a sine function of time as

$$T(0,t) = T_a + A \sin (\omega t + \phi), \qquad\qquad 2.8$$

where: T_a = mean soil temperature (assumed constant with depth),

Table 2.6
TYPICAL THERMAL PROPERTIES OF SOILS AND SNOW (adapted from Van Wijk, 1966).

Soil Type	Volumetric Water Content θ_w[a]	Thermal Conductivity k W/(°C·m)	Volumetric Heat Capacity $C_v = \rho C$ $10^6 J/(m^3 \cdot °C)$	Diffusivity $k/\rho C$ $10^{-6} m^2/sec$	Damping Depth D[b] m
Sand	0.0	0.29	1.3	0.22	1.5
	0.2	1.8	2.1	0.81	2.9
	0.4	2.2	2.9	0.76	2.7
Clay	0.0	0.25	1.3	0.19	1.4
	0.2	1.2	2.1	0.57	2.4
	0.4	1.6	2.9	0.55	2.3
Peat	0.0	0.059	0.50	0.12	1.1
	0.4	0.29	2.2	0.13	1.2
	0.8	0.50	3.9	0.13	1.2
Snow	-	0.063	0.11	0.56	2.5
	-	0.13	0.38	0.34	1.9
	-	0.71	1.1	0.65	2.6

[a] θ_w is the volume of water per unit volume of soil.
[b] D is the damping depth for the annual variation.

A = amplitude of the annual variation of the temperature at the surface,

$\omega = 2\pi/P$; where P is the period of the cycle (12 months), and

ϕ = phase constant whose value depends on the time (starting point) when T reaches a certain characteristic value.

Using this relationship as the surface boundary condition and assuming no heat is gained or lost by vaporization or freezing, Eq. 2.7 can be solved to yield:

$$T(z,t) = T_a + A \exp(-z/D) \sin[\omega t - (z/D) + \phi], \qquad 2.9$$

where $T(z,t)$ = temperature at depth, z, and

$D = \sqrt{2k/\omega\rho C} = \sqrt{2\kappa/\omega}$ (sometimes called the damping depth).

Using Eq. 2.9 it can be shown that the amplitude of the temperature wave at the depth z = D is less than that at the surface by a factor of $1/e (= 0.37)$. Note: the damping depth is not only a property of the conducting medium but also depends on the period of the temperature variation.

Rarely do natural soils possess the homogeneous properties assumed in the derivation of Eq. 2.9. Nevertheless this equation represents well the annual variation in soil temperature, whose amplitude decreases exponentially with depth, at a rate which depends on D, which is proportional to $\sqrt{\kappa}$. The lag between cycles of surface and subsurface temperatures is also determined by the magnitude of D and increases with depth. Figure 2.7, constructed from soil temperature measurements made at Normandin, P.Q., illustrates the effect of depth on the shape of the annual temperature wave. The decrease in the amplitude of the annual oscillation and the increase in lag with depth are clearly evident; the maximum temperature at a depth of 3 m occurs about 3 months later than that at 1 cm. In spring and summer there is a net flow of heat from the atmosphere to the ground resulting in a gradual downward warming. In fall and winter the process is reversed, with cooling beginning first at the surface.

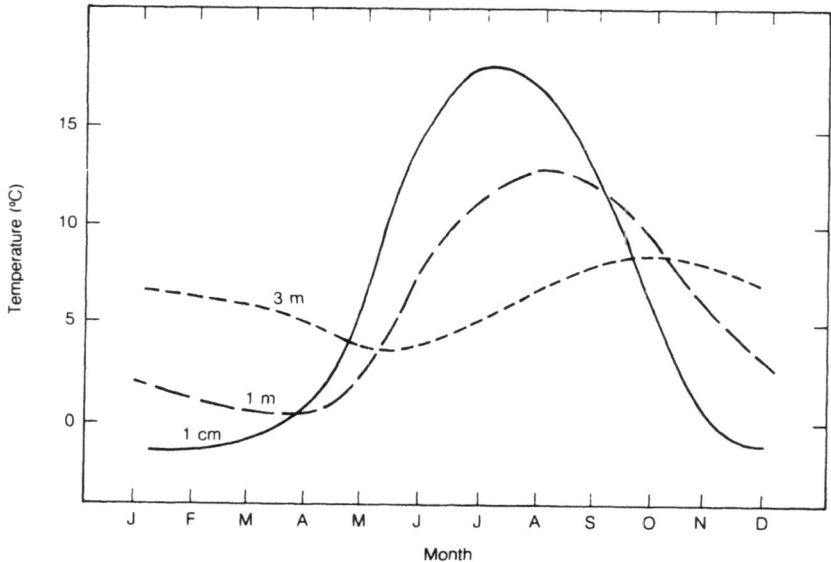

Fig. 2.7 Annual variation of soil temperature at 3 depths, Normandin, P.Q., (48°51'N, 72°31'W) (Aston, 1973).

Table 2.6 lists some typical values of the damping depth D for sand, clay and peat (each with different water contents) and snow. These data allow a comparison to be made of the relative ability of each material to damp surface variations; the larger the damping depth, the smaller the change in amplitude of the temperature wave at a given depth. While the heat flow through materials with low thermal conductivities is relatively small, it does not necessarily follow that surface temperature variations are propagated downwards less effectively, since the propagation also depends on heat capacity. For example, the thermal conductivity of new snow (k = 0.063 $W/(°C \cdot m)$) may be more than an order of magnitude less than that of clay with a volumetric water content of 0.2 (k = 1.2 $W/(°C \cdot m)$); however their volumetric heat capacities differ by the same order of magnitude (see Table 2.6), so that their corresponding ratios, the thermal diffusivities, are approximately the same: 0.56 versus 0.57 x $10^{-6} m^2/s$. Hence, for a common period, P, the two materials have approximately the same damping depth.

Equation 2.9 can also be applied to describe the cyclic variation in temperature at different frequencies. The relationship between damping depths for cycles with an annual period P and an arbitrary period P_1 is

$$D_1 = D\sqrt{P_1/P},\qquad\qquad 2.10$$

where D and D_1 are the corresponding damping depths. Equation 2.10 is valid only for material(s) having identical and constant thermal diffusivities. If the diurnal variation can be treated as periodic, its corresponding damping depth would be about 1/19 of that for the annual variation. This reduction in D means that the magnitude of a diurnal temperature variation decreases much more rapidly with depth than that of the annual cycle.

Although a snowcover reduces the available energy at its surface because of its high albedo to solar radiation and high emissivity in the spectral range of most terrestrial radiation, its insulative properties exert the greatest influence on the soil temperature regime. The thermal conductivity of new snow is roughly an order of magnitude less than that of most soils (see Table 2.6). As snow "ages" its thermal conductivity increases, but generally remains less than that of most soils. Snowcover acts as an insulating layer which reduces the upward flux of heat through the ground and its subsequent loss to the atmosphere, thereby resulting in higher ground temperatures than would occur if the ground were bare. The effect of snowcover on the ground temperature regime is greatest at the surface, becoming progressively less pronounced with depth. On an annual cycle the normal effect of snowcover is to decrease the amplitude of the soil temperature wave and to increase the mean temperature. The magnitude of the changes to the amplitude and mean temperature varies widely from one location to another because of variations in such factors as snow depth and duration, soil physical characteristics, aspect, and air temperature. Also, the thermal conductivity of a snowcover may exhibit wide spatial variability, depending in a complex way on snow depth and meteorological parameters. Williams and Gold (1958), estimated that the mean thermal conductivity at twelve stations in various parts of

Canada ranged between 0.15 and 0.39 W/(°C · m).

Because of the spatial variability in the snowcover properties affecting the ground temperature regime, soil temperature measurements made at a particular site can be assumed to be representative only for that location. Data obtained from point studies are valuable, however, since they provide some insight into the snowcover-ground temperature interaction. Figure 2.8 illustrates the association between air temperature, the temperature within the snowcover (10 cm below the snow surface) and the temperature 1 cm above the ground surface at Fort Greely, Alaska. Except for isolated intervals the temperatures near the ground remained substantially above the air temperature (mean difference 12.3°C) for most of the winter. The data reveal strong damping of the day-to-day air temperature fluctuations by the snowcover, and a lag between high and low temperature amplitudes in the air and the snow, this lag being as much as 26 to 36 hours near the ground (see interval Mar. 23-26, Fig. 2.8). Figure 2.9 shows the average air temperatures, the average ground temperatures at 1 cm, and the average snow depths

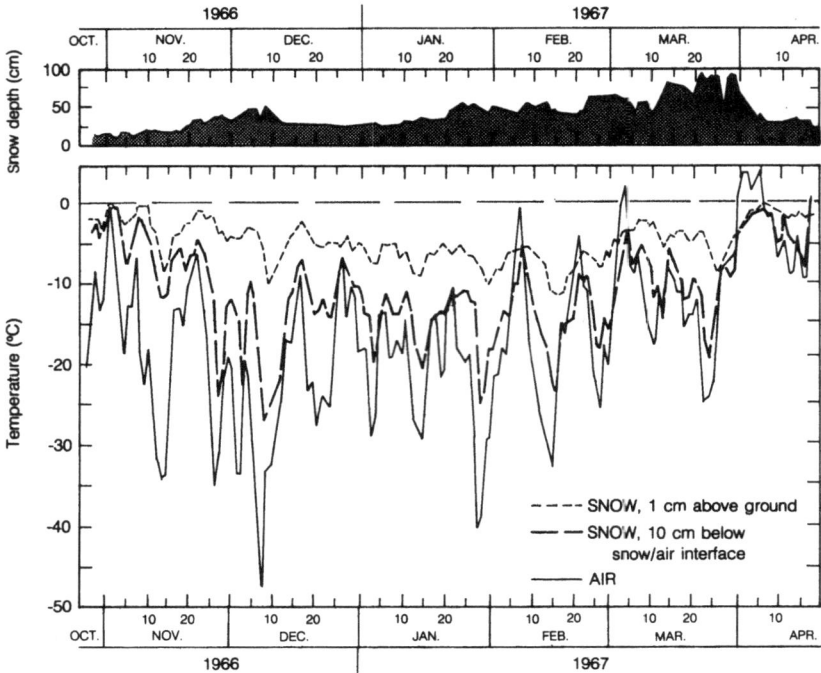

Fig. 2.8 Snow and air temperatures, Fort Greely, Alaska (Bilello et al., 1970).

measured at three locations: Fort Simpson, NWT., where snowcover is relatively deep; Ottawa, Ont., where it usually persists for several months (for a shorter period than at Fort Simpson) but is shallower; and Vancouver, B.C., where it normally remains for only a few days each winter. At Vancouver the

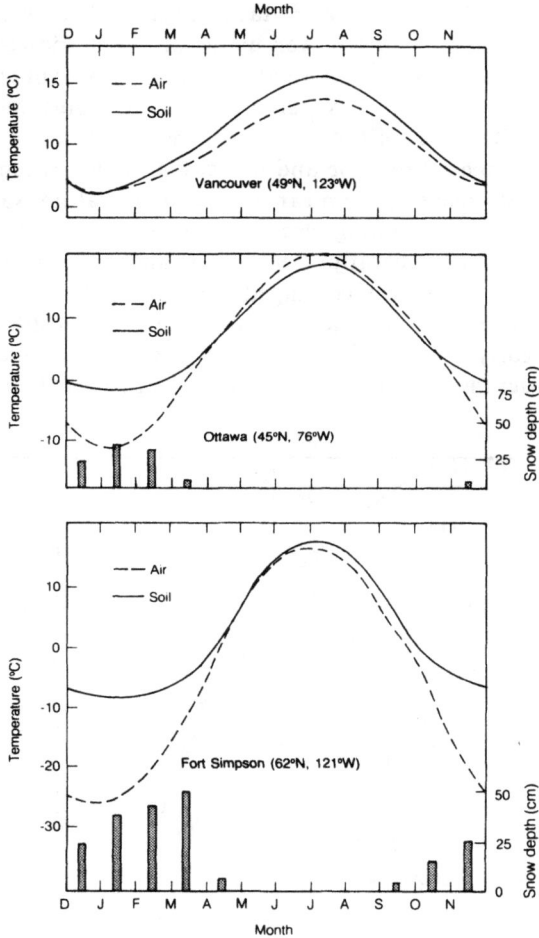

Fig. 2.9 Air (broken lines) and soil temperatures at 1 cm below the ground surface at Vancouver, B.C., Ottawa, Ont. and Fort Simpson, N.W.T. Periods of record vary from 8 to 30 years. Bars indicate average snow depths at end of each month (Potter, 1965).

ground and air temperatures are approximately equal throughout the fall and winter, whereas at the other two stations they differ, the divergence beginning at approximately the time of the first permanent snowcover. The temperature differences at Ottawa and Fort Simpson were greatest in mid-winter, decreasing to their minimum values approximately when snowcover disappears in late winter or spring.

Crawford and Legget (1957) at Ottawa, measured the soil temperature profiles in clay having a grass cover under different snowcovers. At one location the snow was left undisturbed throughout the winter, while at the other it was cleared from a circular area of 2-m radius surrounding the instrument site. Removal of the snowcover increased the amplitude of the annual ground temperature cycle and decreased the mean temperature (see Fig. 2.10). The decrease in the mean annual temperature was found to be 1.8°C near the surface, becoming zero at about 1.8-m depth. In February, the difference between the temperatures of the bare and snowcovered sites within the first 30 cm below the surface ranged from 6 to 8°C.

Figure 2.11 shows the snow and ground temperature profiles measured on one day during early winter, 1973, at two sites near Schefferville, P.Q., having different snow depths: one had a relatively thin cover of 10 cm; the other, a

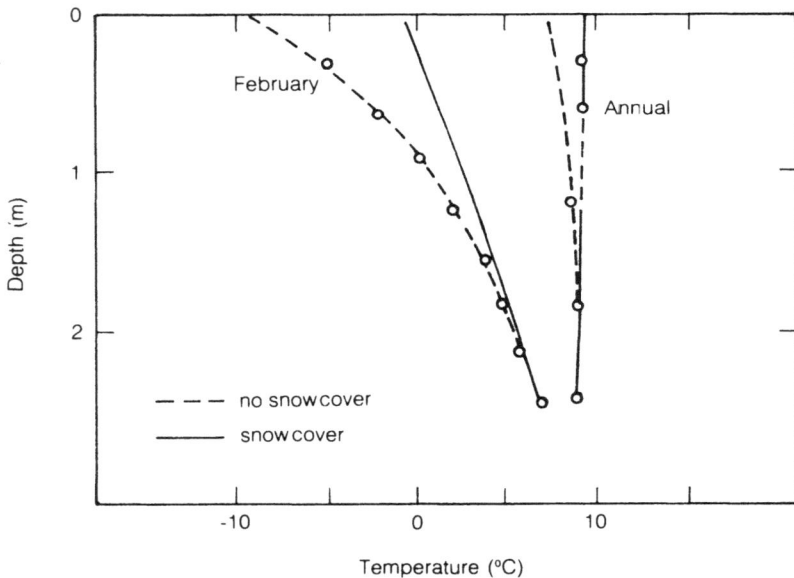

Fig. 2.10 Temperature profiles in clay soil at Ottawa, Ont. from May 1954 to April 1955 (Crawford and Leggett, 1957).

depth of snow greater than 1 m. The data illustrate: (a) a strong temperature gradient immediately below the snow surface, and (b) a pronounced decrease in the temperature gradient in the lower layers when the snow depth exceeds 20 to 50 cm.   Major departures from these profiles do occur, however, particularly in late winter or early spring when the temperature gradient in the pack may become isothermal.

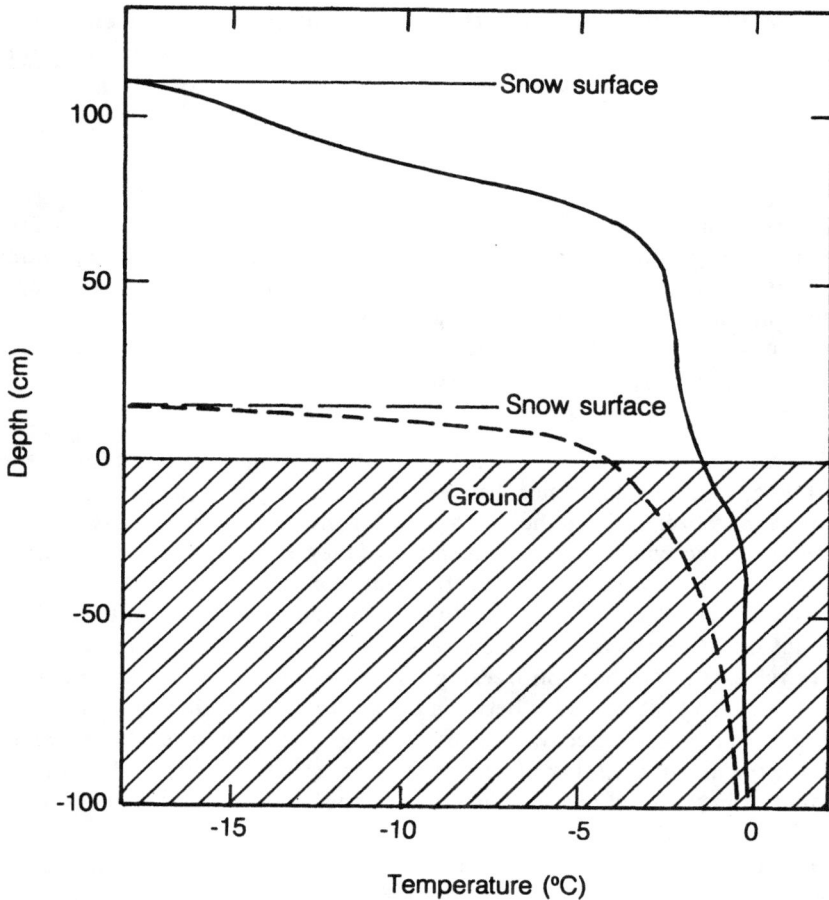

Fig. 2.11 Snow and ground temperatures for a typical day in early winter near Schefferville, P.Q. (55°N 67°W). Profiles are for two sites, one shallow snowcover, the other with relatively deep snow (Nicholson and Grandberg, 1973).

In the Schefferville area the mean annual ground temperature was found to be determined primarily by the snow depth; according to Nicholson and Grandberg (1973) the temperature at depth z had a linear relationship with snow depth ($\leqslant$ 1 m) over a circular area of radius 2 z. The authors suggest that such good correlations between snow depth and ground temperature may be partly due to interdependency between the two variables and other influencing factors. For example, ground water movement may cause soil temperatures to be highest in valleys, where snow tends to accumulate.

Judge (1973) estimated that on the average, ground surface temperatures in Canada are 3.3 °C higher than air temperatures. At least in part, this difference results from the insulative properties of snowcover.

Frost and Permafrost

Information about the depth of penetration of below-freezing temperatures beneath the ground surface, particularly in soils with fairly high water contents, is important to the proper design and construction of buildings, roads, pipelines and other engineering works. It is convenient to consider the effect of snowcover on the extent of below-freezing ground temperatures under non-permafrost and permafrost conditions.

Frost Penetration

The rate of penetration of frost into a soil depends on the net heat flux at the frost line, i.e., the difference between the heat flux propagated upwards through the frozen soil and the flux received from deeper unfrozen layers. The thermal conductivities and heat capacities of the frozen and unfrozen soil, the surface energy budget, and the thermal conductive properties of the snowcover are major factors affecting the magnitudes of these fluxes. The amount of unfrozen water in the soil is another important factor, since the water releases latent heat on freezing. The rate of frost penetration is very sensitive to the heat exchange between the ground and air during the freezing season, and is also affected by the heat exchange during the preceding summer because the net upward flux of heat to the surface in winter depends upon the amount of heat stored in the ground during the warm part of the year (Joynt and Williams, 1973).

Since the theory of frost penetration is extremely complex, in practice the penetration depth is often estimated empirically by relating it to a freezing index, usually taken as the cumulative total of degree-days below the freezing point; one degree-day is a departure of one degree between the freezing point and the mean daily temperature (U.S. Corps of Engineers, 1949). Table 2.7 (cf. Ministry of Transport, 1973), based on data from Canadian sites, gives the ratio of the depth of frost penetration z to the square root of the freezing index F for different snow depths:

Table 2.7
INFLUENCE OF SNOW DEPTH ON THE RATIO OF
THE DEPTH OF FROST PENETRATION z TO THE
SQUARE ROOT OF THE FREEZING INDEX, F -
CANADIAN SITES (Ministry of Transport, 1973).

Snow depth (cm)	5	10	20	30	40	50
Ratio: $z/\sqrt{F}$	1.07	0.95	0.80	0.68	0.55	0.48

These data indicate that for a fixed value of F the frost penetration under 30 cm of snow would be 64 percent of that under 5 cm. Caution should be exercised in using these values since they were derived from data exhibiting wide scatter, and significant factors such as the insulative properties of the snow were not considered. On the average, the snowcover experienced in Canada reduces the depth of frost penetration to about one half of that which would occur if there was no snow (Ministry of Transport, 1973).

Permafrost

Permafrost develops when the ground temperatures are below freezing for periods exceeding the annual cycle of air temperature. It lies beneath a large part of the earth's land surface. In North America this layer is almost entirely continuous throughout the north slope of Alaska, the Northern Yukon, the Arctic Archipelago and much of the District of Keewatin. South of this zone, discontinuous areas of permafrost occur mainly north of 55° N and in alpine regions.

The local snowcover regime can influence the permafrost considerably, particularly in the discontinuous zone. For example, the formation of a thick layer of snow early in the fall inhibits freezing of the soil by reducing the heat loss to the atmosphere. Conversely, its presence until late in the spring can decrease warming. In the Schefferville area of Quebec, permafrost frequently occurs in windswept, treeless, upland areas where snow is not retained to any appreciable depth, but is generally absent in lowland areas, except on those parts associated with shallow snowcover; e.g., muskeg, near hill crests and along lake shores (Grandberg, 1973). Grandberg suggested that, in the Schefferville area a snow depth of 65 to 75 cm is sufficient to prevent the formation of permafrost.

On a larger scale, Brown (1970) estimates that the mean annual ground temperature at the level of zero annual amplitude (5 to 15 m depth) is 3° C higher than the mean annual air temperature in permafrost areas. Wherever the mean annual air temperature is higher than - 4° C, permafrost is usually limited to areas with characteristics favouring its formation, such as peat bogs and north-facing slopes. Snowcover is probably a major cause of the higher ground temperature which limits the extent of permafrost in such areas.

LITERATURE CITED

Aston, D. 1973. *Soil temperature data 1958-72.* Rep. CLI-2-73, Environ. Can., Atmos. Environ. Serv., Toronto, Ont.

Berry, M.O. and R.G. Lawford. 1977. *Low-temperature fog in the Northwest Territories.* Tech. Memo. TEC 850, Atmos. Environ. Serv., Environ. Can., Downsview, Ont.

Bilello, M.A., R.E. Bates and J. Riley. 1970. *Physical characteristics of the snow cover, Ft. Greely, Alaska, 1966-67.* Tech. Rep. 230, U.S. Army Cold Reg. Res. Eng. Lab., Hanover, N.H.

Brown, R.J.E. 1970. *Permafrost in Canada.* Univ. of Toronto Press, Toronto.

Bryson, R.A. and F.K. Hare (eds). 1974. *World Survey of Climatology.* Vol. II, Elsevier Publ. Co., Amsterdam.

Burns, B.M. 1973. *Climate of the Mackenzie Valley-Beaufort Sea.* Clim. Stud. No. 24, Atmos. Environ. Serv., Environ. Can., Downsview, Ont.

Crawford, C.B. and R.F. Leggett. 1957. *Ground temperature investigations in Canada.* Res. Pap. No. 33, Nat. Res. Counc. Can., Div. Build. Res., Ottawa, Ont.

Deacon, E.L. 1953. *Vertical profiles of mean wind in the surface layers of the atmosphere.* Geophys. Memo. No. 91, Meteorol. Office, Great Britain.

DesMarrais, G.A. 1959. *Wind-speed profiles at Brookhaven National Laboratory.* J. Meteorol., Vol. 16, pp. 181-189.

Environment Canada. 1975. *Canadian normals, Vol. 3 - Wind.* Atmos. Environ. Serv., Environ. Can., Downsview, Ont.

Flohn, H. 1974. *Background of a geophysical model of the initiation of the next glaciation.* Quat. Res., Vol. 4, pp. 386-404.

Frost, R. 1948. *Atmospheric turbulence.* Q. J. R. Meteorol. Soc., Vol. 74, pp. 316-338.

Geiger, R. 1961. *Das Klima der bodennahen Luftschicht (The Climate near the Ground.* [English Transl. by Scripta Technica, Inc., Harvard University Press, Cambridge, Mass, 1966].

Grandberg, H.B. 1973. *Indirect mapping of the snow cover for permafrost prediction at Schefferville, Quebec.* North American Contribution to the Second Int. Permafrost Conf., Nat. Acad. Sci., Washington, D.C., pp. 113-120.

Holzworth, G.C. 1974. *Mixing depths, wind speeds and pollution potential for selected locations in the United States.* J. Appl. Meteorol., Vol. 6, pp. 1039-1044.

Joynt, M.I. and P.J. Williams. 1973. *The role of ground heat in limiting frost penetration.* Symp. Frost Action on Roads, Rep. Vol. 1, OECD, pp. 189-203.

Judge, A.S. 1973. *Deep temperature observations in the Canadian North.* North American Contribution to Second Int. Permafrost Conf., Nat. Acad. Sci., Washington, D.C., pp. 35-40.

Kukla, G.J. and H.J. Kukla. 1974. *Increased surface albedo in the Northern Hemisphere.* Science, Vol. 183, pp. 709-714.

Kung, E.C., R.A. Bryson and D.H. Lenschow. 1964. *Study of a continental surface albedo on the basis of flight measurements and structure of the earth's surface cover over North America.* Mon. Weather Rev., Vol. 92, pp. 543-564.

Lamb, H.H. 1955. *Two-way relationships between the snow or ice limit and 100-500 mb thickness in the overlying atmosphere.* Q. J.R. Meteorol. Soc., Vol. 81, pp. 172-189.

Ministry of Transport. 1973. *Estimating the depth of pavement frost and thaw penetrations.* Rep. CBED-6-266, Eng. Design Div., Ottawa, Ont.

Munn, R.E. 1966. *Descriptive Meteorology.* Academic Press, New York.

Namias, J. 1963. *Surface-air interaction as a fundamental cause of drought and other climatic fluctuations.* Proc. Rome Symp. Arid Zone Research, Changes in Climate, Unesco and WMO, Geneva, pp. 345-359.

Nicholson, F.H. and H.B. Grandberg. 1973. *Permafrost and snow cover relationships near Schefferville.* North American Contribution, Second Int. Permafrost Conf., Nat. Acad. Sci., Washington, D.C., pp. 151-158.

Petterssen, S. 1956. *Weather forecasting and analysis.* Vol. 2, McGraw-Hill, New York, Toronto, London.

Portelli, R.V. 1978. *Mixing heights, wind speeds and air pollution potential for Canada.* Clim. Stud. No. 31, Atmos. Environ. Serv., Environ. Can., Downsview, Ont.

Potter, J.G. 1965. *Snow cover.* Clim. Stud. No. 3, Meteorol. Branch, Canada Dept. Transport, Toronto, Ont.

Schneider, S.H. and T. Gal-Chen. 1973. *Numerical experiments in climatic stability.* J. Geophys. Res., Vol. 78, pp. 6182-6194.

Sellers, W.D. 1965. *Physical Climatology.* Univ. of Chicago Press, Chicago.

Stringer, E.T. 1972. *Foundations of Climatology.* W.H. Freeman and Co., San Francisco.

U.S. Army Corps of Engineers. 1949. *Report on frost investigations 1944-45.* Addendum No. 1, 1945-47, Frost Effects Lab., New England Div.

Van Wijk, W.R. (ed). 1966. *Physics of Plant Environment.* North-Holland Publishing Co., Amsterdam.

Vowinckel, E. and S. Orvig. 1965. *The heat budget over the Arctic Ocean.* Publ. in Meteorol. No. 74, McGill Univ. Press, Montreal, P.Q.

Williams, G.P. and L.W. Gold. 1958. *Snow density and climate.* Res. Pap. No. 60, Nat. Res. Counc. Can., Div. Build. Res., Ottawa, Ont.

Williams, J. 1975. *The influence of snowcover on the atmospheric circulation and its role in climatic change.* J. Appl. Meteorol., Vol. 14, pp. 137-152.

3

SNOW AND AGRICULTURE

H. STEPPUHN

*Division of Hydrology, University of Saskatchewan,
Saskatoon, Saskatchewan.*

SNOW AND IRRIGATION

Snow, a Source of Irrigation Water

Snow supplies at least one-third of the water used for irrigation in the world; it is a major source in 28 countries whose total annual irrigated area is approximately 185 million hectares (Table 3.1). Five countries (China, India, USA, Pakistan, and USSR) each annually irrigate a land area in excess of 14 million hectares (FAO, 1979a). Significantly, each also cultivates vast deserts through which flow large rivers originating from snow in the distant, high mountains. Table 3.2 lists some characteristics of four of these rivers.

Snow covering upland watersheds is water in storage which, on melting provides water downstream for irrigation and the growth of crops. Storr (1967), in a study of the precipitation regime of a watershed in the eastern slopes of the Canadian Rockies concluded that snowfall accounted for 70-75% of the annual total. Because snow accumulates during periods of low evapotranspiration, the relative contributions of its melt waters to mountain streamflow often are greater than for rainfall events, e.g., Goodell (1966) estimated that 90% of the annual runoff from the Colorado Rockies above 2740 m originated as snow.

Fortunately the melting of snowcovers and their release of water occur simultaneously with the demands of emerging summer crops. In the western United States, the South Fork of the Ogden River, which originates from a snow-rich watershed, contributes 75% of its average annual volume during the growing season, April to September (Work, 1955). In southern Asia, snowmelt from the Himalayas provides the only significant source of irrigation water to the summer crops of the Indus Valley between seeding and the occurrence of monsoon rains (Cantor, 1967). The hydrograph of Lee Creek in Canada typifies the flow distribution associated with snow-fed

Table 3.1

TOTAL IRRIGATED AREAS FOR EACH COUNTRY WHERE SNOW CONTRIBUTES TO THE WATER SUPPLY (FAO, 1975-78 Production Yearbooks).

Country	Year	Reported Area (x 1000 ha)	Country	Year	Reported Area (x 1000 ha)
China	1977	48,700[a]	Australia	1976	1,475
India	1972	31,590	Chile	1973	1,238
USA	1969	15,832	Bangladesh	1973	1,212[b]
USSR	1976	15,300	Bulgaria	1977	1,149[b]
Pakistan	1973	14,043	Peru	1971	1,110
Iran	1976	5,840	Burma	1976	984
Mexico	1976	4,816	Morocco	1972	850
Iraq	1963	3,695[b]	France	1970	539
Italy	1970	3,345[b]	Canada	1970	421
Spain	1977	2,893	Algeria	1977	285
Afghanistan	1973	2,400	Nepal	1971	181
Turkey	1972	1,939[b]	Yugoslavia	1977	143
Romania	1977	1,854[b]	New Zealand	1972	135
Argentina	1976	1,820[a]	Bolivia	1974	110

[a] FAO Estimate
[b] Total land provided with irrigation facilities

Table 3.2

EXAMPLES OF MOUNTAIN-BORNE, SNOW-FED RIVERS WHICH PROVIDE IRRIGATION WATER TO LARGE DESERT REGIONS (Peterson, 1970).

Country	Mountain Region	River Name	Drainage Area, km^2 (Length, km)	Mean Annual Flow, 10^6m^3	Desert Region
China	Tsing Hai (Kulun)	Hwang Ho	715,000 (4,350)	47,000	Ulan Buh, Hobq, Tengger
Pakistan and India	Kashmir (Himalaya)	Indus	958,000 (2,900)	313,000	Thar (Indus)
USSR	Pamirs (Himalaya)	Amu Dar'ya (Oxus)	326,860 (2,000 est.)	64,900	Kyzl-Kum Kara-Kum
USA and Mexico	San Juan (Rockies)	Rio Grande (Rio Bravo)	471,000 (3,030)	5,600	New Mexican

mountain streams supplying water for irrigation (Fig. 3.1); its major flows occur during the early parts of the growing season between mid-April and mid-June.

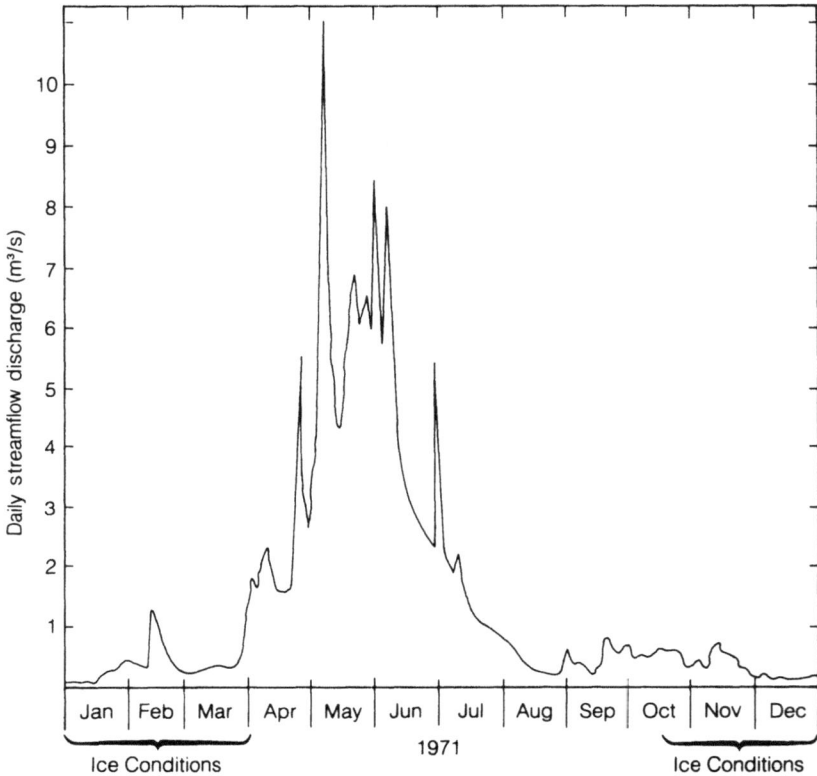

Fig. 3.1 Mean daily streamflow hydrograph for Lee Creek at Cardston, Alberta, drainage area 303 km^2 (Water Survey of Canada, 1972).

Irrigation Water Supply Forecasting

Israelsen and Hanson (1962) noted that the observation of snowcover to index irrigation water supplies is a common, world-wide practice. A variety of techniques that utilize snow survey and related data have evolved to predict the seasonal water volume, peak runoff rate, date of peak, and number of days that flow rates will exceed specified levels. The basic equation used to forecast the runoff volume R from a precipitation index P_i and a loss index, L_i is

$$R = P_i - L_i. \qquad\qquad 3.1$$

The U.S. Soil Conservation Service (1972) have used the following equation in operational practice to forecast runoff:

$$R = a + bB + fF + rP + sS, \qquad\qquad 3.2$$

where   B = base flow,
                     F = fall precipitation,
                     P = spring precipitation,
                     S = snow water equivalent, and
       a, b, f, r, s = regression coefficients.

The time required to develop a satisfactory index is usually 15 years or more.

Many interrelated variables (Table 3.3) have been used in forecasting. The effects of these variables on runoff are complex and distributive in time and space. Snow water equivalent is usually indexed by measuring the snow mass at one or more points with stationary, hydraulically-operated, "snow pillow" balances or by taking snow cores and weighing them with portable scales. A series of observations at one site constitute a snow course and are used to derive a mean value. Groundwater flow into a stream forms the base flow

Table 3.3
VARIABLES USED FOR FORECASTING (U.S. Soil Conservation Service, 1972).

Variable	Used As	Relation to Runoff	Relation to Peak	Significance	%
Snow water equivalent	P_i	Positive	Positive	Very high	60-90
Streamflow (antecedent)	P_i	Positive	Positive	Moderate	5-15
Base flow	P_i	Positive	Positive	Moderate	5-15
Soil moisture	L_i	Positive	Positive	Moderate	5-10
Precipitation					
Fall	P_i	Positive	Positive	Moderate	5-20
Winter	P_i	Positive	Positive	Moderate-high	30-60
Spring	P_i	Positive	Positive	Moderate-high	10-25
Air temperature	L_i	Negative	Positive	Moderate-high	10-25
Wind	L_i	Negative	Negative	Moderate	5-20
Radiation	L_i	Negative	Negative	Moderate	5-15
Relative humidity	L_i	Positive	Positive	Moderate	5-10

Note: This table summarizes the relative significance of each variable. The significance percentages indicate broadly what percentage of the variability accounted for by a multiple regression equation may be attributed to a specific variable.

Year	May 1 Watershed Evap. Index	June 1 Useable Snow Course Index	June 1 Snow Index	$10^6\,m^3$
1948	-2.2	20.3	18.1	136.9
1949	-3.1	20.0	16.9	125.8
1950	-5.0	29.5	24.5	188.7
1951	-6.3	30.2	23.9	183.8
1952	-5.5	25.4	19.9	155.4
1953	-1.6	17.4	15.8	122.1
1954	-8.9	23.3	14.4	109.8
1955	-4.1	16.5	12.4	96.2
1956	-9.6	33.4	23.8	187.5
1957	-0.6	24.7	24.1	182.6
1958	-4.0	22.8	18.8	138.2
1959	-6.7	19.8	13.1	98.7
1960	-4.7	19.0	14.3	102.4
1961	-6.5	15.5	9.0	60.4
1962	-8.0	30.7	22.7	175.2
1963	-3.6	19.0	15.4	117.2

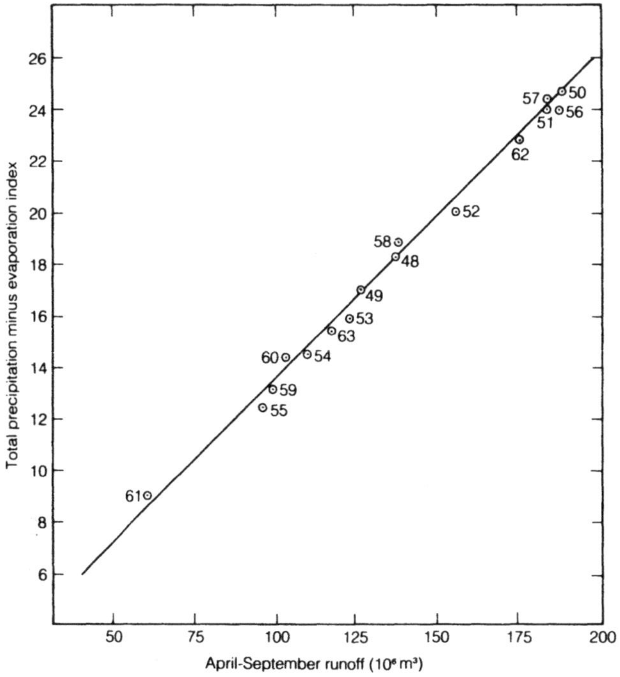

Fig. 3.2 Relation between June 1 watershed snowcourse index minus May 1 evaporation index and April-September runoff for Smith's Fork Watershed, western United States. The numbers beside the points indicate the year. (U.S. Soil Conservation Service, 1972).

which, along with fall, winter and spring rainfalls, contribute to the precipitation index P_i; soil moisture, air temperature, wind, radiation and humidity affect the loss index L_i. Taken together these indices may be used to develop a runoff forecast relationship such as that plotted in Fig. 3.2 (for Smith's Fork watershed).

Steed (1976) correlated mid-winter snow water equivalents obtained in the 303-km^2 Lee Creek Watershed in Alberta with streamflow volumes measured in a section above the Saint Mary's Irrigation Reservoir. He found the mean values from one index snow course statistically accounted for 76% of the yearly variation in the total annual flow volumes available for irrigation. Inclusion of spring and summer precipitation in the correlation increased the index of determination to 86%. Small mountain streams often show close correlations between snow-course measurements and runoff: Golding (1974) reported a mean index of determination of 88% for seven separate snow course regressions relating late March snowcover water equivalent and May-June runoff for the 94-km^2 Marmot Creek watershed in the Canadian Rockies.

The U.S. Soil Conservation Service (1972) issues four types of forecasts for:

(1) Runoff volume for a post-winter period from April through July, based on a snowcover index (Step 1, Fig. 3.3);
(2) Peak flow for one day based on an April to July runoff volume (Step 2, Fig. 3.3);
(3) Number of days that streamflow will exceed a specified rate and the date on which the flow will fall below a specified rate, based on peak flow relationships involving the number of days the flow exceeds a specific value (Step 3, Fig. 3.3) and the number of days between the date of the peak and the date that the flow falls below a specific rate (Step 4, Fig. 3.3);
(4) A hydrograph, based on estimates obtained in (1), (2) and (3).

Preparation of a forecast involves selecting index variables, observation sites and length of record. The number of snow courses required for an index depends on the variability of climate and topography within the forecast region; e.g., mountain basins may require snow courses within each elevation zone on both north and south-facing slopes. Often the choice narrows to one snow course or one precipitation measurement site. An indication of the minimum length of base record required to achieve a stable statistic can be obtained by comparing the annual values of the correlation coefficient for runoff and index snow water equivalent; for example, Fig. 3.4 shows that the statistic is reasonably stable after nine years of record.

Three general assumptions qualify water supply forecasts:

(1) Errors in observation and measurement are small relative to the sampling precision of the index,

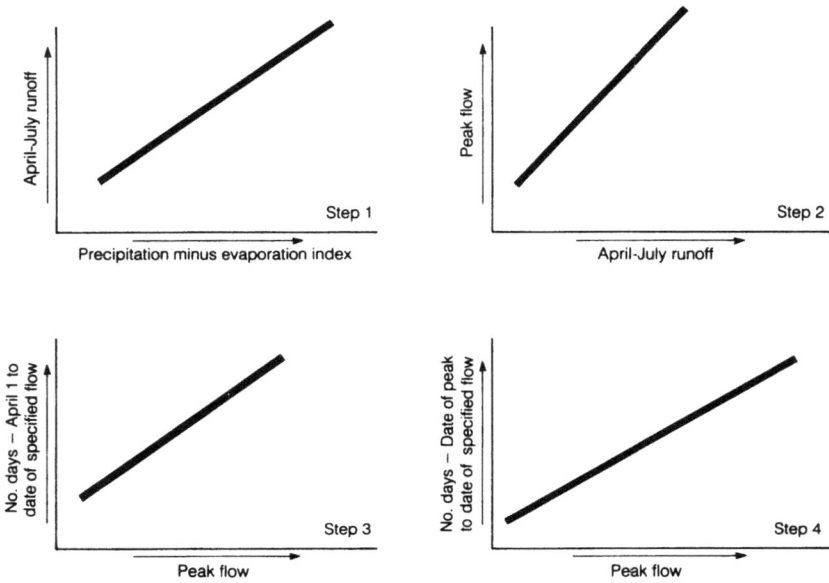

Fig. 3.3 The U.S. Soil Conservation Service (1972) snowmelt runoff forecast procedure for water volume (step 1); the peak one-day flow (step 2); the number of days above a specified flow (step 3) and the number of days between the peak flow and a specified flow (step 4).

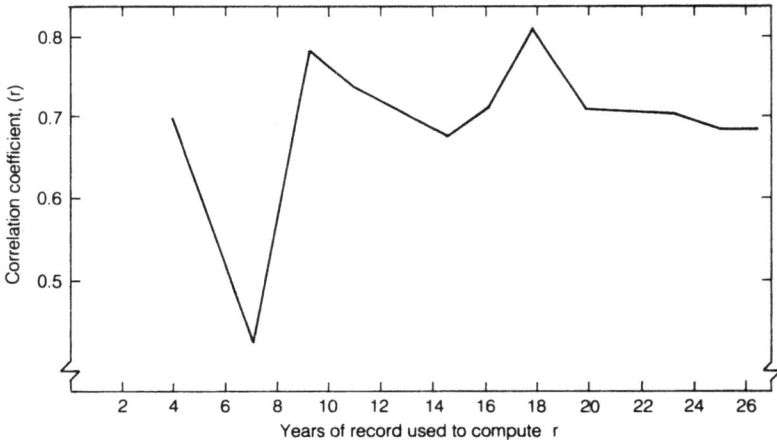

Fig. 3.4 Temporal variability of the correlation coefficient for runoff volume and an index snow water equivalent (U.S. Soil Conservation Service, 1972).

(2) Climatic variations are insignificant within the period of record, and

(3) Future relationships between index and runoff remain about the same as in the past.

The U.S. Soil Conservation Service (1972) estimated their forecast accuracy for runoff volumes during 1922-1963 as:

Date Forecast Issued	Number of Forecasts	Percentage Error[a]
1 February	1593	21.6
1 March	3980	20.0
1 April	6515	18.6
1 May	4928	15.6

[a] % error = 100 (forecast volume - observed volume)/(15-year mean volume).

Snow and Irrigation in Various Climatic Regions

Irrigation systems which depend on snow-fed waters have developed throughout the world under three general climatic environments: arid, seasonal and semi-arid.

Arid climate

The greening of deserts through irrigation with mountain-born snowwater occurs on all continents. One example is the irrigated region in the vast desert stretching across the Kyzl-Kum and the Kara-Kum in the southern Soviet Socialist Republics of Kazakhstan, Tadzhikistan, Turkmeniya and Uzbekistan. Climatic observations at Bairam Ali, Turkmeniya, SSR (Table 3.4) highlight the coincidence of high summer temperatures and scant precipitation characteristic of desert conditions. The snow-clad Pamirs and adjacent mountains supply nearly all the water to irrigate these desert lands which produce such high-value crops as cotton, alfalfa, sugar beets, wheat, fruits and vegetables (Cantor, 1967).

Seasonal climate

Between the Ganga and the Yamuna Rivers in India stretches one of the world's most productive agricultural areas. However, crop success depends heavily on melted snow from the Himalayas as the source of water for irrigation during the critical dry season. The precipitation record for Mainpuri in Uttar Pradesh (Table 3.4) typifies the region's climate and supports Kanwar's (1970) statement that northern India can expect no more than 20% of its annual precipitation to fall during the seven-month interim (November to May) between monsoon seasons. During the dry winters, cereals and vegetables are grown by snow-derived irrigation waters; in summer, millet, rice, cotton, sugar cane and maize are irrigated only when necessary.

Table 3.4

MONTHLY MEAN AIR TEMPERATURE T̄ (°C) AND PRECIPITATION
P̄ (mm) FOR FOUR STATIONS IN IRRIGATED REGIONS (Wernstedt, 1972).

	Bairam Ali Turkmeniya USSR		Mainpuri Uttar Pradesh India		Lyallpur Punjab Pakistan		Lethbridge Alberta Canada	
	37.44N 62.13E 10 year 230 m		27.14N 79.03E 30 year 157 m		31.25N 73.05E 30 year 80 m		49.42N 110.5W 30 year 908 m	
Month	T̄	P̄	T̄	P̄	T̄	P̄	T̄	P̄
Jan	0	17	15	17	12	16	-8	1
Feb	4	19	18	9	15	18	-7	0
Mar	10	29	23	12	20	23	-2	2
Apr	17	22	30	6	26	14	5	17
May	24	8	34	11	32	8	11	47
Jun	28	2	35	64	34	29	15	73
Jul	30	0	31	196	33	96	19	39
Aug	28	0	29	247	32	97	17	40
Sep	22	0	29	152	30	29	15	27
Oct	15	5	26	40	26	5	7	11
Nov	9	9	21	2	19	2	0	2
Dec	4	13	16	6	14	8	-4	1
Mean	16		26		24		5	
Total		124		762		345		260

The Indo-Gangetic Plain fronts the southern flank of the Himalayas for
some 2500 km, exhibiting a very wide range in humidity. Westward across the
Plain, the annual precipitation decreases gradually from a high of 1900 mm in
Bangladesh to 150 mm in Pakistan. As shown by the data, recorded at the
desert station of Lyallpur, Pakistan (Table 3.4) the precipitation regime of the
arid region exhibits the seasonally-skewed distribution characteristic of
monsoon Asia. Consequently, irrigation with water from the Himalayan-
born Indus River is essential during the dry winter, with the demand persisting
well into the "wet" monsoon season. Melt waters from mountain snows
irrigate the winter crops from November through March and initiate the
growth of summer crops from March until the yearly advent of monsoon rains
in June or July (Michel, 1967). The Indus Basin is at once an arid and a
seasonally irrigated desert.

Semi-arid climate

Irrigation in semi-arid climates generally supplements the limited
precipitation falling during the growing season. Most irrigation in Canada

takes place in semi-arid regions. Rapp et al. (1969) reported that the quantity of irrigation used on an average farm in the Vauxhall District of southern Alberta during the years 1958-1968 averaged 43% of the total water available for crops. Semi-arid climates are often relatively cool as shown by the temperatures for Lethbridge, Canada (see Table 3.4). Low air temperatures tend to reduce evapotranspiration, thus compensating for the limited precipitation available for irrigation.

Humid to arid climate

The agriculturally-vital Rio Grande River, which flows through arid zones in Mexico and the USA, originates from many humid mountain regions. From its source in the snow-capped Rocky Mountains to its release into the Gulf of Mexico, the Rio Grande irrigates a diverse array of crops. Mountain forests and pastures receive initial benefit as the melted snow replenishes local soil moisture, and as waters from high elevation streams are diverted into mountain meadows and hayfields to produce forage for livestock. The waters of the upper Rio Grande next flow onto the fields of the extensive, semi-arid San Luis Valley, irrigating potatoes, cereal grains and alfalfa, and supplementing summer rainfall. Some 400-500 km farther downstream, the river greens some 34,000 ha in the arid Middle Valley near Albuquerque, New Mexico. The remaining waters travel another 500 km and irrigate 74,000 ha seeded to small grains, alfalfa, sorghum and vegetables in the deserts surrounding El Paso, Texas. Another 900-km downstream the Rio Grande (called the Rio Bravo del Norte in Mexico) contributes supplemental water to over 40,000 ha of cotton, citrus and winter vegetables on the coastal Gulf Plain of Mexico and the USA (U.S. Bureau of Reclamation, 1961).

The Value of Irrigation

Irrigation has been credited with increasing yield, reducing crop failure, and allowing the cultivation of additional crop varieties (Cantor, 1967). Without it many of the world's most fertile areas would cease to produce, e.g., under natural rainfall the Sonoran Desert of North America will barely produce one cutting of alfalfa hay per year, a sizeable reduction from the six to eight cuttings possible under irrigation. In 1953, irrigated cropland represented 12% of the total area harvested in the USA, but generated 35% of the country's crop revenue (Greenshields, 1955).

The importance of irrigation water for crop production in desert regions is clearly shown by the yield curves of Fig. 3.5. Wheat yields increased quadratically with the amount of water applied, with maxima at about 863 mm/ha, spread over five irrigations. No appreciable precipitation occurred

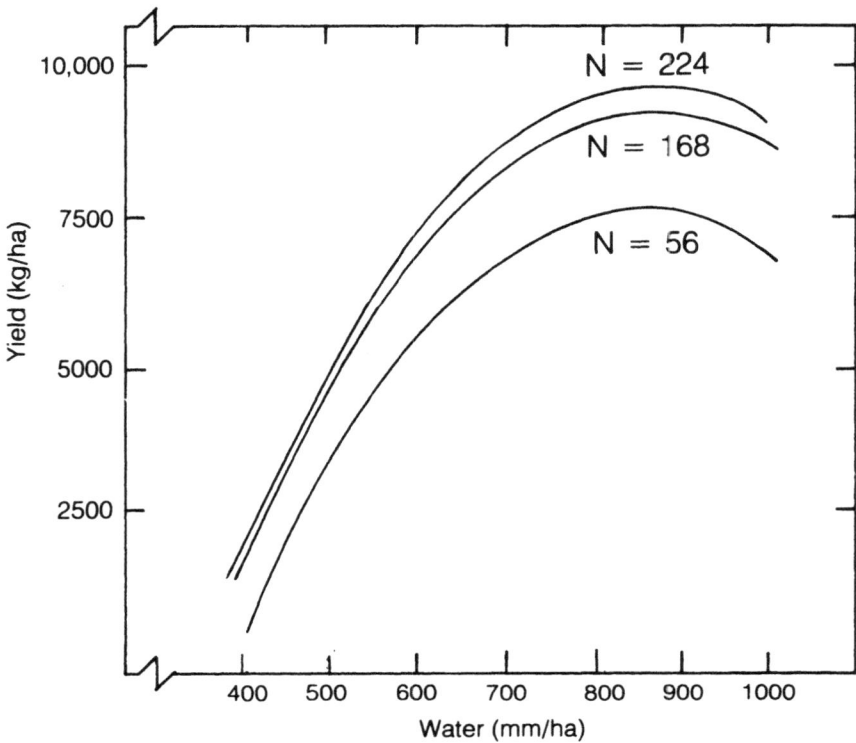

Fig. 3.5 Inia-66 wheat yield response curves to irrigation water inputs for different nitrogen levels of 56, 168 and 224 kg/ha, in the Yuma Valley, Arizona; 1971-72 (Hexem and Heady, 1978). Reprinted by permission from WATER PRODUCTION FUNCTIONS FOR IRRIGATED AGRICULTURE by Roger W. Hexem and Earl O. Heady (c) 1978 by The Iowa State University Press, Ames, Iowa 50010.

during the growing season, emphasizing the complete dependency and survival of the desert wheat crop on irrigation (Hexem and Heady, 1978).

Agricultural production in wetter climates also benefits from irrigation. Tables 3.5 and 3.6 list comparative yields with and without irrigation for selected subhumid Mexican and semi-arid Canadian crops. The respective production changes at the two locations resulting from irrigation ranged from 230 to 494% and from -12 to 120%.

Snow and Water Quality

The most-favoured water for irrigation generally contains the least dissolved solids. The chemical contents of waters from different natural sources listed in Table 3.7 indicate their comparative qualities. Entry of freshly-melted snow into a river system usually contributes the best water for most irrigation schemes (Kovda, 1973).

Table 3.5

AVERAGE YIELDS FOR SELECTED CROPS GROWN IN MEXICO, IRRIGATED AND NON-IRRIGATED LANDS, (United Nations, 1964, p. 162).

Crop	Yield (kg/ha)	
	Irrigated	Non-irrigated
Sugar cane	67,553	51,655
Potatoes	8,170	5,087
Rice	2,839	1,736
Wheat	1,972	1,037
Maize	1,512	913
Beans	1,077	430

Table. 3.6

EFFECT OF IRRIGATION ON THE AVERAGE YIELD OF DIFFERENT CROPS (kg/ha) IN SOUTHERN ALBERTA: 1955-1958 (Sonmor, 1955-58)

Crop (variety)	Non-irrigated		Irrigated	
	Non-fertilized	Fertilized[a]	Non-fertilized	Fertilized[a]
Alfalfa TDM[b] (Ladak)	2,240	2,580	9,860	10,740
Barley (Montcalm)	1,143	1,100	2,782	3,169
Soft wheat (Lemhi)	1,480	1,728	3,429	4,005
Hard wheat (Thatcher)	1,170	1,458	3,193	3,589
Sugar beets	18,200	20,440	44,590	51,740
Field corn[c]	3,230	3,140	6,860	7,840
Potatoes[d] (Netted Gems)	2,750	3,965	17,360	19,580

[a] Fertilizer treatment: 84-112 kg/ha ammonium phosphate (11-48-0)
[b] TDM: total dry matter
[c] 1955, 57 and 58.
[d] 1955, 56 and 58.

Table 3.7
CHEMICAL ANALYSES OF SELECTED WATERS (Hem, 1970).

					River		
Constituent	Snow (1) mg/l	Rain (2) mg/l	Ground water (3) mg/l	Ground water (4) mg/l	Base flow (5) mg/l	Snow-fed (6) mg/l	Rain-fed (7) mg/l
SiO_2	0.0	1.2	23	71	27	22	26
Fe	0.00	--	--	0.0	0.04	0.05	0.08
Ca	0.0	1.2	92	32	51	39	65
Mg	0.2	0.7	38	8.8	10	8.2	12
Na	0.6	0.0	} 110	42	46	25	63
K	0.6	0.0			3.7	3.7	4.3
HCO_3	3	7	153	161	161	115	176
SO_4	1.6	0.7	137	54	102	74	161
Cl	0.2	0.8	205	12	26	11	28
NO_3	0.1	0.2	83	0.6	1.3	1.9	1.7
Total dissolved solids	4.8	8.2	764	310	347	242	448
pH	5.6	6.4	--	7.9	–	--	--
Specific conductance (micromhos/cm at 25°C)			1302	404	526	366	665

(1) Snow, Spooner Summit, Highway 50 east of Lake Tahoe, Nevada, 2164 m, 20 Nov., 1958.
(2) Rain, Menlo Park, California, 90 m, 0800-1400 hr, 10 Jan., 1958.
(3) Irrigation Water Well (84 m) Alluvium, Salt River Valley, Arizona. 14 Aug., 1952.
(4) Domestic, composite from 7 wells (62-218 m) Sante Fe Formation, Albuquerque, New Mexico, 12 Oct., 1951.
(5) River, Rio Grande at San Acacia, baseflow, 1 Nov., 1941-28 Feb., 1942.
(6) River, Rio Grande at San Acacia, snowmelt runoff, 1 Mar.-30 June, 1942.
(7) River, Rio Grande at San Acacia, rain runoff and baseflow, 1 Jul.-31 Oct., 1942.

Snowcover runoff causes rivers to widen, to increase in flow rate, and to decrease in salinity. Unfortunately, the purity of this runoff component, as well as that of the rest of the river water, diminishes during its movement downstream. In spite of the deterioration in chemical quality with length-of-travel the beneficial effects of snowmelt for a river are often realized a great distance from the source. Figure 3.6 shows the decrease in the specific conductance of the Rio Grande at San Acacia, produced by the arrival of spring runoff in May at a location approximately 1200 km downriver of the major snowmelt sources.

In humid regions, snow on cultivated fields can be associated with runoff water of poor quality. Livestock producers often apply manure to the land

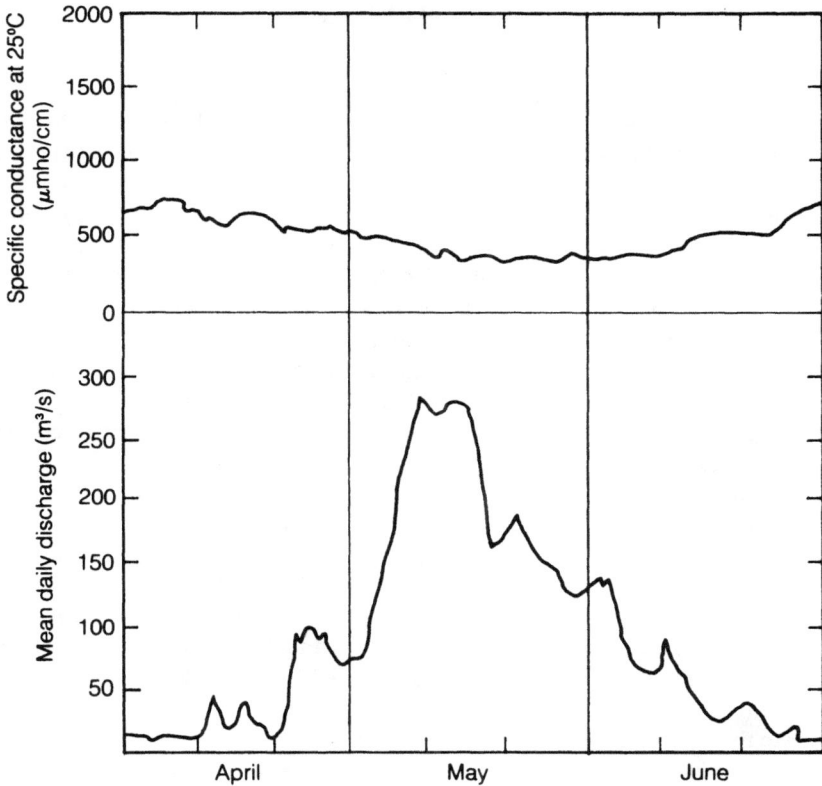

Fig. 3.6 Specific conductance of water samples and mean daily discharge of the Rio Grande at San Acacia, New Mexico, April-June, 1945 (Hem, 1970).

during winter, spreading the nutrient-rich material on top of the snow. If the snow melts rapidly, appreciable amounts of organic residues and solutes move off the fields and into streams. Frozen soil water increases surface runoff by blocking and hindering percolation. Martin et al. (1970) referred to their work and that of Holt's (1969), which indicated that greater amounts of phosphorous appeared in spring runoff from snowmelt than in runoff produced during other times of the year. Presumably, these increases come from the leaching of plant and manure residues that have been frozen over the winter and subsequently are removed in the snowcover runoff. The quantities, though small, were five to six times greater in the surface water than in water percolating through the soil. Hensler et al. (1970) have also detected nitrogen enrichments in waters derived from snow that has been in contact with livestock wastes.

SNOW AND FLOODS

Annual agricultural losses from floods are enormous and often significantly larger than those incurred by urban areas. Warner (1973) estimated that the total deprivation caused by the 1964 May-June flood in southern Alberta, to equal one million Canadian dollars. Ford et al. (1955) estimated the agricultural damage from the spring flood of the Kansas River in 1951 in the USA at 102 million dollars (US) and from the 1952 Missouri flood at 100 million dollars (US). The agricultural losses from floods in the United States in 1975 are estimated to exceed a total of 1,630 million (1967) US dollars (Water Resources Council, 1978). Weinberger (1961) reported that the estimated average annual flood losses in upstream agricultural watersheds in the USA were 513,904,000 in 1960 US dollars. If agricultural damage in the USA is taken as one-tenth of the world total, the annual cost of floods to agriculture exceeds 5,000 million dollars (US) at 1960 value.

In temperate climates major floods occur most frequently in late winter and spring (Harbeck and Langbein, 1949; Kates, 1961) and appear closely associated with the melt of snowcover. Fourteen countries have submitted data to a world catalogue of very large floods and indicated melting snow as a major cause. (Unesco, 1976). In the USA the bulk of the rural flood damage is concentrated in the agricultural interior within and downstream of the annual snowcovered areas (Water Resources Council, 1978). In a plains environment vast areal snowcovers and a wide variety of weather systems combine to increase the region's flood potential. During periods of rapid snowmelt, large quantities of water become available for runoff and downstream flooding. Snow accumulation data are important for flood control operations as shown by the reliance placed on snowcover depth and water equivalent measurements in the preparation of flood forecasts (Martin, 1960).

The consummation of a flood threat from the snowcover over a watershed requires energy fluxes sufficient to melt the snow at a rate exceeding the infiltration rate of the underlying soil. Investigations by Erickson et al. (1978) and others at prairie sites indicated that maximum 24-h snowmelt amounts rarely exceed 55 mm of water equivalent. Normally, most soils in an unfrozen state could infiltrate 55 mm/day (Ayers, 1959; Glymph and Holtan, 1969). When the soil is frozen, reduced infiltration rates can be expected (Gray et al., 1970) along with a substantial increase in flood threat (Bauder et al., 1975). Rostvedt and others (1968) studied a March-April 1962 flood across cultivated lands in South Dakota, and reported the following hydrological sequence:

1) Mid-January 1962:   Snowcover averaged 80-150 mm water equivalent.
2) 24 Jan. to 14 Feb.: 80-130 mm of snow water melted and infiltrated in-
                         to the soil, no runoff;

3) 16 Feb. to 16 Mar.: Air temperature remained below 0° C, freezing the infiltrated water to ice; snowfall added 130-200 mm of water to the snowcover;

4) 23-24 March: Rainfall added 8-13 mm of water, no runoff;

5) 27 March: Snow melts and runoff begins;

6) 29 Mar. thru April: Flood occurs with estimated damage at 3.8 million dollars (US).

It appears to be a principle of nature, that the accumulated snowcover melts when there is a maximum probability of the soil having its lowest infiltration potential.

The extent to which agriculture sustains damage caused by snow-related floods depends on various factors. Lyle (1961) separated factors causing damage into those associated with: the water (e.g., depth, duration, velocity and turbidity); and the land and crop (e.g., soil conservation practices, seed row direction, pasture grazing, treatment of crop residues, variety and vigor of the crop, and stage of plant growth). The different types of loss incurred by agriculture from floods includes:

1) Damage and destruction of the crop directly,
2) Damage to roads, buildings, fences, irrigation systems and drainage facilities,
3) Damage to the land by flood scour, streambank erosion, gullying, and sediment deposition on crops,
4) Increase (or decrease) in salinization of local soil,
5) Loss of stored crops and livestock, and
6) Losses that are indirect, such as delayed field work and reduction in the area cropped.

SNOW AND DRYLAND CROPS

Water Requirements

All agricultural products whether food or fiber, flower or oil, herb or medicine, need water. Pasture and field crops in particular require ample supplies of water derived from soil storage; any scarcity tends to limit production. The amount of soil water required depends on the type of crop, the duration of the growing season, and the evaporative demand of the ambient environment. For example, growth of alfalfa hay used 648 mm in the cool, semi-arid climate of Alberta (Sonmor, 1963), 762 mm in the warm, semi-arid region of Kansas (Hanson and Meyer, 1953), and 1,333 mm in the hot desert of Arizona (Harris, 1951). Potatoes grown adjacent to the alfalfa required about 20% less water at each location.

Much of the world's food, forage and forests are produced in cool, arid and semi-arid environments, such as found in the USSR, China, North America, and the highlands of South America. Thus, the availability of water in these regions during the crop growing season influences the world supply of agricultural commodities. Figure 3.7, indicates a pronounced trend in the relation between wheat production and snowcover in the Northern Hemisphere. Using data reported by Cole (1938) and Staple and Lehane (1954), Willis et al. (1969) estimated that the addition of 10 mm of water above a precipitation base of 200 to 250 mm can increase North American spring wheat yields by 200 kg/ha.

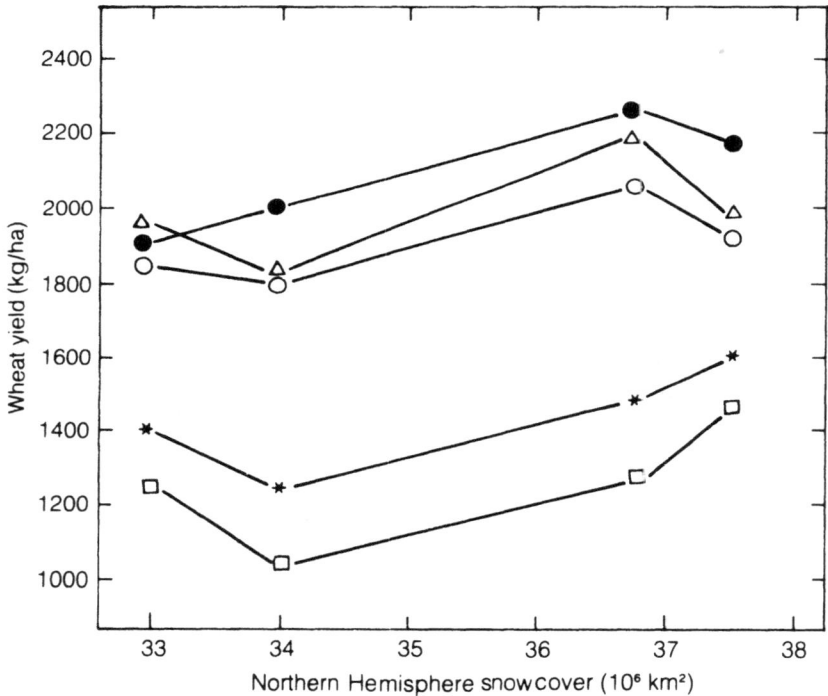

Fig. 3.7 Wheat (spring, winter, durum) produced in various regions as related to snowcover of the Northern Hemisphere during 1968-72, (Willis and Frank, 1975). 1 —•— Wash., Oreg., Idaho, Mont., N. Dak., S. Dak., Minn., Wis.; 2 —o— 1 plus Canada; 3 —△— 1 and 2 plus Wyo., Colo., Nebr., Kans.; 4 —□—U.S.S.R.; 5 —∗— 1, 2, 3 and 4.

Data establishing the field water requirements to grow specific crops under dryland conditions vary regionally, depending on climate; and yearly, depending on weather. Staple and Lehane (1952, 1954) presented 12 years (1938-50, less 1943) of results from field-tests of spring wheat on seven farms in the semi-arid region of southwestern Saskatchewan. Bare survival of the wheat plants, which produced only 60 to 120 kg/ha grain, required 125 to 150 mm of water obtained either from growing-season rainfall (measured by a rain gauge) or from stored soil water (computed as the difference between the soil water content of samples obtained at harvest and seeding). With additional water up to a total of 262 mm, grain yield increased slowly to 942 kg/ha, with each subsequent 25 mm (until a maximum near 412 mm) yielding an additional 230 to 400 kg/ha. de Jong and Rennie (1969) reported that spring wheat yields of 200 to 275 kg/ha were obtained for each additional 25 mm of water (above the long-term precipitation normal) in the more humid east-central region of Saskatchewan where potential evapotranspiration rates are lower. Significant water-to-yield correlations have also been observed for winter wheat and sorghum grown under semi-arid, dryland conditions (Smika and Whitfield, 1966).

The success of fertilizer applications has also been linked to the availability of stored soil water (Warder et al., 1963; Anderson and Read, 1966; de Jong and Rennie, 1969; Read and Warder, 1974). To be effective, nutrients from fertilizers must be dissolved in the soil water so that they may be absorbed by plant roots, with the water being used for transpiration. Consequently, limitations in available soil water also limit nutrient up-take by plants.

A common misconception declares that fertilization leads to larger plants which require more water, thereby resulting in greater crop moisture stress. However, studies in semi-arid environments consistently showed otherwise. For example, Bauer and Young (1966), demonstrated that the water-use efficiency (quantity of grain produced from one unit of water) of spring wheat increased under fertilization. Although driven by the evaporative demand, transpiration processes are controlled by soil water availability. Fertilized plants, grown where soil water is plentiful, use more, simply because it is available. Water-stressed plants use less, whether fertilized or not. According to Bole and Pittman (1978), spring barley in southern Alberta behaved similarly; with an average growing season rainfall of 117 mm and a soil-water storage at seeding time of 100 mm, nitrogen fertilizer, applied at levels of 0, 60, 120 and 180 kg/ha, produced grain yields of 1500, 2100, 2450 and 2500 kg/ha, respectively (Fig. 3.8). For an initial soil-water of 200 mm, yields were 2050, 2900, 3450 and 3650 kg/ha. These values represent comparative water use efficiencies of 6.9, 9.9, 11.3 and 11.5 kg/mm for 100 mm of water storage and 6.4, 9.2, 10.9 and 11.5 kg/mm for 200 mm of storage.

The importance of soil water to agriculture in semi-arid climates can be

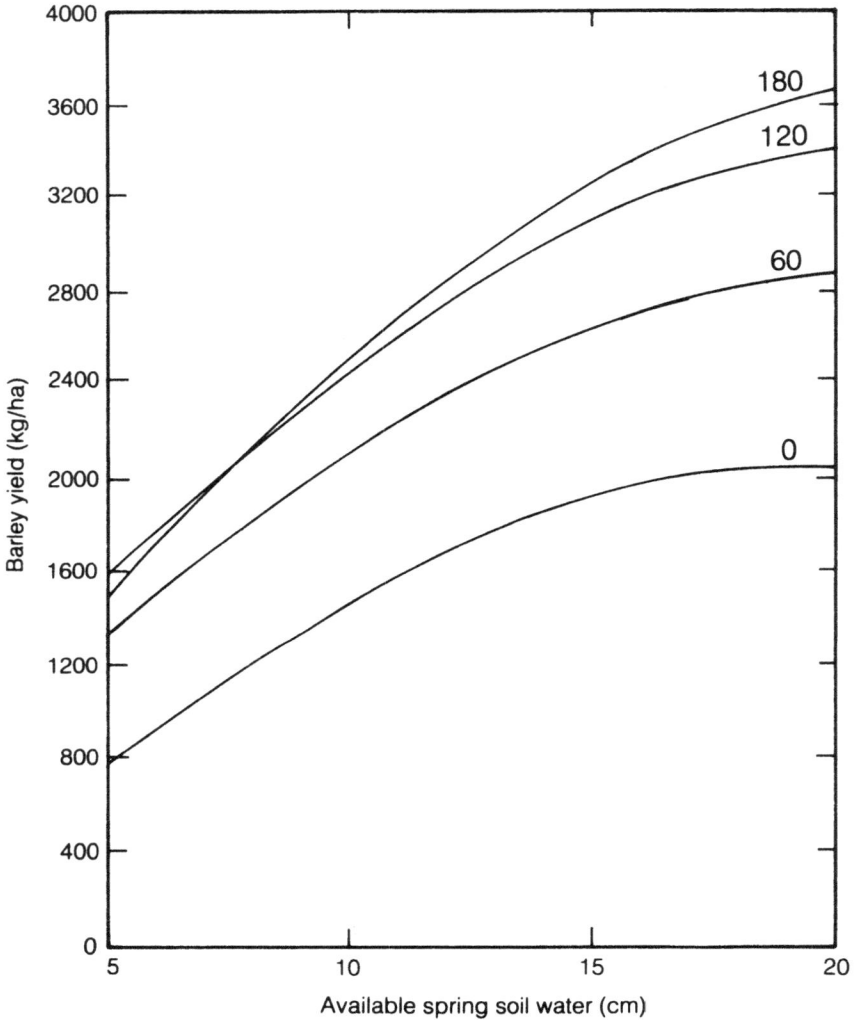

Fig. 3.8 Increase in yield of barley (kg/ha) as a function of available soil water at four nitrogen fertilizer levels: 0, 60, 120 and 180 kg/ha, with an average (117 mm) June - July precipitation (Bole and Pittman, 1978).

detected in precipitation records. Eleven years (1968-78) of data for the semi-arid region of Saskatchewan have an average growing-season (16 May - 15 August) rainfall of 149 mm with 260 mm as a maximum. Obviously, to provide the 262 to 412 mm of water needed (according to Staple and Lehane, 1954) to produce an economical spring wheat crop, requires significant withdrawal of stored soil water supplied by over-winter precipitation. The Saskatchewan Advisory Council on Soils and Agronomy (1978) and the College of Agriculture, University of Saskatchewan (Austenson et al., 1974) recommend that the root-zone soil water supply should be 100 to 125 mm at seeding time to achieve a satisfactory yield.

The production of wood fiber also demands adequate water supplies. Trees large enough to produce lumber-sized material require a growth period of about 150 years in the semi-arid highlands of northern Mexico compared with 60 to 80 years in humid coastal regions. The growth of wood in dry climates depends so closely on the presence of water that a technique has evolved for detecting historical fluctuations in climate, based on the comparative widths of the annual growth rings in tree trunks and limbs. It is more than coincidental that such a dendrochronological dating technique, as pioneered by A.E. Douglass, was developed in the arid regions of the southwestern United States and northern Mexico (Fritts, 1976).

Summerfallow Soil Water from Snow

Perhaps the most widely-used practice for storing adequate soil water for crop production in semi-arid climates is summerfallowing, i.e., one summer crop is sacrificed to conserve soil water for a subsequent growing season. The procedure usually includes three or more summer tillage operations to retard weed growth and to establish a stubble mulch from crop residues. Since weeds transpire the water which summerfallowing attempts to conserve, they must be controlled either mechanically by tillage (Anderson, 1967), or chemically by herbicides (Molberg and Hay, 1968). Janzen et al. (1960) showed (Table 3.8) that delay in controlling early spring weed growth reduces the quantity of soil water conserved by summerfallowing, and decreases the spring wheat yield by ~5% for each week's delay.

The effect of summerfallowing on water storage and utilization has been studied extensively. Staple and Lehane (1952) presented water conservation data for southern Saskatchewan (Fig. 3.9) obtained from studying seven different 21-month periods of summerfallow between 1939 and 1950. They found average soil water storage amounts of 36, 33, 13 and 16% of precipitation for the respective periods covering the first fall, (August-October), the first winter (November-April), the summer and fall (May-October) and the second winter (November-April). The average conserved in 21 months was 21%. They attributed the greater water conservation efficiency of the first winter to an increase in snow retention because of the standing stubble. The data exhibited wide annual variations showing occasional net

Table 3.8

EFFECT OF WEED GROWTH ON MOISTURE CONSERVATION IN SUMMERFALLOW AND ON THE YIELD OF THE FOLLOWING CROP, THREE-YEAR AVERAGE (Janzen et al., 1960). Reproduced by permission of the Minister of Supply and Services Canada.

Treatments	Soil Moisture Stored mm	Comparative Wheat Yields %
Weed growth prevented	130	100
First cultivation May 15	114	88
First cultivation June 15	91	78
First cultivation July 15	48	47

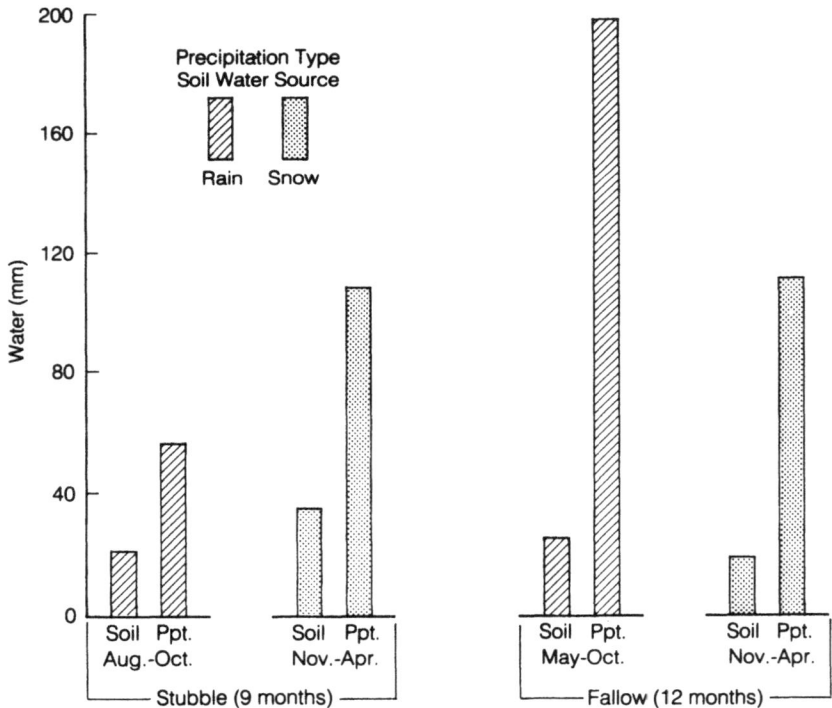

Fig. 3.9 Average depth of soil water stored (soil) and precipitation type and amount (Ppt.) in different intervals during a 21-month summerfallow period; data for seven 21-month periods, 1939-1950, in southern Saskatchewan obtained from Staple and Lehane (1952).

soil water losses during the summer and fall and the second winter, but never during the first winter. In a study covering twenty years of record, Staple et al. (1960) found that the average over-winter gains in root-zone soil water amounted to 37 and 9% of the winter precipitation for first and second fallow-winters. Smika and Whitfield (1966) analyzed soil water data obtained in summer-fallowed fields at North Platte, Nebraska during the years 1907-1941 and 1961-65 and concluded that, "very little moisture was stored in the soil between the first of July and winter-wheat seeding time in September". Snow retained by the standing stubble was cited as the major source of water conserved by summerfallowing.

Water from Snow for Dryland Crops

Dryland crops consist chiefly of cereal grains, oilseeds and forages. Greb (1975) listed eight studies, including his own, in which the recharge of soil water by melted snow was considered highly significant to the overall water economy of these crops. Soil water measurements at the Bad Lake Research Basin in the semi-arid zone of Saskatchewan (see Fig. 3.10) clearly indicate (for a pasture) the annual recharge of water by melted snow and its subsequent withdrawal by evapotranspiration. The quantity of water available from snow to support dryland crops depends primarily on: the frequency, depth, distribution and duration of the local snowcover, the weather conditions fostering snowmelt and the soil properties affecting infiltration and storage.

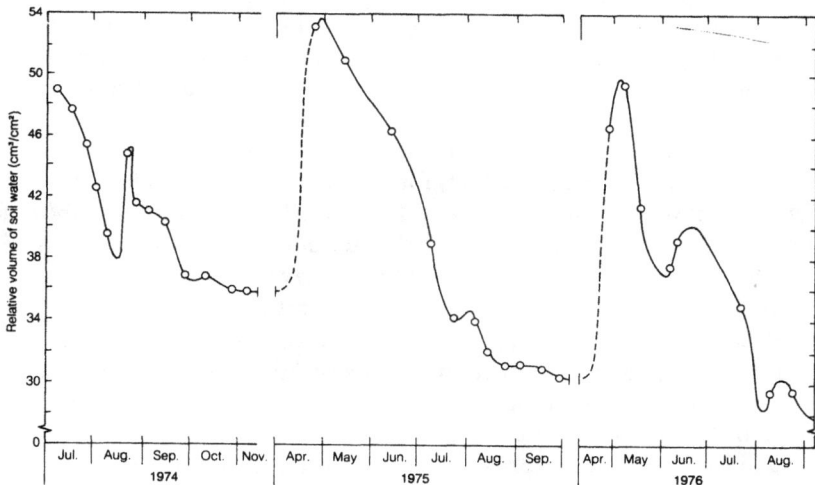

Fig. 3.10 Relative volumes of soil water measured by the neutron back-scatter method in a 122-cm profile of soil sown to perennial grass; Bickleigh, Saskatchewan.

The percentage of snowmelt water entering soil storage generally reflects a region's aridity. Staple et al. (1960) evaluated 20 years of data and related the change in over-winter soil water between harvest and seeding C, to the winter snowfall S and the fall post-harvest rainfall R, for summer-cropped and fallow fields in southwestern Saskatchewan. They derived the following relationships:

stubble: $C = 0.38S + 0.34R + 0.76$,   3.3

fallow: $C = 0.009S + 0.43R - 3.81$,   3.4

where all terms are expressed in millimetres. For stubble with a mean root-zone water content of 33 mm (Eq. 3.3), C was equally affected by S and R (since their corresponding regression coefficients were about the same). Under the wetter fallow (mean root-zone soil water content of 101 mm) the effect of snowfall on C was almost negligible, reflecting the inability of the soil surface to retain a snowcover and/or infiltrate its meltwater. In central North Dakota which is less arid than southwestern Saskatchewan, Willis et al. (1961) found that 48 to 77% of the premelt snowcover on test plots underlain by frozen soils became surface runoff. Bauder et al. (1975) measured spring runoff arising from snowmelt on plots in eastern North Dakota which is more humid, i.e., both the total annual precipitation and winter snowfall are greater than in central and western regions; they reported that 80% or more of the winter precipitation became runoff.

The soil water additions from snowcover appear to be maximized in the dryer fields (Staple et al., 1960). Therefore, snow water contributions to dryland crops tend to be greater for farming operations with less summerfallow and more annual cropping. Black and Siddoway (1975) and others further demonstrated that any operation attempting to optimize the use of snow water to increase yields in a dryland area must also be coupled with an adequate fertilizer program.

The type of crop grown also influences the utilization of the water derived from the snowcover. Hay, pasture, forests, winter wheat and fall rye use the infiltrated snow water immediately upon regrowth in the spring. On the other hand, a spring-seeded crop needs time to develop an extensive root system before it can fully utilize the water; and this delay increases the losses caused by percolation below the root zone and evaporation.

Post-harvest tillage of spring-seeded dryland crops is commonly viewed as a practice which promotes infiltration of over-winter precipitation. Lal and Steppuhn (1980) reviewed the research pertaining to fall tillage in the colder regions of the North American Plains, but failed to discover any study that concluded that fall tillage proved effective in increasing the infiltration of winter precipitation, except for the rare case when a severe early frost froze the soil water in the surface layer upon which rain or snowmelt formed a relatively impervious ice layer. Fall tillage also tends to dislodge standing stubble, reducing snowcover retention.

Snow plays a key role in the concept for minimum tillage cropping. The availability of effective herbicides has reduced the need for tillage to control weeds so that less stubble is disturbed, more snow is trapped and evaporation is reduced. Deibert (1979) compared the fall to spring (non-growing season) soil water gains of four pre-seeding tillage systems in central North Dakota; 48 mm for the summerfallowed field; 97 mm for the spring-cultivated field; 135 mm for the spring-plowed field; and 183 mm for the non-tilled field.

Snow Management in Dryland Areas[1]

The snow covering dryland agricultural areas represents a significant source of manageable water. A 30-cm snowcover with a density of 375 kg/m^3 spread over the northern half of the North American Great Plains (1,865,000 km^2) contains approximately 213.5 x 10^9 m^3 of water, an amount equal to the mean annual volume (1958-73) flowing from the North American Great Lakes into the St. Lawrence River (Water Survey of Canada, 1974). Obviously, a volume of such magnitude presents great potential for increasing the available crop water. In fact, within the hydrological cycle, snow is the only source of water than can be easily manipulated without broad-scale alterations, e.g., by cloud seeding or surface water diversions. The characteristics of snow advantageous to its conservation by management practices include its availability, its susceptibility to redistribution by wind, and its control by intensified agronomic and silvicultural practices.

One of the aims of dryland agronomy is to effect maximum crop yield at the lowest possible unit cost by optimizing the factors affecting production such as land use, soil management, pest control, machinery utilization, and market assessment. The manipulation of snowcovers to direct and accumulate water in desired locations offers agronomy a direct water management opportunity. The widespread practice of summerfallowing as part of dryland cropping rotations attests to its economic success but requires sizable inputs of land, machinery, fuel and labor to effect water conservation. Measurements taken in southwestern Saskatchewan in spring have shown that about an additional 45 mm soil water is available in fallowed fields compared with cropped fields (1971-77 data summarized by the Agrometeorology Research and Service Section, Agriculture Canada, Ottawa). If, as suggested by Willis and Carlson (1962), deJong and Rennie (1969) and others, snow management can increase over-winter soil water in an amount approaching 45 mm (approximately 15 cm of snow), the need for summerfallowing diminishes and more land becomes available for production. An increase in the soil water status also allows: the growth of more water-demanding crops; an increase in the grazing

---

[1]   The reader should refer to material pertinent to this topic contained in Chs. 5, 8, 9 and 16.

capacity of pastures and rangelands (Wight et al., 1975); a more effective utilization of applied fertilizers (Dubetz, 1961); the inclusion of legume or green-manure crops in yearly rotations to reduce costly fertilizer needs (Bowren et al., 1969); and the harvesting of straw for paper and alcohol production or other purposes which otherwise must be left to protect fallowed soil against wind erosion (Coxworth et al., 1978).

Snow management to augment soil water involves two objectives: (1) trapping the snow where it is needed, and (2) holding the melted snow until the soil thaws so that the water can infiltrate. Generally, in dry climates the first objective is more important than the second, while in wetter climates, the second becomes significant.

Management to trap snow on cultivated fields and pastures involves regulating the wind. The forces that keep a particle of snow airborne result from the wind's turbulent structure. Within any horizontally-moving wind stream, turbulence causes individual elements of air to move in all directions, the upward motion imparting the force to buoy the snow particles. An increase in the horizontal speed strengthens the upward-directed force and the capacity of air to transport snow. Any obstacle or barrier protruding into the wind stream produces a local, near-motionless region of air incapable of carrying snow particles, thereby causing their deposition. Thus, a carefully-designed system of regularly-spaced wind barriers can cause and direct the deposition and accumulation of snow wherever desired. Barriers also lessen any subsequent entrainment of the trapped snow by the wind.

To assess the potential of snow management practices for augmenting soil water the surface wind speeds and frequencies of occurrence must be considered. Evidence that sufficient wind occurs in dryland environments to effect snowcover manipulation has been provided by Kuz'min (1960), McKay (1963) and Steppuhn and Dyck (1974). These studies, concerned with measurements of areal snowcover depth and water equivalent, recognized the influence of landscape features (e.g., terrain, vegetative cover and land use) on the distribution patterns of snow in windy regions. The snowcover depth statistics given in Table 5.7 (Ch. 5) attest to the non-uniformity of a prairie snowcover.

Consideration must also be given to available snowfall and snowcover. Willis (1979) cited data from the "Climatic Atlas of the United States" (U.S. Department of Commerce, 1968) in reporting mean annual snowfall totals ranging from 60 to 100 cm over the northern Great Plains in the USA. Although these totals vary widely, they averaged 20 to 25% of the annual precipitation totals. Records from six climatological stations across the Canadian Prairies during the period from September 1974 to the end of August 1975 revealed seasonal snowfall water equivalents between 78 and 137 mm, representing 20 to 33% of the total precipitation during the period (Atmospheric Environment Service, Monthly Records, Sept. 1974 - Aug. 1975).

The duration of a snowcover varies depending on the mass accumulated, the frequency and duration of melt and its susceptibility to wind transport. To examine the snow management potential of a region it may be advantageous to divide it according to snowcover persistence. McKay (1964) mapped two types of regional snowcover that occur over the Canadian Prairies; to which a third can be added:

(1) Western region snowcover: responds to mid-winter melts, frequently disappearing and forming throughout the winter; commonly patchy or discontinuous,

(2) Central region snowcover: experiences only occasional mid-winter thaws, so that mid-winter disappearances and formations are infrequent; often very thin and variable, and

(3) Northern and eastern region snowcover: rarely subject to winter thaws; generally remains continuous throughout the winter.

The amount of snow available for management depends on the amount that sublimates during wind transport (Dyunin, 1959; Tabler, 1971). Tabler (1973) estimated that under favorable conditions 52% of the annual snowfall in a wind-prone environment could sublimate before melting. Observations in Kazakhstan, SSR led Rylov (1969) to suggest that sublimation of snow is related to land use; as shown by the increased loss from cultivated dryland compared to that from virgin prairie. Steppuhn and Gray (1977) applied a theoretical sublimation model to hourly meteorological observations collected during the 1974-75 winter at six stations on the Canadian Prairies. The total average winter snowfall was 105.8 mm; estimates of the average maximum seasonal amounts of snow that could have been transported by wind and sublimated were 10.2 tonnes per metre width and 235 mm, respectively (Table 3.9). Potential sublimation at four of the six stations exceeded the measured snowfall; however, the actual amounts sublimated were less because the potential values assume that an unlimited supply of snow moves in a saturated air stream.

In order for the "managed" snow to be utilized by crops its meltwater must infiltrate and be stored in the root zone until needed. Many investigators have recognized that soil water content is an important factor governing infiltration in frozen soils. Larkin (1962) indicated that if a soil is frozen at a water content greater than field capacity, its infiltration rate will be very low, while if it is saturated, the intake rate will approach zero. Mosienko (1958) observed good infiltration into a soil when its water storage was less than about 60% of field capacity. In studying water entry to frozen prairie soils, Gillies (1968) found that the infiltration rate varied as an inverse exponential function of the initial soil water content of the surface layer (0 to 5 cm depth). Gray et al. (1970) presented several possible shapes (Refer to Fig. 9.12, in Ch. 9) describing the time variation in the infiltration rate of frozen soil based on

Table 3.9
ACCUMULATED SNOWFALL, HOURLY 10-m AVERAGED WINDSPEED, NUMBERS OF STORMS WITH WIND SPEEDS ABOVE THRESHOLD FOR DRIFTING (7m/s), TOTAL DURATION OF STORM (h), ESTIMATED SNOW TRANSPORT, AND POTENTIAL SUBLIMATION AT SIX CANADIAN STATIONS DURING WINTER, 1974-75.
(Steppuhn and Gray, 1977).

Canadian Station	Accumulated snow water equivalent mm	Average hourly wind speed m/s	No. of storms	Duration of Storms h	Est. of max. possible wind-borne snow transport tonnes/m width	Potential sublimation mm
Winnipeg, Manitoba	129.3	9.08	41	900	9.18	109
Regina, Saskatchewan	82.0	9.35	48	1115	13.01	185
Saskatoon, Saskatchewan	78.0	8.80	40	686	5.88	122
Bad Lake, Saskatchewan	101.2	9.38	33	763	8.95	162
Edmonton, Alberta	108.0	8.73	27	329	2.62	64
Lethbridge, Alberta	136.6	10.5	39	1087	21.74	765

its water content and temperature. Some infiltration may be expected under most conditions except perhaps when a saturated soil remains frozen at a very low temperature.

Agricultural practices have potential for controlling or directing snowmelt infiltration. Erickson et al. (1978) monitored snowcover runoff rates from fields which had been summer-cropped and fallowed and found that they were on average in the ratio 1:3. Willis et al. (1969) showed that snowmelt occurred earlier in tall crop stubble and was slightly faster than that in a short stubble. In general, the more rapidly and the earlier the surface becomes bare, the greater the capacity of the soil to infiltrate and absorb melt water.

Snow Management Practices

Snow management to augment soil water for dryland crops is currently practiced on a limited scale in the USSR, USA and Canada. Any practices that induce snow to accumulate preferentially or to melt so as to infiltrate and remain stored in the soil until utilized by a crop, qualify as snow management practices. These are discussed under several categories below.

Field fences. One of the earliest tests to manage snow was initiated in 1937 at Scott, Saskatchewan, Canada. According to Matthews (1940), the investigators reasoned that wind-blown snow would accumulate behind field-sited barrier fences constructed of brush or wood-on-wire. The best fences for snowdrift control have porosities of 50% or more to allow significant snow-laden air to flow through them. Fences have disadvantages: they are expensive and they accumulate snow in a non-uniform areal pattern, resulting in alternate strips of wet and dry soil that prove difficult to cultivate.

Tests on fences and brush piles situated in fields to accumulate snow have been conducted in the steppe and forest-steppe regions of the Kazakhstan, SSR (Kibasov, 1955). In one test during the 1953-54 winter, brush and fence systems were constructed in a chess-board pattern with barriers spaced at intervals of 6 to 8 m. Although the barriers were moved several times during the winter, snow still accumulated unevenly, being 20 to 40 cm deep in the ridges or mounds formed behind the fences, and 10 to 15 cm between them. In the USA Greb and Black (1971) and Willis (1979) both concluded that fences are not practical for extensive use on cultivated lands.

The non-uniform collection of snow behind field-sited fences is less of a problem in pastures and open rangeland (Berndt, 1964; Tabler, 1968). Greb and Black (1971) and Greb (1975) evaluated the performance of vertically-slatted snow fences of 72% porosity in pastures seeded to crested wheatgrass, intermediate wheatgrass and Russian wild rye. Over five years of tests the following results were found within areas 16.8 m to the leeward of the fences: the net over-winter soil water gains averaged 60.1 mm greater than those in adjacent non-fenced pastures; forage yields for the three species averaged 2236, 1900 and 1446 kg/ha, respectively and the volume of stored water, correlated closely with the volume of snow deposited (see Fig. 3.11).

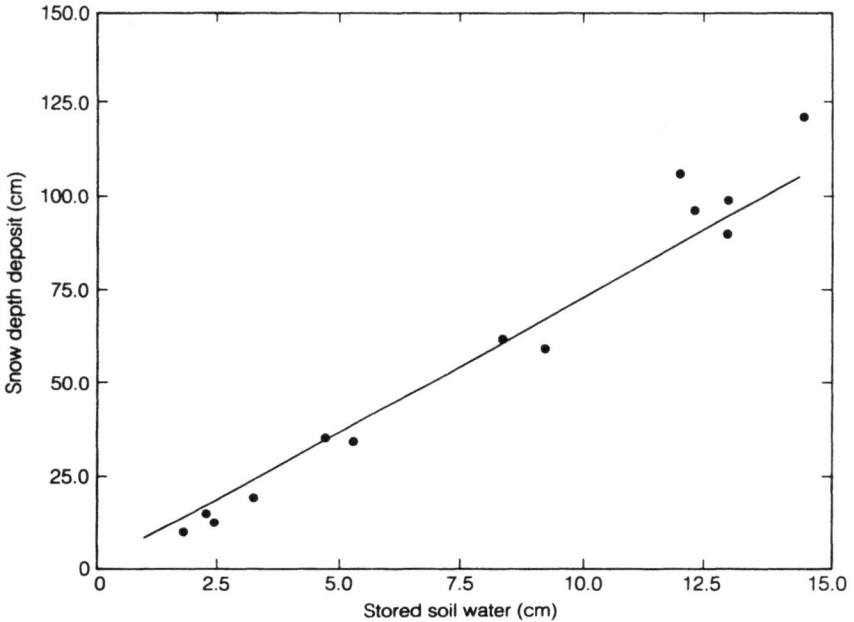

Fig. 3.11 Association between the depth of snow deposit leeward of experimental snow fences and stored soil water from snowmelt, Akron, Colorado, 1964-1968 (Greb, 1975).

Vegetative Barriers. The practice of growing tall, woody windbreaks adjacent to and within cultivated fields to curtail wind erosion, trap snow, reduce evaporation, and increase crop yield has been strongly promoted in the USSR, USA and Canada where vast regions contain numerous fields that are sheltered by live trees and shrubs arranged in variously-spaced rows. In the Astrakhan Steppe, USSR, in 1946-47 snow accumulated to an average depth of 21.2 cm on wind-protected fields compared with 12 cm on open fields (Vasil'yev, 1956). In Canada, Matthews (1949) lauded the water conservative merits of woody windbreaks especially surrounding gardens and vegetable fields. Others (George, 1943; George et al., 1963; Frank and Willis, 1972; Frank et al., 1974) recognized the snow-trapping ability of live woody barriers, but stressed that the ideal windbreak should distribute the snow uniformly over the adjoining cultivated and pasture areas. Frank and George

(1975) show that greater spreading of accumulated snow can be achieved by keeping the windbreak well pruned and porous (Fig. 3.12).

Staple and Lehane (1955) evaluated the influence of windbreak shelters on crop yields in Saskatchewan. The shelterbelts consisted of three continuous rows of caragana (2.5 m in height), ash (7 m) and maple (7 m), spaced 137 m apart, and extending the length of the field. Soil water and wheat yields were sampled at various distances perpendicular to the barriers. As shown in Fig. 3.13, both decreased from the barriers; soil water removal by the trees was not evident beyond distances equal to the tree heights. If the area occupied by the barriers is taken into account, the net increase in spring wheat yield was 47 kg/ha within the sheltered fields during the five-year study period.

Recently, the merits of barriers consisting of live, non-woody vegetation have been investigated, e.g., in Colorado, Greb and Black (1971) grew double rows of sudangrass permanently spaced 11.6 to 18.3 m apart. Within this barrier system they found that: (1) snow averaged 15.2 cm deeper; (2) overwinter soil reserves gained 38 mm more water; and (3) wheat yield was 269

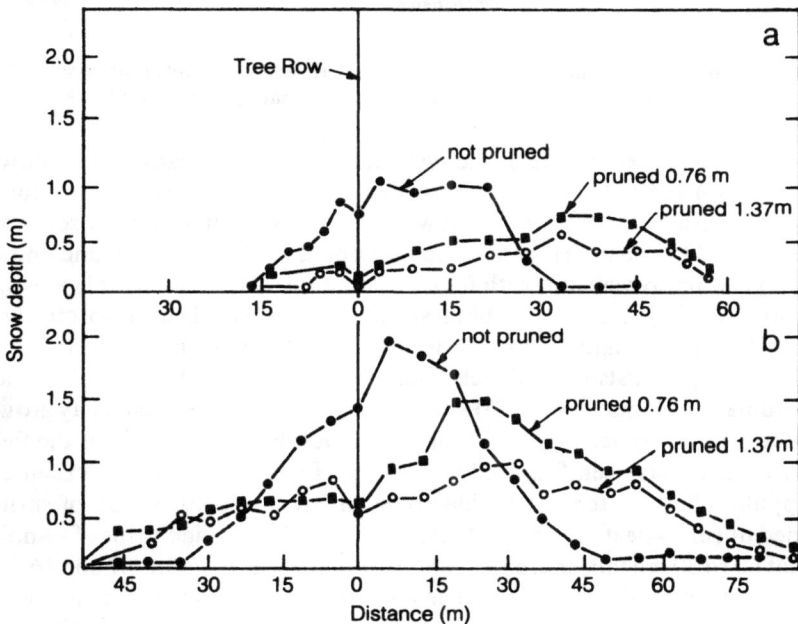

Fig. 3.12 Snowcover depth in 1967 (A), and 1969 (B), adjacent to test sections of a 6.1-m single-row Siberian elm windbreak in its natural growth state, but with branches pruned below 0.76 and 1.37 m (Frank and George, 1975).

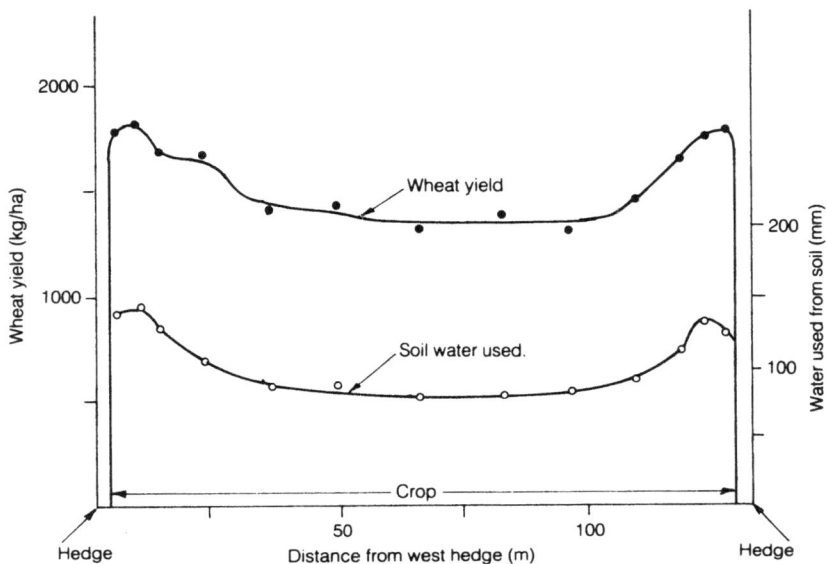

Fig. 3.13 Wheat yield and stored soil water used at various distances between hedges, Aneroid, Saskatchewan, 1950-54 (Staple and Lehane, 1955).

kg/ha greater than outside of the system. In Montana, Black and Siddoway (1975, 1976) evaluated tall wheatgrass as a snow-control barrier and found that the average yearly gain in soil water resulting from its use for seven test years was 43 mm under continuous cropping (Table 3.10). Concomitant grain-yield improvements with 67.2 kg/ha of nitrogen (Table 3.11) equalled 471 and 628 kg/ha for continuous spring and winter wheat, respectively.

Field-sited vegetative barriers need not be permanent to be effective. Extension specialists in Saskatchewan (Saskatchewan Agricultural Services Coordinating Committee, 1975) suggested that ample late-season hay growth or uncut strips, spaced the width of the mower, be left throughout the field. This practice has been followed in the USA (de Vore, 1976). Erickson and Steppuhn (1976) extended the idea of such strips to a 1975 crop of spring-seeded durum wheat. During fall harvest, 30- to 40-cm wide strips of standing wheat 60 to 80 cm tall, spaced 1, 2 and 3 combine widths (5.3, 10.7 and 16.0 m) apart, were left unharvested to act as snow-trap barriers on fields scheduled to be sown the following spring. Snow depths at the time of maximum accumulation measured on the non-stripped areas averaged 21.2 cm; on the stripped areas average snow depths were increased by 88, 82 and 70 percent for corresponding spacings of 1, 2 and 3 combine swaths. With a 15-m spacing, the total snow management investment (non-harvested area) would equal 2%

Table 3.10

SEVEN-YEAR AVERAGE SOIL WATER GAIN AND STORAGE EFFICIENCY[a], WITH AND WITHOUT TALL WHEATGRASS BARRIERS, FOR CONTINUOUS-CROPPING AND CROP-FALLOW SYSTEMS, MONTANA (Black and Siddoway, 1975).

Cropping Sequence	Precipitation mm	Inside Grass Barriers		Outside Grass Barriers	
		Soil Water Gain mm	Storage Efficiency %	Soil Water Gain mm	Storage Efficiency %
Continuous Cropping					
First winter (9 mon.)	176.3	96.5	54.8	53.3	30.3
Crop-Fallow					
First winter (9 mon.)	176.3	99.1	56.2	53.3	30.3
Summerfallow (5 mon.)	196.8	15.2	7.7	15.2	7.7
Second winter (7 mon.)	176.5	25.4	14.4	22.9	12.9
Total gain (21 mon.)	556.3	139.7	25.1	91.4	16.4

[a] Storage efficiency is the percentage of precipitation stored in the soil profile during each period.

Table 3.11

AVERAGE GRAIN PRODUCTION (kg/ha) WITHIN AND OUTSIDE GRASS BARRIERS, AS INFLUENCED BY CROPPING SEQUENCE AND NITROGEN FERTILIZATION, 1968-74, MONTANA, (Black and Siddoway, 1975).

Cropping Sequence	Nitrogen Added (kg/ha)		
	0	33.6	67.2
Inside Grass Barriers			
Continuous spring wheat	874	1244	1373
Spring wheat after fallow	1602	1905	1905
Continuous winter wheat	1468	1855	1883
Winter wheat after spring wheat	1222	1967	2074
Winter wheat after fallow	2062	2342	2365
Outside Grass Barriers			
Continuous spring wheat	863	--	902
Spring wheat after fallow	1244	--	1356
Continuous winter wheat	1098	--	1255
Winter wheat after fallow	1513	--	1670

of the harvested crop, which in 1975-76 amounted to about 4 Canadian dollars per hectare.

Snowplowing. One method of inducing deposition of snow on cultivated fields and pastures involves plowing freshly-fallen snow into parallel ridges, which protrude into the horizontal wind stream causing air-borne snow to be deposited in the furrows between ridges. The most effective ridges are tall and oriented perpendicular to the dominant wind flow.

On the Canadian Prairies, snow ridging, as an agronomic practice to augment soil water for crops, has received periodic attention. Matthews (1940) reported on tests conducted at Scott, Saskatchewan during 1937-39 with ridges spaced 2.5 m apart. Snowcover depths measured during 1939 at the time of maximum accumulation in natural snowdrifts and in and between the ridges are given in Table 3.12. Snowplowing increased the natural accumulation of snow by 100% in the ridges and 30% between them. However, gains in soil moisture (Table 3.13), except in grass pastures, were small which probably explains the rather modest improvements in crop yield (Table 3.14). Matthews concluded that snowplowing would definitely increase yields for some crops and generally reduce the soil erosion potential by maintaining a wetter surface longer into the summer.

In 1962, Keys (1961-72) revived snowplowing tests at Scott and initiated tests at Loverna near the Alberta-Saskatchewan border. He encountered difficulty in establishing ridges in some years, owing to a lack of snow, and therefore recorded a wide yearly variation in results. Nevertheless, ridging was considered to be responsible for an average increase in spring wheat yield of 2% at Scott and 10% at Loverna.

Working in the steppe region of Siberia, Kibasov (1955) tested various snow retention techniques; the results, summarized in Table 3.15, show sunflower fences to be very effective. However, since Kibasov recognized the practical limitations on cultivating sunflowers in the region he favoured snow ridging, because of the 58% increase in snow depth which he obtained with two ridging treatments. Referring to other ridging trials in northern Kazakhstan, USSR, Hockensmith and Harrison (1964) reported that soil water gains amounted to 1000 t/ha, the equivalent to the complete infiltration of a 100 mm rain.

Recent snowplowing trials in Saskatchewan (Dyck et al., 1979) gave varying results for the effectiveness of the practice. In 1972-73 an exceptionally strong wind levelled the ridges, sweeping the snow from the fields; in 1975-76 a mid-winter thaw melted many of the ridges, severely reducing their trapping efficiency; in 1976-77 there was insufficient snowcover to form suitable ridges. However, in 1973-74, one of the effective years, an average of 43% more snow water accumulated while 70% more soil water was stored on wheat stubble fields with ridges than on the test control fields. Also, water from the ridged snow penetrated deeper into the soil. The increases in soil water attributed to ridging at depths of 15, 30.5 and 61 cm were 15, 39 and

Table 3.12
DISAPPEARANCE OF SNOW IN SPRING,
SCOTT, SASKATCHEWAN (Matthews, 1940).
Reproduced by permission of the Minister of Supply
and Services Canada.

Date	Natural snow-drifts cm	Snow in ridges cm	Snow between ridges cm
March 17, 1939	35.56	53.34	54.61
18	35.56	53.34	54.61
20	34.29	52.07	53.34
21	33.02	50.80	50.80
22	30.48	49.53	48.26
23	25.40	44.45	44.45
24	22.86	40.64	43.18
25	22.86	40.64	43.18
27	21.59	39.37	41.91
28	16.51	35.56	40.64
29	15.24	33.02	38.10
30	8.89	24.13	33.02
31	0.0	17.78	27.94
April 1, 1939	-	8.89	22.86
3	-	0.0	16.51
4	-	-	13.97
5	-	-	13.97
6	-	-	13.97
7	-	-	-
8	-	-	-
10	-	-	6.35
11	-	-	6.35
12	-	-	5.08
13	-	-	2.54
14	-	-	0.0

226%, respectively. Dyck et al. concluded that to achieve success with this snow retention method one must be prepared to plow fields immediately following snowfall, to assume the risk of failure with each plowing, and to realize that added benefits will not be produced in some years.

Stubble Management. The merits of leaving standing crop stubble to trap wind-blown snow are well known and documented (Staple and Lehane, 1952; Willis and Haas, 1969; Greb et al., 1970; Rodenko, 1970). Tall stubble and stubble from tall crops, such as sunflowers, grain corn and sorghum, retain

Table 3.13
SOIL WATER CONSERVATION BY SNOW RIDGING, SCOTT, SASKATCHEWAN; EXPRESSED AS A PERCENTAGE INCREASE OVERWINTER WITHIN SELECTED DEPTH LAYERS AND THE 0-91 cm SOIL PROFILE (Matthews, 1940). Reproduced by permission of the Minister of Supply and Services Canada.

| Crop | Year | Land Use | Soil Depth Layers | | | | |
			0-15 cm	15-31 cm	31-61 cm	61-91 cm	0-91 cm
Wheat	1938	Summerfallow	22.8	16.7	22.9	22.2	21.2
Wheat	1939	Summerfallow	12.5	14.3	10.5	12.8	12.5
Oats	1939	Stubble	17.7	11.9	11.2	0.0	10.2
Barley	1939	Stubble	27.0	7.8	32.3	35.6	25.7
Crested wheatgrass	1939	Sod	45.9	25.2	54.0	10.4	33.9

Table 3.14
INFLUENCE OF SNOW RIDGING ON CROP YIELD, SCOTT, SASKATCHEWAN (Matthews, 1940). Reproduced by permission of the Minister of Supply and Services Canada.

| Crop | Year | Land Use | Yield (kg/ha) | | |
			Snow Plowed	Not Snow Plowed	Increase or Decrease
Wheat	1938	Summerfallow	1204	1049	155
Wheat	1939	Summerfallow	1049	1036	13
Oats	1939	Stubble	1653	1557	96
Barley	1939	Stubble	1377	1592	-215
Crested wheatgrass	1939	Sod	4610	2823	1786

more snow than the short stubble of wheat, barley or flax (Keys, 1961-72; Bakeav, 1970; Smika and Whitfield, 1966). In March of 1976 the author measured snowcover depths averaging 9.5 cm in spring wheat stubble and 21.8 cm in sunflower stubble in southwestern Saskatchewan; in central Saskatchewan, corresponding depths were 15 and 32 cm, respectively. Smika and Whitfield (1966) studied the effect of surface residues on the conservation of water from snow at North Platte, Nebraska. They compared the over-winter changes in soil water in fields where the winter wheat stubble was left standing with those in fields where the stubble had been incorporated into the soil by fall tillage (Table 3.16). The soil water enrichments averaged 52 mm for the standing stubble and 9 mm for the incorporated stubble.

Various methods of harvesting a crop have been tried with the purpose of leaving a stubble barrier to trap wind-transported snow. Willis (1979) presented snowcover data measured at Mandan, North Dakota (Table 3.17) on sites where the crop was cut uniformly (as a control), and alternately in

Table 3.15

COMPARISONS OF SNOW RETENTION METHODS, USSR (Kibasov, 1955).

Method of snow retention	Average depth of snowcover (cm)				
	Mar. 10	Mar. 16	Mar. 22	Mar. 27	Apr. 6
Control (natural accumulations)	15.3	14.0	no data		7
Brushwood	22.0	21.9	18.9	14.7	7
Brushwood with subsequent snow-plowing	27.0	24.8	22.0	17.8	8
Snowplowing	24.0	21.0	16.0	7.0	2
Fences (shields)	28.0	27.7	23.7	15.2	8
Sunflower fences	42.0	42.0	36.0	32.0	19

Table 3.16

EFFECT OF SURFACE RESIDUE ON THE STORAGE OF SOIL WATER FROM SNOW (NOV. 15 THROUGH MAR. 31) NORTH PLATTE, NEBRASKA (Smika and Whitfield, 1966).

Over-winter period	Residue	Available soil moisture to 1.83-m depth mm		Precipitation mm	Storage Efficiency %
		Nov. 15	Mar. 31		
1961-62	Standing	94	145	58	88.2
	Fall incorporated	99	105		10.0
1962-63	Standing	139	188	58	83.3
	Fall incorporated	153	168		25.3
1963-64	Standing	112	168	67	83.6
	Fall incorporated	99	105		8.8
1964-65	Standing	142	193	36	140.6
	Fall incorporated	142	104		-105.6

strips 1.5-m wide, separated by 3-m wide strips on which the crop was cut to the ground surface (stubble strips). Snowcover depths over the sites did not differ appreciably; however, because snow densities in the low-cut strips were approximately twice those in the standing stubble, the average water equivalent was 108.6 mm on the sites with stubble strips compared with 62.9 mm on the uniform stubble.

Table 3.17
AVERAGE SNOW DEPTH, DENSITY AND WATER CONTENT ON SITES[a]
WITH STANDING STUBBLE OR WITH STUBBLE STRIPS (Willis, 1979).

| | Control | | | Stubble Strips | | | | | |
| | Standing Stubble | | | Standing | | | Cut | | |
Site	Depth cm	Density kg/m³	Water Content mm	Depth cm	Density kg/m³	Water Content mm	Depth cm	Density kg/m³	Water Content mm
1	48.4	130	62.9	50.8	120	61.0	48.3	221	106.7
2				48.3	137	66.0	48.3	310	149.9
3				50.8	155	78.7	50.8	255	129.5
Ave.	48.4	130	62.9	50.0	137	68.6	49.1	262	128.7

[a] Date from five randomly selected sites in a standing stubble field were composited for the control. Three samples at each of the middle and halfway between the middle and the strip edges were composited for the standing and cut strips.

Combine threshing of light grain crops often involves cutting and feeding two swaths of standing grain into the combine. This operation allows the two harvested swaths (each 4-10 m wide) to be cut at alternate heights, one low ($\leq$ 20 cm above the ground surface), and one high ($\geq$ 40 cm). Three years (1973-75) of water equivalent comparisons showed successive gains in soil water of 7.1, 16.2 and 5.3 mm, on alternately-cut fields compared with uniformly-cut fields (Nicholaichuk and Norum, 1975). Soil water additions measured under alternate-height swathed fields near Swift Current, Saskatchewan by D.W.L. Read (personal communication) exceeded those under uniform stubble by 25 mm or 57%.

One form of stubble management specifically intended for retaining snow involves seeding rows of tall plants in bands as companions to the main crop, e.g., multi-row bands of mustard or sunflowers seeded with cereal grains. Rodenko (1970) applied the practice to the culture of spring wheat in the Altai Region of the Russian Federated Republic, finding that sufficient snow water was retained to allow cropping for a second or third year without summer-fallow. Initially, wheat is seeded so that a 1.8-m wide strip is left unseeded every 30.6 m, which amounts to 6% of the seeded area. About a month after the wheat has emerged the unseeded strips are cultivated and seeded to sunflowers in three rows 15 cm apart. In 1959-60 fields with bands yielded 2570 kg/ha of wheat; those without bands, 2070 kg/ha. For the second crop, the yields were 2370 and 1930 kg/ha, respectively. In addition the practice has been used to establish forage and increase its production.

Tall stubble bands have also been used on summerfallow. Seeded to sunflowers, corn, rapeseed or mustard, these bands provide a supplemental crop as well as act as barriers to retain snow. Bakeav (1970) described the use of bands of mustard with wheat in northern Kazakhstan. He found 4-6 rows of mustard per band more effective in holding snow than 1-2 rows; the bands are usually spaced 12 m apart and seeded early in July. Yields from subsequent spring wheat crops in 1968 and 1969 averaged 2214 and 1807 kg/ha on fields with bands compared with 1895 and 1497 kg/ha on those without. Keys (1961-72), testing single-row bands of sunflowers (sunfallow) for nine years, measured average increases in spring-wheat yields of 9 and 1.3% at two dryland locations in western Canada. Keys also used corn as the vegetative barrier (cornfallow), however, the growing seasons proved to be too short to mature the corn grain. In the USA a soil erosion abatement program has promoted the banding of 2-6 rows of flax on summerfallow. Usually being seeded in August, they rarely attain sufficient height to provide an effective winter snow trap for the 15-m spacing commonly used (Willis, 1979).

Attempts have been made to manage residues so as to increase snow retention on livestock pastures and open prairie rangelands. Galbraith (1971) in Colorado noted that over the winter of 1969-70 the average soil water recharge was 30 mm greater in ungrazed plots compared with that in the surrounding heavily grazed pasture. He surmised that the difference was due to greater snow retention by the ungrazed pasture. Wight et al. (1975) reported the results from a 1971-72 study in northeastern Montana for evaluating the effects of grazing intensity on snow accumulation. After the first storm the snow depths under grazing intensities of zero, moderate and heavy (simulated by clipping), averaged 15.2, 10.2 and 7.6 cm, respectively; after the second storm the average depths were 20.3, 7.6 and 2.5 cm with corresponding coverages of 100, 70 and 20% of complete snowcover. Measurements by Van Haveren and Striffler (1976) in Colorado, during the 1970-71 winter indicated a mean recharge of 15% more soil water in lightly grazed prairie pastures than heavily grazed ranges. In explanation, they cited the taller average height of the lightly grazed plants which trapped more wind-transported snow.

The species composition of range and pasture lands also effects a type of snowcover control. Working in the extensive native rangeland of Wyoming, Hutchison (1965) showed definite preferential accumulations of snow behind tall range plants such as big sage, *Artemisia tridentata*. His work suggests that a rangeland may be managed to include tall, less palatable species which trap more winter snow to augment the soil water. One difficulty may result from the trampling of the tall plants by the grazing animals.

Surface Modifications. Modifications of soil surfaces to effect snow management include plowing, diking, pitting, furrowing, rotary subsoiling and terracing. To varying degrees these practices have advantages; not only by trapping and holding snow where it is needed, but retaining the melt-waters in preferred positions for infiltration. But they also have disadvantages, including high construction costs, and the resultant rough surfaces presented to field machinery. Because of the latter, permanent surface modifications have been mainly used with pasture and range improvements.

Many of the early tests of surface modifications, however, were initiated in cultivated fields principally to control soil erosion, with the secondary objective of holding snow and promoting infiltration of meltwater. Fall plowing with damming listers on summerfallow was tested in 1938 and 1939 at ten sites in southwestern Saskatchewan (Staple et al., 1960). No significant overwinter enrichments of soil water were detected, giving little promise for the practice. The merits of diked, narrow terraces had also been tested in the same dryland region of Canada a few years earlier in 1931-34; Barnes (1938) reported average respective spring wheat yields for diked and non-diked plots of 1770 and 1410 kg/ha on fallow, and 700 and 470 kg/ha on stubble. Hubbard and Smoliak (1953) reported that dikes were effective in spreading and holding snowmelt waters for infiltration and in increasing forage production on Canadian rangelands. They also found that contour furrows, 10.2 to 12.7 cm deep, filled with ice during the winter and held little water during snowmelt.

Neff (1980) also evaluated the snow management potential of contour furrows in the slightly warmer climate of southeastern Montana. The furrows were 15 to 25 cm deep, 50 cm wide and placed with 1.5-m centres on level contours. They were carefully constructed and included small intrafurrow dams every 5 m. In nine years (1969-74, 1976-78) of measurement a yearly average of 22 mm more snowwater accumulated on the furrowed pastures than on the untreated areas. The furrows also retained surface water for longer periods following snowmelt allowing ample time for infiltration into the soil. As a result, over-winter soil-water recharge and herbage production was increased (Neff and Wight, 1977).

Rotary subsoiling for range renovation was studied in eastern Montana by Siddoway and Ford (Wight et al., 1975). A subsoiler was used to punch holes on a one-metre grid pattern so as to rupture the subsoil. The practice increased the soil moisture storage capacity by about 7.6 cm and increased infiltration. Similar claims for shallower treatment (pitting) are probably related to the increase in standing stubble which traps more snow; thereby increasing the soil water reserves more by augmentation than by increasing the rate of infiltration (Rauzi and Lang, 1957).

Level-bench terraces are one of the most effective surface modifications for snow management. They collect wind-carried snow, retain the melted snow water until the ground thaws allowing infiltration, reduce soil erosion, and increase crop yield. Following the design of Zingg and Hauser (1959), bench terraces are constructed level from end to end along the contour and from edge to edge across each terrace; a 30-cm dike on the downslope edge confines the water. Bench width depends on slope; the steeper the slope the narrower the bench, with 9 m being common. Vegetative barriers may also be left standing on the dikes for additional snow collection. As shown in Fig. 3.14, terraces may include upslope contributing areas, but these generally have a minor influence on water storage and crop yield (Willis and Haas, 1971). In North Dakota, the use of level-bench terraces increased spring wheat yields about 370 kg/ha, corn, very little, alfalfa from 2.2 to 4.5 t/ha, and bromegrass from 2.1 to 3.6 t/ha relative to those obtained on non-terraced slopes (Haas and Willis, 1968). Rauzi (1975) obtained 700 kg/ha more alfalfa on level benches than on the non-terraced slopes in dryer Wyoming. Shallower-rooted crested and intermediate wheatgrasses, however, showed no improvements in yield.

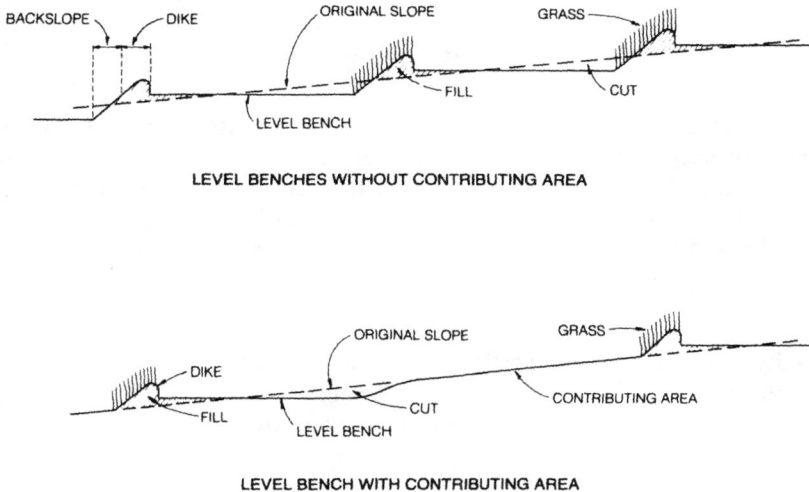

LEVEL BENCHES WITHOUT CONTRIBUTING AREA

LEVEL BENCH WITH CONTRIBUTING AREA

Fig. 3.14 Diagrammatic cross-section of level benches without and with contributing areas (Willis and Haas, 1971).

SNOW AND THE CROP ENVIRONMENT

The accumulation of snow produces significant changes in the ambient environment. Snowcover generally absorbs and transmits less solar radiation, reduces air movement, conducts less heat and increases the water vapor content in the layer above the ground surface. The amount that the ambient environment changes depends largely on snowcover properties such as composition of the solid phase or matrix, liquid water content, impurities, albedo, depth and density. Environmental changes also affect the plant life engulfed by the snow and the soil beneath the cover.

The protection provided by snow to plants completely buried under it is well known. However, if shrubs only are partially covered, they often suffer severe damage just above the snow surface (Van Wijk and Derksen, 1963). The breeding of cold hardy varieties and strains, and the establishing of a snowcover are recognized as the principal means for protecting agricultural crops from damage associated with low temperatures (Vasil'yev, 1956). The application of snowcover protection measures have not only been applied to wheat, oats, barley, rye, hay-cut forages, rapeseed, and pastures, but to berry vines, orchards, lawns, nursery trees and pharmaceutical plants; snow is especially valuable for plants in the seedling stage.

Heat Transfer

The insulative value of a snowcover is related to its poor heat transfer characteristics. Usually, in winter when a snowcover is at a temperature below freezing, heat flows by conduction from the soil to the atmosphere. The change of temperature T with time t at any point in the snowcover can be approximated by the expression

$$\partial T / \partial t = k(\partial^2 T / \partial z^2) / C_v, \qquad 3.5$$

where k and C_v are the effective thermal conductivity and volumetric heat capacity of the snow in the vertical direction z, respectively. de Vries (1975) suggested that the volumetric heat capacity of any composite medium can be formed from the individual capacities of its various phases,

$$C_v = V_a\,C_{va} + V_w\,C_{vw} + V_i\,C_{vi} \qquad 3.6$$

where V and C_v are the volumetric fraction and heat capacity for each component phase of air, water and solid, designated by subscripts a, w, and i. The solid phase in snow is generally taken as ice, although it includes impurities such as soil (dust), organics and dissolved constituents. The heat capacity of each phase is the product of its density ρ and specific heat C. Similarly, if vapor movement is ignored (see Ch. 7) the composite thermal conductivity k of the medium can be obtained from the expression:

$$k = V_a\,k_a + V_w\,k_w + V_i\,k_i. \qquad 3.7$$

Approximate values of ρ, C_v, and k for the components which form snow (assuming air and water at 10° and ice at 0° C) are:

Component	ρ kg/m^3	C_v J/(m^{3}K)	k W/(m K)
air	1.25	1.25 x 10^3	0.025
water	1.0 x 10^3	4.2 x 10^6	0.57
ice	0.92 x 10^3	1.9 x 10^6	2.2

Geiger (1961) calculated the composite thermal conductivity to lie in the range from 0.025 to 1.61 W/(m K) for snow densities between 100 and 800 kg/m^3; assuming k $\propto \rho^2$. Van Wijk and Derksen (1963) reported a thermal conductivity of 0.046 W/(m K) for fresh light snow and 0.326 W/(m K) for old dense snow. The relatively small value of thermal conductivity for snow can be attributed to its large porosity, e.g., snow with a density of 320 kg/m^3 and a 2% liquid water volume has a porosity of about 64% (by volume). Even dense snow (450 kg/m^3) is quite porous (50% by volume).

Some heat may be transferred through a snowcover by convection, however, the amount is usually small. Most values of thermal conductivity listed in the literature include heat transfer by both conduction and convection (see also Ch. 7 - effective or apparent conductivity). In a melting snowcover the heat transfer processes are further complicated by the movement of water.

Soil Temperature

A snowcover reduces the soil heat loss and modifies the ground temperature regime. Usually, a layer of snow increases soil temperatures, especially the daily minima and seasonal averages, reduces the diurnal and annual amplitudes and causes soil temperatures to lag fluctuations of air temperature. Figure 3.15, presented by Ylimäki (1962) shows the effect of snow depth on ground surface temperature in Finland. The warmer ground surface under snowcover acts as a heat source. The effects of snowcover on temperature decrease with increasing soil depths (Table 3.18).

The annual periodicity of temperature is illustrated by data reported by Toperczer (1947) for a 33-year period in eastern Austria (see Fig. 3.16; reproduced from Geiger, 1961). Dingman (1975) reviewed the literature pertaining to frozen ground and concluded that most authors accepted the mean annual ground temperature to be equal to the value occurring at the level of zero annual amplitude, i.e., where summer and winter temperatures are equal. Generally, this temperature is from 2 to 7° C warmer than the mean air temperature, largely because of the insulating effect of a winter snowcover. According to Dingman (1975), Brown (1968) had suggested the following relationship between the ground temperature T_g, air temperature T_a, and snowcover thickness z:

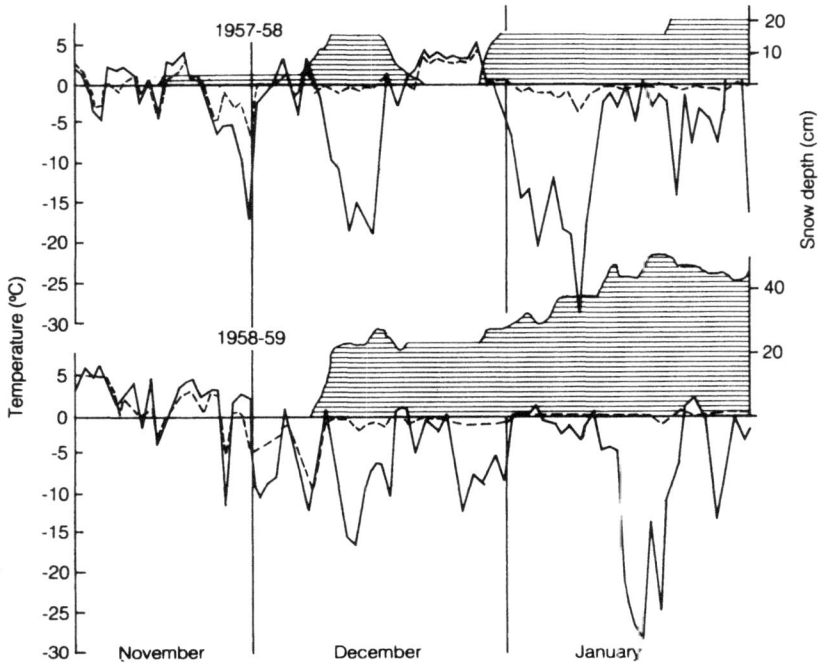

Fig. 3.15 The effect of snow depth on the temperature at the ground surface during two winters with different snow conditions, 1957 and 1958; — minimum air temperature, --- minimum temperature at the ground surface (Ylimäki, 1962).

$$T_g = T_a + \{1 - [1/\exp(z\sqrt{\pi C_v/k\tau})]\}/2, \qquad\qquad 3.8$$

where: A = annual amplitude of mean monthly air temperature,
k = composite thermal conductivity,
C_v = volumetric heat capacity, and
τ = period of oscillation.

The extent to which a snowcover affects the penetration of a freezing front into a soil depends on the snow thickness, the soil water content, the vegetative cover and the temperature history. In Europe, a general rule of thumb is that each centimetre of snowcover produces at least a 0.1° C increase in root-zone soil temperature. According to eight years of data from eastern Europe given by Shul'gin (1957) (Table 3.19), the temperature-modulating effect per unit increase in snow depth on soil temperature decreases with thickness. The insulative effects of snowcovers under similar winter and soil conditions have

Table 3.18
**MEAN MONTHLY TEMPERATURE (°C) OF
BARE SOIL AND OF SNOW-COVERED SOIL,
USSR (Shul'gin, 1957). Reproduced by permission
of Keter Publishing House Jerusalem Ltd.**

Month	Plots	Depth (cm)		
		40	80	160
November	Snow-covered	-2.0	3.8	6.2
	Bare	-3.2	1.0	5.4
	Difference	1.2	2.8	0.8
December	Snow-covered	0.3	1.8	4.2
	Bare	-9.7	-4.7	2.1
	Difference	10.0	6.5	2.1
January	Snow-covered	-1.6	0.2	2.7
	Bare	-13.9	-9.3	-0.6
	Difference	12.3	9.5	3.3
February	Snow-covered	-3.0	-0.8	1.6
	Bare	-13.6	-9.9	-2.2
	Difference	10.6	.1	3.8
March	Snow-covered	-1.6	-0.4	1.7
	Bare	-8.1	-6.7	-2.3
	Difference	6.5	6.3	3.0

Table 3.19
**DIFFERENCE OF TEMPERATURE (°C) BETWEEN
AIR AND SOIL, AS A FUNCTION OF SNOWCOVER
THICKNESS, USSR (Shul'gin, 1957). Reproduced by
permission of Keter Publishing House Jerusalem Ltd.**

Thickness of snowcover cm	Difference between air and soil temperatures per cm of snow °C
0-10	1.1
11-12	0.7
21-30	0.6
31-40	0.4
41-50	0.3
51-60	0.2
61-70	0.1
71-80	0.1

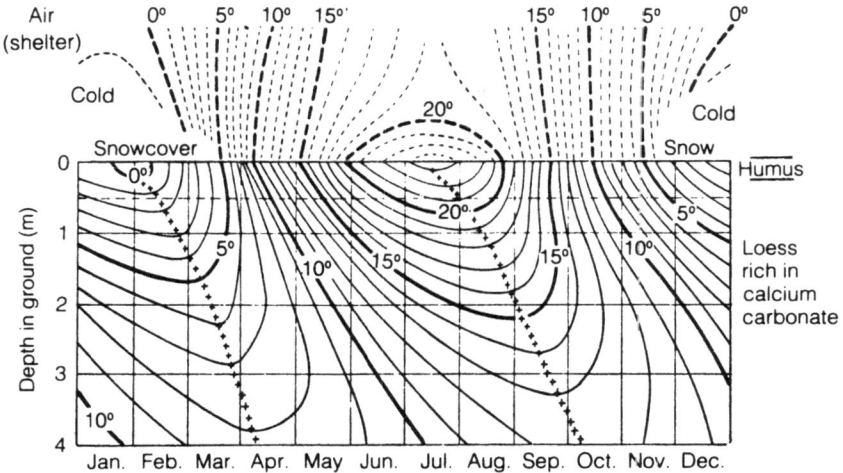

Fig. 3.16 Annual sequence of soil temperature in the gardens of the Vienna Central Institute (after Toperczer, 1947; reproduced from Geiger, 1961). Reproduced by permission of Harvard University Press.

also been studied in Finland by Ylimäki (1962). During three winters (1951-54) he found that the frost depths ranged from 55 to 64 cm on plots kept snow-free, 12 to 22 cm where depths were normal, and 9 to 17 cm under plots receiving above-normal accumulations of snow.

Frost depths are also governed by soil wetness and vegetative cover. A dry soil tends to freeze deeper and faster (but thaws more rapidly) than a wet soil, primarily owing to differences in heat capacity. Figure 3.17, illustrates these trends. Figure 3.18 demonstrates the effect of vegetative cover on frost depth. Heavy-littered woods and grass maintained shallower frost depths than annually-cultivated lands; the insulative effect of the vegetative covers appears to dominate that of the snowcovers. When assessing the interaction between snowcover depth and frost penetration it is essential to consider the duration of snowcover also.

The tendency for snowcovers to moderate a decrease in soil temperature has also been associated with duration of the cover. Shul'gin (1957) showed that the minimum 3-cm soil temperature under a snowcover 25 cm deep was -8.1°C, compared with -17.6°C when the surface had been bare for 30 days (Table 3.20). He also suggested a series of relationships which model the influence of snow depth and the minimum air temperature on the soil temperature at 3-cm depth (location of the winter-wheat tillering nodes) (Fig. 3.19).

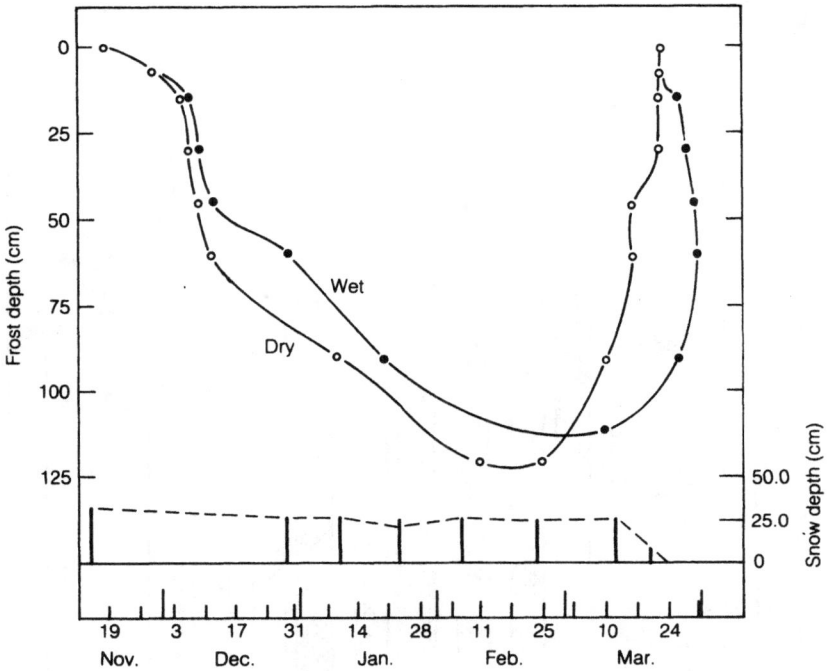

Fig. 3.17 Depth of freezing temperature as a function of time for wet and dry soils during the 1958-59 winter, Mandan, North Dakota (Willis et al., 1961).

Table 3.20
SOIL TEMPERATURE AT 3-cm DEPTH AS A FUNCTION OF DURATION OF A SNOWCOVER, 25 cm THICK, IN EARLY WINTER; USSR, 7 JANUARY 1943 (Shul'gin, 1957).

Reproduced by permission of Keter Publishing House Jerusalem Ltd.

Number of days during which the plot remained bare of snow, Days from 5 December 1942	Minimum soil temperature at depth of 3 cm °C
30	-17.6
20	-13.7
10	-11.8
0	-8.1

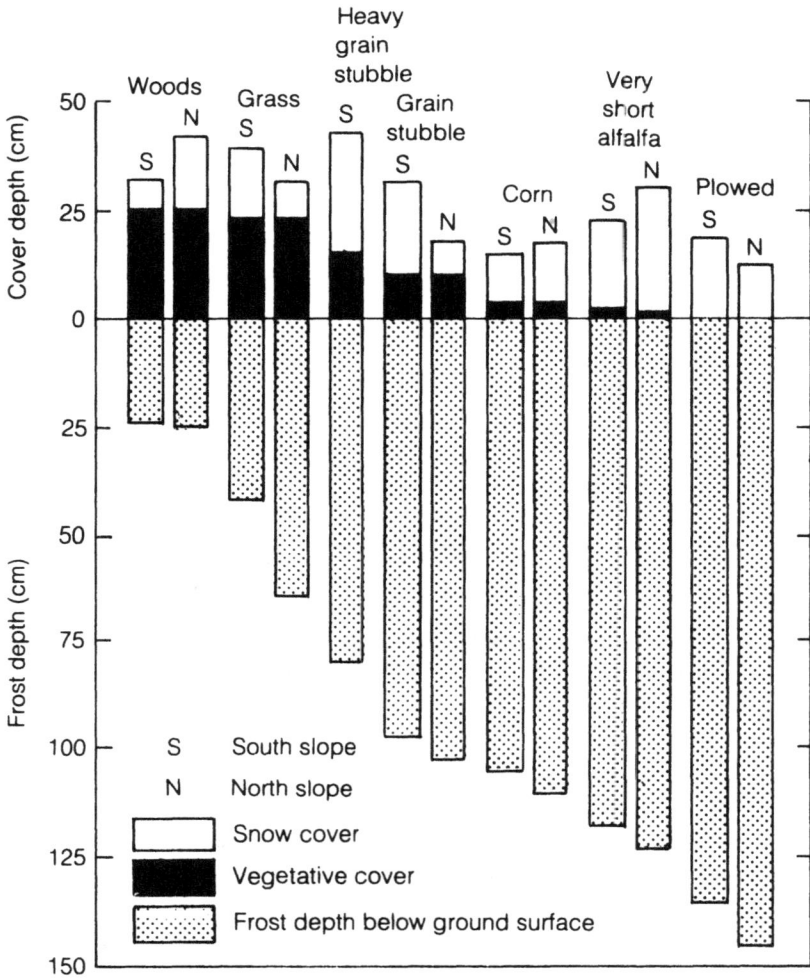

Fig. 3.18 Snow depth, frost depth and vegetative cover at 13 locations in Wisconsin (Atkinson and Bay, 1940).

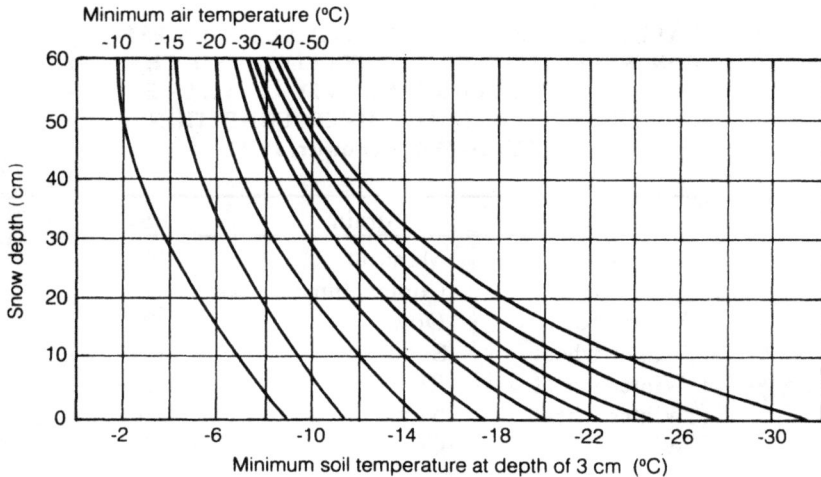

Fig. 3.19 Relationship between soil temperature at 3-cm depth and minimum air temperature, as a function of snow depth in eastern Europe (Shul'gin, 1957). Reproduced by permission of Keter Publishing House Jerusalem Ltd.

Snow and Crop Survival

The FAO (1979b) estimated the 1978 world wheat production at 441.5×10^6 tonnes of which at least half was winter wheat. Some perspective of the economic value of snow as an insulator for this crop can be obtained using the Field Survival Index FSI. Fowler (1978), defined the FSI as that numerical value associated with a specific crop variety or strain such that the difference between any two values equals the difference in crop survival (expressed in percent) during an average winter on the Canadian Prairies in Saskatchewan. Table 3.21 lists some typical values of FSI for different species and strains of wheat, oats, barley and rye, along with the minimal FSI required for low-temperature survival for various snowcover conditions experienced in Saskatchewan.

It is evident that the larger the FSI value the hardier the plant and the harsher the environment that it can survive. As an example of its use, consider the winter wheats Norstar and Kharkov whose index values are 516 and 501, respectively (cf. Table 3.21). Therefore, the FSI is 516 - 501 = 15%, which suggests 15% better survival for Norstar relative to Kharkov. Each centimetre of snow on a bare summerfallow decreases the FSI by 22%. Thus, on bare summerfallow under average winter conditions one would expect that Norstar would be completely winterkilled (650 - 516 = 134%); under 5 cm of

Table 3.21
VALUES OF THE FIELD SURVIVAL INDEX FSI FOR SELECTED CEREAL STRAINS AND THE LOW-TEMPERATURE SURVIVAL VALUES UNDER SPECIFIC SNOWCOVER CONDITIONS, SASKATCHEWAN, (Fowler, 1978).

Species	Strain	FSI
Spring wheat	Manitou	160
Winter oats	Compactum	290
Winter barley	Dover	300
Winter wheat	Sundance	494
Winter wheat	Kharkov	501
Winter wheat	Norstar	516
Winter rye	Cougar	620
	Cover Condition	
	Bare summerfallow	>650
	5 cm snowcover	540
	10 cm snowcover	430
	>15 cm snowcover	<320

snowcover a 24% (540 - 516 = 24) mortality; and under 10 cm of snow complete survival (430 - 516 < 0). If the average winter climate in Saskatchewan is assumed to be colder than most of the world's winter-wheat-growing regions, then a 5-10 cm layer of snowcover might protect, say only 5% of the world crop from winterkill. Thus, from 1978 data (FAO, 1979) each centimetre of snow in the depth 5-10 cm would have been worth one-fifth the multiple of crop survival, production and revenue, i.e.,

$$0.20 \text{cm}^{-1} \times 0.05 \times (220 \times 10^6 \text{t}) \times 135/\text{t} = \$297 \times 10^6/\text{cm (US)}.$$

The hardiness of a plant to direct killing by low temperature varies throughout the winter. The minimum survival temperatures for winter wheat and rye for September to May (Fig. 3.20) reveal an initial sharp decline reaching a minimum near mid-November, a leveling and gradual increase through the winter months, reaching relatively high values in April. These changes relate to the plant's physiology, state of health, and ability to withstand late-winter, early-spring desiccation because of transpiration. Winter desiccation affects a number of crops, especially woody plant species (Vasil'yev, 1956) and lawns (Beard, 1973). Both researchers suggest maintaining an ample uniform snowcover for protection, e.g., by using windbreaks.

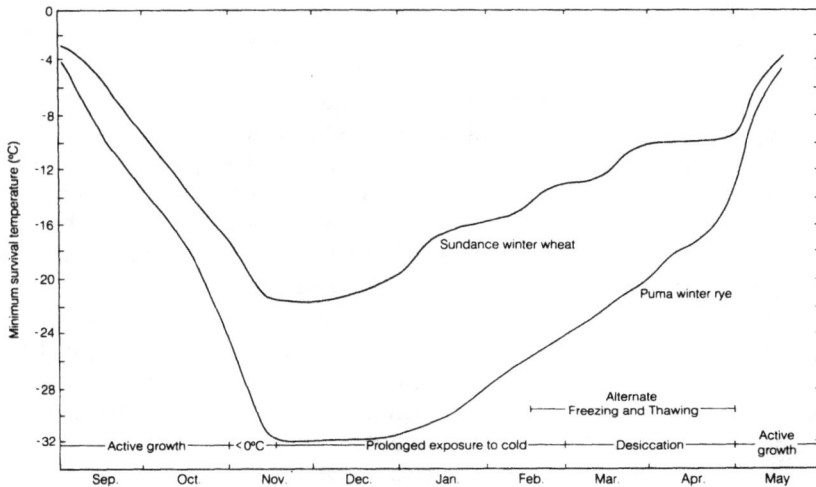

Fig. 3.20 Changes in cold hardiness of winter wheat and rye for the period September to May. The primary factors responsible for these changes are shown at the bottom of the graph (Fowler, 1978).

The growth crown in winter cereals is located below ground within 5 cm of the surface. At this depth, in the absence of a snowcover, temperature changes can be quite abrupt. Levitt (1972) noted that plants may suffer "winter-kill", although not frozen, when subjected to rapid cooling of $\sim 5°\,C/h$, especially if this occurs in the range -10 to -20°C. Direct damage occurs when the intracellular water freezes and expands, rupturing cell walls and destroying tissue.

Winter crops are also subject to frost heaving. The volumetric expansion caused by freezing of soil water lifts crowns and roots exposing tissue to atmospheric desiccation. According to Beard (1973) a lack of snowcover promotes frost heave.

If winter annuals and perennials are covered by ice and liquid water for extended periods they may die. Sprague and Graber (1940) reported injury to alfalfa after 7-10 days under ice, with mortality beginning after 20 days. Most crops, however, can withstand an ice cap of much longer duration, e.g., many turfgrasses can tolerate continuous ice coverage for 60 days or more (Beard, 1973). Even though an ice cover restricts the exchange of gases between plants and atmosphere, oxygen suffocation likely does not occur; rather mortality appears to be caused by the accumulation of toxic gases (Levitt, 1972). Very few workers believe that a deep snowcover inhibits gas exchange. Vasil'yev (1956) reports on successful winter wheat production in the Soviet Union even where 2 m of snow are common, while Beard (1973) indicates there is a reduced potential for ice damage if a snow layer lies between a turfgrass and the ice. Much of the low-temperature kill associated with an ice cover occurs after melting when water submerges the crop (Beard, 1973).

Winter crops must combat various pathogens and pests. A snowcover favours plant survival, but may also foster the over-wintering of beneficial and injurious insects, fungi and weeds. Perennial and annual species of weeds are well known to farmers, and are controlled by tillage operations or herbicide applications. Insects generally over-winter as eggs; however, not all do. For example, the Hessian Fly, a pest of winter wheat, normally winters as a pupa in the plant leaf sheath or as a larva if the ambient temperature remains mild (Dahms, 1967). A snowcover protects over-wintering pupa or larva.

A number of low-temperature fungi affect winter cereals. The stripe rust of wheat, a fungal parasite, has adapted preferentially to the cool regions occurring at high altitudes or in sub-polar latitudes of more than 80 countries (Loegering et al., 1967). These authors described the organism's epidemiology, which follows a sequence common to many winter cereal fungi. Spore innoculation of healthy seedlings flourishes in the fall with rust lesions appearing on leaves which either sporulate or function over-winter but at a reduced rate. The leaves affected may persist throughout the winter if protected by snow, with host and parasite about equally tolerant of low temperatures. Another group of fungal pathogens have been termed snow molds because their effects have often been noted following the disappearance of the winter snowcover. The group includes many species of the genera Fusarium, Sclerotinia and Typhula, which parasite crowns and roots as well as leaves. Although a snowcover benefits wintering of the pathogens, most evidence linking the severity of the disease with thickness of the snowcover remains circumstantial; deep snowcover does not assure snow mold damage; nor is the mold restricted to areas with deep snow. Vasil'yev (1956) claimed that fungi only invade plants injured by early fall frost before snow accumulation begins or only enter leaf stomata under warm, humid conditions in the spring. Britton (1969) emphasized that the growth of pink snow mold occurred primarily under a melting snow layer. Bruehl (1967) associated snow molds of winter cereals with persistent deep snow layers covering unfrozen soil. Failure of the soil to freeze before accumulation of a lasting snowcover was also cited by Ylimäki (1962) as favourable for growth of low-temperature fungi.

Winterkill in most winter cereal regions results principally from direct freezing, desiccation and frost heaving, rather than from fungi or other causes. Thus, the maintenance and enhancement of a snowcover generally favours plant survival and increases crop production. Ylimäki (1962) rated numerically the winter survival of various legumes, cereals and grasses (Table 3.22). Plants survived better on all snowcovered plots than on sites kept free of snow.

Table 3.22

OVERWINTERING OF CEREALS, GRASSES AND LEGUMES IN SNOWCOVER TRIALS 1952-53, TIKKURILA, FINLAND (Ylimäki, 1962).

Species and Strain	Winter Survival[a]	
	Snowcovered	Free of Snow
Winter rye, Toivo	9.0	6.7
Winter wheat, Varma	6.6	1.3
Timothy, Ta 01	5.3	4.7
Meadow fescue, No. 601/50	8.5	2.0
Perennial ryegrass, No. 644/50	3.1	0.1
Italian ryegrass, No. 408/50	0.3	0.0
Red clover, Tammisto	7.2	0.4
Alsike clover, Tetra Weibull	8.0	0.0
Alfalfa, Flammande	6.0	0.0

[a] Winter survival = $10 \times \dfrac{\text{plant density in spring}}{\text{plant density in autumn}}$

Specific management practices to increase snow depth and spread the snow uniformly serve to moderate the ambient overwinter environment and to enhance crop survival. To retain a protective snowcover in windy areas, Vasil'yev (1956) recommended seeding winter wheat directly into the standing stubble of a spring crop wherever soil water reserves permit. He also reported success in sowing winter crops alternately with rapidly-growing, high-stalked mustard or sunflowers; although the tall plants die during the winter, their stalks serve as snow fences. Vasil'yev also cited farmers in the Sverdlovsk region of the USSR who hill the base of young fruit trees and berry vines with snow as a traditional cultivation practice. Beard (1973) suggested using fences or brush to trap snow over temperature-critical turfgrass areas such as golf greens and tees.

SNOW AND LIVESTOCK

Snowstorms and Livestock on Pasture and Range

An Associated Press release (1980) exemplifies the impact of snow on livestock: *"Still digging out after 60 centimetres of snow, concerned Nebraska and Montana ranchers are rounding up new-born calves and taking extra hay to their cattle as a new snowstorm bears down on them from the Colorado Rockies. The heavy wet snow has caught Nebraska cattlemen at the peak of* *calving season, and officials fear a new storm could devastate the calf crop. A 1975 storm killed 56,000 calves, costing cattlemen $4 million. Many ranchers in Nebraska and Montana are gathering cattle into protected areas, feeding them extra hay and keeping a close watch on them and the weather."*

Despite modern technology, snowstorms in agricultural districts still pose a threat to livestock. Concern of the producers is shared by meteorologists who have instituted livestock (stockmen's) warnings as a component of their weather forecasts. These warnings alert ranchers and farmers that livestock on pastures and ranges will require protection from a large accumulation of snow or ice, a rapid drop in temperature, or strong winds. Often a warning is combined with a forecast of an impending blizzard. Blizzards, the most dramatic and perilous of winter storms, are characterized by low air temperatures and strong winds bearing large amounts of fine snow which can drastically reduce visibility. The Weather Almanac (Ruffner and Bair, 1974) contains a special advisory for stockmen to provide extra shelter, water and feed for exposed animals when warned of an approaching storm.

Extra livestock production costs resulting from snowstorms can be high. According to Stoddard et al. (1975) these costs are due to: (1) emergency feeding, especially when aircraft must be used; (2) animal weight losses and deaths; (3) losses of new-born stock, particularly when storms occur during calving and lambing periods. Snowfalls over the western USA during the winter of 1948-49 were double their normal amounts, while the low air temperatures allowed snowcover to accumulate from four to ten times normal depths, severely hindering normal winter grazing. Despite the tonnes of hay and feed concentrate carried over snow-bound roads and dropped from aircraft, 25% of the livestock were lost, representing a value of $2.5 million (U.S. Department of Commerce, 1949).

The literature occasionally refers to the interaction of snowcover with winter grazing, e.g., Koeppe (1931), in describing the Canadian climate, mentioned the warm, desiccating chinook winds which melt the snowcover other southern Alberta just often enough to expose forage and permit year-round grazing. East of this zone, where the chinook influence diminishes, livestock cannot be wintered on pastures and ranges, requiring instead the protection of sheltered feedlots.

Livestock Shelters and Feedlots

Snow may present difficulties to livestock wintering in sheltered quarters. Sheep, as well as beef and dairy cattle, are often housed in open-front sheds. Air flowing over these structures reduces the air pressure within and in front of them favouring deposition of wind-transported snow. McPherson (1962) and Renwick (1962) tested various structural shapes, roof slopes and rear-wall venting to control accumulations in and around open-front buildings. They obtained the best control with 12 to 25% of the rear wall open to the outside. Air currents through the openings were sufficient to maintain positive internal air pressures and reduce snow deposition.

Numerous model studies (Theakston, 1961, 1967; Bellman and Theakston, 1965) have been undertaken in attempts at solving snow accumulation problems and at optimizing the layout of farmsteads. For example, water flowing in a flume around plastic models has been used to simulate wind flowing past farm buildings and structures; fine sand is added to the water to simulate wind-entrained snow. Observations of the sand accumulations around the models were used as guides for the design and layout of the prototype structures.

While fences around feedlots shelter animals, they may also trap large quantities of snow. Moysey and McPherson (1966) measured wind velocities down-wind of fences and found that windbreak fences with 15 to 30% porosity provide the best shelter (the zone of protection extends over a greater area); however, they cautioned that upwind snowtraps would be necessary when using such porous fences.

SNOW AND MECHANICAL DAMAGE

Orchards of apple, cherry, peach and other fruit trees often cover a complete hillslope, from crest to valley floor. As early as 1836, Willis Gaylord, an editor of the "Genesee Farmer", who was also a farmer, wrote a detailed account of the weather associated with an early October snowstorm which wrought considerable damage to fruit orchards in New York State (Ludlum, 1968). He observed that snow falling below the 61 m elevation quickly melted upon hitting the trees which were kept warm by the ambient air whose temperature was above 0°C. Between elevations of 61 and 122 m, the snow was wet and damp, loading trees sufficiently to break limbs and bend trunks. Above 122 m, air temperatures were below freezing and the dry snow blew from the trees causing little damage.

The extent of mechanical damage suffered by vegetation from a load of snow depends on snow mass, vegetation strength, root anchorage and wind shear stress. The damage increases with the length of time that plants must bear the weight of the snow, which causes irrecoverable strain on supporting tissues. Miller (1966) associated wind speed, air temperature, vapor pressure, and insolation with the various ways that snow naturally unloads from trees by falling, melting and evaporating (Table 3.23).

In regions prone to strong winter winds, plants may be damaged by snow abrasion. Winter crops in shallow snowcovers often protrude above a smooth, icy snow surface, across which winds may easily drive loose ice

Table 3.23

PROCESSES TRANSPORTING INTERCEPTED SNOW DURING STORMS (Miller, 1966).

Weather element	Transport by				
	Falling or blowing of dry snow	Sliding or falling of partly-melted bodies of snow	Dripping or flowing of melt water	Vapor flux from melt-water film	Vapor flux from snow
Wind speed	++a	+b	+b, c	+d	+d
Air temperature		++c	+c	+d	+d
Vapor pressure			++	-	--
Insolation		++	+	+	+d

Symbols:

+   indicates an element of storm weather that favors a mode of transport from the crowns;

++ indicates strongly favours;

-   indicates an element of storm weather that discourages a mode of transport from the crowns;

-- indicates strongly discourages.

a = Effect of wind is conditioned by the rate at which masses on intercepted snow are streamlined and wind packed, or develop internal cohesion.

b = Conditional on air temperature being above 0°C.

c = Conditional on air being near saturation.

d = Conditional on low vapor pressure in air.

particles which abrade the exposed plant tissue. Vasil'yev (1956) has observed this kind of damage in winter rapeseed in the USSR.

Early snowfalls can greatly affect standing field crops primarily through lodging. Pendleton (1954) reported that lodging in cereals resulted in lower yields than those for upright plants; also, leaning and bending crops are difficult to harvest.

Occasionally early snowfalls will delay harvesting until fall, winter or early spring (Schlehuber and Tucker, 1967). One example is the cold, wet, snowy weather in September 1965 on the Canadian Prairies (Meteorological Branch, 1965) which brought harvest operations to a standstill. Lethbridge, Alberta, received a record September snowfall of 43 cm; Dodds (1967) estimated the wheat loss at 673 kg/ha. Again in 1968, a snowstorm from 18 to 21 September (Meteorological Branch, 1968) deposited 56 cm of snow on Lethbridge and delayed harvesting. Mechanization of harvest operations has decreased the incidence of crops left in the field because of early snow.

Root crops are also vulnerable to damage if snow delays harvest. To melt an autumn snowcover on fields still containing sugar beets in Colorado, J. Meiman and C. Slaughter (personal communication) spread coal dust on the surface to reopen the harvest season. The dust particles absorb incoming radiation, transfer heat to the snow and accelerate melt. Beard (1973) described the use of lampblack or dark fertilizer to dissipate ice and snow covering turfgrasses.

Forests may incur severe mechanical damage resulting from snow load; evergreens are particularly susceptible (Toumey and Korstian, 1947). Ruffner and Bair (1974) reported that an evergreen, 15-m tall and 6-m wide may be coated with 4.5 t of ice and snow during a severe storm. The extent of injury differs with tree species, age and canopy density. In many regions, damage may occur with such a regularity and a severity that these characteristics are considered in selecting species for planting and in the management of the crop. The amount of snow damage has been linked directly to stand density, i.e., the more trees per unit area, the greater the potential for mechanical harm, especially in mono-cultures of species such as ponderosa pine (Powers and Oliver, 1970), jack pine (Godman and Olmstead, 1962), and red pine (Schantz-Hansen, 1939). Hardwoods may also sustain damage, but apparently have the capacity to recover rapidly (Blum, 1966). On the Canadian Prairies Currie (1953) observed quick recovery in arborescent shelterbelts which had been injured by the above-normal snowcovers of the 1946-47 winter.

LITERATURE CITED

Anderson, C.H. and D.W.L. Read. 1966. *Water use efficiency of some varieties of wheat, oats, barley and flax grown in the greenhouse*. Can. J. Plant Sci., Vol. 46, pp. 375 and 378.

Anderson, D.T. 1967. *The cultivation of wheat*. Proc. Can. Centennial Wheat Symp. (K.F. Nielsen, ed.), Western Fertilizer Ltd., Modern Press, Saskatoon, Sask., pp. 333-335.

Associated Press. 1980. *Ranchers hit hard by winter storm*. News release printed in Star-Phoenix News, 31 March Edition, Armadale Publishers, 204-5th Ave., Saskatoon, Sask., Can., p. 2.

Atkinson, H.B. and C.E. Bay. 1940. *Some factors affecting frost penetration*. EOS: Trans. Am. Geophys. Union, pp. 935-948.

Atmospheric Environment Service. Sept. 1974 - Aug. 1975. *Monthly Meteorological Observations in Canada Records*. Can. Dept. Environ., Downsview, Ont.

Austenson, H., G. Storey, K. MacDonald, K. Kirkland and G. Lee. 1974. *Stubble cropping recommendations for 1974*. Agric. Sci. Bull., Publ. No. 242, Univ. Sask., Saskatoon.

Ayers, H.D. 1959. *Influence of soil profile and vegetation characteristics on net supply to runoff*. Proc. Symp. No. 1, Spillway Design Floods., Queen's Printer, Ottawa, pp. 198-205.

Bakeav, N. 1970. *Banded cover crops for fallow in Northern Kazakhstan*. Zemledeliya, No. 10, pp. 19-21. [English Transl. by Res. Branch, Can. Dept. Agric., Ottawa].

Barnes, S. 1938. *Soil moisture and crop production under dryland conditions in western Canada*. Publ. 595, Can. Dept. Agric., Ottawa, Ont.

Bauder, J.W., L.J. Brun and T.H. Krueger. 1975. *The relationship of soil freezing to snowmelt runoff*. North Dak. Farm Res., Vol. 32, No. 6, pp. 10-13.

Bauer, A. and R.A. Young. 1966. *Fertilized wheat uses water more efficiently*. North Dak. Farm Res., Vol. 24, No. 3, pp. 4-11.

Beard, J.B. 1973. *Turfgrass: Science and Culture*. Prentice-Hall, Inc., Englewood Cliffs, N.J.

Bellman, E. and F.H. Theakston. 1965. *Artificial snow and wind barriers around open-front livestock buildings*. Can. J. Agric. Eng., Vol. 7, pp. 1-4.

Berndt, H.W. 1964. *Inducing snow accumulation on mountain grassland watersheds*. J. Soil and Water Conserv., Vol. 19, pp. 196-198.

Black, A.L. and F.H. Siddoway. 1975. *Snow trapping and crop management with wheatgrass barriers in Montana*. Proc. Symp. on Snow Management on the Great Plains, Publ. No. 73, Great Plains Agric. Counc. and Univ. Nebr. Agric. Exp. Stn., Lincoln, pp. 128-137.

Black, A.L. and F.H. Siddoway. 1976. *Dryland cropping sequences within a tall wheatgrass barrier system*. J. Soil and Water Conserv., Vol. 31, pp. 101-105.

Blum, B.M. 1966. *Snow damage in young northern hardwoods*. Am. J. For., Vol. 64, pp. 16-18.

Bole, J.B. and U.J. Pittman. 1978. *The effect of fertilizer N, spring moisture, and rainfall on yield and protein content of barley in Alberta.* Proc. 1978 Workshop on Soils and Crops, Extension Div., Publ. No. 390, Univ. Sask., Saskatoon, pp. 114-121.

Bowren, K.E., D.A. Cooke and R.K. Downey. 1969. *Yield of dry matter and nitrogen from tops and roots of sweetclover, alfalfa and redclover at five stages of growth.* Can. J. Plant Sci., Vol. 49, pp. 61-68.

Britton, M.P. 1969. *Turfgrass diseases.* Ch. 10, Turfgrass Science (A.A. Hanson and F.V. Juska, eds.), Agron. Ser. No. 14, Am. Soc. Agron., Madison, Wisc.

Brown, J. 1968. *Geocryological considerations of soils.* Unpubl. Tech. Note, U.S. Army Cold Reg. Res. Eng. Lab., Hanover, N.H.

Bruehl, G.W. 1967. *Diseases other than rust, smut, and virus.* Ch. 11, In Wheat and Wheat Improvement (K. Quisenberry and L. Reitz, eds.), Agron. Ser. No. 13, Am. Soc. Agron., Madison, Wisc., pp. 375-410.

Cantor, L.M. 1967. *A World Geography of Irrigation.* Oliver and Boyd, R. Cunningham and Son Ltd., Alva, Great Britain.

Cole, J.S. 1938. *Correlations between annual precipitation and yield of spring wheat in the Great Plains.* Tech. Bull. No. 636. U.S. Dept. Agric., Washington, D.C.

Coxworth, E., D. Thompson, and M. Gimby. 1978. *Energy, agriculture and the food system.* Rep. C78-9 for D.S.S. Contract No. O7SZ.01843-7-0750, Sask. Res. Counc., Saskatoon. (unpublished).

Currie, B.W. 1953. *Prairie provinces and Northwest Territories - snowfall.* Monograph on Climates of the Prairie Provinces and Northwest Territories., Physics Dept., Univ. Sask., Saskatoon.

Dahms, R.G. 1967. *Insects attacking wheat.* Ch. 12, In Wheat and Wheat Improvement. (K. Quisenberry and L. Reitz, eds.), Agron. Ser., No. 13, Am. Soc. Agron., Madison, Wisc., pp. 411-443.

Deibert, E.J. 1979. *No-til influence on soil temperature, moisture and compaction.* Proc. Manitoba-North Dakota Workshop, Agric. Extension Centre, Brandon, Man., pp. 1-13.

de Jong, E. and D.A. Rennie. 1969. *Effect of soil profile type and fertilizer on moisture use by wheat grown on fallow or stubble land.* Can. J. Soil Sci., Vol. 49, pp. 189-197.

de Vore, N. 1976. *Growing up in Montana - a picture essay.* J. Nat. Geogr. Soc., Vol. 149, pp. 650-657.

de Vries, D.A. 1975. *Heat transfer in soils.* Ch. 1, In Heat and Mass Transfer in the Biosphere, Part 1, Transfer Processes in the Plant Environment (D.A. de Vries and N.H. Afgan, eds.), Scripta Book Co., Washington, D.C.

Dingman, S.L. 1975. *Hydrologic effects of frozen ground, literature review and systhesis.* Spec. Rep. 218, U.S. Army Cold Reg. Res. Eng. Lab., Hanover, N.H.

Dodds, M.E. 1967. *Wheat harvesting-machines and methods.* Proc. Can. Centennial Wheat Symp. (K.F. Nielsen, ed.), Western Co-operative Fertilizer Ltd., Modern Press, Saskatoon, Sask., pp. 357-371.

Dubetz, S. 1961. *Effect of soil type, soil moisture, and nitrogen fertilizer on the growth of spring wheat.* Can. J. Soil Sci., Vol. 41, pp. 44-51.

Dyck, G.E., D. Erickson and H. Steppuhn. 1979. *Snow ridging to increase soil water.* Proc. Soils and Crops Workshop. Extension Div., Publ. 403, Univ. Sask., Saskatoon, pp. 1-12.

Dyunin, A.K. 1959. *Fundamentals of the theory of snow drifting.* Izvest. Sibirsk. Otdel. Akad. Nauk. U.S.S.R., No. 12, pp. 11-24. [English Transl. by G. Belkov, Nat. Res. Counc. Can., Ottawa, Tech. Transl. 952, 1961].

Erickson, D. and H. Steppuhn. 1976. *Snow trapping with unharvested strips of grain.* Unpubl. Rep., Div. Hydrol., Univ. Sask., Saskatoon.

Erickson, D., W. Lin and H. Steppuhn. 1978. *Indices for prairie runoff from snowmelt.* Proc. 7th Symp. Water Studies Inst., Appl. Prairie Hydrol., Saskatoon, Sask.

FAO. 1976. *Irrigation.* 1975 Production Yearbook, Vol. 29, FAO Statistics Ser. No. 2. United Nations Food and Agriculture Organization, Rome, pp. 24-25.

FAO. 1977. *Irrigation.* 1976 Production Yearbook, Vol. 30, FAO Statistics Ser. No. 7, United Nations Food and Agriculture Organization, Rome, p. 57.

FAO. 1978. 1977 Production Yearbook, Vol. 31, FAO Statistics Ser. No. 15, United Nations Food and Agriculture Organization, Rome, p. 57.

FAO. 1979a. *Irrigation.* 1978 Production Yearbook, Vol. 32, FAO Statistics Ser. No. 15, United Nations Food and Agriculture Organization, Rome, p. 57.

FAO. 1979b. *Commodity Production.* 1978 Production Yearbook, Vol. 32, FAO Statistics Ser. No. 22, United Nations Food and Agriculture Organization, Rome, p. 96.

FAO. 1979c. *Commodity Prices.* 1978 Production Yearbook, Vol. 32, FAO Statistics Ser. No. 22, United Nations Food and Agriculture Organization, Rome, p. 275.

Ford, E.C., W.L. Cowan and H.N. Holtan. 1955. *Floods - and a program to alleviate them.* In Water, 1955 Yearbook of Agriculture, U.S. Dept. Agric., U.S. Gov. Print. Off., Washington, D.C., pp. 172-174.

Fowler, D.B. 1978. *Winter cereal survival in Saskatchewan.* Proc. 1978 Soils and Crops Workshop, Extension Div., Publ. 390, Univ. Sask.. Saskatoon, pp. 1-13.

Frank, A.B.; D.G. Harris and W.O. Willis. 1974. *Windbreak influence on water relations, growth, and yield of soybeans.* Crop Sci., Vol. 14, pp. 761-765.

Frank, A.B. and E.J. George. 1975. *Windbreaks for snow management in North Dakota.* Proc. Sym. Snow Management on the Great Plains. Publ. 73, Great Plains Agric. Counc. and Univ. Nebraska Agric. Exp. Stn., Lincoln, Nebr. pp. 144-153.

Frank, A.B. and W.O. Willis. 1972. *Influence of windbreaks on leaf water status in spring wheat.* Crop Sci., Vol. 12, pp. 668-672.

Fritts, H.C. 1976. *Tree Rings and Climate.* Academic Press, New York, N.Y.

Galbraith, A.F. 1971. *The soil water regime of a shortgrass prairie ecosystem.* Unpubl. Ph.D. Dissertation, Colo. State Univ., Ft. Collins.

Geiger, R. 1961. *Das Klima der bodennahen Luftschicht (The Climate near the Ground).* [English Transl. by Scripta Technica, Harvard Univ. Press, Cambridge, Mass., 1966].

George, E.J. 1943. *Effects of cultivation and number of rows on survival and growth of trees in farm windbreaks on the northern Great Plains.* J. For., Vol. 41, pp. 820-828.

George, E.J., D. Broberg and E.L. Worthington. 1963. *Influence of various types of windbreaks on reducing wind velocities and depositing snow.* J. For., Vol. 61, pp. 345-349.

Gillies, J.A. 1968. *Infiltration in frozen soils.* M.Sc. Thesis, Univ. Sask., Saskatoon.

Glymph, L.M. and H.N. Holtan. 1969. *Land treatment in agricultural watershed hydrology research.* In Water Resources Symp., No. 2, Effects of Watershed Changes on Streamflow (W.L. Moore and C.W. Morgan, eds.), Univ. Texas Press, Austin, pp. 44-68.

Godman, R.M. and R.L. Olmstead. 1962. *Snow damage is correlated with stand density in recently thinned jack pine plantations.* Tech. Note 625, U.S. Dept. Agric., For. Serv., Lake States For. Exp. Stn., St. Paul, Minn.

Golding, D.L. 1974. *The correlation of snowpack with topography and snowmelt runoff on Marmot Creek Basin, Alberta.* Atmosphere, Vol. 12, pp. 31-38.

Goodell, B.C. 1966. *Snowpack management for optimum water benefits.* Preprint 379, Conf. Water Res. Eng., Am. Soc. Civil Eng., Denver, Colo.

Gray, D.M., D.I. Norum and J.M. Wigham. 1970. *Infiltration and physics of flow of water through porous media.* Handbook on the Principles of Hydrology (D.M. Gray, editor-in-chief), Secretariat, Can. Nat. Comm. Int. Hydrol. Decade, Ottawa, pp. 5.1-5.58.

Greb, B.W. 1975. *Snowfall characteristics and snowmelt storage at Akron, Colorado.* Proc. Symp. on Snow Management on the Great Plains., Publ. 73, Great Plains Agric. Counc. and Univ. Nebr. Agric. Exp. Stn., Lincoln, pp. 45-64.

Greb, B.W. and A.L. Black. 1971. *Vegetative barriers and artificial fences for managing snow in the Central and Northern Plains.* Proc. Symp. on Snow and Ice in Relation to Wildlife and Recreation (A.O. Haugen, ed.), Iowa Coop. Wildlife Res. Unit, Iowa State Univ., Ames, pp. 96-111.

Greb, B.W., D.E. Smika and A.L. Black. 1970. *Water conservation with stubble mulch fallow.* J. Soil and Water Conserv., Vol. 25, pp. 58-62.

Greenshields, E.L. 1955. *The expansion of irrigation in the West.* In Water, 1955 Yearbook of Agriculture, U.S. Dept. Agric., U.S. Gov. Print. Off., Washington, D.C., pp. 247-251.

Haas, H.J. and W.O. Willis. 1968. *Conservation bench terraces in North Dakota.* Trans. Am. Soc. Agric. Eng., Vol. 11, No. 3, pp. 396-398, 402.

Hanson, R.E. and W.R. Meyer. 1953. *Irrigation requirements for Kansas.* Wheat and Wheat Improvement (K. Quisenberry and L. Reitz, eds.), Agron. Ser. No. 13, Am. Soc. Agron., Madison, Wisc., p. 155.

Harbeck, G.E. and W.B. Langbein. 1949. *Normals and variations in runoff 1921-1945.* Water Resour. Rev. Suppl. No. 2, U.S. Geol. Survey, Washington, D.C.

Harris, K. 1951. *Consumptive use and irrigation requirements of crops in Arizona.* Unpubl. Rep., U.S. Soil Conserv. Serv. Cited by H.F. Blaney. 1959. Monthly consumptive use requirements for irrigated crops. J. Irrig. Drain. Div., Am. Soc. Civil Eng., Vol. 85, No. IR1, Part 1, pp. 1-12.

Hem, J.D. 1970. *Study and interpretation of the chemical characteristics of natural water.* Water-Supply-Paper 1473, (2nd Ed.), U.S. Geol. Surv., Washington, D.C.

Hensler, R.F., R.J. Olson, S.A. Witzel, O.J. Attol, W.H. Paulson and R.F. Johannes. 1970. *Effect of method of manure handling on crop yields, nutrient recovery and runoff losses.* Trans. Am. Soc. Agric. Eng., Vol. 13, pp. 726-731.

Hexem, R.W. and E.O. Heady. 1978. *Analysis of wheat experiments.* Ch. 7, In Water Production Functions for Irrigated Agriculture. Iowa State Univ. Press, Ames, pp. 106-119.

Hockensmith, R.D. and P. Harrison. 1964. *Soil conservation, a world movement.* In Farmer's World, 1964 Yearbook of Agriculture. U.S. Dept. Agric., U.S. Gov. Print. Off., Washington, D.C., pp. 69-75.

Holt, F.G. 1969. *Runoff and sediment as nutrient sources.* Bull. 13, Water Resour. Res. Cent., Univ. Man., Winnipeg, pp. 35-38.

Hubbard, W.A. and S. Smoliak. 1953. *Effect of contour dykes and furrows on short-grass prairie.* J. Range Manage., Vol. 6, No. 1, pp. 55-62.

Hutchison, B.A. 1965. *Snow accumulation and disappearance influenced by big sagebrush.* Res. Note RM-46, U.S. Dept. Agric., Rocky Mountain For. and Range Exp. Stn., Fort Collins, Colo.

Israelsen, O.W. and V.E. Hanson. 1962. *Irrigation Principles and Practices.* John Wiley and Sons, Inc., New York, N.Y.

Janzen, P.J., N.A. Korven, G.K. Harris and J.J. Lehane. 1960. *Influence of depth of moist soil at seeding time and of seasonal rainfall on wheat yields in southwestern Saskatchewan,* Publ. No. 1090, Can. Dept. Agric., Res. Branch, Ottawa, Ont.

Kanwar, J.S. 1970. *Agricultural development in India.* In Arid Lands in Transition (H.E. Dregne, ed.), Publ. No. 90, Am. Assoc. Adv. Sci., Washington, D.C.

Kates, R.W. 1961. *Seasonality.* In Papers on Flood Problems (G.F. White, ed.), Dept. Geogr. Res. Pap. No. 70, Univ. of Chicago Press, Chicago, Ill., pp. 114-131.

Keys, C.H. 1961-72. *Moisture conservation and utilization with particular emphasis on winter precipitation.* Unpubl. Annu. Rep., Project 05.01.06, Can. Dept. Agric. Res. Stn., Saskatoon, Sask.

Kibasov, P. 1955. *Effectiveness of various methods of snow retention in Siberia.* Zemledelie, Vol. 3, pp. 60-61. [English Transl. by Res. Branch, U.S. For. Serv., U.S. Gov. Print. Off.]

Koeppe, C.E. 1931. *The Canadian Climate.* McKnight and McKnight, Bloomington, Ill.

Kovda, V.A. 1973. *Quality of irrigation water.* In Irrigation, Drainage and Salinity. An International Source Book, FAO/Unesco, Hutchinson & Co. Ltd., London.

Kuz'min, P.P. 1960. *Snowcover and snow reserves.* Gidrometeorologicheskoe Izdatel'skvo, Leningrad. [English Transl. by U.S. Nat. Sci. Found., Washington, D.C., pp. 99-105.]

Lal, R. and H. Steppuhn. 1980. *Minimizing fall tillage on the Canadian Prairies - a review.* J. Can. Agric. Eng. (In Press).

Larkin, P.A. 1962. *Permeability of frozen soils as a function of their moisture content and fall tillage.* Sov. Hydrol.: Sel. Pap., Am. Geophys. Union, Vol. 4, pp. 445-460.

Levitt, J. 1972. *Responses of Plants to Environmental Stress.* Academic Press, New York, N.Y.

Loegering, W.Q., C.O. Johnson and J.W. Hendrix. 1967. *Culture of wheat.* Ch. 4, Wheat and Wheat Improvement, (K.S. Quisenberry and L.P. Reitz, eds.), Agron. Ser. No. 13, Am. Soc. Agron., Madison, Wisc.

Ludlum, D.M. 1968. *Early American winters II, 1821-1870.* Am. Meteorol. Soc., Boston, Mass.

Lyle, C.V. 1961. *The state of the economic data.* In Economics of Watershed Planning (G.S. Trolley and F.E. Riggs, eds.), Iowa State Univ. Press, Ames, pp. 119-120.

Martin, J.T. 1960. *Use of snow-melt forecasts by the Corps of Engineers for flood control operations on the Rio Grande.* Proc. 28th Annu. Meet., Western Snow Conf., pp. 5-8.

Martin, W.P.; W.E. Fenster and L.D. Hanson. 1970. *Fertilizer management for pollution control.* Ch. 9., In Agricultural Practices and Water Quality (T.L. Willrich and G.E. Smith, eds.), Iowa State Univ. Press, Ames, pp. 142-158.

Matthews, G.D. 1940. *Snow utilization in Prairie agriculture.* Publ. 696, Farmer's Bull. 95, Can. Dept. Agric., Res. Branch, Ottawa, Ont.

Matthews, G.D. 1949. *Scott Dominion Experimental Station, progress report 1937-47.* Can. Dept. Agric., Exp. Farms Serv., King's printer, Ottawa, Ont.

McKay, G.A. 1963. *Relationships between snow survey and climatological measurements.* Int. Union Geod. Geophys., Gen. Assem. Berkeley. [Surface Water]. Int. Assoc. Sci. Hydrol., Publ. No. 63, pp. 214-227.

McKay, G.A. 1964. *Relationships between snow survey and climatological measurements for the Canadian Great Plains.* Proc. 32nd Annu. Meet., Western Snow Conf., pp. 9-18.

McPherson, F.B. 1962. *Movement of air through porous livestock shelters.* M. Sc. Thesis, Dept. Agric. Eng., Univ. Sask., Saskatoon.

Meteorological Branch. 1965. *Canadian weather review.* Can. Dept. Transp., Ottawa, Vol. 3, No. 9.

Meteorological Branch. 1968. *Canadian weather review.* Can. Dept. Transp., Ottawa, Vol. 6, No. 9.

Michel, A.A. 1967. *The Indus River.* Yale Univ. Press, New Haven, Conn.

Miller, D.H. 1966. *Transport of intercepted snow from trees during snow storms.* Res. Pap. PSW-33, U.S. Dept. Agric., For. Serv., Berkeley, Calif.

Molberg, E.S. and J.R. Hay. 1968. *Chemical weed control on summerfallow.* Can. J. Soil Sci., Vol. 48, pp. 255-263.

Mosienko, N.A. 1958. *The water permeability of frozen soil in the Kulundin Steppe.* (In Russian). Pochvovdenie, Vol. 9, pp. 122-127.

Moysey, E.B. and F.B. McPherson. 1966. *Effect of porosity on performance of windbreaks.* Trans. Am. Soc. Agric. Eng., Vol. 9, pp. 74-76.

Neff, E.L. 1980. *Snow trapping by contour furrows in southeastern Montana.* Submitted to J. Range Manage. (Unpubl.).

Neff, E.L. and J.R. Wight. 1977. *Overwinter soil water recharge and herbage production as influenced by contour furrowing on eastern Montana rangelands.* J. Range Manage., Vol. 30, pp. 193-195.

Nicholaichuk, W. and D.I. Norum. 1975. *Snow management on the Canadian Prairies.* Proc. Symp. on Snow Management on the Great Plains, Publ. No. 73, Great Plains Agric. Counc. and Univ. Nebr. Agric. Exp. Stn., Lincoln, pp. 118-127.

Pendleton, J.W. 1954. *The effect of lodging on spring oat yields test weights.* Am. Agron. J., Vol. 46, pp. 265-267.

Peterson, D.F. 1970. *Water in the deserts.* Arid Lands in Transition (H.E. Dregne, ed.), Publ. No. 90, Am. Assoc. Adv. Sci., Washington, D.C.

Powers, R.F. and W.W. Oliver. 1970. *Snow breakage in a pole-sized ponderosa pine plantation*. Res. Note PSW-218, U.S. Dept. Agric. For. Serv., Berkeley, Calif.

Rapp, E.; J.C. Van Schaik and N.N. Khanal. 1969. *Growing irrigated crops in southern Alberta*. Publ No. 1152, Can. Dept. Agric., Res. Branch, Ottawa, Ont.

Rauzi, F. 1975. *Snow management for water conservation in Wyoming*. Proc. Symp. on Snow Management on the Great Plains, Publ. No. 73, Great Plains Agric. Counc. and Univ. Nebr. Agric. Exp. Stn., Lincoln, pp. 180-186.

Rauzi, F. and R.L. Lang. 1957. *Range pitting*. What's New in Crops and Soils, Vol. 9, No. 9.

Read, D.W.L. and F.G. Warder. 1974. *Influence of soil climate factors on fertilizer response of wheat grown on stubble land in southwestern Saskatchewan*. Agron. J., Vol. 66, pp. 245-248.

Renwick, B. 1962. *Air movement around open-front sheds*. B.Sc. Thesis, Dept. Agric. Eng., Univ. Sask., Saskatoon.

Rodenko, G. 1970. *Protecting crop bands*. Zemledeliya, No. 11. [English Transl. by R.P. Knowles, Can. Dept. of Agric. Res. Branch, Saskatoon, Sask.].

Rostvedt, J.O. and others. 1968. *Summary of floods in the United States during 1962*. Water Supply Pap. 1820, U.S. Geol. Surv., U.S. Gov. Print. Off., Washington, D.C.

Ruffner, J.A. and F.E. Bair. 1974. *The Weather Almanac*. Gale Research Co., Book Tower Publ. Co., Detroit, Mich.

Rylov, S.P. 1969. *Snowcover evaporation in the semidesert zone of Kazakhstan*. Trans. Kazakh Hydrometeorol. Sci. Res. Inst., (Trudy Kaz NIGMI), No. 32, pp. 64-77. [English transl., Sov. Hydrol. Sel. Pap., Issue No. 3, 1969, pp. 258-270].

Saskatchewan Advisory Council on Soils and Agronomy. 1978. *Saskatchewan fertilizer and cropping practices 1978-79*. Sask. Dept. Agric. and Extension Div., Univ. Sask., Saskatoon.

Saskatchewan Agricultural Services Co-ordinating Committee. 1975. *Guide to farm practice in Saskatchewan 1975*. Univ. Sask and Sask. Dept. Agric. and Can. Dept. Agric., Edited and produced by Extension Div., Univ. Sask., Saskatoon.

Schantz-Hansen, T. 1939. *Ten-year observations on thinning of 15-year old red pine*. J. For., Vol. 37, pp. 963-966.

Schlehuber, A.M. and B.B. Tucker. 1967. *Culture of wheat*. Wheat and Wheat Improvement (K. Quisenberry and L. Reitz, eds.), Agron. Ser. No. 13, Am. Soc. Agron., Madison, Wisc., pp. 117-179.

Shul'gin, A.M. 1957. *Temperaturni rezhim pochvy (The temperature regime of soils)*. GIMZ Gidrometeorologischeskoe Izdatel'stvo, Leningrad, USSR. [English Transl. by A. Gourevitch, Israel Prog. Sci. Transl., 1965, Jerusalem].

Smika, D.E. and C.J. Whitfield. 1966. *Effect of standing wheat stubble on storage of winter precipitation*. J. Soil and Water Conserv., Vol. 21, pp. 138-141.

Sonmore, L.G. 1955-58. Annual reports, irrigation substation, Vauxhall, Alberta. Can. Dept. Agric., Lethbridge Res. Stn., Lethbridge, Alta.

Sonmor, L.G. 1963. *Seasonal consumptive use of water by crops grown in southern Alberta and its relationship to evaporation*. Can. J. Soil Sci., Vol. 43, pp. 287-297.

Sprague, V.G. and L.F. Graber. 1940. *Physiological factors operative in ice sheet injury of alfalfa*. Plant Physiol., Vol. 15, pp. 661-673.

Staple, W.J. and J.J. Lehane. 1952. *The conservation of soil moisture in southern Saskatchewan.* Sci. Agric., Vol. 32, pp. 36-47.

Staple, W.J. and J.J. Lehane. 1954. *Wheat yield and use of moisture on substations in southern Saskatchewan.* Can. J. Agric. Sci., Vol. 34, pp. 460-468.

Staple, W.J. and J.J. Lehane. 1955. *The influence of field shelterbelts on wind velocity, evaporation, soil moisture and crop yield.* Can. J. Agric. Sci., Vol. 35, pp. 440-453.

Staple, W.J., J.J. Lehane and A. Wenhardt. 1960. *Conservation of soil moisture from fall and winter precipitation.* Can. J. Soil Sci., Vol. 40, pp. 80-88.

Steed, G.L. 1976. *Snowmelt runoff from Lee Creek Basin, Alberta.* Unpubl. Tech. Rep., Alberta Dept. Environ., Agro-Hydrological Branch, Lethbridge, Alta.

Steppuhn, H. and D.M. Gray. 1977. *Potential sublimation on the Canadian Prairies.* Unpubl. Rep., Div. Hydrol., Univ. Sask., Saskatoon.

Steppuhn, H. and G.E. Dyck. 1974. *Estimating true basin snowcover.* Adv. Concepts Tech. Study Snow Ice Resour., Inter. Discip. Symp. U.S. Nat. Acad. Sci., Washington, D.C., pp. 314-324.

Stoddart, L.A., A.D. Smith and T.W. Box. 1975. *Range Management.* (3rd Ed.), McGraw-Hill, Inc., New York, N.Y.

Storr, D. 1967. *Precipitation variations in a small forested watershed.* Proc. 35th Annu. Meet., Western Snow Conf., pp. 11-16.

Tabler, R.D. 1968. *Physical and economic design criteria for induced snow accumulation projects.* Water Resour. Res., Vol. 4, pp. 513-519.

Tabler, R.D. 1971. *Snow fences for watershed management.* Proc. Symp. on Snow and Ice in Relation to Wildlife and Recreation, (A.O. Haugen, ed), Iowa Coop. Wildlife Res. Unit, Iowa State Univ., Ames, pp. 116-121.

Tabler, R.D. 1973. *Evaporation losses of wind blown snow, and the potential for recovery.* Proc. 41st Annu. Meet., Western Snow Conf., pp. 75-79.

Theakston, F.H. 1961. *Snow accumulation around open-front buildings.* Can. J. Agric. Eng., Vol. 3, No. 1, pp. 19 and 32.

Theakston, F.H. 1967. *Advances in the use of models to predict behavior of snow and wind.* Pap. No. 67-433, 60th Annu. Meet. Am. Soc. Agric. Eng. co-sponsored with the Can. Soc. Agric. Eng., Saskatoon, Sask.

Toperczer, M. 1947. *Die Bodentemperaturen in Wien 1911-1944. (The Soil Temperature of Vienna 1911-1944).* Jahrbuch der Zentralanstalt für Meteorol. und Geodyn. (1946), Anhang 6, Wien.

Toumey, J.W. and Korstian. 1947. *Foundations of Silviculture upon an Ecological Basis.* (2nd Ed.) John Wiley and Sons, Inc., New York, N.Y.

United Nations. 1964. *Economic Bulletin for Latin America, VIII,* No. 2, (as of Oct. 1963), New York, p. 162.

Unesco. 1976. *World catalogue of very large floods.* Vol. 21, In Studies and Reports in Hydrology (H. Kikkaw and V. Stanescu, eds.), Unesco Press, Paris.

U.S. Bureau of Reclamation. 1961. *Reclamation project data.* U.S. Gov. Print. Off., Washington, D.C.

U.S. Department of Commerce. 1949. *Climatological data and national summary.* U.S. Gov. Print. Off., Washington, D.C.

U.S. Department of Commerce. 1968. *Climate atlas of the United States.* Nat. Weather Serv., NOAA, ESSA and EDS, Silver Springs, Md.

U.S. Soil Conservation Service. 1972. *Snow survey and water supply forecasting.* Section 22, SCS Nat. Eng. Handbook, U.S. Dept. Agric., U.S. Gov. Print. Off., Washington, D.C., pp. 1.1-9.35.

Van Haveren, B.P. and W.D. Striffler. 1976. *Snowmelt recharge on a shortgrass prairie site.* Proc. 44th Annu. Meet., Western Snow Conf., pp. 56-62.

Van Wijk, W.R. and W.J. Derksen. 1963. *Sinusoidal temperature variation in a layered soil.* Ch. 6, In Physics of Plant Environment (W.R. Van Wijk, ed.), North-Holland Publ. Co., Amsterdam.

Vasil'yev, I.M. 1956. *Wintering of plants.* [English Transl. from Russian by Am. Inst. Biol. Sci., 1961, Washington, D.C.]

Warder, F.G., J.J. Lehane, W.C. Hinman and W.J. Staple. 1963. *The effect of fertilizer on growth, nutrient uptake and moisture use of wheat on two soils in southwestern Saskatchewan.* Can. J. Soil. Sci., Vol. 43, pp. 107-116.

Warner, L.A. 1973. *Flood of June 1964 in the Oldman and Milk River Basins, Alberta.* Can. Dept. Energy, Mines and Resour., Water Resour. Branch, Inland Waters Direct., Ottawa, Ont.

Water Resources Council. 1978. *Second U.S. national water assessment.* Nat. Tech. Inf. Serv., Washington, D.C.

Water Survey of Canada. 1972. *Surface water data - Alberta: Lee Creek at Cardston, Alberta, Station No. 05AE002.* Water Manage. Serv., Can. Dept. Environ., Ottawa, Ont.

Water Survey of Canada. 1974. *Historical streamflow - Ontario Summary: St. Lawrence River at Cornwall, Ontario, Station No. 02MC002.* Water Manage. Serv., Can. Dept. of Environ., Ottawa, Ont.

Weinberger, M.L. 1961. *Potential investment in small watersheds.* In Economics of Watershed Planning (G.S. Tolley and F.E. Riggs, eds.), Iowa State Univ. Press, Ames, pp. 42-68.

Wernstedt, F.L. 1972. *World Climate Data.* Climate Data Press, Penn. State Univ., Lemont.

Wight, J.R., E.L. Neff and F.H. Siddoway. 1975. *Snow management in eastern Montana rangelands.* Proc. Symp. on Snow Management on the Great Plains. Publ. No. 73, Great Plains Agric. Counc. and Univ. Nebr. Agric. Exp. Stn., Lincoln, pp. 138-143.

Willis, W.O. 1979. *Snow on the Great Plains.* Modeling of Snow Cover Runoff (S. Colbeck and M. Ray, eds.), U.S. Army Cold Reg. Res. Eng. Lab., Hanover, N.H., pp. 56-62.

Willis, W.O. and C.W. Carlson. 1962. *Conservation of winter precipitation in the Northern Plains.* J. Soil and Water Conserv., Vol. 17, pp. 122-123.

Willis, W.O., C.W. Carlson, J. Alessi and H.J. Haas. 1961. *Depth of freezing and spring runoff as related to fall soil moisture level.* Can. J. Soil Sci., Vol. 41, pp. 115-123.

Willis, W.O. and A.B. Frank. 1975. *Water conservation by snow management in North Dakota.* Proc. Symp. on Snow Management on the Great Plains, Publ. No. 73, Great Plains Agric. Counc. and Univ. Nebr. Agric. Exp. Stn., Lincoln, pp. 155-162.

Willis, W.O. and H.J. Haas. 1969. *Water conservation overwinter in the northern plains.* J. Soil Water Conserv., Vol. 24, pp. 184-186.

Willis, W.O. and H.J. Haas. 1971. *Snow and snowmelt management with level benches, small grain stubble and windbreaks.* Proc. Symp. on Snow and Ice in Relation to Wildlife and Recreation, (A.O. Haugen, ed.), Iowa Coop. Wildlife Unit, Iowa State Univ., Ames, pp. 86-95.

Willis, W.O., H.J. Haas and C.W. Carlson. 1969. *Snowpack runoff as affected by stubble height.* Soil Sci., Vol. 107, pp. 256-259.

Work, R.A. 1955. *Measuring snow to forecast water supplies.* In Water, 1955 Yearbook of Agriculture. U.S. Dept. Agric., U.S. Gov. Print. Off., Washington, D.C., pp. 95-102.

Ylimäki, A. 1962. *The effect of snow cover on temperature conditions in the soil and overwintering of field crops.* Annales Agriculturae Fenniae, Vol. 1, pp. 192-216. (Seria Phytopathologia No. 3, Sarja Kasvitaudit 3).

Zingg, A.W. and V.L. Hauser. 1959. *Terrace benching to save potential runoff for semiarid land.* Agron. J., Vol. 51, pp. 289-292.

PART II

SNOWFALL AND SNOWCOVER

127

4

SNOWFALL FORMATION

R. S. SCHEMENAUER

Atmospheric Research Directorate, Atmospheric Environment Service, Downsview, Ontario.

M. O. BERRY

Canadian Climate Centre, Atmospheric Environment Service, Downsview, Ontario.

J. B. MAXWELL

Canadian Climate Centre, Atmospheric Environment Service, Downsview, Ontario.

INTRODUCTION

The physics of snowfall is a special case of the broader subject, the physics of precipitation, since the meteorological conditions producing snowfall are the same as those generating other forms of precipitation. In general, the occurrence of precipitation is determined by the availability of atmospheric moisture and the presence of mechanisms which can convert this moisture into precipitation. This conversion is mostly associated with cooling resulting from the vertical motion of the air. For snow, the temperature must be at or below the freezing point.

During its "formative" stage in the atmosphere snow can be defined simply as particles of ice formed in a cloud which have grown large enough to fall with a measurable velocity and reach the ground. In practice, certain restrictions must be placed on this definition; for example, ice pellets and hailstones are not considered "snow"; also rime or hoarfrost are not included because they are formed at the point of deposition.

GENERAL PRINCIPLES OF THE METEOROLOGY OF PRECIPITATION

Moisture Sources

The formation and distribution of moisture in the atmosphere is basic for development of precipitation. Moisture is acquired by evaporation from oceans, rivers, soils, and plants. The rate and direction of moisture exchange between the air and the underlying surface depends on several factors. Over water, evaporation occurs when heat is supplied directly from the sun or air, and the vapor pressure gradient is positive; i.e., when the vapor pressure of an air layer is less than the saturation vapor pressure of the underlying air layer adjacent to the surface. On land, the availability of water, which depends on climate and characteristics of the soil and plant regimes, has an important effect on this exchange process.

During winter in the Northern Hemisphere the oceans, particularly the Pacific and Atlantic north of 40° N, and the Gulf of Mexico, are the main sources of atmospheric moisture. A continental air mass moving out over the ocean usually has a much lower temperature and vapor pressure than the underlying water surface, resulting in large upward fluxes of both heat and moisture to the lower layers of the atmosphere. The height to which this moisture is transported depends on the surface exchange processes as well as the large-scale vertical atmospheric motion, which is influenced by the upper flow pattern.

Although the oceans are the primary source of moisture on a hemispheric scale in winter, other bodies of water can be important on a smaller scale. For example, in Canada before ice cover forms in the early part of winter, Hudson Bay and the Great Lakes can be important sources.

Atmospheric Moisture Distribution

The amount of water vapor contained in the atmosphere in a particular region is influenced by both the temperature of the air, which determines the upper limit of its vapor capacity, and the accessibility of the region to air flow from moisture sources. The pronounced decrease in the moisture holding capacity of air (increase in relative humidity) with lower temperature is shown in Table 4.1.

For North America, the North Pacific is a major source region of moisture which is carried eastward by the prevailing westerly circulation. In winter, the maximum moisture transport across the Pacific coast probably occurs between latitudes 50° and 55° N with the mean flow decreasing almost to zero

Table 4.1
THE CAPACITY OF THE ATMOSPHERE TO RETAIN MOISTURE
AS A FUNCTION OF TEMPERATURE (McKay, 1970)

Air Temperature °C	Relative Humidity %					
30	16	24	31	45	57	100
20	28	42	54	79	100	
16	36	53	77	100		
10	52	77	100			
6	67	100				
0	100					
Water Vapor content (g/m^3)	4.85	7.27	9.41	13.65	17.31	30.4

between 30° and 35° N (Rasmusson, 1967). A portion of this moisture is lost as precipitation while air crosses the Cordillera; that which is retained is carried across the plains. In eastern Canada and the northeastern United States, the Atlantic Ocean and the Gulf of Mexico also serve as source regions.

When considering the distribution of atmospheric moisture associated with transport patterns, it is useful to use the concept of precipitable water; i.e., the depth to which liquid water would stand if all water vapor were condensed from a vertical atmospheric column of uniform cross-section. Hay (1970) computed the distribution of mean precipitable water for Canada; the January distribution is shown in Fig. 4.1. The largest values (8 to 12 mm) occurring along the British Columbia coast can be attributed to warm moist air from the Pacific. A second maximum (4 to 7 mm) occurring over the Atlantic provinces reflects the influence of the Atlantic Ocean and, to a lesser extent, the Gulf of Mexico on the atmospheric moisture supply. Over much of Ontario, Quebec and the Prairie provinces, values of the mean precipitable water range from 2 to 4 mm. The minimum values of 2 mm or less occur over Keewatin, N.W.T. and the Arctic Islands, since the air temperatures during January are very low over these regions which are also distant from the major moisture sources.

Vertical Motion

Precipitation normally forms when the upward vertical motion of air causes the atmosphere to cool by adiabatic expansion (an expansion process is adiabatic when there is no heat exchange between a mass of air and its

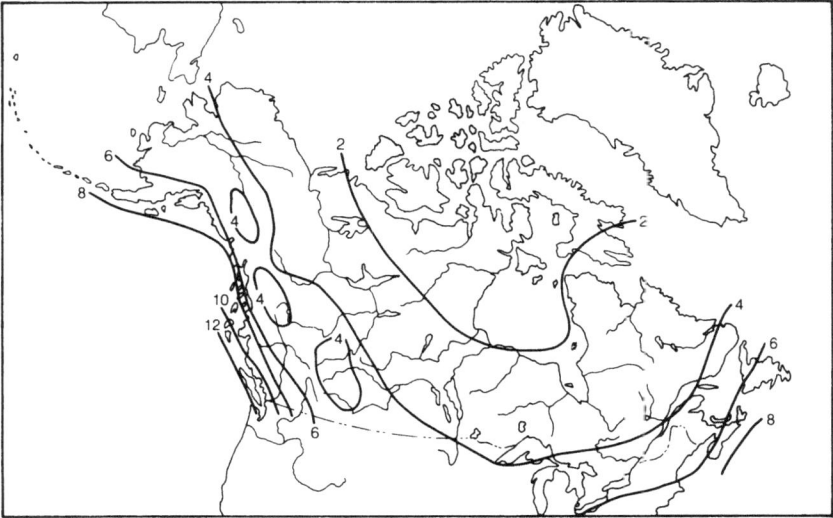

Fig. 4.1 Distribution of the mean depth of precipitable water (mm) over Canada in January (Hay, 1970).

surroundings). Even though this cooling is rare in the atmosphere, the non-adiabatic effects are small enough to be ignored for most applications. Four types of vertical motion associated with the formation of significant amounts of precipitation can be identified.

Horizontal convergence. The wind field directs the flow of air into a particular area, e.g., a surface low pressure area, causing the air to lift.

Orographic lift. When air flows against a mountain barrier, a ridge, or a group of hills, some is forced to rise, causing cooling, extensive condensation and precipitation. This is common along the west slopes of the Rocky Mountains in British Columbia whenever strong upslope winds bring moist air into the area.

Convective lift. Differential heating or advection results in some of the atmosphere becoming more buoyant than its environment, e.g., when air flows over a warmer body of water. The resultant vertical motion can cause the formation of cumuliform clouds which can produce heavy showery snowfall under freezing conditions. The snowbelt areas south and east of Lake Huron and Lake Ontario receive major accumulations of snow because of convective activity.

Frontal lift. Fronts are the zones which separate large masses of air having significantly different physical properties (namely temperature and moisture). Differences in the density and motion of the two air masses associated with a front frequently result in large-scale lifting causing extensive cloud systems and precipitation. Where a warm air mass displaces a cold air mass (warm front), the warmer air, being of lower density, is forced to rise above the heavier, cooler air. For the other case, (cold front), the warm air is lifted vigorously by the advancing wedge of cold air. Frontal storms tend to give widespread precipitation with the heaviest amounts often occurring in the northeast quadrant of a surface low pressure area where the combined lifting due to convergence and frontal over-running is the greatest.

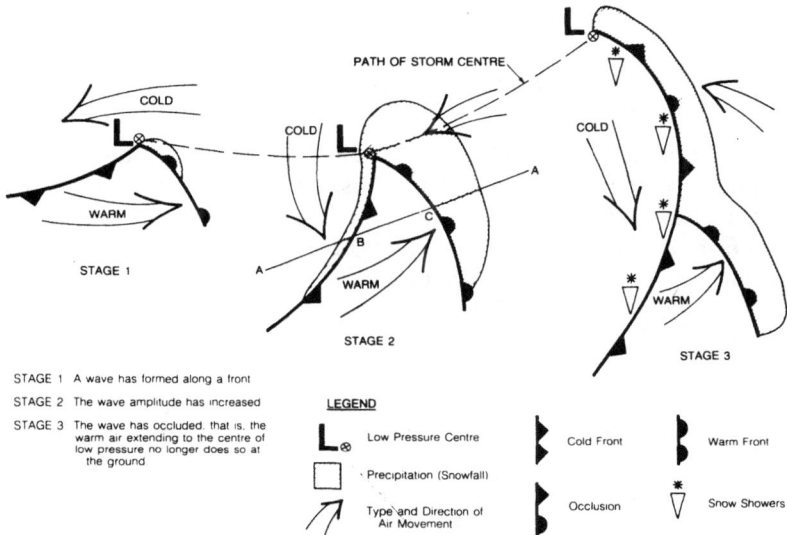

Fig. 4.2 Stages in the development of a frontal system.

Synoptic Weather Systems

The amount of cloud and precipitation (snow) associated with fronts in major synoptic weather systems depends on the rate of lifting of one air mass above another. This is usually greatest near the waves which form on a front.

Figure 4.2 illustrates the development of a frontal wave. Three stages in its life cycle may be identified:

(1) Formation on a front,
(2) Increase in amplitude, and
(3) Occlusion, i.e., the warm air extends to the center of the low pressure region aloft but not at the ground.

In winter, an average frontal wave moves at ~11 m/s. Its precipitation pattern (in the absence of instability) usually shows steady precipitation extending 250 km ahead of the warm front and within 40 to 80 km of the cold front. Precipitation rates are generally heaviest within 80 km of the low pressure area in the northeast quadrant, and tend to decrease with distance away from both the front and the low pressure center.

However, in any given storm, conditions may vary markedly depending on the moisture supply and the vigour of the wave. Strong convective development, indicative of instability in the frontal system, further complicates the precipitation pattern by producing pockets of heavy precipitation.

Topographic and Orographic Factors

Topographic and orographic factors significantly affect the cloud and precipitation patterns of synoptic systems. When a warm frontal surface crosses a mountain range, the orographic upglide superimposed upon the general frontal upglide increases the rates of condensation and precipitation. Since the mountains also retard the front's motion the duration of the precipitation is prolonged. The topographic and orographic factors produce effects on the formation and character of precipitation similar to those associated with the passage of a cold front or an occlusion.

In winter, coastal valleys, fjords, and lowlands are frequently filled with stagnant cold air which, with the onshore movement of a disturbance above, can be accelerated towards the coast. This drainage, acting simultaneously with the inland flow from the sea, establishes an irregular zone of low-level convergence along the coast. As a result, a zone of ascending motion is initiated and maintained producing cloudiness and precipitation along the coast until either the supply of cold air has been exhausted or the disturbance has passed inland.

Cloud and Precipitation Patterns

Figure 4.3 illustrates the typical cloud structure observed along the cross-section AA through the warm and cold fronts of Fig. 4.2. The characteristic snowfall which may be associated with each cloud type is listed in Table 4.2.

The cloud pattern of a typical occluded system (Stage 3 in Fig. 4.2) usually consists of lines of convective cloud embedded in stratiform cloud layers. Unsaturated layers are interspersed, particularly at mid-levels; generally, the snowfall is less intense and less continuous than that in the earlier stages when the mature wave of the system is developing.

LEGEND

Frontal Surface

Air Flow

Precipitation

Cloud Abbreviations are discussed in Table 4-2.

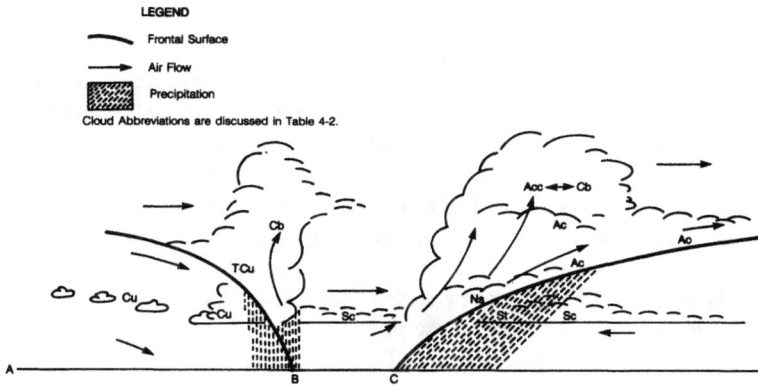

Fig. 4.3 Typical cloud structures associated with warm and cold fronts.

Table 4.2
SNOWFALL ASSOCIATED WITH VARIOUS CLOUD TYPES

Front	Cloud Type	Possible Snowfall
Warm	Cirrus and Derivatives (Ci)	
	Altocumulus (Ac)	Usually virga; light snow showers
	Altocumulus Castellanus (Acc)	may occur from the Acc.
	Altostratus (As)	Light continuous or intermittent snow from As; however, when the snow is heavy, the As has probably graduated to a nimbostratus cloud.
	Nimbostratus (Ns)	Continuous snow. Virga occurs from both Ns and As.
	Stratocumulus (Sc)	Intermittent light powdery snow (fine flakes).
	Stratus (St)	Continuous light powdery snow. (This is the frozen precipitation analogue of warm precipitation drizzle.)
Cold	Cumulus (Cu) and Towering Cumulus (Tcu)	Light snow showers possible; more likely from the Tcu.
	Cumulonimbus (Cb)	Moderate to heavy snow showers.

THE FORMATION OF SNOW

The formation of snow in the atmosphere depends on many variables; the most important being that the ambient temperature must be less than 0° C and that supercooled water must be present. The flow diagram shown in Fig. 4.4 outlines the processes which produce different types of snow. Initially a cloud is formed by condensation of the water vapor in an ascending region of warm moist air. Once the cloud temperatures drop below freezing, conditions are suitable for forming snow. At about -5° C the nuclei present in the atmosphere form tiny crystals through the process of ice nucleation. The ice crystal is the initial stage in growth of the snow particle, having a small size, a diameter usually less than 75μm; a very low fall velocity, typically less than 5 cm/s; and a very simple crystal habit or shape, often a hexagonal plate. If the ice crystal continues to grow by sublimation a snow crystal is formed. This is a large individual particle, often having a very intricate shape, and of such a size that it is readily visible to the naked eye. A snowflake is an aggregation of snow crystals. In some cases, a snow crystal passes through a region within the cloud of higher cloud droplet concentrations where riming occurs if its crystal size is greater than approximately 300 μm, i.e., small (10-40 μm) cloud droplets freeze on contact onto the snow crystal. This process, occurring typically at temperatures from -5 to -20° C, results in the formation of a "rimed crystal" and, in cases of extreme riming, in the formation of a graupel particle (snow pellet). Thus, snow may consist of snow crystals, rimed snow crystals and snowflakes as well as fragments of any of these.

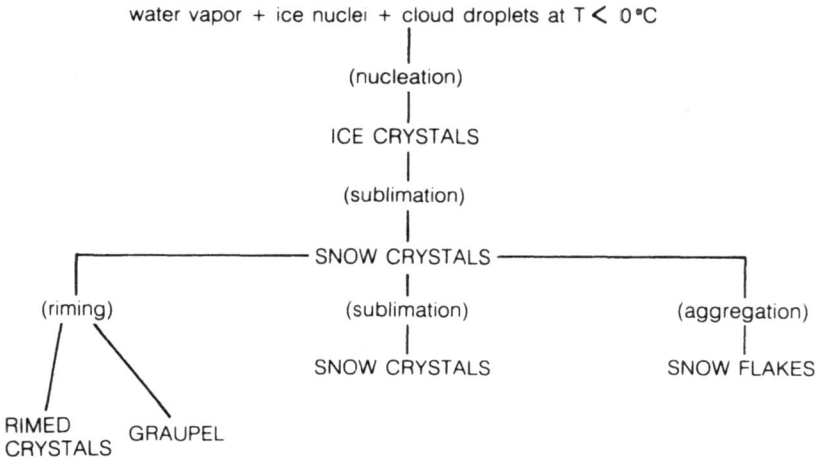

water vapor + ice nuclei + cloud droplets at T < 0 °C

(nucleation)

ICE CRYSTALS

(sublimation)

SNOW CRYSTALS

(riming)   (sublimation)   (aggregation)

SNOW CRYSTALS   SNOW FLAKES

RIMED   GRAUPEL
CRYSTALS

Fig. 4.4 Flow diagram of the formation of different types of snow.

Ice Nucleation

At any given time one cubic centimetre of the lower atmosphere contains from several hundred to many thousands of aerosol particles, most of which have sizes ranging from 0.01 to 1 μm. Particles on which water vapor will condense when the air is supersaturated are called "condensation" nuclei. "Ice" nuclei are particles which cause the formation of ice through either the direct freezing of cloud droplets or the freezing of water deposited as a vapor onto the particle's surface; the nucleation of the ice phase by these particles is a process called "heterogeneous" nucleation. At temperatures below -40° C a minute ice particle may be formed by a chance combination of water molecules, in the process called "homogeneous" nucleation.

Cloud physics is concerned primarily with heterogeneous nucleation, which exists as three basic and extremely complex processes involving ice nuclei. If an ice nucleus is present in a cloud droplet which is cooled below 0° C the droplet will freeze at a temperature which is determined by the nature of the nucleus. In this case the ice nucleus has acted as an "immersion-freezing" nucleus. If an ice nucleus touches the outer surface of a droplet and causes it to freeze, it has served as a "contact" nucleus. The third type is the "deposition" nucleus; it is a particle on which direct deposition of vapor will occur, forming an ice layer on the surface.

Ice nuclei are found with a wide variety of chemical compositions; each is characterized by a particular activation temperature at which ice formation can be initiated. Nuclei active in the range -5 to -10° C would be classed as efficient whereas those active below -20° C would be considered as poor nucleating agents.

The major source of ice nuclei in the atmosphere is the earth's surface from which dust is raised by the wind. Clay-silicate particles are one of the most common and most efficient natural ice-nucleating agents. Other, less important sources are industrial plants and forest fires. Particles of extraterrestrial origin and organic matter from vegetation may also serve as nuclei. In addition, these sources are major producers of other aerosols. Only a very small fraction of aerosols are active as ice nuclei, about one in 10^9, at -10° C. The most probable sizes of ice nuclei are in the range 0.1-1 μm. The concentration (number per unit volume of air) of "active" nuclei increases rapidly with a decrease in air temperature, e.g., between -10 and -30° C it increases approximately by a factor of 10, with each decrease of 4° C. Concentrations also vary greatly in time and space: changes of one or two orders of magnitude or more can occur over periods from a few minutes to a few weeks, and over distances of less than a few decametres.

Growth of Crystals by Vapor Deposition

At a given temperature an ice crystal embedded in a cloud of water droplets will grow at the expense of the droplets because the vapor pressure at the ice surface will be less than that at the water surface. At temperatures conducive to the formation of snow a cloud may only be slightly supersaturated with respect to water but 10 to 20% supersaturated with respect to an ice surface. The net result is a transfer of water vapor from the droplets to the surface of the ice crystal, which grows by absorbing water molecules into the crystal faces. This preferential growth of ice crystals in a cloud is the basis of the Bergeron mechanism (1935) to explain precipitation development and formation especially in temperate latitudes. The crystals rapidly reach a size when they attain a downward fall velocity relative to the cloud droplets. Each crystal falls and sweeps out the cloud droplets, and thus grows predominantly by a riming process. If, during its fall, it passes through the $0°$ C level of the cloud it may melt completely to fall as rain.

The basic habit (shape) of an ice crystal is determined by the temperature at which it grows whereas its rate of growth and secondary crystal features are determined by the degree of supersaturation. Habit changes are produced by the different relative growth rates of the crystal faces; this leads in extreme cases to prism-like or plate-like crystals which result from more rapid growth of the prism or the basal (base or end) faces, respectively. In the temperature range 0 to $-4°$ C, plates predominate; -4 to $-10°$ C, prism-like crystals, scrolls, sheaths and needles; -10 to $-20°$ C, thick plates, dendrites and sector plates; and from -20 to $-35°$ C, sheaths and hollow columns (see Fig. 4.5 for crystal types). The basic transitions in crystal habit occur approximately at $-4°$ C, $-10°$ C, and from -20 to $-22°$ C; at these temperatures the curves representing the rates of growth of the basal and prism faces cross.

The rate of increase with time t of a mass m of a crystal through diffusion of water vapor onto its surface is given by

$$dm/dt = 4\pi CDFA_c (\rho_\infty - \rho_0),   \qquad 4.1$$

where: C = shape factor,
          D = diffusivity of water vapor in air,
          F = ventilation factor depending on the relative motion of the crystal with respect to the air,
          A_c = a function of the crystal size and the accommodation co-efficient, which depends on the temperature and the nature of the crystal's surface,
          ρ_∞ = vapor density (the mass of water vapor per unit volume of moist air) in the stream at a large distance from the crystal, and
          ρ_0 = vapor density at the surface of the crystal.

Equation 4.1 is only valid if the crystal diameter is less than a few hundred micrometres, when its growth by diffusion becomes relatively less important than that by capture of cloud droplets through collisions. At larger diameters the accretion of cloud droplets becomes the dominant factor in its mass growth.

Riming of Snow Crystals

For a snow crystal to grow by accretion the cloud droplets must collide with and adhere to the crystal surface. The collision efficiency is defined as the ratio of the number of droplets that actually impact on the crystal to the number of droplets that are swept out by it inside a column of air enclosed by the crystal's cross sectional area and fall distance. Much experimental and theoretical work has been done to determine the collision efficiencies for the different crystal types but their values are still not well defined. The adhesion efficiency is defined as the ratio of the number of droplets that freeze and remain attached to the crystal to the total number of droplets that impact or graze the surface. Its value is very close to unity and therefore it is often ignored in calculations of crystal growth by riming. The onset of riming occurs at different sizes for different crystals: e.g., hexagonal plates and dendrites with dimensions greater than 200 to 300 μm; columns with minor axes longer than 50 μm.

Study of the riming process is important because riming affects the fall velocity and the motion of snow crystals, and through this process graupel forms and most hailstones originate. Recent evidence indicates that secondary ice particles are ejected during the freezing of a cloud droplet onto a snow crystal. Thus, riming may lead to a "multiplication" of ice particles in a cloud.

The mass growth rate due to riming (dm/dt) of a disc-like crystal falling in a supercooled cloud can be expressed as

$$dm/dt = \pi r^2 abwW, \qquad\qquad 4.2$$

where: r = radius of the crystal,
             a = adhesion efficiency,
             b = collision efficiency,
             w = crystal's fall velocity relative to the droplets, and
             W = liquid water content of the cloud.

This equation shows that the mass growth rate increases rapidly as the particle size increases; similar equations predict that a crystal with a radius of 250 μm may grow to a 1 to 2 mm graupel within 10 to 20 min.

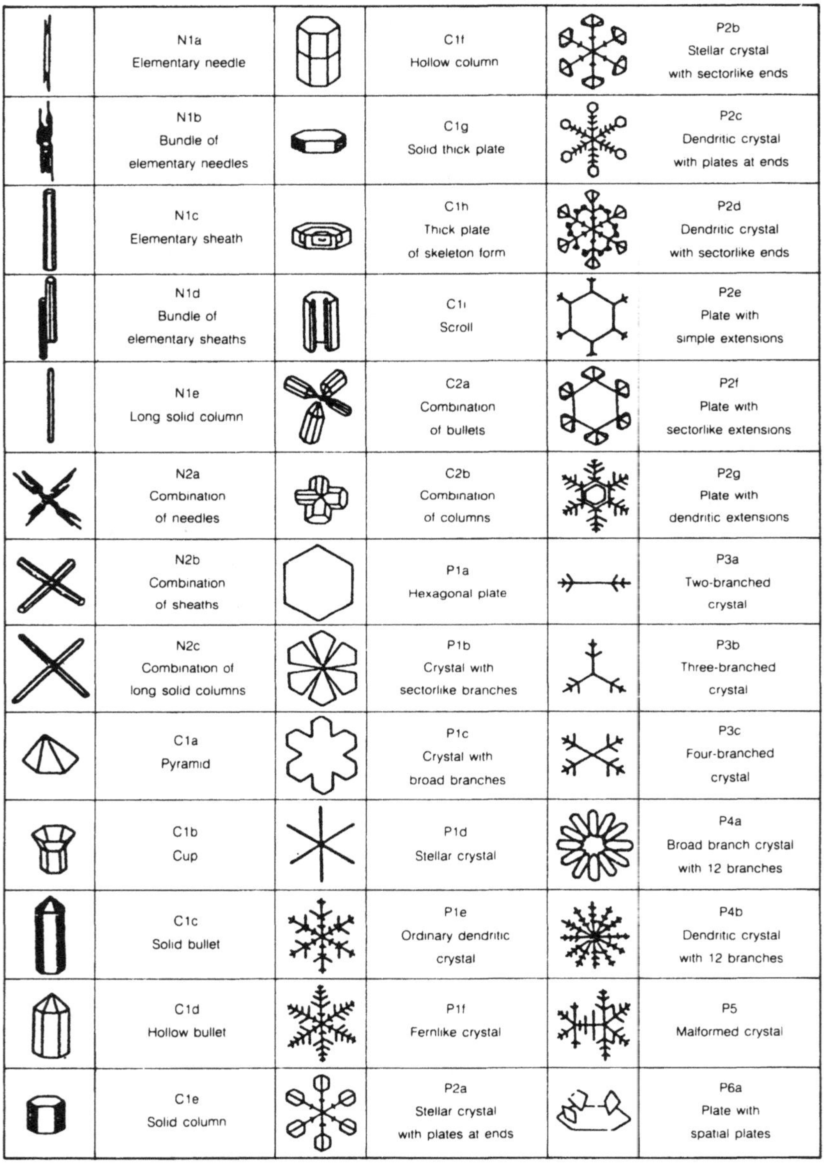

	N1a Elementary needle		C1f Hollow column		P2b Stellar crystal with sectorlike ends
	N1b Bundle of elementary needles		C1g Solid thick plate		P2c Dendritic crystal with plates at ends
	N1c Elementary sheath		C1h Thick plate of skeleton form		P2d Dendritic crystal with sectorlike ends
	N1d Bundle of elementary sheaths		C1i Scroll		P2e Plate with simple extensions
	N1e Long solid column		C2a Combination of bullets		P2f Plate with sectorlike extensions
	N2a Combination of needles		C2b Combination of columns		P2g Plate with dendritic extensions
	N2b Combination of sheaths		P1a Hexagonal plate		P3a Two-branched crystal
	N2c Combination of long solid columns		P1b Crystal with sectorlike branches		P3b Three-branched crystal
	C1a Pyramid		P1c Crystal with broad branches		P3c Four-branched crystal
	C1b Cup		P1d Stellar crystal		P4a Broad branch crystal with 12 branches
	C1c Solid bullet		P1e Ordinary dendritic crystal		P4b Dendritic crystal with 12 branches
	C1d Hollow bullet		P1f Fernlike crystal		P5 Malformed crystal
	C1e Solid column		P2a Stellar crystal with plates at ends		P6a Plate with spatial plates

Fig. 4.5a Classification of natural snow crystals (Magono and Lee, 1966).

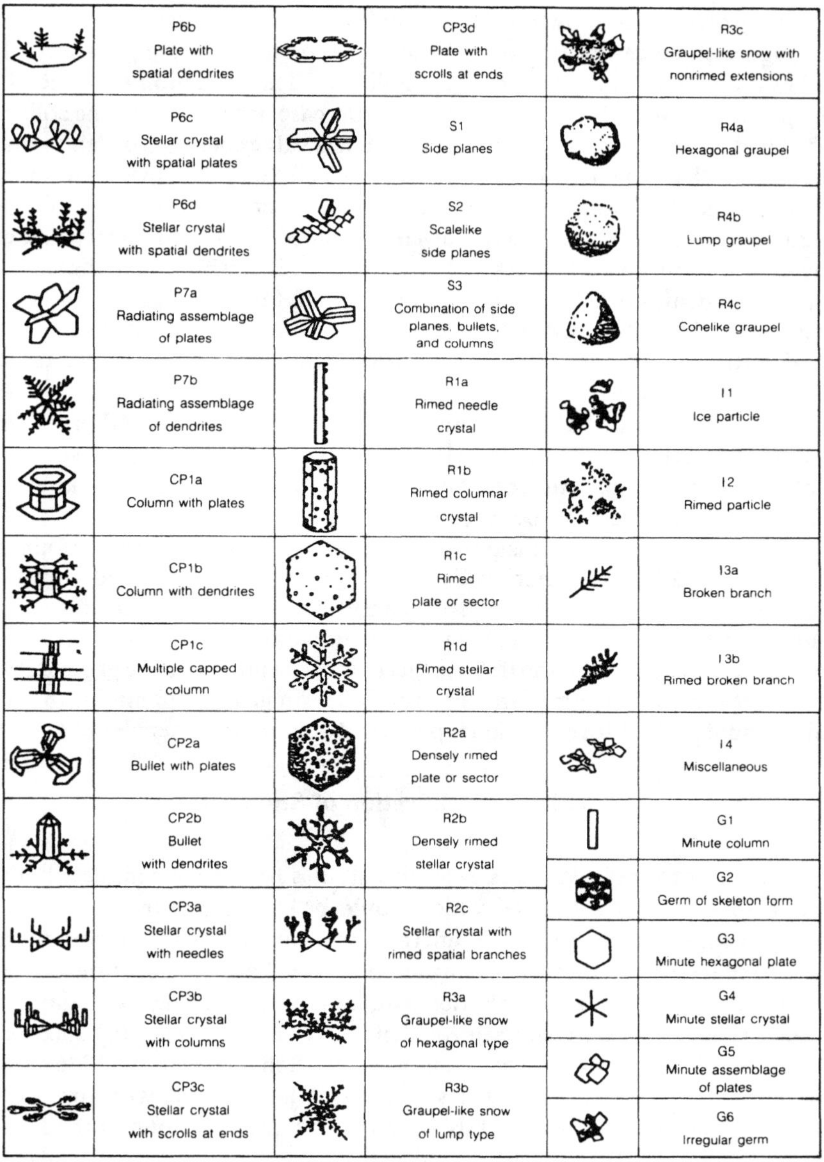

	P6b Plate with spatial dendrites		CP3d Plate with scrolls at ends		R3c Graupel-like snow with nonrimed extensions
	P6c Stellar crystal with spatial plates		S1 Side planes		R4a Hexagonal graupel
	P6d Stellar crystal with spatial dendrites		S2 Scalelike side planes		R4b Lump graupel
	P7a Radiating assemblage of plates		S3 Combination of side planes, bullets, and columns		R4c Conelike graupel
	P7b Radiating assemblage of dendrites		R1a Rimed needle crystal		I1 Ice particle
	CP1a Column with plates		R1b Rimed columnar crystal		I2 Rimed particle
	CP1b Column with dendrites		R1c Rimed plate or sector		I3a Broken branch
	CP1c Multiple capped column		R1d Rimed stellar crystal		I3b Rimed broken branch
	CP2a Bullet with plates		R2a Densely rimed plate or sector		I4 Miscellaneous
	CP2b Bullet with dendrites		R2b Densely rimed stellar crystal		G1 Minute column
					G2 Germ of skeleton form
	CP3a Stellar crystal with needles		R2c Stellar crystal with rimed spatial branches		G3 Minute hexagonal plate
	CP3b Stellar crystal with columns		R3a Graupel-like snow of hexagonal type		G4 Minute stellar crystal
					G5 Minute assemblage of plates
	CP3c Stellar crystal with scrolls at ends		R3b Graupel-like snow of lump type		G6 Irregular germ

Fig. 4.5b Classification of natural snow crystals (Magono and Lee, 1966).

Aggregation of Snow Crystals

The growth of a snowflake by sweeping out snow crystals while falling through a cloud can be described by an equation like 4.2. Aggregation is the adhesion of snow crystals after their collision. The collection efficiency E is the product of the adhesion efficiency (in this case, ice to ice) and the collision efficiency. Most studies have found that E has values between 0.1 and 1 and a magnitude that depends on such parameters as the cloud temperature and the strength and direction of the electric field at the crystal.

Jayaweera and Mason (1965) showed that the wake of a moving particle can increase its collision efficiency. If two similar discs which have dimensions and fall velocities simulating those of real crystals fall one behind the other the attraction of the leading disc for the other can be measured at distances of up to 40 particle diameters, therefore collisions can occur. Similar results were found for cylinders.

For natural snowflakes, adhesion can result from several processes, including interlocking of the crystals, riming, vapor deposition, and sintering. The latter is the evaporation of material from one part of the snowflake and its condensation at the neck between two crystals.

Measurements made inside clouds generally show that maximum snowflake sizes occur near $0°C$, resulting from the increased forces of adhesion at these temperatures. In addition, the greater the distance a snowflake falls the larger it becomes. At temperatures above $0°C$ they melt rapidly and fall as rain if the $0°C$ isotherm is high enough above ground level. In a typical cloud a 1-mm diameter snowflake can grow to 10 mm in about 20 min., and if melted form a raindrop about 1 mm in diameter (Hobbs, 1974).

Artificial Modification of Snowfall

In the past twenty-five years extensive studies have been conducted in the search for artificial ice nuclei which would be effective in initiating the ice phase in supercooled clouds. Numerous organic and inorganic compounds have been tested and their activating temperatures have been catalogued (e.g., see Mason, 1971). Two of the most widely used seeding agents today were among those first discovered: dry ice and silver iodide. The first successful cloud seeding experiment was conducted by Schaefer during 1946 in the United States. After seeding a supercooled layer cloud with dry ice he observed that many snow crystals were formed and the cloud in the seeded region cleared shortly afterward. Project Cirrus, conducted during 1948 in the United States was the first operational test of seeding with silver iodide. A layer cloud composed of supercooled cloud droplets was converted into ice crystals by dropping pieces of burning charcoal impregnated with silver iodide through the cloud.

The redistribution and the initiation and increase of snowfall from a cloud system have been the two main objectives of winter weather modification activities. Both objectives are accomplished by essentially the same means. Heavy seeding produces large numbers of ice crystals which grow by vapor deposition while being carried downwind. If seeding had not been done the smaller number of crystals would normally rime and fall out of the cloud upwind. The net effect of the seeding operation is to increase snowfall farther downwind. Experiments, using this technique over Lake Erie, conducted by the Cornell Aeronautical Laboratory in the late 1960's (Eadie, 1970) were somewhat successful in distributing the snowfall from winter-time lake-effect storms farther inland. The Cloud Physics Group at the University of Washington began their Cascade Project in 1969 to study winter clouds in the Cascades and the effects of seeding these clouds with silver iodide and dry ice. Hobbs et al. (1974) reported that snowfall was successfully targeted onto a predetermined area on the crests of the Cascade range in Washington.

Changes in the microphysical structure of winter-time clouds can be achieved by seeding with artificial nuclei. However, it remains to be proven that snowfall can be targeted consistently onto an area of interest.

PHYSICAL CHARACTERISTICS OF SNOW CRYSTALS

Classification of Snow Crystals

"Snow" and "snow crystals" are terms that are much too general for describing the different precipitation types and processes in detail. A good classification system is necessary so that scientists or laymen carry on meaningful discussions. This is particularly important for scientists in the fields of meteorology and glaciology as well as for skiers and general outdoorsmen. It is difficult to imagine such a classification system not being created since man, when faced with the inherent beauty and variety of the different types of snow would try to communicate this information with his fellow men.

The habit of an ice crystal at its origin in the cloud depends largely on the same atmospheric factors which determine its growth, namely, the temperature and the water vapor available. Hence, the snow crystal that arrives at the ground may be a simple product of ice crystal formation or it may be the complex result of a convoluted life history during which its original physical character was greatly modified and changed. Several systems have been devised to classify ice particles in the atmosphere; they are based on the different shapes and growth processes of the crystal types. Sometimes subtle distinctions among these types will reflect widely different atmospheric conditions. This is somewhat analagous to the manner in which different snow types can affect the degree and rate of metamorphism that the snowpack undergoes which, in turn, influences its strength and stability, an important aspect for avalanche warning procedures.

One system that has been used frequently to classify snow in the atmosphere was proposed by the International Association of Hydrology, Commission on Snow and Ice in 1951 (National Research Council, 1954). Ice particles were classified into ten major categories: plate, stellar crystal, column, needle, spatial dendrite, capped column, irregular crystal, graupel, ice pellet and hail. Within each category the precipitation type may be distinguished according to: broken crystals, rimed particles, clusters, wet or melted, and the maximum particle dimension. Unfortunately, the International Snow Classification System permits snow crystals to be classified only in a broad sense, since irregular forms cannot readily be identified separately. Because of this and other deficiencies this system has fallen into disuse in cloud physics. A more precise classification scheme developed by Magono and Lee (1966) contains sufficient detail to describe almost all crystal types and is now routinely used (see Fig. 4.5a and b). Other particles, such as ice pellets and hail, which are not included in the figure, are covered by standard definitions.

Sizes, Masses, Densities, Terminal Velocities and Concentrations of Snow Crystals

Individual snow crystals observed at the earth's surface range in maximum dimension from $\sim 50\mu$m to ~ 5 mm - a difference of two orders of magnitude. Studies of natural snow crystals, snow flakes and graupel have resulted in mass-diameter relationships describing the different particle types. These expressions are necessarily approximate since each represents the mean of numerous measurements on many particles which are similar but not identical. A further complication arises with measurements made at different locations or times. These may not agree exactly because of differing meteorological conditions. Despite such limitations the information obtained is useful for calculating growth rates or precipitation rates. In these relationships the diameter is that of the smallest circle circumscribing the particle. Table 4.3 gives typical relationships for some common snow types. If more precise estimates of the terminal velocities of snow crystals are desired they are generally calculated directly using an experimentally-determined drag coefficient (e.g., see Hobbs, 1974).

In most cases the density of a given type of snow crystal is difficult to specify because of the difficulties inherent in determining a representative volume.

Table 4.3
MASS-DIAMETER-VELOCITY RELATIONSHIPS FOR
DIFFERENT TYPES OF SNOW CRYSTALS

Snow Crystal Type	Mass-Diameter[a] Relationship	Velocity-Diameter[b] Relationship	Reference
Dendrites	$m = 0.0038D^2$		Nakaya and Tereda (1935)
Needles	$m = 0.0029D$		"
Rimed plates and stellar dendrites	$m = 0.027D^2$		"
Powder snow and spatial dendrites	$m = 0.010D^2$		"
Conical graupel	$m = 0.073D^{2.6}$	$u = 1.2D^{0.65}$	Locatelli and Hobbs (1974)
Hexagonal graupel	$m = 0.044D^{2.9}$	$u = 1.1D^{0.57}$	"
Unrimed combinations of plates, sideplanes, bullets and columns	$m = 0.037D^{1.9}$	$u = 0.69D^{0.41}$	"
Unrimed radiating assemblages of dendrites	$m = 0.073D^{1.4}$	$u = 0.8D^{0.16}$	"
Densely-rimed radiating assemblages of dendrites or dendrites	$m = 0.037D^{1.9}$	$u = 0.79D^{0.27}$	"
Densely rimed columns	$m = 0.033L^{2.3}$	$u = 1.1L^{0.56}$	"

[a, b] m = mass of crystal (mg), D = diameter of crystal (mm), L = length of crystal (mm), u = velocity (m/s)

The particle densities for heavily-rimed crystals and for graupel, whose volumes can be measured with reasonable accuracy, range from approximately 100 to 700 kg/m^3. Also, typical concentrations of snow crystals and snowflakes in a snowfall are very difficult to specify because they may vary widely from 0.1 to 100 particles/l. The concentrations are related to such cloud factors as top temperature, thickness and the height of base above ground. Hobbs et al. (1974) suggest that the ice crystal concentration in a layered, convective cloud top may be expressed by

$$\log_{10} C = -1.11 - 0.13T_c, \qquad\qquad 4.3$$

where: C = crystal concentration (particles/l), and
         T_c = cloud top temperature ($^\circ$C).

Equation 4.3 can be applied for determining concentrations above 0.1 particles/l in clouds with tops colder than -8°C.

PHYSICAL CHARACTERISTICS OF SNOWFLAKES

Snowflakes may consist of from two to several hundred snow crystals joined together. Generally, for snowflakes to form, an array of crystals should be moving at different velocities at air temperatures slightly lower than 0° C. Because they have abundant radiating arms, dendritic crystals (Ple-P2d, see Fig. 4.5) tend to aggregate more readily than other types and are often found as the constituent crystals in snowflakes. However, other crystal types such as needles (Nla-Nle), are also common.

Like a snow crystal, a snowflake is a particle type whose size, fall velocity and other pertinent parameters are difficult to specify, since an infinite variety of possible crystal combinations can produce a snowflake. Their maximum diameter may range from one tenth of a millimetre to several centimetres. The largest occur at temperatures near 0° C with the size decreasing with decreasing temperature (Hobbs, 1973). Results indicate that aggregation ceases at temperatures below -20° C. The fall speed of a snowflake depends on such factors as size, type of constituent crystal and degree of riming. Some fall-speed relationships for snow crystals are given in Table 4.3. Snowflake densities are also quite variable; however, the work of Magono and Nakamura (1965) showed that the density of a snowflake decreased with size according to the equation

$$\rho d^2 = 2 \times 10^{-2}, \qquad\qquad 4.4$$

where ρ is the density in g/cm^3 and d is the diameter in cm.

ROLE OF SNOW IN THE REMOVAL OF PARTICULATE MATTER FROM THE ATMOSPHERE

In addition to transferring moisture from the atmosphere to the earth's surface snow transfers large quantities of particulate matter, especially in and near cities and near large industrial plants. Aerosols are removed from the atmosphere by four mechanisms, which in order of importance are: "washout", which is the scavenging of particles by rain or snow below the cloud base; "rainout" or "snowout", a process in which particles become incorporated in, or attached to, cloud droplets which subsequently grow to raindrops or become rimed on snow crystals before falling out of the cloud; "impaction on objects on the earth's surface"; and "sedimentation", which only is active for particles with dimensions greater than 10 to 20 μm.

A wide variety of material can be removed from the atmosphere by snow e.g., aerosols, ions and trace gases. Therefore snow, like rain, provides a natural mechanism for sampling the atmosphere. The concentration of material embedded in snow also reflects its concentration in the atmosphere, providing an indication of the production mechanisms or the aerosols occurring in the vicinity. Since a snowflake or snow crystal will fall slower than a raindrop of equal mass and sweep out a larger area, it will have a greater exposure to some pollutants and therefore be a better indicator of their presence.

Sood and Jackson (1970) studied the effects of crystal habit, dimensions and particle size on the scavenging (collection) efficiency of natural snow crystals (1 to 10 mm) for submicron-sized particles. Scavenging efficiency was found to be independent of crystal habit, increased with decreasing crystal diameter, and did not fall below 0.1 for particles in the range 0.3 to 0.5 μm for the size combinations investigated. These results illustrate the efficiency with which snow can remove aerosols from the atmosphere and deposit them in the snowcover.   When the snowcover melts the aerosols contribute to the contamination of surface and groundwater supplies, and significantly change the chemistry of the surface water in regions with air pollution.

AREAL DISTRIBUTION OF SNOWFALL

Relationship Between Atmospheric Processes and Occurrence of Snow

As stated above, the following atmospheric conditions are important in determining the occurrence of significant amounts of snowfall:

(1) Sufficient moisture and active nuclei at a temperature suitable for the formation and growth of ice crystals,

(2) Sufficient depth of cloud to permit growth of snow crystals by aggregation or accretion,

(3) Temperatures below 0° C in most of the layer through which the snow falls, and

(4) Sufficient moisture and nuclei to replace losses caused by precipitation.

Usually, there is an adequate supply of available active nuclei for snow formation when the other conditions are satisfied.

In large-scale weather systems, from which most snow falls, extensive areas of ascending air are cooled by adiabatic expansion. If enough moisture is available, extensive altostratus-nimbostratus type clouds are formed. After these clouds have formed, most of the water vapor which has been converted into ice or water reaches the ground as precipitation, i.e., the precipitation rate is approximately equal to the rate of condensation and sublimation. The primary factors determining the rate of condensation and sublimation are the vertical velocity, temperature, and pressure of the air. Fulkes (1935) derived an approximate relationship for the precipitation rate P of an ascending layer of saturated air of unit cross section:

$$P = bw\Delta z, \qquad\qquad 4.5$$

where: b = coefficient whose value depends on the temperature
                 and pressure of the layer,
           w = vertical air velocity, and
           Δz = thickness of the layer.

The variation of b with temperature and pressure (height) is indicated by the curves of Fig. 4.6 showing the precipitation rates in mm/h for a layer of

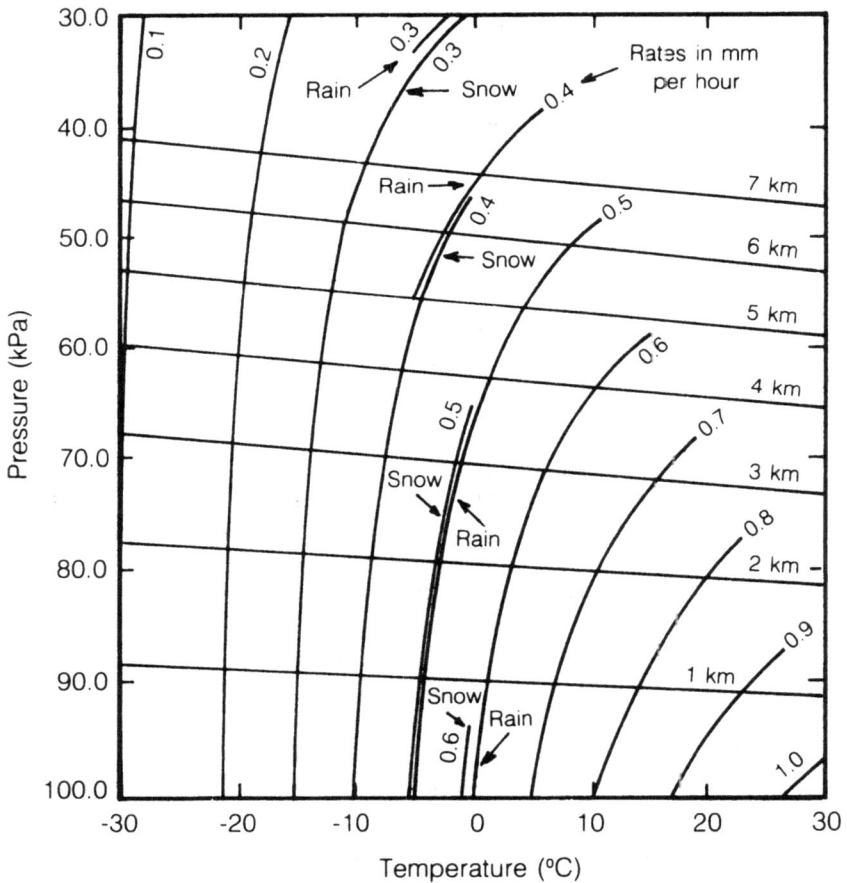

Fig. 4.6 Rates of precipitation from adiabatically ascending air for a 100-m layer with a vertical velocity of 1 m/s (Fulkes, 1935).

thickness Δz = 100 m ascending with a velocity w = 1.0 m/s. Under conditions when most snow forms (temperatures below 0° C and heights below 6 km) the rate of precipitation decreases rapidly with decreasing temperature, and, to a much lesser degree, with decreasing pressure (increasing height). The vertical velocity is determined mainly by the characteristics of individual weather systems and the extent to which terrain features affect the airflow, for example, by orographic lift.

Another important aspect of snowfall is its duration. For a synoptic system, this depends on several factors including its speed and track and the size and shape of the associated snowfall area. The development of a frontal storm is shown in Fig. 4.2. Complete assessment of seasonal snowfall patterns also requires information about the frequency of occurrence of snow-producing systems.

The study of snowfall only as a product of large-scale vertical motion associated with major synoptic systems is incomplete. Convective processes can also produce significant amounts of snow. The snow-formation processes are similar for both lifting mechanisms in the sense that precipitation results from the cooling by expansion of ascending air. However, with convective lifting, the upward motion occurs because of the differential heating of air over localized areas while the rate of snow formation and the resultant amount of precipitation depend on the available moisture and the degree of instability of the air. Convection often occurs with the movement of cold air over relatively warm bodies of water; the amount of modification of atmospheric moisture and stability depends on such factors as the initial temperature and moisture distribution of the air, the temperature of the water, and the length of over-water trajectory.

Instability can also be produced by the large-scale upward motion associated with frontal systems. If an atmospheric layer whose vapor content decreases sufficiently with height is lifted to a height where condensation occurs, it becomes unstable, and convective cloud and precipitation may result (see Fig. 4.3).

Effects of Moisture and Precipitation Mechanisms on Snowfall Distribution

The distribution of the mean annual snowfall amounts over Canada and the United States is shown in Fig. 4.7, in which there are two areas of relatively heavy snowfall. Certain areas of western British Columbia, the Yukon and Alaska adjacent to the mountain ranges which parallel the Pacific Coast, receive seasonal values exceeding 400 cm. These areas are directly exposed to moisture-laden disturbances moving eastward from the Pacific. The vertical motion associated with a low-pressure system is enhanced by the coastal terrain so that the snowfall can be very heavy, in some locations exceeding

Fig. 4.7 Mean annual snowfall (cm) over Canada and the United States (1941-1970).

1000 cm annually. However, amounts are highly variable; e.g., along the southern coast of British Columbia, near sea level, the air temperatures are normally above freezing so that most of the winter precipitation is rain, and the seasonal average snowfall is less than 60 cm. Snowfalls are also relatively light in areas to the lee of the mountains.

Widespread heavy snowfall also occurs in eastern Canada throughout central Ontario, southern Quebec, much of the Atlantic provinces, Labrador, and the east coast of Baffin Island where the seasonal amounts range from 250 to 400 cm. These parts of Canada lie on or near several principal tracks of transient low pressure systems which are frequently vigorous and well developed. Varying amounts of moisture are supplied to these areas from the Pacific and Atlantic Oceans and the Gulf of Mexico. In addition, the Great

Lakes serve as an important moisture source for local precipitation, e.g., average seasonal snowfalls greater than 250 cm occur southeast of Lake Huron. Snowfall amounts decrease rapidly in the southward direction from the eastern Ontario-northern New England area to the southeastern United States. This is mostly a result of increasing temperatures, as opposed to decreasing precipitation.

Over the prairie provinces of Alberta, Saskatchewan, and Manitoba the seasonal snowfall is considerably lower than in the eastern or western regions of Canada, averaging between 75 and 140 cm. The small amounts of snowfall over these regions can be attributed, in part, to the infrequent occurrence of vigorous weather systems. Also, the relatively flat terrain is not conducive to snowfall formation since the Pacific air moving inland subsides because of the downward slope in topography from the Rocky Mountains.

The western half of the Arctic Islands receives less snow ($<$ 80 cm) than most other parts of Canada. Although this area experiences long winters, it is remote from major moisture sources; the extremely low temperatures over the region reduce the moisture holding capacity of the air to extremely low values thereby reducing snowfall amounts.

LITERATURE CITED

Bergeron, T. 1935. *On the physics of cloud and precipitation.* Int. Union Geod. Geophys. Gen. Assem. Lisbon., Int. Assoc. Hydrol. Sci. Publ. 20, p. 156.

Eadie, E.J. 1970. *The experimental modification of lake-effect weather.* Rep. VC-2898-P-1, Cornell Aeronaut. Lab., Buffalo, N.Y.

Fulkes, J.R. 1935. *Rate of precipitation from adiabatically ascending air.* Mon. Weather Rev., Vol. 63, pp. 291-294.

Hay, J.E. 1970. *Aspects of the heat and moisture balance of Canada.* Ph.D. Thesis, Univ. London, London.

Hobbs, P.V. 1973. *Ice in the atmosphere: a review of the present position.* Physics and Chemistry of Ice (E. Whalley, S.J. Jones and L.W. Gold, eds.), R. Soc. Can., Ottawa, pp. 308-319.

Hobbs, P.V. 1974. *Ice Physics.* Clarendon Press, Oxford.

Hobbs, P.V., L.F. Radke, R.R. Weiss, D.G. Atkinson, J.D. Locatelli, K.R. Biswas, F.M. Turner and C.E. Robertson. 1974. *The structure of clouds and precipitation over the Cascade Mountains and their modification by artificial seeding (1972-73).* Contributions from the Cloud Physics Group, Res. Rep. VIII, Dept. Atmos. Sci., Univ. Wash., Seattle.

Jayaweera, K.O.L.F., and B.J. Mason. 1965. *The behaviour of freely falling cylinders and cones in a viscous fluid.* J. Fluid Mech., Vol. 22, pp. 709-720.

Locatelli, J.D. and Hobbs, P.V. 1974. *Fall speeds and masses of solid precipitation particles.* J. Geophys. Res., Vol. 79, pp. 2185-2197.

Magono, C. and C. Lee. 1966. *Meteorological classification of natural snow crystals.* J. Fac. Sci., Hokkaido Univ., Ser. VII, Vol. 2, pp. 321-335.

Magono, C. and T. Nakamura. 1965. *Aerodynamic studies of falling snowflakes*. J. Meteorol. Soc. Jpn., Vol. 43, pp. 139-147.

Mason, B.J. 1971. *The Physics of Clouds*. 2nd Ed., Clarendon Press, Oxford.

McKay, G.A. 1970. *Precipitation*. Handbook on the Principles of Hydrology (D.M. Gray, ed.), Nat. Res. Counc. Can., Ottawa, pp. 2.1-2.111.

Nakaya, U. and T. Tereda. 1935. *Simultaneous observations of the mass, falling velocity and form of individual snow crystals*. J. Fac. Sci., Hokkaido Univ., Ser II, Vol. 1, pp. 191-201.

National Research Council, 1954. *The international classification for snow*. Int. Assoc. Sci. Hydrol., Tech. Memo. 31, Nat. Res. Counc. Can., Ottawa, Ont.

Rasmusson, E.M. 1967. *Atmospheric water vapor transport and the water balance of North America: Part 1*. Mon. Weather Rev., Vol. 95, pp. 403-426.

Schaefer, V.J. 1946. *The production of ice crystals in a cloud of supercooled water droplets*. Science, Vol. 104. pp. 457-459.

Sood, S.K. and M.R. Jackson. 1970. *Scavenging by snow and ice crystals*. Proc. of Symp. on Precipitation Scavenging (1970), Richland, Washington, June 2-4, 1970. Div. Tech. Inf., USAEC, Washington, D.C., pp. 121-136.

5

THE DISTRIBUTION OF SNOWCOVER

G.A. McKAY

Canadian Climate Centre, Atmospheric Environment Service, Downsview, Ontario.

D.M. GRAY

Division of Hydrology, University of Saskatchewan, Saskatoon, Saskatchewan.

INTRODUCTION

Snowcover comprises the net accumulation of snow on the ground resulting from precipitation deposited as snowfall, ice pellets, hoar frost and glaze ice, and water from rainfall, much of which subsequently has frozen, and contaminants. Its structure and dimensions are complex and highly variable both in space and time. This variability depends on many factors: the variability of the "parent" weather (in particular, atmospheric wind, temperature and moisture of the air during precipitation and immediately after deposition); the nature and frequency of the parent storms; the weather conditions during periods between storms when radiative exchanges may alter the structure, density and optical properties of the snow and wind action may promote scour and redeposition as well as modification of snow density and crystalline structure; the process of metamorphism and ablation which can alter the physical characteristics of the snowcover so that it hardly resembles the freshly-fallen snow; and surface topography, physiography and vegetative cover. Being the end product of both accumulation and ablation, snowcover is the product of complex factors that affect accumulation and loss.

The areal variability of snowcover is commonly considered on three geometric scales:

(1) Macroscale or Regional scale: areas up to 10^6 km^2 with characteristic linear distances of 10^4 to 10^5 m depending on latitude, elevation and orography, in which the dynamic meteorological effects such as standing waves, the directional flow of wind around barriers and lake effects are important,

(2) Mesoscale or Local (within region) scale: characteristic linear distances of 10^2 to 10^3 m in which redistribution along meso-relief features may occur because of wind or avalanches and deposition and accumulation may be related to the elevation, slope and aspect of the terrain and to the canopy and crop density, tree species or crop type, height, extent and completeness of the vegetative cover.

(3) Microscale: characteristic distances of 10 to 10^2 m over which major differences occur and the accumulation patterns result from numerous interactions, but primarily between surface roughness and transport phenomena.

FACTORS CONTROLLING SNOWCOVER DISTRIBUTION AND CHARACTERISTICS

Snow accumulation and loss are controlled primarily by atmospheric conditions and the "state" of the land surface. The governing atmospheric processes are precipitation, deposition, condensation, turbulent transfer of heat and moisture, radiative exchange and air movement. The major land features to be considered are those which affect the atmospheric processes and the retention characteristics of the ground surface.

Several of the atmospheric processes and land factors to which snowcover responds are also discussed in Chs. 2, 3, 4, 8, and 16.

Temperature

Snowcover is a residual product of snowfall and has characteristics quite different from those of the parent snowfall. The temperature at the time of snowfall, however, controls the dryness, hardness and crystalline form of the new snow and thereby its erodability by wind. The importance of temperature is apparent on mountain slopes, where the increase in snowcover depth can be closely associated with the temperature decrease with increasing elevation. Wet snow, which is heavy and generally not susceptible to movement by wind action, falls when air temperatures are near the melting point; this commonly occurs when air flows off large bodies of water. Within continental interiors where colder temperatures often prevail the snowfall is usually relatively dry and light.

Wind

The roughness of the land surface affects the structure of wind and hence its velocity distribution. Because of the frictional drag exerted on the air by the earth's surface the wind flow near the ground is normally turbulent and snowcover patterns reflect a resulting turbulent structure. Also, the wind moves snow crystals, changing their physical shape and properties, and redepositing them either into drifts or banks of greater density than the parent material. For example, Church (1941) found that fresh snow with densities of 36 and 56 kg/m^3 increased in density to 176 kg/m^3 within 24 hours after being subjected to wind action. Gray et al. (1971) found similar results for the Prairies; the density increased from 45 to 230 kg/m^3 within 24 hours due to wind action. Although initiated by wind action this time-densification of snow is also influenced by condensation, melting, and other processes. Table 5.1 lists the densities of snowcover subjected to different levels of wind action.

Table 5.1
DENSITIES OF SNOWCOVER

Snow Type	Density (kg/m^3)
Wild snow	10 to 30
Ordinary new snow immediately after falling in the still air	50 to 65
Settling snow	70 to 90
Very slightly toughened by wind immediately after falling	63 to 80
Average wind-toughened snow	280
Hard wind slab	350
New firn[a] snow	400 to 550
Advanced firn snow	550 to 650
Thawing firn snow	600 to 700

[a] snow consolidated partly into ice (after Seligman, 1962)

Wind transports loose snow (analagous to the movement of sediment by streamflow in a river channel) causing erosion of the snowcover, packing it into windslab and crust, and forming drifts and banks. A loose or friable snowcover composed of dry crystals, 1-2 mm in diameter, is readily picked up even by fairly light winds with speeds $\sim$ 10 km/h. The formation of a glaze by the freezing of condensation or surface melt may inhibit transport; however, under very strong winds even large sheets of glazed snow may move. Erosion prevails at locations where the wind accelerates (at the crest of a ridge), and deposition from a fully-laden air stream occurs where the wind velocity decreases (along the edges of forests and cities).

The rate of transport is greatest over flat, extensive open areas, free of obstructions to the airflow, and is least in areas such as cities and forests having great resistance to flow. Table 5.2 (Miller, 1976), summarizes the mean winter transport flux rates for different physiographic and climatic regions. These data show that the transport rates in the highly exposed Arctic Coast and Tundra regions are substantially greater than those in more sheltered regions, such as the Rocky Mountains.

Table 5.2
MEAN WINTER SEASON TRANSPORT FLUX RATES, Tonne/m,
(adapted from Miller, 1976)

Central European Russia (Mikhel', et al, 1969, p. 54)	91 to 182
Kamennaia Steppe (Mikhel', et al, 1969, p. 54)	210
Tundra (Mikhel', et al, 1969, p. 54)	550 to 907
Arctic Coast (Mikhel', et al, 1969, p. 54)	907
Wyoming (Tabler, 1973)	150
Rocky Mountains (Martinelli, 1973)	136
Northern Idaho ridge: Increase in transport rate into	272
a lee cornice when windward slope was cleared	13 to 109
(Haupt, 1973)	14 to 120

Major deposition occurs in areas located windward of zones with high aerodynamic roughness where the greatest deceleration in wind speed occurs. Drifts are deepest where a long upstream fetch covered with loose snow has sustained strong winds from one direction. The drifts are less pronounced when the winds change direction, especially at low speeds. Very slight perturbations in the airflow, such as produced by tufts of grass, ploughed soil, or fences, may induce drift formation. In areas with no major change in land use, and where the wind distributions are repeated seasonally, the drifts tend to form in approximately the same shapes and locations from year-to-year. The largest drifts are caused by major wind storms such as blizzards which may have speeds exceeding 40 km/h. Such storms occur with unpredictable frequency being absent in some years and recurring in others.

Most snow is transported by saltation or turbulent diffusion (see Ch. 7), two modes of transport that have led to the development of two snowdrift theories, the dynamical and diffusion theories, based on the respective works of Bagnold (1941) and Schmidt (1925). As detailed by Radok (1977) the basic equations describing these theories are of somewhat marginal relevance in practice. In essence, the most telling difference between them results from their dominant processes and vertical scales: the dynamical theory views snow drifting as a near-surface phenomenon caused by small eddies in the lowest 10 cm which mainly produce saltation; the diffusion theory (relating to conditions on polar ice sheets) emphasizes the larger eddies in the free air stream extending to tens or even hundreds of metres above the surface. In evaluating the theories Radok (1977) stresses that the real power of the diffusion theory comes from its detailed predictions of drift concentrations and velocity profiles and a greater understanding of the snow drift process.

An important aspect to consider in the redistribution of snowcover by wind is the mass change of a snow crystal, while it is being transported, resulting from its exchange of vapor with the surrounding air. Schmidt (1972) modelled the evaporation or sublimation of a single particle during transport finding that: the sublimation rate appeared to double for each 10-degree temperature rise in the range from -20° to 0° C, and more than doubled when the particle diameter doubled; and the percentage mass loss in unit time increased markedly as particle size decreased. Schmidt assumed that all the sublimated vapor is transferred vertically upward by turbulent exchange; therefore it would be hypothesized that a substantial amount of water to the ground surface may be lost because of this process. Equations for estimating the condensation/sublimation loss of a mass of blowing snow have been proposed by Dyunin (1961) and Tabler (1971), who made use of the concept of a "transport distance", Ru, the distance over which the average sized drift particle is completely evaporated. Obviously, Ru frequently differs from the contributing distance (fetch) so that the amount of blown snow collected by some barrier or obstacle must be adjusted by the amount which has evaporated. Tabler (1975) suggests that the percentage of relocated snow water lost to evaporation should be an exponential function of the ratio of the fetch to the transport distance. For example, in Wyoming, in the case where environmental conditions may be constant over the fetch distance downwind of a boundary, the percentages of relocated snow evaporating in transit would be 37, 57, 75 and 83 for fetches equal to 0.5, 1.0, 2, and 3 Ru, respectively.

Tabler (1975) developed a combined transport and evaporation equation:

$$Q = \sum_{i=1}^{n} \int_{R_{i-1}}^{R_i} (P_r)_i (0.14)^{R/R_u} dr, \qquad 5.1$$

where: Q = total water equivalent volume of snow arriving at a point after transport over a distance, R, (m^3/m),

          n = number of increments of ΔR within R where P_r and evaporation can be assumed constant,

     $(P_r)_i$ = the water equivalent of the precipitation swept off the ith ΔR interval (m) - for annual transport calculations P_r is the total amount of winter precipitation, less the amounts held by the vegetation and terrain irregularities, plus the amount of melt,

Ru = the transport distance required for complete evaporation of the average size particle (evaluated from field measurements, (m)), and

dr = incremental length (m).

Tabler demonstrates the application of the model for determining transport volumes and evaporation losses for different combinations of vegetation and terrain features.

Table 5.3 presents data, from several studies (summarized by Miller, 1976) on the average distances of snow transport over different types of terrain.

Table 5.3
AVERAGE DISTANCES OF SNOW TRANSPORT IN DIFFERENT TERRAIN, km, (Miller, 1976).

In mountainous dissected topography (Kotlyakov, 1973)	0.1 to 0.5
On plains of western Siberia (Dyunin et al, 1973)	1 to 3
On plateaus of southeastern Wyoming (Tabler and Schmidt, 1973)	
--small particles (0.1 mm diameter)	0.5
--larger particles (0.2 mm diameter)	1.4
On ice domes (Kotliakov, 1968)	Up to 5
On polar ice caps (Kotliakov, 1968)	To 30-50

Interaction in a forest environment

Maximum accumulations of snow often occur at the edges of a forest as a result of snow being blown in from adjacent areas, but depend very highly on the porosity of the stand borders. Within the stand accumulations may not be uniform, however, generally the snowcover distribution is more uniform within hardwoods than within coniferous forests whose dense crowns create a "ridge and hollow" effect. Further, most studies have reported that more snow is found within forest openings than within the stand.

A unique phenomenon influencing snowcover distribution within a forest (which is affected primarily by wind) is the transport of intercepted snow. Wind causes a tree to vibrate, resulting in a loosening and erosion of intercepted snow, and the transport of fragments downwind. Miller (1966) summarized the results from different studies of the transport process. His review shows that a physical understanding of the process is complicated not only by the complexity of the air flow patterns and velocity distributions within different forest covers but also by other factors including the manner in which snow accumulates, collects on and adheres to vegetation types, and the cohesion and adhesion properties of the snow. Hoover and Leaf (1967) in

Colorado concluded from timed-sequence photographic measurements of snow interception that mechanical removal probably predominated over vaporization in removing snow from tree crowns.

At present, because of the lack of field measurements, the amount of snow transported by wind in a forest environment and the manner in which this snow affects the snowcover distribution is largely unknown.

Energy and Moisture Transfer

During the winter months energy and moisture transfers to and from the snowcover are significant in changing its state. Prior to the period of continuous snowmelt the radiative fluxes are dominant in determining changes in depth and density. The underlying surface, the physical properties of the snowcover and trees, buildings, roads or other features, and activities which interrupt the snowcover or alter its optical properties, affect the net radiative flux to the snow. Such factors, therefore, influence how the snowcover is modified by the different radiative fluxes to change its erodability, mass and state. One property of the snowcover surface which directly affects the solar energy absorbed by the snow is its albedo, the ratio of the reflected to the incident short-wave radiation. Some typical values of the surface albedo of different snow surfaces are given in Table 5.4.

The spatial changes in albedo of a snowcover relate at least to the snow depth, which is a regional characteristic. Kung et al. (1964) found that the space-averaged surface albedo remained high when the depth exceeded 12 cm but decreased with depth for values less than this (see Fig. 5.1). This is explained by the increase in both the exposed patches of ground and the transparency of the snowcover with decreasing depth: the latter causes the reflective properties of the underlying ground to affect the albedo.

Table 5.4
ALBEDO OF SNOWCOVER (K. Kondratyev, 1954 quoting P.P. Kuz'min).

Surface	Albedo in %
Exposed with continuous snow-cover	80
Exposed with changing (melting) snowcover	55
Wet, after snowmelt	15
Coniferous forest with snowcover	12
Deciduous forest	18

Heat and mass transfers from the air and ground lead to changes in the crystal structure within the snowcover and to loss of mass as melt or water vapor. The turbulent transfer of heat and moisture, which occurs with föhn winds, can lead to evaporation, melting, the formation of glaze, and general physical alterations within the snowcover. Some föhn winds do not surface at low elevations so that their spectacular effects are confined to intermediate elevations of a mountain.

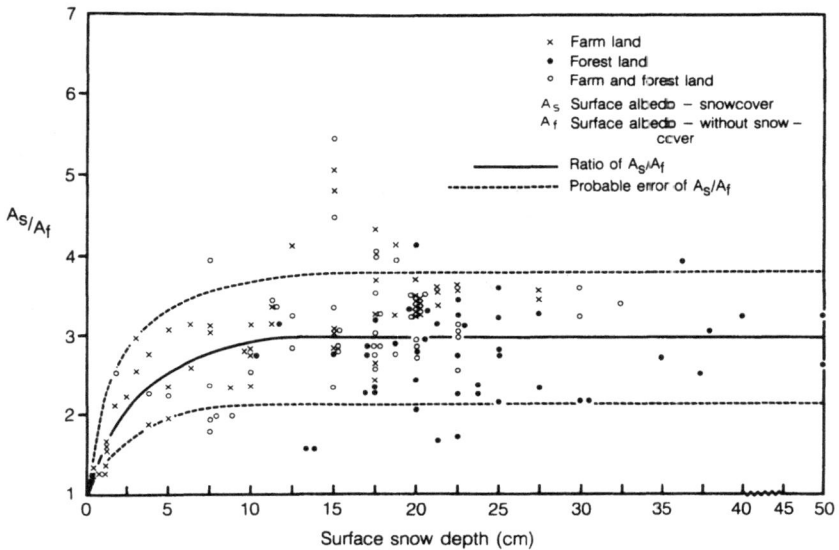

Fig. 5.1 Surface albedo related to depth of snowcover (Kung et al., 1964).

Physiography

Landform and the juxtaposition of surfaces with different thermal and roughness properties are major factors governing snowcover characteristics. Winter snowcover reaches the greatest depths in snowbelt areas to the lee of open water areas, and on windward slopes which stimulate the precipitation process. Shallow depths occur on sheltered slopes, particularly those with sunny exposures and at lower elevations where melt losses are more probable. The usual wind patterns and slides occurring in rugged terrain may result in extremely varied depths.

The physiographic features which rationally and demonstrably relate to snowcover variations are elevation, slope, aspect, roughness and the optical and thermal properties of the underlying materials.

Elevation

Normally, in mountainous regions elevation is presumed to be the most important factor affecting snowcover distribution. Often a linear association between snow accumulation and elevation can be found within a given elevation interval at a specific location (U.S. Army Corps of Engineers, 1956). Using these relationships for other sites is highly suspect, since the influence of elevation alone is indeterminate because it depends on climate and slope. Figure 5.2, constructed from data reported by Meiman (1970) shows the increase in snow water accumulation with elevation as measured in eleven separate investigations undertaken in Alberta, California, Colorado, Idaho and New Mexico. In one study (Colorado) measurements were made within selected elevation bands for three consecutive years. The information reflects the large variation in snow water increases encountered between major physiographic areas as well as the spatial-temporal variations within a given area. Meiman cites that numerous workers have substantially improved the correlations by including other land surface features. However, as Peck (1964) and others have emphasized, the influence of climatic factors or elements of the parent weather system must be recognized to interpret snow distribution and accumulation patterns adequately, e.g., the normal decrease in temperature with elevation results in a decrease in melt losses. Also, since the moisture available for the precipitation process decreases with elevation it can be argued that the increases observed with elevation reflect the combined influence of slope and elevation on the efficiency of the precipitation mechanism.

To the hydrologist, the snowcover density is equally as important as depth. Storr and Golding (1974) found a linear association between snow water equivalent and depth for the Marmot Basin in Alberta. Grant and Rhea (1974) reported that snow density varied greatly in the mountain passes of the central Colorado Rockies. Greater densities observed at lower elevations were attributed to a relatively higher frequency and greater amount of riming of ice crystals.

Slope

Mathematically, the orographic precipitation rate is predominantly related to terrain slope and windflow rather than elevation. That is, if the air is saturated the rate at which precipitation is produced is directly proportional to the ascent rate of the air mass and, over unsloping terrain this rate is directly proportional to the product of the wind speed and the slope angle.

Rhea and Grant (1974) provided a general mathematical expression for orographic mountain snowfall including the effects of large-scale vertical air mass movement, convective activity and orographic lift; air mass movement over upwind mountain barriers; interception; and gauge exposure. The orographic component depends on the specific humidity gradient, the vertical

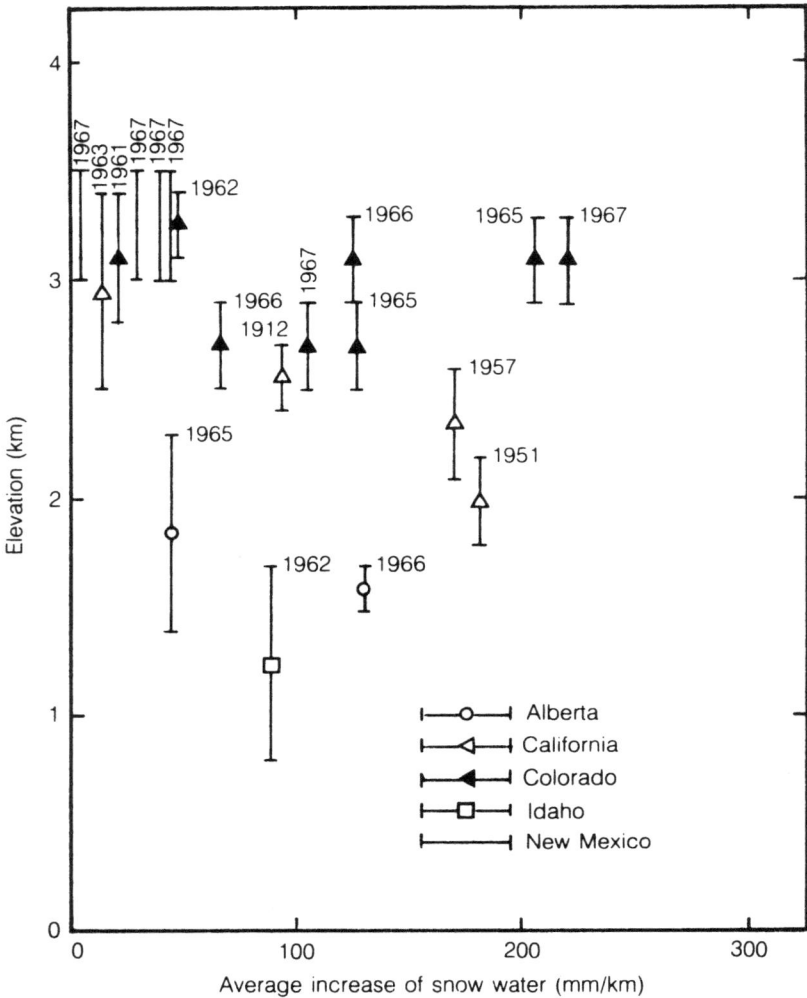

Fig. 5.2 Average increase in snow water with elevation at different locations (data from Meiman, 1970).

velocity component (a function of slope and time), the density and the air mass, the precipitation efficiency and the elevation of the site. They found from analyzing Colorado winter precipitation data that the long-term average at a point was strongly correlated (positively) with the topographic slope of the fetch, 20 km upwind. Calculations were based on slope values obtained from 30° directional classes weighted according to the relative frequency of precipitation days, the mean 700 mb specific humidity gradient and an estimated average potential precipitating cloud depth with a base at 725 mb. The correlation coefficient r between measured winter precipitation and computed "orographic" precipitation was slightly improved (r = 0.82 to r = 0.90) by incorporating a factor to account for the partial depletion of available condensate because of precipitation induced by upstream barriers. They concluded that long-term average precipitation is not well correlated to station elevation except for points on the same ridge.

Even where orography is the principal lifting mechanism and snowfall may be expected to increase with elevation, the depth of accumulation or deposition may not exhibit this trend. Besides the many factors affecting distribution, winds of high speed and long duration at the higher elevations are more frequent causing transport and redistribution. Table 5.5 reported by Miller (1976) shows the high variability in the relative amounts of snow deposited on different topographic facets in the Ural Mountains in the USSR.

In areas topographically-similar to the Prairies, where snow is primarily due to frontal activity and the exposed snowcover is subjected to high wind shear forces, slope and aspect are important terrain variables affecting the snowcover distribution. Snow depth along a slope oriented in the direction of the prevailing wind tends to decrease with distance. Gray et al. (1979) showed that the ratios of snow amounts retained by level plains, gradual slopes and

Table 5.5

RELATIVE SNOW DEPOSITION ON TOPOGRAPHIC FACETS ALONG A TRAVERSE ACROSS THE URAL MOUNTAINS (Kotliakov, 1968, p. 161 reported by Miller, 1976).

Topographic Facet	Length, km	Precipitation (Relative)	Accumulation (Relative)	Eddy
Flat	10	1.0	1.0	
Lower slope	2	1.75	1.2	
Upper slope	1.	1.25	0.25	V
Ridge	1.3	1.4	2.1	H
Lee of main second ridge	1.	0.5	1.1	
Lee of second ridge	0.8	0.6	0.6	V
Downwind plateau	0.7	0.8	0.9	
Valley of Igan River	0.7	0.7	1.3	
Windward slope	0.5	0.9	0.3	V
Ridge and its lee slope	1.3	0.6	0.7	H
Lower lee slope	0.5	1.0	0.9	
Standard deviation		0.43	0.52	

Symbols: H-Horizontal-axis eddy, V-Vertical axis eddy.

hill tops, all in summer fallow, to that measured by a Nipher snow gauge are ~ 0.60, 0.70 and 0.2 respectively. During many years on the Prairies, hilltops may be free of snow at the times of maximum accumulation on other landscape types. In this region the lee of steep slopes, drainageways and gullies are major collection areas. Gray et al. also provide examples that suggest that the amounts retained in small drainageways and on steep hill and valley slopes may be ~ 1.5 to 2.4 and 3 to 4.5, respectively, times those on gradual slopes. McKay (1970) cites a case where the snow water equivalent remaining in a channel of a prairie valley during spring after the adjacent plains were free of snow amounted to 14,000 m^3 per km.

Aspect

The importance of aspect on accumulation is shown by the large differences between snowcover amounts found on windward and leeward slopes of coastal mountain ranges. In these regions the major influences of aspect contributing to these differences are assumed to be related to: the directional flow of snowfall-producing air masses; the frequency of snowfall; and the energy exchange processes influencing snowmelt and ablation. Meiman (1970) suggests the effect of aspect appears to be predominantly a melt effect. Similarly, the results of Wilm and Collet, (1940), Goodell (1952) and Stanton (1966) indicate that aspect did not affect the maximum snowcover in natural forest conditions if the melt opportunity is minimized; however, its effect increased in areas where winter melt is common. Meiman (1970) reported that multifactor studies of snow accumulation indicate that the effect of aspect on accumulation in most areas tends to be considerably less than that of elevation. The results of a study reported by Stanton (1966) provide an example of the importance of aspect, along the eastern slopes of the Canadian Rockies (see Table 5.6).

Within the Prairie environment it is accepted that the influence of aspect on accumulation is outweighed by the snow transport phenomenon and to a lesser extent by local energy exchange.

Table 5.6
MEAN ACCUMULATION OF SNOW (cm) UNDER
DIFFERENT COVER CONDITIONS IN THE EASTERN
ROCKIES AS RELATED TO ASPECT (Stanton, 1966).

Cover	Aspect		
	N	S	E
Forest	41	41	39
Cut Forest	45	53	65

Vegetative Cover

Vegetation influences the surface roughness and wind velocity thereby affecting the erosional, transport and depositional characteristics of the surface. If the biomass extends above the snowcover it affects the energy exchange processes, the magnitudes of the energy terms and the position (height) of the most active exchange surface. Also, a vegetative canopy affects the amount of snow reaching the ground. Most studies of the interaction between vegetation and snow accumulation can be divided into separate investigations of forest and non-forest (short vegetative cover) ecosystems.

Forest

A forest differs from other vegetative covers mainly in providing a large intercepting and radiating biomass above the snowcover surface. Kitteredge (1953) demonstrated the effects of a forest canopy interception on local snowcover distributions. Numerous other studies on this topic were summarized by Miller (1966) and Meiman (1970); most have concentrated on measuring differences between snowcover amounts under different species of forest and in openings. Though more snow is consistently found in forest openings than within the stand, absolute estimates of the interception amounts cannot be made because of the lack of information about the interception and loss processes.

According to Kuz'min (1960) the amount of snow accumulated in a forest clearing depends on the size of the wooded area, e.g., the average depths in clearings were 53 cm and 36 cm in areas of 100 m x 200 m and 1000 m x 2000 m; respectively. Also the accumulation depends on the species of tree. Deciduous forests have greater accumulations than pine or fir forests; e.g., the latter have respective values of 82 and 63 percent of that measured in a birch forest. Multifactor analysis disclose that snowcover amounts increase from 8 to 56 mm of water equivalent for a 10% decrease in canopy density depending on tree species (Meiman, 1970).

Many investigators (e.g., Goodell, 1959; Molchanov, 1963) have found snow accumulation to be inversely related to canopy density. Therefore forest density could likely be used as an index of the amount of interception. Kuz'min (1960) reports that the snowcover water equivalents in a fir forest WEP_f and in a clearing WEP_c can be related to tree density p (expressed as a fraction) as follows:

$$WEP_f = WEP_c (1 - 0.37p).$$ 5.2

In addition to affecting the wind velocity distribution and interception, which influence snow accumulation and distribution, a forest modifies the energy flux exchange processes which change snowcover erodability, mass and state. Investigations by the U.S. Army Corps of Engineers (1956) and Reifsnyder and Lull (1965) provide excellent guidance on the effects of a forest canopy on the radiative exchange components. By means of a theoretical analysis Bohren (1973) suggested that the total surface area of the canopy was a relevant parameter determining the amount of radiant energy incident on a snowcover. The turbulent transfers of the sensible and latent heat between the snowcover and the atmosphere in a forest are largely unknown.

Figure 5.3 shows the association between the snowcover depth and vegetation in a forest environment, as found by Adams and Findlay (1966).

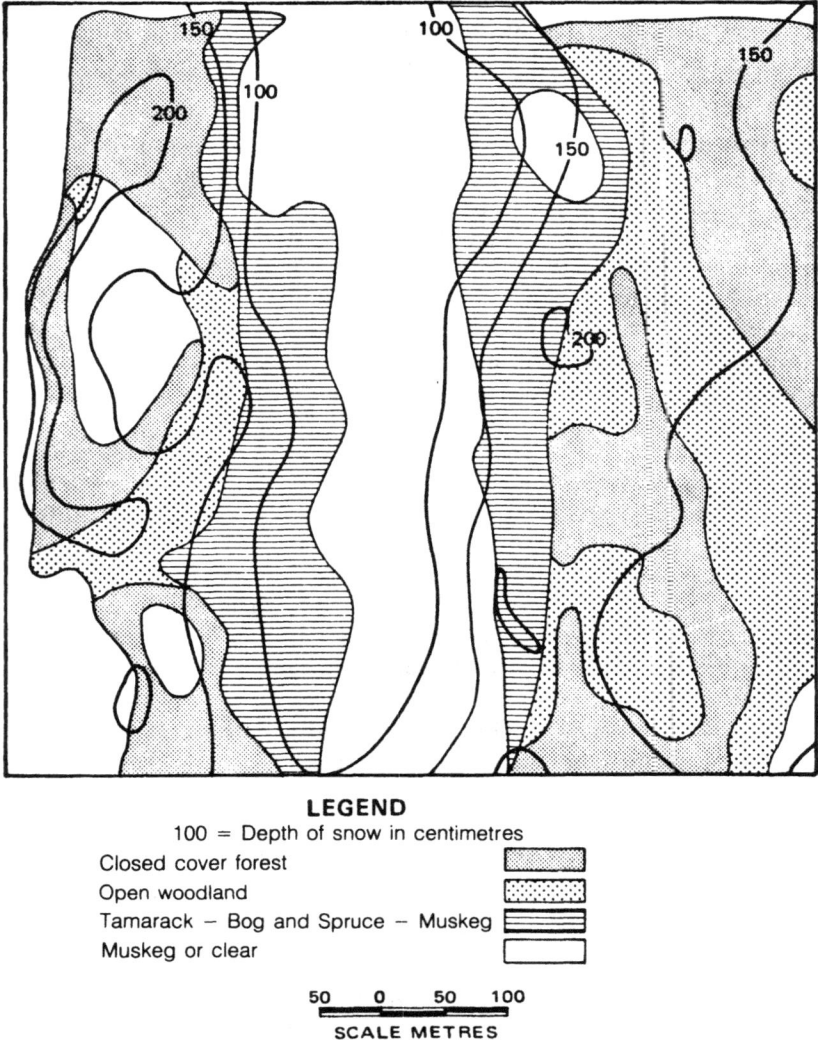

LEGEND
100 = Depth of snow in centimetres
Closed cover forest
Open woodland
Tamarack — Bog and Spruce — Muskeg
Muskeg or clear

50   0   50   100
SCALE METRES

Fig. 5.3 Relationship between the depth of snowcover and vegetation in a forest environment (Adams and Findlay, 1966). Reprinted by permission of the Minister of Supply and Services Canada.

Prairie (Grassland) and Steppes

Kuz'min (1960) and McKay (1963) stressed the importance of terrain and wind for establishing snowcover patterns on the Prairies. Over the highly-exposed, relatively flat or moderately-undulating terrain, the increased aerodynamic roughness resulting from meso- and microscale differences in vegetation may produce wide variations in accumulation patterns. Accumulations are most pronounced where sustained strong winds from one direction act on a long upstream fetch of loose snow and less pronounced when winds frequently change direction, especially for low speeds.

Lakshman (1973) cites examples of areas, stratified according to land use and topography, for which snow water equivalent and snow depth are linearly related. According to Steppuhn and Dyck (1974) consistent similarities exist in the areal variation of snowcovers within areal units having similar landscape features. That is, forests, pastures, cultivated fields, ponds, etc., within the same climatic region tend to accumulate snow in recurring patterns unique to specific terrain features and land use. Table 5.7, taken from Steppuhn (1976) shows the snowcover depth statistics by landscape type for west central Saskatchewan. Several aspects of the data are noteworthy:

(1) The depth of snow collected by scrub brush is consistently higher than that collected on fallow, stubble or pasture, independent of the terrain features.

(2) A strong dependency exists between vegetation and terrain in relation to the comparative amounts of snow retained by fallow, stubble and pasture.

(3) The number of observations required to obtain comparable values of the coefficient of variation varies widely with landscape type.

It should be noted that the measurements were taken in a year with above normal snowfall.

In addition to affecting the depth of a Prairie snowcover vegetative patterns also affect its average densities, and consequently hardnesses. The lowest densities occur over forested areas that are least subject to wind action and thaws; the highest densities occur on non-vegetative, exposed areas.

The retention coefficients in Table 5.8 (Kuz'min, 1960) summarize the effects of the physiographic-vegetation interaction on snowcover depth. The data are for landscape units of a regional scale; the magnitude of the coefficient for each physiographic, land-use unit depends on wind speed and other factors.

Table 5.7
SNOWCOVER DEPTH STATISTICS BY LANDSCAPE
TYPE, 1974, BAD LAKE, SASK. (Steppuhn, 1976).

Landscape Type	No. Obs.	Mean Depth cm	Coeff. of Variation
Plain			
fallow	360	41.5	.155
stubble	216	46.4	.133
Rolling Plain			
fallow	668	49.4	.151
stubble	180	58.8	.083
pasture	578	56.2	.174
Gradual slope			
fallow	324	50.4	.202
stubble	180	47.4	.147
scrub	183	65.6	.240
Sharp Slope (Ice)			
pasture	400	111.5	.199
scrub	869	126.5	.239
Broad Lowland			
fallow	395	101.1	.188
stubble	435	95.2	.084
pasture	219	97.0	.114
scrub	537	112.3	.183
Topland			
fallow	507	22.9	.277
stubble	218	37.6	.136
pasture	181	21.2	.434

SNOWCOVER PATTERNS

Global

Snowcover commonly occurs at the higher latitudes since near or below freezing temperatures affect both the frequency of occurrence of snowfall and the probability of snowmelt. These characteristics are evident in Fig. 5.4

Table 5.8
**SNOW RETENTION COEFFICIENTS FOR VARIOUS
SURFACES RELATIVE TO VIRGIN SOIL (Kuz'min, 1960).
Reproduced by permission of Keter Publishing House Jerusalem
Ltd.**

Virgin soil	1.0
Open ice	0.4 to 0.5
Arable land	0.9
Hilly land	1.2
Large forest tracts	1.3 to 1.4
River beds	3.0
Rushes	3.0
Forest cutting 100 to 200 m radius and edges of forests	3.2 to 3.4

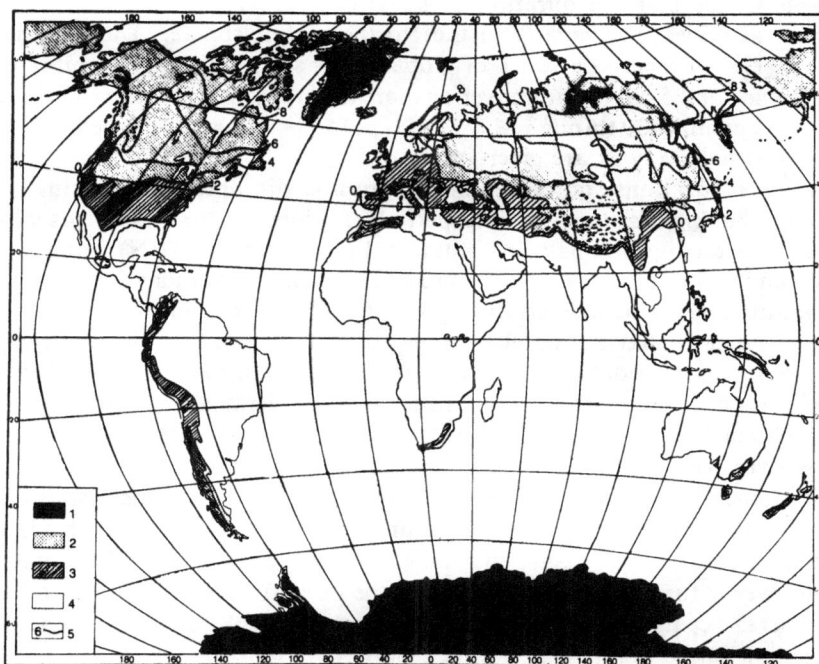

Fig. 5.4 World distribution of snowcover (Rikhter, 1945).
1. Permanent cover of snow and ice. 2. Stable snowcover of varying duration forms every year. 3. Snowcover forms almost every year, but is not stable. 4. No snowcover. 5. Duration of snowcover (months).

showing the world distribution of snowcover; the snowcover duration is longest near the poles and on high mountains.

On the global scale the snowcover on Antarctica is a dominant feature. Its mass and extent, covering an area of approximately 1.3×10^9 ha, makes it a major factor in the thermal and moisture balance of the earth. The formation of hoar frost and the deposition of ice crystals are considered to contribute significantly to the annual accumulation. Orvig (1970) reports no-cloud precipitation is almost a continuous phenomenon on the plateau area. Similar findings have been reported for the Greenland Ice Cap.

Generally the depths of snowcover in continental interiors north of 50° N and in polar areas range between 40 to 80 cm. However, this is not universal since snowcover depends on topography and geographical location, e.g., in mountainous regions of Norway and Southern Alaska average depths exceed 120 and 180 cm, respectively.

In addition to depth and density, the dates (time) of occurrence and disappearance and the duration of snowcover are important factors for human activities, particularly in the Northern Hemisphere. Dickson and Posey (1967) prepared a series of global maps showing the probability of occurrence of 2.5 cm of snowcover or more at the end of each month for the period September 30 through May 31. Maps for September 30, December 31, March 31 and May 31 are given in Fig. 5.5 a, b.

At most locations snowcover may form and disappear several times a season. At high latitudes a long period of winter snowcover is virtually assured (exceeding 182 days in continental areas north of 60° N), but at the onset and end of winter, and even into summer, snowcover may form briefly before melting. This indeterminate period occurs throughout the winter in milder climates having small seasonal accumulations. Towards lower latitudes, (e.g. Florida), even a single occurrence of snowcover is unlikely.

Because early and late seasonal snows are ephemeral, analysts often have difficulty in deciding exactly what criteria determines the length of the winter snowcover period. The illustrations in this chapter used several different definitions, e.g., the snowcover may be considered to form on the first day in autumn when the snow depth is 2 cm or more. In many instances the data would be misleading because the snow may disappear before a permanent snowcover is formed. This problem may be partially overcome by rejecting ephemeral occurrences. McKay and Thompson (1968) define the date of formation as the first day after which snowcover remains for at least seven days. Different assumptions of this kind make it difficult to compare many of the maps of snowcover distribution appearing in the literature.

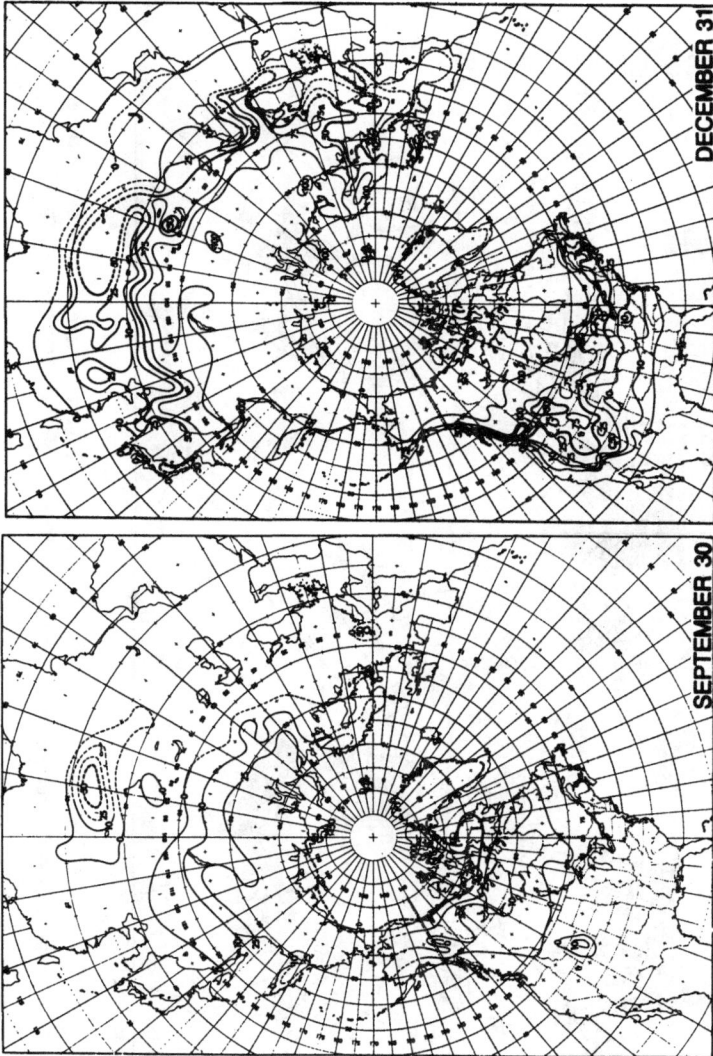

Fig. 5.5a Isolines showing the probability of occurrence of 2.5 cm or more of snowcover at the end of the months of September and December, for the Northern Hemisphere (Dickson and Posey, 1967).

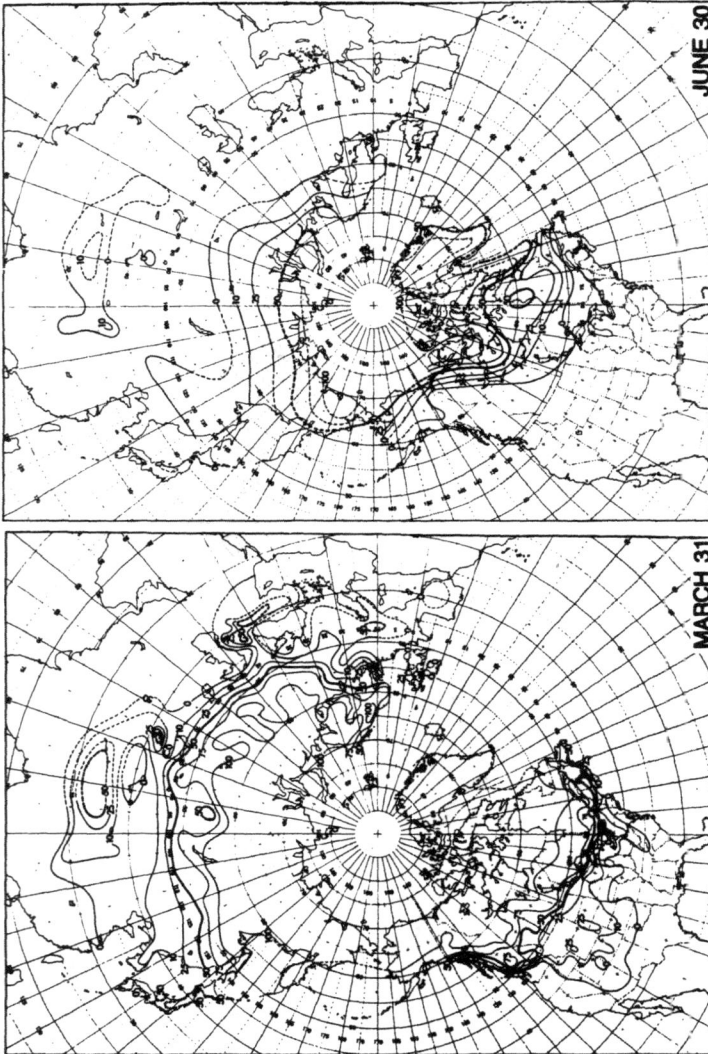

Fig. 5.5b Isolines showing the probability of occurrence of 2.5 cm or more of snowcover at the ends of the months of March and May for the Northern Hemisphere (Dickson and Posey, 1967).

Zonal

Climate, physiography and vegetation interact in a complex manner to govern snowcover accumulation and distribution. Many investigators (Rikhter, 1945; McKay and Findlay, 1971; Khodakov, 1975) found it possible to classify and map some general characteristics of snowcover according to vegetation zones. This procedure is often effective because the type of vegetation is frequently indicative of climate. Simple zonation by vegetation is often useful for interpreting snowcover maps and data. Figure 5.6 presents the major vegetation zones of the Northern Hemisphere.

Tundra

In the tundra zone, with the exception of permanent snow fields, the average period of snowcover on land surfaces ranges from eight months to over ten months in high latitude regions of Greenland and Ellesmere Island (see Figs. 5.7 and 5.8). The actual mean dates of occurrence and disappearance are difficult to define and measure and are highly variable from year to year. Generally, snowcover forms in late September or early October and disappears by June while the maximum accumulation occurs in February, March or April (see Fig. 5.9).

In Canada, the mean maximum depths of snowcover range from 30 cm on the arid areas to 300 cm on mountain slopes near open seas (see Fig. 5.10). Although the snowfall distribution throughout the tundra zone may be considered as being regionally uniform, the snow is quickly redistributed by wind. Scour and sedimentation of graupel and ice needles produces a cover which may be highly variable in density and depth with numerous exposed areas, drifts, dunes and sastrugi. The eroded snow accumulates along the upwind sides of valleys and along the edges of airflow obstacles, such as shrubs or rocks.

The average density of snowcover in this zone is usually taken to be 300 kg/m^3 (or greater) over most of the season. The cold dry snow forms a finely-structured surface having a high bearing capacity. Major changes (increases) in density do not occur until the time of active melt in late May or June (see Fig. 5.11).

Taiga and Boreal forest

During October, the area of snowcover moves rapidly southward to include most of the Taiga and Boreal Forest except those parts inflenced by a maritime climate. For example, in Eurasia the snowcover is widespread throughout Siberia in October, but does not extend to the Baltic until December. In North America, the average duration of snowcover in these zones is 200 to 240 days in the far north and 120 days at their southern limits with the maximum depths occurring in February and March. However, the

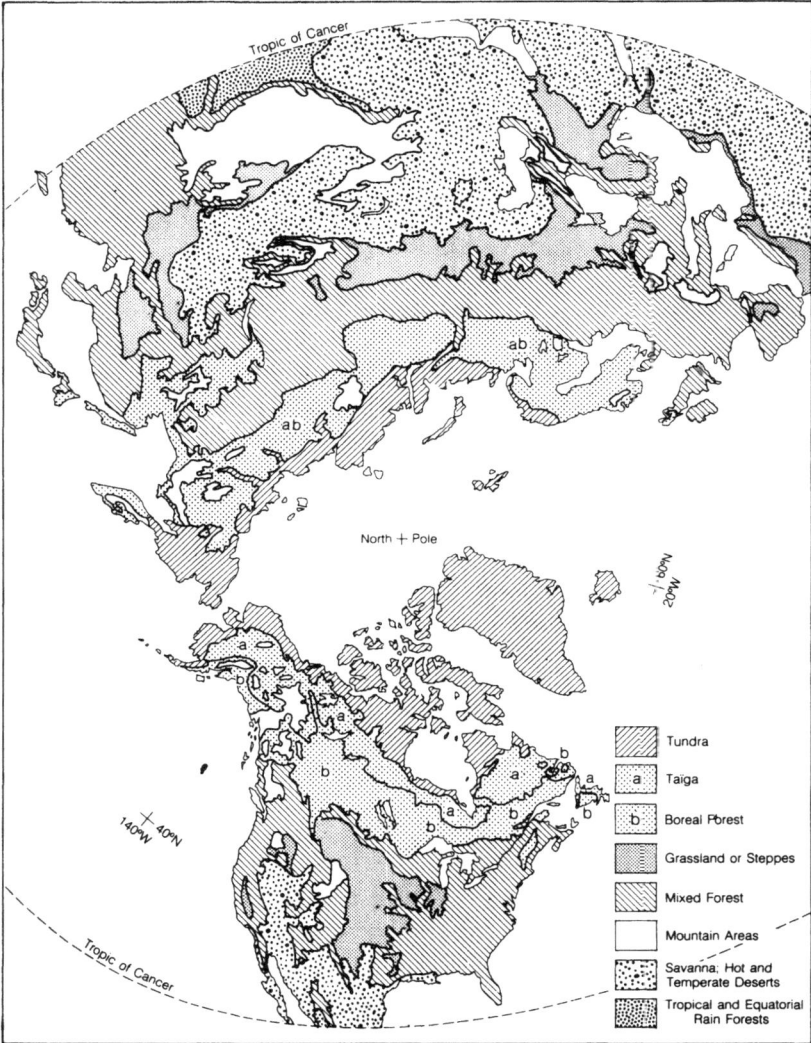

Fig. 5.6 Major vegetation zones in the Northern Hemisphere used in studies of snowcover.

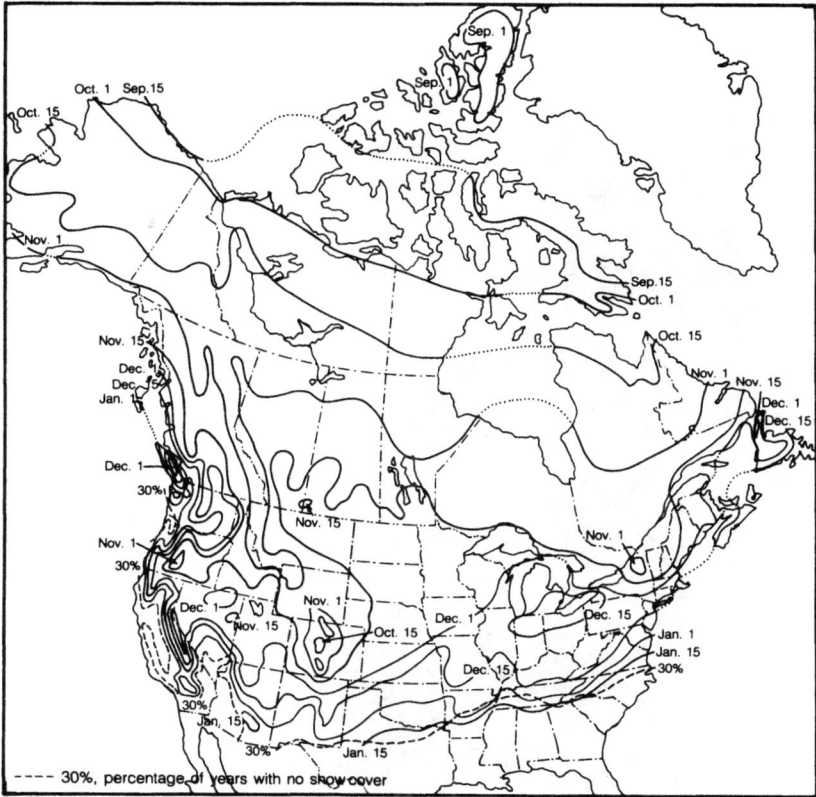

Fig. 5.7 Mean date of snowcover formation (North America).

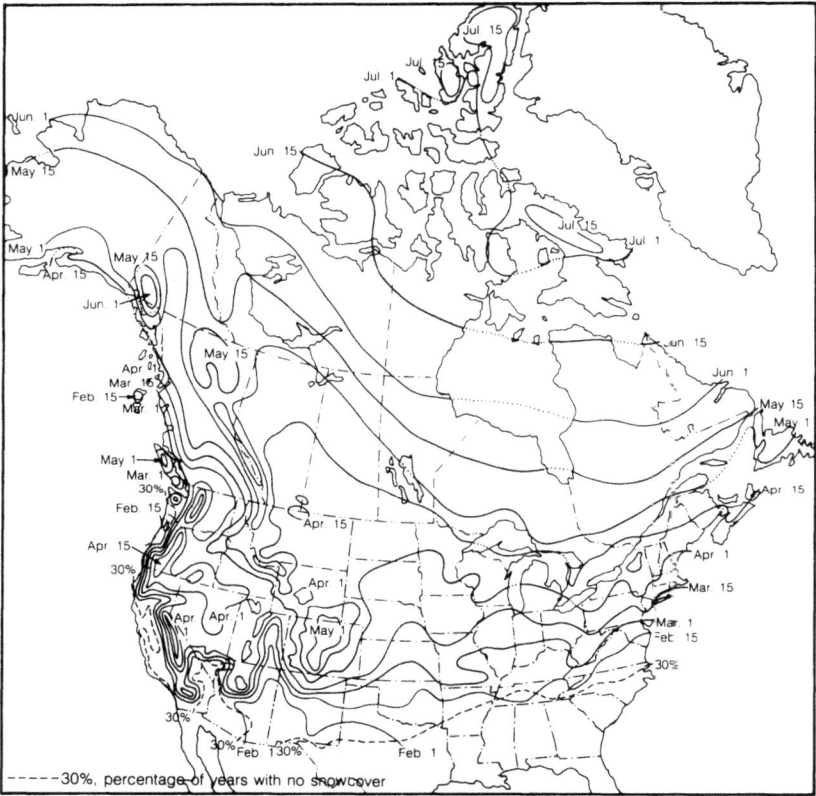

Fig. 5.8 Mean date of snowcover disappearance (North America).

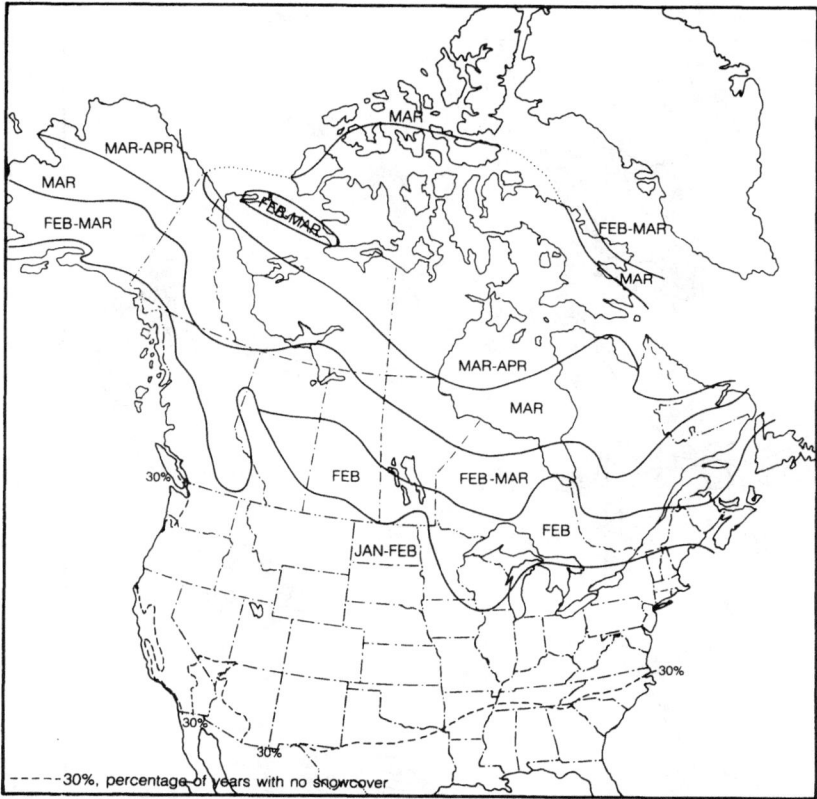

Fig. 5.9 Month of maximum snowcover (North America).

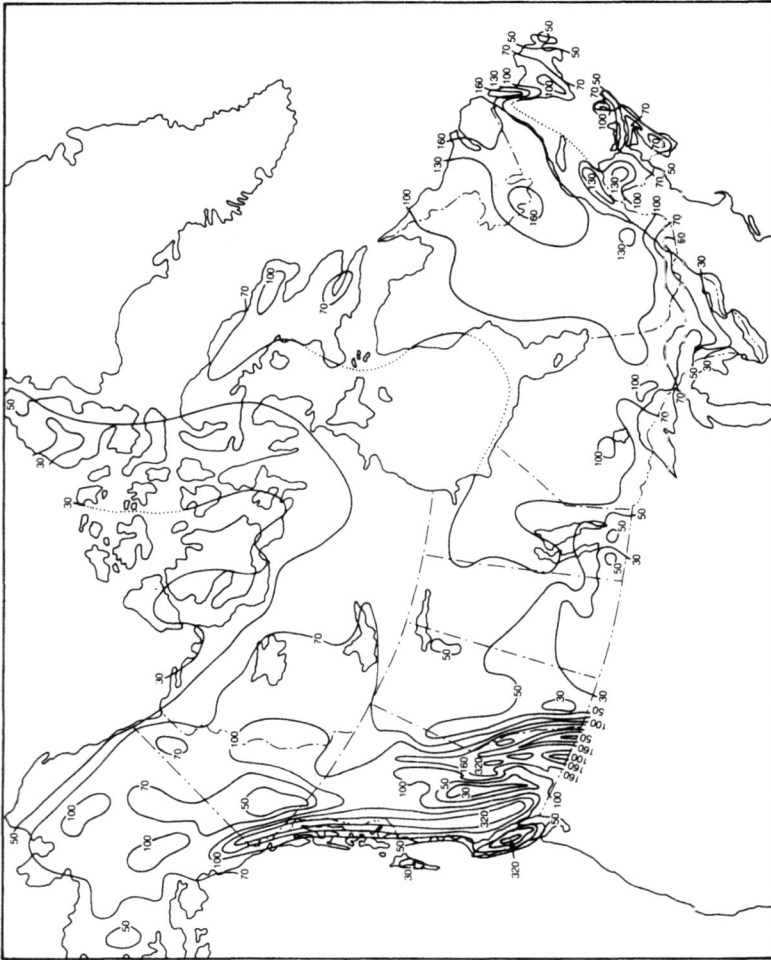

Fig. 5.10 Mean maximum depth of snow - cm (Canada) (Fisheries and Environment Canada, 1978). Reproduced by permission of the Minister of Supply and Services Canada.

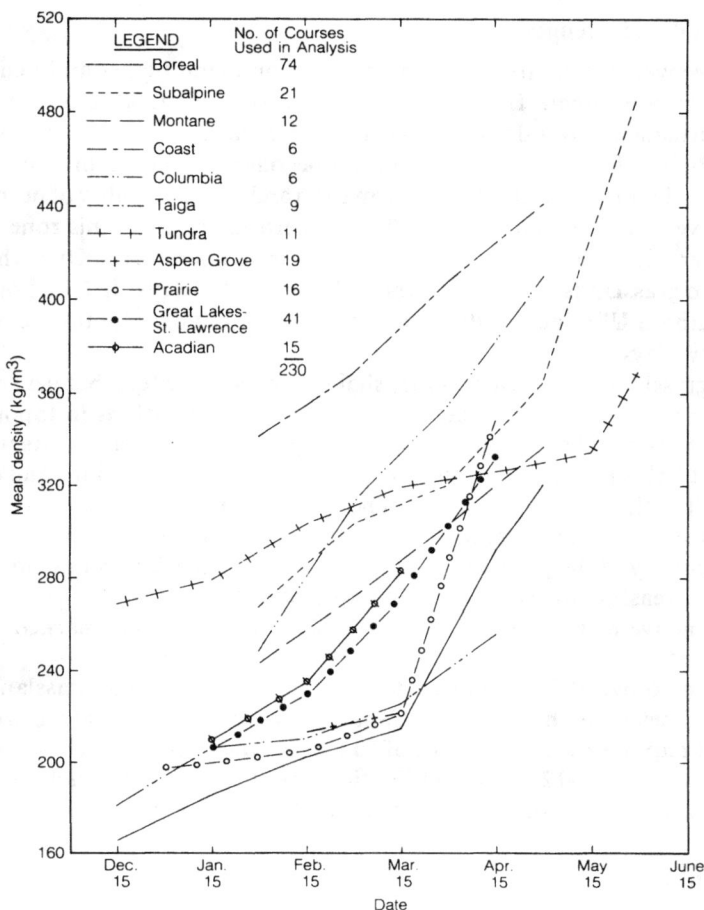

Fig. 5.11 Time-density variation of the snowcover by vegetation zone.

dates of occurrence of maximum depths throughout the region vary widely from season to season and between clearings and forests. Since most weather stations are located in cleared areas, their reports probably provide early estimates. In Canada the mean maximum depths of accumulation range from 50 to 150 cm depending on the land form and proximity to the open sea.

The snowcover within the evergreen forest differs from that in the tundra and grasslands. Forest cover intercepts falling snow, serves as a wind break and shelters the snowcover from solar radiation, thereby extending its duration, and resulting in less compacted formations. The average densities of snowcover in these regions are much lower than that in tundra, viz. 170 to 210 kg/m^3 during most of the non-melt period.

Grassland and Steppes

Snowcover usually forms on the colder, continental grassland plains and steppes in November. In the Soviet Union east of the Urals a permanent or semi-permanent cover does not form until December, whereas in the southern Great Plains of North America, the cover becomes permanent in December or January. Estimates of the date of formation and the probability of permanent snowcover are quite unreliable in the southern extremity of this zone. These characteristics are largely dependent on air temperature. Over the cold northern grasslands snowcover persists for 120 to 160 days. In Oklahoma and the Southern Ukraine it lasts from 30 to 60 days, but over central Texas only for a few days.

The grassland snowcover is fairly shallow and well-drifted, but nevertheless is rather uniform spatially, being broken by local variations in topography and vegetation. Shelter belts and buildings may cause massive drifts whereas adjacent fields may be relatively free of snow. Also, depressions and eroded areas fill with well-packed snow. The mean annual accumulated depths of snowcover in this zone mainly fall in the range from 20 to 50 cm.

The density of the prairie snowcover is $\sim 200 \text{ kg/m}^3$ throughout most of the winter increasing slowly with time because of metamorphic changes up to the time of active melt (about March 15) when the changes are marked as melt progresses.

A climatological feature of particular interest in both the grasslands and tundra zones is the blizzard. In Canada, for a storm to be categorized as a blizzard requires the co-existence of 40 km/h winds, an air temperature equal to or lower than -12°C and visibilities less than 0.8 km which last for a duration of at least six hours. Such conditions can lead to massive, hard drifts.

Mixed forest

The most extensive mixed forest zones are in eastern North America, south of 35°N; along the Pacific Coast; and in regions of Europe west of 10°E with low elevations. Snowcover usually forms in late November and December and recedes in two directions; from the south in mid-February, and from the north in late March or early April. Predictions of snowcover duration are unreliable since the cover does not remain on the ground for long periods.

In this zone, because of its mild climate and sheltering effect against wind action, the snowcover tends to be moist, except at high elevations and during periods when cold outbreaks follow snow storms. The average snowcover density is $\sim 200 \text{ kg/m}^3$ in mid-winter increasing progressively over the winter to $\sim 300 \text{ kg/m}^3$ by late March. Freeze-thaw cycling of air temperature, freezing rain and drizzle are common and produce hard crusts, ice layers and crystalline changes within the snowcover.

One feature of the Great Lakes Forest climate is the "lake effect" storm which deposits massive amounts of snow over fairly small areas downwind of a lake. For example, the areas near Buffalo and Oswego, N.Y. are seriously affected by such storms which originate from Lake Ontario and are also marked by gale force winds and blowing snow (see Fig. 5.12 for a typical deposition pattern). According to Sykes (1965) 257 cm of snow which accumulated over a five-day period in 1966 at Oswego was due to lake effect. The snowbelts which occur to the south and east of each of the Great Lakes can be largely attributed to lake effects.

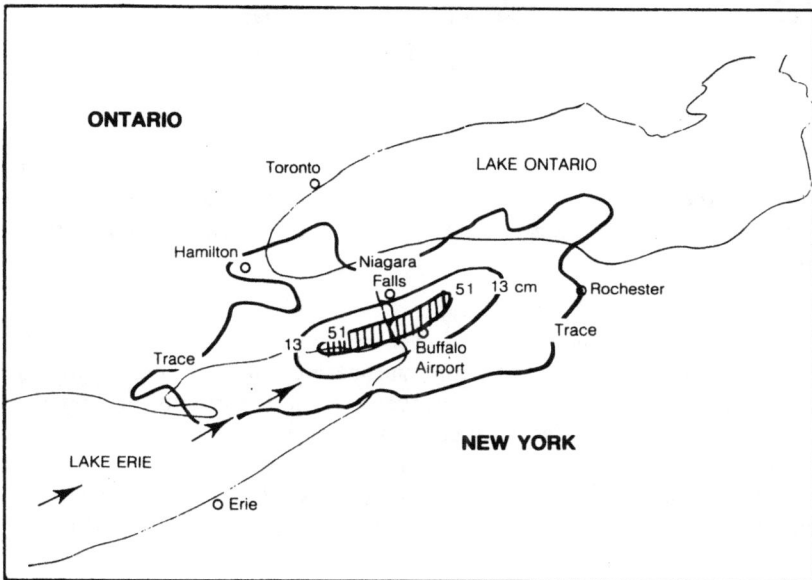

Fig. 5.12 Typical snowfall deposition pattern produced by a "lake effect" storm in the Great Lakes region of Canada and the United States; Nov. 15, 1974.

Mountain areas

Snow exists on high mountain tops every month of the year, at elevations which vary with latitude and climate. Table 5.9 (Geiger, 1961) shows the increase of snowcover duration with altitude in the Swiss Alps.

The snowcover on rugged terrain above the tree line is highly heterogeneous because of its exposure to slides and wind action. At the higher elevations tundra conditions prevail and the snowcover undergoes severe erosion and wind packing which may result in the formation of slabs. With decreasing

Table 5.9
ELEVATION AND SNOWCOVER CHARACTERISTICS IN THE EASTERN
ALPS (Geiger, 1961). Reproduced by permission of Harvard University Press.

Elevation m	% frequency of snow in winter	No. of days with snowfall per year	Maximum depth of snow on ground cm	Annual No. of days with snow on ground
200	49	27	20	38
400	61	32	31	55
800	79	45	73	109
1200	90	62	120	138
1600	96	85	142	169
2000	98	113	199	212
2400	100	143	296	270
3000	100	188	545	354

elevation the type of forest changes gradually from coniferous to deciduous in phase with the climate. The snowcover pattern in forested areas is governed by the species of trees, stand density, land features and topography (e.g., steppes).

Other zones

The other major zones where snowcover occurs include the mid-latitude deserts, the Mediterranean area and the warmer climatic regions of North America and Asia. In these, snowcover does not normally present major problems to man's activities because it is shallow and has a short duration and low frequency of occurrence. The characteristics of these snowcovers can generally be determined by analogy with the snowcover in adjacent areas although allowances must be made for the continentality of the climate.

SIMULATION AND SYNTHESIS

Snow transport and accumulation phenomena are highly complex, so that the exact mechanisms of movement are difficult to explain even for the simplest situation of a smooth, flat infinite plane. Some of the physical, mathematical and empirical approaches for describing areal accumulation and distribution of snowcover are discussed below.

The approach adopted for investigating the distribution depends mainly on such factors as: study objectives, available resources and intended use of the results. For example, an engineer primarily interested in distribution patterns around buildings, air terminals and runways, roadways and other engineering structures resulting from a single storm may use physical models. On the other hand a hydrologist mainly interested in the areal distribution of the snow water equivalent at the time of maximum accumulation, would reject the physical model as being completely impractical - not only because it is physically large and costly but because most natural systems are sufficiently complex to defy construction of a model that will satisfy the geometric, kinematic and dynamic similitude requirements.

Before attempting to apply simulation techniques to the snowcover distribution it is important to determine whether improvements in estimating spatial accumulation patterns would necessarily improve the final result for which these data are used, whether for forecasting, engineering design or other purposes. Frequently, a one-dimensional model of the snowcover will serve equally as well as a spatially-varied model when used as input to a system whose response is governed by other factors.

Physical Models

Mellor (1965) has pointed out that the value of wind-tunnel testing in snow-drifting studies has been widely recognized for about 40 years although relatively few studies were conducted during that period. By using sawdust and flake mica to simulate snow in a wind tunnel, Finney (1939) was able to propose improved methods for controlling snowdrift along highways. Theakston (1962) studied snow accumulation around structures first by using wind tunnels and later, water flumes. Quite extensive studies on drifting snow have been conducted at New York University by Gerdel and Strom (1961). In the early investigations, modelling criteria were apparently not explicitly considered. It is impractical to satisfy the large number of similarity criteria yielded by a simple dimensional analysis of the system. However, a similarity analysis which draws more heavily on physical analysis and experimental evidence promises to be more fruitful. Strom et al. (1962), Odar (1962, 1965), Isyumov (1971) and Kind (1976) have considered this approach. The reader is referred to Ch. 7 for a comprehensive review of the application of physical models to the study of snow drifting.

Mathematical, Empirical and Other Approaches

The complex nature of the transport and deposition processes and the large number of factors affecting them defies development of a generalized, physically-based mathematical model for describing areal snowcover accumulation and distribution. However, the works of Tabler (1975) on a combined transport and evaporation model and Rhea and Grant (1974) on a method to account for the effects of orographic lift on snowfall are creditable contributions in providing a physical basis toward a better understanding of the accumulation process. Such models are recognized to involve a number of simplifying assumptions and to require field data in their application.

If sufficient observations have been obtained over an area, maps of such variables as depth, density and water equivalent can be drawn and used to estimate their distributions. Such maps provide grossly simplified snowcover distributions, however, if complimented with data from aerial photographs or satellite imagery they may serve a useful operational purpose. Another method of obtaining areal averages from point observations utilizes the Thiessen polygon approach (U.S. Army Corps of Engineers, 1956).

Many users, lacking an understanding of the physical processes governing snowcover distribution, of necessity, have opted to use empirical relation-ships between snowcover properties and topographical, vegetative and land use features to define their spatial distributions. To do this it is assumed that within the same climatic region consistent similarities in the areal variations exist within areal units having similar landscape features. The extrapolation, use and application of these empirical relationships outside of the region for which they were developed is highly questionable. In many cases, the associations are established through regression techniques; e.g., the U.S. Corps of Engineers (1956) and Meiman (1970) provide excellent summaries of the relationships developed in forested, mountain regions between snowcover depth and such factors as elevation, slope and aspect. Much work in these regions has been directed to studying the effects of clearing size on accumulation. In the less-densely forested regions of rolling topography in Eastern Canada, Adams and Rogerson (1968) and Adams (1976) focussed on the different snowcover characteristics of the vegetation zones. For the prairies, Steppuhn and Dyck (1974) demonstrated that sampling of watershed areas with similar land use, terrain and vegetal cover is valuable to obtain an accurate estimate of the basin mean water equivalent. Broader scale studies (Bilello, 1969, and McKay, 1972) show that different snowcover properties may be expected in different climatic and vegetation regions. Finally, Kuz'min (1960) studied the interaction of snowcover accumulation patterns and vegetation and topography and obtained "snow retention coefficients" for a range of landscape classes (see Table 5.8).

The fact that accumulation patterns may be explained in terms of landscape features serves as a useful tool for gaining an insight into areal snowcover. The amounts accumulated by different features may differ appreciably from year to year because of variabilities in the aerodynamic processes affecting accumulation and changes in land use. The transport-condensation model proposed by Tabler (1975) explained above in this chapter delineates some of these factors. Adams (1976) suggests that significant differences can be expected (e.g., in the maximum depths of snowcover on specific vegetation types in individual years) and that the differences between strata (vegetative types) may not be consistent between years. He points out that the consistency of the differences between the depth, density and water equivalent of different vegetative-based snowcovers will depend largely on the consistency in their evolution from initiation to peak accumulation. In a region where snowcover is subject to periods of melt or rain the changes in the physical properties caused by these factors may mask the changes caused by vegetation. Adams findings suggest that a colder environment is more conducive to the preservation of different seasonal depositional patterns on different landscapes.

In principle, the application of landscape-based classes to differentiate areal snowcover distribution or to determine the mean basin water equivalent is fairly straightforward. By using aerial photographs and vegetative (land use) maps the investigator can identify and determine the areas of each landscape type. Selected areas of each type are sampled to obtain statistically-valid estimates of the mean depth, density or water equivalent for each class. These values are then used for mapping and calculations. In many cases this procedure may not find immediate use for operational forecasting purposes by management agencies because:

(1) The division of a basin into different landscape types requires prior knowledge of snow accumulation facets; the preparation of maps is time consuming, independent of any computer mapping facilities which may be available. New maps may need to be prepared each year to account for land use or vegetative cover changes caused, for example, by farming practices, forest manipulations, fires or other natural or artificial practices.

(2) Comprehensive snowcover data are required from representative landscape types to establish reliable statistical estimates of the snowcover properties.

(3) Provision must be made for updating the snowcover conditions to account for precipitation, sublimation/condensation and melt occurrence after the last snow sampling.

Young (1974) described a method of obtaining snow accumulation maps for glaciers by using data from irregularly-spaced sampling points. A grid-square technique is used to describe terrain characteristics based on altitude, slope, azimuth and local relief, and to establish field sampling locations. The method assumes that a strong linear association between snow depth (or water equivalent) and altitude exists, the total accumulation on the glacier can be calculated by such a relationship and local patterns are fairly constant from year to year and closely associated with the local configuration of the glacier geometry. The depth of the snow surface at altitude z is given by:

$$d = a + bz + cR + eS \qquad\qquad 5.3$$

where: R = local relief - indexed by the extent a grid point is above
                  or below a plane fitted to the altitude of the point itself and
                  the altitudes of the four nearest neighbour points of the grid,
              S = surface slope angle - the maximum slope of the plane,
                  and
          a,b,c,e = empirically - derived coefficients.

The magnitudes of the coefficients c and e are mean values describing the
persistence of the local snow distribution patterns on the glacier. Values of the
coefficients a and b are established each year from depth and density samples
taken at selected points at different altitudes. Using a linear relationship
between water equivalent with altitude, the values are applied to all points on
the grid to obtain an estimate of the snow water equivalent for the entire
glacier. The water equivalent values so produced are weighted in accordance
with the local shape factors; i.e. local relief and surface slope, to obtain the
final water equivalent distribution. Maps are then prepared from the adjusted
values. A standardized data bank coupled with high speed data processing
systems would make the method suitable for real-time studies to predict melt.

The grid-square technique has become popular for the spatial extrapolation
of hydrometeorological data because it is adaptable for computer operations.
Solomon et al. (1968), for example, have applied this procedure to estimate
precipitation, temperature and runoff for Newfoundland.

Even in the more sophisticated mathematical models with distributed
parameters, e.g., those used in streamflow forecasting, it is evident that the
snowcover input data will be spatially-averaged either by smoothing data
obtained from randomly selected points or stratified according to selected
landscape units. Stratification, simply by providing improved areal
representativeness, reduces not only the systematic errors in estimating areal
snowcover but also the interpolation error inherent with the use of point data
to construct maps for determining an average value. Steppuhn and Dyck
(1974) have proven that with stratification significant improvements in
estimates of the "true" areal mean value are obtained.

Landscape-course snow surveys are used operationally in the USSR.
Uryvaev and Vershinina (1971) report that these surveys can be applied to
determine the water equivalent for areas of 250-1000 km^2 with an error within
5-10% of the estimates obtained from comprehensive snow-survey data.

Vershinina (1971) presents a comprehensive review, including both
theoretical considerations and field results, of the accuracy of determining the
snowcover water equivalent at a point, over a course and for an area (average,
based on landscape snow courses). His findings indicate that the root mean
square error decreases exponentially as area increases for a constant network
density, but increases as network density decreases.

A complete discussion of snowfall and snowcover measurement is
presented in Ch. 6.

LITERATURE CITED

Adams, W.P. 1976. *Areal differentiation of snowcover in East-Central Ontario.* Water Resour. Res., Vol. 12, No. 6, pp. 1226-1234.

Adams, W.P. and B.F. Findlay. 1966. *Snow Measurement in the vicinity of Knob Lake, Central Labrador-Ungava, Winter 1964-65.* Proc. 23rd Annu. Meet. East. Snow Conf. pp. 26-40

Adams, W.P. and R.J. Rogerson. 1968. *Snowfall and snowcover at Knob Lake, Central Labrador-Ungava.* Proc. 25th Annu. Meet. East. Snow Conf., pp. 110-139.

Bagnold, R.A. 1941. *The Physics of Blown Sand and Desert Dunes.* Methuen and Co., London.

Bilello, M.A. 1969. *Relationship between climate and regional variations in snow-cover density in North America.* Res. Rep. 267, U.S. Army Cold Reg. Res. Eng. Lab., Hanover, N.H.

Bohren, C.F. 1973. *Theory of radiation heat transfer between forest canopy and snowpacks.* The Role of Snow and Ice in Hydrology: Proc. Banff Symp., Sept. 1972, Unesco-WMO-IAHS, Geneva-Budapest-Paris, Vol. 1, pp. 165-175.

Church, J.E. 1941. *The melting of snow.* Proc. Central Snow Conf., East Lansing, Mich., pp. 21-32.

Dickson, R.R. and J. Posey. 1967. *Maps of snow-cover probability for the Northern Hemisphere.* Mon. Weather Rev., Vol. 95, pp. 347-355.

Dyunin, A.K. 1961. *Ispareniye snega (Evaporation of snow).* Novosib. Izv. Sib. Otd. Akad. Nauk, SSSR.

Finney, E.A. 1939. *Snowdrift control by highway design.* Bull. No. 86, Mich. Eng. Expt. Sta., East Lansing.

Fisheries and Environment Canada, 1978. *Hydrological atlas of Canada.* Cat. No. EN 37-26/1978, Supply Serv. Can., Ottawa, Ont.

Geiger, R. 1961. *Das Klima der bodennahen Luftschicht (The Climate near the Ground).* [English Transl. by Scripta Technica, Inc., Harvard University Press, Cambridge, Mass., 1966].

Gerdel, R.W. and G.H. Strom. 1961. *Wind tunnel studies with scale model simulated snow.* Int. Union Geod. Geophys., Gen. Assem. Helsinki , [Snow and Ice], Int. Assoc. Hydrol. Sci. Publ. No. 54., pp. 80-88.

Goodell, B.C. 1952. *Watershed management aspects of thinned young lodgepole stands.* J. For., Vol. 50, pp. 374-378.

Goodell, B.C. 1959. *Management of forest stands in western United States to influence the flow of snow-fed streams.* Water and Wood Aids, Int. Assoc. Sci. Hydrol. Symp., Hannoversch-Munden, Pub. No. 48, pp. 49-58.

Grant, L.O. and J.O. Rhea. 1974. *Elevation and meteorological controls on the density of new snow.* Adv. Concepts Tech. Study Snow Ice Resour., Interdiscip. Symp., U.S. Nat. Acad. Sci., Washington, D.C., pp. 169-181.

Gray, D.M., D.I. Norum and G.E. Dyck. 1970. *Densities of prairie snowpacks.* Proc. 38th Annu. Meet. West. Snow Conf., pp. 24-30.

Gray, D.M., H.W. Steppuhn and F.L. Abbey. 1979. *Estimating the areal snow water equivalent in the Prairie environment.* Proc. Can. Hydrol. Symp.: 79 - Cold Climate Hydrol., Nat. Res. Counc. Can., Ottawa, pp. 302-332.

Haupt, H.F. 1973. *Relation of wind exposure and forest cutting to changes in snow accumulation.* The Role of Snow and Ice in Hydrology: Proc. Banff Symp., Sept. 1972, Unesco-WMO-IAHS, Geneva-Budapest-Paris, Vol. 2, pp. 1399-1405.

Hoover, M.D. and C.F. Leaf. 1967. *Process and significance of interception in Colorado subalpine forest.* In Proc. Int. Symp. For. Hydrol., Pergamon Press, pp. 213-222.

Isyumov, N. 1971. *An approach to the prediction of snow loads.* Ph.D. Thesis, Univ. Western Ont., London.

Khodakov, V.G. 1975. *Role of the snowcover in the nature of northern landscapes and its physical properties.* Sov. Hydrol.: Sel. Pap. Issue No. 2, pp. 60-65. (English edition published July 1976).

Kind, R.J. 1976. *A critical examination of the requirements for model simulation of wind-induced erosion/deposition phenomena such as snow drifting.* Atmos. Environ., Vol. 10, pp. 219-227.

Kittredge, J. 1953. *Influences of forests on snow in the ponderosa-sugar pine-fir zone of the Central Sierra Nevada.* Hilgardia, Vol. 22, No. 1, pp. 1-96.

Kondratyev, K. Ya. 1954. *Luchistaya Energiya Solntsa (The Radiant Energy of the Sun).* Gidrometeoizdat, Leningrad.

Kotliakov, V.M. 1968. *Snowcover of Earth Glaciers.* Gidrometeoizdat, Leningrad.

Kotliakov, V.M. 1973. *Snow accumulation on mountain glaciers.* The Role of Snow and Ice in Hydrology: Proc. Banff Symp., Sept. 1972, Unesco-WMO-IASH, Geneva-Budapest-Paris, Vol. 1, pp. 394-400.

Kung, E.C., R.A. Bryson and D.J. Lenschow. 1964. *Study of a continental surface albedo on the basis of flight measurements and structure of the earth's surface cover over North America.* Mon. Weather Rev., Vol. 92, pp. 543-564.

Kuz'min, P.P. 1960. *Formirovanie Snezhnogo Pokrova imetody opredeleniya snegozapasov (Snowcover and snow reserves).* Gidrometeorologicheskoe, Izdatelitvo, Leningrad. [English Transl. by Israel Prog. Sci. Transl., Transl. 828].

Lakshman, G. 1963. *Drainage basin study.* Prog. Rep. 8, No. E73-6, Eng. Div., Sask. Res. Counc., Saskatoon.

Martinelli, M. Jr. 1973. *Snow fences for influencing snow accumulation.* The Role of Snow and Ice in Hydrology: Proc. Banff Symp., Sept. 1972, Unesco-WMO-IASH, Geneva-Budapest-Paris, Vol. 2, pp. 1394-1398.

McKay, G.A. 1963. *Relationships between snow survey and climatological measurements.* Int. Union Geod. Geophys. Gen. Assem. Berkeley, [Surface Water], Int. Assoc. Hydrol. Sci. Publ. No. 63, pp. 214-227.

McKay, G.A. 1970. *Precipitation.* Handbook on the Principles of Hydrology. (D.M. Gray, ed.), The Secretariat, Can. Nat. Comm. Int. Hydrol. Decade, Ottawa. pp. 2.1-2.111.

McKay, G.A. 1972. *The mapping of snowfall and snowcover.* Proc. 29th Annu. Meet. East. Snow Conf., pp. 98-110.

McKay, G.A. and B. Findlay. 1971. *Variations of snow resources with climate and vegetation in Canada*. Proc. 39th Annu. Meet. West. Snow Conf., pp. 17-25.

McKay, G.A. and H. Thompson, 1968. *Snowcover in the Prairie Provinces of Canada*. Trans. Am. Soc. Agric. Eng., Vol. 11, No. 6, pp. 812-815.

Meiman, J.R. 1970. *Snow accumulation related to elevation, aspect and forest canopy*. Proc. Workshop Semin. Snow Hydrol. Queen's Printer of Canada, Ottawa. pp. 35-47.

Mellor, M. 1965. *Blowing snow*. Mono. III-A3c, U.S. Army Cold Reg. Res. Eng. Lab., Hanover, N.H.

Mikhel', V.M., A.V. Rudneva and V.I. Lipöuskaya. 1971. *Snowfall and snow transport during snowstorms over the USSR.* [Englisk Transl. by Israel Prog. Sci. Transl., Transl. 5909].

Miller, D.H. 1966. *Transport of intercepted snow from trees during snowstorms*. Res. Pap. PSW-33, USDA For. Serv., Berkeley, Calif., pp. 1-30.

Miller, D.H. 1976. *Spatial interactions produced by meso-scale transports of water in the atmospheric boundary layer*. Pap. presented to the Annu. Meet., Assoc. Am. Geogr., New York, N.Y. April.

Molchanov, A.A. 1963. *The hydrologic role of forests*. [English Transl. by Israel Prog. Sci. Transl., Transl. 870].

Odar, F. 1962. *Scale factors for simulation of drifting snow*. Proc. Am. Soc. Civ. Eng., Eng. Mech. Div., Vol. 88, No. EM2, pp. 1-16.

Odar, F. 1965. *Simulation of drifting snow*. Res. Rep. 174, U.S. Army Cold Reg. Res. Eng. Lab., Hanover, N.H.

Orvig, S. (ed.). 1970. *Climates of Polar Regions*. In World Survey of Climatology, Vol. 14. (H.E. Landsberg, Editor-in-Chief), Elsevier, Amsterdam.

Peck, E.L. 1964. *The little used third dimension*. Proc. 32nd Annu. Meet. West. Snow Conf., pp. 33-40.

Radok, U. 1977. *Snow drift*. J. Glaciol., Vol. 19, pp. 123-129.

Reifsnyder, W.E. and H.W. Lull. 1965. *Radiant energy in relation to forests*. Tech. Bull. 1344, U.S. Dept. Agric., Washington, D.C.

Rhea, J.O. and L.O. Grant. 1974. *Topographic influences on snowfall patterns in mountainous terrain*. Adv. Concepts Tech. Study Snow Ice Resour. Interdiscip. Symp., U.S. Nat. Acad. Sci., Washington, D.C., pp. 182-192.

Rikhter, G.D. 1945. *Snezhyi Pokrov, Ego Gormirovanie; Svoistva. (Snow cover, its Formation and Properties)*. Izv. Akad. Nauk. SSSR, Moscow. [English Transl. by U.S. Army Snow, Ice Permafrost Res. Estab., Transl. 6].

Schmidt, W. 1925. *Der Massenaustrausch in freier Luft und verwandte Erscheinungen*. Hamburg, Grand.

Schmidt, R.A. Jr. 1972. *Sublimation of wind-transported snow - A model*. Res. Pap. RM-90, USDA For. Serv., Rocky Mtn. For. and Range Expt. Stn., Fort Collins, Colo.

Seligman, G. 1936. *Snow Structure and Ski fields*. MacMillan and Co. London.

Solomon, S.I., J.P. Denouvilliez, E.J. Chart, J.A. Woolley and C. Cadou. 1968. *The use of a square grid system for computer estimation of precipitation, temperature and run-off*. Water Resour. Res., Vol. 4, pp. 919-929.

Stanton, C.R. 1966. *Preliminary investigation of snow accumulation and melting in forested and cut-over areas of the Crowsnest Forest.* Proc. 34th Annu. Meet. West. Snow Conf., pp. 7-12.

Steppuhn, H. 1976. *Areal water equivalents for prairie snowcovers by centralized sampling.* Proc. 44th Annu. Meet. West. Snow Conf., pp. 63-68.

Steppuhn, H.W. and G.E. Dyck. 1974. *Estimating true basin snowcover.* Adv. Concepts Tech. Study Snow Ice Resour. Interdiscip. Symp., U.S. Nat. Acad. Sci., Washington, D.C., pp. 314-328.

Storr, D. and D.L. Golding. 1974. *A preliminary water balance evaluation of an intensive snow survey in a mountainous watershed.* Adv. Concepts Tech. Study Snow Ice Resour. Interdiscip. Symp., U.S. Nat. Acad. Sci., Washington, D.C., pp. 294-303.

Strom, G., G.R. Kelly, E.L. Keitz and R.F. Weiss, 1962. *Scale model studies on snow drifting.* Res. Rep. 73, U.S. Army Snow, Ice Permafrost Res. Estab.

Sykes, R.B. 1966. *The blizzard of '66 in central New York state - legend in its time.* Weatherwise, Vol. 19, No. 6, pp. 240-247.

Tabler, R.D. 1971. *Design of a watershed snow fence system, and first-year snow accumulation.* Proc. 39th Annu. Meet. West. Snow Conf. pp. 50-55.

Tabler, R.D. 1975. *Estimating the transport of blowing snow.* Snow Management on the Great Plains. Publ. 73, Res. Comm. Great Plains Agric. Counc. and Univ. Nebr. Agric. Exp. Stn., Lincoln, pp. 85-117.

Tabler, R.D. and R.A. Schmidt. 1973. *Weather conditions that determine snow transport distances at a site in Wyoming.* The Role of Snow and Ice in Hydrology: Proc. Banff Symp., Sept. 1972, Unesco-WMO-IASH, Geneva-Budapest-Paris, Vol. 1, pp. 118-127.

Theakston, F.H. 1962. *Snow accumulations about farm structures.* Agric. Eng., Vol. 43, pp. 139-141, 161.

Uryvaev, V.A. and L.K. Vershinana. 1971. *Accuracy of landscape-course snow surveys.* In Issledovaniya Metodov, Apparatuny i Tochnosti Opredeleniya Zapasov Vody v Snezhnom Pokrove (Determination of the Water Equivalent of Snow Cover, Methods and Equipment), (L.K. Vershinina and A.M. Dimaksyan, eds.), [English Transl. by D. Lederman, Israel Prog. Sci. Transl., Keter Press, Jerusalem, pp. 94-99].

U.S. Army Corps of Engineers. 1956. *Snow Hydrology.* U.S. Dept. Commer., Washington, D.C.

Vershinina, L.K. 1971. *Assessment of the accuracy in determining water equivalent of snowcover.* In Issledovaniya Metodov, Apparatury i Tochnosti Opredeleniya Zapasov Vody v Snezhnom Pokrove (Determination of the Water Equivalent of Snow Cover, Methods and Equipment), (L.K. Vershinina and A.M. Dinaksyan, eds.), [English Transl. by D. Lederman, Israel Prog. Sci. Transl., Keter Press Jerusalem], pp. 100-127.

Wilm, H.G. and M.H. Collet. 1940. *The influence of lodgepole pine forest on storage and melting of snow.* Trans. Am. Geophys. Union, Vol. 21, pp. 505-508.

Young, G.J. 1974. *A data collection and reduction system for snow accumulation studies.* Adv. Concepts Tech. Study Snow Ice Resour., Interdiscip. Symp., U.S. Nat. Acad. Sci., Washington, D.C., pp. 304-313.

6

MEASUREMENT AND DATA ANALYSIS

B. E. GOODISON

*Canadian Climate Centre, Atmospheric Environment Service,
Downsview, Ontario.*

H. L. FERGUSON

*Air Quality and Inter-Environmental Research Branch,
Atmospheric Environment Service,
Downsview, Ontario.*

G. A. McKAY

*Canadian Climate Centre, Atmospheric Environment Service,
Downsview, Ontario.*

INTRODUCTION

Snowfall is the depth of fresh snow which falls during a given "recent" period. Snowfall measurements are summed to determine the total for any time period: a single storm, a day, a month or a year. Total precipitation is the sum of the vertical depth of all liquid precipitation and of the water equivalent (depth) of all forms of solid precipitation, including snowfall.   Snowcover refers to the amount of snow on the ground at the time of an observation; the ground may be either completely or partly covered.

Proper utilization of snowfall-snowcover data depends on the user's requirements, assuming that the user understands how the data were obtained, realizes the problems of measuring and processing the data and is aware that errors may exist in the data.

POINT SNOWFALL MEASUREMENT

Snow Ruler, Snow Boards

A graduated ruler inserted vertically into the snow is the most direct method for measuring the depth of freshly-fallen snow. Where the snow has not drifted the mean depth of snowfall is determined from measurements made at several points. To ensure that "old" snow is not sampled, the measurement is made on a patch or a snow board whose surface has been kept free of snow before the snowfall. A snow board is a piece of plywood or lightweight metal at least 40 cm by 40 cm, painted white or covered with white flannel, which provides a reference level for the measurement.

To obtain a representative 'mean' depth of new snow under drifting conditions requires careful judgement by the observer. A large number of measurements must be taken in both drifted and exposed areas.

The water equivalent of fresh snowfall from ruler measurements may be estimated by using an approximate relation between depth and water equivalent. Commonly, the average density of newly-fallen snow is accepted as 100 kg/m³, that is, 1 cm of snow is taken equal to 1 mm of snow water equivalent. In Canada, the Atmospheric Environment Service uses this method to estimate snow water equivalent for more than 85% of the observing stations. In reality, the density of newly-fallen snow varies with region, with individual storm events and often throughout the duration of a storm. If more precise data about the water equivalent are required these are obtained by melting samples of the snow taken from the boards.

Limited-Capacity, Non-Recording Snow Gauges

Snow gauges measure snowfall water equivalent directly. Essentially, any open cylinder in which snow can accumulate and be measured can serve as a snow gauge. The cylinder is generally shielded to reduce wind turbulence around the orifice and is mounted far enough above the snow surface to minimize the accumulation of blowing snow in the gauge. The water equivalent is determined by weighing the contents or by melting the snow and measuring the liquid volume. Gauges with limited capacity must be emptied frequently, usually once a day.

In Canada, beginning with the winter of 1960-61, the MSC Nipher shielded snow gauge (Fig. 6.1a) was designated as the official Canadian instrument for measuring snowfall water equivalent. The shield used on Canadian gauges is a modification of that originally designed by Nipher (1878). It has the shape of an inverted bell, and is usually constructed of spun aluminum or fiberglass.

Fig. 6.1 Types of gauges used to measure the water equivalent of snowfall in different countries. (a) MSC Nipher Shielded Gauge (Canada). (b) Swedish SMHI Precipitation Gauge. (c) USSR Tretyakov Precipitation Gauge.

Wind tunnel tests conducted by the National Research Council of Canada indicated that this shield design is effective in minimizing disturbances to the airflow over the gauge orifice (Potter, 1965). The collector is a hollow, metal cylinder about 52 cm long, open at one end, and 12.7 cm in diameter. It is placed inside the gauge so that its top rim is level with the top edge of the shield. The gauge and shield usually are mounted on an adjustable stand, so that the lip of the shield can be maintained approximately 1.5 m above the snow surface (Meteorological Branch, 1965). During periods of light winds snow may accumulate on the solid shield, while some may be blown into the collector by subsequent gusts of wind. Hence, the shield is cleared of snow after each observation. Snow caught by this gauge is melted and measured in a special glass graduate to obtain the water equivalent.

At present about 14% of the 2500 precipitation stations in Canada are equipped with an MSC Nipher shielded snow gauge. At these stations total winter precipitation is the sum of any rainfall and the snowfall water equivalent measured by the Nipher. These data are published in the Monthly Record, Meteorological Observations in Canada (Atmospheric Environment Service, 1980). Detailed observational procedures are given in publications by the Atmospheric Environment Service (1973a, 1977).

Unlike Canada, some countries such as Sweden, the USA and the USSR have one national gauge for measuring both rainfall and snowfall. In winter the rainfall funnel of the gauge is removed so that snow enters directly into the collector. Figures 6.1b,c, respectively, illustrate the standard Swedish SMHI precipitation gauge with solid windshield and the USSR's Tretyakov shielded gauge. The Tretyakov gauge includes a collector, a metal cylinder 41 cm long and 16 cm in diameter, surrounded by a flexible shield of 15 steel slats which are fastened together at the top and bottom by steel wire to provide an upper diameter of 1.04 m and a lower diameter of 0.44 m. Snow accumulated in the collector is melted then measured in a glass graduate to obtain the water equivalent.

Errors in measurement associated with limited capacity, non-recording gauges are attributable to the effect of wind speed on gauge catch and to evaporation, spillage, wetting and retention losses. The last type of loss results from water retained on the walls of the measuring cylinder by surface tension. Soviet results (Struzer, 1965) for the Tretyakov gauge indicates this loss to be about 0.25 mm for each measurement. Goodison (1978b) reports a mean retention loss of 0.15 ± 0.02 mm for each MSC Nipher shielded gauge measurement. Evaporation loss depends on the time intervals and climatic conditions between observations. For a gauge read daily, this loss, plus any that occurs during melting of the snow, should be less than 0.25 mm.

Weighing-Type Precipitation Gauges

Weighing-type precipitation gauges measure all forms of precipitation. They use the principle of a simple spring balance. Precipitation is collected in a catch bucket mounted on a spring, which becomes compressed and activates a recording mechanism, usually a pen, to produce a trace on a chart. The mechanical displacement of the spring can be converted to a digital output signal which can be recorded *in situ* or telemetered by radio, satellite or telephone to a centrally-located facility. The capacities of weighing-type gauges range from 300 to 600 mm water equivalent. Some can operate unattended for up to one year; their time resolution capabilities can vary from 5 min to several hours. Capacity, time resolution and the duration of operation are closely interrelated characteristics which must be considered when selecting a gauge to satisfy specific study requirements.

Weighing-type gauges should be equipped with shields to help reduce wind turbulence over the gauge orifice. One type, the Alter shield, (see Fig. 6.2) is adaptable to several models of long-duration gauges; it is flexible, consisting of a number of slats mounted to surround the orifice of the gauge. When the shield is used in very windy, unprotected sites the slats are sometimes bridled or joined to each other by a light chain to prevent them "looping" on the ring thereby increasing turbulence. In sheltered sites the slats are left "free-swinging" because if they are bridled snow may collect between the gauge and the lower end of the slats and then accumulate on this base to cover the orifice. The shield should be attached to extend 1.25 cm above the rim of the orifice. The height at which the orifice is mounted differs from country to country, the most common being 2 m above ground level. In regions of high snowfall it is recommended that the gauge be mounted so that its orifice remains at least 1 m above the surface of the maximum expected snowpack.

In snowy climates, long-duration gauges require an antifreeze charge to prevent freezing of precipitation in the collector. The amount of antifreeze required depends on the expected amount of precipitation and the minimum temperature expected at the time of maximum dilution. An ethylene glycol-water solution is commonly used as the antifreeze, but since glycol is denser than water, a problem may arise with the mixture if it is not stirred periodically to prevent freezing. All new snow must be completely melted by the charge. To prevent evaporation from the catch bucket a thin layer of light oil, such as transformer oil is added. Mayo (1972) recommends the use of an antifreeze solution of 40% ethylene glycol and 60% methyl alcohol by volume. Because water and ice are denser than the glycol-alcohol solution, they induce mixing on being added. However, this antifreeze must be covered with a thicker layer of heavier oil to minimize evaporation of the methyl alcohol.

Fig. 6.2 Fischer and Porter Precipitation Gauge with Alter Shield.

The specifics for the two most common long-duration recording gauges used in North America are given below (trade names are given for reference purposes only).

Fischer and Porter. A Fischer and Porter, automatic, long-duration, recording precipitation gauge with free-swinging Alter shield (Fig. 6.2) is particularly suited for use in remote locations where regular observations are not made. This gauge collects all forms of precipitation and records the accumulated total on punched paper tape. The standard gauge has an orifice diameter of 20.3 cm, a collector capacity of 630 mm water equivalent and a recording range from 0 to 500 mm in increments of 2.54 mm. Improved resolution can be obtained by using an optical absolute position encoder instead of the mechanical encoder. The timing readout options range from 5-min to 12-h intervals with a 15-min interval being the most common. Power may be supplied by line or battery. The maximum duration of operation of

Fig. 6.3 Universal Precipitation Gauge.

the gauge without attendance depends on the readout interval, precipitation capacity and associated power requirements. These gauges have been left unattended in remote locations for up to one year; however, to ensure reliable continuous operation they should be serviced at least every three months. The signal from the Fischer and Porter gauge is readily adaptable for telemetry via land line or satellite. Pollock et al. (1973), Ferguson et al. (1974) and Goodison and McKay (1978) have discussed the problems most commonly encountered with the Fischer and Porter gauge and its resulting data.

Universal Recording Gauge. The Universal recording precipitation gauge (see Fig. 6.3) collects all forms of precipitation and continuously records the accumulated total. Its orifice is 20.3 cm in diameter and it has a standard maximum capacity (including antifreeze) of about 300 mm of liquid. A high capacity (760 mm liquid) version of the gauge is also available. The chart drive is spring wound or battery operated and provides cylinder rotations ranging from 6 h to 36 days per revolution; the time resolution is about 1 h with an 8-day revolution. Use of the potentiometer and a resistance-to-voltage transmitter allows data to be transmitted from remote stations.

Precipitation Storage Gauges

Storage gauges have large capacities, up to 2540 mm of water equivalent, and are used to measure seasonal precipitation totals at remote or unattended sites. With these gauges, as with recording gauges, an Alter shield (or equivalent), an antifreeze solution, and a light oil are necessary accessories since only the depth of liquid is measured. In a storage gauge it is important that a slush layer, which may solidify in very cold weather, not be permitted to form in the collector. Poor mixing of the precipitation and antifreeze, leading to the formation of slush layers, can be prevented by using inexpensive nitrogen gas bubblers or a self-mixing antifreeze solution.

Storage gauges are monitored manually at regular or irregular intervals depending on the accessibility of the site. The usual practice is to read the vertical depth of fluid with a ruler or with a T-stick to measure the distance from the top of the gauge to the surface of the liquid. The water equivalent in the gauge is calculated from the liquid depth and the density of the antifreeze mixture. A storage gauge can be modified for telemetering by connecting it to a stilling well enclosed in an adjacent shelter. A float-activated, parallel,

Fig. 6.4 Sacramento Storage Gauge.

digital recorder may be used to monitor the depth of fluid in the standpipe and provide input to a telemetry system. This type of gauge is referred to as a float-type gauge.

The Sacramento storage gauge (see Fig. 6.4) is commonly used to measure precipitation in mountainous regions characterized by high precipitation. It is cone-shaped with a length of about 120 cm and has a 20.3 cm orifice and a maximum capacity of 2540 mm water equivalent.

In "low" snowfall regions smaller standpipe storage gauges may be used. They are usually straight-sided plastic pipes of varying length, constructed specifically to suit measurement requirements.

Visibility Attenuation

One method of measuring snowfall is based on the attenuation of a light beam by snowflakes passing between a pulsed light source and a light sensitive detector (Warner and Gunn, 1967). The relation between attenuation and rate of snowfall can be affected by drifting snow, atmospheric pollutants, fog, and liquid precipitation contained in the snow. Wasserman and Monte (1972) have shown that the visibility attenuation method can be used to estimate hourly snow accumulations. They state that for a given storm such information could be used to forecast additional snow accumulations, to estimate snowfall amounts when direct measurements are missing or to estimate the frequency of selected hourly rates (since manual measurements are made only every six hours).

Telemetry of Point Precipitation Measurements

Modern electronics have led to rapid advances in the development of telemetry systems suited to remote operation in harsh environments. Continuing development of new communication techniques permits access to precipitation and other hydrometeorological data (e.g., temperature, wind, or snowpack water equivalent) from remote stations on a real- or near-real-time basis. Data transmission systems include telephone telemetry (the least expensive method if there is a line available at the site), radio systems (FM, VHF, UHF), and satellite retransmission (Landsat, GOES, ARGOS). Depending on the system used, hydrometeorological information from remote stations is transmitted at a time varying from a few seconds to a few hours after the observation is made. Some systems monitor and transmit data continuously; some monitor only when interrogated; others monitor and store data which are later transmitted on request. Data on winter snow accumulation and meteorological conditions during the spring melt runoff period can now be obtained from remote areas, and are used to improve streamflow flood forecasting and reservoir regulation.

Chadwick (1972) reported the development of an operational, battery-operated, radio telemetry system having high resolution and accuracy for transmitting precipitation data from remote mountain sites in Utah. Barnes (1974) described a hydrometeorological network operating in California with VHF radio and microwave UHF telemetry systems to transmit data to a central base station. The successful operation of an Alberta network of automatic remote precipitation stations linked to a central computer facility by telephone lines is described by Graham et al. (1977).

More recently, the transmission of hydrometeorological data via satellites has been shown to be an excellent alternative (Halliday, 1975; Schumann, 1975; Flanders, 1979 and Coles and Graham, 1980) and many agencies are planning to use this system. A Data Collection Platform (DCP) which accepts data inputs in analogue, parallel digital or serial digital format is used to transmit data from a precipitation gauge, such as a Fischer and Porter, to the satellite. These battery-powered DCP units are designed to function in severe temperatures and high humidities.

DCP technology is advancing rapidly and several types of platforms are now available. New systems will continue to develop; e.g. a near-real-time operational data acquisition system (SNOTEL) uses the reflection of VHF signals from ionized meteor trails to relay data from remote sites in the Arctic to base stations. This system is being used by the United States Soil Conservation Service (Barton and Burke, 1977; Farnes, 1978; Woodward et al., 1980).

COMPARISON OF POINT SNOWFALL MEASUREMENT TECHNIQUES

The ruler provides a measurement of depth from which the snowfall water equivalent can be estimated, whereas a snow gauge provides a measurement of snowfall water equivalent from which the depth can be estimated. The causes of errors in these point measurements and estimates are known; the magnitudes of the errors or the differences between "measured" and "true" catches are not well known, largely because of the difficulty in determining "true snowfall".

Limitation of Ruler Measurements

Errors in ruler measurements of snowfall depth mainly originate from poor siting (e.g., an open site susceptible to drifting) and from observer bias. The type and magnitude of the errors vary from storm to storm, observer to observer, and station to station.   Hence it is difficult to apply universal corrections to daily depth data.

Appreciable error in snowfall water equivalent estimates can also result by assuming that snow has a density of 100 kg/m^3 (Table 6.1); Canadian experiments to determine the 10 to 1 ratio were originally performed in Toronto.   However, the density of new snow exhibits wide temporal and spatial variations, and is primarily controlled by the amount of air in the interstices between individual snow crystals.   The air space is a function of the type of snowfall and its crystal structure, the meteorological conditions (especially upper-air temperature and wind speed and direction near the surface) prevailing during and immediately following the snowfall, the elapsed time of measurement of snowfall after the beginning or end of the storm, the siting of the measurement station and observer bias.

The data listed in Tables 6.2 and 6.3 illustrate the temporal variations in snowfall density at a sheltered site.   Densification of a fresh snowfall may occur rapidly such that water equivalent estimates, based on an initial density for the precipitate and depth measurements taken only a few hours after a storm, can be quite erroneous.   Snowfalls in different geographic and/or climatic regions also have different densities.   Potter (1965) showed that the mean values of the ratio of water equivalent to depth had regional variations throughout Canada from 0.07 to 0.11, with extremes from 0.02 to 0.28.   On the Canadian Prairies, Gray et al. (1970) measured average densities for freshly-fallen snow of 45 kg/m^3 during non-drifting conditions, and 230 kg/m^3 for

Table 6.1
COMPARISON OF MEASURED SNOWFALL WATER EQUIVALENT WITH THAT ESTIMATED USING A MEAN DENSITY OF 100 kg/m^3.

Hour	Accumulated Storm Board Depth cm	Measured Water Equivalent mm	Estimated Water Equivalent mm	(Estimated - Measured) as a Percentage of Measured %
1	1.5	0.89	1.5	+68.5
2	3.3	2.26	3.3	+46.0
3	4.4	3.86	4.4	+14.0
4	5.2	5.69	5.2	- 8.6
5	7.2	8.73	7.2	-17.5
6	9.5	11.00	9.5	-13.6
7	9.8	12.57	9.8	-22.0
8	11.8	16.13	11.8	-26.8
9	12.0	17.09	12.0	-29.8

Table 6.2

VARIATION IN DENSITY (kg/m³) OF FRESHLY-FALLEN SNOW AND CHANGES IN DENSITY (kg/m³) OF FRESH SNOW ON THE GROUND OCCURRING DURING STORMS. MEASUREMENTS MADE AT SHELTERED SITE (NON-DRIFTING CONDITIONS) IN SOUTHERN ONTARIO

Date d/mos./y	SWE[a] mm	Measure-ment[b]	Density as a Function of Time After the Beginning of Storm (h)											
			1	2	3	4	5	6	7	8	9	10	11	12
28/2/75	1.53	1	89	108	78									
		2	89	102	99									
7/3/75	16.73	1	59	71	179	200	108	112	189	168	139			
		2	59	68	88	110	121	116	128	137	143			
27/11/75	11.71	1	100	138	145	140	212	353						
		2	100	125	132	128	141	141						
19-20/1/76	2.58	1	78	98	101	124	119	128						
		2	78	91	105	110	120	121						
1-2/3/76	25.4	1	105			205		199		191		210		168
		2	105			200		192		196		243		220

[a] SWE: snowfall water equivalent.
[b] 1. Density of snowfall during hour as measured on a snow board.
2. Density of snow accumulated on snow board from the time of beginning of storm.

Table 6.3

CHANGES IN DENSITY OF FRESHLY FALLEN SNOW ON THE GROUND AFTER CESSATION OF SNOWFALL.

Time[a]	Measured Depth cm	Density kg/m³	Measured Water Equivalent[b] mm	Water Equivalent Using 100 kg/m³ mm
1030	8.7	104	9.06	8.70
1315	7.0	131	9.14	7.00
1700	6.0	152	9.09	6.00

[a] General climatic conditions during the day: cloudy, maximum temperature 1°C, wind speed ~ 6 m/s.
[b] Determined by clearing an individual 930 cm² snow board.

drifted snow a day later, during blizzard conditions; a six-fold increase in density caused by drifting had occurred within a period of less than 24 hours.

In the Canadian Rockies a depth to water equivalent ratio of 12 to 1 rather than 10 to 1 has been found to yield a better estimate of snowfall water equivalent (P. Schaerer, personal communication). Also, in the Colorado

Rockies, Grant and Rhea (1974) reported mean new snow densities of 77 to 101 kg/m^3, with 58% of their stations recording mean values from 80 to 90 kg/m^3. Regions surrounding the Great Lakes are often characterized by lake effect snow storms. The density of newly fallen snow from such storms has commonly been measured at 35 to 50 kg/m^3; therefore, a mean density factor of 0.10 would overestimate the snowfall water equivalent in many regions.

Comparison of Snow Gauge Measurements

In order to obtain the best measurements from snow gauges, each instrument must be serviced, levelled, and maintained on a regular basis. The measurement accuracy of each gauge depends on type, siting or exposure conditions, shielding, the height of the orifice and the relevant storm characteristics (notably wind speed). Most precipitation gauges catch a snowfall amount which is less than the true amount.

Long-term observations of the catches of different gauge types at the same site provide basic comparative data. From a 15-year study of United States standard gauges Allis et al. (1963) found that they had greater variations in their catches of snow, as compared to rain, and increased catches when shielded. Harris and Carder (1974), in a ten-year study conducted at Beaverlodge, Alberta, found that over a wide range of snowfall amounts the snow water equivalent totals recorded by the MSC Nipher gauge were consistently larger than those measured by a United States Weather Bureau (USWB) gauge equipped with an Alter Shield: in 567 comparisons, the USWB gauge recorded smaller amounts in 55% of the cases; the same in 40%; and larger amounts in only 5%.

Data in Table 6.4, (Goodison, 1977), provide comparisons of the catches of different gauges to the "ground" snowfall water equivalents as measured on a snow board. It can be noted that on the average the Canadian MSC Nipher shielded snow gauge gave better estimates of snowfall water equivalent than the other gauges. The results also confirm the well-known facts: shielded gauges have higher catch efficiencies than unshielded gauges, gauges located in a sheltered site catch more snow than those located in a windy open site, and the gauge and shield design influence gauge catch.

It is difficult to correlate the absolute value of the snowfall totals measured by a gauge at principal stations with those measured by ruler at the same stations or nearby climatological stations. In Canada, direct measurements of the density of new snow at observing stations is not standard procedure, so that only comparative Nipher gauge/ruler ratios can be developed. Ferguson and Pollock (1971) calculated such monthly mean ratios for geographic regions of Canada (Table 6.5). In any year, the application of a mean ratio for estimating snowfall water equivalent may result in considerable error. For

example, observations taken from 1960 to 1974 at the Toronto International
Airport and Thunder Bay indicated that the annual Nipher gauge/ruler ratios
for these stations ranged from 0.072 to 0.114 and from 0.074 to 0.096,
respectively.

Table 6.4
**MEAN RATIO OF GAUGE CATCH TO GROUND "TRUE" MEASUREMENT
1974-1976 FOR MSC NIPHER SHIELDED, FISCHER AND PORTER
RECORDING, UNIVERSAL RECORDING, AND USSR TRETYAKOV
SHIELDED GAUGES (Goodison, 1977)[a].**

Gauge Description[b]	Site Description	Mean Ratio[c] (gauge/snow board)	Mean wind speed[d] m/s
Snow Board on Ground	Sheltered	1.00	---
MSC Nipher Shielded	Bush-sheltered	1.01	0.48
MSC Nipher Shielded	Valley-sheltered	1.05	1.74
MSC Nipher Shielded	Open-plateau	0.94	3.80
MSC Nipher Shielded	Open-flat	0.91	3.95
12.5 - cm snow collector (no Nipher shield)	Open-flat	0.48	3.95
Fischer and Porter c/w[e] Free Swinging Alter Shield	Bush	0.96	0.48
Fischer and Porter c/w Free Swinging Alter Shield	Valley	0.86	1.74
Fischer and Porter c/w Free Swinging Alter Shield	Plateau	0.51	3.80
Fischer and Porter c/w Free Swinging Alter Shield	Flat	0.47	3.95
Fischer and Porter - unshielded	Plateau	0.32	3.80
USSR Tretyakov Shielded	Flat	0.57	4.24
Universal Recording c/w Free Swinging Alter Shield	Flat	0.58	3.95
Universal Recording - unshielded	Flat	0.37	3.95

[a] Results are means of values obtained from data collected during snowstorms, irrespective
of wind speed or temperature, as measured at Cold Creek Hydrometeorological Research
Station, Bolton, Ontario.
[b] All gauges mounted at 2 m.
[c] Mean ratio of gauge catch to ground catch measured on a snow board at a sheltered site.
[d] Mean wind during snowstorms, measured at each site at 2 m above ground.
[e] c/w: complete with.

Table 6.5

REGIONAL SNOW RULER DEPTHS AND NIPHER GAUGE WATER EQUIVALENT RATIOS (Ferguson and Pollock, 1971). Reproduced by permission of the Minister of Supply and Services Canada.

Region	Station Months of Data	Monthly Mean Depth from Snow Ruler cm	Ratio
Eastern Arctic	222	12.5	0.087
Western Arctic	292	18.3	0.077
B.C. Interior	120	43.7	0.075
Alberta	200	29.2	0.077
Saskatchewan	125	23.6	0.076
Manitoba	160	25.9	0.079
Northern Ontario	85	38.4	0.081
Southern Ontario	71	47.0	0.082
Quebec	235	60.2	0.089
Nova Scotia New Brunswick }	75	66.3	0.094
Newfoundland	112	68.6	0.095

Effect of Wind on Gauge Catch

Many investigators (Weiss and Wilson, 1957; Struzer, 1965; Hamon, 1973; Larson and Peck, 1974; and Goodison, 1978a) indicate that wind is the major cause of error in precipitation gauge measurements. The undercatch by any gauge because of wind is related to three factors: the wind speed and its vertical profile, the buoyancy of the particle, and the obstruction of the gauge to the airflow. Figure 6.5 illustrates the airflow over an unshielded cylinder of a Nipher gauge — eddying or turbulence is absent over the orifice while the flow paths are compressed indicating acceleration of the air flow.

Proper siting of a gauge minimizes the effect of wind on catch; the best sites are situated in natural "well-protected" locations. Peck (1972) recommends that a gauge in a coniferous forest should be sheltered in all directions by vegetation below an imaginary line subtending angles of 30 and 45° to the horizontal from the gauge orifice; the forest should be deep enough to minimize eddy effects. In practice, this requirement seldom can be fulfilled and hence gauges are often located in exposed windy areas.

Procedures are available for adjusting the measured gauge catch to allow for the effects of different meteorological variables, e.g., Struzer (1969) reported on the adjustment of Soviet Tretyakov measurements for wind velocity and air temperature. Kuz'min (1975) reported that, on the average,

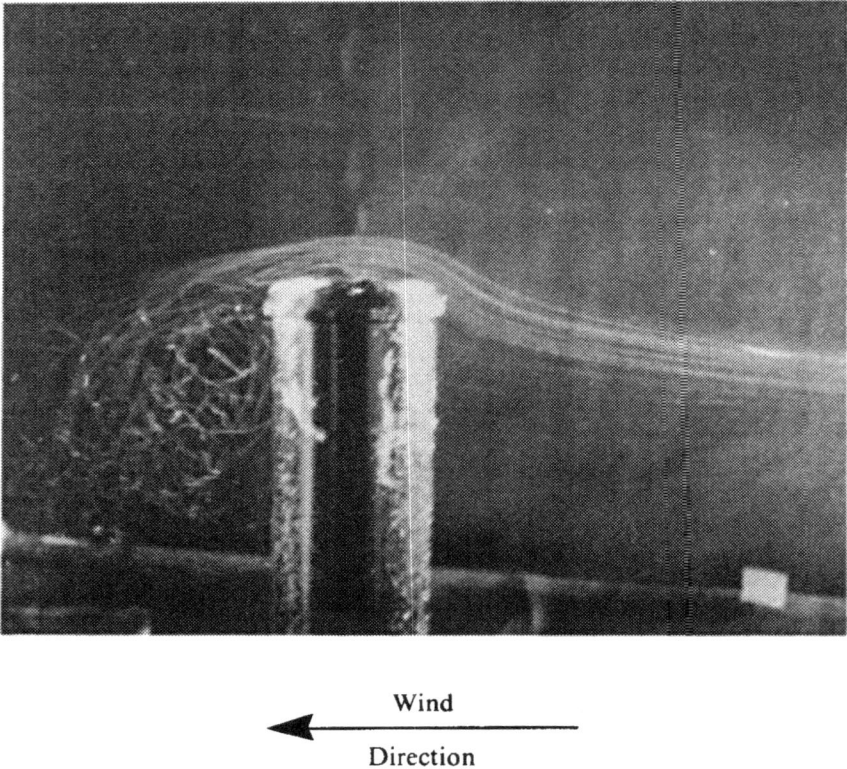

Wind
Direction

Fig. 6.5 Airflow pattern over an open cylinder.

the Tretyakov precipitation gauge catches about 70, 50 and 35% of solid precipitation at wind speeds of 3, 5 and 7 m/s, respectively, at the orifice height.

Larson and Peck (1974) prepared catch deficiency curves (Fig. 6.6) for shielded and unshielded precipitation gauges. Their values for "true catch" were determined from measurements by similar gauges located in a nearby sheltered site. The data show that the deficiency increases non linearly with wind speed and that shielded gauges catch more than unshielded gauges. Also, this relation is a function of location and measurement period; the differences reflect the integrated effect of gauge exposure and other meteorological elements on gauge catch.

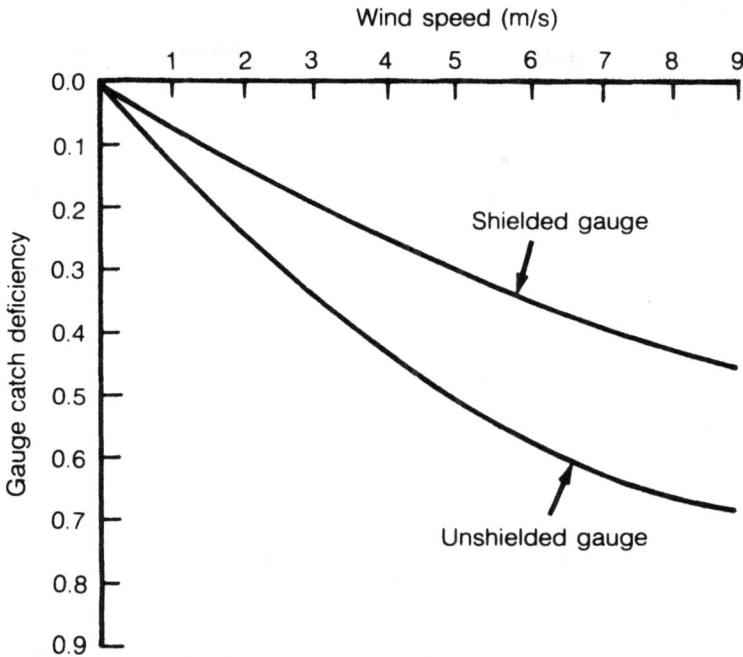

Fig. 6.6 Mean gauge catch deficiency of shielded and unshielded United States gauges for snow as a function of wind speed (Larson and Peck, 1974).

Figure 6.7 shows the mean ratio of gauge catch to ground true value (as measured on snowboards in a sheltered site) as a function of wind speed for selected shielded gauges. All ratios, except that for the MSC Nipher shielded gauge, decrease with increasing wind speed. The different shape of curve for the Nipher shielded snow gauge is attributed to its shield design. The Nipher shield minimizes disturbance of the airflow over the gauge and eliminates updrafts over the orifice which result in an improved catch.

Care and caution must be exercised in applying data obtained from a mean gauge deficiency curve to adjust gauge readings at a specific site. Furthermore it should be recognized that a correction factor for gauge catch is only applicable to point measurements of precipitation. The reliablility of areal estimates obtained from these point values is a function of the representativeness of the site and the variability of the precipitation in the region.

New methods for improving gauge catch at windy, exposed sites are being tested and evaluated in the USA and USSR, e.g., the use of dual snow fence

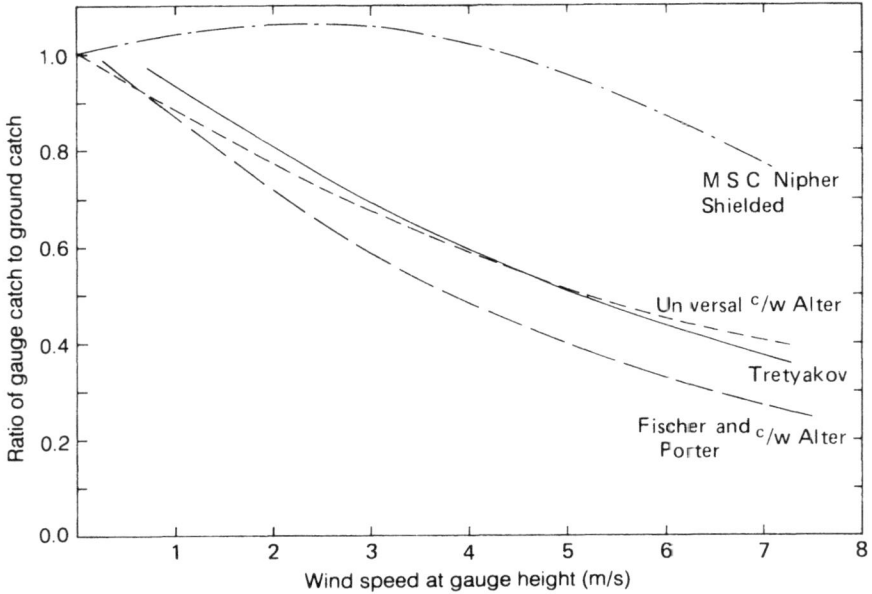

Fig. 6.7 Relationship between gauge catch and ground catch as a function of wind speed for different types of gauges (Goodison, 1978b).

shields to screen the gauge from the wind. In the USSR the installation (Fig. 6.8) is equipped with two fences, each having 1.5-m slats at a porosity of 50%, mounted vertically at radii of 2 m and 6 m around a Tretyakov gauge. The orifice of the gauge is set at a height of 3 m, the bottom of the outer fence at a height of 2 m, and of the inner fence at a height of 1.5 m above the ground. This arrangement of fences prevents accumulation of drifts. The results obtained from tests conducted at the Valdai test site in the USSR show that this method of shielding significantly increased the catch of the gauge. For wind speeds between 3 and 6 m/s, the catch ratio of the shielded gauge ranged from 99 to 92% compared with 70 to 40% for a standard installation.

In the USA, the "Wyoming shield" has been developed for use in windy, exposed locations where a natural, well-protected site is not available. This shield (Fig. 6.9) consists of two slatted fences having porosities of 50%, placed as concentric rings with diameters of 3 and 6 m around a shielded snow gauge. The inner fence is inclined outward at 45° to the horizontal and the outer fence at 60°; both fences tend to deflect the airflow downward. Rechard et al. (1974) reported that on the average the catch of a gauge equipped with a Wyoming Shield should be within 10% of that recorded by a standard gauge placed in a small forest opening.

Fig. 6.8 Dual snow fence used with the Tretyakov gauge (USSR).

Fig. 6.9 Wyoming Shield (USA).

An empirical method for calculating the true amount of precipitation using measurements from a dual-gauge arrangement (shielded and unshielded gauges) has been suggested by Hamon (1973). The procedure assumes that the following relationship exists between the catches of an unshielded gauge P_u and a shielded gauge P_s, and the actual snowfall, P_a :

$$\ln (P_u/P_a) = B \ln (P_u/P_s) \qquad\qquad 6.1$$

where B is a calibration coefficient whose magnitude is thought to be dependent on gauge type, but essentially independent of wind speed and type of precipitation. Canadian research has indicated that B also depends on air temperature. Results from a study conducted in Southwestern Idaho showed that the value of B for the Universal gauge using a rigid, bridled Alter shield was 1.70. Rawls et al. (1975) found that measurements from a gauge using a free-swinging Alter shield could also be used to compute "actual" precipitation by this method.

AREAL MEASUREMENT OF SNOWFALL

Point measurements of precipitation serve as the primary data base for an areal analysis. Even if the absolute value of the depth of snowfall at a point could be obtained the value would be representative of the depth on a very limited area, whose size is a function of the length of accumulation period, the physiographic homogeneity of the region, local topography, and the precipitation-producing process. Point snowfall measurements may be used to estimate depths over relatively large areas when the accumulation period is long, the terrain is flat, the snowfall is steady and major redistribution by wind does not occur. Conversely, when the accumulation period is short, the terrain is fairly rugged and the snowfall is showery (e.g., lake effect snowfall) point measurements will provide reliable estimates of depth over only relatively small areas.

Radar offers one possibility for measuring the spatial distribution of snowfall. Wilson (1974) indicates that radar can be used for estimating the snowfall water equivalent with approximately the same accuracy as that for measuring rainfall provided an empirically derived range correction is applied to the radar estimates. A study conducted as part of the International Field Year for the Great Lakes (IFYGL) when radar was used to measure snowfall showed that for radar ranges of less than 50 km, 80 to 90% of the radar estimates (which were range-adjusted for each storm type) were within a factor of two of the gauge measurements (Wilson, 1975).

The measurement of snowfall by radar is more difficult than the measurement of rainfall. Ohtake and Henmi (1970) report that the Z-R relation (between the reflectivity of the snowflakes Z measured by the radar and the rate of snowfall R) varies with crystal type (see Table 6.6). When information about the snowflake crystals in a particular storm is unavailable the relation, $Z = 1780\ R^{2.21}$, has been found to be applicable (Sekhon and Srivastava, 1970). Puhakka (1975) suggests the mean surface temperature is also a factor affecting the reliability of radar measurements for estimating snowfall (see Table 6.6).

Table 6.6
Z-R RELATIONS FOR DIFFERENT CRYSTAL FORMS.

Hail	$Z = 320\ R^{1.6}$
Graupel	$Z = 900\ R^{1.6}$
Snowflakes (Ohtake and Henmi, 1970)	
plates and columns	$Z = 400\ R^{1.6}$
needle crystals	$Z = 930\ R^{1.9}$
stellar crystals	$Z = 1800\ R^{1.5}$
spatial dendrites	$Z = 3300\ R^{1.7}$
Snow (Puhakka, 1975)	
Dry $(\overline{T} < 0°C)^a$	$Z = 1050\ R^{2}$
Wet $(\overline{T} > 0°C)$	$Z = 1600\ R^{2}$

[a] $\overline{T}$ = mean air temperature.

The results from the study conducted during IFYGL suggest that radar was more accurate for defining snowfall in large-scale storms where the primary snowfall occurred at elevations above 2 km than in small-scale, "lake-effect" snowstorms having relatively-small horizontal and vertical extents. The low beam elevation of the radar scanning results in poor backscatter signals in areas of strong relief because hills or mountains block the view and produce ground clutter. This problem severely restricts the use of the method for estimating snowfall in such areas.

Empirical range corrections must be applied to the radar estimates to adjust for different storm types and non-uniform beam filling. Also, it has been found that measurements from reference gauges can be used to adjust the radar estimates to improve their accuracy (Peck et al., 1974; Wilson and Pollock, 1977). Generally, radar is used to define the spatial distribution of precipitation while the gauges are used to provide depth measurements. The error reduction in the radar estimates obtained with this approach depends on the storm type, the distance between the radar facility and each reference gauge and geographical location.

Precipitation Network Design

Until new techniques for measuring areal precipitation are fully developed and tested point measurements will remain the primary source of data. So that these "point" measurements may provide reliable estimates of the spatial distribution of precipitation, consideration must be given to network design, that is, the number and spacing of the gauges. This design must be consistent with the study objectives and requirements. In Canada, as in most countries, national networks are multipurpose and, as such, are not intended to supply detailed information to satisfy "special" user-requirements. These networks developed mainly in response to public and air transport requirements and their growth has been influenced by observer availability (or population density), site access and economics (or government financial policy) (Stark, 1974). The snowfall statistics provided by the national network serve as useful indices, but they are representative of only a limited area surrounding the station.

Snowfall data obtained from regional networks are used on a real-time basis for preparing weather forecasts and warnings, for predicting avalanches, for planning transportation operations and recreational activities, and for making hydrological forecasts. These data are also used, on a seasonal basis, in broad-based regional studies concerned with agriculture, ecology, climatology and hydrometeorology. The design of a regional network varies with the scope and purpose of the individual study, and depends greatly on the regional physiography. For example, in the USSR, Gandin (1970) showed that on two areas of equal size where the gauges were uniformly distributed, five stations in the Valdai region gave about the same relative error in the arithmetic mean precipitation as twenty stations in the Ukraine region. Gandin found that the error in determining the average areal precipitation increased relatively slowly as the size of the area increased, but decreased relatively rapidly with an increase in gauge density within a specified area.

The accuracy with which the mean areal snowfall can be determined is also a function of the period of accumulation. Figure 6.10 shows the change in the absolute error of areal winter precipitation over southern Saskatchewan with period of accumulation (30 to 90 days) for different network densities expressed as a percentage of the value obtained with a station density of 8 gauges per 25,000 km^2; the absolute error decreases with an increase in network density or period of accumulation. The national average station density for precipitation in Canada is only 7 stations per 25,000 km^2, about 1/6 that of the United States network; for over half of Canada the density is less than 2 stations per 25,000 km^2.

Another method of areal analysis is the grid-square technique (Solomon et al., 1968; Solomon, 1972; Ferguson and Louie, 1974; den Hartog and

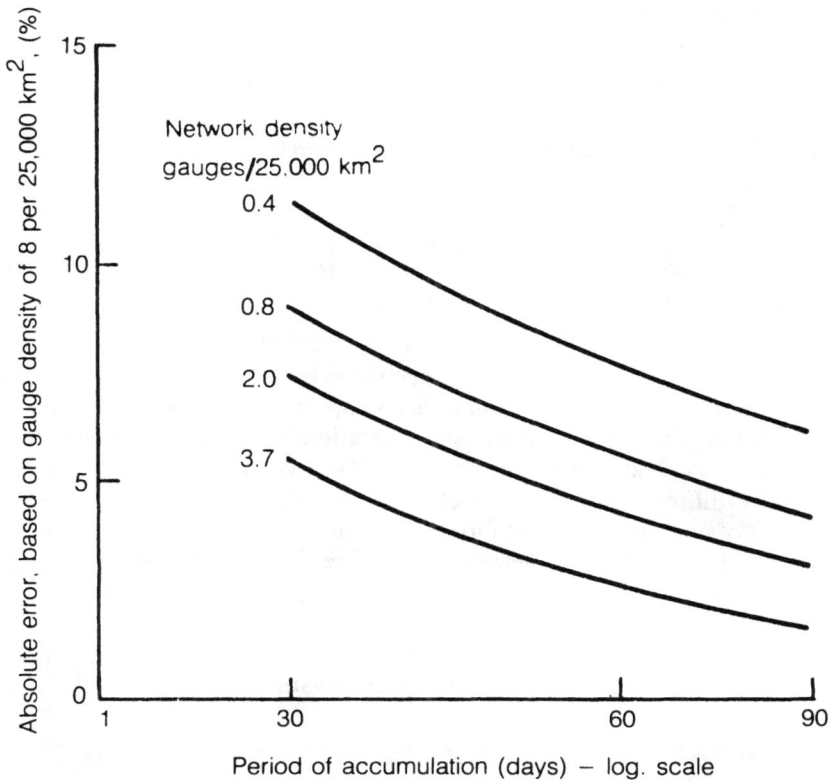

Fig. 6.10 Average error of estimated areal winter precipitation over southern Saskatchewan, expressed as a percentage of the value obtained with a network density of 8 gauges per 25,000 km^2.

Ferguson, 1975) which allows for storing, processing, estimating and plotting of information on physical, meteorological, hydrological and other characteristics of the study area. Precipitation data (usually monthly, annual or normal values) are listed for each grid square together with pertinent physiographic information such as mean elevation, mean slope, mean aspect, distance from a major topographic barrier and the height of the barrier. Regression equations are derived relating precipitation and the physiographic factors for those squares having complete data sets. Then these equations are used to estimate the depth of precipitation for those squares having no data (i.e., no observing stations).

On a watershed or sub-regional scale, the researcher may have to either develop a new network or modify an existing one to solve a specific problem.

Ferguson and Storr (1974) suggest that design of an efficient precipitation network requires answers to the following questions:

1. What is the objective(s) of the study? For what purposes are the data to be used?
2. What data are required to achieve the objective? What time resolution and accuracy are required?
3. What is the most suitable instrumentation available to provide the required data? How accurate are the point measurements?
4. What density of network is required? How representative of the surrounding terrain are parameter networks?
5. What financial and manpower resources are available?

The answers to these questions vary from basin to basin and region to region. In those regions where snow is an important climatological variable, network planning, development and operation is difficult (Findlay et al., 1972; Otnes, 1972; Ferguson et al., 1974). Most areas have common problems: difficult access, severe climate, less than ideal siting conditions, high cost of operation, the inability of instruments to measure accurately and efficiently for an extended period under extreme weather conditions, and a small number of existing network precipitation stations. Because of these problems, some compromise must be made, for example, in mountainous regions stations are usually located at accessible valley locations, even though most of the land and snowfall is at higher elevations.

POINT MEASUREMENTS OF SNOWCOVER DEPTH

Most simply, depth measurements of snowcover (snow accumulated on the ground) are made with a snow ruler or similar graduated rod which is pushed through the snow to the ground surface. Representative depth measurements by this method may be difficult to obtain in open areas since the snowcover undergoes drifting and redistribution by the wind, and may have embedded ice layers that limit penetration with a ruler. At each observing station a number of measurements are made and averaged.

In remote regions, graduated snow stakes or aerial snow markers may be used. The depth of snow at the stake or marker is observed from distant ground points or from aircraft by means of binoculars or telescopes. Aerial snow depth markers are vertical poles (up to 8 m in length, depending on the maximum snow depth) with horizontal cross arms mounted at fixed heights on the poles and oriented with reference to the point of observation. Miller (1962) and the U.S. Soil Conservation Service (1972) provide specific details on the construction and installation of aerial markers.

Rulers, stakes and aerial markers do not provide information on snowcover water equivalent.

MEASUREMENT OF SNOW WATER EQUIVALENT

Gravimetric Measurement

The water equivalent of a snowcover is the vertical depth of water which would be obtained by melting it. The standard method of obtaining this is by gravimetric measurement using a snow tube to obtain a sample core. This method serves as the basis for snow surveys, a common procedure in many countries for obtaining a measure of water equivalent.

Snow Pillows

The pressure snow pillow gauge is similar to a large air mattress filled with an antifreeze fluid. The fluid pressure responds to changes in the weight of snow on the pillow, and is measured with a manometer or pressure transducer. It is a non-destructive sampling technique.

Pillows come in various shapes, sizes and materials. Octagonal or circular pillows are most common; however, shape does not affect the measurement accuracy. Pillows may be fabricated from butyl rubber, neoprene rubber, sheet metal or stainless steel, and are filled with an antifreeze mixture of methyl alcohol and water or a methanol-glycol-water solution having a specific gravity of 1.0 (Mayo, 1972).

Barton (1974) recommends the following minimum pillow sizes (m^2) for specific snow water equivalents (mm): 3.7 m^2 for a water equivalent less than 750 mm; 5.6 m^2 for 750-1270 mm; 7.4 m^2 for 1270-1900 mm, and 11.2 m^2 for > 1900 mm. A 3.66-m diameter pillow generally is large enough for most snowcover depths; however, a small pillow used in a deep snow will give a pressure reading indicating a higher snow water equivalent than that of the snow on the pillow (Beaumont, 1965).

Accessibility of a site also influences the type of installation. Large pillows are bulky and require a large volume of antifreeze that is heavy and difficult to transport. As an alternative, small, easily-transported metallic pillows (1.22 m by 1.52 m), which may be combined to obtain a larger surface area (Washichek, 1973), are now in common use.

Figure 6.11 shows a typical snow pillow installation. Davis (1973) and Cox et al. (1978) discuss in detail the relative merits of different types of pillows with respect to their design, materials and fluids and summarize the correct installation, operation and maintenance procedures.

The most common readout device for a pillow installation is a manometer — usually a clear plastic tube fastened vertically on a scale from which the fluid level can be read. Onsite and/or telemetry data acquisition systems can

Fig. 6.11 Snow pressure pillow installation (photograph courtesy of British Columbia Ministry of Environment).

be installed to provide continuous measurements of depth through the use of a standpipe and float actuated charts or digital recorders. However, corrections must be made to float measurements, to account for changes in the density of the fluid. Alternatively, the pressure changes in the pillow can be monitored by a transducer, whose electrical output can be interfaced into a telemetry system. The transducer must be sensitive, free from inertial effects and able to endure harsh environments.

The snow pillow provides a "point" value of the average water equivalent of cover accumulated on its surface. Use of this value as an absolute quantity depends on the representativeness of the site. Snowfall or snowmelt events are identifiable from the chart record, but anomalies do occur. Smaller pillows take longer to respond to the added weight of a heavy snowfall on a deep pack than larger pillows; this delay can produce an erroneous depth-time distribution for a storm. Storr (personal communication) and Tarble (1968) report delayed responses of 5 h (3.66-m pillows) to 10 d (1.5-m pillows) under certain conditions. Intermittent freeze-thaw periods or rain-on-snow events may lead to the formation of ice layers within the pack, resulting in "bridging". When this occurs the recorded water equivalent is less than the true value. Recovery to the true weight may or may not occur with time. Bridging may also occur if there is separation of the pillow surface from the overlying snow.

Pillows have been used with some success in measuring the water equivalent of a shallow snowcover where bridging did not occur (Kerr, 1976). It was also reported that pillows provided rough estimates of daily snowmelt losses from the snowcover. In shallow snowcover, diurnal temperature changes may cause expansion or contraction of the fluid in the pillow giving spurious indications of snowfall or snowmelt. In deep mountain packs, diurnal temperature fluctuations are unimportant except at the beginning and end of the snow season.

Radioisotope Snow Gauges

Attempts to use radioisotopes for snowcover measurement have continued throughout the past twenty-five years with recent advances in technology permitting many refinements in gauge construction and operation. Nuclear gauges measure the total water equivalent of the snowcover and/or provide a density profile. They are a non-destructive method of sampling and are adaptable to on-site recording and/or telemetry systems. Most, particularly those employing an artificial source of radiation, are used at a fixed location to yield measurements only for that site. Portable gauges are being developed and field tested (Young, 1976). Nearly all systems operate on the principle that water, snow or ice attenuates radiation; samples are situated between a radiation source and a detector. As with other methods of point measurement, siting in a representative location is critical for interpreting and applying point measurements as areal indices.

The earliest gauges, which measured only total water content, consisted of a radiation source (^{60}Co or ^{137}Cs) placed on the ground with a Geiger-Muller (G-M) tube suspended above it at a height greater than the maximum expected snow depth. As snow accumulated, the count rate decreased in proportion to the water equivalent of the snowpack. Modifications to the system have involved reversing the positions of the source and detector and using a scintillation detector instead of a G-M tube. More recently, a system using naturally occurring uranium as a ring source around a single pole detector has been successfully used to measure packs up to 500 mm water equivalent, or 150 cm depth (Morrison, 1976).

A profiling radioactive snow gauge at a fixed location provides data on total snow water equivalent and density and permits an accurate study of the water movements and density changes that occur with time in a snowpack (Smith et al., 1972). A profiling gauge consists of two parallel vertical access tubes, spaced about 66 cm apart, which extend from a cement base in the ground to a height above the maximum expected depth of snow. A gamma ray source (commonly 5 mCi ^{137}Cs) is suspended in one tube and a scintillation gamma-ray detector (sodium iodide crystal), attached to a photomultiplier tube, in the other. The source and detector are set at equal depths within the snowcover and a measurement made. Vertical density profiles of the snowcover are obtained by taking measurements at about 2-cm increments of depth.

The profiling gauge described by Smith et al. (1972) can profile a snowpack at a speed of 30 cm/min providing density measurements of the horizontal snow layer (between the source and the detector) in increments of approximately 1.3 cm with a precision of ± 1%. Thin ice lenses or rapid density changes are readily detected by the system (Fig. 6.12). Changes in the density profile of the snowpack in response to snowmelt and surface lowering can also be detected. Limpert and Smith (1974) have shown the system to be useful in supplying data on the amount and intensity of rainfall on the snow

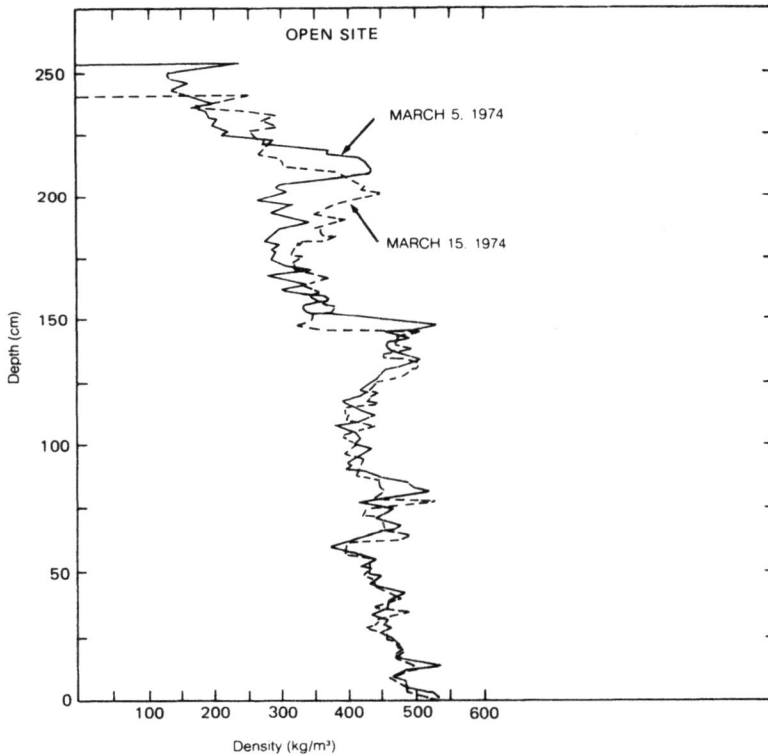

Fig. 6.12 Profiles obtained from nonforested site, Central Sierra Snow Laboratory with profiling snow gauge (U.S. Department of Agriculture Forest Service, 1974).

before the release of water from the pack, the amount of new snow, the melt rate and, as well, changes in soil moisture content. Data on the density profile and on the amount and rate of new snow accumulation are particularly important in avalanche studies (Armstrong, 1976).

Several variations in the design and operation of a profiling gauge have been tested. One modification utilizes a ^{60}Co source, which can penetrate water better than ^{137}Cs, equipped with a Geiger-Muller tube as the detector, to provide good temperature stability (Armstrong, 1976).

Another design employs modular construction that measures water content in discrete zones of 75- or 150-cm increments. It provides a measure of the total water equivalent of the snowcover and a coarse measurement of the vertical density distribution (Shreve and Brown, 1974; Morrison, 1976). The advantages of this type are that it contains no moving parts, has low power requirements, and can be installed to meet local snow conditions.

Blincow and Dominay (1974) and Young (1976) reported on the development and testing of a portable profiling gauge, a single probe instrument containing a Kr-85 gaseous source and a G-M tube detector. The probe is lowered down a tube into the snow and measures the density of the snowcover by backscatter rather than transmission of the gamma rays. Although the gauge yields very good values of the average density through the pack, its resolution (5 cm) does not permit narrow ice lenses to be identified. However, this method is a practical alternative to digging deep snow pits, while instrument portability allows assessment of areal variations of density and water equivalent. Modifications to improve gauge ruggedness during field operations are continuing to be made.

Natural Gamma Radiation: Terrestrial Survey

Snow, water and ice attenuate gamma radiation, whether emitted from an artificial source or from natural radioactive elements in the soil. Terrestrial gamma surveys can consist of a point measurement at a remote location (Bissell and Peck, 1973; 1974), a series of point measurements, or a selected traverse over a region (Loijens, 1975). The equipment includes a portable gamma-ray spectrometer which utilizes a small scintillation crystal to measure the rays in a wide spectrum and in three spectral windows (reflecting potassium, uranium and thorium emissions). With this system, pre-snow or no-snow measurements of gamma levels are required at the point or along the traverse being studied. To obtain absolute estimates of the snow water equivalent it is necessary to correct the readings for soil moisture changes in the upper 10 to 20 cm of soil for variations in background radiation resulting from cosmic rays, instrument drift, and the washout of radon gas (which is a source of gamma rays) in precipitation with subsequent buildup in the soil or snow. To formulate the relationship between spectrometer count rates and water equivalent, supplemental snow water equivalent measurements are initially required. Snow tube measurements are the common reference standard.

The natural gamma method can be used for snowpacks having up to 300 mm water equivalent; with appropriate corrections its precision is ± 20 mm. The advantage of this method over the use of artificial radiation sources is the lack of a radiation hazard.

SNOW SURVEYING

Snow surveys are made at regular intervals at designated stations throughout the winter to determine the depth, vertically-integrated density and water equivalent. A snow course is a permanently marked traverse where snow surveys are conducted. In 1977 there were over 1500 "network" snow courses throughout Canada, the same number as in 11 western States in the USA.

Purpose of Snow Surveys

The development of a snow survey network and the frequency and accuracy of associated measurements must be related to its purpose. Most snow courses are established to obtain an index of the snow water equivalent over an area for use in predicting spring runoff volumes. However, the data is also extensively used in other studies concerned with hydrology, as well as in agriculture, ecology, transportation, recreation and engineering. Before establishing a snow course it should be decided whether an index or an absolute estimate of snow water equivalent is required. If another agency's data are used, their applicability to the solution of a specific problem should be determined. When the data are used as an index, McKay (1970b) emphasizes that it is important that equipment, procedures and siting of the course remain consistent over time. If absolute snow water equivalent values are required, such as in watershed research studies, the representativeness of the terrain and landscape is an important factor in network design since it affects the representativeness of the snowcover. Statistically valid estimates of the absolute areal water equivalent should include an estimate of the probable error, or associated confidence levels.

Snow Course Selection and Network Design

The snow survey course

The conventional course is a selected line of marked sampling points along which depth and density measurements are made. The length of the course and the distance between sampling points vary depending on site conditions and the uniformity of the snowcover. For example, in hilly terrain a snow course is generally 120 to 270 m in length along which observations are taken at 30-m intervals; on the Prairies it may be longer with density measurements taken 100 to 500 m apart and depth measurements made at about five, equally-spaced locations between the sampling points. The course may take a shape other than a straight line. The first and last points of a course are generally permanently marked, with the intermediate points clearly indicated so that succeeding samples can be taken within a radius of 1 m of each point. In establishing a snow course on an area for which little statistical data about the snowcover exists the best practice is to "over-sample"; i.e., use long lines and a high sampling density. On reappraisal of the data collected, the length of the course and the number of its measurement points may be reduced depending on some accepted level of measurement precision.

Site selection

The selection of the site for the snow course depends on the purpose the course is to serve. Data that are to be used merely as an index of a parameter need only show a satisfactory correlation with the physical parameter. However, attempts are frequently made to associate the mean of the measurements with the absolute areal snowcover values for the immediate (and often large-scale) region or for water-producing areas exhibiting similar terrain and vegetation. A few general rules for establishing a snow course are listed below:

1. The site should be selected so as to yield consistent results over time. Areas where land use manipulations may take place, such as by logging operations, should be avoided. The permanence of the site and the continuity of record are key factors in site selection.
2. The site should be accessible by foot, skis or vehicle.
3. Individual sampling points should be located remote from ground irregularities such as boulders, fallen logs or shrubs, and never be located in areas subject to ponding or flooding during snowmelt periods.
4. A slightly sloping terrain is preferred to a flat area; however, very steep slopes should be avoided as the movement of the snowcover destroys the integrity of the measurements.
5. Areas affected by snow removal operations should be avoided.

After a site has been selected, it should be properly marked, mapped and documented. Figure 6.13 and Table 6.7 exemplify the basic maps and documentation used for courses operated by the British Columbia Water Resources Service. In some regions different agencies operate snow courses and it is important to have good documentation to allow data to be

intercompared. Ferguson and Goodison (1974), Goodison (1975) and
Findlay and Goodison (1978) suggest methods of standardizing these
procedures.

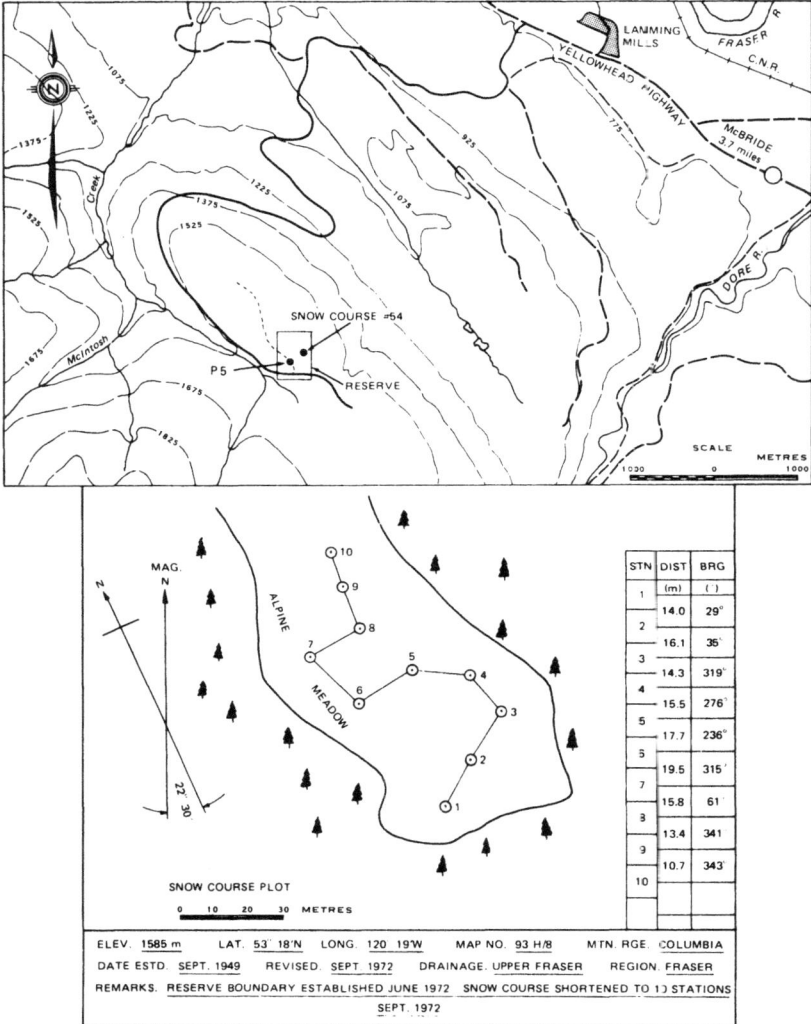

Fig. 6.13 Example of basic maps used to define a snow course.

Table 6.7
DOCUMENTATION OF SNOW COURSE INFORMATION INCLUDED WITH FIG. 6.13.

McBRIDE   Snow Course No. 54

GENERAL

Region: *Fraser*

Drainage: *Upper Fraser*

Maps: *93 H/8 1:50,000*

Climate Station: *McBride*

Stream Gauging Station: *Fraser R. at McBride   #8KA005*

Mountain Range: *Columbia*

Range Faces: *North-easterly*

AT SNOW COURSE SITE

Elevation: *1615 m*   Latitude: 53° 18' N   Longitude: *120° 20' W*

Reference Markers: *2 red plates on trees for each station*

Exposure Aspect: *flat*

Canopy Cover: *50%*

Ground Slope: *0 - 5% to South-east*

Tree Types: *Spruce - Balsam 30.5-m tall*

Undergrowth Density: *light in alpine meadow*

OTHER INFORMATION

Access: *By logging road 12.9 km from Highway 16 to Pillow station*
P5 then 230 m by trail.

Cabins: *none*

Co-operator: *---------*

Sampler: *Mr. Norman Lamming*

Address: *Box 481*
McBride, B.C.   Phone No.: *569-2485*

Date Established: *Sept. 27, 1949*   By: *D.A. McLean & V. Raudsepp*

Remarks: *Snow course reduced to 10 stations from 12 stations Sept. 1972. Map reserve*
established around snow course and snow pillow station June, 1972; revised
December, 1979.

Network and sampling design criteria

Data from a network of snow courses or from a special snowcover survey provide estimates of the mean depth and water equivalent over a selected area. The accuracy of these estimates depends on the representativeness of the sample, whose determination requires a minimum number of observations, which varies for each survey.

The design of a snow survey depends on whether absolute water equivalent values are required throughout a basin or whether indices are satisfactory. In either case, a reasonable estimate of the areal water equivalent over a basin should be provided.

When only an index of the amount of snow water available for runoff is required snow courses are generally located in areas of high accumulation to indicate changes in snow water equivalent, particularly throughout the accumulation and ablation periods. Different areas of the basin are sampled: forest and non-forest regions, north-facing and south-facing slopes, and individual elevation zones, as each of these areas may contribute runoff in a different manner. When the data are used as indices, it is most important that biases introduced by equipment or the method of sampling remain consistent with time.

Absolute estimates of the areal water equivalent can be obtained only after allowances are made for instrumental, observational and siting biases, and after the sampling network has been designed to reflect areal snowcover variability. Snowcover is continuously modified by wind transport, resulting in zones of erosion and deposition. Terrain, land use and vegetative cover are important factors affecting snowcover distribution. River banks, gullies, shelterbelts, and the edges of forests are common zones for accumulation, whereas hilltops, ridges and smooth level surfaces are favoured zones for erosion.

The design and implementation of a sampling system based on terrain and vegetative features provides the best estimate of snow water equivalent over an area (Dickinson and Whitely, 1972; Steppuhn and Dyck, 1974; Young, 1974; Steppuhn 1976; Adams, 1976; Goodison, 1978c). Snow water equivalent at any point is the product of the depth and the vertically-integrated density. In shallow snowcovers these have been shown to be essentially independent. Separation of the sampling of depth and density can speed field surveys and still provide statistically-valid areal water equivalents in shallow snowpack areas, such as the Prairies, since there is less temporal and spatial sample variability in density than in depth. Dickinson and Whitely (1972) showed that the standard error of the water equivalent was least when the same number of depth and density samples were taken, while it increases as the number of density samples decreases relative to the number of depth samples. For a population of 41 depth and density samples they found that the standard error of estimate of the snow water equivalent was increased by a factor of 1.5 when the number of density samples was reduced to 10. However, the increase in the standard error represented only a 10% increase in the coefficient of variation. The results of these and other studies suggest that the ratio of density to depth samples may be substantially reduced from 1:1 while retaining statistically-valid estimates of the areal snow water equivalent.

This procedure is particularly attractive because the density is more difficult, more time-consuming and more expensive to sample than the depth.

Steppuhn and Dyck (1974) showed that reliable estimates of areal snow water equivalent can be obtained from snow surveys if basins have been stratified according to landscape features, namely, terrain and land use. Table 5.7, Ch. 5 illustrates the effect that separation of depth data by landscape types can have on the mean depths and coefficients of variation relative to that obtained without stratification. Steppuhn (1976) shows that central depth sampling in conjunction with landscape stratification is an even more efficient sampling technique for determining areal water equivalents of snowcover on the Prairies. This procedure involves an initial depth survey on a given landscape type from which a central depth statistic, such as the median depth, is determined; a density survey follows with samples being taken at locations having the central depth value. Steppuhn emphasizes that the validity of such a sampling system depends on: (a) measurement errors being small relative to sampling errors, (b) the frequency distribution of water equivalents about the regression line for the central depth being similar to those for all depths, and (c) the depth survey being representative of the snowcover of the area investigated.

In most countries the stratified sampling procedure is at present only used by researchers in special surveys designed to obtain an absolute estimate of snow water equivalent. Agencies operating snow courses for the purpose of runoff prediction usually operate a standard snow survey of 5-10 points; the same numbers of depth and density samples are taken. Since the sampling points are at fixed locations, repeatability and compatibility of successive samples are possible.

Snow Survey Measurement Techniques

Equipment

The basic snow sampling equipment consists of a graduated tube with a cutter fixed to its lower end to permit easy penetration of the snow, and a spring balance (reading directly in water-equivalent units) to weigh the tube and its contents. Most snow tubes are made of aluminum, others are fiberglass or plastic. The large-diameter shallow snowcover samplers are a single tube; the smaller diameter samplers (for deeper snowpacks) are in sections for easier portability. Besides the spring balance, accessories include a driving wrench or turning handle, a weighing cradle, and spanner wrenches for uncoupling sections of tube. Figure 6.14a illustrates the Canadian MSC shallow snow sampler and accessories; Fig. 6.14b shows the standard Federal snow sampler, currently the most widely used sampler in North America.

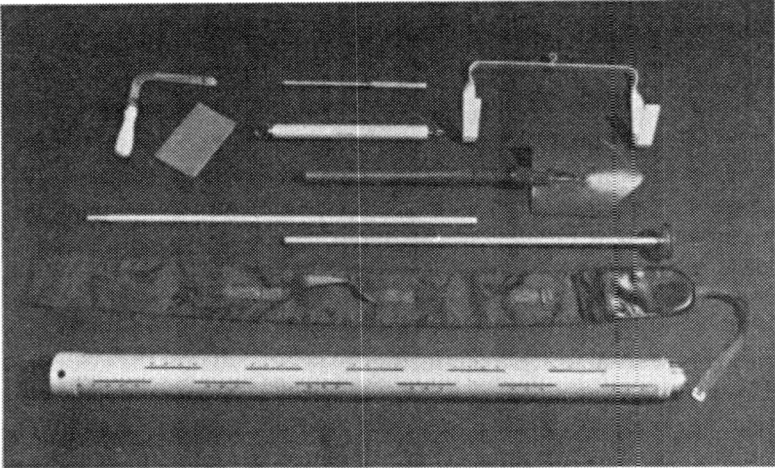

(a) Canadian MSC snow sampler.

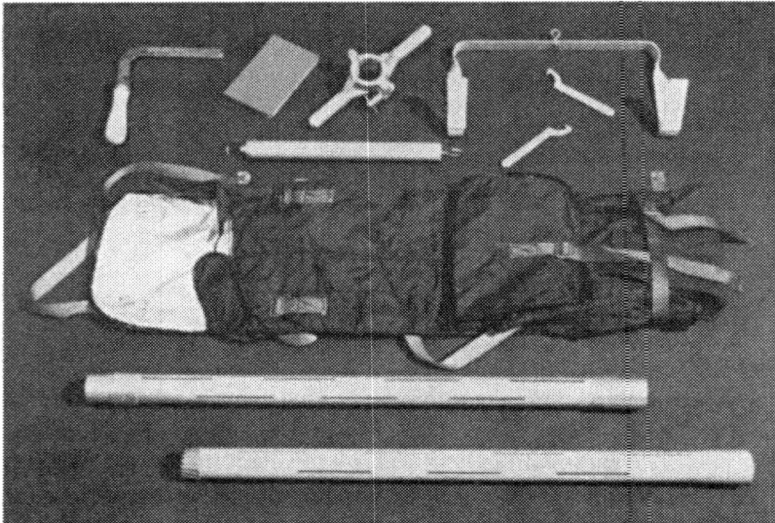

(b) Standard Federal snow sampler.

Fig. 6.14 Snow samplers used in North America.

Table 6.8 lists the basic characteristics of these and some other snow samplers used in North America.

Table 6.8
SNOW SAMPLER PROPERTIES

	Standard[a] Federal	Federal[b]	Bowman[c] L-S	McCall[d]	Canadian[e] MSC	Adirondack[f]
Material	Aluminum	Aluminum	Plastic or Aluminum	Heavy Gauge Aluminum	Aluminum	Glass Fiber
Length of Tube[g] (cm)	76.2	76.2	76.2	76.2	109.2	153.7
Theoretical ID of cutter (cm)	3.772	3.772	3.772	3.772	7.051	6.744
Number of teeth	16	8	16	16	16	None[h]
Depth of snow that can be sampled (m)	> 5	> 5	> 3.5	> 5	1.0	1.5
Retains snow cores readily	Yes	Yes	Yes	Yes	No	No

[a] Standard sampler used in the Western United States and Canada.
[b] Identical to "Standard Federal" but has an 8-tooth cutter.
[c] Cutter has alternate cutter and raker teeth and may be mounted on plastic or standard aluminum tubing. It is more an experimental rather than operational sampler.
[d] Used in dense snow or ice. It is a heavy gauge aluminum tube with 5-cm cutter with straight flukes. It may be driven into the pack with a small slide drop hammer producing an ice-pick effect.
[e] Atmospheric Environment Service large diameter sampler used in shallow snowcover.
[f] Large diameter fiberglass sampler commonly used in Eastern United States.
[g] Most snow samplers in North America use inches and tenths as their basic units of measurement. Values in this table are corresponding metric equivalents.
[h] Stainless steel circular cutter edge or small teeth.

In every snow sampler's design the inside diameter of the tube is slightly larger than the inside diameter of the cutter to allow the snow core to move freely up the tube. The practice of waxing and polishing both the inside and outside of the tube helps to minimize friction between the snow core and the inner wall of the tube and to prevent the tube from sticking in the snow; however, recent tests have shown that a baked silicone coating on the tubes is more effective and durable than wax for these purposes.

Snow tubes made of plastic, fiberglass, or heavy gauge aluminum have the advantage over thinner aluminum tubes of minimizing sticking problems since they do not conduct heat as readily under conditions with marked differences in air and snow temperatures. Most tubes have narrow slots or holes which allow the length of the snow core to be observed, thus aiding in the detection of any snow plugging the tube, and which provide entrances for a cleaning tool to remove the snow core. However, repeated turning of the sampler while using the cutter to penetrate dense snow can cause the snow to enter the tube through the slots.

Snow sampler cutters are designed especially to penetrate various layered snows which may contain crusty snow, ice lenses, or solid ground ice. The cutters must be sharp to penetrate ice layers with a minimum force, not to compact snow and force excessive amounts into the tube, and to seize a ground plug to retain the snow core. Smaller diameter samplers retain cores better than the larger ones.

When the densities of various snow strata in a snowcover are to be measured, rather than just the vertical mean density, a CRREL snow tube can be used. It is a stainless-steel, 500-cm^3 cylinder with a sharpened edge. After the sample has been obtained rubber caps are attached to each end to retain the sample for weighing (see Fig. 6.15).

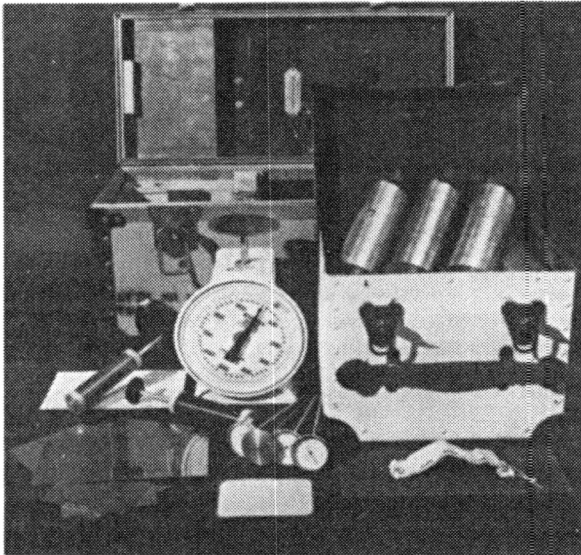

Fig. 6.15 CRREL snow tube.

Testing of modified versions of the standard snow samplers and cutters is continuing in order to improve their performance under different snowcover conditions (Farnes et al. 1980).

Snow sampling procedure

The snow sampler is lowered vertically into the snowpack with a steady thrust downward. A small amount of twisting aids in driving the tube and cutting thin ice layers; however, considerable force and twisting of the sampler with the driving wrench may be required to penetrate hard layers of ground ice. Penetration to extract a soil plug helps to prevent the loss of the snow core from the tube, a trace of soil or litter in the cutter indicating no loss has occurred. Observation of the length of the snow core permits a quick assessment of whether a complete core is obtained; the depth of snow having been measured previously. Compaction of the snow core will occur during sampling, the amount depending on snow conditions. Blocking of the core or freezing of the snow may also occur in the tube, preventing snow from entering. In such cases the core should be discarded and another sample taken. Freezing in the tube is a problem occurring when the snow temperature is below 0° C and the air temperature above 0° C. When a good snow core is obtained, the sample is weighed in the tube and the combined weight (in water equivalent units) is read directly with the spring balance. The tare weight of the tube is subtracted to obtain the snow water equivalent.

When sampling shallow snowcover it may be easier and more accurate to put each snow core into a plastic bag to be weighed later indoors on a balance (Bray, 1973; Goodison, 1978d), or to combine all samples to form a bulk sample (Atmospheric Environment Service, 1973b). The bulk sampling technique provides mean values of water equivalent and density for the course.

The density of the snow is determined by dividing the water equivalent by the depth of the snow. Of these three parameters, density will generally show the least areal variability.

Table 6.9 illustrates a convenient format for recording snow survey information in the field. Such a form also provides for the documentation of any problems encountered while surveying that may affect the accuracy of the survey and interpretation of the results.

More detailed information on suggested snow survey procedures is available in *Snow survey and water supply forecasting* (U.S. Soil Conservation Service, 1972), *Snow surveying* (Atmospheric Environment Service, 1973b) and *Guide to hydrological practices* (World Meteorological Organization, 1974).

Accuracy of measurement of snow samplers

Snow course data may be used for preparing quantitative stream runoff

Table 6.9
FORMAT FOR RECORDING SNOW SURVEY INFORMATION

Snow Course No. _____
Name _____
Sampler _____ Date_____

Station No.	Snow Depth cm	Weight Tube & Core	Wt. Tube Only Before Sampling	Water Equivalent cm	Core Length cm
Total					
Average					

Checked _____   Date _____

SNOW SAMPLING: Began _____ a.m. p.m.   Ended _____ a.m. p.m.

Sampling Conditions
(Please *check* items descriptive of present conditions)
Weather at time of sampling: Temp. _____°C

_____Clear   _____Snowing
_____Partly Cloudy   _____Blowing
_____Overcast   _____Freezing
_____Raining   _____Thawing

Snow Conditions at Snow Course
_____Crusted-supports man on skis/snowshoes.
_____Breakable crust-breaks under man on skis/snowshoes.
_____Snow soft and powdery-not sticky.
_____Snow soft and wet-sticky.
_____Snow samples obtained easily.
_____Snow samples obtained with moderate difficulty.
_____Snow samples obtained with extreme difficulty.
_____Ice layer on ground. How thick?   _____cm
_____Ground frozen under snow.
_____Ground not frozen under snow.
_____Ground dry under snow.
_____Ground damp under snow.
_____Ground wet (saturated) under snow.

General Snow Conditions
What elevation is snow-line generally?   _____m
Is snow melting on north and east slopes?   _____
Is snow melting on south and west slopes?   _____
How many centimetres of fresh snow at snow course?   _____cm
Is there evidence of snow-slides?   _____
Weather conditions of past month
_____generally overcast and stormy.
_____generally clear and cold
_____generally clear and melting.

REMARKS: _____

forecasts and for calculating indices of snowcover depth and water equivalent. The accuracy of other snowcover measurement techniques, both ground based and remote sensors, are commonly judged against snow sampler measurements which are still regarded as the best approximation of the true snow water equivalent. It is important, therefore, to know the types of equipment used and the errors associated with the measurements from different samplers.

Work et al. (1965) carried out extensive tests on different snow samplers and their associated equipment. Surveys performed with a Federal sampler in deep snow (> 250-mm water equivalent) showed that the percentage errors produced by the following factors were small: an erratic inner cutter point diameter; changes to scale precision caused by air temperature fluctuation; and scale readings by different observers. For the same absolute deviations the percentage errors in shallow snowcovers can be larger than those in deep snowpacks so that greater care is required in maintaining proper calibration of the spring balance and in reading the scale. For shallow snowcovers, a large diameter sampler is recommended to obtain a larger core to reduce the measurement error.

Tables 6.10a, b and c (after Work et al., 1965; Goodison, 1978d) give the measurement errors with which different samplers can estimate snow water equivalent under various depth and density conditions relative to that determined by weighing. The procedure followed in obtaining these results involved taking a sample of snow from a test plot then removing and weighing all snow from the plot to determine its snow water equivalent. The results in the Tables show that, on the average, most snow samplers give a positively-biased estimate of the true snow water equivalent. The accuracy of the sample is largely determined by the ability of the sampler to cut and retain the snow core. Work et al. (1965) concluded that the overestimate by the Federal sampler originates from the design of the cutter point which tends to force more snow into the tube than the inside diameter of the cutter would suggest. Bindon (1964) emphasized the importance of sharp teeth on a sampler to ensure a clean separation of the snow at the inner wall of the cutter.

Tests suggest that a sharp Federal sampler is suitable for use in all types and depths of snowcover. Similarly, the Adirondack sampler provides consistent, accurate samples in snows where ice layers do not restrict its ability to obtain a core. The Canadian MSC sampler has found satisfactory use in the shallow snowcovers of the prairie and tundra regions. Cooperative testing by North American agencies through the Western Snow Conference is continuing in an attempt to develop a standard metric sampler for deep and shallow snowcovers that will provide consistent, accurate and repeatable measurements (Farnes et al., 1980).

Table 6.10

SNOW SAMPLER COMPARISONS

(a) Goodison (1978d).

  Location: Ontario.
  Range in Snow Water Equivalent: 12 to 180 mm.
  Range in Snow Density: 140 to 408 kg/m³.
  Number of tests: 11.

Snow Sampler[a]	Error Range %	Mean %
Standard Federal	-0.3 to 12.4	4.6
Federal (L-S)	-2.0 to 18.5	10.3
Bowman Cutter[b]	-5.0 to 10.3	3.5
McCall	-3.0 to 5.8	0.3
Adirondack[c]	-9.1 to 4.1	0.0
Canadian MSC	-1.8 to 13.0	6.0

  [a] All samples bagged and weighed on a precision balance indoors.
  [b] Mounted on standard aluminum tube.
  [c] Seven samples.

(b) Work et al (1965).

  Location: Alaska (different sites).
  Range in Snow Water Equivalent: 66.8 to 82.3 mm.
  Range in Snow Density: 151 to 171 kg/m³.
  Number of tests: 5.

Snow Sampler	Error Range %	Mean %
Standard Federal	3.0 to 11.4	8.2
Federal (Slotless)	7.5 to 13.9	10.0
Bowman[a]	-1.5 to 7.1	4.1
Adirondack	-3.9 to 8.0	1.9
CRREL	-0.6 to 13.3	7.1

  [a] Plastic tube used.

(c) Work et al. (1965).

  Location: Mt. Head, Oregon
  Range in Snow Water Equivalent: 214.1 to 2169.2 mm

Snow Sampler	Number of Tests	Error Range %	Mean %
Standard Federal	7	6.8 to 12.3	10.5
Federal (L-S)	1	8.0	
Federal Sharpened[a]	2	3.7 to 4.6	
Bowman[b]	2	-0.6 to 1.8	
Rosen	5	0.6 to 6.4	3.8
Adirondack[c]	3	-0.4 to 2.8	

  [a] Standard 16-tooth Federal Sampler with teeth sharpened.
  [b] Plastic tube.
  [c] Similar to the McCall sampler, but has 8-tooth cultivator.

COMPARISON OF GROUND BASED SNOWCOVER
MEASUREMENT TECHNIQUES

In any study there is always a question concerning the best measurement method for determining point or areal snow water equivalent. The method selected depends on the purpose of the study and the manner in which the data are used, the accessibility of the site, the time resolution required, and the capital and operating costs of the measurement system. Snow survey data are the ultimate base of comparison for other snowcover system measurements.

Figure 6.16 compares three years of records obtained from a 3.66-m diameter snow pillow and a standard 10-point snow course located at a sheltered mountain site in British Columbia. It is apparent from these data that there is reasonable agreement between the two methods, the agreement in 1976 being very good. A distinct advantage of the pillow is that it provides a continuous record.

Figure 6.17 illustrates the results from tests conducted in Idaho (Morrison, 1976) in which appreciable differences were found between the measurements made with a snow pillow, nuclear gauge and snow tube. The reason that the water content of the snowcover measured by the snow pillow (after February 14, 1974) is consistently less than the corresponding values obtained by the other methods is attributed to bridging. In regions where winter thaws are infrequent, accumulated precipitation totals may provide reliable estimates of

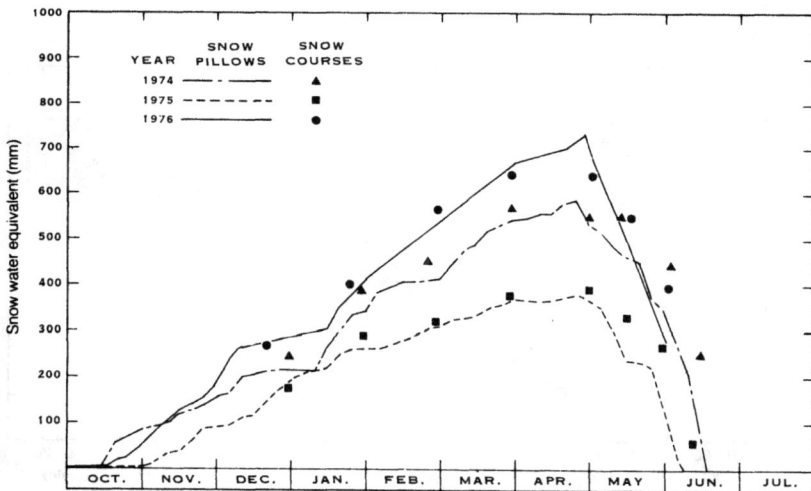

Fig. 6.16 Comparison of snow pillow and snow course measurements near McBride, B.C.

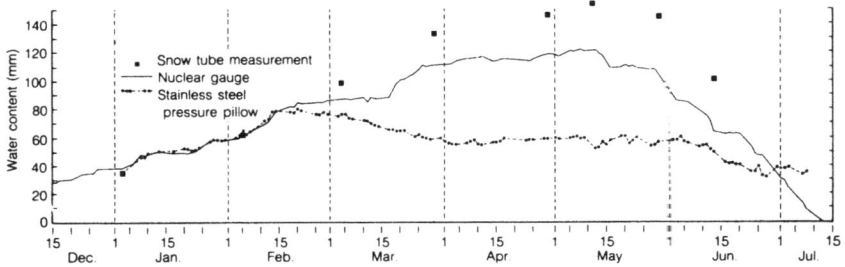

Fig. 6.17 Comparison of snow pillow, nuclear gauge and snow tube measurements at Boise River Drainage Basin, Idaho (Morrison, 1976).

the accumulated water equivalent of the snowcover. Figure 6.18 shows the accumulated snowfall water equivalent with time measured by different precipitation gauges (MSC Nipher and Fischer and Porter), a snow pillow and a snow course at a site in the Spring Creek Watershed in Northern Alberta. The values obtained by the different methods are in reasonably close agreement throughout the accumulation period.

These examples demonstrate the differences between instruments in snow water equivalent measurements. Obviously, these differences may be amplified when attempts are made to extrapolate the results spatially to obtain areal estimates of the extent of snowcover and its depth and water equivalent.

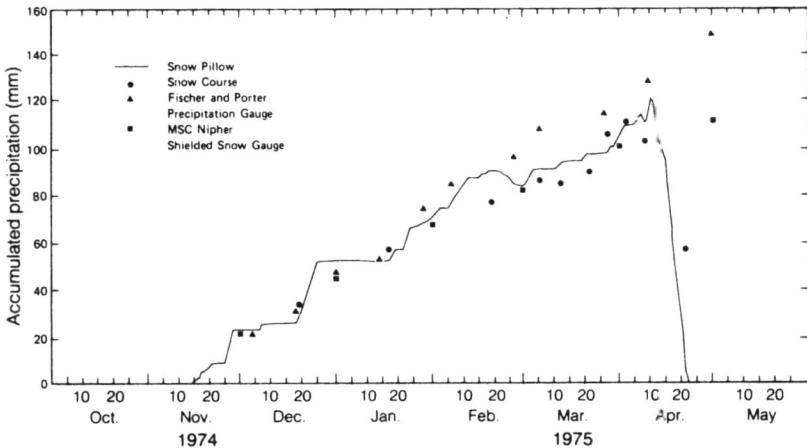

Fig. 6.18 Comparison of snow pillow, snow course and accumulated precipitation gauge measurements from Spring Creek Watershed, Alberta (Kerr, 1976).

Goodison (1977, 1978c) discusses the compatibility of snowfall and snowcover measurements in a shallow snowcover. Figure 6.19 shows measured and calculated net snowcover accumulation up to the time of maximum accumulation during two winters. The calculated values were obtained by correcting and adjusting MSC Nipher gauge data for gauge catch and for snowmelt and evaporation losses. In the same figure the mean areal accumulated snowcover are also plotted, as determined from snow survey measurements weighted according to land use. Error limits on these mean values are defined at the 1% confidence intervals. It is clearly demonstrated that the accumulations calculated from the gauge measurements by a mass balance method are in close agreement with snow course measurements stratified according to land use.

REMOTE SENSING OF SNOWCOVER

Ground-based measurement techniques will not always detect significant areal snowcover variations. In mountainous regions large variations in snow depth and water equivalent occur because of variations in slope, aspect, elevation, exposure and surface cover. In specific areas, accessibility can limit both the number and representativeness of ground measurements. In plains regions, variations in snowcover are dominated by local land use and topographic variations, in addition to the spatial variations in initial snowfall distribution.

For many studies the percentage snowcover in a basin is an important areal parameter which ground measurements alone may not provide with sufficient accuracy, especially in sparsely-instrumented regions. However, the rapid development during the last 20 years of remote-sensing techniques for snowcover applications has provided new methods for observation, measurement and analysis. The successful application of these methods relies on accurate "ground truth" data for calibration and verification.

With improvements in both photographic methods and aircraft technology during the past 60 years, aerial photography has developed into a major field of investigation primarily aimed at photographing, interpreting, and mapping topography and other landscape features. However, to take vertical imagery over a large area is both time consuming and expensive. In Canada, as in most other countries, because air photos were used for topographic mapping the photography was preferably done in late spring after the snowcover was gone but before the leaves were on the trees. Consequently, for most of the country, the most readily available photography was not applicable for the analysis of snowcover. Instead, special flights were required to obtain these data. However, the rapid development of satellite remote sensing has provided a promising alternative for the areal analysis of snowcover over large regions. Aircraft observations can be used to supplement large-scale data on selected areas of interest.

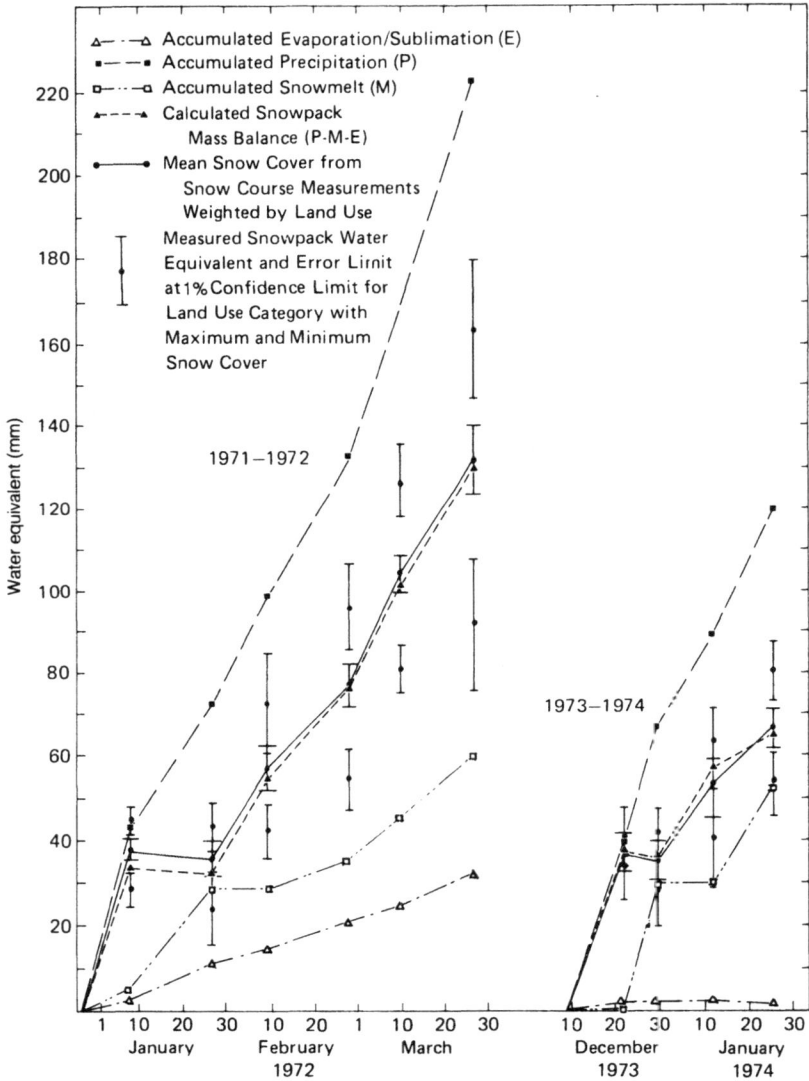

Fig. 6.19 Calculated and measured snowcover mass balance, Cold Creek Basin, Ontario: 1971-1972, 1973-1974.

Aerial Markers, Snowline Flights, Aerial Surveys

In mountainous or remote regions where adequate snow course networks may be very expensive and difficult to service, observations from aircraft may be a reasonable alternative. Snow depth at selected locations can be obtained by aircraft observations of aerial markers. An estimate of water equivalent for these sites is possible by using density data from nearby (within 40 km) snow courses located at similar elevations. A network of markers can provide areal snowcover data which cannot be obtained from point measurements.

Snowmelt runoff from mountain basins is very important for flood prediction, reservoir regulation, and hydroelectric production. Canadian and U.S. agencies use daily runoff simulation models to predict runoff for selected mountainous drainage basins. An important parameter in some models is the snowline elevation; that is, the boundary between the lower snow-free area and the higher snowcovered area. This elevation, in conjunction with the hypsometric curve for the drainage basin, is used to calculate the percent of snowcovered area for the basin. Field observations of the snowline from aircraft in British Columbia have been shown to greatly improve the prediction of a runoff simulation model (Sporns, 1976). For the Columbia River Basin both the United States Army Corps of Engineers and British Columbia Hydro and Power Authority conduct "snowline flights" to determine the snowline elevation. Personnel, flying in small aircraft at the altitude of lowest snowcover, estimate the snowline elevation for different aspects at pre-selected points marked on a topographical map. The aircraft altimeter and a hand level aid the observer. Two to four flights are made each year for each area being studied, depending on the type of season and the flying conditions. In Canada, flights are made in May and June when the snowline is most distinct.

Snowline flights are effective, yet are relatively expensive and time consuming. They may present a safety problem and, as well, a reliability problem since they are subject to suitable flying conditions, while the accuracy depends somewhat on the experience of the observer. As an alternative, efforts are being made to abstract snowlines from satellite imagery. Not only can a large area be mapped, but also more frequent observations are possible. Rapid changes in available sensor packages have occurred in the past few years. Sensors that measure radiation reflected and emitted by the earth's surface within different spectral bands are common. For example, a radiometer which is spatially and spectrally matched to the four spectral bands of Landsat imagery is available for airborne use. The evaluation of satellite imagery for snowcover mapping may thus be supplemented with imagery from an aircraft in flight in addition to ground snow survey measurements obtained at the same time as the satellite passes.

Natural Gamma Radiation: Airborne Survey

The aerial gamma survey was developed in the USSR in the 1960's (Kogan et al., 1965) and since then experimental aerial surveys have been carried out in Norway, the USA and Canada to measure snowpack water equivalent. Natural gamma radiation is emitted from the ground by potassium-40, bismuth-214 and thallium-208, and is attenuated by material lying between the earth's surface and the sensor or detector. Since water, in either the solid or liquid phase, attenuates gamma radiation, its amount can be calculated from the reduction in the intensity of photons reaching the detector. The basic theory of the method and the processing procedures for Canadian airborne measurements are summarized by Loijens and Grasty (1973). For snow water equivalent measurement, the potassium and integral (total count) data have been shown to be the most useful (Grasty et al., 1974; Jones et al., 1974; Peck et al., 1974).

As in terrestrial surveys, a no-snow measurement of the gamma levels along the flight line is needed, while adjustments to the data are required for changes in soil moisture in the top 15 cm of the ground. The aerial survey method presents other problems for which corrections must be made. For example, the mass of air between the aircraft and the ground attenuates gamma radiation, but its mass can be calculated from measurements of atmospheric temperature and pressure at flight level (typically 150 m) to compute the correction factor. Radon gas in the atmosphere also emits gamma radiation which particularly affects computations involving the total count data. Entrapment of radon gas by an inversion layer near the earth's surface causes special measurement problems because of changes to the background radiation emission. Background radiation is most easily measured by flying over a lake, since the water attenuates all gamma radiation emitted by the underlying ground. In other words, the reading over the lake establishes the "zero" value for the survey. If this method for establishing the zero value is not possible, other methods are available, but have greater uncertainty (Jones et al., 1974). Finally, deviation from the pre-snow flight track, especially over areas of variable ground radioactivity, may introduce significant error in the snow water equivalent calculation. The errors caused by variations in the count rate resulting from changes in the background radiation, the mass of air below the aircraft, and variations in the signal response of the counting equipment may be twice as large as those resulting from changes in soil moisture.

Pilot projects using gamma-ray spectrometer surveys of snowcover in Ontario have given encouraging results (Grasty et al., 1974; Liojens, 1979), e.g., the average snowcover water equivalent over 16-km sections could be measured to a precision of 12 mm using the potassium count and to 17 mm using the total count (Grasty et al., 1974). Snowcover water equivalents up to 140 mm were sampled; comparative ground based information was obtained from snow surveys along the flight track.

The United States National Weather Service has conducted operational tests of a gamma system over the northern plains states, particularly the Souris River Basin at the time of peak snow accumulation. The aim of such surveys was to provide near real-time snow water equivalent data as an aid in snowmelt forecasting. Good results are reported; the estimated errors are of the same order as those found in the Canadian experiment (Loijens, 1975; Peck et al., 1979).

In contrast to snow course measurements, the gamma survey measures the mean snowcover water equivalent over an area. Use of fixed wing aircraft which requires low-level flight tracks restricts the technique to relatively flat areas; helicopters can be used in mountainous terrain. Precise navigation equipment is needed to ensure that the flight tracks are the same as those used to obtain the ground truth data. The great advantage of the aerial gamma technique is its large area coverage. By averaging the water equivalent measurements in sections along the line, the high local variations in snowcover water equivalent, which commonly affect snow survey measurements, can be minimized.

Work by Grasty (1979) indicates that a single flight technique which monitors variations in the scattered and direct gamma-ray fluxes from potassium-40 can be used to measure snow water equivalent to an accuracy of better than 16 mm. This technique has the advantages of being insensible to variations in soil moisture, and of eliminating navigational problems arising from duplicate flights. Although further work is required to assess and evaluate its limitations, development of such a method would make the aerial gamma survey an even more useful operational technique for measuring snow water equivalent.

Microwave Sensing of Snowcover from Aircraft

Both radar and passive microwave systems operate in the electromagnetic spectrum between 0.1 and 100 cm. Passive systems sense some of the natural radiation emitted by objects; active systems, e.g., radar, emit radiation and measure the reflected and backscattered return signals. For snowcover applications both techniques are still in the research and development stage. Microwave emission (or brightness temperature) has been shown to change with the depth of accumulation and the wetness of the snowcover (Meier, 1975; Hall et al., 1978). Studies are aimed at assessing the utility of passive microwave data for determining the position of the snow line, determining the beginning of snowmelt, and estimating the liquid water content of the snowpack (Hall et al., 1979; Rango et al., 1979; Foster et al., 1980). The great advantages of microwave systems are their capabilities of operating and sensing through cloud cover. Studies are currently being conducted on the usefulness of synthetic aperature radar (SAR) for measuring snowcover water equivalent (Goodison et al., 1980). Side-looking airborne radar (SLAR) has been used for observing ice cover on the Great Lakes and ocean areas and can provide high resolution imagery (tens of metres) over sizable swath widths (tens of kilometres) with no scale variation across the image.

Satellite Observations of Snowcover

Studies of the application of satellite imagery to snowcover analysis began in the 1960's. Their current status is briefly reviewed here, though technological changes will rapidly render this summary obsolete.

In 1979 the two principal satellite systems still used for snowcover studies and operational analysis in North America were the polar orbiting Landsat and NOAA systems. The geostationary SMS/GOES satellites provide useful data for snowcover analysis south of about 50° N with a time resolution of 30 min., but, the viewing angle causes distortion problems at higher latitudes. The resolution of a few kilometres over southern Canada is not as good as that provided by Landsat or NOAA.

The Landsat and NOAA satellites

The first Earth Resources Technology Satellite (ERTS) was launched in 1972 (and later renamed Landsat 1). Landsat scans a surface swath width of 185 km with a resolution of about 90 m at nadir. The Multispectral Scanner Subsystem (MSS) provides data in four spectral bands:

MSS 4: 0.5-0.6 μm - green,
MSS 5: 0.6-0.7 μm - red (strong reflectance of dry and melting snow; similar to NOAA-VHRR visible band),
MSS 6: 0.7-0.8 μm - near infrared, and
MSS 7: 0.8-1.1 μm - near infrared (least water penetration, low reflectance from melting or metemorphosed snow).

This multi-spectral imagery can be used separately to obtain representative signatures in each of the four bands, or combined to give black and white or false colour composites. Landsat provides data for a given area at mid-latitudes once every 18 days. Convergence of swaths at high latitudes provides for more frequent coverage. When two Landsats are in orbit a temporal resolution of 9 days is obtained, but even coverage at nine-day intervals may be insufficient for many applications. Furthermore, because of cloud cover, a significant proportion of this imagery (typically 50% or more) cannot be used for snowcover studies nor in a real time mode. Usually "quick-look" imagery can be obtained within about ten days, while corrected and annotated images are available within two or three months. In general, Landsat data are being used for research purposes and to supplement NOAA-VHRR and conventional ground data for operational snowcover analyses.

Landsat also has a data retransmission capability: it can collect surface observation data, for example, from precipitation gauges or snow pillows, and from automatic telemetering stations, and transmit them to distant data collection centres. This capability is useful in areas of significant snow storage that are remote or relatively inaccessible, where it may not be practical to install manned observing stations. With the Landsat system, remotely-sensed imagery and ground-based data can be collected simultaneously, thereby eliminating any significant time lag errors in correlating observations.

The NOAA series of satellites provides twice-daily coverage (mid-morning and late evening over southern Canada) of all areas in a swath width of about 1900 km, with distortion because of the earth's curvature that is acceptable for snowcover analysis. The NOAA-5 satellite measures radiation in the visible (0.6 - 0.7 μm) and thermal infrared (10.5 - 12. 5 μm) bands. TIROS-N, the third generation polar orbiting satellite, was launched in late 1978, and carries the Advanced Very High Resolution Radiometer (AVHRR), providing twice daily thermal infrared (IR) and daily visible and near-IR data.

Recent snow studies and current applications

In 1968 the World Meteorological Organization initiated a project on "snow studies by satellites", and subsequently published a status report and a report on an international seminar (World Meterological Organization, 1973, 1976) which incorporated the results from several countries including Canada and the United States. The main emphasis of this project was the determination of the areal extent of snowcover, expressed as a fraction of each total basin area that was covered to a depth of at least 5 cm. In a few cases the feasibility of determining other parameters, such as snow depth, was also examined. Canadian studies in support of the WMO project were conducted on the Saint John, Souris, Lake-of-the-Woods and Columbia basins. In general these studies were primarily based on NOAA-VHRR data (Ferguson and Lapczak, 1977) in conjunction with limited "ground truth" data. Landsat imagery supplemented the analysis while image analyzers (density slicers) were used to provide grey scale analyses of the images. Optical equipment superimposed original or density sliced images onto base maps. The method was shown to be feasibile for determining the areal extent of snowcover. Its accuracy is highest over non-forested terrain and lowest over dense coniferous forests where the snowcover may only be visible in large clearings. However,

Hogg and Hanssen (1979) report that computer enhancement of digital infrared satellite data has worked well to produce snowcover maps for the coniferous forests of the Saint John Basin. Snowcover maps and areal estimates are produced without any need to draw a snowline manually.

In some situations correlations have been found between image brightness and snow depth (McGinnis et al., 1975a, b; Ferguson and Lapczak, 1977) but the interpretation problems are formidable so that a general operational technique employing this type of correlation has not been developed. Because attenuation limits the penetration of visible light into the snowpack, it seems likely that the remote sensing of snow depth will rely on radiation measurements within other wavelength bands, particularly the microwave band.

Operational snowcover mapping is being carried out by the U.S. National Environment Satellite Service. As of June, 1976, a service has been provided for 23 basins, including five extending partially into Canada (NOAA, 1977); for each basin a percentage snowcover message is transmitted via teletype and a snowcover map is transmitted via telecopier, usually once or twice a week during the snow season. An accuracy within ± 5% is claimed for areas greater than 5000 km^2 (Wiesnet, 1974).

It has been recommended (World Meteorological Organization, 1976) that for operational hydrological purposes Landsat data, NOAA-VHRR imagery and NOAA-SR data be applied to basins with areas exceeding 10, 1,000 and 10,000 km^2, respectively. For hydrological applications it would be desirable that forecasting models be adapted to accept remote-sensing inputs.

Satellite snowcover applications also include regional or hemispheric delineations of snow and ice cover (Wiesnet et al., 1978). The National Oceanic and Atmospheric Administration distributes hemispheric maps of snowcover weekly (NOAA, 1977) and also prepares 10-day composite hemispheric minimum brightness charts for detecting snowcover under changeable cloud conditions. These charts are updated daily, recorded on tape and displayed as photographic images.

A summary of operational applications of snowcover observations is given by Rango and Peterson (1980).

Future possibilities

At present, fairly simple optical-electrical interpretation techniques are being used for operational snowcover mapping using satellite data. These methods involve the skill and subjectivity of the operator. More sophisticated and objective methods of analyzing digital data have the potential to produce more accurate results, but are still in the research and development stage. Suitable cost-effective analytical routines (based on the raw data with appropriate correction factors) will likely be developed over the next few years

(Hogg and Hanssen, 1979). However, ground-based data will still be needed for calibration and verification and for filling gaps in the satellite data set. Satellites are thus perceived as one component of a total observing system which will continue to include surface observations.

Wiesnet et al. (1978) provides information on future U.S. satellite systems relevant to snow studies. It has been recommended (WMO, 1976) that a channel in the 1.55 - 1.75 μm band be installed for distinguishing between cloud and snowcover, that a multispectral imaging microwave radiometer be included in future programs, and that future Landsat multispectral scanners and similar sensors be designed to measure snow radiance through its entire range of expected values and not "cut-off" or "saturate", as they do on Landsat 1 and 2.

DATA ANALYSIS

The following discussion is limited to frequently-used, standard, methods-of-analysis of snow data. For further information the reader is referred to specialized texts dealing with the analysis of hydrometeorological data, such as, *Statistical Methods in Hydrology* by Haan (1977), the WMO *Guide to climatological practices*, (World Meteorological Organization, 1974) and *Climatological Statistics* by Brooks and Carruthers (1953).

Definition of the data requirements is a first step to problem solving, but is only possible after the problem has been defined and the variability of each parameter required is understood. In practice, most snow studies are approached using data acquired for other purposes - not necessarily with the user's needs in mind. For some research purposes such data are unacceptable, but for most engineering purposes they are usually the prime basis for decision making. Their limitations and the manipulative techniques for optimizng their information content are extremely important.

Data Sources

Major data sources are the national or regional archives and libraries of the agencies responsible for making snow measurements. Many agencies may be involved in data acquisition, e.g., those concerned with climate, streamflow forecasting, transportation, construction and agriculture. Some other data sources are project and research reports, field survey documents, newspaper archives and published maps. Unconventional or indirect sources, such as remote sensing or tree ring data banks, can be extremely useful, depending on the type of information required.

Data series should generally be of sufficient length, quality and homogeneity for statistical treatment and of sufficient areal coverage to show the features of interest. Until recently only climatological data have approached these requirements, at least on a national scale, e.g., snow depth has been measured at least monthly over the past fifty years at most Canadian climatological stations. Starting in 1941 all principal weather stations in North America reported snow depth daily; after 1952 many first-order climatological stations in the United States made measurements of the snowcover water equivalent.

Snow survey records for some areas have equally long records, but survey networks have often been developed to serve the specific needs of watershed operations rather than those of a region or the nation. Beginning in 1953, Canada, the U.S. Weather Bureau, the U.S. Cold Regions Research and Engineering Laboratory and the U.S. Soil Conservation Service cooperated in measuring the snowcover and ice cover, thereby creating a network which extended across most of North America (Bilello, 1969). In 1954-55 Canada started publishing snow survey data for eastern Canada and in 1962-63 the Canadian Meteorological Service initiated a national snow survey program at its own stations and converted *Snow Cover in Canada* to a national publication that contained observational data submitted by all organizations. In 1975, mean snow depth and water equivalent data measured on different survey dates for about 1300 snow courses were included in this publication.

Individual agencies may publish snow course and snow pillow measurements on a close to "real-time" basis. For example, British Columbia Water Resources Service publish monthly snow survey bulletins which have the added advantage of including comparative historical data.

In Canada daily measurements of snow depth are available from synoptic observing weather stations for the 1200 GMT observations. Depth of snow on the ground at the end of each month is recorded at all climate stations.

Because of the deficiencies of climatological and snow survey networks, supplementary data, such as measurements obtained by aerial photography, lysimeters, snow stakes, and gamma radiation counters can be used to great advantage. Another interesting supplementary source is satellite pictures whose use has been described by Barnes and Bowley (1969), and McGinnis et al., (1975a).

The use of more than one data source may pose major problems of data consistency and data reconstruction. Data usually are valid for specific points and times, which affects their bias, error and statistical distribution. Furthermore, the normal variability of climate may cause short-duration samples to be unrepresentative. For planning and design purposes it is advisable to assess all data for their quality and representativeness in time and space, and to standardize them, for example, to common instrument and exposure conditions, before analysis and interpretation.

Data Quality

The number and quality of data, as well as their statistical nature, impose limitations on the information that can be usefully deduced. All measurements are inaccurate to some degree. The observer and his procedures, the instruments and their maintenance, the data transmission and transcription, may each contribute individually or collectively to errors in the published values. Errors tend to be mainly systematic for data obtained at a given site or station but tend to be highly random if the data from many stations are analyzed collectively. Archived data reflect agency objectives. Many observational programs are confined to depth and water equivalent, although some provide more physical data about the snowcover.

Consistency

For comparability, measurements should be consistent. Public service agencies attempt to maintain consistency, but the snowfall and accumulation characteristics of a site can be greatly altered by natural events such as forest fires or by man-induced actions such as land-use modificiations and road construction. With snowcover, the exposure may change as the depth increases, so that the accumulation rate is a variable related to wind and snowfall. Inconsistencies are common when several organizations are collecting data, using different observational practices and instruments.

Frequently, tests for consistency amount to comparing values at a selected site with those from one or more nearby locations whose consistency has already been established, for example, by simple graphical analysis of plotted points identified by year. Changes in grouping of the points indicate periods during which inconsistency may have occurred. Regression lines fitted to the groupings may be used to reconstruct any data deemed to be inconsistent.

Double mass analysis has been commonly used by hydrologists to detect changes in data series. This technique involves plotting the time accumulated values of a control variable (commonly the mean of measured values from several sites) against those for the site being examined. Significant changes in the slope of the resulting curve should be checked since they may possibly be due to changes in data consistency. However, these changes might be due to the normal variability of climate, changes in measurement practices, or instrument malfunction. Accordingly, a decision to reconstruct data should be undertaken only following a careful review of all evidence relating to the observing program and the regional climate. Standard statistical tests performed on variances of the slope or on deviations of the individual points from the fitted line may also be used to assist in making this decision. Figure 6.20 illustrates the application of the double mass curve procedure to data on the depth and water equivalent of annual snowfall for Thunder Bay and Kakebeka Falls, Ontario. Prior to 1960, the snow water equivalent at both stations was calculated from ruler measurements assuming the ratio of water equivalent to depth to be 0.10. After 1960, the snow water equivalent at Thunder Bay was measured with an MSC Nipher shielded gauge but was estimated at Kakebeka Falls from the snow depth. At both stations during the entire period snow depth was measured by ruler. The plotted data show that the relationship between snow depths at the two stations remained constant. However, after the Nipher gauge was introduced at Thunder Bay the relationship between their snow water equivalents changed markedly, the snow water estimated at Kakebeka Falls being much greater than that measured at Thunder Bay.

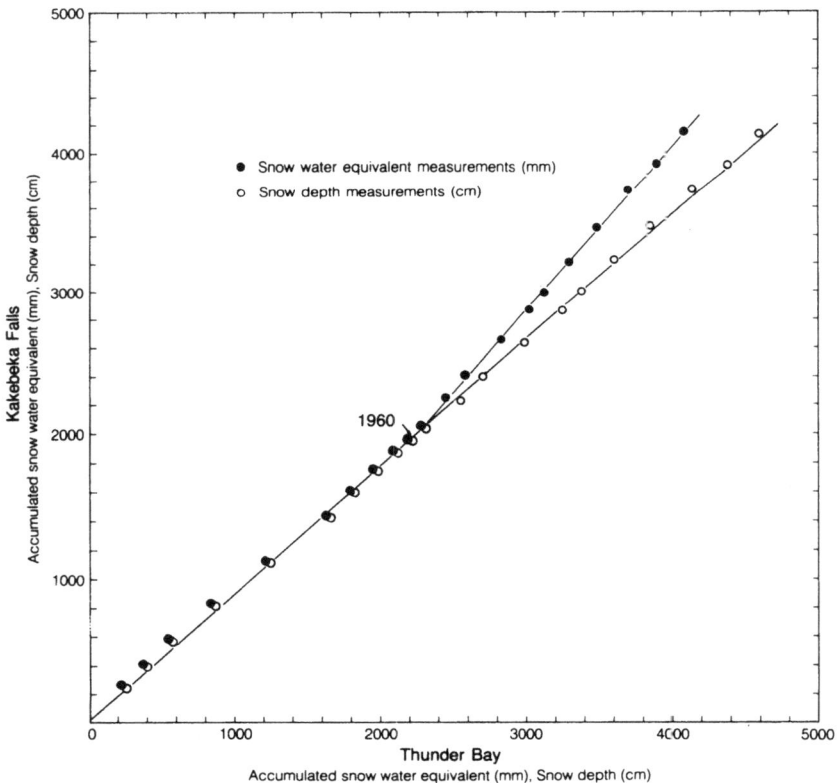

Fig. 6.20 Accumulated depth and water equivalent of annual snowfall (1951-1974) at Thunder Bay and Kakebeka Falls, Ontario.

Errors

Errors may be random, such as mistakes in transcribing numbers, or systematic, such as a bias introduced by an observer or an instrument. Random errors tend to cluster around the mean value and are generally both positive and negative so that "Normal" or "Gaussian" statistics apply. Occasionally these type of errors include major mistakes such as those which may arise in reading a scale; these are usually detected in the process of quality control. Some "obvious" errors can be easily explained and corrected; others must be rejected if they lack a sound physical explanation, but should not be discarded, since later evidence may provide an explanation. Errors that fall within reasonable limits of possibility are the most insidious since they are virtually impossible to detect. A mean value of several measurements is a better estimate of the true value, provided systematic errors are negligible. Similarly, the average of a time series may give a superior measure of its true (normal) value.

Systematic errors may either be constant or proportional to the magnitude of the variable, and appear only under specific environmental conditions, e.g, when snow is wet and adhesive rather than dry and easily transported by the wind. Such errors are minor in data for indices, but are serious in data required for quantitative values. Adjustment factors can be determined to compensate for exposure bias but are somewhat subjective and cannot be freely transposed to other seasons or sites.

Snow scientists have reduced the measurement errors and increased data consistency using standard instruments, instrument exposures and operational procedures. While these techniques provide consistent data, it is well-recognized that measurements of snowfall and snowcover provide better indices than quantitative estimates of areal amounts. Accordingly, attempts have been made to adjust the measurements to obtain more precise estimates. Soviet Union hydrometeorologists have recommended corrections of 40 to 60% for data from gauges with a 500 cm^2 orifice exposed 2 m above the snow surface (Nordenson, 1968) to overcome the deficiency caused by wind. Tollan (1970) corrected snow gauge catches for the Filefjell Basin in Norway by 50% for the same reason. Hare and Hay (1971) consider that precipitation estimates for the Canadian Arctic may be deficient by 40%, based on water and energy balance calculations. Ample evidence clearly shows that the present snow gauges cannot offset the effects of air turbulence on catch induced by the gauge and its surroundings.

As discussed earlier in this Chapter (see Table 6.10) snow tubes also have their limitations. Freeman (1965) and Goodison (1978d) found positive biases of about 9 to 11% for the Federal sampler, and 7% for the Canadian MSC sampler. Considering that the variety of samplers is large and that the sampler bias depends on the condition of the cutter, the analyst may be hard pressed to evaluate the correction that should be applied to get more accurate estimates.

Representativeness of the Measurement Area

Most rainfall measurement procedures have been standarized and designed to achieve representativeness. These procedures are extremely difficult to apply to snow since frequently two sets of data are available, one from climatological stations, the other from snow survey sites, neither of which are collocated. These data may show major dissimilarities even when the two groups are collocated because of differences in measurement procedures and siting. For example, grassland snow courses catch and retain only about 60% of the accumulated snowfall that is reported at adjacent climatological stations whereas forest snow courses report approximately equal amounts (McKay, 1963).

Most snow courses have been established to aid in predicting runoff volumes and peaks. Their data measurements are used as indices so that the measured values need not be representative of a large area. Preferably they indicate snowcover amounts over areas that contribute substantially to runoff.

A simple comparison with values in the same general area will often indicate the extent to which exposures are comparable. Major differences should be explainable in terms of elevation, land form, vegetation, or other climatic or physical features. The catch and retention characteristics of sites can be determined and then compared to those of the surrounding region (McKay, 1963).

Frequency Distributions

Snowfall and snowcover data are highly amenable to statistical analyses and probablistic statistical association. Therefore regression and other statistical techniques are potent aids to the snow scientist. Statistical methods are the subject of many excellent books and are not discussed here in detail. Rather, an attempt is made to describe any peculiarities in snowfall and snowcover data which might be relevant in applying the statistical methods.

These peculiarities must always be kept in mind, e.g., the data must be examined critically for the occurrence of "zero" values and for the frequency distribution most appropriate for analysis. Furthermore, the varied sources of snowfall, the topographic controls and climatic variability may complicate the analysis for specific locations. Allowances for these differences may be made by stratifying the data, but generally speaking, the analytical approaches are more or less similar to those used for precipitation data.

The choice of the most suitable theoretical distribution to be fitted depends on each data set. For example, a normal distribution can be fitted to the dates of formation and disappearance of a Prairie snowcover (McKay and Thompson, 1967); an extreme value or log-normal distribution to maximum annual 24-h snowfall amounts (Thom, 1966; Nkemdirim and Benoit, 1975; Tobiasson and Redfield, 1976); and the incomplete gamma function to many other snowfall and snowcover variables (Crutcher et al., 1977).

For a discussion of the characteristics of commonly-used distributions and the use of normalizing transformations, the reader is referred to the works of Brooks and Carruthers (1953), Stidd (1953), Chow (1954), Gumbel (1958), McGuinness and Brakensiek (1964), and the National Research Council of Canada (1966). Because of inconsistencies and errors commonly found in snow data, an initial graphical plotting is recommended to check the kind of distribution; for example, Fig. 6.21, shows the annual maximum 24-h snowfall amounts recorded at Winnipeg, Manitoba plotted on extreme-value or Gumbel paper. The distribution of a single variable can frequently be expressed in a linear form:

$$X(F) = \overline{X} + s \cdot k(F) \qquad\qquad 6.2$$

where $\qquad$ $X(F)$ = expected value of the variable whose probability of not being exceeded is F,

$\overline{X}$ = estimated mean of the population,

s = estimated standard deviation, and

$k(F)$ = frequency factor which is chosen to correspond to a given probability level, F, and whose magnitude depends on the frequency distribution and sample size.

Table 6.11 shows the return periods $[t_r = 1/(1 - F)]$ derived from Eq. 6.2 for the maximum annual 24-h snowfall at selected Canadian cities, assuming the events follow an extreme value distribution.

One problem in analysing snow data arises from the occurrence of "zero" values. For example, snowcover does not form every year in some areas, so that the time series of data for such parameters as dates of formation and disappearance of snowcover and maximum depths contain zero values. This kind of data can be handled by dividing the data series into populations, one containing zero values, and the other the remaining values (greater than zero). Thom, (1966) treated the problem in the following manner while studying the annual snow water equivalent maxima, WE. He assumed the water equivalent maxima follow a log-normal distribution such that the transformed value, u = ln WE, will be normally distributed with the distribution function

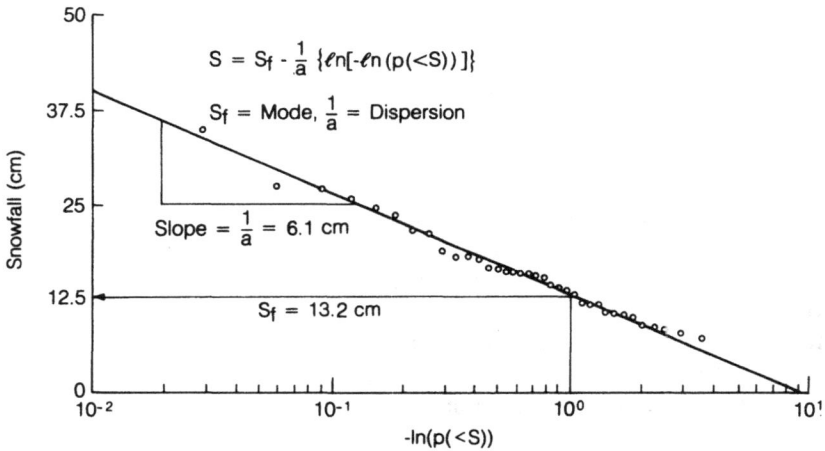

Fig. 6.21 Distribution of annual maximum 24-h snowfall amounts at Winnipeg, Manitoba (Jan. 1938 to Dec. 1972) plotted by the Gumbel method (Isyumov and Mikitiuk, 1976).

Table 6.11

COMPUTED ESTIMATES ON MAXIMUM ANNUAL 24-h SNOWFALL (cm) FOR SELECTED CANADIAN CITIES, EXPRESSED IN TERMS OF RETURN PERIODS AND PERCENTAGE PROBABILITIES OF AMOUNTS EXCEEDING 20 TO 40 cm[a].

Location	Return Period years			Amount cm	
	5	20	25	20	40
Vancouver	20	34	48	20	3.4
Calgary	23	36	47	29	3.7
Winnipeg	20	34	45	20	2.8
Toronto	25	41	55	37	5.5
Montreal	30	41	51	70	5.5
Halifax	34	48	59	90	10.0

[a] Based on observed 24-h snowfall amounts at International Airports, except Halifax, where observations were made at a site in the city.

N(u). If q is the probability of a year with no snow, then p = 1 - q lies at the probability of a year with snow. Since the zero component with probability q is the start of the distribution, the mixed distribution function G(u) may be expressed as

$$G(u) = q + pN(u). \quad\quad 6.3$$

Solving Eq. 6.3 for N(u) and transforming it to the unit normal distribution gives

$$[G(u) - q]/p = N |(u - \mu)/\sigma|, \quad\quad 6.4$$

where μ and σ are the population mean and standard deviation of the variable, u, respectively. Rearranging Eq. 6.4 and solving for u the following expression is obtained:

$$u(G) = sN^{-1}|[(G(u)-q)/p]| + \overline{\ln WE} \quad\quad 6.5$$

where: u(G) = specific value of u corresponding to a probability G(u),

s = standard deviation of the sample, an estimate of the population standard deviation, σ, and

$\overline{\ln WE}$ = sample mean, an estimate of the population mean, μ.

Equation 6.5 can be used to derive the design quartiles u(G) for specific values of 1 - G(u), i.e., the probabilities of exceeding a design value of u(G). For example, the distribution of the annual maxima of the snow water equivalent WE can be expressed as,

$$p (>\ln WE) = 1 - \{sN^{-1}[G(\ln WE)-q]/p] + \overline{\ln WE} \}, \quad\quad 6.6$$

which gives the probability that the annual snow water equivalent exceeds a given value of WE. If the events do not contain zero values the distribution reverts to a log-normal distribution.

Isyumov and Davenport (1974) studied distributions of snowfall depth at several Canadian cities. Assuming snowfall events on successive days were independent, they suggested that the probability of occurrence of a snowfall of depth S on any day t may be expressed as

$$p (> S, t) = p (S \neq 0, t) \cdot p' [> S, t | (S \neq 0)] \quad\quad 6.7$$

where: p(> S,t) = probability of snowfall exceeding S on day t,

p(S $\neq$ 0, t) = probability of snow falling on day t, and

p'[> S,t | (S $\neq$ 0)] = probability of the snowfall exceeding a depth S on snowy days.

The authors found that snowfall depths in temperate climates are generally skewed and may be suitably fitted to a Weibull distribution. Hence, Eq. 6.7 can be written as:

$$p(> S, t) = p (S \neq 0, t) \cdot \exp \{-[S|c(t)]^{k(t)}\} \qquad\qquad 6.8$$

where c(t) and k(t) are Weibull parameters which provide measures of the prevalent snowfall and skewness respectively. Isyumov and Davenport present data for several Canadian locations to exemplify the fit of daily snowfall amounts to the Weibull distribution, and the seasonal variations in c and k.

Equations 6.6 and 6.8 are not valid unless the data originate from a single population (homogeneous) and are mutually independent. For a series of annual maximum values, such as snow water equivalent, independence needs to be established, e.g., by testing the correlation of snowfall on successive days. Furthermore, caution must be taken as to the population sampled, e.g., daily snowfall regimes in December and February will usually differ. In other words, probabilities such as those given by Eq. 6.8 vary from one time period to another (see Isyumov and Davenport, 1974). Monthly values can usually be assumed to comprise a distinct, homogeneous population.

Variability in Time

In climatology a 30-y normal period has been adopted and used extensively as a basis for design. The use of normals has led to the impression that climate is stable, but it is actually highly variable, so that time variations should be considered in planning, design and operation. The 30-y period was selected because the mean values of series tend to be stable when calculated over this duration. Hershfield (1961) has shown that the standard deviation of 24-h extremes tends to stabilize when the sample size exceeds 50 years. However, the whole concept should be examined because the means and variances are continually changing. A recurrence of the snowfall cycle for the period 1865-1875 (see Fig. 6.22) in present day Toronto would have interesting implications for snow removal budgets and strategies. Ratio techniques can usually be used to adjust the means of short duration series to match those of longer series, but natural temporal changes in the ratios should not be overlooked. Court (1967) has recommended that 15-year median values computed every five years be used as predictors.

Because of the variability of data, for example snowfall amount, with time it is possible to obtain a short duration time series that has statistical characteristics which differ substantially from those of the long duration series from which it is obtained. Generally a search of the historical records will reveal long duration series that closely resemble the parent population, and against which those of shorter durations can be judged and tested. This somewhat spurious feature of the short duration record is normal in climatic data, and does not justify their rejection, but rather demands great caution in their use. An example of the effects of the length of the time series on sample statistics is illustrated by the data given in Table 6.12.

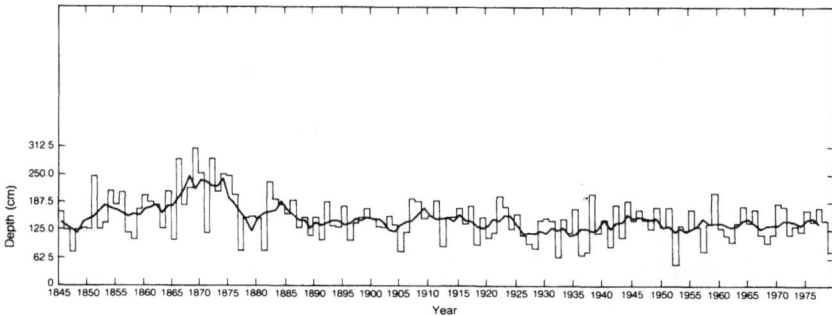

Fig. 6.22 Annual snowfall at Toronto, Canada (1845-1975): annual values are given by vertical bars, the meandering line is the 5-y moving mean plotted on the median years.

Table 6.12
VARIATIONS IN THE MEAN DEPTH AND STANDARD DEVIATION OF SNOWCOVER AT SHAWINIGAN FALLS, QUEBEC, DETERMINED FROM DIFFERENT PERIODS - OF - RECORD.

Period	Mean cm	Standard Deviation cm
1967-76	92.7	24.9
1928-76	73.1	23.6
1929-38	65.5	13.0

Seasonal analyses disclose other features. Because of its extreme cold the far North has relatively light, dry snowfall in regions remote from the open oceans. Continental areas that are frequented by arctic air tend to have light midwinter snowfalls, but have the heaviest, more humid snowfalls at the onset and cessation of winter. Further to the south or near the oceans, the wetter snow, and freezing rains that create a glaze within the snowcover, frequently (but not always) occur in conjunction with early- or late-winter storms (Fig. 6.23a,b). Changes in the physical characteristics of snowfall occur within a storm, and even hourly, as the temperature and moisture regimes change. Where quality of snow is a factor these changes may have to be considered.

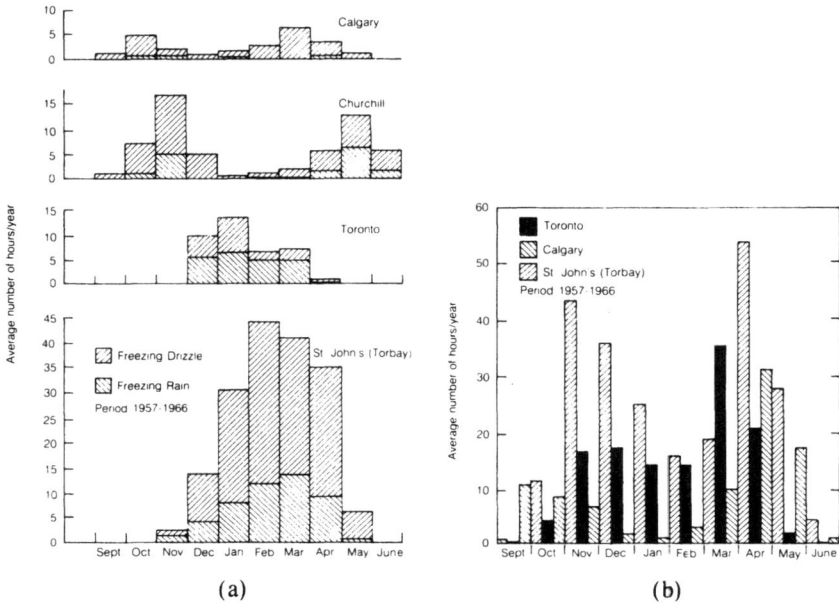

Fig. 6.23 (a) Average number of hours of freezing precipitation at selected stations in Canada. (b) Average number of hours of snow when air temperatures are greater than 0.56° C at selected stations in Canada.

Their effects are manifest in the stratigraphy of the snowcover in colder regimes, but may vanish as the snowcover commences to melt.

Cyclical patterns are sometimes evident in snowfall and snowcover records (Thomas, 1975, - see Fig. 6.24). The temptation always arises to use cycles for prediction purposes, but apart from the diurnal and annual cycles, most cycles are ephemeral, some times showing complete reversals in phase. Where they do appear to have validity they account for a very small fraction of the total variability which is dominated by other meteorological processes. Snow data should be carefully examined for the times of the cycles and their amplitudes and for changes in variability. Too simplistic approaches to snowfall cycles should be treated with caution.

Areal Variability

Some of the deficiencies that are claimed in gauge-catch of snowfall may be due to orographic effects, since snowfall amounts are highly related to land form, vegetative cover, extensive fetches of wind over water areas, as well as

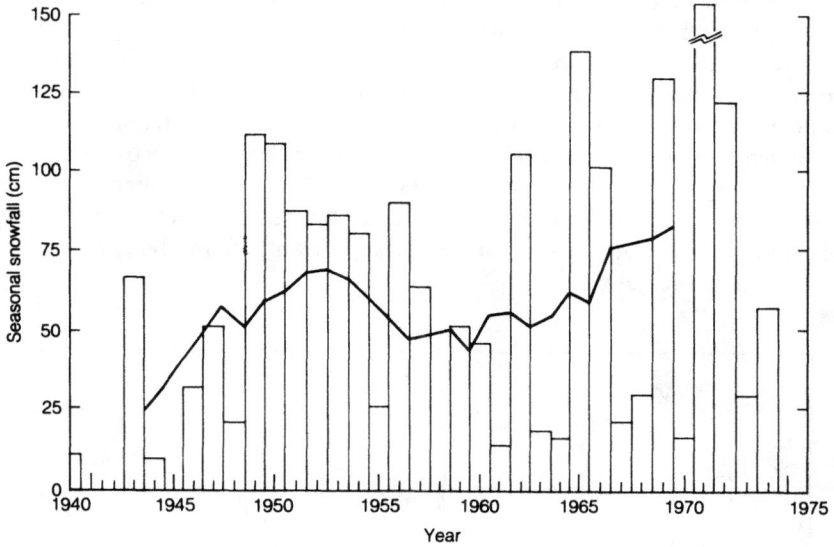

(a)

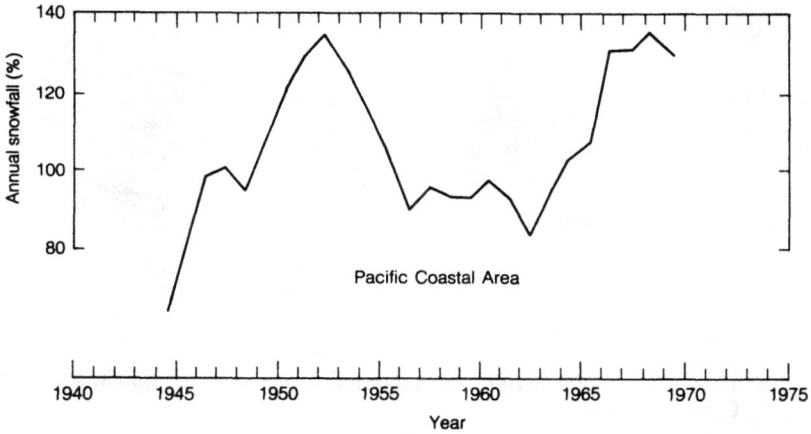

(b)

Fig. 6.24 Cyclic patterns of snowfall (Thomas, 1975).
  (a) Seasonal snowfall: Vancouver.
  (b) Annual snowfall: Pacific Coast Area expressed as a percentage of the
  1940 to 1974 average value.

general variations in climate. Figure 6.25 gives the 30-y return period
maximum winter snow depth for Canada which exhibits an areal variability
corresponding to general physiographic and climatic regions. Many
climatological stations are in sheltered locations to avoid inclement weather,
so that the extrapolation of data from one site to others having different
exposures is a very difficult problem. Those who have evaluated snowfall data
have generally compared it with measurements either of snow on the ground
near the gauge area, or of runoff. These "true" estimates probably compensate for topographic influences as well as instrumental deficiencies, i.e., they
provide areal averages which can differ significantly from the "point" value
measured at the gauge.

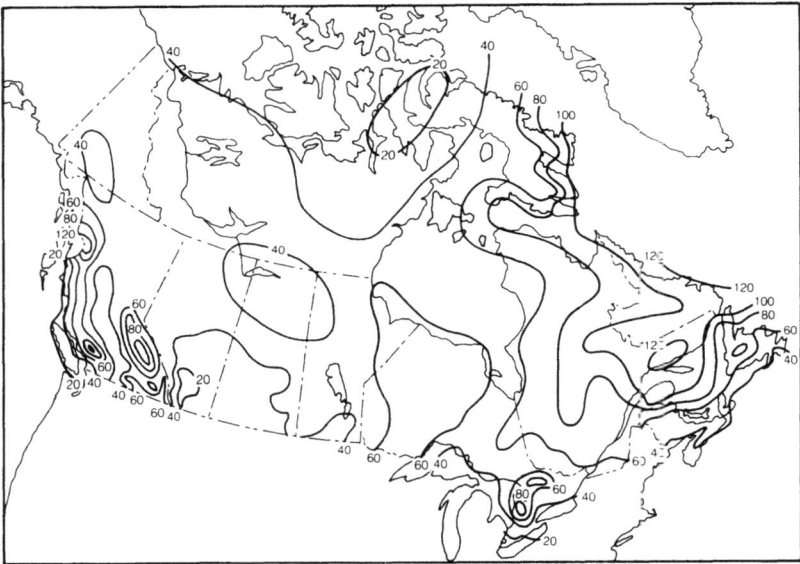

Fig. 6.25 Thirty-year return periods of maximum winter snow depth (Boyd, 1961).
Values in centimetres.

Expressing point climatological measurements in terms of areal snowfall
amounts is a major challenge. Relationships have been developed which aid
in doing this, e.g., see Rikhter (1945) and Kuz'min (1960), who provide
comprehensive reviews of the differences in snow accumulations that occur
over different types of terrain (see Ch. 5). The transposability of such results is
valid only qualitatively, however, and it is evident that much work remains to
be done to establish similar relationships for other areas before snowfall or

snowcover estimates of a desired precision can be provided for most snowfall regions. Survey courses are usually intentionally placed in locations which have abnormal accumulations to ensure not only continuity of measurement through the years, but also a high correlation with water yield. This bias may not exist when the networks have been established for other purposes, such as to study tractionability or snow hardness.

Too often the analyst accepts dogmatically that point values are absolute. This error is prevalent in snowcover mapping where isolines used to depict the areal variability frequently give a false impression of accuracy. The risk of misinterpretation may be reduced by using a map scale commensurate with the accuracy of the data, and a meaningful isoline interval (Espenshade and Schytt, 1956). Small-scale maps reduce the risk of misinterpretation by effectively broadening the zone covered by an isoline. Whenever users insist on larger detailed maps, the analyst must determine the sampling error in the most convenient manner possible and select an isoline interval that adequately conveys the degree of reliability of the data. The resultant map will probably not have the desired "eye appeal", but is less likely to induce significant errors of estimation. Physical relationships, e.g., those established by a regression analysis, are frequently used to introduce further interpretative skill, however, the reliability of the product is usually speculative.

If statistically-valid estimates of the mean areal value of a snowcover property are required a sufficient number of samples, obtained with calibrated instruments, should be taken to establish the population variance. For example, within an area of relatively homogeneous snowcover, the true mean depth lies between $\bar{S}_c \pm (s/\sqrt{N})t$ where N is the number of samples, $\bar{S}_c$ is the sample mean, s is the sample standard deviation, and t is the "Student's t" which is available in tabular form for given probability levels and sample sizes in most books on statistics. For the true mean to lie within $\bar{S}_c \pm \Delta\bar{S}_c$ of the sample mean at a specified confidence level then $\Delta\bar{S}_c < (s/\sqrt{N})t$. For example, if 20 snow depth measurements have a standard deviation of 2.0 cm, then $t_{0.05} = 2.09$, so that the mean snow depth has greater than a 95% probability of lying within 1 cm of the true value. The number of observations required to obtain an estimate of the mean for any selected level of confidence can be calculated by the preceding method when the standard deviation is known.

Table 6.13, reported by Steppuhn and Dyck (1974), compares the statistics for the mean snowcover depth $\bar{S}_c$ and standard deviations obtained from repeated samples taken on an upland pasture in Saskatchewan; over the same course the samples had different numbers of observations. It is evident from these data, assuming $\bar{S}_c$ and s calculated from 64 samples to be estimates of the true population statistics, that a progressive increase in the standard deviation occurs with decreasing sample size. If it can be assumed that the optimum number of observations should yield sample statistics with values within about 5% and 10% of the population mean and standard deviation, respectively, the number of observations to be taken should be between 26 and 32.

Stratification of an area according to catch and retention characteristics, and the use of the above statistical criteria provides for a more accurate quantitative estimate of the mean areal snowcover. The reduction of the sample variance obtained with stratification has merit: it increases confidence in the sample, reduces the number of samples required, increases the confidence that snowcover is uniformly distributed over a specified landscape area, and aids in extrapolating data.

Steppuhn and Dyck also point out that the areal frequency distributions of snowcover data for a given landscape class are often non-symmetric; the depth distribution often shows greater skew than the density distribution. These characteristics do not preclude the use of normal distribution statistics, since the sample means tend toward a normal distribution as the number of samples

Table 6.13
STATISTICS FROM REPEATED MEASUREMENTS OF SNOW DEPTH TAKEN OVER AN AREAL UNIT OF UPLAND-PASTURE CLASS, BAD LAKE, SASKATCHEWAN, 1972 (Steppuhn and Dyck, 1974). Reproduced from "Advanced Concepts and Techniques in the Study of Snow and Ice Resources, 1974" with the permission of the National Academy of Sciences, Washington, D.C.

No. of Observations N	Mean Depth $\bar{S}_c$ cm	Relative Deviation in $\bar{S}_c$ %	Standard Deviation s cm	Relative Deviation in s %
64	13.85	0	8.82	0
57	13.97	1	8.93	1
54	14.10	2	8.99	2
51	13.31	4	7.84	11
48	14.31	3	9.46	7
46	13.56	2	7.96	10
43	14.37	4	9.29	5
38	13.33	4	8.09	8
32	13.26	4	8.05	9
26	14.61	5	9.90	12
22	14.61	5	10.24	16
18	14.59	5	10.94	24
16	12.47	10	6.58	25
13	16.02	16	12.14	38
11	12.00	13	6.76	23
8	16.12	16	14.37	63

increase. Separate sampling of depth and density is highly advantageous when the areal variabilities of the two variables differ appreciably. Frequently an investigator will find that depth influences the value of the water equivalent much more than density. In this situation the number of depth samples should be increased relative to the number of density samples.

Estimation

A wide variety of estimation techniques can be used to repair incomplete series as well as to interpolate or extrapolate data, varying from simple visual or mathematical adjustments to complex models with built-in consistency checks.

A procedure commonly used to repair gaps in precipitation records makes use of the normal ratio involving records for nearby locations; an example for four stations is,

$$P_4 = N_4 \left[(P_1/N_1) + (P_2/N_2) + (P_3/N_3)\right]/3, \qquad 6.9$$

where N_i are the normal values of the data, and P_i their corresponding actual values for the locations, $i = 1, 2, 3, 4$. More precision can usually be obtained from an estimate based on isolines on a map drawn so as to fit measured values and known topographic relationships. Extrapolative procedures are more hazardous and are very suspect when based on transposed physical relationships, or on regression lines extended beyond the values for which they were derived.

Ratio techniques are commonly employed to adjust short-term records to the normal period, or to estimate winter season snowfall from annual precipitation averages, and to estimate values for a specific return period, using a well-known field of data and well-established frequency distributions as a basis for prediction. For example, a curve that envelops the ratio of extreme-to-mean values is sometimes used to estimate extremes for specific locations.

Whenever possible, to estimate values for data sets, the relationships between snowcover and physiographic and climatic factors should be used. Relationships have been developed to aid in this process; see the comprehensive reviews of Rikhter (1945), Kuz'min (1960) and Miller (1976).

Regression is used extensively for extrapolating and interpolating, e.g., to estimate changes in snowfall or snowcover water equivalent with elevation (Fig. 6.26; see also Ch. 5). The empirical nature of the regression must be kept in mind, for example, the increase of snowfall with elevation is often attributed uniquely to elevation, whereas it is actually due to the compound effect of many factors, including air temperature, the reduction in melt and the spillover of snowfall across ridges. The introduction of all of these variables might improve the estimate, but would also greatly reduce the number of degrees of freedom and affect the significance of the results. The transposition of regression equations used for estimation must always be questioned. The equations reflect both seasonal and local topographic influences that may change significantly even over brief periods of time or short distances.

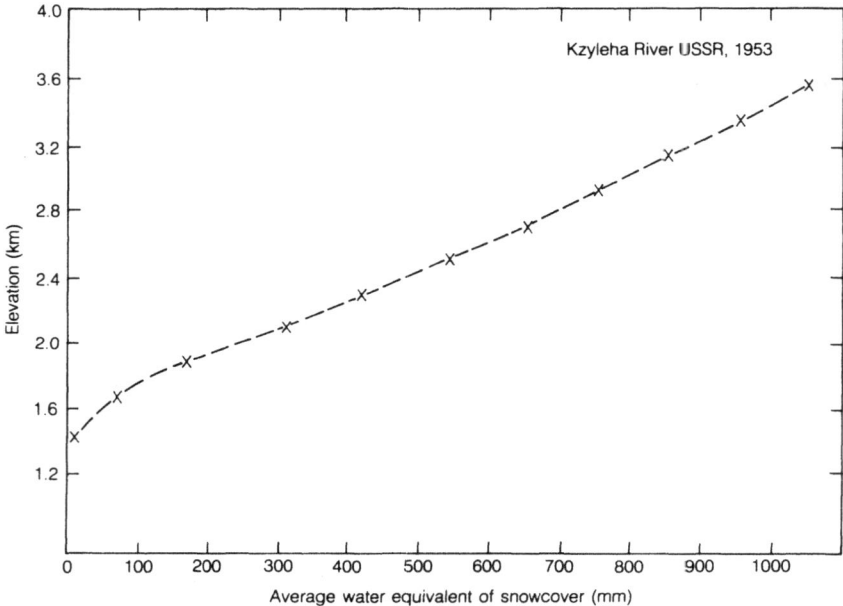

Fig. 6.26 Height dependency curve for snowcover water equivalent (based on data reported by Kuz'min, 1960).

Design Values

Chapter 13 presents the rationale and methods used to estimate snow loads on buidings. Often it is necessary to estimate the snow resources for other design purposes. Frequency distributions are commonly used to do this, e.g., in estimating a 1:100 return-period snowfall amount.

Other techniques have been employed, particularly in hydrology. To obtain a "lower estimate" of the extreme, envelopes of measured extremes over an area are sometimes used. The length of the time series used in the analysis must be considered, since in general, the longer the series, the greater the extreme that is likely to be recorded. Very long series for the area selected should be examined and reference made to newspaper archives and to other records to ensure that the envelopes derived are appropriate. A "systematic" approach to snowcover maxima may also be used. For example, the maximum precipitation of record for November, December and other months may be summed to obtain an estimate of a seasonal extreme. Meteorological persistence tends to support this technique, however, it has many weaknesses: the extreme value, so estimated, is influenced by the sampling period, e.g., the summation of daily extremes equals or exceeds the summation of weekly and monthly extremes. Discretion must be used to select a time interval that reflects the natural storm frequency.

Both snowcover and snowfall data may be used in the above. Since snowcover survey data include the effects of ablation, their use may help achieve a more realistic estimate of the snowcover extreme, as compared with the use of snowfall data for this purpose. Topographic variations in snowcover and snowfall should also be taken into account when interpolation or transposition techniques are being employed.

The maximization of storm snowfall is achieved by assuming that each contributing storm could have occurred with the maximum atmospheric water vapor possible for the specific area and time of year. In practice, measured snowfall values are increased by the ratio of the potential to actual precipitable moisture of the source air mass. Comparing the two procedures for basins in Quebec, Bruce and Clark (1966) found that the sum of the maximized storm values were consistent with "systematic" estimates for a four-day interval (see Fig. 6.27).

Presentation of Data

Snow data are presented in numerical, graphical and cartographical forms, usually for the following general categories:

1. Days of formation, disappearance, or duration, sometimes combined with other characteristics,
2. Accumulated amounts and depths,
3. Physical characteristics: density, hardness, wetness, etc., and,
4. Functional relationships between snowcover properties and other elements, e.g., depth versus elevation.

The data are usually expressed as means, medians, normals, departures, variances, frequencies, rates, extremes or durations for a specified time or period and location.

Tabular presentation is the easiest to prepare and produce and is usually the best way of displaying large volumes of precise data values. However, for many purposes, graphs and maps are superior since they enable anomalous values and mathematical and physical relationships to be easily recognized, and thereby introduced in estimation and transposition.

In preparing maps and diagrams it is usually advantageous to follow conventions such as those of the World Meteorological Organization as described in the *Guide to climatological practices* (World Meteorological Organization, 1960). When they are to be used in design and for operational purposes, however, the specific requirements of the project dictate both format and content. In such instances the analyst must make careful decisions to ensure that the diagram does not convey more accuracy than the data

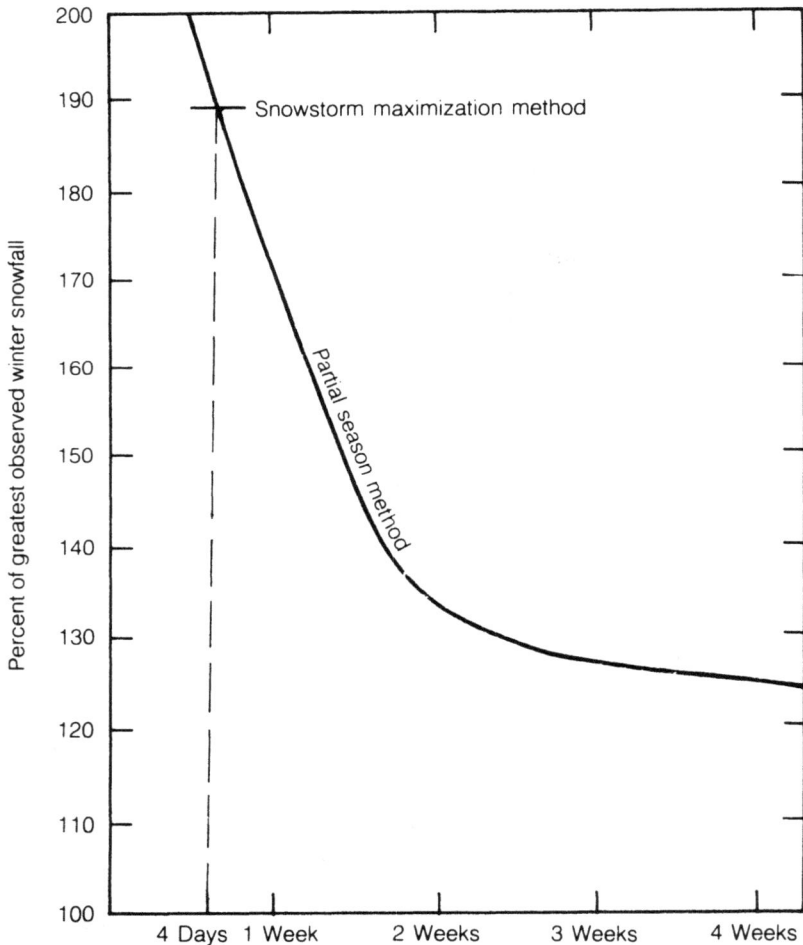

Fig. 6.27 Estimates of maximum seasonal snowfalls for Manicouagan and Outardes basins, Quebec (Bruce and Clark, 1966).

possesses. The accompanying legend or text for each map or diagram should identify the data base, and the estimation techniques used for preparing each map and their limitations.

Formation, duration and ending of snowcover

Generally speaking, the date of formation or disappearance of the snowcover is readily definable for high latitudes, but not for more southerly latitudes where there are frequent winter thaws. To overcome this difficulty, criteria can be set up to suit the user's needs, e.g., McKay and Thompson (1967) postulated that the seasonal snowcover begins on the first day of the first seven-day period in winter when the ground remains covered. A similar assumption was made for the termination of snowcover. The durations are simply derived from the dates of formation and termination.

The frequency distributions for dates of formation and termination and for durations are approximately Gaussian so that normal probability paper can be used to obtain return period values.

Accumulated depth

The analysis of accumulated depth data pose few problems. However, in regions with mild winters, zero values may occur, becoming more common as the climate becomes warmer. The occurrence of zero values may necessitate using the median instead of the mean, and complicates the use of frequency distributions.

The distributions of snowfall and snowcover change with the length of the measurement period. Most attention is given to seasonal or maximum accumulations. In statistical analysis, as shown earlier, their values or their logarithms can often be fitted to normal (Gaussian) or an extreme value (Gumbel) distribution. For sites at which "zero" values occur, these distributions cannot be applied to the complete data set without some modification. McKay and Thompson (1967) found that for the Canadian Prairies the slopes of lines fitted to non-zero values obtained from a station during winters when the snowcover was interrupted were conservative and consistent with those at adjacent stations where the snowcover was not interrupted. They used a partial series, and synthetic mean values (having a return period of 2.12 years - extreme value distribution) to estimate values for higher return periods.

Snowcover is not always measured on a synoptic basis so that for some purposes depth measurements must be evaluated for a specific time. This is usually achieved by using daily meteorological measurements of snowfall and snowcover. The adjustment of mass is simple when snowmelt has not occurred; and may be estimated by water balance techniques when it has (McKay, 1963).

The snow environment

The properties of snowfall and snowcover and their environment, are highly dynamic. Descriptions should carefully identify the controlling environment including its heat and moisture balances, and the physical objects that may control air flow and drifting. Just as the existence of a snowcover and its physical nature are related to atmospheric conditions, an extensive snowcover limits or modifies the nature of the surrounding environment. Recognition of such interactions is often the key to meaningful analysis of snow data.

Accordingly, in using other climatological data (e.g., temperature) it must be remembered that their monthly means may incorporate values for days without snowcover. Careful stratification of data may be necessary when attempting to establish their functional relationships with snowcover. The occurrence of a very high air temperature may provide a valuable clue to the absence of extensive snowcover. Atmospheric dew point temperatures quickly approach 0°C when the air has an extensive trajectory over snow; therefore this process must be considered when using snowmelt relationships. The amount of free water, the density of the snowcover and its adhesion or compaction are highly related to atmospheric conditions both present and past.

The physical characteristics of snowfall and snowcover change continuously in response to changing atmospheric conditions and gradients of temperature and vapor pressure. Density and hardness vary with such factors as crystal structure, packing and free-water content. Presentation of data representing these features may generally be enhanced by stating the context in which measurements were made. Density and wetness provide an example. During periods of active melt the flow of meltwater through the snowcover may increase the density to as high as 500 kg/m^3. If a period of non melt follows, the density may decrease to about 370 kg/m^3 owing to drainage of the water. This "cycling" of density or wetness may be significant for forecasting tractionability of a winter road, skiing conditions and other applications, and shows the need for prudence when comparing data observed on different hours and days.

Relative exposure to wind is another important factor which must be considered. Bilello (1967) found that average snowcover density (kg/m^3) was related on a seasonal basis with average air temperature T (°C) and wind speed U (m/s) by the following equation:

$$\rho = 152 - 0.31T + 1.9U. \qquad\qquad 6.10$$

Since T and U are partially controlled by topography large spatial variations in ρ are normal.

Rapid changes occur in snow density with time even in the absence of melting, and even from the moment the snow crystal reaches the ground. Church (1943) observed a density increase from 36-56 kg/m³ to 176 kg/m³ after 24 hours of drifting. The frequency of occurrence of wet snow can be indicated simply by contingency statistics, e.g., most snowfall is wet when air temperatures are greater than or equal to -1°C.

Within the snowcover, and particularly near the ground-snow interface, the energy balance assumes major importance. For example, if solar radiation is strongly absorbed snow may melt at ground level with air temperatures of -5°C at the screen level. On the other hand, it may not melt at the air-snow interface even with air temperatures above 0°C if radiation hardly penetrates into the snow. Under these circumstances glazing may occur at the interface, but the energy transfer is insufficient to melt the snow at depth. The radiative balance may be strongly affected by vegetative cover, by underlying materials when the snowcover is shallow, and by admixtures such as dust (see Chs. 7 and 9).

These complexities and the dynamic characteristics of snow must be kept continually in mind in order to make rational comparisons and to extract meaningful information from snowfall and snowcover data.

LITERATURE CITED

Adams, W.P. 1976. *Areal differentiation of snowcover in East Central Ontario.* Water Resour. Res., Vol. 12, No. 6, pp. 1226-1234.

Allis, J.A., B. Harris and A.L. Sharp. 1963. *A comparison of performance of five rain gauge installations.* J. Geophys. Res., Vol. 68, pp. 4723-4729.

Armstrong, R.L. 1976. *The application of isotopic profiling snow gauge data to avalanche research.* Proc. 44th Annu. Meet. West. Snow Conf., pp. 12-19.

Atmospheric Environment Service. 1973a. *Precipitation.* Environ. Can., Downsview, Ont.

Atmospheric Environment Service. 1973b. *Snow surveying. 2nd Ed.* Environ. Can., Downsview, Ont.

Atmospheric Environment Service. 1980. *Monthly record, meteorological observations in Canada.* Environ. Can., Downsview, Ont.

Atmospheric Environment Service. 1977. *Manual of surface weather observations - 7th Ed.,* Environ. Can., Downsview, Ont.

Barnes, G.W., Jr. 1974. *A new California Department of Water Resources telemetry system.* Adv. Concepts Tech. Study Snow Ice Resour. Interdisc. Symp., U.S. Nat. Acad. Sci., Washington, D.C. pp. 329-338.

Barnes, J.C. and C.J. Bowley. 1969. *Operational snow mapping from satellite photography.* Proc. 26th Annu. Meet. East. Snow Conf., pp. 79-103.

Barnes, J.C., C.J. Bowley and J.L. Cogan. 1974. *Snow mapping applications of thermal infrared data from the satellite very high resolution radiometer (VHRR).* Final Rep., Doc. 0438-F, Environ. Res. Tech., Inc.

Barton, M. 1974. *New concepts in snow surveying to meet expanding needs.* Adv.
    Concepts Tech. Study Snow Ice Resour. Interdiscip. Symp., U.S. Nat. Acad.
    Aci., Washington, D.C. pp. 39-46.
Barton, M. and M. Burke. 1977. *SNOTEL: An operational data acquisition system
    using meteor burst technology.* Proc. 45th Annu. Meet. West. Snow Conf., pp.
    82-87.
Beaumont, R.T. 1965. *Mt. Hood pressure pillow snow gauge.* J. Appl. Meteorol.,
    Vol. 4, pp. 626-631.
Bilello, M.A. 1967. *Relationships between climate and regional variations in snow-
    cover density in North America.* Physics of Snow and Ice (H. Oura, ed.), Proc.
    Int. Conf. Low Temp. Sci., Hokkaido Univ., Sapporo, Vol. 1, Part 2, pp. 1015-
    1028.
Bilello, M.A. 1969. *Surface measurements of snow and ice for correlation with aircraft
    and satellite observations.* Sp. Rep. No. 127, U.S. Army Cold Reg. Res. Eng.
    Lab., Hanover, N.H.
Bindon, H.H. 1964. *The design of snow samplers for Canadian snow surveys.* Proc.
    21st Annu. Meet. East. Snow Conf., pp. 23-28.
Bissell, V.C. and E.L. Peck. 1973. *Monitoring snow water equivalent by using natural
    soil radioactivity.* Water Resour. Res., Vol. 9, pp. 885-890.
Bissell, V.C. and E.L. Peck. 1974. *Measurement of snow at a remote site:
    natural radioactivity technique.* Adv. Concepts Tech. Study Snow Ice Resour.
    Interdiscip. Symp., U.S. Nat. Acad. Sci., Washington, D.C. pp. 604-613.
Blincow, D.W. and S.C. Dominay. 1974. *A portable profiling snow gage.* Proc. 42nd
    Annu. Meet. West. Snow Conf., pp. 53-57.
Boyd, D. 1961. *Maximum snow depths and snow loads on roofs in Canada.* Proc.
    29th Annu. Meet. West. Snow Conf., pp. 6-16.
Bray, D.I. 1973. *A report on the variability of snow water equivalent measurements at
    a site.* Proc. 30th Annu. Meet. East. Snow Conf., pp. 20-31.
Brooks, C.E.P. and N. Carruthers. 1953. *Handbook of Statistical Methods in
    Meteorology.* Her Majesty's Sta. Off., London.
Bruce, J.P. and R.H. Clark. 1966. *Introduction to Hydrometeorology.* Pergamon
    Press, London.
Chadwick, D.G. 1972. *Precipitation telemetry in mountainous areas.* Water Resour.
    Res., Vol. 8, pp. 255-258.
Chow, V.Te. 1954. *The log-probability law and its engineering applications.* Am.
    Soc. Civ. Eng. Proc., Vol. 80, Separate 536.
Church, J.E. 1943. *Perennial snow and glaciers.* Scientific Monthly, Vol. 56, pp. 211-
    231.
Coles, G.A. and D.R. Graham. 1980. *Alberta Environment's real-time hydrometeoro-
    logical network.* Proc. Data Collection Platform Networks Workshop, Can.
    Remote Sensing Soc., 29-30 Sept. 1979, Quebec City, P.Q.
Court, A. 1967. *Climate normals as predictors.* Rep. No. 67-824, Vol. 1, San
    Fernando Valley State College Foundation, Northridge, Los Angeles, Calif.
Cox, L.M., L.D. Bartee, A.G. Crook, P.E. Farnes and J.L. Smith. 1978. *The care and
    feeding of snow pillows.* Proc. 46th Annu. Meet. West. Snow Conf., pp. 40-47.

Crutcher, H.L., G.F. McKay and D.C. Fulbright. 1977. *A note on a gamma distribution computer program and computer-produced graphs.* NOAA Tech. Rep. EDS, Nat. Tech. Inf. Serv., Springfield, Va.

Davis, R.T. 1973. *Operational snow sensors.* Proc. 30th Annu. Meet. East. Snow Conf., pp. 57-70.

den Hartog, G. and H.L. Ferguson. 1975. *National water balance maps of evapotranspiration and precipitation.* Proc. Can. Hydrol. Symp.-75, Winnipeg, Aug. 11-14. pp. 511-525.

Dickinson, W.T. and H.R. Whiteley. 1972. *A sampling scheme for shallow snowpacks.* Bull. Int. Assoc. Hydrol. Sci., Vol. 17, pp. 247-258.

Espenshade, E.B. and S.V. Schytt. 1956. *Problems in mapping snowcover.* Rep. 27, U.S. Snow, Ice and Permafrost Res. Estab.

Farnes, P.E., 1978. *Future snow survey operations with SNOTEL.* Proc. 46th Annu. Meet. West. Snow Conf., pp. 15-20.

Farnes, P.E., B.E. Goodison, N.R. Peterson and R.P. Richards. 1980. *Proposed metric snow samplers.* Proc. 48th Annu. Meet., West. Snow Conf. (in press).

Ferguson, H.L. 1973. *Precipitation network design for large mountainuous areas.* WMO/OMM No. 326, World Meteorological Organization, Geneva, pp. 85-105.

Ferguson, H.L. and B.E. Goodison. 1974. *Mean snowpack water equivalent maps and snow course data problems over southern Ontario.* Proc. 31st Annu. Meet. East. Snow Conf., pp. 91-111.

Ferguson, H.L., H.I. Hunter and D.G. Schaefer. 1974. *The IHD mountain transects project, Part II - Instrumentation problems and data record.* Can. Meteorol. Res. Rep. CMRR1/75, Atmos. Environ. Serv., Downsview, Ont.

Ferguson, H.L. and S. Lapczak. 1977. *Satellite imagery analysis of snow cover in the Saint John and Souris river basins.* Proc. Fourth Can. Symp. Remote Sensing, Quebec City, Can. Aeronaut. and Space Inst., Ottawa, Ont., pp. 126-142.

Ferguson, H.L. and P.Y.T. Louie. 1974. *Precipitation and evapotranspiration models by the Atmospheric Environment Service.* Tech. Suppl. II, Ch. 6, Final Rep., Canada-British Columbia Okanagan Basin Agreement, Off. Study Director, Penticton, B.C. pp. 82-116.

Ferguson, H.L. and D. Storr. 1974. *Hydrometeorological network design for basin studies.* IHD Workshop Seminar on Hydrologic Data Collection, Oct. 21-22, Can. IHD Secretariat, Ottawa.

Ferguson, H.L. and D.M. Pollock. 1971. *Estimating snowpack accumulation for runoff prediction.* Can. Hydrol. Symp., No. 8, Runoff from Snow and Ice, Queen's Printer, Ottawa, Vol. 1, pp. 7-27.

Findlay, B.F., D. Gray, G.A. McKay and H.A. Thompson. 1972. *Networks in Arctic areas.* Casebook on Hydrological Network Design Practice, WMO No. 324, World Meteorological Organization, Geneva, pp. V5.1-V5.1-6.

Findlay, B.F. and B.E. Goodison. 1978. *Archiving and mapping of Canadian snow cover data.* Presented at Workshop on Mapping and Archiving of Data on Snowcover and Sea Ice Limits, Inst. Arct. Alp. Res., Univ. Colo., Boulder, 2-3 November.

Flanders, A.F. 1979. *Design of automatic telemetering and satellite transmission systems: hydrological data transmission.* Rep. WMO Working Group on Hydrol. Data Transmission, Processing and Retrieval, World Meteorological Organization, Geneva.

Foster, J.L., D.K. Hall, A.T.C. Chang, A. Rango, L.J. Allison and B.C. Diesen, III. 1980. *The influence of snow depth and surface air temperature on satellite - derived microwave brightness temperature.* NASA Tech. Memo. TM-80695, Goodard Space Flight Center, Greenbelt, Md.

Freeman, T.G. 1965. *Snow survey samplers and their accuracy.* Proc. 22nd Annu. Meet. East. Snow Conf., pp. 1-10.

Gandin, L.S. 1970. *The planning of meteorological station networks.* WMO Tech. Note No. 111, World Meteorological Organization, Geneva.

Goodison, B.E. 1975. *Standardization of snow course data: reporting and publishing.* Proc. 32nd Annu. Meet. East. Snow Conf., pp. 12-23.

Goodison, B.E. 1977. *Snowfall and snow cover in Southern Ontario: principles and techniques of assessment.* Ph.D. Thesis, Univ. Toronto, Toronto, Ont.

Goodison, B.E. 1978a. *Canadian snow gauge measurements: accuracy, implications, alternatives, needs.* Proc. Seventh Symp. Appl. Prairie Hydrol., Water Studies Inst., May 9-11, Saskatoon, Sask., pp. 7-15.

Goodison, B.E. 1978b. *Accuracy of Canadian snow gage measurements.* J. Appl. Meteorol., Vol. 27, pp. 1542-1548.

Goodison, B.E. 1978c. *Comparability of snowfall and snowcover in a Southern Ontario basin.* Proc., Modeling Snow Cover Runoff (S.C. Colbeck and M. Ray, eds.), U.S. Army Cold Reg. Res. Eng. Lab., Hanover, N.H., pp. 34-43.

Goodison, B.E. 1978d. *Accuracy of snow samplers for measuring shallow snowpacks: an update.* Proc. 35th Annu. Meet. East. Snow Conf., pp. 36-49.

Goodison, B.E. and D.J. McKay. 1978. *Canadian snowfall measurements: some implications for the collection and analysis of data from remote stations.* Proc. 46th Annu. Meet. West. Snow Conf., pp. 48-57.

Goodison, B.E., S.E. Waterman and E.J. Langham. 1980. *Application of synthetic aperture radar data to snow cover monitoring.* Proc. 6th Remote Sensing Symp., 21-23 May, Halifax, N.S.

Graham, D.R., W.E. Kerr and R.J. Grauman. 1977. *Automatic hydrometeorological telemetry network of Alberta.* Proc. Can. Hydrol. Symp.-77, August 29-31. Edmonton, Alta., Nat. Res. Counc. Can., Ottawa, Ont., pp. 136-143.

Grant, L.O. and J.O. Rhea. 1974. *Elevation and meteorological controls on the density of new snow.* Adv. Concepts Tech. Study Snow Ice Resour. Interdiscip. Symp., U.S. Nat. Acad. Sci., Washington, D.C., pp. 169-181.

Grasty, R.L. 1979. *One flight snow-water equivalent measurement by airborne gamma-ray spectrometry.* Paper presented at WMO Workshop on Remote Sensing of Snow and Soil Moisture by Nuclear Techniques, Voss. Norway, 23-27 April.

Grasty, R.L., H.S. Loijens and H.L. Ferguson. 1974. *An experimental gamma-ray spectrometer snow survey over southern Ontario.* Adv. Concepts Tech. Study Snow Ice Resour. Interdiscip. Symp., U.S. Nat. Acad. Sci., Washington, D.C., pp. 579-593.

Gray, D.M., D.I. Norum, G.E. Dyck. 1970. *Densities of Prairie snowpacks*. Proc. 38th Annu. Meet. West. Snow Conf., pp. 24-30.

Gumbel, E.J. 1958. *Statistics of Extremes*. Columbia Univ. Press, New York.

Haan, C.T. 1977. *Statistical Methods of Hydrology*. The Iowa State Univ. Press, Ames.

Hall, D.K., A. Chang, J.L. Foster, A. Rango and T. Schmugge. 1978. *Passive microwave studies of snowpack properties*. Proc. 46th Annu. Meet. West. Snow Conf., pp. 33-39.

Hall, D.K., J.L. Foster, A.T.C. Chang and A. Rango. 1979. *Passive microwave applications to snowpack monitoring using satellite data*. NASA Tech. Memo. TM-80310, Goodard Space Flight Center, Greenbelt, Md.

Halliday, R.A. 1975. *Data retransmission by satellite for operational purposes*. Proc. Int. Semin. on Modern Developments in Hydrometry, Sept. 8-13, Padua, Italy.

Hamon, W.R. 1973. *Computing actual precipitation in mountainous areas*. WMO No. 326. World Meteorological Organization, Geneva, pp. 159-173.

Hare, F.K. and J.E. Hay. 1971. *Anomalies in the large-scale annual water balance over Northern North America*. Can. Geogr., Vol. XV, No. 2, pp. 79-94.

Harris, R.E. and A.C. Carder. 1974. *Rain and snow gauge comparisons*. Can. J. Earth Sci., Vol. 11, pp. 557-564.

Hershfield, D.M. 1961. *Estimating the probable maximum precipitation*. J. Hydraulics Div., Proc. Am. Soc. Civil Eng., Vol. 87, HY5, pp. 99-116.

Hogg, W.D. and A.J. Hanssen. 1979. *Computer enhanced snow cover analysis of satellite data*. Can. Hydrol. Symp.:79 - Cold Climate Hydrology, Nat. Res. Counc., Can., Ottawa, Ont., pp. 414-423.

Isyumov, N. and A.G. Davenport. 1974. *A probablistic approach to the prediction of snow loads*. Can. J. Civil Eng., Vol. 1, pp. 28-49.

Isyumov, N. and M. Mikitiuk, 1976. *Climatology of snowfall and related meteorological variables with application to roof snow load specifications*. Proc. 33rd Annu. Meet. East. Snow Conf., pp. 41-69.

Jones, E.B., A.E. Fritzsche, Z.G. Burnson, and D.L. Burge. 1974. *Areal snowpack water-equivalent determinations using airborne measurements of passive terrestrial gamma radiation*. Adv. Concepts Tech. Study Snow Ice Resour. Interdisc. Symp., U.S. Nat. Acad. Sci., Washington, D.C., pp. 594-603.

Kerr, W.E. 1976. *Snow pillow experiences in a prairie (Alberta) environment*. Proc. 44th Annu. Meet. West. Snow Conf., pp. 39-47.

Kogan, R.M., M.V. Nikiforov, V.P. Chirkov, A.F. Yakovlev and Sh. D. Fridman. 1965. *Determination of water equivalent of snow cover by method of aerial gamma survey*. Sov. Hydrol. Sel. Pap., No. 2, pp. 183-187.

Kuz'min, P.P. 1960. *Snowcover and snow reserves*. Gidrometeorologicheskoe Izdatelsko Leningrad. [English Transl. by U.S. Nat. Sci. Found., Washington, D.C., 1963].

Kuz'min, P.P. 1975. *Development of a reference precipitation gauge for the measurement of snow*. Unpubl. written comminication with the Chairman of the WMO/CIMO/WG on Measurement of Precipitation, Evaporation and Soil Moisture, Geneva, Oct. 1975.

Larson, L.W. and E.L. Peck. 1974. *Accuracy of precipitation measurements for hydrologic modelling*. Water Resour. Res., Vol. 10, pp. 857-863.

Limpert, F.A. and J.L. Smith. 1974. *Utility of isotope profiling snow gauge for water management*. Adv. Concepts Tech. Study Snow Ice Resour. Interdiscip. Symp., U.S. Nat. Acad. Sci., Washington, D.C., pp. 625- 631.

Loijens, H.S. 1975. *Measurements of snow water equivalent and soil moisture by natural gamma radiation*. Proc. Can. Hydrol. Symp.-75, Aug. 11-14, Winnipeg, pp. 43-50.

Loijens, H.S. 1979. *Measurement of snow water storage in the Lake Superior basin using aerial gamma-ray spectrometry*. Paper presented at WMO Workshop on Remote Sensing of Snow and Soil Moisture by Nuclear Techniques, 23-27 April, Voss, Norway.

Loijens, H.S. and R.L. Grasty. 1973. *Airborne measurement of snow-water equivalent using natural gamma radiation over southern Ontario, 1972-1973*. Sci. Ser. No. 34, Inland Waters Directorate, Environ. Can., Ottawa, Ont.

Mayo, L.R., 1972. *Self-mixing antifreeze solution for precipitation gauges*. J. Appl. Meteorol., Vol. 11, pp. 400-404.

McGinnis, D.F., J.A. Pritchard and D.R. Wiesnet. 1975a. *Determination of snow depth and snow extent from NOAA-2 satellite very high resolution radiometer data*. Water Resour. Res., Vol. 11, pp. 897-902.

McGinnis, D.F., J.A. Pritchard and D.R. Weisnet. 1975b. *Snow depth and extent using VHRR data from the NOAA-2 satellite*. Tech. Memo. NESS-63, Washington, D.C.

McGuinness, J.L. and D.L. Brakensiek. 1964. *Simplified techniques for fitting frequency distributions to hydraulic data*. U.S. Dept. Agric. and Water Cons. Res. Div., (unnumbered), Washington, D.C.

McKay, G.A. 1963. *Relationships between snow survey and climatological measurements*. Int. Union Geod. Geophys., Gen. Assem. Berkeley [Surface Water]. Int. Assoc. Sci. Hydrol. Publ. 63, pp. 214-227.

McKay, G.A. 1970a. *Precipitation*. Handbook on the Principles of Hydrology (D. M. Gray, ed.), Water Information Centre, Inc., Port Washington, N.Y., pp. 2.1-2.111.

McKay, G.A. 1970b. *Problems of measuring and evaluating snowcover*. Snow Hydrology, Proc. of CNC/IHD Workshop Seminar, Feb. 28-29, 1968, Univ. New Brunswick. Can. Nat. Comm. Int. Hydrol. Decade, Ottawa, Ont., pp. 48-62.

McKay, G.A. and H.A. Thompson. 1967. *Snow cover in the prairie provinces of Canada*. Trans. Am. Soc. Agric. Eng., Vol. 11, pp. 812-815.

Meier, M.F. 1975. *Application of remote-sensing techniques to the study of seasonal snow cover*. J. Glaciol., Vol. 15, pp. 251-265.

Meteorological Branch. 1965. *The nipher shielded snow gauge*. Meteorol. Branch, Can. Dept. Transport, Toronto, CIR-4353, INS-154.

Miller, D.H. 1976. *Spatial interactions produced by meso-scale transports of water in the atmospheric boundary layer*. Paper presented to Assoc. Am. Geogr., New York, N.Y.

Miller, R.W. 1962. *Aerial snow depth marker configuration and installation consideration*. Proc. 30th Annu. Meet. West. Snow Conf., pp. 1-5.

Morrison, R.G. 1976. *Nuclear techniques applied to hydrology.* Proc. 44th Annu. Meet. West. Snow Conf., pp. 1-6.

National Research Council of Canada. 1966. *Statistical methods in hydrology.* Proc. Hydrol. Symp. No. 5, Queens Printer, Ottawa, Ont.

Nipher, F.E. 1878. *On the determination of the true rainfall in elevated gauges.* Am. Assoc. Advancement Sci., Vol. 27, pp. 103-108.

Nkemdirim, L.C. and P.W. Benoit. 1975. *Heavy snowfall expectations for Alberta.* Can. Geogr., Vol. 19, pp. 60-72.

NOAA (National Oceanic and Atmospheric Administration). 1977. *National environmental satellite service catalog of products.* NOAA Tech. Memor. NESS88, U.S. Dept. Commerce, Washington, D.C., pp. 83-89.

Nordenson, T. 1968. *Preparation of coordinated precipitation, runoff and evaporation maps.* IHD Rep. No. 6, World Meteorological Organization, Geneva.

Ohtake, T. and T. Henmi. 1970. *Radar reflectivity of aggregated snowflakes.* 14th Radar Meteorol. Conf., Tucson, Ariz., Am. Meteorol. Soc., Boston Mass. pp. 209-210.

Otnes, J. 1972. *Network in mountainous areas.* Casebook on Hydrological Network Design Practice, WMO No. 324, World Meteorological Organization, Geneva, pp. V.1.1-V.1.10.

Peck, E.L. 1972. *Snow measurement predicament.* Water Resour. Res., Vol. 8, pp. 244-248.

Peck, E.L., V.C. Bissell, E.B. Jones and D.L. Burge. 1971. *Evaluation of snow water equivalent by airborne measurement of passive terrestrial gamma radiation.* Water Resour. Res., Vol. 7, pp. 1151-1159.

Peck, E.L., L.W. Larson and J.W. Wilson. 1974. *Lake Ontario snowfall observation network for calibrating radar measurements.* Adv. Concepts Tech.Study Snow Ice Resour. Interdiscip. Symp., Nat. Acad. Sci., Washington, D.C., pp. 412-421.

Peck, E.L., T.R. Carroll and S.C. Van Demark. 1979. *Operational aerial snow surveying in the United States.* Paper presented at WMO Workshop on Remote Sensing of Snow and Soil Moisture by Nuclear Techniques, 23-27 April, Voss, Norway.

Pollock, D.M., J. Rogalsky and D.A. Carr. 1973. *A processing system for Fischer and Porter precipitation gauge data.* CL1-6-71, Atmos. Environ. Serv., Environ. Canada, Downsview, Ont.

Potter, J.G. 1965. *Water content of freshly fallen snow.* CIR-4232, TEC-569, Meteorol. Branch, Dept. of Transport, Toronto, Ont.

Puhakka, T. 1975. *On the dependence of the Z-R relation on the temperature of snowfall.* Proc. 16th Radar Meteorol. Conf., 22-24 Apr., Houston, Tex., Am. Meteorol. Soc., Boston, Mass., pp. 504-507.

Rango, A. and R. Peterson (eds.), 1980. *Operational applications of satellite snowcover observations.* Proc. of Final Workshop, 16-17 April, 1979, Sparks, nev., NASA Conf. Pub. 2116, Goodard Space Flight Center, Greenbelt, Md.

Rango, A., A.T.C. Chang and J.L. Foster. 1979. *The utilization of spaceborne nicrowave radiometers for monitoring snowpack properties.* Nordic Hydrol., Vol. 10, pp. 25-40.

Rawls, W.J., D.C. Robertson, J.F. Zuzel and W.R. Hamon. 1975. *Comparison of precipitation gage catches with a modified Alter and rigid Alter type wind shield.* Water Resour. Res., Vol. 11, pp. 415-417.

Rechard, P.A., R.D. Brewer and A. Sullivan. 1974. *Measuring snowfall, a critical factor for snow resource management.* Adv. Concepts Tech. Study Snow Ice Resour. Interdiscip. Symp., U.S. Nat. Acad. Sci., Washington, D.C., pp. 706-715.

Rikhter, G.D. 1945. *Snezhnyi Pokrov, Ego Gormirovanie i Svoistva (Snow cover its formation and properties).* Izv. Akad. Nauk., SSSR, Moscow. [English Transl. by U.S. Army Snow, Ice and Permafrost Res. Estab., Transl. No. 6.]

Schumann, H.H. 1975. *Operational applications of satellite snowcover observations and LANDSAT data collection systems operations in central Arizona.* Proc. Operational Applications of Satellite Snowcover Observations, 18-20 August, South Lake Tahoe, Calif., NASA, Washington, D.C., pp. 13-28.

Sekhon, R.S. and R.C. Srivastava. 1970. *Snow size spectra and radar reflectivity.* J. Atmos. Sci., Vol. 27, pp. 299-307.

Shreve, D.C. and A.J. Brown. 1974. *Development and field testing of a remote radio-isotopic snow gauge.* Adv. Concepts Tech. Study Snow Ice Resour. Interdiscip. Symp., U.S. Nat. Acad. Sci., Washington, D.C., pp. 661-673.

Smith, J.L., H.G., Halverson and R.A. Jones. 1972. *Central Sierra profiling snow-gauge: a guide to fabrication and operation.* USAEC Rep. TID-25986, Nat. Tech. Inf. Serv., U.S. Dept. Commer., Washington, D.C.

Solomon, S.I., J.P. Denouvilliez, E.J. Chart, J.A. Woolley and C. Cadou. 1968. *The use of a square grid system for computer estimation of precipitation, temperature and runoff.* Water Resour. Res., Vol. 4, pp. 919-929.

Solomon, S.I. 1972. *Joint mapping.* Casebook on Hydrological Network Design Practice. WMO No. 324, World Meteorological Organization, Geneva, pp. III 2.1-III 2.16.

Sporns, U. 1976. *Snow cover mapping in mountainous areas - Columbia river drainage above Mica Dam.* Unpubl. Rep., B.C. Hydro., Vancouver, B.C..

Stark, R.G. 1974. *Precipitation, evaporation and snow cover data collection system: Part 1: National scale.* IHD Workshop Seminar on Hydrologic Data Collection, 21-22 Oct., Can. IHD Secretariat, Ottawa, Ont.

Steppuhn, H. and G.E. Dyck. 1974. *Estimating true basin snowcover.* Adv. Concepts Tech. Study Snow Ice Resour. Interdiscip. Symp., U.S. Nat. Acad. Sci., Washington, D.C., pp. 314-324.

Steppuhn, H. 1976. *Areal water equivalents for prairie snowcovers by centralized sampling.* Proc. 44th Annu. Meet. West. Snow Conf., pp. 63-68.

Stidd, C.K. 1953. *Cube-root-normal precipitation distributions.* Trans. Am. Geophys. Union., Vol. 34, pp. 31-35.

Struzer, L.R. 1965. *Principal shortcomings of methods of measuring atmospheric precipitation and means of improving them.* Sov. Hydrol.: Sel. Pap. No. 1, pp. 21-35.

Struzer, L.R. 1969. *Method of measuring the correct values of solid atmospheric precipitation.* Sov. Hydrol.: Sel. Pap. No. 6, pp. 560-565.

Tarble, R.D. 1968. *California federal-state snow sensor investigations, problems and rewards.* Proc. 36th Annu. Meet. West. Snow Conf., pp. 106-109.

Thom, H.C.S. 1966. *Distribution of maximum annual water equivalent of snow on the ground.* Mon. Weather Rev., Vol. 94, pp. 265-271.

Thomas, M.K. 1975. *Recent climatic fluctuations in Canada.* Atmos. Environ. Serv. Environ. Can., Downsview, Ont.

Tobiasson, W. and R. Refield. 1976. *CRREL is developing new snow load design criteria for the United States.* (Abstract), Proc. 33rd Annu. Meet. East. Snow Conf., pp. 70-72.

Tollan, A. 1970. *Determination of areal values of water equivalent of snow in a representative basin.* Nordisk Hydrologisk K. Konferens, Stockholm, Vol. 11, pp. 97-105.

U.S. Department of Agriculture Forest Service. 1974. *Snow lab notes.* Pacific Southwest For. Range Expt. Sta., Berkely, Calif.

U.S. Soil Conservation Service, 1972. *Snow survey and water supply forecasting.* Section 22, SCS Nat. Eng. Handb., U.S. Dept. Agric., Washington, D.C.

Warner, C. and K.L.S. Gunn. 1967. *Measurement of snowfall by optical attenuation.* J. Appl. Meteorol., Vol. 8, No. 1, pp. 110-121.

Washichek, J. 1973. *Collection of atmospheric data for project skywater.* The Role of Snow and Ice in Hydrology, Proc. Banff Symp., Sept. 1972, Unesco-WMO-IASH, Geneva-Budapest-Paris, Vol. 1, pp. 644-655.

Wasserman, S.E. and D.J. Monte. 1972. *A relationship between snow accumulation and snow intensity as determined from visibility.* J. Appl. Meteorol., Vol. 11, pp. 385-388.

Weiss, L.L. and W.T. Wilson. 1957. *Precipitation gauge shields.* Int. Union Geod. Geophs. Gen. Assem. Toronto [Trans. Vol. 1, Land Erosion, Instruments, Precipitatin] Int. Assoc. Sci. Hydrol. Publ. 43, pp. 462-484.

Wiesnet, D.R. 1974. *The role of satellites in snow and ice measurements.* NOAA Tech. Memo. NESS58, Nat. Oceanic Atmos. Admin., U.S. Dept. Commer., Washington, D.C.

Wiesnet, D.R. 1978. *Future U.S. satellite programs of interest to snow scientists.* Proc. 35th Annu. Meet. East. Snow Conf., pp. 81-88.

Wiesnet, D.R., M. Matson and D.F. McGinnis. 1978. *NOAA satellite monitoring of snowcover in the northern hemisphere during the winter of 1977.* Proc. 35th Annu. Meet. East. Snow Conf., pp. 63-80.

Wilson, J.W. 1974. *Measurement of snowfall by radar.* Adv. Concepts Tech. Study Snow Ice Resour. Interdiscip. Symp., U.S. Nat. Acad. Sci., Washington, D.C., pp. 391-401.

Wilson, J.W. 1975. *Measurement of snowfall by radar during the IFYGL.* Proc. 16th Radar Meteorol. Conf., 22-24 April, Houston, Tex., pp. 508-513.

Wilson, J.W. and D.M. Pollock. 1977. *Precipitation (radar) project of the IFYGL lake meteorology program.* IFYGL Spec. Bull. No. 20, NOAA, Washington, D.C.

Work, R.A.; H.J. Stockwell, T.G. Freeman and R.T. Beaumont. 1965. *Accuracy of field snow surveys, western United States, including Alaska.* Tech. Rep. 163, U.S. Army Cold Reg. Res. Eng. Lab., Hanover, N.H.

World Meteorological Organization. 1960. *Guide to climatological practices.* Rep. No. 100, TP.44, Secretariat, World Meteorological Organization, Geneva.

World Meteorological Organization. 1973. *Snow survey from earth satellites, a technical review of methods.* WMO/IHD Rep. No. 19, WMO. No. 353, World Meteorological Organization, Geneva.

World Meteorological Organization. 1974. *Guide to hydrological practices.* 3rd Ed. Ch. 2, Instruments and Methods of Observation. WMO No. 168, World Meteorological Organization, Geneva, pp. 2.1-2.90.

World Meteorological Organization. 1976. *International working seminar on snow studies by satellites.* Final Rep., WMO, Geneva.

Young, G.J. 1974. *A stratified sampling design for snow surveys based on terrain shape.* Proc. 42nd Annu. Meet. West. Snow Conf., pp. 14-22.

Young, G.J. 1976. *A portable profiling snow gauges - results of field tests on glaciers.* Proc. 44th Annu. Meet. West. Snow Conf., pp. 7-11.

7

PHYSICS AND PROPERTIES OF SNOWCOVER[1]

E.J. LANGHAM

*Snow and ice Division, National Hydrology Research
Institute, Environment Canada, Ottawa, Ontario*

PHYSICS

Snow Deposition

The purpose of this chapter is to describe the physical properties of snow on the ground and the processes that take place within the snowcover to change these properties. Ice particles that form in the atmosphere have a large variety of crystal habits and sizes (see Ch. 4). By the time they reach the ground they have already undergone a number of transformations resulting from growth, disintegration or agglomeration. For example, if dendritic crystals form floccular aggregates in light winds they may be deposited in a layer of extremely low density (~ 10 kg/m^3). On the other hand graupel (coalesced frozen water droplets) in the same wind conditions may be deposited in a layer of high density (~ 500 kg/m^3). Between these extremes, any intermediate density is possible so that the density of the surface deposition may vary by a factor of 50 or more.

Wind speed near the surface also determines how crystals arriving at the surface are packed. At high speeds, crystals are broken first in the extremely turbulent boundary layer in the lowest few metres above the surface; further comminution occurs as the crystals are bounced and dragged over the snow surface (saltation) either during a snow storm or afterwards as blowing snow. After being reduced in size and shaped more symmetrically the crystals can be packed much more closely to produce a much denser surface layer than would otherwise occur.

One important structural characteristic of snowcover is its ice layers (see Fig. 7.1) which affect not only the rate of transmission of air and water

---

[1] The reader should note that this chapter does not include a complete tabulation of all the properties of snow. Additional information is given in other chapters: e.g., snow density and albedo, Chs. 5 and 9; hardness, Chs. 11 and 19.

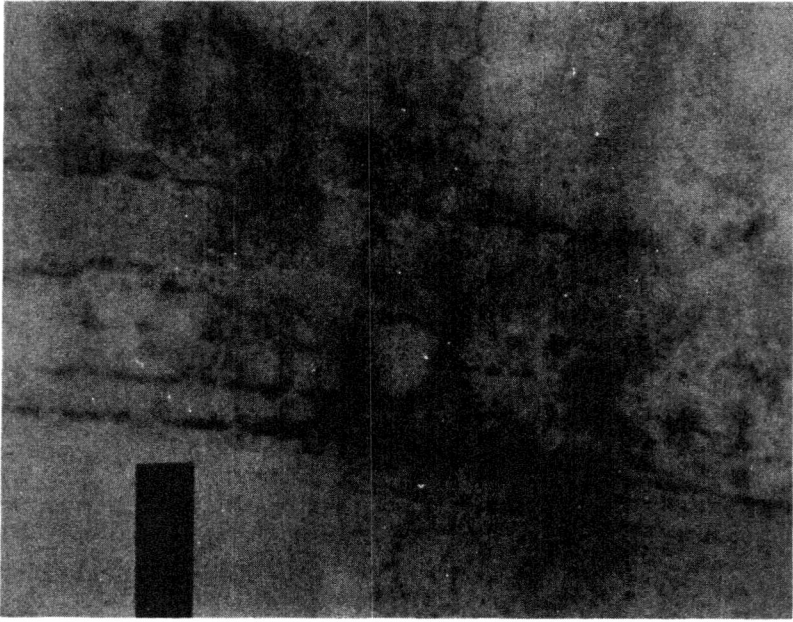

Fig. 7.1 Stratification of a natural snowpack. The dark, nearly horizontal, lines are icy layers containing dyed water (photo courtesy of A.C. Wankiewicz).

through the snow but also the trafficability over snow. Preceding the period of active melt, ice layers are formed by: (a) surface melt (primarily the result of radiation) which is refrozen and covered by deposits of new snow, and (b) the addition of freezing rain. Rain can form in the atmosphere by the coalescence of liquid water droplets; in the absence of nucleation of the ice phase (refer to Ch. 4), this process can occur at temperatures down to -10°C or lower. When droplets of freezing rain contact the snowcover, freezing commences immediately and part or all of the liquid turns to ice. The thickness and continuity of the resulting ice sheet depends on such factors as the temperatures of the supercooled droplets, the ambient air and the upper layer of the snowcover. If the ice sheet becomes continuous and impermeable, pools of water may collect on its surface. If these pools subsequently freeze, they produce thick ice lenses that become part of the snowcover. Otherwise some water drains into the snow altering the internal structure. In each case the layers may later be covered by new snow.

A snowcover that has formed from a number of snowfalls, each occurring with different meteorological conditions, will be highly stratified. The deposition of the snow is not horizontally uniform since obstacles affect the boundary-layer turbulence causing the snow to accumulate in drifts. A flat, rough surface under conditions of saltation produces similar aerodynamic effects which result in dunes over open ground or frozen lakes. Both drifts and dunes have distribution patterns that depend on wind direction. Since the wind direction and the type of snow reaching the ground may vary during each snowfall, the stratification of the snowpack is spatially very heterogeneous. This presents considerable problems in making reliable estimates of the water equivalent of the snowpack (see Ch. 5).

Evolution of the Snowcover

After the snow is deposited the particle shapes are modified by a process known as metamorphism. Thus dendritic crystals decompose into fragments and the larger fragments grow at the expense of the smaller ones. This process continues until the fragments have been reduced to more or less rounded grains[1] of ice or until a significant temperature gradient develops within the pack. The mechanisms which cause these initial changes in shape are not completely understood; however, considerable experimental evidence suggests that the movement of molecules through the vapor phase dominates the process. Thermodynamically, the snow crystals are moving to an equilibrium state and the thermodynamic property which determines this state is the free energy[2] which for snow crystals implies minimizing the ratio of surface area to volume.

Simultaneous with the breakup of the dendritic assemblies of newly-deposited crystals is the formation of bonds at the points of contact between grains. This process, known as sintering, increases the strength of the snowpack[3]. Two grains of ice in point contact form a system which is not in thermodynamic equilibrium because the total surface free energy is not a minimum. A neck will form between the grains to decrease the total surface area. Vapor transfer plays a dominant role in this process as well.

Once a significant temperature difference is established across any layer within the snowcover[3] the process of metamorphism is completely altered.

---

[1] A grain is a single crystal, in that all its molecules lie in the same three dimensional array, but whose surface is irregular in shape and does not display a habit, which is a set of plane outer surfaces whose orientation is related to the symmetry of the molecular array.

[2] Throughout this chapter the term "free energy" is used in the thermodynamic sense; it is energy free to do useful work. The reader is referred to a fundamental text regarding this subject (e.g., Denbigh, 1966), for a more precise definition as applied to closed or open systems.

[3] Throughout this and other chapters the terms "snowpack" and "snowcover" are used interchangeably to refer to the snow on the ground. Snowcover is used in the more generaly context; snowpack in reference to deep accumulations.

Such differences are associated with heat transfer taking place at the upper and lower surfaces of the pack. At the upper surface, radiative, sensible and latent heat transfer occurs while at the lower surface the important process is conduction.

Since vapor pressure is temperature dependent, temperature gradients produce associated vapor pressure gradients, which cause water vapor to diffuse from warmer to colder parts of the snowpack. Yoshida and others (1955) suggest that the water vapor is transferred from one ice grain composing the snow layer to the next in a "hand-to-hand" transfer process. Thus, there is a series of transfers involving movement from the solid phase to the vapor phase and back to solid; a significant portion of the snow will pass through the vapor phase and be deposited as part of a new crystal. The newly formed crystals (known as depth hoar) take many shapes but have a characteristic layered structure resulting in a stepped or ribbed surface on some crystal facets. These crystals are only weakly bonded together. The low shear strength of such layers is a primary cause of slab avalanches (see Ch. 11).

Metamorphism can also result from compaction caused by the pressure of overlying layers of snow. This process is responsible for transforming snow into glacial ice whose crystals sometimes attain sizes of the order of 10 cm. During its early stages, the refreezing of melt water can accelerate the densification process.

All the phenomena mentioned above are discussed in greater detail below. However, it should be clear that the variation of snowcover structure, because of its mode of deposition and subsequent evolution, produces an extra-ordinarily complex medium. Moreover, although the properties of snow are discussed separately in this chapter for the sake of an orderly presentation, changes in them are closely related to the metamorphic processes.

The Role of Free Energy in the Metamorphism of Snow

The principles of thermodynamics are often used to explain the observed changes resulting from the metamorphism of snow crystals. If a volume containing several snow crystals is assumed to be a thermodynamic system, then differential changes in their state may be described by the first law of thermodynamics (or conservation of energy) according to the expression:

$$dU = \delta Q - \delta W, \qquad\qquad 7.1$$

where: dU = the change in internal energy, U,
            δQ = heat added to the system, and
            δW = work done by the system.

If it is assumed that work done on the system of snow crystals will only result in changes in volume or surface area of the ice then:

$$\delta W = pdV + \xi dA \qquad 7.2$$

where: p = pressure,
                V = volume,
                ξ = surface energy/unit area, and
                A = surface area.

The second law of thermodynamics for a system of fixed mass is:

$$dS > \delta Q/T \qquad 7.3$$

where: dS = change in entropy, and
                T = temperature of the system at the point of
                    heat transfer.

If the heat transfer takes place reversibly and temperature differences are kept infinitely small or if the only process which the system undergoes is the addition or removal of heat while being maintained at a constant temperature then:

$$dS = \delta Q/T. \qquad 7.4$$

For a completely isolated system the second law states that all changes occur such that

$$\Sigma dS_i > 0, \qquad 7.5$$

where the summation extends over the i phases of the system. For a given mass of snow this implies that the summation should extend over all crystals plus the air or water vapor likely to influence the system during the period when the differential change of state takes place.

The equilibrium criterion may be stated in terms of the Gibbs Free Energy, G, as

$$dG = \Sigma dG_i = 0, \qquad 7.6$$

where $G = U - TS + pV + \xi A.$ \qquad 7.7

The derivation of the criterion in terms of G is found in standard works on thermodynamics (e.g., Hatsopoulos and Keenan, 1965). Expanding Eq. 7.6 in terms of Eq. 7.7 gives

$$dG = dU - TdS - SdT + pdV + \xi dA + Ad\xi = 0. \qquad 7.8$$

Under conditions of constant temperature, pressure and surface energy the Gibb's free energy may decrease because of decreases in surface area or volume of one of the phases. Thus, when their temperatures are uniform the snow crystals tend to change their shape so that the ratio of surface area to volume approaches a minimum. Equilibrium is attained when changes produce no further decrease in the Gibb's free energy (Eq. 7.6. i.e. dG = 0).

While surface effects are important in a consideration of snow crystals some of the better known properties of snow can be explained without reference to surface conditions. In this case it is convenient to consider a thermodynamic system consisting of pure water in the bulk state. Equations 7.1, 7.2 and 7.3 still apply to the bulk state although Eq. 7.2 and Eq. 7.8 take correspondingly simpler forms;

$$\delta W = pdV, \qquad\qquad 7.9$$

and

$$dG = dU - TdS - SdT + pdV + Vdp. \qquad\qquad 7.10$$

If Eq. 7.9 and Eq. 7.4 are substituted into Eq. 7.1 and the resultant expression is used to eliminate dU from Eq. 7.10 then the following relation is obtained for changes in free energy,

$$dG = - SdT + Vdp. \qquad\qquad 7.11$$

For a system with different phases (solid, liquid, vapor) a condition of equilibrium requires that they have the same free energy per unit mass (see, for example, Hatsopoulos and Keenan, 1965). It follows that changes in free energy per unit mass of each phase must also be equal if they are to remain in equilibrium. Thus,

$$dG_v = dG_l = dG_s, \qquad\qquad 7.12$$

where

$$dG_v = - S_v dT + V_v dp \text{ (vapor)}, \qquad\qquad 7.12a$$

$$dG_l = - S_l dT + V_l dp \text{ (liquid), and} \qquad\qquad 7.12b$$

$$dG_s = - S_s dT + V_s dp \text{ (solid)}. \qquad\qquad 7.12c$$

The elimination of dG from any two of the Eqs. 7.12a, b and c gives

$$dp/dT = (S_1 - S_2)/(V_1 - V_2),$$

in which the subscripts 1 and 2 refer to any two phases. At the point of equilibrium where $G_1 = G_2$ the transition between the two phases is reversible so that Eq. 7.2 applies. Therefore,

$$\Delta S = S_1 - S_2 = Q/T = L_{12}/T, \qquad\qquad 7.13$$

where L_{12} is the latent heat for the phase transition. Hence:

$$dp/dT = L_{12}/(T\Delta V_{12}), \qquad\qquad 7.14$$

where ΔV_{12} is the increase in volume during the phase change. This equation, known as the Clausius-Clapeyron equation, determines the pressure increase dp needed to maintain phase equilibrium for a temperature increase dT. The derivation of the three equations of this type does not depend on any assumptions about the phases, although here the discussion refers to a single-component system (i.e. pure water). The graphical representation of Eq. 7.14 is called a phase diagram; e.g., Fig. 7.2 is for pure water. When all three phases are present in a system there is a unique solution of Eq. 7.14 which is the

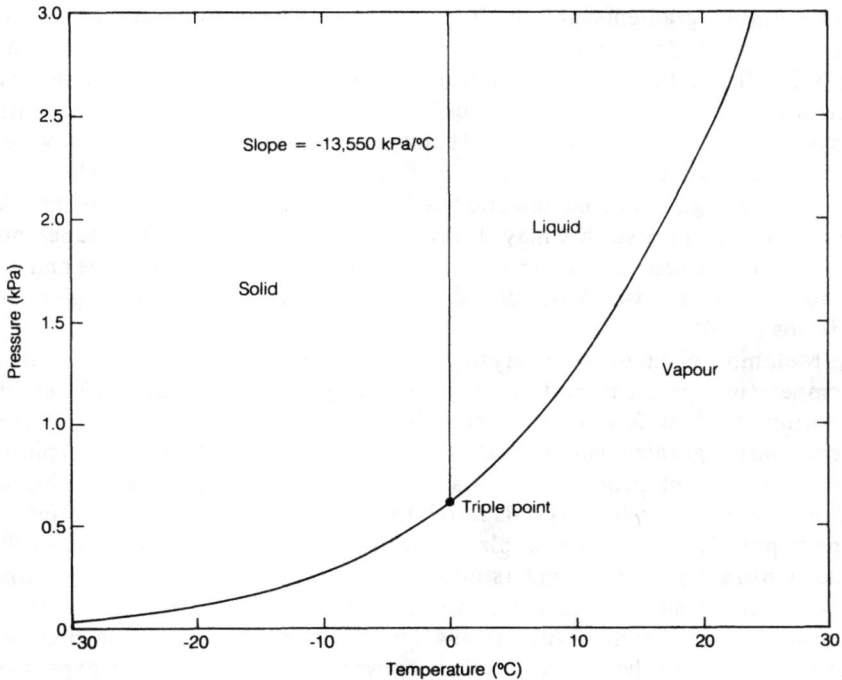

Fig. 7.2 Phase diagram for pure water.

meeting point of the three boundary curves at the triple point (+ 0.0099° C, 0.615 kPa, for pure water).

In the presence of air at atmospheric pressure the positions of these curves are modified very slightly so that they intersect at a new point, known as the ice point (0° C and 0.6095 kPa). The free energy of ice crystals in air depends on the air pressure. Since this effect is extremely small and is the same for all crystals it will be ignored in the subsequent discussion. The ice point is used for reference in tabulations of the thermodynamic properties of snow.

In a dry snowpack deposited over a period of several days, temperature variations with depth are quite common. The resulting temperature differences are associated with vapor pressure differences which can be calculated using an integrated form of Eq. 7.14 (see, for example, Denbigh, 1966; p. 203). Since vapor diffuses towards lower pressures there is a net transfer of material from the warmer part of the snowpack where the snow crystals have a higher vapor pressure to the colder part where they have a lower vapor pressure. Thus, colder crystals grow at the expense of the warmer ones. This process is referred to as temperature gradient metamorphism, a phrase which is often used only to discuss crystal growth. When the temperature gradients are small, the snow grains grow large but do not develop a well-defined crystal habit. When the gradients are large, the crystals develop flat surfaces parallel to internal crystal planes related to the molecular arrangement. These crystals are called depth hoar because they frequently grow in the lowest layers of a cold snowpack as a result of diffusion of vapor from the relatively warmer soil below. They may also occur elsewhere in a snowpack, e.g., after cold new snow is deposited on a relatively warmer pack. Frost on the upper surface may also consist of crystals with similar shapes, but they form because of nocturnal radiative cooling of the snow surface and are known as hoar frost. More details of crystal growth dynamics are given by Hobbs (1974).

Metamorphism of snow crystals takes place even in the absence of a temperature gradient and its corresponding vapor pressure differences. During the first few hours after snow is deposited (before a significant temperature gradient can be established within it), destructive metamorphism is the dominant process. Its first effect is to reduce snow consisting of dendritic or needle-like crystals to relatively round grains. Mass transfer via the vapor phase is known to play an important role in this process but the mechanism is not clearly understood. On the other hand, it is well known that the vapor pressure can vary over an irregular surface such as that of a snow crystal. For example, Adam (1941) showed that the resulting pressure difference across the curved interface between two phases may be expressed as:

$$dp_s - dp_v = \xi \, (1/r_1 + 1/r_2) \qquad\qquad 7.15$$

where: dp_s = difference between the vapor pressure of the solid at a curved surface and that of an infinite plane,

dp_v = corresponding pressure difference for the vapor phase,

ξ = surface energy per unit area, and

r_1 and r_2 = principle radii of curvature of the ice surface.

In general, Eq. 7.15 suggests that convex surfaces have a higher vapor pressure and concave surfaces a lower vapor pressure than plane surfaces.

In cases where the surface is anticlastic a net increase in vapor pressure may still occur. For example, although the radius of curvature of the narrowest part of a crystal may be infinite or even large and negative in one direction, its equilibrium vapor pressure may still be greater than that of a flat surface because the radius of curvature in the plane perpendicular to its length is small. On the other hand, a relatively flat, plate-shaped crystal where both radii of curvature are large produces a very small increase in the vapor pressure. If such high and low vapor pressure areas exist near each other then diffusion towards lower pressures can transfer ice by sublimation from one area to the other.

When the ice grains are subject to a local stress such that dp_s is positive, then, for the same radii of curvature dp_s increases. Although dp_s is defined herein as a compressive stress, any shear stress, or combined stresses such as those for bending, will also increase the free energy and thus also the vapor pressure. The experiments of Yoshida and others (1955) support this theory. They made a detailed study of the thinning process at the root of a dendritic crystal and found that the rate of reduction of the root radius takes place several orders of magnitude faster than can be accounted for by curvature alone. In addition, for thin rod-like ice grains, elastic stresses caused by the weight of adjoining crystals could significantly speed up the rate of thinning.

Once the dendritic or needle-like crystals have been reduced to more or less round grains equi-temperature metamorphism can still continue through sublimation from the small grains to the larger ones. This reduces the net surface area and consequently the free energy. Sintering also reduces the air/ice surface area by increasing the areas of contact between the larger crystals. This process has been studied by a number of workers in recent years (Ramseier and Keeler, 1966; Jellinek and Ibrahim, 1967; Hobbs, 1968; Gow, 1974). Hobbs and Mason (1964) showed that vapor transport is the dominant mechanism by which the area of contact or neck between two sintering spheres grows. The gradient that causes vapor movement results from the difference in curvatures of the ice surface at the neck and the adjoining ice particles. This produces densification and increasing mechanical continuity. It is the increased continuity which is responsible for the increased strength and hardness of the snow.

Any water in the snowpack collects at points of contact between grains. Colbeck (1973) distinguishes two saturation regimes in wet snow: the pendular regime — saturations $< \sim 14\%$ of the pore volume, with air in more or less continuous paths throughout the snow matrix; the funicular regime — saturations $> 14\%$ of the pore volume with air in distinct bubbles. According to Colbeck (1973) the metamorphism of wet snow is best considered by looking at local temperature differences in the vicinity of the grain boundaries. These differences exist as a result of the radii of curvature of the solid-liquid, liquid-vapor or solid-vapor interfaces. He showed that in the funicular regime heat flows from the larger to the smaller grains causing them to dissipate while the larger grains increase in size. In the pendular regime the radii of curvature of the ice grains have less effect on the transfer process so that capillary effects dominate, resulting in a reduction of temperature differences between particles. Also, the amount of liquid through which heat may flow is reduced. Hence much lower rates of grain growth are observed. Colbeck states that the snow strength in the pendular regime is quite high, since no melting occurs at the grain contacts, whereas in the funicular regime it is relatively low because very little bond-to-bond strength exists. On refreezing, the snowpack becomes very strong because of the resulting continuity between grains. This process is melt-freeze metamorphism, also referred to as firnification in perennial snowpacks.

Before leaving the theory of metamorphism, recrystallization must be mentioned. Although the process is somewhat peripheral to snow studies, since it relates mainly to crystal growth in polycrystalline ice, it is also partly responsible for transforming snow into ice. Variations in stress, and consequently free energy, occur in the matrix of crystals within a layer of deposited snow. Such variations act to transfer material between crystals at snowpack densities exceeding $\sim$ 580 kg/m^3, corresponding to close random packing of loose grains. Hobbs and Radke (1967) showed that the transfer mechanism at these densities is volume diffusion through the ice lattice to the highly stressed zone at the area of contact between crystals. They suggest that the rate of compaction is independent of pressure at least for times up to 15 hours. However, as the pressure of the overlying snow increases, viscoplastic flow may occur and dominate the densification process.

During viscoplastic flow, the crystals are permanently deformed. To explain the relationship between this flow and grain growth the crystal structure of ice must be considered. The oxygen atoms, and hence molecules of ice, are arranged on the regular three-dimensional lattice shown in Fig. 7.3. In the direction of the c-axis the lattice has hexagonal symmetry, i.e., a rotation of 60 degrees (and a displacement in the c-direction) brings the lattice into coincidence with the original molecular arrangement. This symmetry is responsible for the hexagonal crystal habits of plates, stars and prisms of snow. In an ice crystal lattice the molecules lie in planes whose separations are greatest perpendicular to the c-direction. Therefore, the molecular bonds between such planes are the weakest so that the resistance of ice to shear is least for stresses parallel to them. This plane of least resistance is called the basal plane. For an elaboration of the role of the crystal lattice in the structure of ice see Fletcher (1970).

In snow subject to high stresses resulting from the weight of the overlying layers or the slope of the snowcover, permanent deformation takes place in those crystal lattices oriented so that the shear stress is parallel to the basal planes. Other crystals withstand higher stresses before deforming permanently, but gain free energy. This increase acts so as to transfer material to those crystals which have undergone plastic deformation.

Also, the number of defects in the crystal lattice affect the free energy, and hence the growth of crystals. The lattice arrangement is usually not perfect, and can have various kinds of dislocation in the regular array of molecules. These constitute a form of internal free energy and are more numerous in damaged crystals. Dislocations move most easily in the basal plane, thus further reducing the resistance to shear. In other directions the stresses tend to increase the concentration of dislocations and therefore their free energy. Thus, in addition to the elastic energy, the dislocation free energy caused by inelastic deformation is conducive to preferential grain growth.

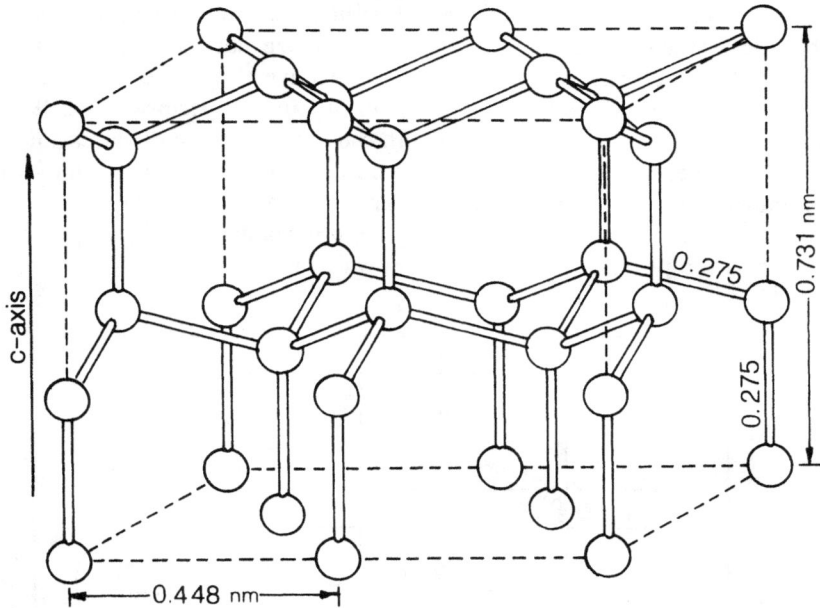

Fig. 7.3 Oxygen atom positions in an ice structure. The oxygen atom of each water molecule is shown by a ball, and hydrogen bonds by rods. The hydrogen atoms are omitted but lie approximately on the hydrogen bond lines. Dimensions shown are for a temperature of 77K (Kamb, 1968). [From Structural Chemistry and Molecular Biology (A. Rich and N. Davidson eds.). Freeman and Co. San Francisco, Copyright 1968].

CLASSIFICATION OF THE CRYSTALLINE NATURE OF THE SNOWPACK

Each of the metamorphic processes tends to produce a distinctive snow crystal. The need for classifying snow has led to several systems suited for different purposes. The earliest systems date from the nineteenth century; since they were proposed by mountaineers and skiers, they emphasize superficial appearance (powder snow, fluffy snow), and superficial properties (e.g., skiing characteristics) and their mutual relationships. The best descriptive system is given by Seligman (1936) who illustrates many of the snow conditions by photographs. Although he puts emphasis on nomenclature, he offers many suggestions for the causes of interesting features. He discusses metamorphism under the title of firnification and also describes the mechanical effects of wind packing.

Bader et al. (1939) attempted a field classification system based on physical properties of the snowcover, using data on grain size, hardness and amount of liquid. Although the system was very subjective they tried to introduce quantitative descriptions by measuring density and air permeability. While the results of their laboratory experiments did not allow clear demarcations to be established, they generally show how these parameters vary with metamorphic state. Their data are summarized in Fig. 7.4. The work also marked the beginning of an interest in the importance of metamorphism in relation to the physical properties of snow.

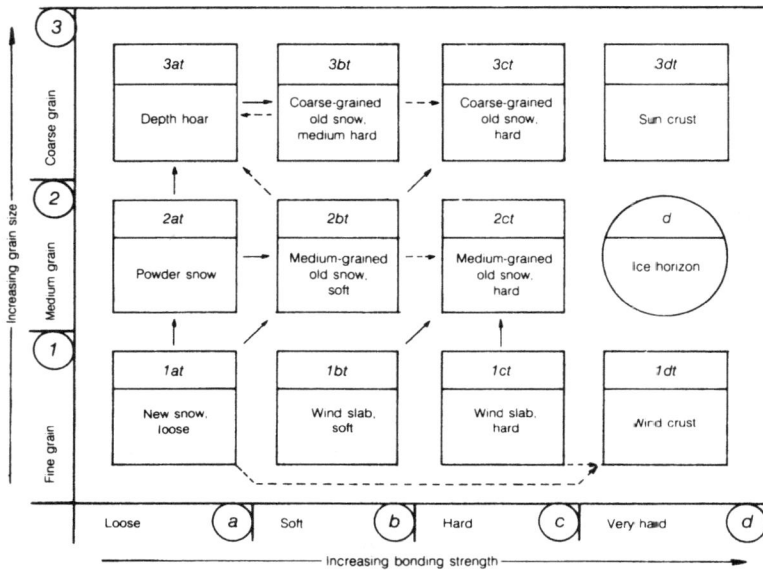

Fig. 7.4 Classification of snowcover based on bonding strength and grain size. (Bader et al., 1939).

1 = Fine grain	a = Loose or soft	d = Very hard
2 = Medium grain	b = Medium hard	t = Dry
3 = Coarse grain	c = Hard	

Subsequent efforts to measure the physical properties of the snowcover led to the inclusion of such parameters as density and hardness in the description (Klein, 1950), and later to the formal adoption by the International Commission on Snow and Ice of the (then) International Association of Scientific Hydrology (National Research Council, 1954) of an international classification for snow (Klein et al., 1950) embracing both qualitative descriptions and quantitive values of a few physical and mechanical properties. Symbols are given for describing the vertical profiles in the snow and for identifying the proposed physical measurements. Tables 7.1 and 7.2 illustrate the approach. The physical properties are treated in a similar way. Table 7.3 summarizes this classification while Fig. 7.5 illustrates one application to a snowpack.

Shumskii (1964) discusses metamorphism quantitatively in terms of the crystallographic processes involved, using the concept of minimum free energy which includes both surface free energy and internal free energy due to stresses on the ice crystals. The consideration of strain in crystals led him to include recrystallization in his classification. His system is shown in Table 7.4. The reader is referred to Shumskii's book for definitions of unfamiliar crystallographic terms.

During the decade 1960-1970, interest in the physical processes occurring in the snowpack resulted in an emphasis on classification based on the state of metamorphism. The best classification of this type was presented by Sommerfeld and LaChapelle in 1970 (see Table 7.5). *The Field Guide for Snow Crystals* (LaChapelle, 1969) provides lucid descriptions and numerous excellent illustrations to assist in using this system.

Table 7.1
PRIMARY FEATURES OF DEPOSITED SNOW (National Research Council, 1954). Reproduced by permission of the National Research Council of Canada.

	Feature	Units	Symbol
	Specific gravity, or Density	non-dimensional g/m^3, or kg/m^3	G
	Free water content	% by weight, or see table of descriptions	W
	Impurities	% by weight	J
	Grain shape	see table of descriptions (Table 7.2)	F
	Grain size	millimetres	D
Structure	Strength represented by: Compressive yield strength Tensile strength Shear strength at zero normal strength or Hardness	g/cm^2 g/cm^2 g/cm^2 according to instrument	Kp Kz Ks R
	Snow temperature	degress Centigrade	T

Table 7.2

DESCRIPTIONS AND SYMBOLS FOR CLASSIFICATION OF GRAIN SHAPE (National Research Council, 1954). Reproduced by permission of the National Research Council of Canada.

	Description	Symbol	Graphic Symbol
Class "a"	Class "a" refers to freshly deposited snow composed of crystals, or parts of broken crystals of types F1 to F7 (classification of falling snow). Snow which has lost its crystalline character while falling to earth, and graupel, ice pellets and hail do not belong to this class. Class "a" snow is generally very soft.	a	+ + +
Class "b"	This class refers to snow during its initial stage of settling. It has not reached the very fine grain-size condition which is generally regarded as the conclusion of the initial stage of transformation. Although it has lost a great deal of its crystalline character, some crystalline features can be observed. Class "b" snow is usually fairly soft.	b	> > >
Class "c"	When snow is transformed by melting, or melting followed by freezing, it completely loses all crystalline features and its grains become irregular and more or less rounded in form. This is Class "c" snow. It has no sparkle effect even in bright sunlight and can be readily recognized by its dull appearance. It is usually fairly soft when wet, but can be very hard when frozen. Class "c" snow may have any size of grains from very fine to very coarse.	c	: : :
Class "d"	At temperatures well below freezing and without any apparent melting, snow is transformed in Class "d" by the process of sublimation which produces irregular grains with flat facets. These facets give the snow a distinct sparkle effect in bright sunlight. In the Arctic, where temperatures are low and persistent winds accelerate the sublimation practically all of the settled snow is Class "d" and has almost as much sparkle as a deposit of F1 crystals. Class "d" snow is usually fairly hard.	d	□ □ □
Depth hoar	Depth hoar is characterized by its hollow cup-shaped crystals. These crystals are produced by a very low rate of sublimation during a long uninterrupted cold period and are most frequently found below a more or less impermeable crust in the lower part of the snowcover. The strength of a layer of depth hoar is very low.	e	Λ Λ Λ

Table 7.3
SUMMARY OF CLASSIFICATIONS AND SYMBOLS FOR MEASUREMENTS AND SURFACE CONDITIONS (National Research Council, 1954). Reproduced by permission of the National Research Council of Canada.

DEPOSITED SNOW

FEATURE	SYMBOL	SUBCLASSIFICATION				
		a	b	c	d	e
SPECIFIC GRAVITY	G					
FREE WATER (%)	W	dry	moist	wet	very wet	slush
GRAIN SHAPE	F	F1-F7 crystals	partly settled	rounded grains	grains with facets	depth hoar
GRAIN SIZE (mm)	D	<0.5	0.5 - 1	1 - 2	2 - 4	>4
COMPRESSIVE YIELD STRENGTH (g/cm²)	K	0 –10	10 –10²	10² –10³	10³ –10⁴	>10⁴
SNOW TEMPERATURE, T (°C)	ICE LAYER, i			IMPURITIES, J (%)		

SNOWCOVER MEASUREMENTS

	VERTICAL	⊥ to inclined surfaces	
CO-ORDINATE (cm)	H	M	INCLINATION OF SURFACE, N (degrees)
TOTAL DEPTH (cm)	HS	MS	WATER EQUIVALENT OF COVER, HW (mm)
DAILY NEW SNOWFALL (cm)	HN	MN	SNOWCOVERED AREA / TOTAL AREA Q (tenths)
			AGE OF DEPOSIT, A (h, days, etc.)

SNOW SURFACE CONDITIONS

SURFACE DEPOSIT	SURFACE HOAR	SOFT RIME	HARD RIME	GLAZED FROST
SYMBOL	V1	V2	V3	V4
GRAPHIC SYMBOL	⌊_⌋	∨	▼	∾

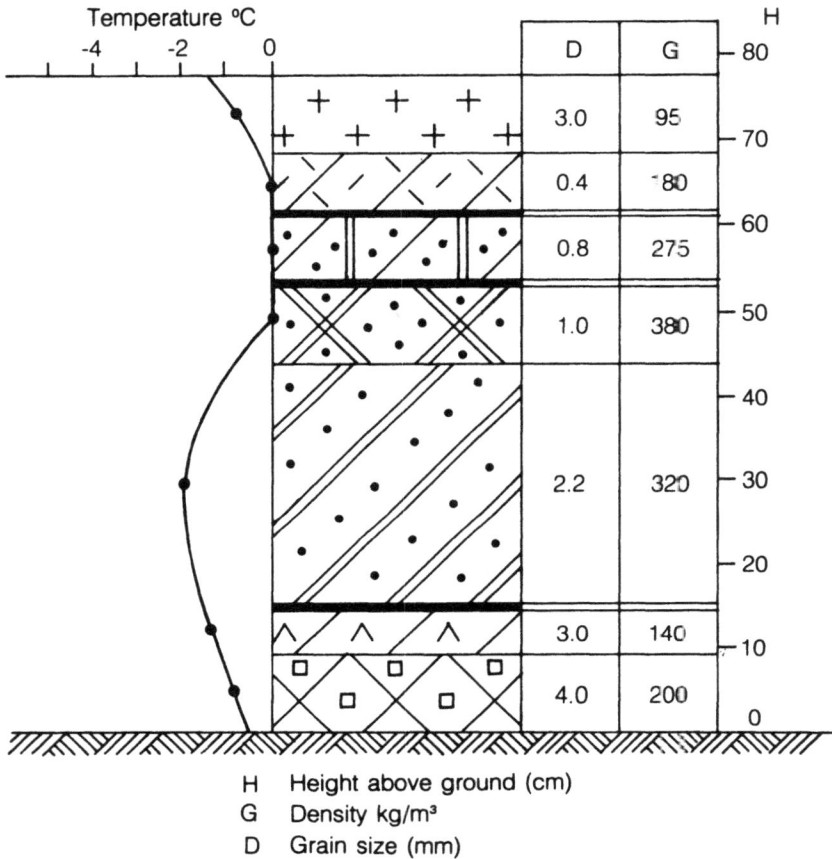

Fig. 7.5 Illustration of the use of symbols to describe a snowpack (National Research Council, 1954). Reproduced by permission of the National Research Council of Canada.

Table 7.4
CLASSIFICATION OF PROCESSES OF METAMORPHISM (Shumskii, 1964).

Energy Source	General characteristics	Types of Metamorphism		
		Recrystallization and dislocation metamorphism (in the solid phase)	Regelation metamorphism (through the liquid phase)	Sublimation metamorphism (through the vapor phase with surface migration)
Internal energy of the rock — Free surface energy of the crystals	Rounding of the crystals (approximation of equilibrium form)	Recrystallization rounding	Regelation rounding	Sublimation rounding
	Collective pere-crystallization	Collective recrystallization	Collective regelation pere-crystallization	Collective sublimation pere-crystallization
Free internal energy of the stressed state	Paratectonic pere-crystallization	Migratory recrystallization / Polygonization (recrystallization diaphthoresis)	Paratectonic regelation pere-crystallization (according to Riecke's principle)	Paratectonic sublimation pere-crystallization (according to Riecke's principle)
External energy	Dislocation metamorphism	Cataclasis and Mylonitization	Regelation dynamometamorphism of friction and pressure	Sublimation diaphthoresis (growth of depth hoar)

Dynamic metamorphism

Table 7.5

CLASSIFICATION OF SNOW BY METAMORPHIC STATE (Sommerfeld and LaChapelle, 1970; reprinted from the *Journal of Glaciology*).

I   **Unmetamorphosed snow**
   A. No wind action: Many fragile snow crystal forms easily distinguishable; little difference from snow in air. *We recommend that subclassification in accordance with the system of Magono and Lee (1966).*
   B. Wind blown: Shards and splinters of original snow crystals; parts of original forms may be recognizable but whole forms very uncommon. *Magono and Lee's (1966) classifications.*
   C. Surface hoar.

II **Equi-temperature metamorphism**
   A. Decreasing grain size
      1. Beginning: original snow crystal shapes recognizable, but corners show rounding and fine structure has disappeared.
      2. Advanced: very few indistinct plates or fragments recognizable; grains show distinct rounding.
   B. Increasing grain size
      1. Beginning: no original snow crystal shapes recognizable; grains show a distinct equi-dimensional tendency; a few, indistinct facets may be visible.
      2. Advanced: larger equi-dimensional grains present; a strong tendency toward uniform grain size; faceting generally absent.

III **Temperature-gradient metamorphism**
   A. Early: the result of a strong thermal gradient on new-fallen snow; associated with the first snowfalls of the season.
      1. Beginning: angular or faceted grains common; stepped surfaces not visible.
      2. Partial: medium-sized angular grains predominate; poorly formed steps visible.
      3. Advanced: medium to large angular grains predominate; well-developed facets and steps visible; a few filled or hollow cups may be found.
   B. Late: The result of a strong thermal gradient acting on snow in the later stages of equi-temperature metamorphism.
      1. Beginning: medium to large angular or faceted grains predominate; some stepped surfaces visible.
      2. Advanced: large grains predominate; many very fragile hollow cups or lattice grains; very deep steps.

IV **Firnication**
   A. Melt-freeze metamorphism
      1. Limited: single thaw-freeze cycle and limited gain in ice density.
      2. Advanced: repeated thaw-freeze cycles and appreciable gain in density and mechanical strength; density range 600 to 700 kg/m^3.
   B. Pressure metamorphism
      1. Beginning: grains deformed and rearranged by pressure; density range 700 to 800 kg/m^3.
      2. Advanced: pore spaces become non-communicating; permeability zero; density range 800 to 830 kg/m^3.

There is no doubt that this last classification is based on a more complete physical knowledge than the earlier attempts. Some of the speculative explanations of the older works are in error and perhaps the same will be true for assertions made in more recent proposals concerning matters that are presently either unstudied or disputed. Certainly the definitive classification system for a snowcover has not been constructed; it is likely that current interest in contamination of the snowpack and in remote-sensing techniques will lead to classifications which include trace contaminants or electromagnetic properties.

However, in spite of subsequent changes of approach, the older systems retain their value and are certainly worth the attention of the reader, since each provides descriptions of the different physical features of snow, and is accompanied by a glossary. These glossaries differ but together provide an extensive vocabulary which is used in the scientific discussion of snow properties.

PROPERTIES OF DRY SNOW

Metamorphism and Bulk Properties

The discussion in the preceding sections has emphasized the various processes of metamorphism and the classifications based upon them. These processes control the bulk properties of snow. Thermal properties that depend only on density (specific heat, latent heat) are well defined. However, those that depend on conductivity or permeability of the snowpack are affected by sintering, particle size, ice layers and depth hoar. Mechanical properties are affected mainly by sintering; electrical and optical properties by sintering and particle size. Unfortunately, measurements of bulk properties are not usually accompanied by a careful documentation of the snow condition so that the disparities between many of the published measurements cannot be explained.

The compilation of data in this section is selective and gives an indication of the range of variability of bulk properties. Most of these data result from measurements of homogeneous snow on a scale of tens of centimetres. At this macroscopic scale, equations are sought linking the response of the snow to an imposed stress (for example, compression to pressure). In recent years some attention has been given to the problem of relating these phenomenological equations to the crystalline structure of the snow. Development of macroscopic theory from microscopic theory is called macroscopisation. It is important both for understanding the causes of the bulk phenomena and for understanding certain phenomenological laws (for example, Darcy's Law defining the flow of fluids through porous media) when applied to a porous medium such as snow. There are two approaches to this problem. One is to develop a general theory of macroscopisation (e.g., Coudert, 1973), an approach used by Male and Norum (1971) who incorporated the theory of Mokadam (1961) to apply irreversible thermodynamics to the study of a wet snowpack. The other approach relies on a conceptual model of the microstructure of the snowpack that is simple enough to be modelled mathematically (de Quervain, 1973).

Thermal Properties

The specific and latent heats of snow are the simplest thermal properties to determine since the contributions from air and water vapor can be discounted; each property is simply the product of the snow density and the corresponding property for ice. The temperature dependence of the specific heat of ice given by Dorsey (1940) is:

$$C = 2.115 + 0.00779 \ T, \qquad\qquad 7.16$$

where: $\quad$ C = specific heat (kJ/(kg $\cdot$ °C), and
$\qquad\qquad\quad$ T = temperature (°C).

The latent heat of melting of ice at 0°C and standard atmospheric pressure is 333.66 kJ/kg (Dorsey, 1940). The latent heat of sublimation is temperature dependent as shown in Table 7.6.

For one-dimensional, steady-state heat flow by conduction in a solid the thermal conductivity is the proportionality constant of the Fourier equation:

$$q = -k \ dT/dz, \qquad\qquad 7.17$$

where: $\quad$ q = heat flux, and
$\qquad\qquad\quad$ dT/dz = the temperature gradient.

Table 7.6
**LATENT HEAT (ENTHALPY) OF SUBLIMATION
FOR ICE (Keenan et al, 1969). Copyright © 1969 by John
Wiley and Sons, Inc. Reproduced by permission of John
Wiley and Sons, Inc.**

0	2834.8
-10	2837.0
-20	2838.4
-30	2839.0
-40	2838.9

The thermal conductivity of snow is a more complex property than specific heat because its magnitude depends on such factors as the density, temperature and the microstructure of the snow. The thermal conductivity of ice varies inversely with temperature by $\sim 0.17\%/°C$; the same may be expected for snow. A temperature gradient could induce a transfer of vapor and the subsequent release of the latent heat of vaporization, thereby changing the thermal conductivity value. Mellor (1977) points out that in non-aspirated dry snow the heat transfer process involves: conduction of heat in the network of ice grains and bonds, conduction across air spaces or pores, convection and radiation across pores (probably negligible) and vapor diffusion through the pores. Yoshida and others (1955) suggest that the movement of water vapor contributes 37% to the apparent thermal conductivity of snow at a density of 100 kg/m^3, but only 8% at a density of 500 kg/m^3. Because of the complexity of the heat transfer processes, the thermal conductivity of snow is generally taken to be an "apparent" or "effective" conductivity k_e to embrace all the heat transfer processes. It should be recognized that "measured" values of k_e reported in the literature depend on the method of measurement (transient and steady-state) in addition to the factors discussed. Considerable scatter can be expected in the data, e.g., Fig. 7.6, plotted from data compiled by Mellor (1977), shows the variability in the relationship between k_e and snow density as reported by different scientists. Using these data, an approximate relationship between k_e of the snowcover and density can be developed which is useful in engineering design (see Fig. 7.7). Considerable care should be exercised in applying this relationship, since it is difficult to determine whether the thermal regime encountered in the field problem will correspond to the regime under which the curve was developed.

The degree of surface packing (for example, hardness) also affects the flow of heat through snow, probably because a surface crust of low air permeability inhibits ventilation in the upper snow layer. Even though ventilation is not heat conduction, it is conveniently included in this discussion since the two processes interact. Ventilation may result from natural convection or from variations in atmospheric pressure which force air through the snowpack. The influence of ventilation on vapor transport was studied by Yen (1962) who showed that it can change k_e by a few percent. However, to apply his results requires data about the air flow which are not normally available. Shallow snowcovers often develop breathing holes early in the winter indicating that air movement is not uniform. This further complicates the application of Yen's results to a shallow snowcover.

It should be noted that the thermal conductivity of snow, even when dense, is very low compared to that of ice or liquid water; therefore snow is a good insulator. This is an important factor affecting heat loss from buildings and the rate of freezing of lake and river ice.

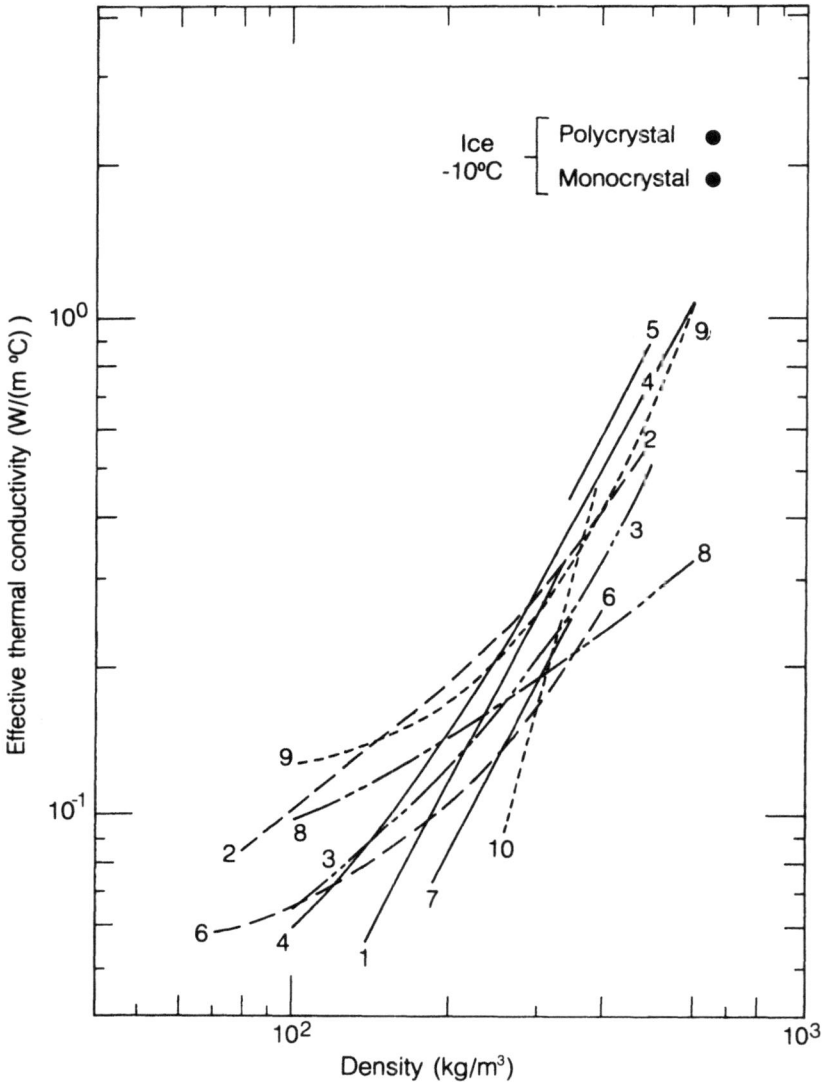

Fig. 7.6 Effective thermal conductivity of snow as a function of density. Data from 1 Abels (1892); 2 Jansson (1901); 3 Van Dusen (1929); 4 Devaux (1933); 5 Kondrat'eva (1945); 6 Yosida and others (1955); 7 Bracht (1949); 8 Sulakvelidze (1959) (-20°C); 9 Pitman and Zuckerman (1967) (-5°C); 10 Jaafar and Pigot (1970). The data for ice is from Glen (1974). (Adapted from Mellor, 1977).

Thermal diffusivity κ is, strictly speaking, a derived property being defined by the expression;

$$\kappa = k/\rho C, \qquad\qquad 7.18$$

where ρ = density, and
                      C = heat capacity.

However, many values of conductivity, k, are actually derived from a determination of diffusivity. Values of κ for snow lie between 0.0025 and

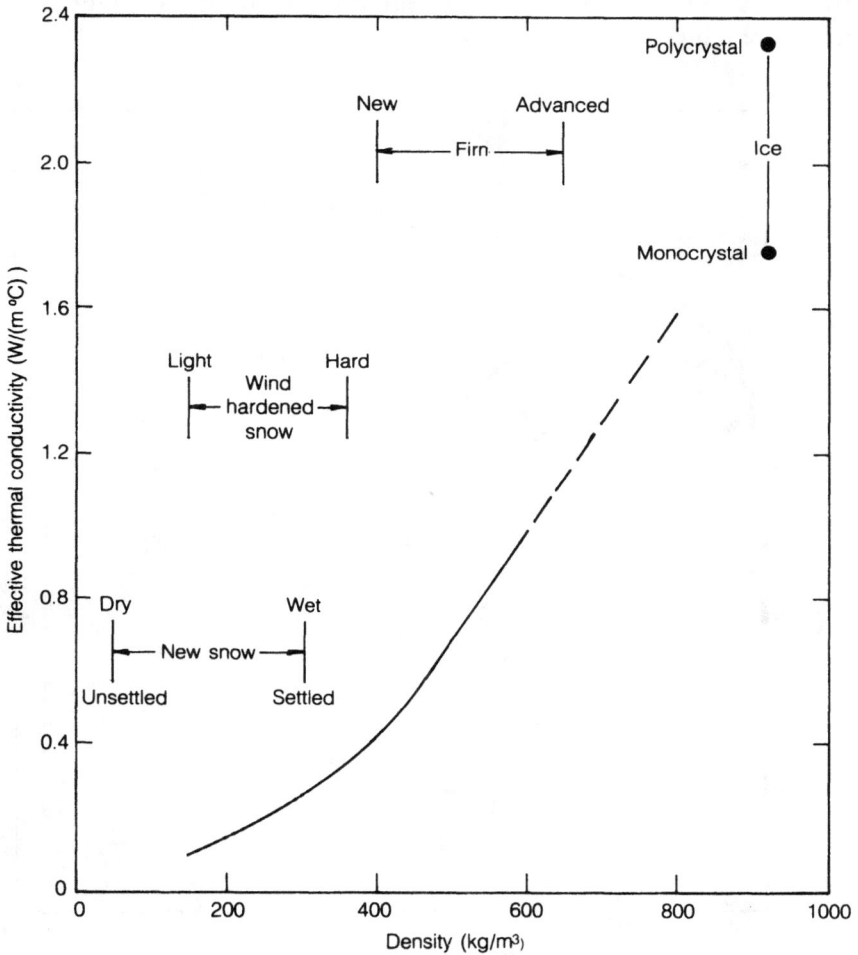

Fig. 7.7 Approximate relation between effective thermal conductivity of snow and ice and density.

0.005 cm^2/s (Dorsey, 1940, p. 483). The thermal diffusivity is important because it governs the speed and attenuation of temperature waves propagating into the snow from the surface. Figure 7.8 illustrates the attenuation with depth of a diurnal surface temperature wave which is assumed to follow a cosine function at different depths within a snowcover assuming a constant κ. The temperature regime within a layered snowcover where κ is not constant (anisotropic condition) can be obtained by solving the diffusion equation (see Carslaw and Jaeger, 1959). However, since the surface temperature of a snowcover cannot exceed 0° C (because of the phase change) it is more difficult to solve the diffusion equation for melting snow.

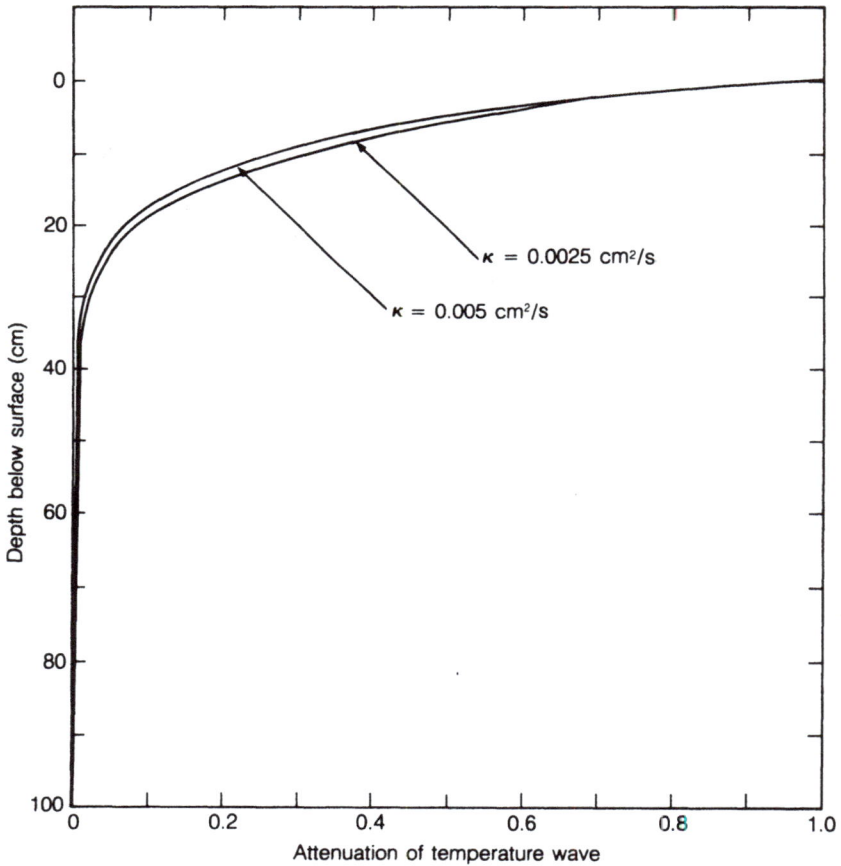

Fig. 7.8 Attenuation of the amplitude of a diurnal temperature wave propagating into snow from the upper surface.

Mechanical Properties

Snow may be deformed elastically when subjected to a small load applied for a short period of time. Under these conditions the strains are small enough not to disrupt the grain structure and are recoverable once the stress is removed. Snow also deforms continuously and permanently if a sustained load is applied; this is referred to as creep or viscous or plastic flow. Strictly speaking, plastic flow requires a threshold stress to be reached before flow can start. However, for snow this stress is so small that it cannot be measured and so snow is referred to as a visco-plastic material.

The behaviour of snow under stress is better understood by comparing it to a mechanical system of springs and dashpots known as a Burger's body (see Fig. 7.9). When such a system is compressed by applying a load at point A or B there is an immediate elastic response and spring (1) compresses. After this, creep begins and is governed by dashpot (4), which slides until stopped by the compression of spring (3). Throughout this entire period dashpot (2) has been moving slowly and, during the final stages of compression, governs the creep since springs (1) and (3) and dashpot (4) are immobile. The elastic constants of the springs and the time constants of the dashpots are functions of the temperature, density, metamorphic state and the stress/strain history of the snow. When stresses are high enough failure occurs, presenting even greater difficulties to the attempts at describing the snow's behaviour.

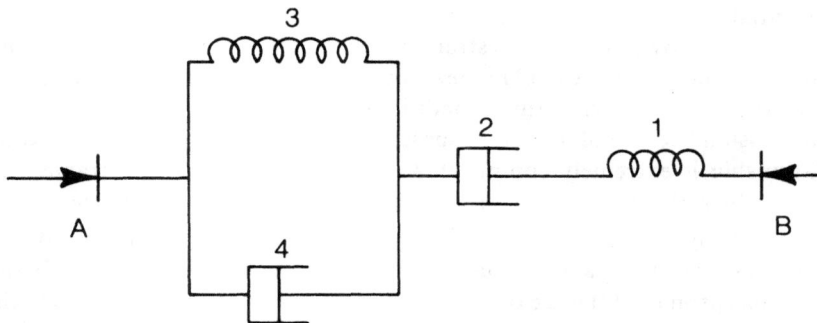

Fig. 7.9 Burger's body model of visco-elastic behaviour of snow. Compressive forces are applied at A and B.

Mellor (1975) states "there is no material of engineering significance that displays the bewildering complexities of snow". His opinion is confirmed by the papers presented at an international conference devoted solely to snow mechanics held in Grindelwald, Switzerland in 1974. Therefore the following account only summarizes the most important properties; much remains to be learned.

The best measurements of elastic properties have been made by subjecting samples of snow to vibrations at acoustic or ultra-sonic frequencies so that permanent deformation does not occur. Figure 7.10 (from Mellor, 1975) shows data for dry snow that has undergone appreciable sintering and is well bonded. The influence of temperature on Young's Modulus of snow E (the ratio of the longitudinal stress to longitudinal strain; or the internal force per unit cross sectional area divided by the change in length per unit length) is shown in the inset to the figure. It should be noted that E is plotted on a logarithmic scale since there is a factor of nearly 10^5 between its smallest and largest measured values. Granular snow of low cohesion or snow without appreciable sintering has a much lower value of E. Nakaya (1961) shows that E increases with time during sintering, e.g., see Fig. 7.11. Although sintering affects all the physical properties of snow, its most important effect is on the mechanical properties.

Under compression, the stress/strain rate of snow can vary by six or seven orders of magnitude, partly because of the effect of compression itself on the strain rate; this can be minimized by making measurements over small ranges of strain. However, since the strain history of a sample is also important, such data have limited use and a graph of all the data shows much scatter.

Because of these difficulties, the strain is resolved into two components in the theory of elasticity (see Westergaard, 1964): the deviatoric (constant volume) and volumetric strains. The former is the pure shear strain, defined as the shear angle ϕ (rad), measured under conditions of pure shear stress (the tangential force per unit area). Thus the problem of compression is avoided. Figure 7.12 shows such stress/strain relationships interpolated for various densities. The data are useful in the study of snow flowing down a steep slope. Since shear flow under natural conditions is normally slow, the short-term compression because of its weight has less effect. The volumetric stress/strain relationship is extremely complicated and a suitable analytical framework within which these measurements can be made is lacking at present.

A relatively simple situation is the uniaxial strain occurring naturally in a horizontal dry snowpack because of the weight of the overburden. Bader (1953) has promoted the use of the term "compactive viscosity" defined as the ratio of the overburden pressure to the vertical strain rate for this situation. This quantity, being shear free, is almost linear with density for pressures less than 80 kPa exerted over a short period. Figure 7.13 shows, however, that even under such conditions the data has considerable dispersion.

The failure limits of dry, coherent snow under tension or compression, as measured by the fracture or collapse of a test sample are shown in Fig. 7.14 as a function of density. Similar data for snow under shear stress are shown in Fig. 7.15. Shear failure is important for the development of avalanches, but for slab avalanches (see Ch. 11) the bulk shear properties of the snowpack are

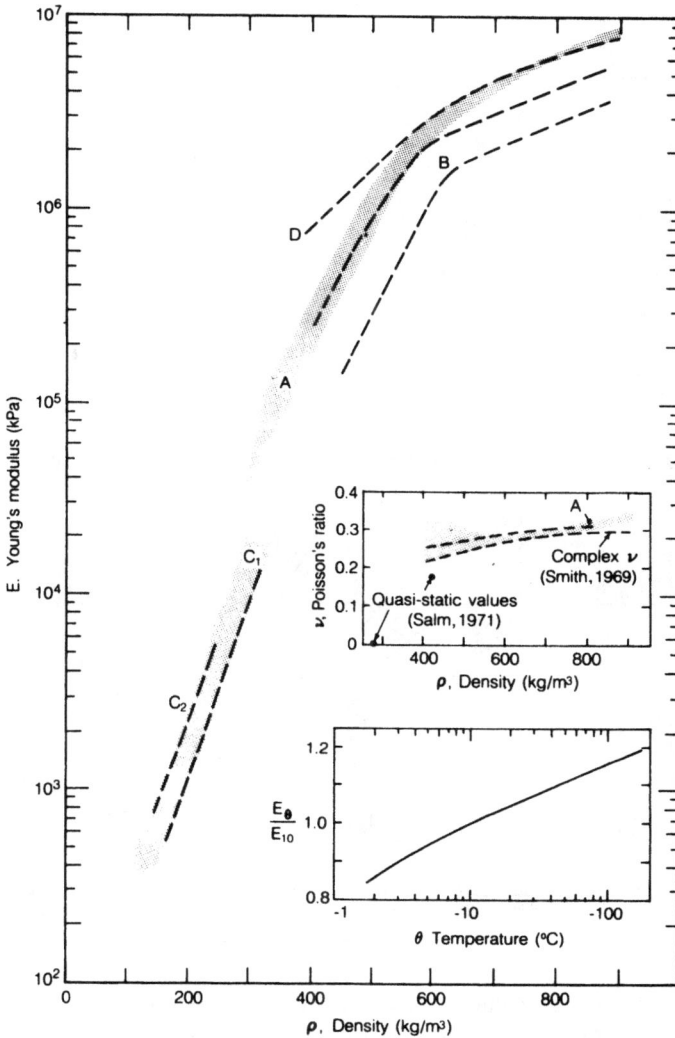

Fig. 7.10 Young's modulus for sintered snow as a function of density (Mellor, 1975). A: pulse propagation or flexural vibration at high frequencies, -10 to -25°C (Bentley *et al.*, 1957; Nakaya, 1959a, b; Lee, 1961; Crary et al, 1962; Ramseier, 1963; Smith, 1965). B. Uniaxial compression, strain rate approximately 3×10^{-3} to 2×10^{-2} sec^{-1}, -25°C (Kovacs et al., 1969). C_1: Uniaxial compression and tension strain rate approximately 8×10^{-6} to 4×10^{-4} sec^{-1}, -12°C to -25°C. C_2: Static creep test, -6.5 to -19°C (Kojima, 1954). D: Complex modulus, 10^3 Hz, -14°C (N. Smith, 1969).

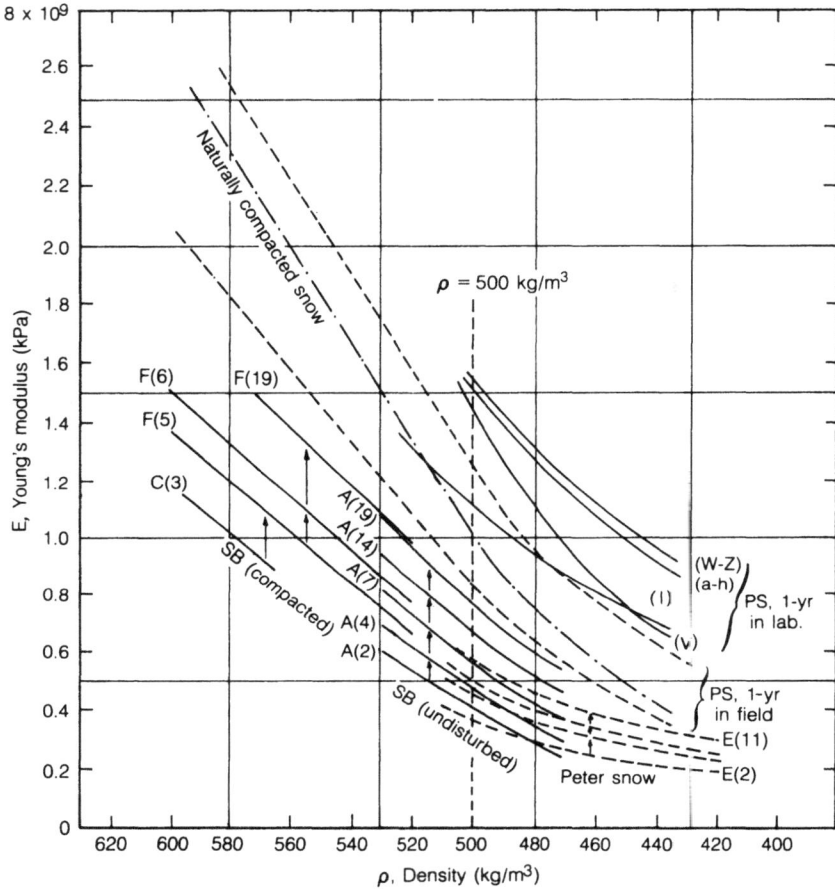

Fig. 7.11 Young's modulus of snow as a function of density. Figures in parentheses are days after snow was passed through a snow blower (from Nakaya, 1961). The letters represent different series of samples according to the following arrangement:

PROCESSED	UNDISTURBED	COMPACTED
SB: new deposit	A,D	C,F
PS: new deposit	E,P,I,N	-----
PS: 1 year in the	V,W-Z	-----
laboratory	a-h,l	

Note -SB signifies snow which has been passed through a Snowblast miller while PS signifies the use of a Peter miller.

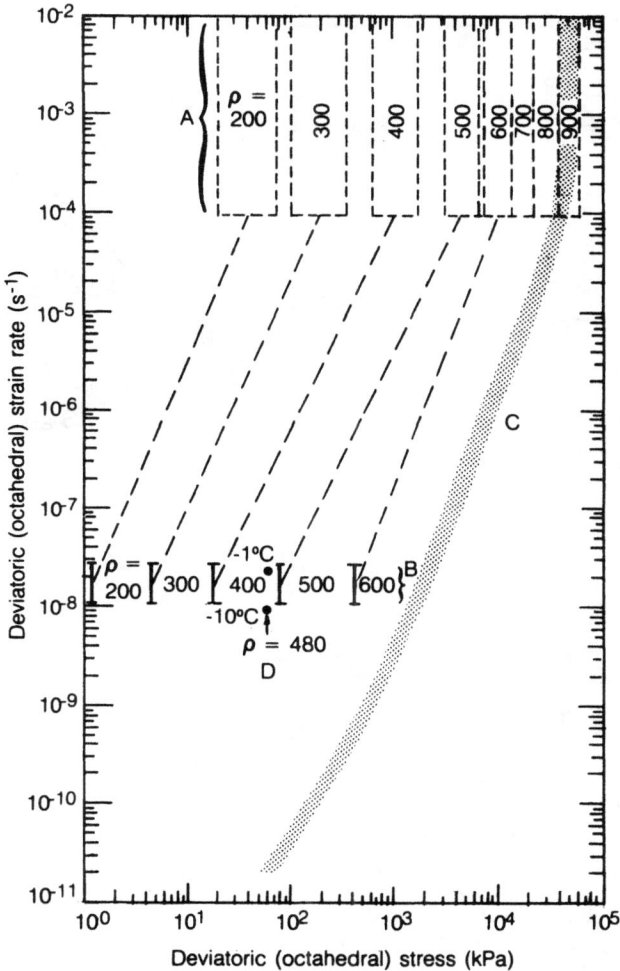

Fig. 7.12 Deviatoric stress/strain rate relationships deduced indirectly from various data sources (Mellor 1975). The slanting dashed lines connect data for corresponding densities and may be used for interpolation. A: snow, $\simeq 0°$C (data from Bucher and Roch, 1948; Butkovich, 1956; Smith, 1965; Mellor and Smith, 1966; Kovacs *et al.*, 1969). B: snow, $\simeq 0°$C to $-7°$C (calculated and extrapolated from data by Haefeli, 1939; Martinelli, 1960; Frutiger and Martinelli, 1966; Judson, as communicated to Mellor). C: ice, $\simeq 900$ kg/m³, -2 to $-10°$C (data from Mellor and Smith, 1967; Mellor and Testa, 1969; Hawkes and Mellor, 1972). D: snow, 480 kg/m³, -1 to $-10°$C (Mellor and Smith, 1967).

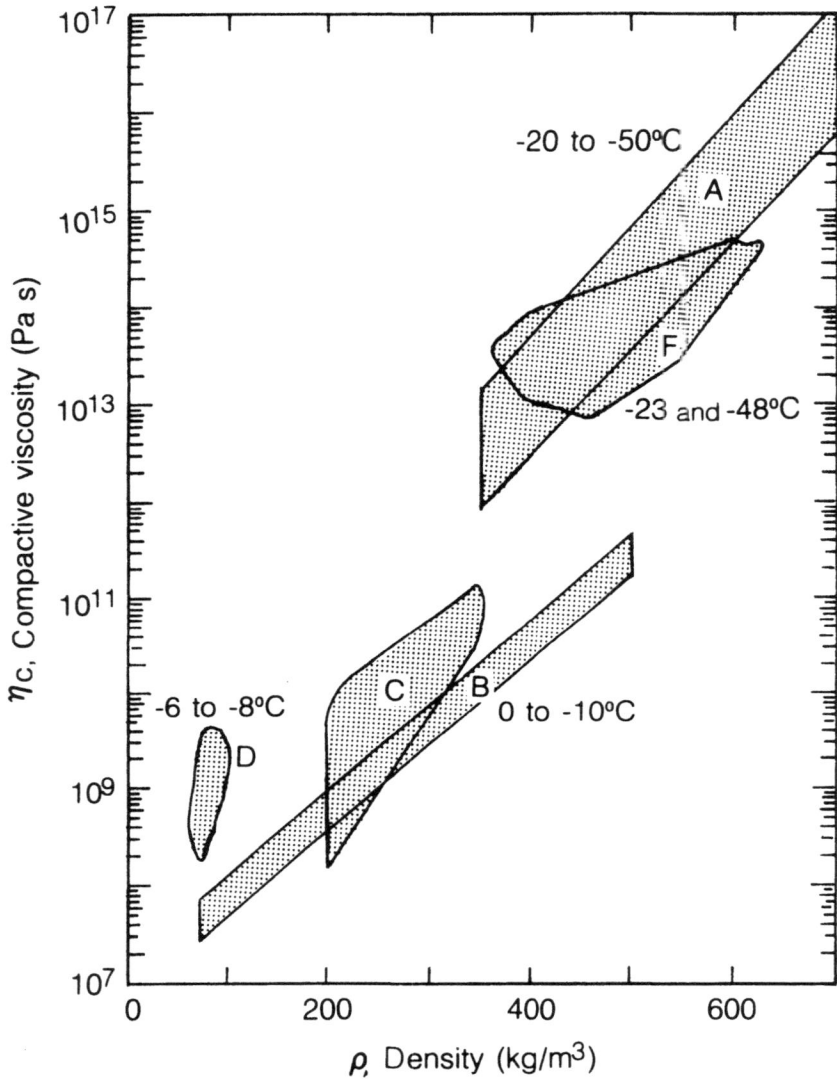

Fig. 7.13 Compactive viscosity of snow as a function of density for various snow types (Mellor, 1975).

A: Greenland and Antarctica, -20 to -50°C (Bader, 1960, 1962). B: Seasonal snow, Japan, 0- to -10°C (Kojima, 1967). C: Alps and Rocky Mountains (Keeler, 1969). D: Uniaxial-strain creep tests, -6 to -8°C (Keeler, 1969). F: Uniaxial strain creep tests, -23 and -48°C (Mellor and Hendrickson, 1965).

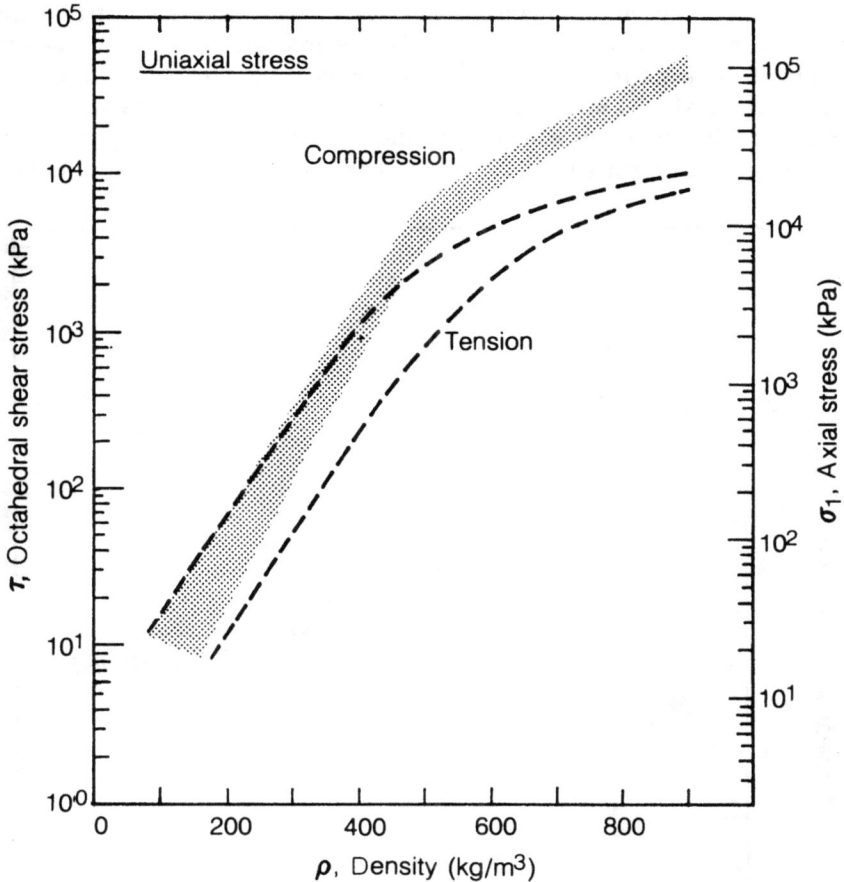

Fig. 7.14 Failure limits for uniaxial compression and tension of dry, coherent snow under rapid loading in uniaxial stress states (Mellor, 1975). (Data from Haefeli, 1939; Bucher, 1948; Butkovich, 1956; Ramseier, 1963; Smith, 1963; Smith, 1965; Mellor and Smith, 1967; Keeler and Weeks, 1967; Keeler, 1969, and Kovacs et al., 1969).

usually less important than those of thin layers that have very low shear resistances (Smith and Curtis, 1975), such as layers of depth hoar, ice or wet snow. In these layers the grain boundaries lose their strength. The actual release of a slab avalanche is more dependent on the tensile failure along the upper boundary; before the release, the stress concentration at the end of a propagating crack is the important factor, rather than the stress in the continuous snow. Perla (1975) suggests that for slab avalanches, the shear and tension failures occur simultaneously at the upper (crown) wall; then the shear fracture front advances across the avalanche area with the tension fracture. An increase in shear stress results from the tensile failure, which causes the shear failure also to propagate down the slope. Lang et al. (1974) support Perla's explanation, but add that the slide of a snow layer also gives rise to buckling which further weakens the resistance of the snow to shear failure.

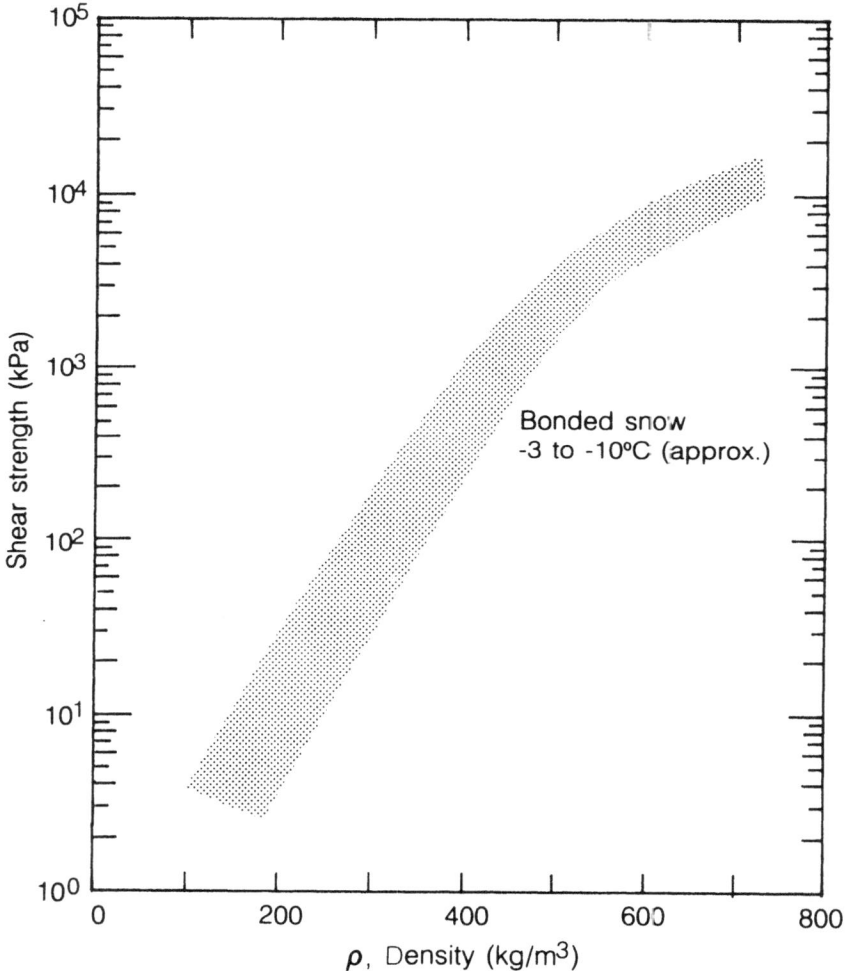

Fig. 7.15 Failure limits of dry coherent snow under shear stress (Mellor, 1975). (Data from Haefeli, 1939; Butkovich, 1956; Ballard et al., 1965; Keeler and Weeks, 1967, and Keeler, 1969). Also avalanche release data from various sources summarized by Keeler, 1969. Some values from in situ shear vanes were rejected.

Frictional Properties

The frictional properties of snow have considerable practical importance for designing and using snow-covered airfields and for hauling materials with sleds. They are also of interest for skiing (see Ch. 19).

The sliding or kinetic friction of snow is extremely low. Bowden (1955) showed that this is due to the presence of a liquid film as revealed by measurements of the electrical conductivity of the snow in contact with a runner. Near 0°C this film is probably produced by pressure at the points of contact between the snow grains and the runner, while at lower temperatures it is likely generated by frictional heating. This conjecture is supported by the fact that the coefficient of friction increases at lower temperatures, which is

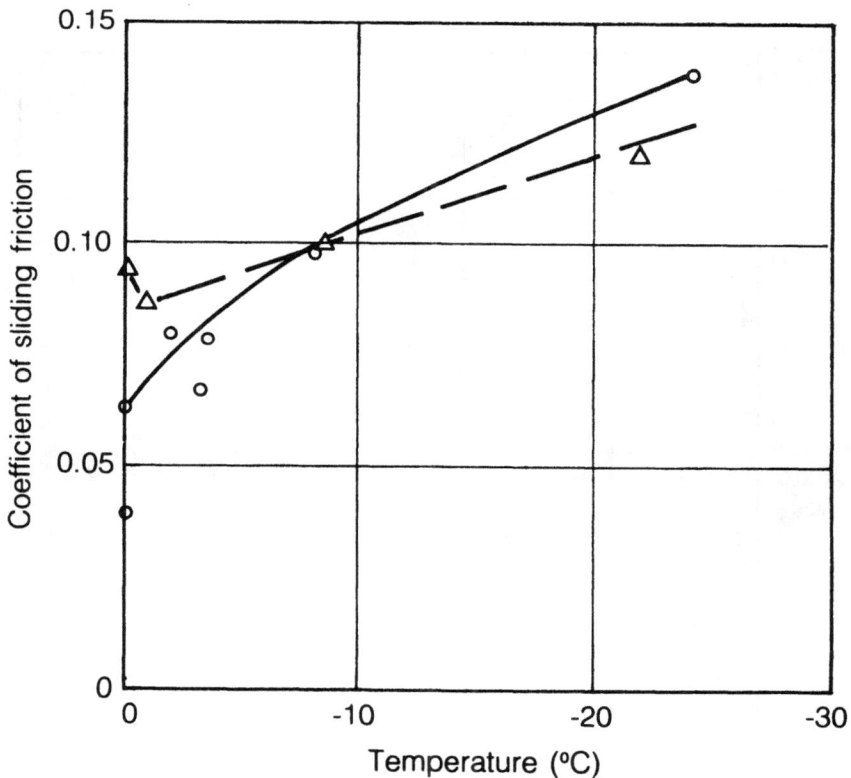

Fig. 7.16 Coefficient of sliding friction vs. temperature of new snow. Circles and solid line are for steel runners: total surface 60 cm², load $\simeq$ 20 kg. The triangles and broken line are for paraffin-covered wooden runners: total surface = 150 cm²; load $\simeq$ 5 kg. Sliding speed = 2.5 m/s (redrawn from Eriksson, 1949).

expected since more work has to be done at such temperatures to provide the
energy to maintain the contact surface temperature at 0°C. This effect is
shown by the data in Fig. 7.16 (Eriksson, 1949): for steel runners the increase
in sliding friction with decreasing temperature is greater than that for wood
runners. This result is attributed to the higher thermal conductivity of the
metal which tends to remove heat more rapidly from the contact surface.
Close to 0°C the coefficient of sliding friction decreases as the load increases
(see Fig. 7.17) because of increased pressure melting. Eriksson also found that
below -3°C the coefficient of sliding friction was less for rough than for
smooth steel surfaces (see Fig. 7.18).

Table 7.7 lists the lower and upper limits of the sliding friction coefficient
for a number of common materials on snow. The lower values are for dry
snow at 0°C. For wetable runner materials the friction increases rapidly as

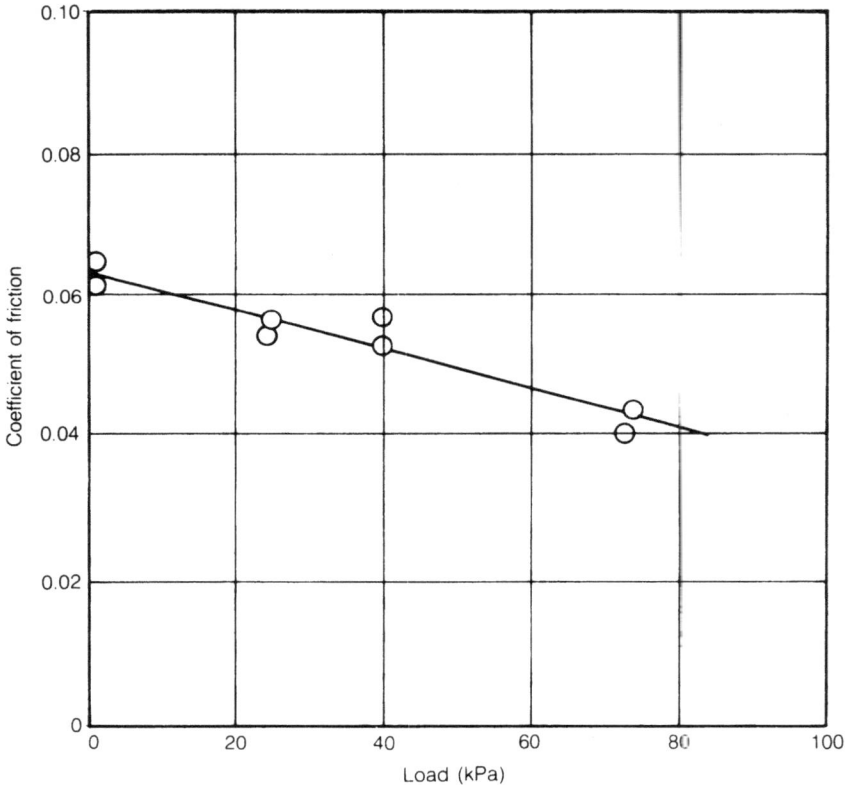

Fig. 7.17 Coefficient of friction vs. load on new snow. Steel runners. Temp. = ± 0°C.
Sliding speed ≅ 2.5 m/s (Eriksson, 1949).

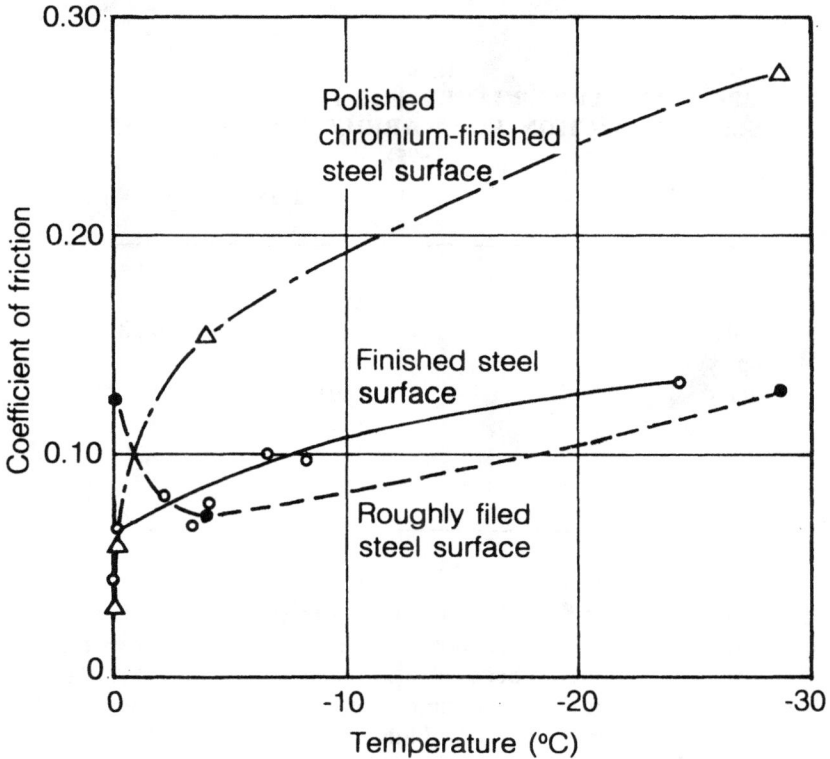

Fig. 7.18 Effect of roughness on sliding friction of steel as a function of temperature. Note that the coefficient of friction is *lower* for the rougher surfaces. (Eriksson, 1949).

soon as the snow contains liquid water, and is probably due to work expended against surface tension forces as the contact meniscii are dragged over the sliding surface.

Starting friction must be separated into two types: static friction and adhesion friction. Static friction is defined in the conventional way as a tangential force exerted by one solid surface on another, both being at rest. However, it can only be measured when adhesion has been eliminated. At temperatures of 0°C static friction is of the same order as the sliding friction, while at low temperatures it exceeds the kinetic friction because no liquid layer is present to lubricate the surface of contact. Adhesion, which usually masks static friction, is due to refreezing of the liquid layer as soon as the sliding surface comes to rest. The pressure of the sliding surface promotes rapid sintering in the snow immediately underneath, especially at temperatures near

Table 7.7
UPPER AND LOWER LIMITS FOR THE COEFFICIENTS OF
SLIDING FRICTION FOR VARIOUS MATERIALS (Mellor,
1975).

Material	Lower Value (dry snow)	Upper Value
poly tetraflorethelene	0.02	0.04
poly ethylene	0.02	0.05
poly vinyl chloride	0.03	0.08
epoxy resin	0.03	0.07
glass	0.015	0.02
aluminium	0.04	0.05
iron	0.015	0.20
wood	0.06	0.18
beeswax	0.04	0.28

$0°$ C, so that a layer of snow becomes frozen to the material. The adhesion friction therefore is determined by the force required to shear this layer.

Another type of cohesion between snow particles occurs in the absence of any imposed stress. It develops immediately, even during snowfall in calm conditions, and is the cause of the high angle of repose of the snow accumulating on obstacles. This angle depends on crystal shape and particularly on temperature (Kuroiwa et al., 1967); e.g., see Fig. 7.19 for dendritic snow. The size of the angle of repose depends on the shape and velocity of impact of the particles. This form of cohesion probably starts as a result of surface tension effects in a liquid like surface layer. Nakaya and Matsumoto (1954) and others have investigated this using small spheres of ice which were allowed to come into contact briefly. They found that the ice spheres could rotate freely relative to each other, even though the force of adhesion varied from 5 x 10^{-5}N at $-2°$ C to 1 x 10^{-5}N at $-15°$ C. This variation of the force of adhesion with temperature explains the dependence of the angle of repose. Of course, if the snow particles are left in contact long enough, sintering will occur.

Many of the rheological properties of snow are attributable to the fact that it is relatively close to its melting point. As a consequence, the mobility of the crystal defects is high and they can diffuse rapidly through the crystal lattice. This is a characteristic of ductile materials and is exploited, for example, in annealing and in the production of preferred crystal orientations in metals. Such processes are similar to many of those that occur in snow.

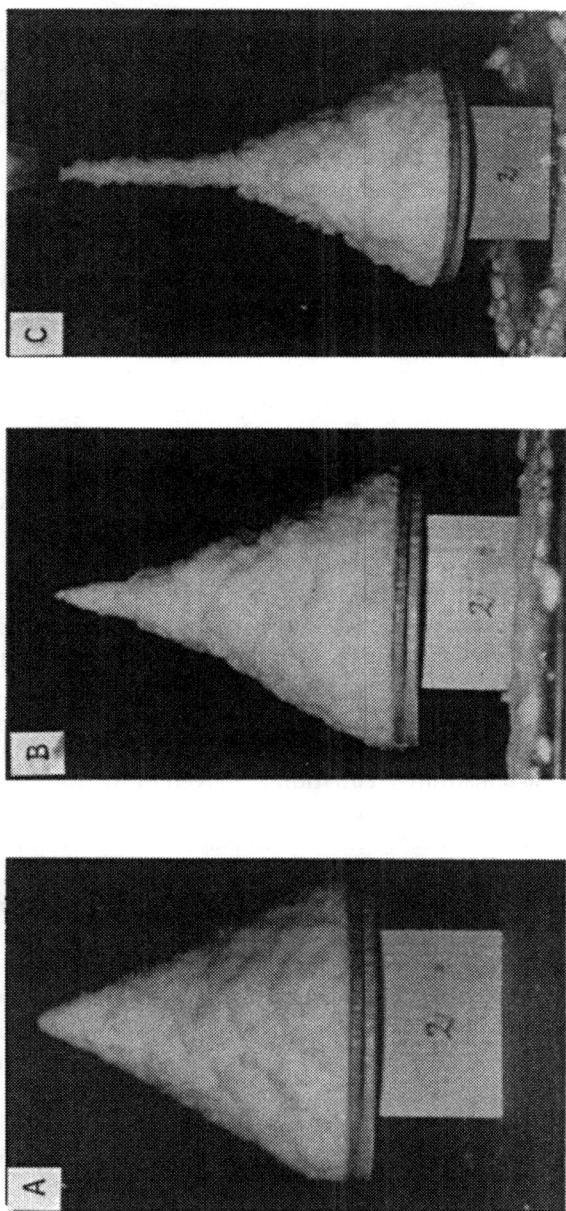

Fig. 7.19 Temperature dependence of angle of repose of dendritic snow crystals A: −35°C. B: −12°C. C: −4°C. (Kuroiwa et al., 1967).

Constitutive Equations for Snow and the Prediction of Failure

The dispersion in the published data relating to the various mechanical properties of snow is due partly to unspecified differences in its metamorphic state, but also to its complex visco-elastic properties. Recently much work has been done and considerable progress has been made towards formulating constitutive equations which describe the visco-elastic properties and relate the kinematic response to the imposed stresses.

These constitutive equations are part of the set of equations required to describe snow behaviour using continuum mechanics. The remaining equations are the kinematic relationships which describe the motion of the continuum, the continuity equation (conservation of mass), the momentum and angular momentum equations, and the energy equation. Zeigler (1975) reviews this approach to visco-elastic deformations. A constitutive equation may be formulated in many ways depending on its underlying assumptions.

Salm (1975) derived an equation for uniaxial compression (or tension) of snow based on a non-linear Burger's body model (see Fig. 7.9). He expressed the dissipation function for irreversible work done in the dashpots into terms up to the sixth degree in the stress tensor in order to fit calculated and measured curves relating the deformation rate to the stress. Salm's work did not, however, include the thermodynamic equations. By introducing continuum thermodynamics to the study of visco-elastic properties of snow Brown and Lang (1975) were able to include the effect of strain history into its behaviour under stress. They derived a constitutive equation containing seven memory functions. Once these are evaluated, the Helmholtz free energy and the dissipation could be calculated. More details of the mathematical development of the constitutive equation are given in Brown (1976a). Some theoretical results have been shown to agree very well with measured values. The stress curves for various sequences of constant strains computed for snow samples classified as Type III Al according to the Sommerfeld-LaChapelle classification system and having a density of 335 kg/m^3 were found to agree very well with the measured stresses. These experiments were continued to failure of the snow samples and therefore, are particularly interesting. At the start of a test the samples behaved elastically, that is, the free energy increased rapidly with little dissipation of energy. However, when the stress reached about 250 kPa, the energy dissipation increased, at first rapidly then approached a lower steady rate of increase. Samples strained at a rate of 0.0068/min or less had a high energy dissipation and fractured at a free energy of 70 J/m^3; those, 0.0135/min or greater, had low dissipation and fractured at only 35 J/m^3. Brown and Lang (1975) suggest this difference is due to modifications of the elastic and fracture properties by the dissipated energy. This interpretation is supported by experiments in which the snow samples were initially strained at a low rate until the dissipation mechanism was fully developed, then the strain was increased to cause fracture. Under these conditions, the free energy at fracture increased by an order of magnitude, that is, to about 700 J/m^3. In a third series of experiments, the snow samples were strained at a low initial rate and then the stress removed. After a period of relaxation the samples were stressed to fracture. These tests showed a progressive recovery of the snow toward its original state.

Brown (1976b) extended this analysis to develop snow failure criteria which depend on free energy and dissipation. Tension, compression and shear failure tests showed that the deviatoric stored energy was related to the volume stored energy for high strain rates. When the initial rates were low enough for dissipation to occur, the failure critera could be expressed in terms of the deviatoric and volumetric energies and the dissipation.

These studies mark a significant advance in understanding the mechanics of snow. They are supported by Bradley and St. Lawrence (1975) who measured the acoustic emission of snow under stress, and found that during the initial period of strain, acoustic emissions are low but increase with strain. They suggest that during the quiet period the crystals flow and grain boundaries deform plastically. These results correspond to those of Brown (1976b). According to Bradley and St. Lawrence (1975) the acoustic emissions are associated with fracture of grain boundaries which ultimately leads to fracture of the sample; they suggest that the time for fracture to occur after the onset of this process is related to the time it takes for grain-boundary fractures to become sufficiently numerous so as to form a continuous fracture. These authors also suggest that these sound emissions can give forewarning of avalanches. In laboratory tests the maximum time from the beginning of the acoustic emissions to failure was 42 min. Experiments with geophones installed in avalanche slopes, however, show greater delays. Both Sommerfeld (1977) and St. Lawrence and Williams (1976) have recorded strong bursts of low frequency (0-30 Hz) noise during the two-day period preceding an avalanche. Sommerfeld also showed that a seemingly potential avalanche that had not produced such emissions could not be released with explosives.

The work of Bradley and St. Lawrence (1975) further connects the acoustic properties of snow with its rheological memory. They found that acoustic emissions stopped when the stress was removed and did not recommence when the stress was reimposed until the stress exceeded the original stress. This behaviour, which they refer to as the Kaiser effect, agrees with results of the free energy/dissipation experiments of Brown (1976b).

Electrical, Electromagnetic and Optical Properties[1]

In describing its electrical properties, dry snow may be considered a heterogeneous system of air and ice, which does not, however, exhibit the polarization problems of ice in direct conduction. Measurements of the complex impedance Z of both snow and pure ice as a function of frequency indicate that they each behave electrically like the circuit shown in Fig. 7.20. For this circuit Z is given by

$$1/Z = 1/R_0 + 1/[R_1 + 1/(i\omega C_1)] + i\omega C_\infty, \qquad\qquad 7.19$$

where: R_0, R_1 represent resistors and their corresponding
                         resistances,

           C_1, C_∞ represent capacitors and their corresponding
                         capacitances,

           $i = \sqrt{(-1)}$, and

           ω = the angular velocity of the electrical signal.

At low frequencies the direct current resistance R_0 dominates. The d.c. conductivity (inverse of R_0) was measured by Kopp (1962) under various conditions; see Figs. 7.21 and 7.22 which show the marked effects of density,

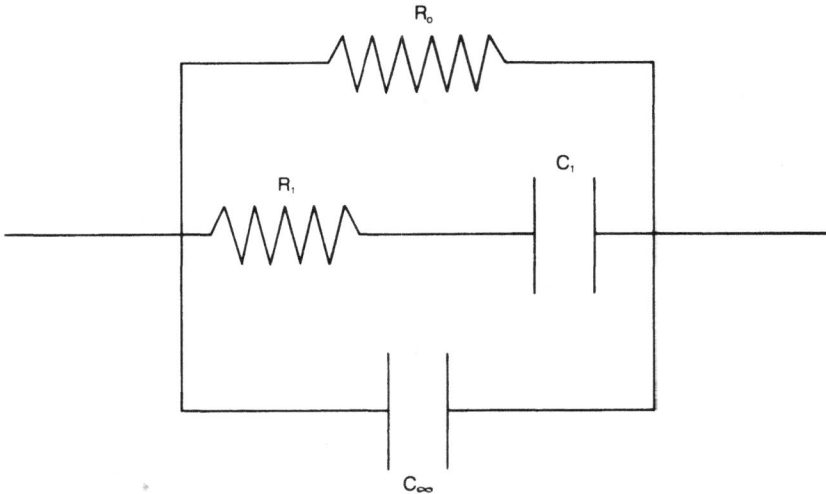

Fig. 7.20 Equivalent electrical circuit for ice and snow.

---

[1]   Throughout this section the properties are expressed in terms of frequency (f) or wavelength (λ) which are related by the expression f (Hz) $\cdot$ λ (m) = 2.9979 x 10^3 (m/s).

temperature and microstructure on electrical conductivity. The conductivity of a snow sample that had previously reached the melting point was found to be significantly lower than one that had not. This was attributed to trace quantities of contaminants having been washed out of the sample rather than to metamorphism. Kuroiwa (1954) gives values of $\sim 4 \times 10^{-4}$ mho/cm for snow containing 20 ppm of chloride ions.

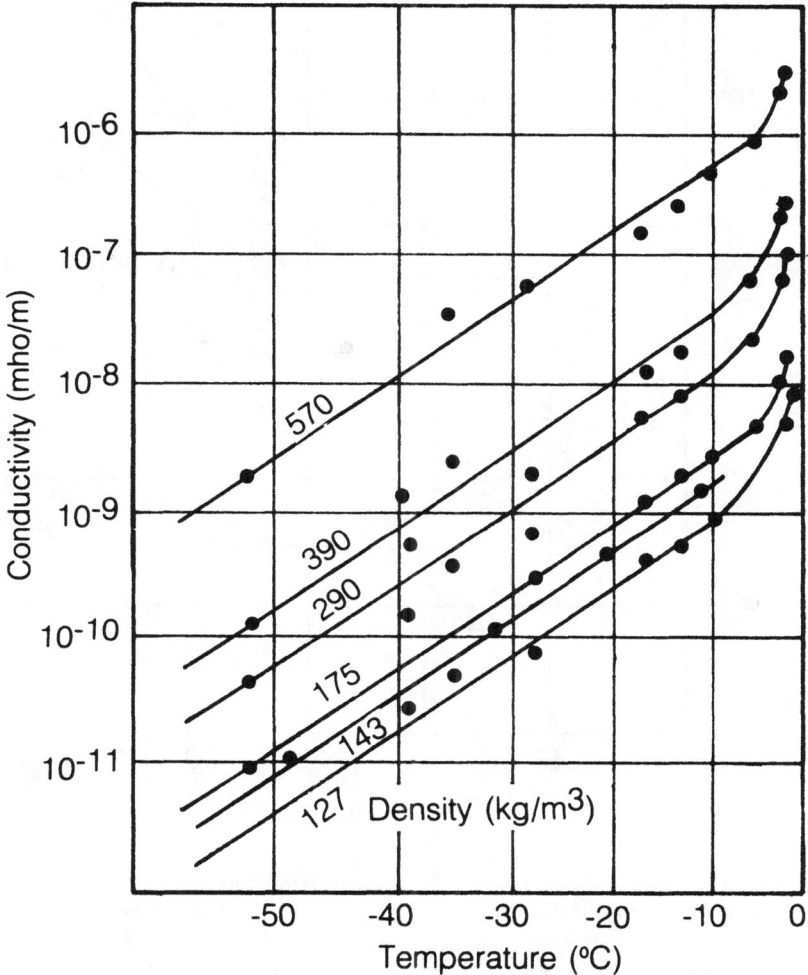

Fig. 7.21 Electrical conductivity as a function of temperature for fresh snow compressed to various densities (Kopp, 1962). Densities (kg/m^3) are noted beside each curve.

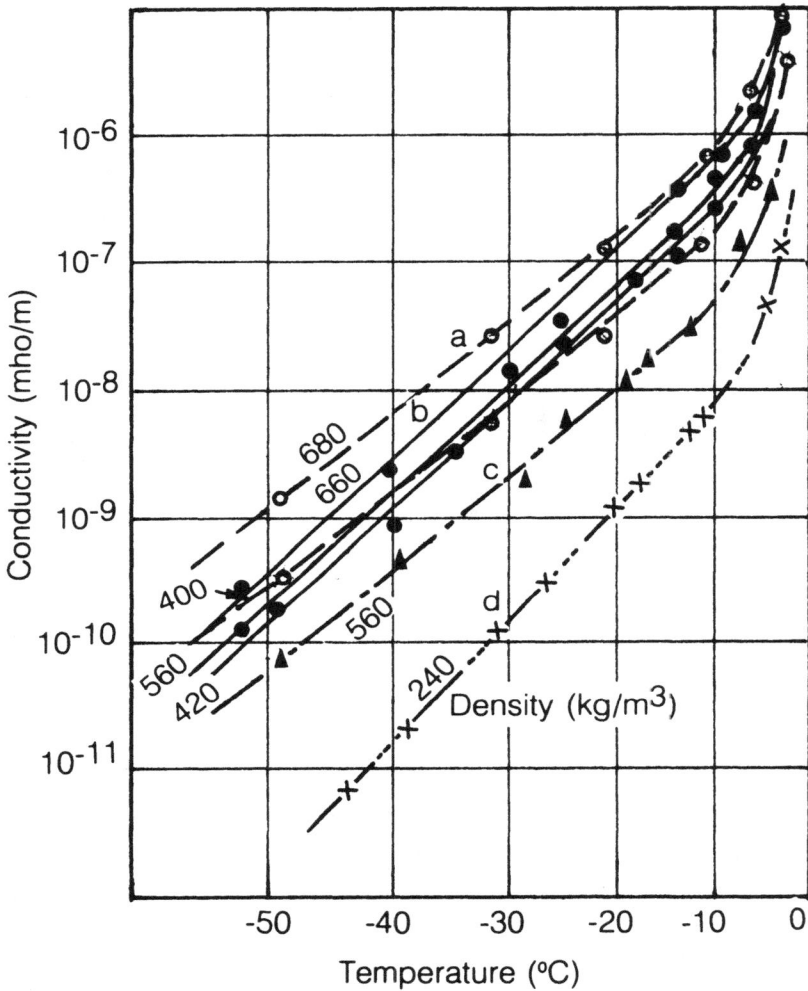

Fig. 7.22 Electrical conductivity as a function of temperature for various states of metamorphism: a) untransformed snow, o---o b) after temperature gradient metamorphism, ●——● c) large grained neve (firn), ▲— - —▲ d) fresh snow equitemperature metamorphosed to granular form x—— -- ——x. Densities (kg/m³) are noted beside each curve. As in Fig. 7.21, different densities in one snow type produce systematic differences in electrical conductivity. However, different metamorphic states have different conductivities with similar densities (Kopp, 1962).

The a.c. dielectric effects of ice or snow can be considered if the term R_0 is omitted from Eq. 7.19. Thus the a.c. impedance Z_c becomes:

$$1/Z_c = 1/[R_1 + 1/(i\omega C_1)] + i\omega C_\infty. \qquad 7.20$$

The dielectric constant (the ratio of the capacitance of a capacitor with material between its plates to its capacitance with a vacuum between the plates) or the relative permittivity ϵ of both ice and snow have been measured by several investigators and, following Hobbs (1974), can be related to the components of the equivalent circuit (Fig. 7.20) in the following manner. The electric field F in the ice or snow sample is given by $F = V/L$ where V is the potential difference across the sample of thickness L. F may be related to ϵ using Gauss's theorem as $F = Q/(A\epsilon_0\epsilon)$ where Q is the total charge in coulombs on one end of the sample of surface area A, and ϵ_0 is the relative permittivity of free space. If the sample of snow or ice can be characterized by a capacitance C, then $C = Q/V = A\epsilon_0\epsilon/L$.

Also

$$1/Z_c = i\omega C = i\omega A\epsilon_0\epsilon/L. \qquad 7.21$$

Comparison of Eq. 7.20 and 7.21 shows that

$$i\omega A\epsilon_0\epsilon/L = 1/[R_1 + 1/(i\omega C_1)] + i\omega C_\infty, \qquad 7.22$$

so that

$$\epsilon = LC_1/[\epsilon_0 A (1 + i\omega R_1 C_1)] + L C_\infty/(\epsilon_0 A). \qquad 7.23$$

The product R_1C_1 in Eq. 7.23 is known as the dielectric time constant τ of the circuit. Equation 7.23 is frequently written in the form of the well-known Debye dispersion formula (Debye, 1929) as:

$$\epsilon = \epsilon_\infty + (\epsilon_s - \epsilon_\infty)/(1 + i\omega\tau) = \epsilon' - \epsilon'', \qquad 7.24$$

where:
$$\epsilon_s = L(C_1 + C_\infty)/(\epsilon_0 A),$$
$$\epsilon_\infty = LC_\infty/(\epsilon_0 A),$$
$$\epsilon' = \epsilon_\infty + (\epsilon_s - \epsilon_\infty)/(1 + \omega^2\tau^2) = (L/\epsilon_0 A)[C_1/(1 + \omega^2\tau^2) + C_\infty]$$
$$\epsilon'' = (\epsilon_s - \epsilon_\infty)\omega\tau/(1 + \omega^2\tau^2) = L \omega\tau C_1/[\epsilon_0 A (1 + \omega^2\tau^2)],$$

in which;
ϵ_s = static relative permittivity,
ϵ_∞ = high frequency relative permittivity,
ϵ' = ordinary relative permittivity or dielectric constant of the ice, and
ϵ'' = dielectric loss factor or conductivity.

For a fixed sample temperature a plot of ϵ'' against ϵ' for different frequencies yields a semicircle having a diameter $(\epsilon_s - \epsilon_\infty)$ with the centre of the circle lying on the ϵ' axis at a distance of $(\epsilon_s + \epsilon_\infty)/2$ from the origin. Such plots are called Cole-Cole plots (Cole and Cole, 1941), or Argand diagrams, e.g., see Fig. 7.23 for ice free of voids and conducting impurities: curves (a) and (b) are at -10.8

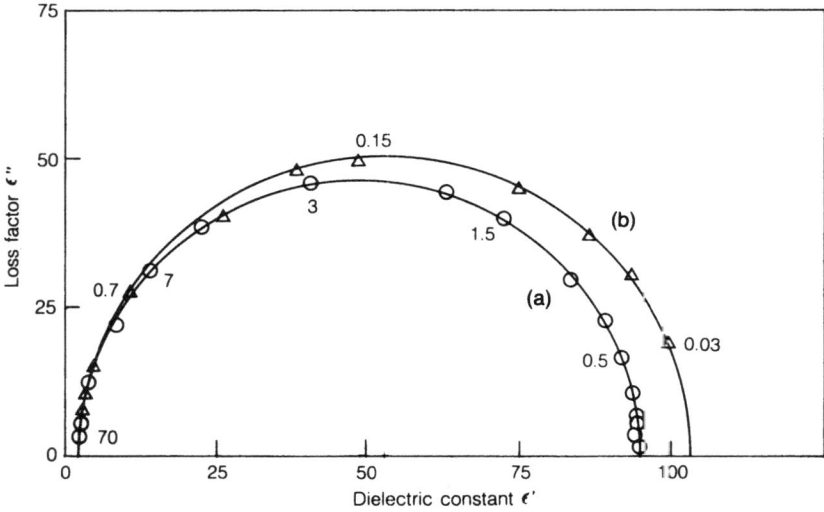

Fig. 7.23 Cole-Cole plot of the dielectric constant and loss factor for pure ice. Curves are for ice (a) at -10.8°C and (b) at -40°C. Numbers beside points are for frequencies in kilohertz (Auty and Cole, 1952).

and -40°C, respectively. Note that the static relative permittivity is approximately 100 and the high frequency permittivity is approximtely 3 (at 70,000 Hz). Similar plots for snow are shown in Fig. 7.24. The departure from the semi-circular shape is apparent in the low frequency range where the d.c. contribution due to trace contaminants is significant.

The permittivity of snow is different from that of pure ice, mainly because there are air gaps in the snow. Snow can be treated as a heterogeneous mixture of air and ice so that dielectric properties of dry snow may be derived from those of ice. Glen and Paren (1975) discuss three equations which can be used for this purpose and suggest that Looyenga's (1965) equation is the most suitable for the air-ice mixture. Assuming $\epsilon = 1$ for air this equation has the form

$$\epsilon^{1/3} - 1 = v(\epsilon_i^{1/3} - 1), \qquad\qquad 7.25$$

where: ϵ = relative permittivity of snow,
               v = proportion of ice by volume, and
               ϵ_i = relative permittivity of ice.

In Fig. 7.25, $(\epsilon^{1/3} - 1)$ is plotted as a function of snow density using the data of Cummings (1952) measured at 9.57×10^9 Hz. The data in this figure show that the high frequency relative permittivity is extremely dependent on density. The Looyenga equation predicts a straight line through the origin, and a value of ϵ equal to 3.17 at a density of 920 kg/m³.

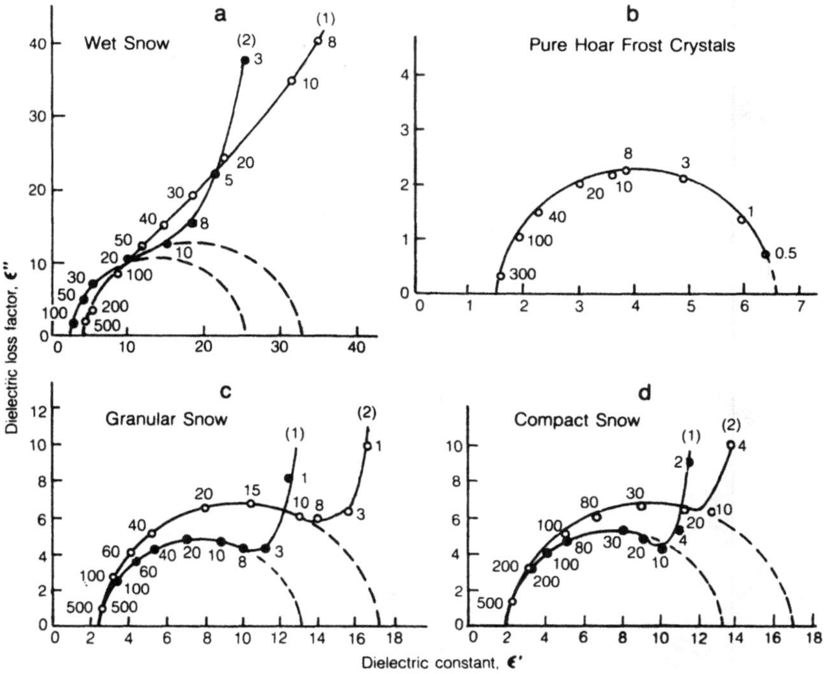

Fig. 7.24 Cole-Cole plots of the dielectric constant and loss factor for various kinds of snow (Kuroiwa, 1954). ϵ'' is the loss factor; ϵ' is the dielectric constant. Numbers beside individual points represent frequencies in kHz. (a) wet snow, average density 380 kg/m^3 curve 1 at 0°C, curve 2 is derived from the same sample at -9°C. (b) pure hoar frost crystals prepared in a cold chamber. Average density 260 kg/m^3, temperature -8.5°C. (c) granular snow, density 410 kg/m^3, temperature -4°C. (d) compact snow, density 400 kg/m^3, temperature -3°C. For figures (c) and (d) curve 1 is from measurements obtained immediately after the preparation of the sample while curve 2 was obtained approximately 1/2 day later.

Figure 7.24 shows a marked increase in the dielectric constant of wet snow compared with that of dry snow, a difference caused by free water in the wet snow. The dielectric properties of liquid water and ice exhibit similar relaxation effects; however, the value of the dielectric constant of water changes from 80 at frequencies below 3 x 10^9 Hz to 3 at frequencies above 3 x 10^{10} Hz. Thus, at lower frequencies this high value causes the increase in the permittivity of wet snow.

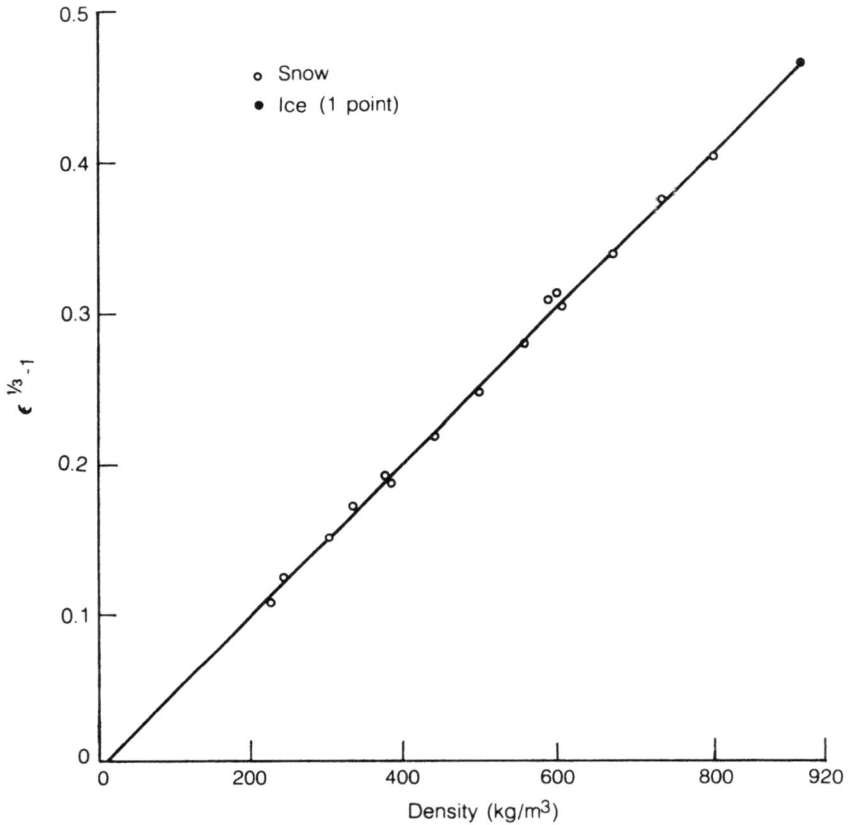

Fig. 7.25 Plot of $\epsilon^{1/3} - 1$ for snow at 9.57 GHz as a function of density (Glen and Paren, 1975). Snow samples at -18°C. Reprinted from the *Journal of Glaciology*.

Above the ice relaxation frequency[1] (approximately 10^5 Hz) dry snow shows no appreciable changes in dielectric properties until about 3×10^7 Hz. Between these two frequencies the attenuation of radio waves transmitted into snow is exponential and may be expressed by the equation:

$$I/I_o = e^{-bz}$$
                                                                    7.26

where: I = field strength at distance z from the source,
           I_o = field strength at the source, and
           b = the extinction coefficient $(\omega\sqrt{\epsilon'}/2c) \tan \delta$ (δ is the loss angle such that $\tan \delta = \epsilon''/\epsilon'$, and c is the velocity of light).

_____

[1] The relaxation frequency is that rate of reversal of an electric field beyond which the molecular dipoles can no longer rotate rapidly enough to align themselves continuously with the field (see Debye, 1929).

For ice Fig. 7.26 shows that the product, f tan δ, where f is the frequency in MHz, is almost constant at a given temperature for f < ~ 10^3MHz. Thus, since ε' is constant, there is little variation in b. Beyond 10^{11} Hz the effects of infrared absorption bands become apparent in the wavelength interval from 20 to 1 μm (5×10^{11} to 3×10^{12} Hz). Along with this "optical" variation in dielectric properties the relative permittivity decreases to a value of 1.72. The

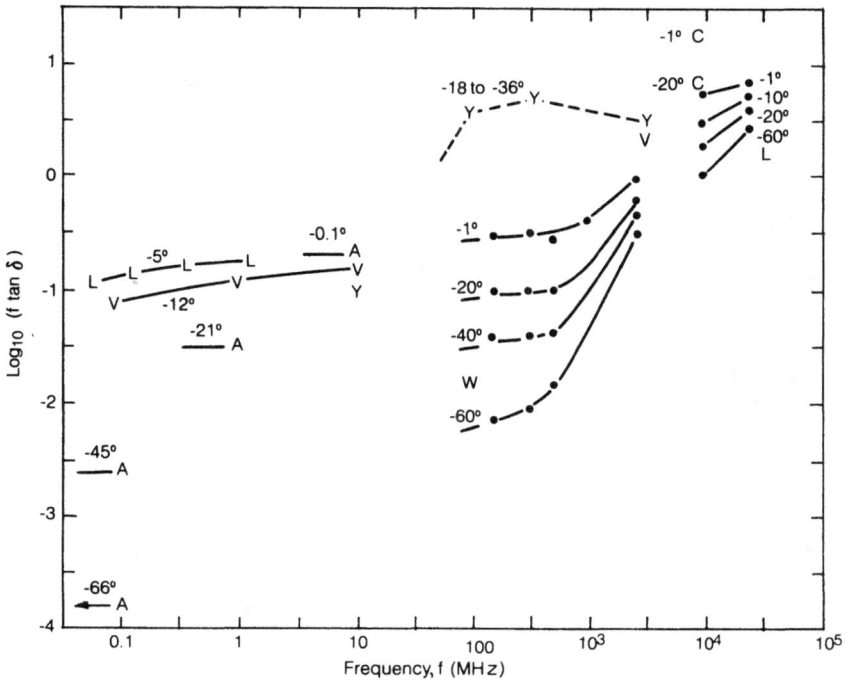

Fig. 7.26 Loss tangent of ice as a function of frequency, where f is the frequency in MHz. Temperatures are indicated in °C. Note that for a given temperature $\log_{10}$ (f tanδ) does not vary appreciably until the frequency increases above 10^3 MHz. A: Auty and Cole (1952). Conductivity water, ice free from stress. Limiting values plotted at 1000 times the relaxation frequency. C: Cummings (1952). Distilled water, tap water, and melted snow (no observable difference). L: Lamb (1946) and Lamb and Turney (1949). Distilled water, ice not annealed, V: Von Hippel (1954). Conductivity water, ice not annealed. W: Westphal, as communicated to Evans (1968). Greenland ice, annealed, density 900 kg/m³. Y: Yoshino (1961). Antarctic ice core samples, not annealed, density 910 kg/m³. (Evans, 1968; reprinted from the *Journal of Glaciology* by permission of the International Glaciological Society).

absorption spectrum for ice in the frequency range from 10^{12} to 2.1×10^{13} Hz is shown in Fig. 7.27 (Bertie et al., 1969). The absorptivity K represents the fraction of the incident radiation absorbed by ice. Within the range of wave numbers shown, the absorption by ice is so high that for all practical purposes it can be considered a black body and therefore can be assumed to absorb all radiation incident on it.

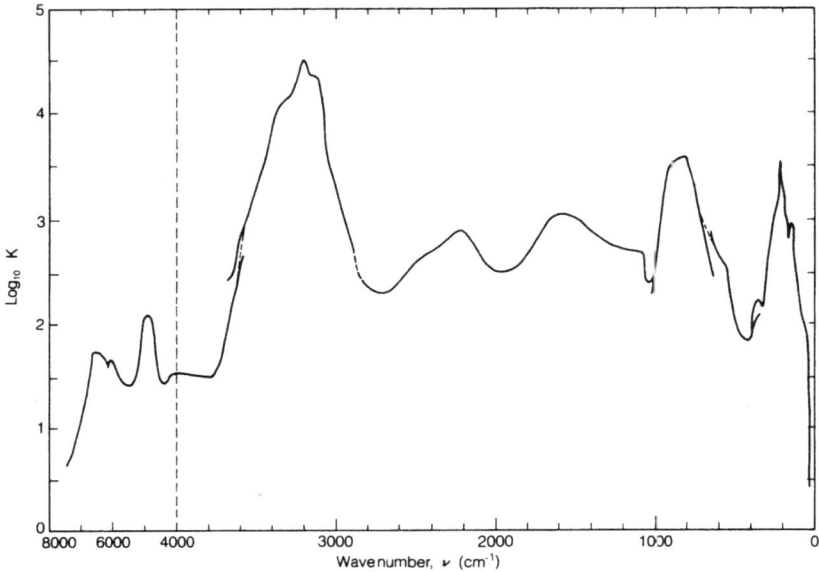

Fig. 7.27 Infra-red absorption spectrum of ice at -173°C. K is the absorptivity (cm^{-1}) (Bertie, et al., 1969).

In the visible spectrum (7.5×10^{14} to 4.3×10^{14} Hz) ice is very transparent; its slightly lower absorptivity in the blue region of the spectrum is responsible for the blue colouration of snow and ice.

The attenuation of light passing through snow, called extinction, is due to both absorption and diffusion. Theoretical studies of the extinction process have been made by Bohren and Barkstrom (1974), who show that diffusion is by far the greater of the two attenuating processes and is due to refraction and diffraction; contrary to general belief, reflection at the surface of snow grains plays little part in the extinction process. These results apply only to unsintered snow; the effect of sintering has yet to be investigated.

Measurements of the extinction coefficient are conflicting, partly because the studies have been conducted with different boundary conditions at the upper surface. The conditions of the upper surface can influence extinction to depths of 20 cm or more.   If radiation attenuation is assumed to be exponential with depth (Eq. 7.26) the coefficient of extinction lies between 0.07 cm^{-1} (Bergen, 1971) and 1.5 cm^{-1} (Mellor, 1965).

In calculating radiative energy transfer to the snowpack, its albedo is most often used. It is the ratio of the intensities of reflected radiation to incident radiation calculated using radiation intensities averaged over the short-wave radiation spectrum (frequently defined by the method of measurement but always including visible radiation) and sometimes encompassing the wavelengths from 0.2 to 3.2 μm. The albedo gives a good measure of the ability of the snow to absorb radiative energy. It is influenced not only by structural and geometrical modifications resulting from metamorphism and densification, but also by the effect of contaminants on the snow surface. These may originate either from direct deposits or from previously buried deposits exposed at the surface during melting.

Clearly, the albedo varies widely with snowpack conditions and time since the last snowfall. In particular, the proximity to industrial aerosol sources and to blowing soil is important. The albedo of fresh snow ranges from 0.92 to 0.98; and its change with time for a deep pack is shown in Fig. 7.28 (see also Ch. 9; Fig. 9.6).

Because albedo governs the capability of snow to absorb solar radiation, attempts were made by numerous workers during the last one hundred years to increase the melting rate of snow by artificially decreasing its albedo. This is achieved by scattering coal dust, soot or similar material onto the snow

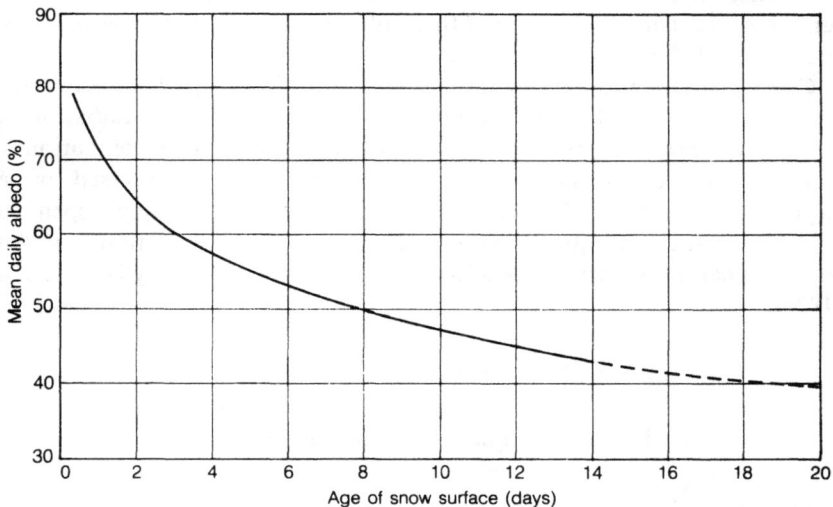

Fig. 7.28 Variation of snow albedo with age during the melt season. (U.S. Army Corps of Engineers, 1956).

surface. Unfortunately, the experiments are not well documented in terms of radiation measurements, comparative melt rates and other factors. Nevertheless, there is no doubt that the materials have a pronounced effect on the melt rate. Slaughter (1969) reviews a large number of such experiments. It appears that by using additives the disappearance of a snowpack can be advanced by one to two weeks depending on the depth of the snow.

Enthusiastic proponents of albedo modification have proposed grandiose schemes for controlling meltwater supply to power stations, for lengthening the growing season in cold climates and so forth. While some of these may look interesting, their economic and ecological costs eliminate them as serious undertakings. There are, however, some minor but nevertheless interesting applications. For example, in the Prairies, the snow that has accumulated in drainage ditches is deeper than the snowcover over the adjacent land areas so that the ditches are usually still blocked after snow on the land has melted. The resulting floods not only make roads impassable but frequently cause washouts. Under certain conditions the use of coal dust or similar material over the limited surface areas of the ditches could melt enough snow to eliminate this problem. Frequently, however, the effects of the treatment on albedo modification are nullified by the redeposition of a shallow depth of wind-blown snow onto the treated area thereby increasing the albedo to about its former value.

At shorter wavelengths than those in the visible spectrum a strong absorption band occurs at about 0.15 μm (2 x 10^{15} Hz) where snow is also black (Dressler and Schnepp, 1960). At still shorter wavelengths, its spectral properties and others such as fluorescence, phosphorescence, triboluminescence, triboelectricity and thermoluminescence, may be derived from those of ice (Hobbs, 1974).

One important property of snow is its absorption of γ-rays ($>$ 10^{18} Hz). Methods based on the attenuation of natural or artificial γ-radiation are being developed to measure the snow water equivalent. The attenuation is a function of the thickness of absorbing material, as expressed by an exponential integral. The coefficients of absorption strongly depend on wavelength and consequently also on the γ-ray source. Equations and data for some naturally occurring radioactive elements are given by Grasty, et al. (1974).

PROPERTIES OF WET SNOW

Effects of Liquid Water in the Snowpack

As indicated above, small differences in surface energy influence the processes that occur in dry snow. If liquid water is present, its free energy is still related to the curvature of the liquid surface but the equilibrium temperature is no longer fixed at the ice point (Colbeck, 1973; Langham, 1974a). Below - 0.01°C the effects of liquid water on the free energy are negligible. Minute depressions of the equilibrium temperature caused by liquid water are important since they result in the production of strong gradients of capillary tension which in turn cause water to migrate through the snowpack; they also dramatically influence the permeability of the ice layers (Langham, 1975).

When the liquid water content of the snowpack is $\sim 1\%$ by volume, capillary suction is no longer important; then water is transmitted by gravitational percolation. Percolation in homogeneous snow can be shown to follow Darcian flow (Colbeck, 1971). According to Darcy's law the vertical volume flux of water V_w can be described by the equation:

$$V_w = -(k_l/\mu_l)\,(\partial p/\partial z + \rho g), \qquad\qquad 7.27$$

where:
$\quad k_l$ = permeability,
$\quad \mu_l$ = dynamic viscosity of water,
$\quad \partial p/\partial z$ = pressure gradient,
$\quad \rho$ = density of water, and
$\quad g$ = acceleration of gravity

k_l can be related to the water content of snow by the following expression (Kuriowa, 1968):

$$k_l = c \exp (b\phi) \qquad\qquad 7.28$$

where:
$\quad \phi$ = liquid water content, and
$\quad c, b$ = constants.

Thus a change in the surface melt rate, because it determines the amount of water available and thus the magnitude of k_l (Eq. 7.28), makes Eq. 7.27 nonlinear in space and time. There is reason to believe, however, that the flow may be so modified by the effects of snow structure and by instability in the flow that Eq. 7.28 is of limited use (Wooding and Morel-Seytoux, 1976; Langham, 1973).

For water to pass through an isothermal snowpack at 0°C calculations based on the assumption of homogeneous flow give travel times of about one to several hours, depending on the melt rate and the depth of snow. These times may be reduced in the case of unstable flow, particularly during heavy rain. Impermeable ice layers may considerably increase the time and even bring the flow to a halt (Langham, 1973, 1974a); the retained water is subsequently released, often rather suddenly. Each one of these conditions presents important problems in forecasting snowmelt floods from calculated melt rates. Colbeck (1976) has provided an excellent review of experiments and theories on the flow of water through snow.

Thermal, Mechanical and Electrical Properties

Since wet snow cannot sustain temperature gradients in the same way as dry snow, it cannot conduct heat. Melting takes place at the upper surface because of absorbed solar radiation and heat transferred from the air. Warm rain can carry energy a short distance into the snow but the melting that occurs is very near the upper surface. The absorption of radiation can also cause melting within the snow (Langham, 1974b), but again because of the large extinction coefficient, this occurs near the surface. However, meltwater or rain entering the snowpack can raise the temperature of cold snow to 0° C by releasing latent heat as it freezes.

The freezing of wet snow, on the other hand, does occur via conduction of heat towards the cold surface of dry snow. At the moving boundary between the wet and dry snow, the latent heat that is liberated complicates the heat diffusion equation. This is known as the Stefan problem which requires special methods for its solution (Ockendon and Hodgkins, 1975).

During daily nocturnal refreezing of a snowpack the freezing front generally advances only some 10 cm into the snow, forming a strong crust. When the melting and refreezing occur in association with changing atmospheric air masses, the cycle time may be of the order of a week and temperature variations may affect the snow to a depth of one metre or more.

The process of melting and refreezing changes the microstructure of the snow and is known as melt-freeze metamorphism. The grains become rounded during the melting process, some smaller grains disappearing completely. Water is held between the grains and there is a general compaction of the snowpack. When the water refreezes the snow has an increased density; the greater continuity of the resulting ice matrix increases its strength and causes changes in associated properties (e.g., increased strength and hardness).

The effect of liquid water on the dielectric properties of the snowpack is of particular interest. Since the relaxation frequency for water is about 10^{10} Hz, any frequency below this value and above 10^5 Hz (the relaxation frequency for ice) may be used to measure the liquid water content. This property was extensively studied by Ambach and Denoth (1974), who used it to investigate the relationship between water content and the incidence of wet avalanches. In the microwave region, it has been used to study the liquid water distribution in snow (Linlor et al., 1974) and to search for bodies buried in avalanches (Fritzsche, 1974). It also offers prospects of obtaining synoptic data on snowpack condition by means of remote sensing techniques with X-band microwaves (both passive and active) and of studying vertically distributed properties and processes using time-domain reflectometry in the same wave band.

The presence of water also modifies the albedo of snow. Unfortunately, little quantitative data are available about this, but the effect has been found to be pronounced for near-infrared radiation as revealed in aerial photographs and satellite images. The reduction in albedo because of liquid water is attributed to the reduction in the surface area in contact with the air which, in turn, reduces the refractive diffusion.

CONTAMINATION OF THE SNOWPACK

Snow on the ground contains many impurities. Dust has already been mentioned as an important impurity through its effect on the albedo. However, a snowpack may also contain many trace elements which, because of their toxicity, contribute to environmental pollution.

In a study of ice cores from the Greenland Ice Cap, Murozumi et al. (1969) showed that the concentration of lead in snow reaching the ground has increased from 0.001 μg/kg in 1940 to 0.2 μg/kg in 1968. This is for an area with no local urban influence. They ascribe the increase to the global pollution of the atmosphere resulting from the use of lead tetra-alkyls in gasoline. In urban areas, the concentrations are extremely high locally and may reach 10^5 μg/kg in snowpacks adjacent to major highways (LaBarre et al., 1973). Snow clearing procedures may increase lead concentrations in rivers if the melted snow is dumped directly into them or if the meltwater released from snow dumps runs off or percolates towards them. The lead does not appear to be very soluble, however, possibly because of adsorption on solid particles or the low solubility of the lead chlorides and bromides formed by its reaction with the salt used to clear streets. LaBarre et al. (1973) have measured concentrations of lead of 100 μg/kg in snow dump meltwater. However, it should be pointed out that this is released continuously during the summer until the snow is completely melted.

In addition to the pollutants that enter the snowpack when it is on the ground, a large number of contaminants are also carried by the falling snow. These include natural aerosol particles, which act as nuclei for the formation of ice particles in the atmosphere, and dust particles collected by the crystals while falling through the air. Likewise, industrial aerosol pollutants may be swept out by the falling snow so that an extensive area of the snowcover may contain high levels of lead and zinc metals highly toxic to aquatic life (Hagen and Langeland, 1973).

Gases are also adsorbed on the surface of snow crystals. Sulpher dioxide and oxides of nitrogen are of particular importance; they are also oxidized in the atmosphere and produce sulphuric and nitric acid. High sulphur contents in a snowpack are due to industrial sources. For example, the smelting furnaces at Sudbury, Ontario release more than 2.27 x 10^9 kg of sulphur dioxide into the atmosphere per year; snow more than 40 km away has been found to have a pH of less than 3 (Beamish and Harvey, 1972). This level of acidity cannot be attributed to any other source.

Hagen and Langeland (1973) consider that industrial pollutants can be deposited in high concentrations hundreds of kilometres from their sources as a result of meteorological factors that cause the air pollutants to be transported and then to be deposited by adhering to the falling snow. They found that this process produces highly variable concentrations in successive layers of the snowpack but similar values in corresponding layers at two sites located 48 km apart (see Table 7.8).

Table 7.8

CONTAMINATION OF SNOWPACK BY CHEMICAL FALLOUT AT TWO SITES 48 km APART (Hagen and Langeland, 1973).

	Sulphate	Calcium	Zinc	Lead	Iron	Cadmium	Copper
Fyresdal site	10.6	0.9	85.	67.	80.	2.	15.
Tokke site	7.4	0.4	85.	60.	180.	4.	20.
	mg/l			μg/l			

During winter, snow removal procedures and the application of sand, cinders and other abrasives are widely used to improve highway trafficability. The use of these methods in highway snow and ice removal and control programs is decreasing partly because they may cause blockage of sewer systems. As an alternative, increasing use is being made of salts such as sodium and calcium chlorides, e.g., some cities use as much as 7,000 kg per lane kilometre (Hawkins and Judd, 1972). The brine generated from the melted snow may enter streams in the runoff from streets or snow disposal sites. Not only is the brine often colder and therefore denser than the water with which it mixes, but it is also marginally denser because of the dissolved salts. As a result, it tends to sink to the bottom layers of lakes, so stabilizing the water that normal spring overturn does not take place. This prevents oxygenation of the water which, together with the high salinity, kills fish, other animals and plants living near the bottom.

Other chemicals frequently occur in the salt used for road clearing, either naturally or as additives. The mixture of nutrients and toxic materials is alarming and includes poly-phosphates, nitrates, urea, nitrites, ferrocyanides, sulphates, chromium, borax, ammonium compounds and other organic substances (Boies and Bortz, 1965).

It is difficult to be precise about the general effects of salts in increasing toxicity levels, since the rate of release of salts depends on snow disposal practice and the water bodies' ability to dilute the snow. Clearly, near sewer outfalls, local concentrations of salts can be very high, especially shortly after application when they may reach 0.25% by weight (American Public Works Association, 1969; Van Loon, 1972).

If radioactive elements are present they are usually not dangerous, but rather are of interest for studying the processes occurring within the snowpack. Radon, thoron and actinon are released as products of the decay of radium in the soil. The half-life of radon is 3.8 days so that variations in its concentration can be used to trace the diffusion processes between the soil and the snow (Kovach, 1945). Radon can be identified by its γ-radiation peak at 2.62 Mev (Bissell, 1974).

In atmospheric fallout radioactive elements such as strontium and tritium also collect in the snow. Tritium has been used for hydrologic studies of snow accumulation and melting of glaciers (Prantl and Loijens, 1975). Naturally-occurring stable isotopes of hydrogen and oxygen provide other means of studying snowmelt dynamics (Krouse et al., 1975).

An immense variety of particles from the atmosphere accumulate in the snow: airborne pollens, fungal spores and spore phases of bacteria. Most of these remain inert in the snow, but certain algae, fungi and microbes are cryophilic or at least are adapted to survive in snow at temperatures near the melting point.

Stein and Amundsen (1967) describe seven genera of algae belonging to Chlorophyta Volvocales and Chlorophyta Chlorococcales. Of fifteen or so species identified, many are flagellates which require melting snow for part of their life cycle. Since many of these algae are bright red or orange during part of their life cycle (and green during another part) they impart a striking coloration to the snow. The same authors also refer to two species of fungus of genus Chionaster Wille (Eumycophyta) which have only been found in snow.

Visser (1973) reports the presence of many types of bacteria in samples from snow dumps and some in snow samples from the Laurentian forest: e.g. species of monococci, diplococci, streptococci, staphylococci, vibria and streptomyces. Although they can metabolize at $0°C$, they are likely introduced to the snow from soil particles, animal excrement and other sources. Many species of microflora can multiply in snow (Stokes et al., 1962), but they have not as yet been studied very much.

LITERATURE CITED

Abels, H. 1892. *Beobachtungen der iä glichen periode der temperatur im schnee und bestimmun des wärmeleitungsvermögens des schnees als function seiner dichtigkeit.* Rep. Meteorol. Bd. XVI, No. 1, pp. 1-53.

Adam, N.K. 1941. *Physics and Chemistry of Surfaces.* Oxford Univ. Press, London.

Ambach, N. and A. Denoth, 1974. *On the dielectric constant of wet snow.* Proc. Int. Symp. on Snow Mechanics, Grindelwald. Int. Assoc. Hydrol. Sci. Publ. 114, pp. 136-144.

American Public Works Association (APWA) 1969. *Water pollution aspects of urban runoff.* WP20-15, USDI, FWPCA, Washington, D.C.

Auty, R.P. and R.H. Cole, 1952. *Dielectric properties of ice and solid D_2O.* J. Chem. Phys., Vol. 20, pp. 1309-1314.

Bader, H., R. Haefeli, E. Bucher, I. Neher, O. Eckel and Chr. Thams. 1939. *Der Schnee und seine Metamorphose* (Snow and its Metamorphism). Beit. Geol. Schweiz, Geotech. Ser., Hydrol., Lief. 3, Bern. [English Transl. by Snow, Ice Permafrost Res. Estab., Transl. 14].

Bader, H., 1953. *Sorge's law of densification of snow on high polar glaciers.* Res. Pap. 2., U.S. Army Snow, Ice Permafrost Res. Estab.

Bader, H., 1960, 1962. *Theory of densification of dry snow on high polar glaciers.* Pt. I, Res. Rep. 69, Pt. II Res. Rep. 108. U.S. Army Cold Reg. Res. Eng. Lab., Hanover, N.H.

Ballard, G.E.H., E.D. Feldt and S. Toth, 1965. *Direct shear study on snow, procedure and data.* Sp. Rep. 92. U.S. Army Cold Reg. Res. Eng. Lab., Hanover, N.H.

Beamish, R.H. and H.H. Harvey, 1972. *Acidification of the La Cloche mountain lakes, Ontario, and resulting fish mortalities.* J. Fish. Res. Bd. Can., Vol. 29, No. 8, pp. 1131-1143.

Bentley, C.R., P.W. Pomeroy and H.J. Dorman. 1957. *Seismic measurements on the Greenland Ice Cap.* Ann. Geophys., Vol. 13, No. 4, pp. 253-285.

Bergen, J.D. 1971. *The relation of snow transparency to density and air permeability in a natural snow cover.* J. Geophys. Res., Vol. 76, pp. 7385-7388.

Bertie, J.E., H.J. Labbé, and E. Whalley. 1969. *Absorptivity of ice I in the range 4000-30 cm^{-1}.* J. Chem. Phys., Vol. 50, pp. 4501-4520.

Bissell, V.C., 1974. *Natural gamma spectral peak method for snow measurement from aircraft.* Adv. Concepts Tech. Study Snow Ice Resour. Interdiscip. Symp., U.S. Nat. Acad. Sci., Washington, D.C., pp. 614-623.

Bohren, C.F. and B.R. Barkstrom, 1974. *Theory of the optical properties of snow.* J. Geophys. Res., Vol. 79, pp. 4527-4535.

Boies, D.B. and S. Bortz, 1965. *Economical and effective deicing agents for use on highway structures.* Rep. 19, Nat. Coop. Highw. Res. Prog., Highw. Res. Board, NAS-NRC.

Bowden, F.P., 1955. *Friction on snow and ice and the development of some fast running skis.* Nature, Vol. 176, p. 946.

Bracht, J., 1949. *Über die wärmeleitfähigkeit des erdbodens und des schnees und den wärumeumsatz im erdboden* (On the thermal conductivity of soil and snow, and the heat utilization in soil), Veröffentl. Geophysikalischen Inst. Univ. Leipzig, Ser. 2, Vol. 14, No. 3, pp. 147-225.

Bradley, C.C. and W. St. Lawrence, 1975. *Kaiser effect in snow.* Proc. Int. Symp. on Snow Mechanics, Grindelwald, Int. Assoc. Hydrol. Sci. Publ. 114, pp. 145-154.

Brown, R.L., 1976a. *A thermodynamic study of materials representable by integral expansions.* Int. J. Eng. Sci., Vol. 14, pp. 1033-1046.

Brown, R.L., 1976b. *Fracture criterion for snow*. J. Glaciol., Vol. 19, pp. 111-121.

Brown, R.L. and T.E. Lang. 1975. *On the fracture properties of snow*. Proc. Int. Symp. on Snow Mechanics, Grindelwald, Int. Assoc. Hydrol. Sci. Publ. 114, pp. 196-207.

Bucher, E. 1948. *Beiträge zu den theoretischen Grundlagen des Lawinenverbaus* (Contribution to the theoretical foundations of avalanche defence construction). Beit. Geol. Schweiz, Geotech. Ser., Hydrol., Lief. 6. [English Transl. by U.S. Army Snow, Ice Permafrost Res. Estab., Transl. 18].

Bucher, E. and A. Roch. 1948. *Reibungs-und Packungswiderstande bei raschen Schneebewegungen* (Friction and resistance to compaction of snow under rapid motion). Mitteilungen des Eidgenöss, Instituts fur Schnee und Lawinenforschung. Davos-Weissfluhjoch.

Butkovich, T.R., 1956. *Strength studies of high-density snows*. Res. Rep. 18, U.S. Army Snow, Ice Permafrost Res. Estab.

Carslaw, H.S. and Jaeger, J.C. 1959. *Conduction of Heat in Solids*. Oxford Univ. Press, London.

Colbeck, S.C. 1971. *One dimensional water flow through snow*. Res. Rep. 296, U.S. Army Cold Reg. Res. Eng. Lab., Hanover, N.H.

Colbeck, S.C. 1973. *Theory of metamorphism of wet snow*. Res. Rep. 313, U.S. Army Cold Reg. Res. Eng. Lab., Hanover, N.H.

Colbeck, S.C. 1976. *The physical aspects of water flow through snow*. In Advances in Hydroscience (Ven Te Chow, ed.), Academic Press, New York, Vol. 11, pp. 165-200.

Cole, K.S. and R.H. Cole. 1941. *Dispersion and absorption in dielectrics I: alternating current characteristics*. J. Chem. Phys., Vol. 9, pp. 341-351.

Coudert, J.F. 1973. *Théorie macroscopique des écoulements multiphasiques en milieu poreux. Première partie: Approche d'une théorie mathématique pour établir les équations en valeur moyenne décrivant l'ecoulement d'un fluide hétérogène*. Rev. Inst. Fr. Pét., Vol. 28, pp. 171-183. *Deuxième partie: Méchanique*, Vol. 28, pp. 373-398.

Crary, A.P., E.S. Robinson, H.E. Bennett and W.W. Boyd. 1962. *Glaciological studies of the Ross Ice Shelf, Antarctica, 1957-1960*. Int. Geophys. Year, Glaciol. Rep. 6, Am. Geograph. Soc.

Cummings, W.A. 1952. *The dielectric properties of snow and ice at 3.2 centimeters*. J. Appl. Phys., Vol. 23, No. 7, pp. 768-773.

Debye, P. 1929. *Polar Molecules*. Dover Publications Inc., New York, N.Y.

Denbigh, K. 1966. *The Principles of Chemical Equilibrium with Applications in Chemistry and Chemical Engineering*. The Univ. Press, Cambridge.

Devaux, J. 1933. *Ann. de phys.* Ser. 10, Vol. 20, pp. 5-67. (as cited by Dorsey, 1940. p. 483).

de Quervain, M.R. 1973. *Snow structure, heat and mass flux through snow*. The Role of Snow and Ice in Hydrology: Proc. Banff Symp., Sept. 1972, Unesco-WMO-IASH, Geneva-Budapest-Paris, Vol. 2, pp. 203-226.

Dorsey, N.E. 1940. *Properties of Ordinary Water-substance in all its Phases: Water Vapor, Water and all the Ices*. Mono. Ser. No. 8, Am. Chem. Soc. Rheinhold Publ. Corp. (reprinted Hafner Publ. Co., New York, 1968).

Dressler, K. and O. Schnepp. 1960. *Absorption spectra of solid methane, ammonia and ice in the vacuum ultra-violet.* J. Chem. Phys., Vol. 33, pp. 270-274.

Eriksson, R. 1949. *Medens friktion mot snö och is (Friction of runners on snow and ice).* Foren Skogsarbet., Kgl., Domänstyrelsens arbetsstud., Med 34-35, [English Transl. by U.S. Army Snow, Ice Permafrost Res. Estab., Transl. 44].

Evans, S. 1968. *Dielectric properties of ice and snow: a review.* J. Glaciol. Vol. 5, pp. 773-792.

Fletcher, N.H. 1970. *The Chemical Physics of Ice.* Cambridge Univ. Press. Cambridge, U.K.

Fritzsche, W. 1974. *New electronic avalanche rescue devices.* Rep. Inst. High Frequency and Electronics, Tech. Univ. Graz, Austria.

Frutiger, H. and M. Martinelli. 1966. *A manual for planning structural control of avalanches.* Res. Pap. RM-19, USDA Rocky Mountain For. Range Exp. Sta.

Glen, J.W. 1974. *The physics of ice.* Mono II-C2a., U.S. Army Cold Reg. Res. Eng. Lab., Hanover, N.H.

Glen, J.W. and J.G. Paren. 1975. *The electrical properties of snow and ice.* J. Glaciol., Vol. 15, No. 73, pp. 15-38.

Gow, A.J. 1974. *Time-temperature dependence of sintering in perennial isothermal snowpacks.* Proc. Int. Symp. on Snow Mechanics, Grindelwald, Int. Assoc. Hydrol. Sci. Publ. 114, pp. 25-41.

Grasty, R.L., H.S. Loijens, and H.L. Ferguson. 1974. *An experimental gamma-ray spectrometer snow survey over southern Ontario.* Adv. Concepts Tech. Study Snow Ice Resour. Interdiscip. Symp., U.S. Nat. Acad. Sci., Washington, D.C., pp. 579-593.

Haefeli, R. 1939. *Snow mechanics with references to soil mechanics.* In (Bader et al., 1939), *Der Schnee und seine Metamorphose,* Beit. Geol. Schweiz, Geotech. Ser., Hydrol., Lief. 3, Bern. [English Transl. by U.S. Army Snow, Ice Permafrost Res. Estab., Transl. 14].

Hagen, A. and A. Langeland. 1973. *Polluted snow in southern Norway and the effect of the meltwater on freshwater and aquatic organisms.* Environ. Pollut., Vol. 5, pp. 45-57.

Hatsopoulos, G.N. and J.H. Keenan. 1965. *Principles of General Thermodynamics.* John Wiley and Sons, Inc., New York, N.Y.

Hawkes, I., and M. Mellor. 1972. *Deformation and fracture of ice under uniaxial stress.* J. Glaciol., Vol. 11, No. 61, pp. 103-131.

Hawkins, R.H. and J.H. Judd. 1972. *Water pollution as affected by street salting.* Water Resour. Bull., Vol. 8, pp. 1246-1252.

Hobbs, P.V. 1968. *Metamorphism of dry snow at a uniform temperature.* Int. Union Geod. Geophys. Gen. Assem. of Bern, [Trans. Vol. 5, Snow and Ice], Int. Assoc. Sci. Hydrol. Publ. 79. pp. 392-402.

Hobbs, P.V. 1974. *Ice Physics.* Clarendon Press, Oxford.

Hobbs, P.V. and B.J. Mason. 1964. *The sintering and adhesion of ice.* Phil. Mag., Ser. 8, Vol. 9, No. 98, pp. 181-197.

Hobbs, P.V. and L.F. Radke. 1967. *The role of volume diffusion in the metamorphism of snow.* J. Glaciol., Vol. 6, No. 48, pp. 879-891.

Jaafar, H. and J.J.C. Picot. 1970. *Thermal conductivity of snow by a transient state probe method.* Water Resour. Res., Vol. 6, No. 1, pp. 333-335.

Jansson, M. 1901. *Öfvers. K. svenska.* Vet. Akad. Förh Vol. 58, pp. 207-222 (as cited by Dorsey, 1940, p. 483).

Jellinek, H.H.G. and S.H. Ibrahim. 1967. *Sintering of powdered ice.* J. Colloids Interface Sci., Vol. 25, pp. 245-254

Kamb, B. 1968. *Ice polymorphism and the structure of water.* In Structural Chemistry and Molecular Biology (A. Rich and N. Davidson eds.), W.H. Freeman and Co., San Francisco and London, pp. 507-542.

Keeler, C.M. 1969. *Some physical properties of alpine snow.* Res. Rep. 271, U.S. Army Cold Reg. Res. Eng. Lab., Hanover, N.H.

Keeler, C.M. and W.F. Weekes. 1967. *Some mechanical properties of Alpine snow, Montana, 1964-66.* Res. Rep. 227, U.S. Army Cold. Reg. Res. Eng. Lab., Hanover, N.H.

Keenan, J.H., F.G. Keyes, A.G. Hill and J.G. Moore. 1969. *Steam Tables.* John Wiley and Sons, Inc., New York, N.Y.

Klein, G.J. 1950. *Canadian survey of physical characteristics of snow covers.* Tech. Memo. 15, Div. Build. Res., Nat. Res. Counc. Can., Ottawa, Ont.

Klein, G.J., C.D. Pearce and L.W. Gold. 1950. *Method of measuring the significant characteristics of a snow cover.* Tech. Memo 18, Assoc. Comm. Soil Snow Mech., Nat. Res. Counc. Can., Ottawa, Ont.

Kojima, K. 1954. *Visco-elastic property of snow.* Low Temp. Sci. Ser. A, Vol. 12, pp. 1-13.

Kojima, K. 1967. *Densification of seasonal snowcover.* Physics of Snow and Ice, (H. Oura, ed.), Proc. Int. Conf. on Low Temp. Sci., Hokkaido Univ., Sapporo, Vol. 1, pp. 929-952.

Kondrat'eva, A.S. 1945. *Teploprovodnost' snegovogo pokrova i fizicheskie protsessy, prois khodiaschie v nem podvlianiem temperaturnogo gradienta (Thermal Conductivity of the snow cover and physical processes caused by the temperature gradient).* Akad. Nauk. SSSR [English Transl. by U.S. Army Snow, Ice Permafrost Res. Estab., Transl. 22].

Kopp, M. 1962. *Conductivité électrique de la neige, au courant continu.* Z. Math. Phys., Vol. 13, pp. 431-441.

Kovach, E.M. 1945. *Meteorological influences upon the radon-content of soil - gas.* Trans. Am. Geophys. Union, Vol. 26, pp. 241-248.

Kovacs, A., W.F. Weeks and F. Michetti. 1969. *Variation of some mechanical properties of polar snow, Camp Century, Greenland.* Res. Rep. 276., U.S. Army Cold Reg. Res. Eng. Lab., Hanover, N.H.

Krouse, H.R., R. Hislop, H.M. Brown, K. West and J.L. Smith. 1975. *Climate and spatial dependence of the retention of D/H and O^{16}/O^{18} abundances in snow and ice of North America.* Int. Union Geod. Geophys. Gen. Assem. Grenoble [Isotopes and impurities in snow and ice], Int. Assoc. Sci. Hydrol., Publ. 118, pp. 242-247

Kuroiwa, D. 1954. *The dielectric property of snow.* Int. Union Geod. Geophys. Gen. Assem. Rome, [Trans. Vol. 4, Snow and Ice], Int. Assoc. Sci. Hydrol. Publ. 46, pp. 52-63.

Kuroiwa, D. 1968. *Liquid permeability of snow.* Low Temp. Sci. Ser. A, Vol. 26, pp. 87-100 [Japanese with English summary].

Kuroiwa, D., Y. Mizuno and M. Takenuchi. 1967. *Micromeritical properties of snow.* Physics of snow and ice. (H. Oura, ed.), Proc. Int. Conf. on Low Temp. Sci., Hokkaido Univ., Sapporo, Vol. 2, pp. 751-772.

LaBarre, N., J.B. Milne and B.G. Oliver. 1973. *Lead contamination of snow.* Water Resour. Res., Vol. 7, pp. 1215-1218.

LaChapelle, E.R. 1969. *Field Guide to Snow Crystals.* Univ. Washington Press, Seattle.

Lamb, J. 1946. *Measurements of the dielectric properties of ice.* Trans. Faraday Soc., Vol. 42A, pp. 238-44.

Lamb, J. and A. Turney. 1949. *The dielectric properties of ice at 1.25 cm wavelength.* Proc. Physical Soc., Sect. B. Vol. 62, Pt. 4, pp. 272-273.

Lang, T.E., R.L. Brown, W.F. St. Lawrence and C.C. Bradley. 1974. *Buckling characteristics of a sloping snow slab.* J. Geophys. Res., Vol. 78, pp. 339-351.

Langham, E.J. 1973. *Un modèle du manteau nival relié a sa méso-structure.* Bull. Sci. Hydrol., Vol. 18, pp. 33-43.

Langham, E.J. 1974a. *The occurrence and movement of liquid water in the snowpack.* Adv. Concepts Tech. Study Snow Ice Resour. Interdiscip. Symp., U.S. Nat. Acad. Sci., Washington, D.C., pp. 67-75.

Langham, E.J. 1974b. *Phase equilibria of veins in polycrystalline ice.* Can. J. Earth Sci., Vol. 11, pp. 1280-1287.

Langham, E.J. 1975. *The mechanism of rotting of ice layers within a structured snowpack.* Proc. Int. Symp. on Snow Mechanics, Grindelwald, Int. Assoc. Hydrol. Sci. Publ. 114, pp. 73-81.

Lee, T.M. 1961. *Note on Young's Modulus and Poisson's Ratio of naturally compacted snow and processed snow.* Unpubl. Tech. Note, U.S. Army Cold Reg. Res. Eng. Lab., Hanover, N.H.

Linlor, W.I., M.F. Meier and J.L. Smith. 1974. *Microwave profiling of snowpack free water content.* Adv. Concepts Tech. Study Snow Ice Resour., Interdiscip. Symp., U.S. Nat. Acad. Sci., Washington, D.C., pp. 729-736.

Looyenga, H. 1965. *Dielectric constants of heterogeneous mixtures.* Physica, Vol. 31, pp. 401-406.

Magono, C. and C.W. Lee. 1966. *Meteorological classification of natural snow crystals,* J. Faculty Sci., Hokkaido Univ., Ser. VII (Geophysics) Vol. 2, pp. 321-325.

Male, D.H. and D.I. Norum. 1971. *Movement of water through snow and ice: A thermodynamic analysis.* Hydrol. Symp. No. 8, Quebec, Nat. Res. Counc. Can., Queen's Printer, Ottawa. pp. 83-108.

Martinelli, M. 1960. *Creep and settlement in an alpine snowpack.* Res. Note 43, USDA Rocky Mountain For. Range Exp. Sta.

Mellor, M. 1964. *Properties of snow.* Mono. III-A1. U.S. Army Cold Reg. Res. Eng. Lab., Hanover, N.H.

Mellor, M. 1965. *Some optical properties of snow.* Proc. Int. Symp. on Scientific Aspects of Snow and Ice Avalanches, Davos, Switzerland, Int. Assoc. Hydrol. Sci. Publ. 69, pp. 128-140.

Mellor, M. 1975. *A review of basic snow mechanics.* Proc. Int. Symp. on Snow Mechanics, Grindelwald, Int. Assoc. Hydrol. Sci. Publ. 114, pp. 251-291.

Mellor, M. 1977. *Engineering properties of snow.* J. Glaciol., Vol. 19, No. 81, pp. 15-66.

Mellor, M. and G. Hendrickson. 1965. *Confined creep tests on polar snow.* Res. Rep. 138. U.S. Army Cold Reg. Res. Eng. Lab., Hanover, N.H.

Mellor, M. and J.H. Smith. 1966. *Strength studies on snow.* Proc. Int. Symp. on Scientific Aspects of Snow and Ice Avalanches, Davos, Switzerland, Int. Assoc. Hydrol. Sci. Publ. 69, pp. 100-113.

Mellor, M. and J.H. Smith. 1967. *Creep of snow and ice.* Physics of Snow and Ice. (H. Oura, ed.), Proc. Int. Conf. on Low Temp. Sci., Hokkaido Univ., Sapporo, Vol. 1, pp. 843-855.

Mellor, M. and R. Testa. 1969. *Effect of temperature on the creep of ice.* J. Glaciol., Vol. 8, No. 52, pp. 131-152.

Mokadam, R.G. 1961. *Thermodynamic analysis of the Darcy Law.* Trans. Am. Soc. Mech. Eng., Ser. E, J. Appl. Mech., Vol. 28, pp. 208-212.

Murozumi, M., T.S. Chow and C. Patterson. 1969. *Chemical concentration of pollutant lead aerosols, terrestrial dusts and sea salts in Greenland and Antarctic snow strata.* Geochim. Cosmochim. Acta, Vol. 33, pp. 1247-1294.

Nakaya, U. 1959a. *Viscoelastic properties of snow and ice from the Greenland Ice Cap.* Res. Rep. 46. U.S. Army Snow, Ice Permafrost Res. Estab.,

Nakaya, U. 1959b. *Viscoelastic properties of processed snow.* Res. Rep. 58. U.S. Army Snow Ice Permafrost Res. Estab.

Nakaya, U. 1961. *Elastic properties of processed snow with reference to its internal structure.* Res. Rep. 82, U.S. Army Snow, Ice Permafrost Res. Estab.

Nakaya, U., and A. Matsumoto. 1954. *Simple experiment showing the existence of "liquid water" film on the ice surface.* J. Colloid Sci., Vol. 9, pp. 41-49.

National Research Council. 1954. Int. Assoc. Hydrol. Sci., *Classification for Snow.* Tech. Memo No. 31., Nat. Res. Counc. Can., Ottawa, Ont.

Ockendon, J.R. and W.R. Hodgkins, eds. 1975. *Moving Boundary Problems in Heat Flow and Diffusion.* Clarendon Press, Oxford.

Perla, R.I. 1975. *Stress and fracture of snow slabs.* Proc. Int. Symp. on Snow Mechanics, Grindelwald. Int. Assoc. Hydrol. Sci. Publ. 114, pp. 208-221.

Pitman, D. and B. Zuckerman, 1967. *Effective thermal conductivity of snow at $-88°C$, $-27°C$ and $-5°C$.* J. Appl. Phys. Vol. 38, pp. 2698-2699.

Prantl, F.A. and H.S. Loijens. 1975. *Application des méthodes nucléaires aux études glaciologiques au Canada.* Int. Union Geod. Geophys. Gen. Assem. Grenoble [Isotopes and impurities in snow and ice]. Int. Assoc. Hydrol. Sci. Publ. 118, pp. 237-241.

Ramseier, R.O. 1963. *Some physical and mechanical properties of polar snow.* J. Glaciol., Vol. 4, pp. 753-769.

Ramseier, R.O. and C.M. Keeler. 1966. *The sintering process in snow.* J. Glaciol., Vol. 6, pp. 421-424.

St. Lawrence, W. and C.C. Bradley. 1975. *The deformation of snow in terms of a structural mechanism.* Proc. Int. Symp. on Snow Mechanics. Grindelwald. Int. Assoc. Hydrol. Sci. Publ. 114, pp. 155-170.

St. Lawrence, W. and T.R. Williams. 1976. *Seismic signals associated with avalanches.* J. Glaciol., Vol. 17, pp. 521-526.

Salm, B. 1975. *A constitutive equation for creeping snow.* Proc. Int. Symp. on Snow Mechanics, Grindelwald, Int. Assoc. Hydrol. Sci. Publ. 114, pp. 222-235.

Seligman, G. 1936. *Snow Structure and Ski Fields.* Macmillan and Co., London.

Shumskii, P.A. 1964. *Principles of Structural Glaciology. The Petrograph of Freshwater Ice as a Method of Glaciological Investigation.* [Transl. by D. Kraus], Dover Publications Inc., New York, N.Y.

Slaughter, C.W. 1969. *Snow Albedo modificiation: a review of the literature.* Tech. Rep. 217, U.S. Army Cold Reg. Res. Eng. Lab., Hanover, N.H.

Smith, F.W. and J.O. Curtis. 1975. *Stress analysis and failure prediction in avalanche snowpacks.* Proc. Int. Symp. on Snow Mechanics, Grindelwald, Int. Assoc. Hydrol. Sci. Publ. 114, pp. 332-340.

Smith, J.L. 1963. *Crushing strength and longitudinal wave velocity in processed snow.* Tech. Rep. 137, U.S. Army Cold Reg. Res. Eng. Lab., Hanover, N.H.

Smith, J.L. 1965. *The elastic constants, strength, and density of Greenland snow as determined from measurements of sonic wave velocity.* Tech. Rep. 167, U.S. Army Cold Reg. Res. Eng. Lab., Hanover, N.H.

Smith, J.L. 1969. *Shock tube experiments on snow.* Tech. Rep. 218, U.S. Army Cold Reg. Res. Eng. Lab., Hanover, N.H.

Smith, N. 1969. *Determining the dynamic properties of snow and ice by forced vibration.* Tech. Rep. 216, U.S. Army Cold Reg. Res. Eng. Lab., Hanover, N.H.

Sommerfeld, R.A. 1977. *Preliminary observations of acoustic emissions preceding avalanches.* J. Glaciol., Vol. 19, pp. 399-409.

Sommerfeld, R.A. and E. LaChapelle. 1970. *The classification of snow metamorphism.* J. Glaciol., Vol. 9, No. 55, pp. 3-17.

Stein, J.R. and C.C. Amundsen. 1967. *Studies on snow algae and fungi from the front range of Colorado.* Can. J. Bot., Vol. 45, pp. 2033-2045.

Stokes, J.L., A.H. Rose and J.L. Ingraham. 1963. *Psychrophilic microorganisms.* In Recent Prog. Microbiol., Symp. Int. Congr. Microbiol. No. 8 (N.E. Gibbons ed.), Univ. Toronto Press, pp. 181-212.

Sulakvelidze, G.K. 1959. *Thermal conductivity equation for porous media containing saturated vapor, water or ice.* Bull. Acad. Sci. USSR Geophys. Ser., January, pp. 180-188.

U.S. Army Corps of Engineers. 1956. *Snow Hydrology.* Portland, Oreg., North Pacific Div., Corps of Engineers.

Van Dusen, M.S. 1929. *Thermal conductivity of non-metallic solids.* Int. Crit. Tables, Vol. 5, pp. 216-217.

Van Loon, J.C. 1972. *The snow removal controversy.* Water Pollut. Control, Vol. 71, pp. 18-20.

Visser, S.A. 1973. *The microflora of a snow depository in the city of Quebec.* Environ. Lett., Vol. 4, pp. 267-272.

Von Hippel, A. ed. 1954. *Dielectric Materials and Applications.* Technology Press of Massachusetts Institute of Technology; Cambridge, Mass.; John Wiley and Sons, Inc.; Chapman and Hall, Ltd. New York, London.

Westergaard, H.M. 1964. *Theory of Elasticity and Plasticity*. Dover Publications Inc., New York.

Wooding, R.A. and H.J. Morel-Seytoux. *Multiphase fluid flow through porous media*. Annu. Rev. Fluid Mech., Vol. 8, pp. 233-273.

Yen, Y-C. 1962. *Effective thermal conductivity of ventilated snow*. J. Geophys. Res., Vol. 67, pp. 1091-1098.

Yoshino, T. 1961. *Radio wave propagation on the ice cap*. Antarctic Record (Tokyo), No. 11, pp. 228-233.

Yosida, Z. and others. 1955. *Physical studies on deposited snow: Thermal properties*. Inst. of Low Temp. Sci., Hokkaido Univ., Sapporo, Ser. A. No. 27, pp. 19-74.

Zeigler, H. 1975. *Continuum mechanics: a powerful tool in solving ice and snow problems*. Proc. Int. Symp. on Snow Mechanics, Grindelwald. Int. Assoc. Hydrol. Sci. Publ. 114, pp. 185-195.

8

SNOW DRIFTING

R. J. KIND

*Department of Mechanical and Aeronautical Engineering,
Carleton University, Ottawa, Ontario*

INTRODUCTION

Drifting snow has a considerable economic impact in Canada and other countries where significant amounts of snow fall during the winter. Wind-induced drifting often produces snow accumulations many times greater than those due to snowfall alone and large sums of money are spent annually to remove drifted snow from roads, railways, airports and building areas. Snow loads on roof tops can be increased substantially by drifting (see Ch. 13), and this affects the safety and cost of buildings. Also, wind-induced snow distribution patterns greatly affect the design of areal snowcover measurement programs, snowmelt and runoff phenomena, animal migration patterns and agricultural production. In some instances snowdrift formation is encouraged: to increase snow depths on ski slopes, to augment water resource supplies, and to reduce the depth of frost penetration.

In many respects, the process of snow drifting is similar to that of sand blowing in the deserts and soil drifting on the Prairies. Although the mechanisms affecting the transport of the two materials are similar, it should be recognized that during movement the mass, shape and other properties of a snow crystal may change significantly whereas the corresponding properties of a sand particle remain relatively unchanged. Bagnold (1941) has provided an extensive analysis of the mechanics of blowing sand and his work provides an excellent introduction to many aspects of snow drifting. He explains qualitatively how various surface features such as dunes and ripples are formed. Mellor (1965) also extensively reviews many aspects of blowing snow.

338

CHARACTERISTICS OF THE WIND NEAR THE EARTH'S SURFACE

Basic Wind Structure

The characteristics of the wind near the earth's surface are of major importance in determining whether snow drifting occurs and what the drift patterns are. At altitudes greater than about one kilometre the motion of the atmosphere (wind) is such that the pressure-gradient forces are approximately in equilibrium with the Coriolis forces arising from the interaction of atmospheric motion with the earth's rotation. In effect, the winds at these altitudes are governed by the pressure distributions associated with large-scale weather systems and their characteristics are independent of the surface topography. These winds are known as "geostrophic winds".

At the earth's surface, the so-called no-slip condition requires that the wind speed be zero. Therefore there must be a "boundary layer" across which the speed increases from zero at the earth's surface to its geostrophic value at the top of the layer. The thickness of the boundary layer and the distribution of wind velocity with height z above ground level (the "velocity profile") depend on surface roughness. Over relatively rough terrain the boundary layer is thicker and the wind speed increases relatively slowly with height; over flat, open terrain the opposite is true. Figure 8.1 schematically illustrates these differences. The boundary-layer thickness and velocity profile are also influenced by the thermal stratification of the atmosphere e.g., for unstable stratification (potential temperature decreasing with height above the surface) the boundary layer tends to be thicker than for stable stratification.

The wind speed quoted in weather reports is normally measured at a height of 10 m at a local airport site where the terrain is relatively flat and open; it generally differs from the wind speeds at other heights and/or over different terrain.

Velocity gradients in the boundary layer imply the existence of shear stresses in the wind flow. Shear stress is a maximum at the earth's surface and decreases gradually with height, becoming zero in the geostrophic wind above the boundary layer. The shear stress exerted by the wind on the snowcover surface causes the movement of loose snow.

The wind near the earth's surface is always turbulent or gusty i.e., the wind speed fluctuates with time and location. Since the magnitude of the speed fluctuations is proportional to the average or mean speed, the subsequent discussions will refer to the mean speed.

Snow drifting phenomena depend directly on wind characteristics fairly close to the earth's surface (height < 25 m, say); virtually all natural surfaces act as rough surfaces for the wind. If the wind blows over a rough, but

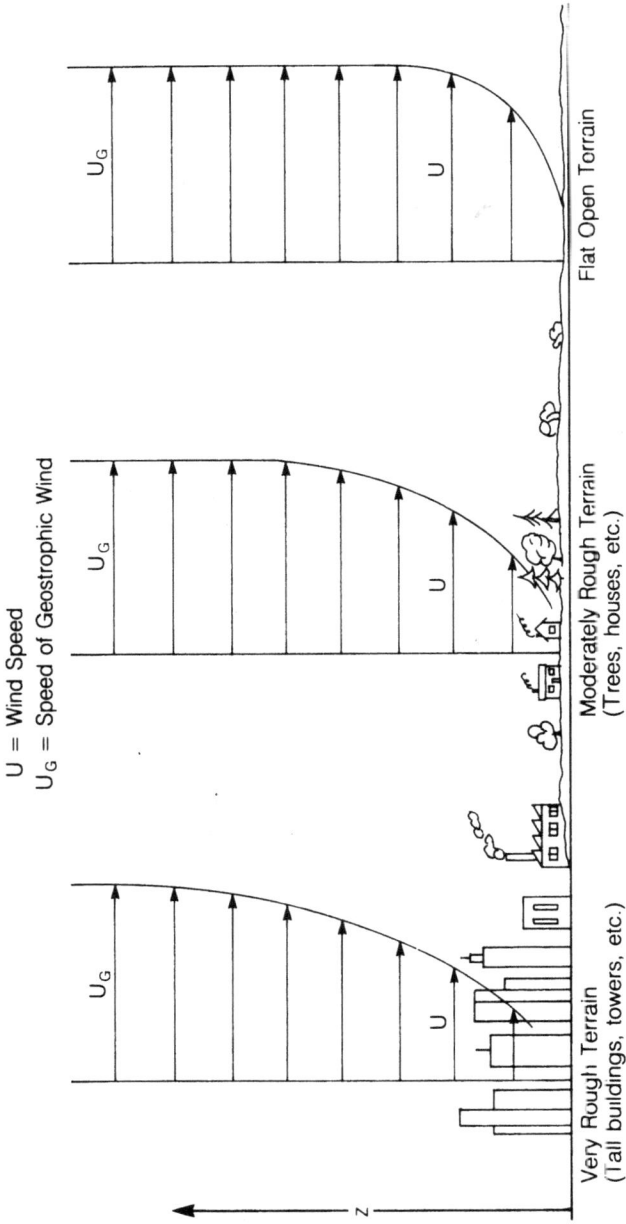

U = Wind Speed
U_G = Speed of Geostrophic Wind

Fig. 8.1 Velocity profiles over terrain with different roughness.

otherwise uniform surface, the mean flow in the lower portion of the atmospheric boundary layer is two-dimensional i.e., the mean wind direction is invariant with height and the flow parameters are invariant in the cross-wind direction. When the wind speed is strong enough to induce snow drifting the flow near the surface is dominated by the shear forces and is not affected by thermal stratification. Under these conditions similarity (dimensional) considerations and field observations both show that the mean wind speed U depends on the height, z, the shear stress at the surface, τ_0, the density of the air, ρ, the mean height of the roughness elements, k_0, and their non-dimensional spacing, λ, according to the following relationship:

$$U/U^* = 2.5 \ln(z/k_0) + 5.5 - C(\lambda) \qquad\qquad 8.1$$

where $U^* = \sqrt{\tau_0/\rho}$, and
          $C(\lambda)$ = a constant depending on λ.

The parameter U^* has dimensions of velocity and is known as the "friction velocity" or "shear velocity". The maximum height above the surface to which this logarithmic velocity-profile equation applies is limited by the upstream fetch over which the surface roughness is reasonably uniform. The equation is appoximately valid for $z <$ fetch/20 (Tani and Makita, 1971) or $z < 50$ m, whichever is least, and for $z > 2k_0$. Thus, it does not describe the velocity profile near or below the tops of the roughness elements where the flow pattern is very complex and often three-dimensional, e.g., near the snow surface where bushes and trees, buildings and other obstacles affect the wind pattern. Alternatively, if the wind flow is not affected by obstructions or roughness elements, Eq. 8.1 adequately describes the velocity profile close to the snow surface as long as the effective roughness height k_0 is properly chosen.

Wind Structure Near Surface Features

The term "surface feature" includes any obstruction or relatively abrupt change of surface geometry, e.g., buildings, snow fences, cliffs, embankments, highway cuts, bushes and trees. If surface features with an average height k_0 are distributed fairly uniformly they act as distributed roughness elements for wind. Equation 8.1 then represents the mean velocity profile above a height of about $2k_0$. However, the flow near and below the tops of the surface features or near relatively-isolated surface features is complex, giving rise to rapid spatial variations of the shear stress τ_0 or the shear velocity U^*. This results in the formation of snowdrifts in some areas and in erosion of loose snow from others.

In steady flow, streamlines represent the paths followed by fluid particles; no flow can cross streamlines. The mass density ρ remains constant when air flows over the surface of the earth. Therefore, if streamlines converge such that the flow between them must pass through a smaller cross-section the flow must accelerate; on the other hand, if the streamlines diverge or spread apart, the flow must decelerate. Because of viscous forces, fluid flows break away or separate from solid surfaces if motion along the surface would cause the flow to decelerate rapidly. For this reason flow over a spherical object separates from the rear of the object resulting in a so-called separation bubble behind it; the air in this bubble is sometimes referred to as "dead air" because its velocity is very low and its motion is turbulent and disorganized. On the other hand, flow over a "tear-drop" shaped or "streamlined" object does not separate and no separation bubble is formed.

As outlined above, separation bubbles commonly form downwind of two-dimensional surface features, but can also occur on the upwind side, usually less frequently and with a smaller size. Since the velocity of the "dead air" region is low, the surface shear stress τ_0 is also very small, encouraging the formation of snow drifts and limiting their subsequent erosion. Figures 8.2a and 8.2b are schematic diagrams of flow over a slatted snow-fence and over an embankment, respectively, showing the extent of each separation bubble.

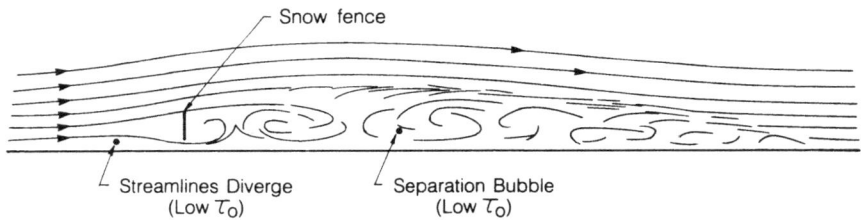

(a) Flow over a Slatted Snow Fence

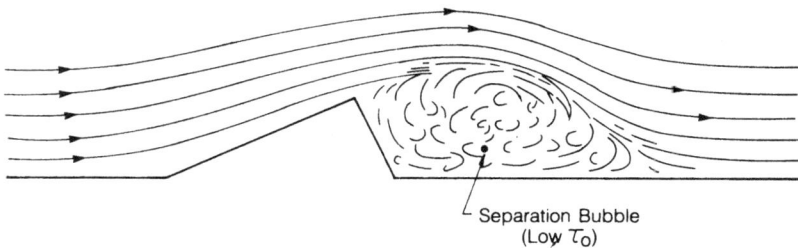

(b) Flow over an Embankment

Fig. 8.2 Wind flow patterns over two-dimensional surface features.

The flow over three-dimensional objects, such as buildings, is even more complex than that over approximately two-dimensional features such as snow fences or embankments. Separation bubbles still occur although their flow is sometimes relatively better organized compared with that in a two-dimensional separation bubble. Three-dimensional separation and re-attachment often differs greatly from the two-dimensional case, giving high, rather than low, surface shear stress in some regions of the flow. A common occurrence found in wind flow over a three-dimensional surface feature is the formation of a horseshoe vortex around its base (Fig. 8.3a shows a flow pattern for a cylinder). The surface shear stress under the vortex tends to be fairly high and usually erodes the snow around the base; see Fig. 8.3b.

The limitless varieties and combinations of surface features encountered in nature make it impossible to discuss the flow for all possible situations. The foregoing discussion should give the reader an appreciation of some of the major effects influencing snowdrift patterns produced when wind blows over surface features.

RELATIONSHIPS BETWEEN WIND AND SNOW MOVEMENT

Threshold Conditions

Individual particles of snow lying on the surface are initially caused to move by the drag force (i.e., downwind force) exerted on them by the moving air. This drag force (per unit plan area) is the shear stress τ_0 between the moving air and the snow surface, it is this force which causes particles to move. Snow particles often tend to stick to one another so that the inter-particle cohesive forces and particle weights oppose motion. Thus, before snow drifting can occur, the shear stress must exceed some "threshold" value sufficient to overcome these resisting forces. In studies of snow drifting the shear velocity is commonly used instead of the shear stress. The shear velocity is a hypothetical velocity, defined as $U^* \equiv \sqrt{\tau_0/\rho}$ (see Eq. 8.1); its "threshold" value U^*_{th} depends on the size, shape and weight of each snow particle and on the cohesive forces, and therefore its magnitude varies widely. At snow temperatures below $\sim$ -2°C the cohesive forces are negligible in newly-fallen snow, and after a little movement the large snow particles are broken into smaller ones. The major portion of the mass in the surface layer of the snowpack then consists of particles having fairly similar shapes and a size (nominal diameter) of about 0.5 mm. Their threshold shear velocities may range from about 0.1 to 0.2 m/s (0.15 m/s, a representative mean). While immobile they shelter the very fine particles (snow dust) from wind action.

The threshold velocity of a snowcover is also affected by the "wetness" and "hardness" of the snow. Because of larger cohesive forces, U^*_{th} values are

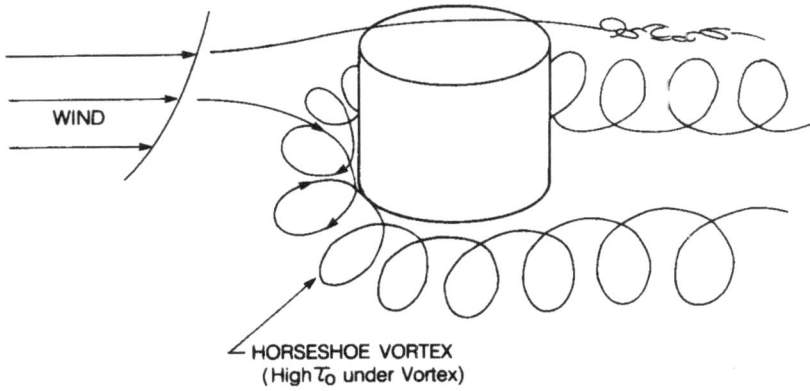

(a) Wind flow patterns around a cylindrical surface feature.

(b) Snowdrift pattern near a three-dimensional surface feature.

Fig. 8.3 Patterns around simple three-dimensional surface features.

much higher for wet and aged snows than for newly-fallen dry snow. The drifting process itself causes compaction and hardening of the snow surface; as a result of wind hardening, the threshold shear velocity increases progressively when drifting continues in the absence of snowfall. Table 8.1 and Fig. 8.4 give approximate values of $U*_{th}$, for different snowcover surface conditions. Snow particles scattered thinly on smooth hard surfaces are more exposed to wind action and their threshold shear velocities are lower than those for continuous layers of similar particles.

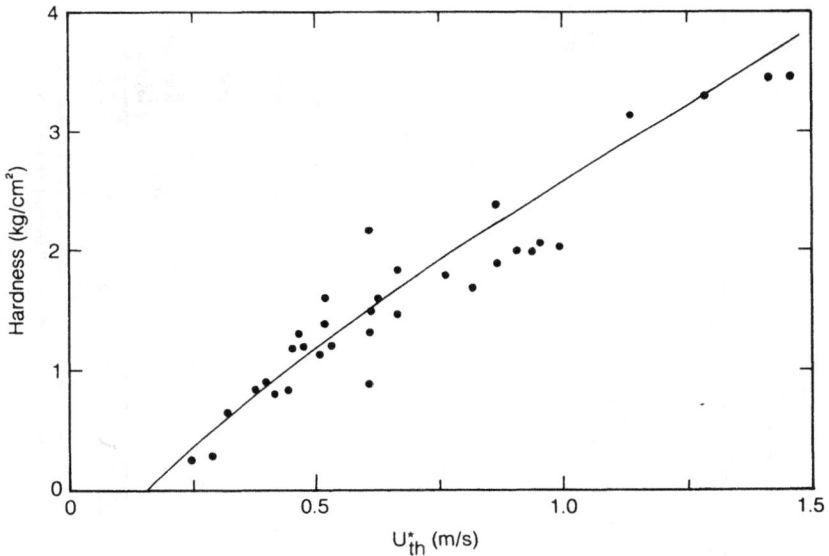

Fig. 8.4 The influence of the hardness of the snow surface at a temperature of $\simeq -15°$ C on threshold shear velocity (based on Antarctic field data reported by Kotlyakov, 1961).

Modes of Transport of Blowing Snow

The average density of a snow particle is approximately that of ice but is much greater than that of air. Owen (1969) shows that the aerodynamic lift forces acting on the particles are negligible compared to the aerodynamic drag forces. Therefore, a snow particle can only rise high above the earth's surface if the vertical velocity components in the air flow are approximately equal to or greater than the terminal fall velocity of the particle. In blowing snow the only significant vertical components are normally those in the turbulent

Table 8.1

THRESHOLD SHEAR VELOCITY FOR VARIOUS SURFACE CONDITIONS

CHARACTERISTICS OF SURFACE LAYER OF SNOWPACK

DESCRIPTION	Surface Hardness[a] kPa	Snow Density kg/m³	Temperature (°C) Snow	Temperature (°C) Air	$U^*{}_{th}$ m/s	SOURCE OF FIELD DATA
Exceptionally light dry snow					0.07	Rikhter (1945) and Kungertsev (1956) as quoted by Isyumov (1971).
Loose fresh dry snow	0.10	50	<-2.5° C		0.15	Kobayashi (1973); Oura et al. (1967); Rikhter (1945) and Kungertsev (1956) as quoted by Isyumov (1971).
Slightly aged (several hours) dry snow				-7	0.22	Oura et al. (1967)
Newly fallen snow at 0° C			0		0.25	Oura et al. (1967)
Slightly packed dry snow	0.9	120		-5.5	0.27	Kobayashi (1973), run 52.
Slightly aged (several hours) snow near 0° C				-1	0.4	Oura et al. (1967)
Old hardened snow					0.4	Kungertsev (1956) quoted by Isyumov (1971).
Wind hardened snow (a)	100	350		-15	0.4	Kotlyakov (1961); see also Fig. 8.4.
(b)	250	400		-15	1.	

[a] Hardness of the snow surface is defined as the pressure required to produce collapse of the surface when exerted by a standard disc.

eddies. The eddy velocities approximate the terminal fall velocities of the particles only when the shear velocity U^* is greater than $\sim 5\, U^*_{th}$ (Bagnold, 1973), e.g., for a snowcover with $U^*_{th} = 0.15$ m/s, U^* would have to be 0.75 m/s for the vertical velocity to be roughly equal to the fall velocity. In nature, wind speeds of ~ 14 m/s at a height of one metre can produce such a high shear velocity but are very rare.

Therefore, the dominant size-group of snow particles, whose characteristics govern the value of U^*_{th}, normally cannot rise to significant heights above the snow surface. When $U^*_{th} < U^* < 5U^*_{th}$ the snow particles bounce along close to the surface, in a mode of motion called "saltation". When U^* is greater than U^*_{th}, the mobility of the dominant size-group of particles exposes the very fine particles to the wind action. Because this snow dust has a very low terminal fall velocity it can rise to great heights in the atmosphere, become suspended and airborne in the turbulent airflow, reduce visibility and may cause plugging of air intakes placed above the ground. Usually, however, the largest portion of the total mass flux of snow in movement is by saltation or "ground drift" of the heavier particles very near the surface. The distribution of the horizontal mass flux of blowing snow with height, as developed from field data reported by Oura (1967) and Kobayashi (1973), is shown in Fig. 8.5. The trajectories of saltating snow particles are shown in Fig. 8.6 (taken from Kobayashi, 1973). These figures support the proposition that most of the mass transport occurs by saltation in a thin layer above the surface and that this is the dominant transport mechanism involved in the erosion or deposition of snow through wind action. Budd (1966) and Budd et al. (1966) present both theoretical and experimental results for the mass flux occurring above the saltation layer.

Wind Velocity Profiles in Blowing-Snow Conditions

As discussed earlier, the velocity of the wind varies with height above the surface, having a profile which depends mainly on surface roughness. Suspended snow dust has a negligible effect on the wind velocity profile near the surface. In contrast, the mass flux of saltating snow particles exerts a pronounced effect on this profile. While rising from the surface the particles are accelerated horizontally by the aerodynamic drag force, and extract kinetic energy from the wind. This energy is subsequently dissipated in the snowcover when each particle strikes the surface at the end of a trajectory. Saltating particles thus exert a drag force on the airflow similar to fixed roughness elements such as vegetation. Therefore saltation increases the effective roughness of the snow surface and modifies the wind velocity profile, as verified by both wind-tunnel and field data. Owen (1964) shows that the

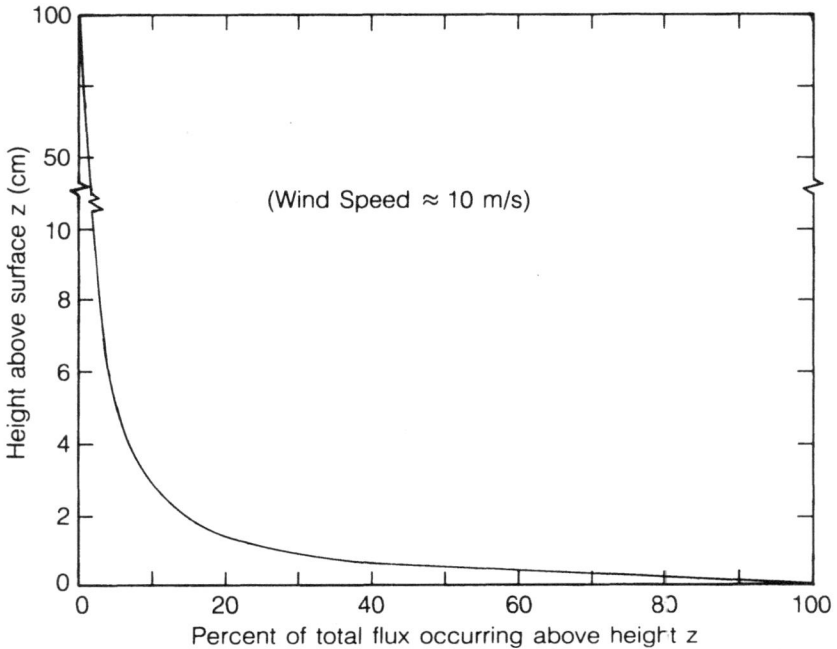

Fig. 8.5 Approximate distribution of horizontal mass flux of blowing snow with height (based on data reported by Oura, 1967 and Kobayashi, 1973).

profile above the saltation layer over a saltating bed of soil or sand of various grain sizes can be described by the equation:

$$U/U^* = 2.5 \ln (2gz/U^{*2}) + 9.7. \qquad\qquad 8.2$$

Since the saltation layer is thin in height, z can be measured with sufficient precision from the snow surface, although, strictly speaking, its zero plane (the height at which the wind velocity given by Eq. 8.2 is zero) is at some unknown height within the saltation layer. Kind (1976) using field data given by Oura et al. (1967) showed that Eq. 8.2 also satisfactorily represents the wind velocity profile over saltating snow. This equation is effectively the same as Eq. 8.1 which describes the wind speed profile over surfaces with fixed roughness elements, provided $U^{*2}/2g$ is interpreted as being proportional to the effective roughness height k_o of the saltating surface. Dimensional analysis of the drifting snow process suggests that the height of the saltation trajectories should be proportional to $U^{*2}/2g$. Equation 8.2 applies in blowing-snow conditions, subject to the same restrictions as those for Eq. 8.1.

The relationship between U^* and the wind velocity U at height z above the snow surface given by Eq. 8.2 can be used to determine the wind velocity

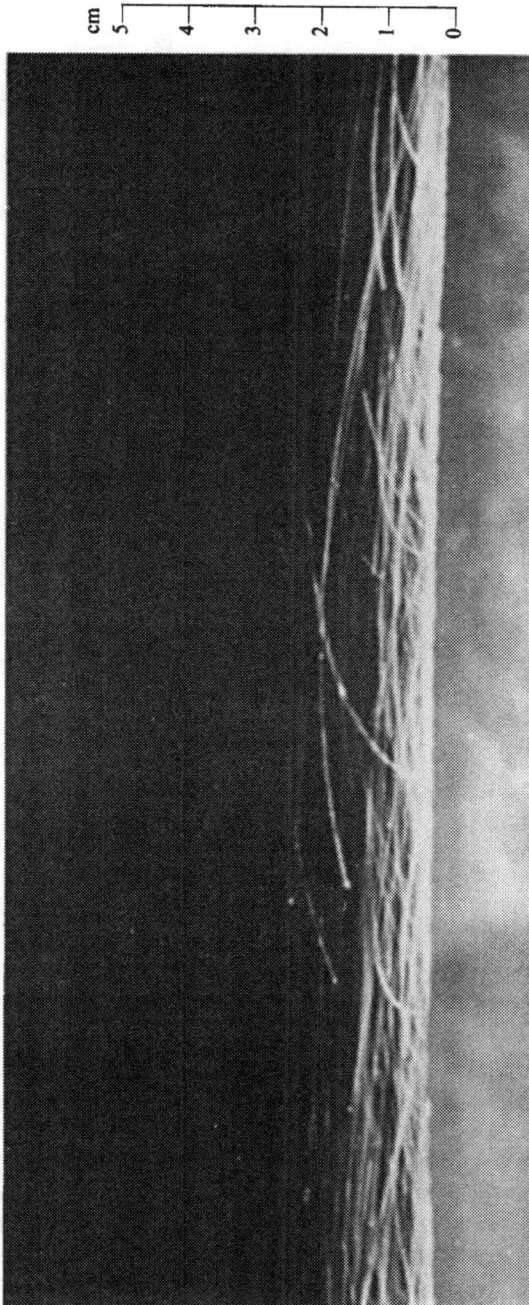

Fig. 8.6 Photograph of trajectories of saltating snow particles; U = 3.9m/s at z = 1 m. (Kobayashi, 1973).

required at any height to cause blowing of snow for a given threshold shear velocity $U*_{th}$. Figure 8.7, which is derived from Eq. 8.2, gives the minimum wind velocity as a function of z for several $U*_{th}$ values; the calculated velocities required to produce threshold conditions for fresh dry snow agree closely with field observations.

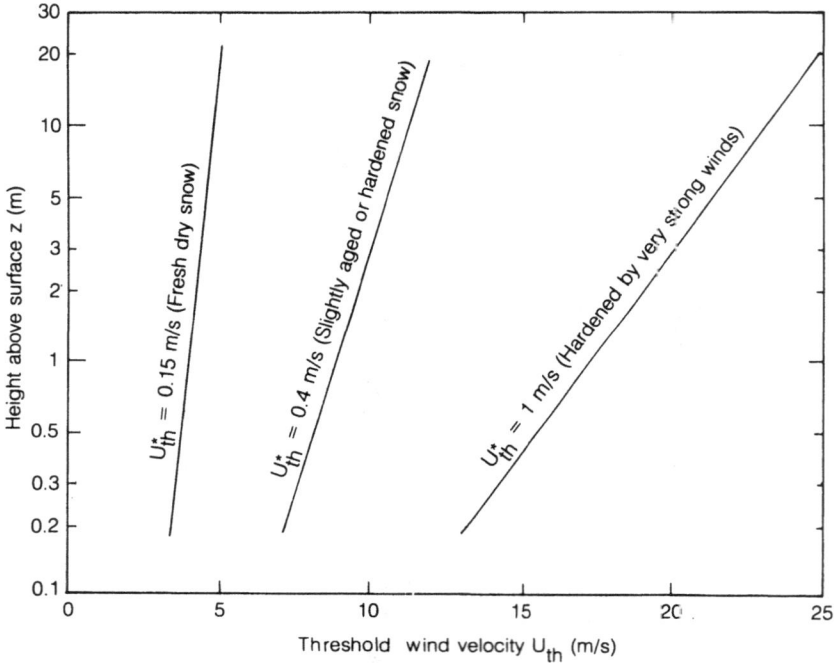

Fig. 8.7 Threshold wind velocity required to cause blowing snow.

Rate of Transport of Snow by Wind

When saltation is the major mechanism causing snow movement most of the mass flux occurs in a thin "saltation layer" about 1 cm thick (see Fig. 8.5). The individual snow particles rise approximately vertically from the snow surface and are accelerated horizontally by aerodynamic drag forces. Obviously, the particles are subject to the force of gravity which tends to move them back to the snow surface, where upon impact, most of their kinetic energy is dissipated by "splashing" into the surface. The force of impact causes other snow particles to be ejected more or less vertically from the surface so that the process is continued. Since the airborne particles are accelerated horizontally by aerodynamic drag and lose their kinetic energy upon impact, they extract horizontal momentum from the air flow. Thus the saltating particles are responsible for a shear stress between the wind and the surface; the ability of the wind to transport snow by saltation depends on the magnitude of the shear stress in the near-surface flow. Bagnold (1941) discusses how the foregoing argument can be used to identify the main parameters controlling the rate of mass transport by saltation. Consider a "lane" aligned with the wind direction in which the mass flow of saltating particles is G (g/s per unit width perpendicular to the wind direction). Similarity considerations suggest that all velocities near the surface must be proportional to the shear velocity $U*$ so that the average horizontal velocity U attained by the particles during a saltation trajectory is proportional to $U*$. Similarity theory and experimental evidence also show that the average height of the saltation trajectory is proportional to $U*^2/g$; hence the average length of the trajectories must also be proportional to $U*^2/g$. The rate at which the saltating particles extract horizontal momentum from the wind per unit surface area is equal to the shear stress τ exerted by the wind on the saltating particles, i.e.,

$$\tau \propto GU*/\text{surface area}. \qquad\qquad 8.3$$

Since the appropriate surface area is equal to the unit width multiplied by the average trajectory length Eq. 8.3 can be written as

$$\tau \propto GU*/(U*^2/g). \qquad\qquad 8.4$$

Owen (1964) argues that the shear stress exerted by the wind on the surface itself must always remain at its threshold value, $\rho U*^2_{th}$, so that τ given by Eq. 8.4 must represent the difference between the total shear stress, $\tau_0 \equiv \rho U*^2$, exerted by the wind and $\rho U*^2_{th}$. Thus Eq. 8.4 can be written;

$$G = \rho U*^3 \alpha \, [1 - (U*_{th}/U*)^2]/g, \qquad\qquad 8.5$$

where α is a constant. Owen (1964) collected experimental evidence which suggests that α takes the form

$$\alpha = 0.25 + (w/3U*), \qquad\qquad 8.6$$

where w is the terminal falling velocity of the particles. For snow, Isyumov (1971) suggests a typical value of w is 0.75 m/s.

Equation 8.5 shows that the mass transport rate G is approximately proportional to $U*^3$. The ratio $U/U*$ at a fixed height varies slowly in response to changes in velocity; therefore the capacity of the wind to transport snow increases approximately as U^3. Figure 8.8 shows graphs of Eq. 8.5 using $U/U*$-values calculated from Eq. 8.2 and α calculated from Eq. 8.6 for representative values of w and $U*_{th}$. Some field data of Kobayashi (1973) have been included for comparison. The curves show a very rapid increase of the drift rate with increasing wind velocity. The calculated drift rates are based on wind velocities calculated for a 1 m height. These velocities can be related to those at other heights using Eq. 8.2 or by referring to Fig. 8.7 which gives, in effect, plots of Eq. 8.2 for several values of the shear velocity.

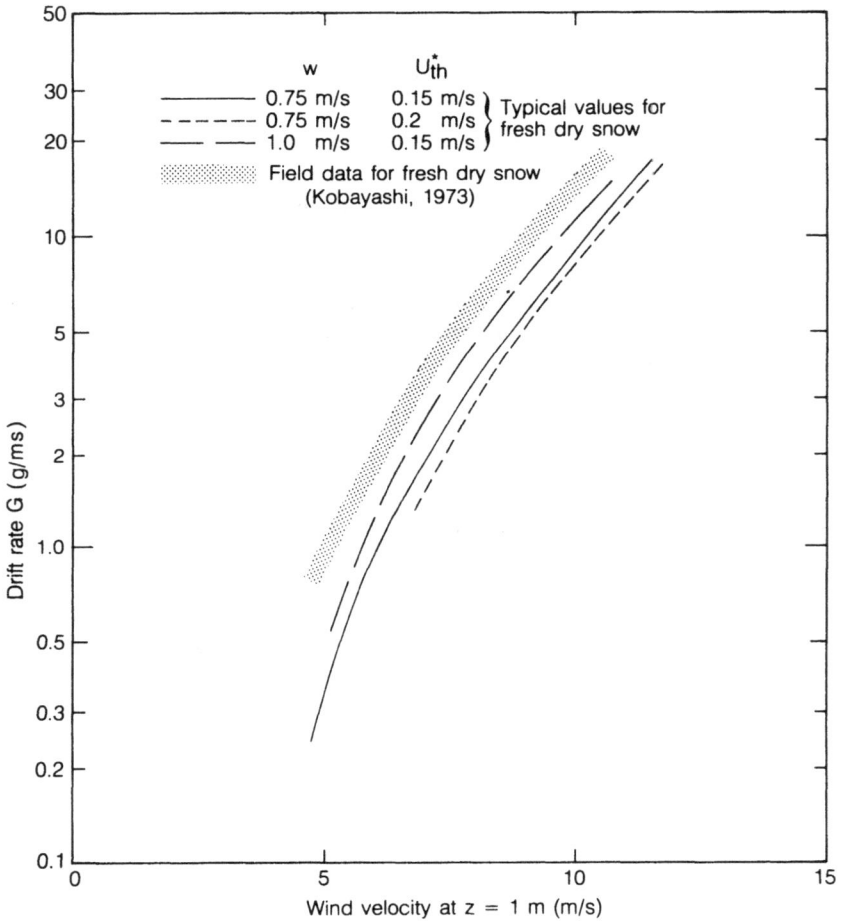

Fig. 8.8 Variation of snow drifting rate with wind speed at z = 1 m (two-dimensional flow).

It should be noted that Eq. 8.5, like Eq. 8.2, only applies when the wind flow is two-dimensional, the shear velocity is not varying rapidly and the wind has passed over a substantial upstream fetch of loose snow so that equilibrium saltation conditions have been established. Such conditions do not prevail, for example, when the wind is blowing around buildings or other surface features. Similarity considerations indicate, however, that for all conditions $Gg/\rho U*^3$ depends only on the location and the ratios $U_{th}*/U*$ and $w/U*$. Thus, even when Eq. 8.5 does not hold quantitatively, the ability of the wind to transport snow is approximately proportional to $U*^3$ at any given site.

Some uncertainty also exists regarding the minimum upstream fetch of loose snow required to establish the equilibrium drift rate. Kobayashi (1973) suggests a fetch of 30 to 60 m. Some unreported wind-tunnel experiments conducted by the author, using silica sand, suggest that equilibrium saltation conditions may develop over fetch distances an order of magnitude less than these, at least when $U*/U*_{th}$ is greater than about 1.5.

Field measurements show that, while drifting occurs, substantial amounts of snow undergo sublimation (Dyunin, 1967; Tabler, 1971, 1975), since all surfaces of the saltating particles are exposed to the moving air. However, sublimation is not expected to significantly affect the drift rate because the wind will simply pick up additional particles from the surface to replace the mass lost by sublimation. Thus sublimation should not affect either the drift rate or snowdrift patterns, but would reduce the total mass of snow deposited in a snowdrift in those cases where the upwind supply of loose snow is limited.

Unlike surfaces consisting of loose snow particles, hard immobile surfaces such as bare roads or bare rooftops absorb little kinetic energy from the impact of saltating particles. Consequently the wind can transport larger fluxes of snow over such surfaces where there is less tendency for snow deposition to occur.

Snowdrift Formation

When snowfall occurs during windy conditions localized snow accumulations much deeper than the average "areal" snowfall depth can form by means of two mechanisms: by focussing of snowflake trajectories; and by non-uniform ground drift. Buildings and other surface features cause local non-uniformities in the flow field near them, i.e., the wind is accelerated in some regions and is slowed down in others, notably in the wake of a surface feature. These velocity changes affect the snow distribution pattern. The horizontal non-uniformity of wind speed causes a certain amount of "focussing" of the trajectories of the falling snowflakes so that snow deposits roughly two or three times deeper than the average snowfall depth can be formed in regions with relatively low wind speeds. Also, the non-uniform flow field exerts a non-uniform shear stress on the snowcover, the shear stress being lower in regions of low wind speed and vice versa. As the drift rate or rate of mass transport by saltation strongly depends on the shear stress, a non-uniform drift rate results, so that snow erodes from some areas and deposits in others. More precisely, in regions where U* and G decrease along the direction of flow (s-direction) the drift rate into the region is greater than the drift rate out, and snow must be deposited. That is,

$$\text{for deposition } dU^*/ds < 0, \text{ and}$$

$$\text{for erosion } dU^*/ds > 0.$$

Of the two mechanisms capable of causing relatively deep snow deposits the "focussing" of snowflake trajectories is less important, except perhaps in special cases such as the formation and distribution of snow loads on rooftops. The focussing can only occur during a snowfall, and moreover can only be appreciable when the obstructions are rather high so that the horizontal non-uniformity of the wind speed extends over a fairly deep vertical layer.

A non-uniform ground drift can occur whenever the wind velocity exceeds the threshold value, regardless of whether snow is falling. Consequently, the depth of deposit produced by this mechanism continuously increases provided sufficient loose snow lies upwind of the obstruction which itself must not be completely buried by snow. For example, an unprotected highway cut in open terrain can easily be completely filled with snow because of ground drifting even after only a light snowfall. Generally the snow deposits commonly termed "snowdrifts" are formed by this ground drift mechanism. Because the shear stress patterns are very complex the corresponding drift patterns produced are equally as complex. Moreover, the stress patterns change as the snow deposits build up and change the structure of the flow. Figure 8.3b shows the drift pattern around a small three-dimensional surface feature, and clearly shows erosion of snow from the region of high shear stress under the horseshoe vortex.

Principles of Drift-Control Devices

Snowdrifts are formed when two conditions exist simultaneously: there is ground drift or saltation; and the shear stress or shear velocity decreases in the flow direction. Two principles can therefore be exploited to control (and prevent) snowdrift formation. The wind velocity must either be reduced so that the blowing snow may be deposited before passing over the area to be protected or it must be maintained so that the shear stress is large enough to transport the snow past the area.

Snow fences are the most common device used to control snowdrift formation. They cause a decrease in the local shear stress or shear velocity at the ground downwind from their location and, to a lesser extent, upwind, resulting in deposition of the ground-drifting snow. Thus, if a snow fence is properly located it will force the snow to be deposited before reaching the area being protected. It must be placed a sufficient distance upwind of the area so that the snowdrift produced does not extend into it. On the other hand the fence must not be placed so far upwind so that the drift rate can recover substantially before reaching the area. Practical rules for positioning snow fences are given in Ch. 16. Snow fences lose their effectiveness when they become "saturated" i.e., when they are nearly buried in the drift which they produce. When saturated they take on a streamlined shape and will not appreciably lower the shear velocity.

Snowdrift formation can also be prevented by ensuring that the shear velocity is not reduced. In practice this problem is solved by designing a structure to have an appropriate shape (geometric) so that the air flow does not separate when passing over it. For example, drift formation on a highway can be prevented by raising its surface slightly above that of the surrounding fields so that the streamlines converge and the flow accelerates. In this regard the production of high snowbanks during snow removal should be avoided.

In some instances it may be desirable to encourage drift formation. The principles to be exploited are unchanged, but the objectives are reversed. If possible, a long unobstructed fetch should be provided upwind of the site so that the drift rate can reach its maximum value there, and a local lowering of the shear velocity should be induced over the site.

MODEL SIMULATION OF SNOW DRIFTING

Snow drifting is clearly a complex phenomenon whose exact dynamics of movement are difficult to explain even in the simplest situation, i.e., for a smooth, flat, infinite plane. The complexity increases greatly when the geometry is affected by hilly terrain, trees, buildings, snow fences, etc. For such cases no theoretical methods are available for predicting the location, size, and depth of snowdrifts. Often this is not a serious handicap to engineering practice as past experience can provide sufficiently accurate and reliable design information. Unusual situations, for which no directly relevant data exist, nevertheless arise fairly frequently. Reliable modelling techniques for predicting significant properties of snowdrifts are therefore desirable.

Mellor (1965) points out that the value of wind-tunnel testing in snowdrifting studies has been widely recognized for about 40 years, but few studies were conducted during that time. By using sawdust and flake mica to simulate snow in a wind tunnel Finney (1939) was able to suggest improved methods for snowdrift control along highways. More recently, Theakston (1962) studied snow accumulation around structures by using first wind tunnels and later water flumes. Quite extensive studies on drifting snow have been conducted at New York University by Gerdel and Strom (1961). In the early tests modelling criteria were apparently not explicitly considered. It is impracticable to satisfy the large number of similarity criteria yielded by simple dimensional analysis of the system. However, similarity analysis which draws more heavily on physical analysis and experimental data should be more fruitful. Strom et al. (1962), Odar (1962, 1965), Isyumov (1971) and Kind (1976) have also studied this problem.

Since the similarity analysis is relatively complex only a brief summary is presented here. The mean and fluctuating velocity fields of the model and prototype flows must be similar. It is assumed that the collisions between saltating snow particles can be neglected so that they are in free flight except at the beginning and end points of their trajectories. Correct modelling of the free-flight portion of a trajectory requires that the ratios: drag force: weight, and drag force: inertia force, be the same for the model and prototype particles. Similarly, saltation must begin at corresponding velocities in the model and prototype. Kind (1976) concluded that these requirements are met when the following design conditions are satisfied:

$$(U^*_{th}/V)_p = (U^*_{th}/V)_m,$$ 8.7

$$(w/V)_p = (w/V)_m,$$ 8.8

$$(V^2/Lg)_p = (V^2/Lg)_m,$$ 8.9

where: V = reference wind velocity, e.g., the nominal wind velocity,
  w = terminal velocity of a saltating particle,
  L = reference dimension, e.g., the dimension of a building around which drift formation is being studied, and
  p, m = subscripts indicating prototype and model systems, respectively.

As discussed earlier the saltating particles introduce an effective roughness to the surface which is proportional to $U^{*2}/2g$. Experiments have shown that the flow of fluids over rough surfaces is independent of the kinematic viscosity of the fluid ν when the Reynolds number N_R (the ratio of inertial to viscous forces calculated using the roughness height as the length dimension) exceeds a certain value. This flow case prevails under field conditions and hence in the model simulation $N_R = U^{*3}/2g\nu$ should exceed 30.

The angle of repose of the model particle material should be as large as possible, although it is almost impossible to model this property of snow accurately. The size and density of the particles are not directly important except in the way they influence U^*_{th} and w. The coefficient of restitution is also not important provided the model particles lie within a fairly narrow range of sizes.

Methods for selecting model particles are outlined by Kind (1976). Similarity of model and prototype velocity fields requires that the model be geometrically similar to the prototype, and the velocity profile of the wind flowing towards the model must be similar to that in the prototype; i.e., both profiles must have the same value of U^*/V. The model must also include a long fetch of correctly modelled terrain, complete with particles, upstream of the region of interest. This is necessary so that saltation may be properly established by the time the flow reaches this region. Unfortunately, the requirement that V^2/Lg be the same in both model and prototype usually conflicts with the requirement that $U^{*3}/2g\nu$ be greater than 30. This problem is discussed by Kind (1976) and Kind and Murray (1980).

If the above requirements are satisfied, snowdrift patterns in the model will develop in the same way as those in the prototype. The non-dimensional drift-rate parameter $Gg/\rho V^3$ will then also have the same value at corresponding locations and non-dimensional times in the model and prototype. However, the mass densities of the "snow" particles will usually differ and this must be taken into account when determining the times required for volume accumulations of snow when using data from the model tests.

The overall simulation requirements include those for correctly simulating the free-flight of falling particles. Snowfall and snowdrifting can therefore be modelled simultaneously without additional difficulty. Simulation of snowfall is accomplished by dropping particles into the flow at an appropriate rate from the roof of the wind tunnel.

Sometimes only the simulation of airborne blowing snow is necessary and ground-drift is irrelevant, e.g., in studies of the ingestion of snow into air intakes installed at about one or several metres above the ground. In such cases the values of U^*_{th}/V and $U^{*3}_{th}/2g\nu$ are of no concern. If the airborne particles are very small they essentially follow the airflow so that smoke or dye tracers could be substituted.

As mentioned earlier the ratio of drag force to lift force of snow particles is large because of the large ratio of particle density to air density. However this would not be true if water were the model fluid; therefore the use of water to model ground drifting is questionable. Nevertheless, model studies in water flumes have been and will continue to be useful (Theakston, 1962; Isyumov, 1971). The flow of water over buildings and other surface features is basically the same as that of air. Saltation is also qualitatively the same in both fluids but there are large differences in magnitudes for certain effects. Drift patterns obtained in a water flume should be reasonably realistic, if the velocity profile in the flow approaching the model is adequately simulated [1], and if the model itself is geometrically similar to the prototype. Such simulations would not, however, be expected to yield reliable quantitative results for rates of snow accumulation or drift depths, and would tend to be somewhat inaccurate for drift shape and location even though usually qualitatively correct. The same would apply to wind-tunnel tests if the conditions imposed by Eqs. 8.7, 8.8 and 8.9 and the Reynolds number criterion ($U*^3_{th}/2g\nu > 30$) were not met. The consequences of relaxing the latter criterion are not clear at the present time and more experimental data are required to clarify this.

LITERATURE CITED

Bagnold, R.A. 1941. *The Physics of Blown Sand and Desert Dunes*. Methuen and Co., London.

Bagnold, R.A. 1973. *The nature of saltation and of bed-load transport in water*. Proc. R. Soc. (London), Ser. A, Vol. 332, pp. 473-504.

Budd, W.F. 1966. *The drifting of non-uniform snow particles*. Studies in Antarctic Meteorol., Am. Geophys. Union, Antarc. Res. Ser., Vol. 9, pp. 59-70.

Budd, W.F., R. Dingle and U. Radok. 1966. *Byrd snowdrift project: outline and basic results*. Studies in Antarctic Meteorology, Am. Geophys. Union, Antarc. Res. Ser., Vol. 9, pp. 71-134.

Dyunin, A.K. 1967. *Fundamentals of the mechanics of snow storms*. Physics of Snow and Ice (H. Oura, ed.), Proc. Int. Conf. on Low Temp. Sci., Hokkaido Univ., Sapporo, Vol. 1, Pt. 2, pp. 1065-1073.

Finney, E.A. 1939. *Snowdrift control by highway design*. Bull. No 86, Mich. Eng. Exp. Sta., East Lansing, Mich., pp. 1-58.

Gerdel, R.W. and G.H. Strom. 1961. *Wind tunnel studies with scale model simulated snow*. Int. Union Geod. Geophys., Gen. Assem. Helsinki, [Snow and Ice], Int. Assoc. Sci. Hydrol. Publ. No. 54, pp. 80-98.

Isyumov, N. 1971. *An approach to the prediction of snow loads*. Ph.D. Thesis, Univ. Western Ont., London.

Kind, R.J. 1976. *A critical examination of the requirements for model simulation of wind-induced erosion/deposition phenomena such as snow drifting*. Atmos. Environ., Vol. 10, pp. 219-227.

Kind, R.J. and S.B. Murray. 1980. *Saltation flow measurements relating to modelling of snowdrifting*. Unpubl. Pap., Dept. Mech. Aeronaut. Eng., Carleton Univ., Ottawa, Ont.

Kobayashi, D. 1973. *Studies of snow transport in low-level drifting snow*. Rep. No. 231, Inst. Low Temp. Sci., Sapporo, pp. 1-58.

---

[1] Workers in the field of wind effects on buildings and structures generally acknowledge that failure to simulate the wind velocity profile and turbulence structure can sometimes lead to unrealistic flow patterns and unrealistic test results (Scruton, 1967).

Kotlyakov, V.M. 1961. *Results of study of the ice sheet in Eastern Antarctica*. Int. Union Geod. Geophys. Gen. Assem. Helsinki [Antarctic Glaciology], Int. Assoc. Hydrol. Sci. Publ. 55, pp. 88-99.

Kungertsev, A.A. 1956. *The transfer and deposit of snow*. Geogr. Inst., Acad. Sci. USSR [English Transl. by Univ. Colo. Eng. Sci. Res. Div., Boulder, Colo.]

Mellor, M. 1965. *Blowing snow*. Mono. III-A3c, U.S. Army Cold Reg. Res. Eng. Lab., Hanover, N.H.

Odar, F. 1962. *Scale factors for simulation of drifting snow*. Proc. Am. Soc. Civ. Eng., Eng. Mech. Div., Vol. 88, pp. 1-16.

Odar, F. 1965. *Simulation of drifting snow*. Res. Rep. 174, U.S. Army Cold Reg. Res. Eng. Lab., Hanover, N.H.

Oura, H. 1967. *Studies of blowing snow I*. Physics of Snow and Ice (H. Oura, ed.), Proc. Int. Conf. on Low Temp. Sci., Hokkaido Univ., Sapporo, Vol. 1, Pt. 2, pp. 1085-1097.

Oura, H., T. Ishida, D. Kobayashi, S. Kobayashi and T. Yamada. 1967. *Studies on blowing snow, II*. Physics of Snow and Ice (H. Oura, ed.), Proc. Int. Conf. on Low Temp. Sci., Hokkaido Univ., Sapporo, Vol. 1, Pt. 2, pp. 1098-1117.

Owen, P.R. 1964. *Saltation of uniform grains in air*. J. Fluid Mech., Vol. 20, Pt. 2, pp. 225-242.

Owen, P.R. 1969. *Pneumatic transport*. J. Fluid Mech., Vol. 39, Pt. 2, pp. 407-432.

Rikhter, G.D. 1945. *Snezhnyi Pokrov, Ego Gormirovanie i Svoistva (Snow cover, its formation and properties)*. Izv. Akad. Nauk. SSSR, Moscow. [English Transl. by U.S. Army Snow, Ice and Permafrost Res. Estab., Transl. No. 6].

Scruton, C. 1967. *Aerodynamics of structures*. Proc. Int. Res. Sem. Wind Effects Build. Struct., Ottawa, Vol. 1, pp. 115-161.

Strom, G., G.R. Kelly, E.L. Keitz and R.F. Weiss. 1962. *Scale model studies on snow drifting*. Res. Rep. 73, U.S. Army Cold Reg. Res. Eng. Lab., Hanover, N.H.

Tabler, R.D. 1971. *Design of a watershed snow fence system, and first-year accumulation*. Proc. 39th Annu. Meet. West. Snow Conf., pp. 50-55.

Tabler, R.D. 1975. *Estimating the transport of blowing snow*. Snow Management on the Great Plains, Res. Comm., Great Plains Agric. Counc. and Univ. Nebr. Agric. Exp. Sta., Lincoln, Publ. 73, pp. 85-117.

Tani, I. and H. Makita, 1971. *Response of a turbulent shear flow to a stepwise change in wall roughness*. Z. Flugwiss, Vol. 19, pp. 335-339.

Theakston, F.H. 1962. *Snow accumulations about farm structures*. Agric. Eng., Vol. 43, pp. 139-141, 161.

9

SNOWCOVER ABLATION
AND RUNOFF

D. H. MALE

*Division of Hydrology, University of Saskatchewan,
Saskatoon, Saskatchewan*

D.M. GRAY

*Division of Hydrology, University of Saskatchewan,
Saskatoon, Saskatchewan*

INTRODUCTION

In many countries snow constitutes a major water resource; its release in the form of melt water can significantly affect agriculture, hydro-electric energy production, urban water supply and flood control. The ablation of a snowcover or the net volumetric decrease in its snow water equivalent is governed by the processes of snowmelt, evaporation and condensation, the vertical and lateral transmission of water within the snowcover and the infiltration of water to the underlying ground. In turn, water yield and streamflow runoff originating from snow are governed by these same processes as well as the storage and the hydraulics of movement of water in channels. In recent years it has become apparent that a better understanding of the physics of the ablation process is central to improving techniques of forecasting the time of melt, the quantity and rate of water released, the volume of water entering the soil and the amount of evaporation.

Regular forecasting of runoff from snowmelt in North America was first attempted at Lake Tahoe, Nevada, in 1909. Engineers of the local power company correlated changes in the lake water levels during the spring with the water content of the snow on Mount Rose (as determined from snow surveys made by Dr. J. E. Church). This correlation allowed the company to regulate releases from the lake to prevent spring flooding and to use the melt water more efficiently for power production. From this early beginning, research into the snow ablation phenomenon has increased significantly. Most studies have been concerned with the prediction of floods and peak discharge rates for designing hydraulic structures and flood control works, and melt water flow to optimize its use for hydro-electric power generation, recreation and irrigation. Recently, there has been a greater emphasis on studies of the impact of snow on the agricultural industry (crop production) and the environment. The interactions between human activities (for example, deforestation, urban development) and the snowmelt regime are receiving increasing emphasis.

The most comprehensive study of snowmelt published to date is *Snow Hydrology* prepared by the U. S. Army Corps of Engineers (1956). Although the investigations summarized in this report were conducted in mountainous regions in the United States their findings have served as a foundation for many subsequent studies undertaken in other parts of North America. In particular, the results and methods described in this publication form the basis for many of the snowmelt runoff models currently used to forecast streamflow. The major Soviet publication on snowmelt is that by Kuz'min (1961) and deals primarily with conditions on the Russian Steppes. This work contains a thorough discussion of the physical processes influencing snowmelt, and an excellent summary of empirical methods for estimating melt rates suitable for incorporation in streamflow forecasting procedures and for other water management purposes.

PHYSICS OF SNOWMELT

General Considerations

Seasonal snowcovers normally develop from a series of winter storms and are modified by the action of freezing rain, wind and diurnal melting and refreezing at the surface. As a result, both deep snowpacks in the mountains and the shallow covers in regions of low relief develop a characteristic layered structure (Gerdel, 1948; Langham, 1974) with "ice" layers or relatively-impermeable, fine-textured high-density layers alternating with coarse-textured, low density and highly permeable layers. Early in the melt sequence vertical drainage channels develop in the snow contributing further to its heterogeneity. The internal structure significantly influences the retention and movement of melt water through the snow, making a detailed analysis of the transmission process extremely difficult. However, during most of the melt period the total melt water produced is governed by the energy exchanges at the upper and lower snow surfaces.

When the pack is primed to produce melt it is at a temperature of $0°C$ throughout and its individual snow crystals are coated with a thin film of water; also, small pockets of water may be found in the angles between contacting grains, normally amounting to 3 to 5% of the snow by weight although some investigators have measured values as high as 25% (de Quervain, 1948). Any additional energy input produces melt water which subsequently drains to the ground. When melt rates are at their highest, 20% (by weight) of the pack or more may be liquid water, most of which is in transit through the snow under the influence of gravity.

The amount of energy available for melting snow is determined from the energy equation. This equation is applied to a volume of snow whose upper and lower surfaces are the snow-air and snow-ground interfaces respectively, and may be written as:

$$Q_m = Q_{sn} + Q_{ln} + Q_h + Q_e + Q_g + Q_p - dU/dt, \qquad 9.1$$

where

Q_m = energy flux available for melt,
Q_{sn} = net short-wave radiation flux absorbed by the snow,
Q_{ln} = net long-wave radiation flux at the snow-air interface,
Q_h = convective or sensible heat flux from the air at the snow-air interface,
Q_e = flux of the latent heat (evaporation, sublimation, condensation) at the snow-air interface,
Q_g = flux of heat from the snow-ground interface by conduction,
Q_p = flux of heat from rain, and
dU/dt = rate of change of internal (or stored) energy per unit area of snowcover.

The net long-wave radiation and convective heat transfer processes are operative *at* the snow-air interface whereas the short-wave radiation exchange is strongest at the surface although limited amounts penetrate into the pack. The ground heat flux, which is usually small, may produce small amounts of melt near the snow-ground interface. Water is released from this lower layer when the snow reaches 0°C and is holding the maximum liquid. Rain may penetrate considerable depths into the snow resulting in a mass transport of heat which is distributed more uniformly throughout the pack than the heat obtained from other sources. However, melt water is generated primarily at the snow-air interface.

The daily amount of melt produced by a given value of Q_m ($kJ/m^2 \cdot d$) (see Eq. 9.1) may be calculated by the expression

$$M = Q_m/(\rho h_f B), \qquad 9.2$$

in which

M = snowmelt water equivalent (cm/d),
h_f = latent heat of fusion, (kJ/kg),
ρ = density of water, (kg/m^3), and

B = thermal quality or the fraction of ice in a unit mass of wet snow.

For normal melt conditions h_f = 333.5 kJ/kg and ρ = 1000 kg/m³. Thus, Eq. 9.2 reduces to

$$M = Q_m/(3335\ B). \hspace{3cm} 9.3$$

As mentioned previously, a melting snowpack generally will retain 3 to 5% water (by weight) against free drainage, corresponding to a thermal quality between 0.95 and 0.97.

Table 9.1 indicates the magnitude of the various energy fluxes during the melt period for clear days in Saskatchewan at a site with no vegetation protruding above the snow surface. The wide range of possible values for each of the fluxes is evident. The presence of a cloud cover would produce an even greater variation. Relative changes in the fluxes can be attributed to changing wind conditions, relative humidity, air temperature and time of year, factors considered in detail below. Table 9.1 also shows the dominant size of the net radiation Q_n during snowmelt in open areas.

Table 9.1
SELECTED DAILY ENERGY FLUX TRANSFER (kJ/m²)[a] DURING THE MELT PERIOD IN THE ABSENCE OF VEGETATION (BAD LAKE, SASKATCHEWAN).

Date (Day/Mon/Yr)	Q_{sn}	Q_{ln}	Q_n[b]	Q_h	Q_e	Q_g
11-4/75	8090	-6320	1770	186	-855	- 45
12-4/75	9620	-8480	1140	782	26	- 22
14-4/75	12290	-9430	2860	13	-395	- 4
17-3/76	4630	-4500	130	1830	-555	64
27-3/76	7200	-7720	- 520	1517	-208	-237
28-3/76	7790	-7120	670	70	-201	-111
29-3/76	9070	-7660	1410	532	- 60	-180
30-3/76	9290	-6040	3250	827	140	-270

[a] positive values indicate an energy gain by the snow.
[b] the daily net radiation flux transfer: $Q_n = Q_{sn} + Q_{ln}$.

Short-wave (Solar) Radiation

Incident radiation

The radiation which influences snowmelt is electromagnetic radiation emitted from a medium by virtue of its temperature and falls in the wavelength range from about 0.2 to 100 μm. Within this range two types of radiation are distinguished. Short-wave radiation (emitted by the sun) is generally considered as that portion falling within the range from 0.2 to 2.2 μm; long-wave radiation (emitted by the atmosphere and the earth) lies between 6.8 and 100 μm.

Radiation from the sun is short-wave radiation falling within a very narrow wavelength band with maximum intensity at 0.47 μm. At the top of the earth's atmosphere extra-terrestrial solar radiation incident on a surface perpendicular to the sun's rays at a mean earth-sun distance of 149.5 x 10^6 km is equal to the solar constant, 1.365 kW/m^2. This value varies by about 7% during a year primarily because of the changing distance between the earth and the sun. The amount of solar radiation penetrating the earth's atmosphere to be received at the surface varies widely depending on latitude, season, time of day, topography (slope and orientation), vegetation, cloud cover and atmospheric turbidity. While passing through the atmosphere radiation is reflected by clouds, scattered diffusely by air molecules, dust and other particles and absorbed by ozone, water vapor, carbon dioxide and nitrogen compounds. The absorbed energy increases the temperature of the air which in turn increases the amount of long-wave radiation emitted to the earth's surface and to outer space.

Short-wave radiation reaching the surface of the earth has two components: a direct beam component along the sun's rays and a diffuse component scattered by the atmosphere but with the greatest flux coming from the direction of the sun. The daily amount of direct beam radiation is a complex function of the factors mentioned above. Garnier and Ohmura (1970) have presented a useful method for analyzing the interaction between these factors. They express the direct clear day radiation I_d falling upon a slope as

$$I_d = (I_0/r^2) \int p^m \cos (X \Lambda S) dH, \qquad\qquad 9.4$$

where

I_0 = solar constant,

r = radius vector of the earth's orbit, (the distance from the centre of the sun expressed in terms of the length of the semi-major axis of the earth's orbit),

p = mean transmissivity of the atmosphere along the zenith path; this is a measure of the fraction of solar radiation which reaches the earth's surface without being scattered or absorbed, m = optical air mass which is the ratio of the distance the sun's rays travel through the atmosphere to the depth of the atmosphere along the zenith path,

$\cos (X \Lambda S)$ = cosine of the angle of incidence of the sun's rays on the slope, (X is a unit normal vector pointing away from the surface and S is a unit vector expressing the sun's position), and

H = hour angle measured from solar noon, the integral being taken over the duration of sunlight on the slope.

Values for I_0, r and m are given by List (1968).

Figure 9.1, which is calculated using Eq. 9.4, shows the annual variation in daily values of solar radiation received by a horizontal surface at several latitudes; p is assumed to be 1.0, implying that all the energy reaches the surface. The influence of transmissivity is illustrated in Fig. 9.2. By multiplying the data in Fig. 9.1 by the ratios of Fig. 9.2, the daily radiation values received by a horizontal surface can be calculated for any atmospheric transmissivity. For example, at 50°N latitude, if p is 0.7, the radiation would be 18 x 0.405 = 7.29 MJ/m^2 on March 1, 13.9 MJ/m^2 on April 1, about twice the value of a month earlier. The time of year obviously is an important factor governing the solar radiation flux incident on the earth's surface. Since this flux is a major component in the total energy flux of the snow, the time of year also has an important influence on the melt rate — a fact well known to hydrologists involved in flood forecasting. As a rule, the longer the spring melt is delayed the greater the danger of flooding. This is due partly to increases in the radiative flux and partly to the increased probability of rain.

The transmissivity p is the highest in winter and lowest in summer because the atmosphere contains more water vapor during summer. It also varies somewhat with latitude, increasing northwards. Kondratyev (1969) reports a mean annual value for p at Pavlovsk, USSR, of 0.745 with a deviation of ± 0.05 during the years 1906-1936. Williams et al. (1972) suggest average monthly values ranging from 0.92 to 0.50 for four locations in middle latitudes (Pavlovsk, USSR; Benson, U.K.; Paris, France; Tacubaya, Mexico). Kuz'min (1961) recommends the use of 0.80 ± 0.05 for the snowmelt period within the European USSR.

It is a common observation that snow on a south-facing slope melts faster than snow on a north-facing slope, the reason being that the orientation of the slope affects the amount of direct beam solar radiation the area receives. The effect of slope and orientation is illustrated in Fig. 9.3. The results are symmetric about a north-south line; as might be expected the influence of orientation diminishes towards the summer solstice. Even on a 10° slope the effect of orientation can be significant; e.g., at 50° N on April 1, a south-facing

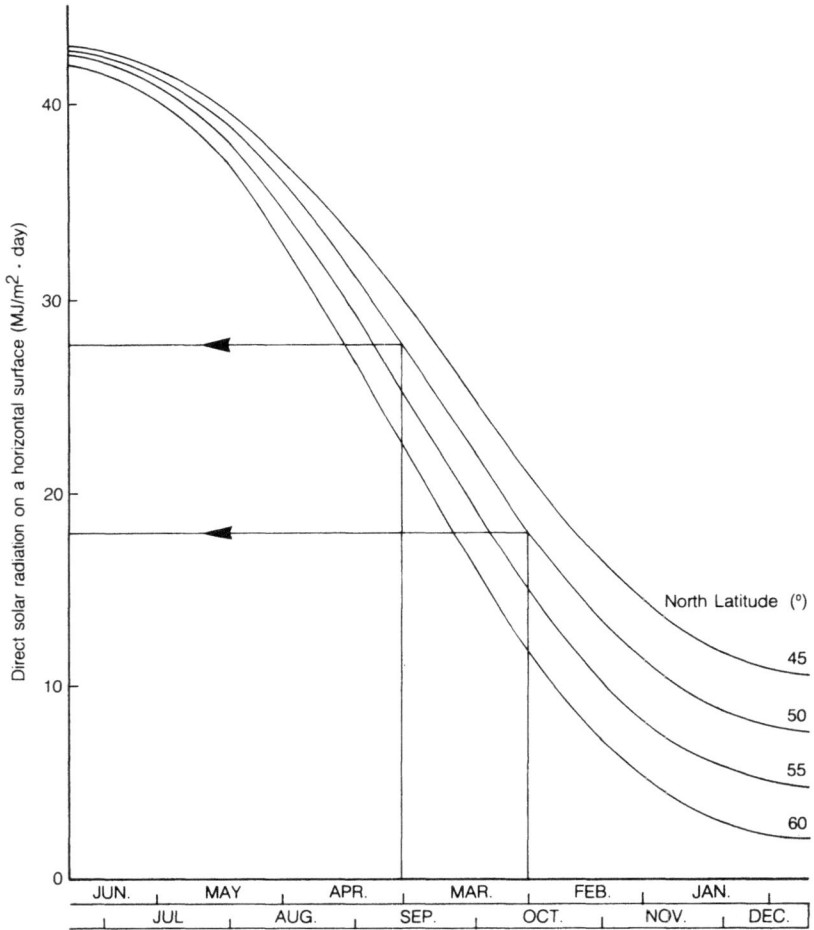

Fig. 9.1 Daily values of direct solar radiation received on a horizontal surface —
transmissivity p = 1.

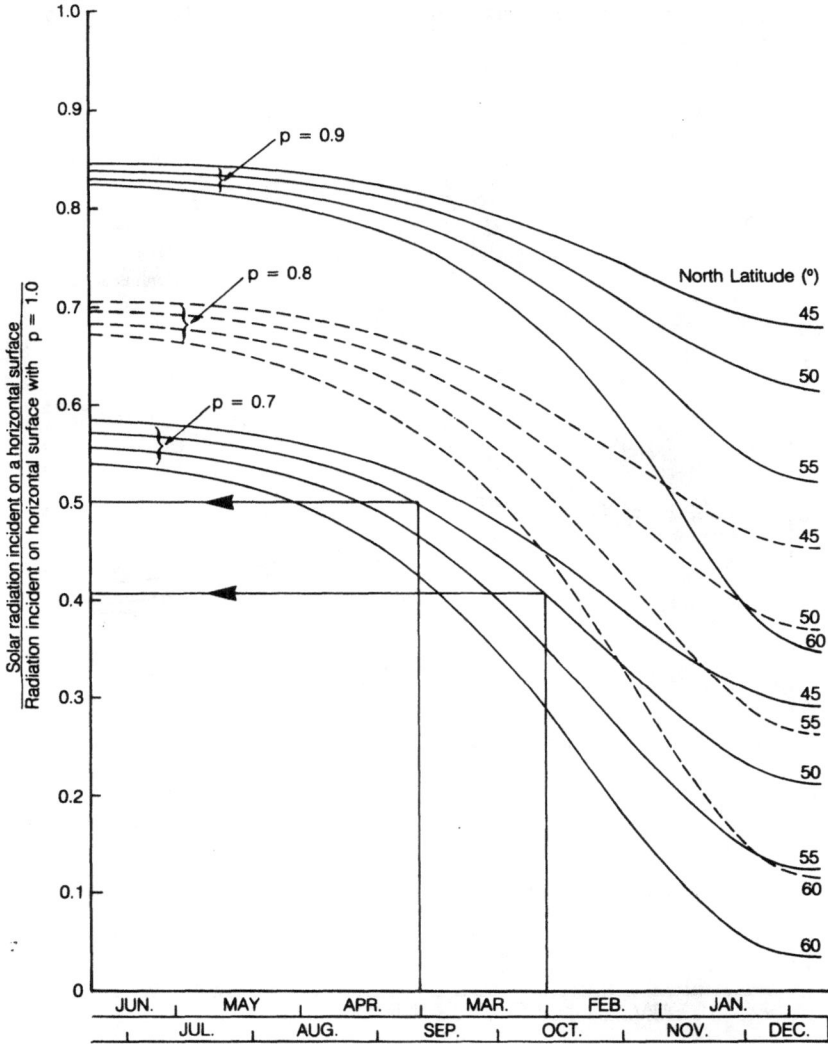

Fig. 9.2 Influence of transmissivity (p) on daily solar radiation received on a horizontal surface.

slope receives approximately 40 percent more direct beam radiation than a north-facing slope (see Fig. 9.3a). The ratios in Fig. 9.3 are calculated assuming p = 1.0; and decrease slowly as p decreases, becoming about 5 percent less for p = 0.7.

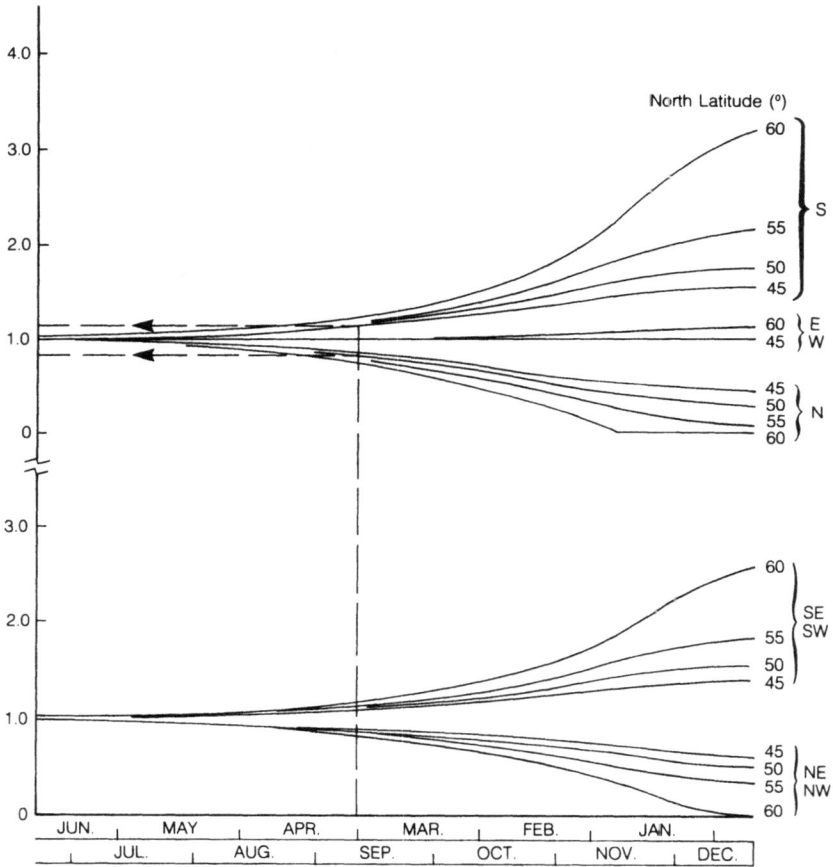

Fig. 9.3a Ratio of direct daily solar radiation received by a 10° slope to that received by a horizontal surface.

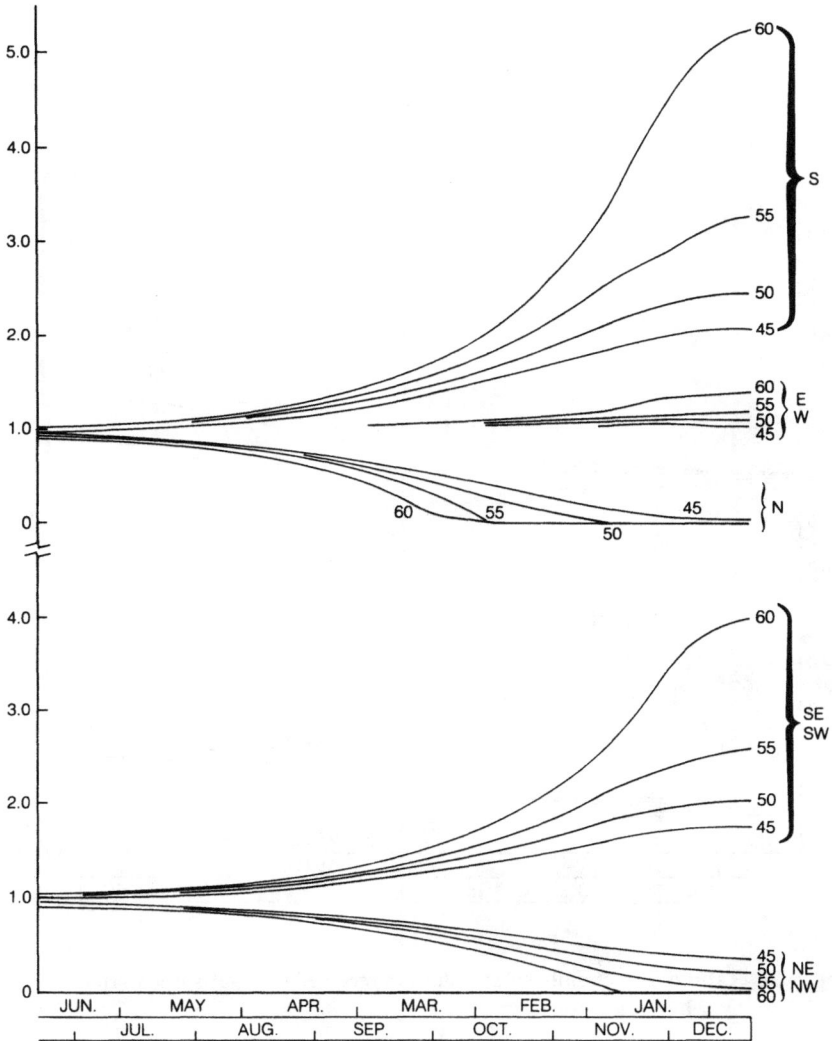

Fig. 9.3b Ratio of direct daily solar radiation received by a 20° slope to that received by a horizontal surface.

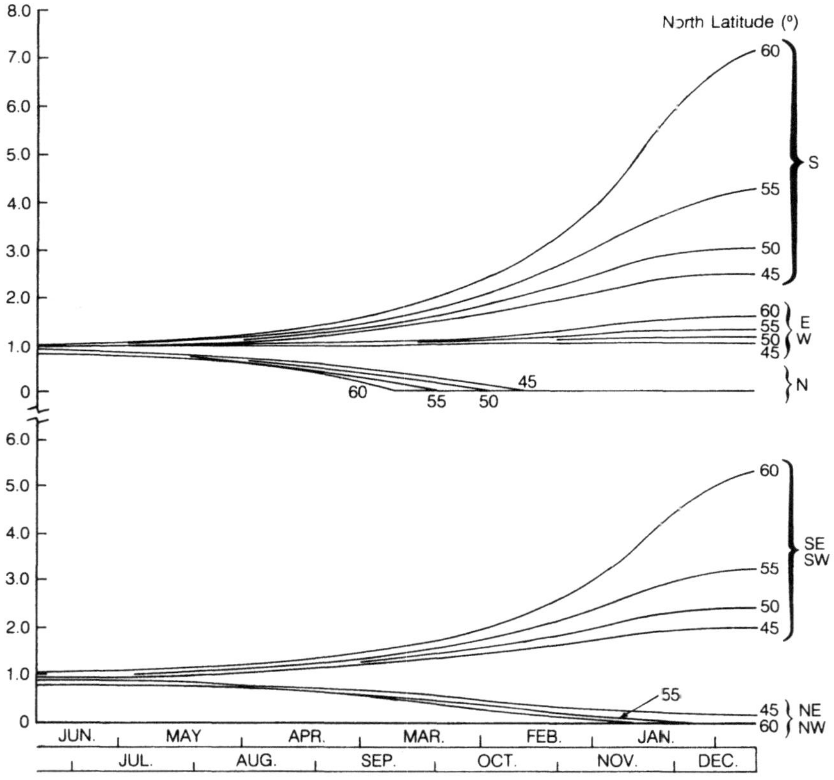

Fig. 9.3c Ratio of direct daily solar radiation received by a 30° slope to that received by a horizontal surface.

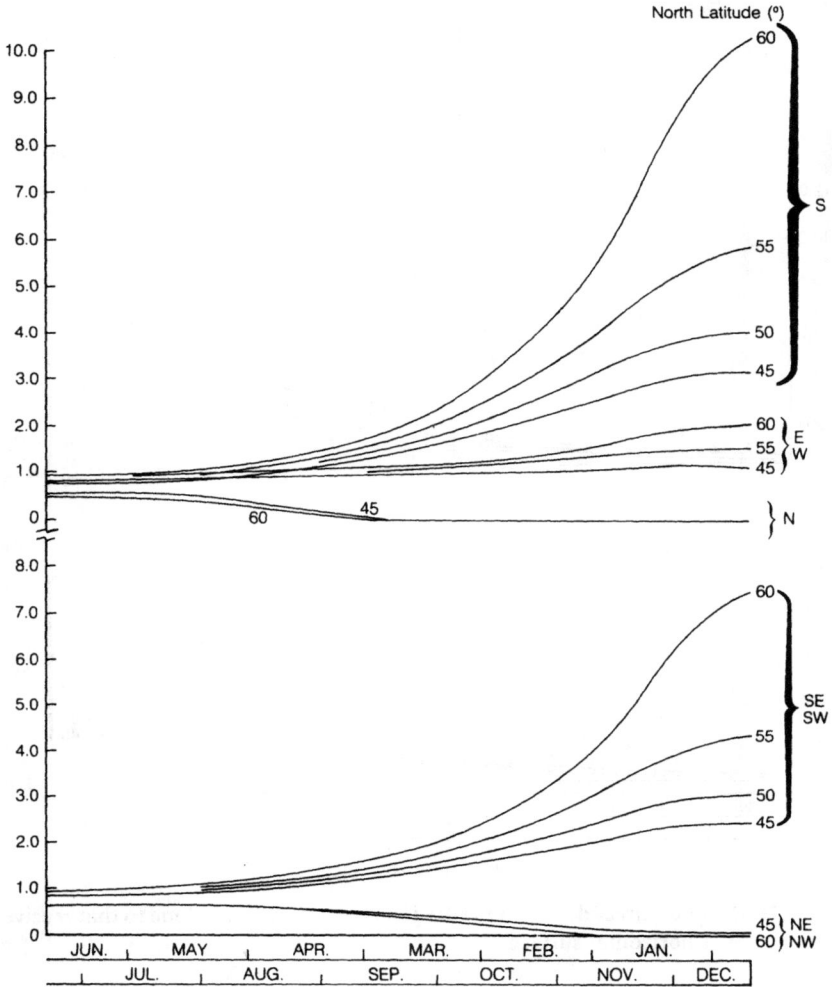

Fig. 9.3d Ratio of direct daily solar radiation received by a 50° slope to that received by a horizontal surface.

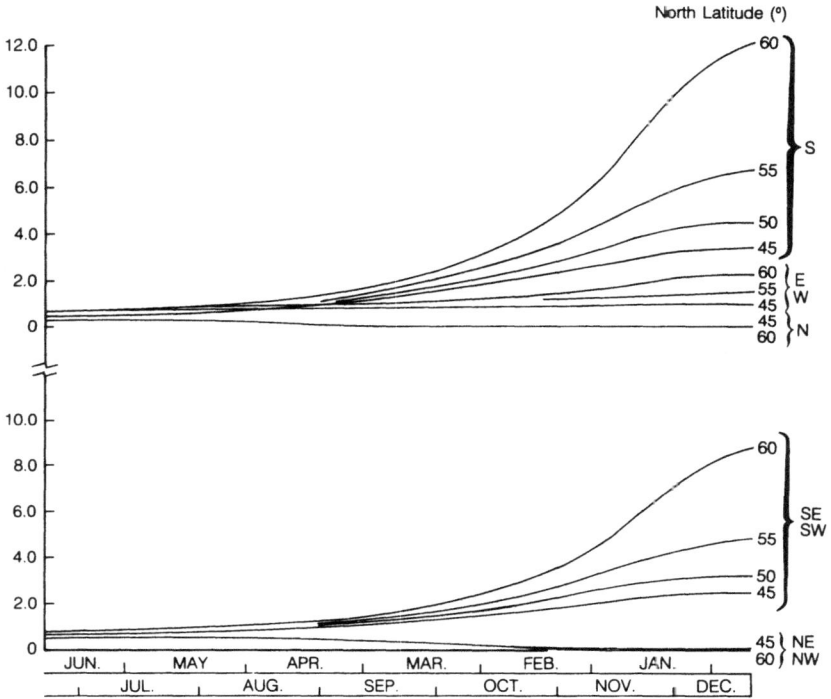

Fig. 9.3e Ratio of direct daily solar radiation received by a 70° slope to that received by a horizontal surface.

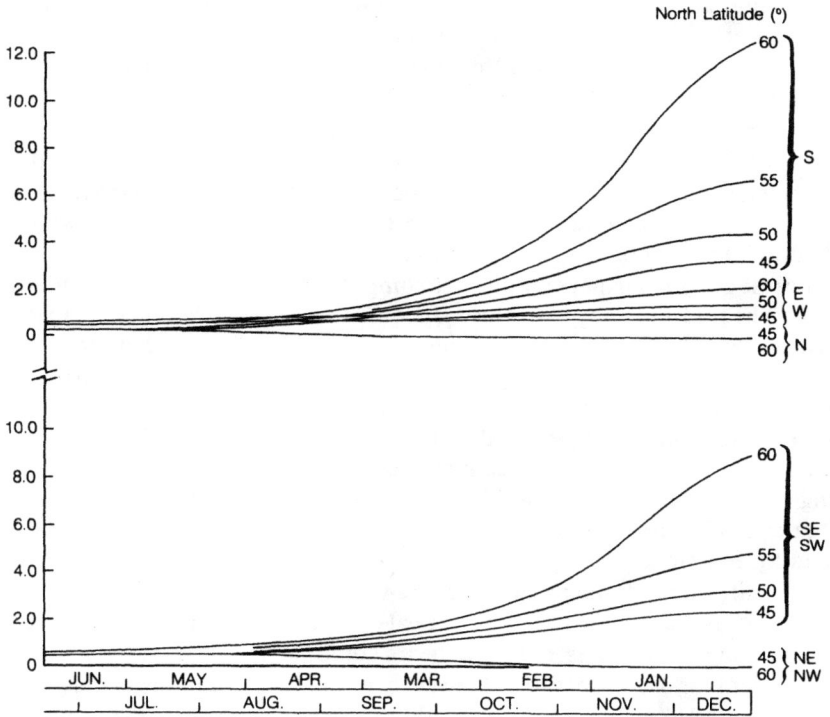

Fig. 9.3f Ratio of direct daily solar radiation received by a 90° slope to that received by a horizontal surface.

In recent years models of solar radiation transfer have been developed which more closely approximate the physical processes occurring in the atmosphere (Suckling and Hay, 1976; Dozier, 1979). In these models separate transmission functions are defined for the attenuation of radiation resulting from absorption by ozone, water vapor, oxygen, carbon dioxide, methane and nitrous oxide, and for scattering by atmospheric molecules (Rayleigh scattering) and aerosols. Each function depends on the optical path length and the attenuation coefficient for each atmospheric constituent.

In mountainous terrain shadows play an important role in determining the amount of direct solar radiation reaching a given point. Methods for considerating this effect have been developed by Williams et al. (1972) and Dozier (1979). The region being analyzed is divided into a uniform grid. By considering the elevation at each grid point and searching along the line of the sun's azimuth for points high enough to block the sun, the grid points in shadow at any hour angle may be found.

Forest canopy further complicates computation of the surface net radiation flux. The forest canopy is a heterogeneous, anisotropic medium that absorbs, scatters and reflects the direct beam solar radiation and emits long-wave radiation. The amount of direct radiation received by the snow surface has been reduced by the shading of the trunks and canopy. To calculate the reduced amount the incoming direct radiation I_d is usually multiplied by the factor (1-V); where V is a beam-shading function depending on the solar angle. V effectively averages the surface in the shade with those in the sun. Dozier (1979) describes a complete solar radiation model which includes a shading function.

The diffuse component of the short-wave radiation reaching the earth's surface is less amenable to mathematical computation and is often ignored in energy balance estimates; the direct beam component is then taken as an index of the total short-wave flux. However, the diffuse component may be the significant energy flux: on clear, bright days it may be 10% of the short-wave radiation received while on partly cloudy days it may be more than 50% and on overcast days it is 100%.

Under clear sky conditions the amount of diffuse radiation depends on the altitude of the sun above the horizon, the atmospheric transmittance and the amount of incident radiation reflected by the underlying surface. Kuz'min (1961) reports that the increase in scattered radiation over snow relative to that over bare ground ranges from 65% (solar elevation ~ 0°) to 12% (solar elevations: 36 to 55°). Various methods have been suggested for calculating the diffuse flux under cloudless sky conditions (Kuz'min, 1961; List, 1968; Kondratyev, 1969) but to obtain precise energy balance data over snow, direct measurements are normally made.

The simplest expression for calculating D_o, the diffuse radiation flux on a horizontal surface under cloudless conditions is given by List (1968, p. 420 attributed to Fritz):

$$D_0 = 0.5 [(1-a_w - a_0)I_t - I_d], \qquad 9.5$$

where a_w = radiation absorbed by water vapor (assumed to be 7%),

a_0 = radiation absorbed by ozone (assumed to be 2%),

I_t = extra-terrestrial radiation on a horizontal surface,

$$I_t = (I_0/r^2) \int \cos z_s \, dH,$$

where r is the radius vector of the earth (see Eq. 9.4) and z_s is the sun's zenith distance (values of I_t are tabulated by List (1968, p. 418)) and,

I_d = direct radiation reaching a horizontal surface of the earth (see Eq. 9.4).

The factor 0.5 applied in Eq. 9.5 expresses the assumption that half of the direct solar beam is scattered toward the surface and half is scattered away from it. Other empirical methods of calculating D_0 are described by Liu and Jordan (1960) and Stanhill (1966). To use these procedures some estimate or measurement of the total (or global) short-wave flux (direct and diffuse components) on a horizontal surface is required. Models have also been developed to calculate the diffuse component by taking the scattering process into account. Dozier (1979) presents a computing scheme in which the amount of radiation initially scattered out of the beam is determined by wavelength-dependent coefficients for aerosol and Rayleigh scattering and for ozone and water vapor absorption. This type of model is complex requiring estimates of the ozone and water vapor distributions with altitude and the change in air pressure with altitude.

An important source of diffuse radiation to the snow surface is the back-scattered short-wave radiation, i.e., that portion of the incoming radiation reflected at the snow surface which is again scattered and reflected downward by the atmosphere. Backscattering models have been developed by Hay (1976) and Dozier (1979). The critical parameter in model calculations is the reflectance or albedo of the snow surface, considered below.

For a sloping surface the diffuse component can be calculated from the expression of Kondratyev (1969), assuming that the diffuse radiation field is isotropic; that is, it has the same properties in all directions:

$$D = D_0 \cos^2 (\theta/2), \qquad 9.6$$

where θ is the angle of inclination of the slope. In mountainous regions any slope is likely to have a sky dome that is restricted by the surrounding topography. The fraction by which the downward scattered radiation is reduced can be calculated by assuming that the radiation intensity is constant from all parts of the sky dome, i.e., isotropic. For a specific location the reduction factor is a constant and need be determined only once. Details about this calculation are given by Obled and Harder (1979) and Dozier (1979).

A further source of diffuse radiation for a slope in mountainous terrain is the direct beam radiation and diffuse radiation reflected by the adjacent terrain. For snow, specular reflection is rare, so that the reflected direct beam radiation is assumed to be isotropic. Dozier (1979) includes this effect in his model, as a special case in the general problem of the reflection of short-wave radiation from snow.

Most calculations of diffuse radiation are made on the assumption that the incoming flux is isotropic. However, Kondratyev (1969) shows that clear sky diffuse radiation is significantly anisotropic, with the greatest intensity in the vicinity of the sun, and near the horizon where the optical thickness of the atmosphere is greatest. The angular distribution of diffuse radiation has been examined by Steven (1977) and Temps and Coulson (1977). The correction factor for angular effects that was developed by Temps and Coulson is used in the computing scheme of Dozier (1979).

Under cloudy sky conditions, the flux of diffuse radiation depends on the sun's elevation angle, the amount of incident radiation reflected by the underlying surface and the amount and types of cloud. Clouds reduce direct solar radiation but increase diffuse radiation; they also reduce the total radiation flux incident on a surface relative to that for clear skies, and the reflected energy lost to space. When clouds are present a large part of the energy reflected by the surface is scattered diffusely by air molecules or reflected by the cloud base downward to the earth's surface. The effect of this exchange process on the incident diffuse radiation flux is more pronounced over a snow surface than over a snow-free surface. Kondratyev (1954) shows that at a fixed solar angle the diffuse radiation flux is approximately 50% greater over snow than over bare ground.

Models have been developed for estimating short-wave radiation under cloudy conditions (e.g., Suckling and Hay, 1977); they require observations of the type of cloud, the number of cloud layers and the fraction of the sky covered by each layer. Unfortunately, these data are not routinely available. For snowmelt studies that warrant a detailed examination of the spatial variation of the short-wave flux the approach of Garnier and Ohmura (1970) is suggested. They use point observations of direct and diffuse radiation on a horizontal surface as input for Eqs. 9.4 and 9.6 to prepare maps of the areal distribution of radiation fluxes.

Mean values of daily global solar radiation measurements (direct plus diffuse components) have been published, e.g., Fig. 9.4 shows the distribution of mean daily global solar radiation for Canada for March (Titus and Truhlar, 1969). Radiation decreases with increasing latitude and isolines tend to be parallel to the latitude lines in northern regions.

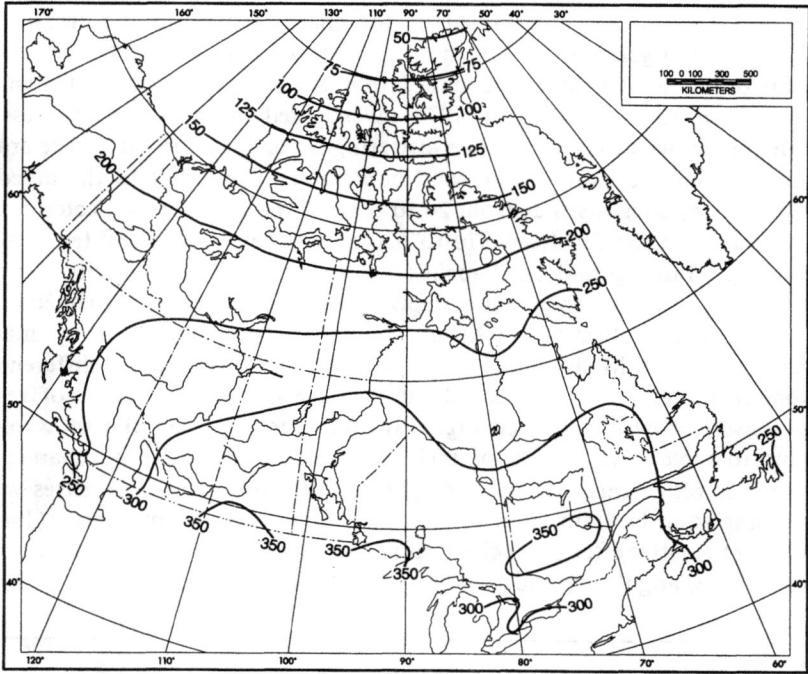

Fig. 9.4 Mean daily global solar radiation in langleys (Titus and Truhlar, 1969) (1 langley = 41.868 kJ/m^2).

Albedo

A large portion of the short-wave radiation incident on the snow surface is reflected. The ratio of the reflected radiation Q_r to the incident radiation Q_{si} is referred to as the albedo, A, i.e.

$$A = Q_r/Q_{si} \qquad \qquad 9.7$$

and is expressed as a decimal fraction or a percentage. Thus, the net short-wave radiation absorbed by a snow surface Q_{sn} may be determined as:

$$Q_{sn} = (1-A)Q_{si}. \qquad \qquad 9.8$$

Albedo may be defined as the integrated reflectance of light over the short wavelength spectrum, i.e.,

$$A = \int_{\lambda_1}^{\lambda_2} r(\lambda)I(\lambda)d\lambda \Big/ \int_{\lambda_1}^{\lambda_2} I(\lambda)d\lambda, \qquad \qquad 9.9$$

where $r(\lambda)$ is the reflectivity at wavelength λ and $r(\lambda)$ and $I(\lambda)$ are the reflected and incident monochromatic intensities, respectively.   The limits of integration λ_1, λ_2 are determined by the characteristics of the measuring instrument and generally fall in the range from 0.3 to 3 μm.

Part of the incident short-wave radiation is transmitted beneath the snow surface.   Bohren and Barkstrom (1974) investigated the attenuation process in detail and showed that subsurface scattering occurs primarily from the grain boundaries of the snow matrix and contributes somewhat to the albedo. Mellor (1966) suggests that from 20 to 60% of the radiation reflected from snow, having a density from 370 to 620 kg/m³ could be attributed to this scattering.   Giddings and LaChapelle (1961) show that the reflectivity of snow at 0.6 μm becomes independent of depth when the depth exceeds 10 to 20 mm.

The spectral reflectivity $r(\lambda)$ of snow has been measured by many investigators but Mellor (1977) shows there are considerable differences among their data sets.   In general the short-wave spectral reflectance of fresh, low-density snow is high (see Fig. 9.5), but decreases once the snowcover begins to melt.   Refreezing of melt water causes virtually no change in snowcover reflectance; Kondratyev (1966) concluded that it decreases with increasing solar elevation and wavelength.   Data reported by Mellor (1966) confirms the latter effect.   Much of the scatter can be attributed to differences in grain size, angularity and bonding of the various samples.

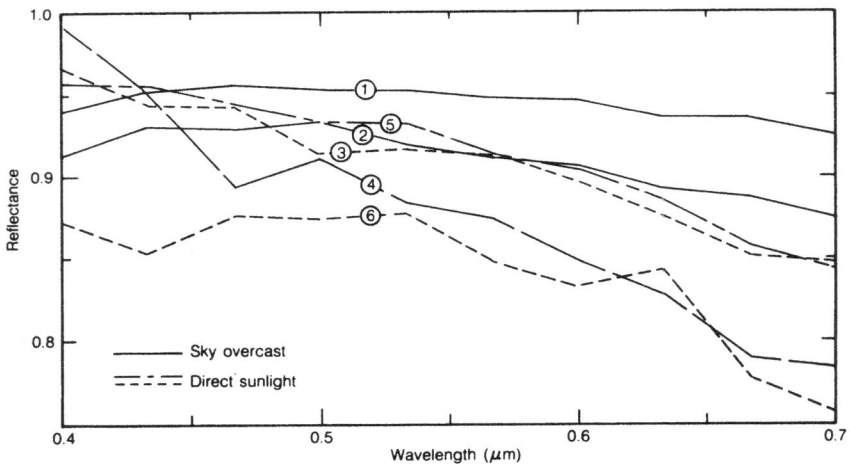

Fig. 9.5 Spectral reflectivity of snow (Mellor, 1966).
1. Fresh Snow (dry), 280 kg/m³, 0°C.
2. 1-2 cm. fresh snow (100 kg/m³) on older snow (400 kg/m³), 0°C.
3. Metamorphosed snow, 430 kg/m³ 0°C.
4. Slightly metamorphosed new snow, 200 kg/m³, 0°C.
5. Wet snow, 2 d old, 400 kg/m³, melting during test.
6. Same as 5 after 4 h more melting.

The mean reflectance or albedo governs the amount of radiant energy absorbed by a snowcover and hence is important for determining the rate of melting. Albedo is known to depend on grain size and density, the method of illuminating the surface (by diffuse or direct radiation), sun angle and surface roughness. Bohren and Barkstrom (1974) calculated that the albedo would vary inversely with the square root of the grain size. The albedo variations with changing elevation of the sun are caused by changes in the direct beam radiation and its spectral composition. The latter is the most important factor causing the diurnal change in albedo of a snow surface. Table 9.2 lists some typical values of albedo for various snowcovers.

Snowcover albedo changes with time during the melt period, e.g., see Fig. 9.6 for different snowcover conditions. The curve for the deep snowpack, given by the U.S. Army Corps of Engineers (1956), is assumed typical of the changes which occur in the albedo of a deep mountainous pack where metamorphic processes acting on the snow crystals produce gradual structural changes which affect the scattering characteristics of the pack. The curve for a shallow snowpack ($<$ 25 cm deep) reported by O'Neill and Gray (1973) is characteristic of the spatial change in albedo with time which Prairie

Table 9.2
ALBEDO OF SNOW (K. Kondratyev (1954) quoting P.P. Kuz'min).

Condition of Snow Surface	Sun Angle degrees	Albedo percent
Compact, dry, clean	30.3	86
	29.7	88
	25.1	95
Clean, wet, fine grain	33.3	64
	34.5	63
	35.3	63
Wet, clean, granular	33.7	61
	32.0	62
Porous, very wet, greyish color	35.3	47
	36.3	46
	37.3	45
Very porous, grey, full of water, sea ice visible	32.8	43
	31.7	43
Very porous, dirty, saturated with water, sea ice visible	37.3	29

snowpacks undergo through the melt sequence. As shown in Fig. 9.6 during melt the temporal rate of change in albedo decreases for a deep mountain snowpack but increases for a shallow snowcover. This result for shallow snow is attributed to the increasing influence of the lower albedo, characteristic of the underlying ground or bare patches, on the radiative reflection and absorption processes as the snowcover ablates (see also Fig. 5.1, Ch. 5).

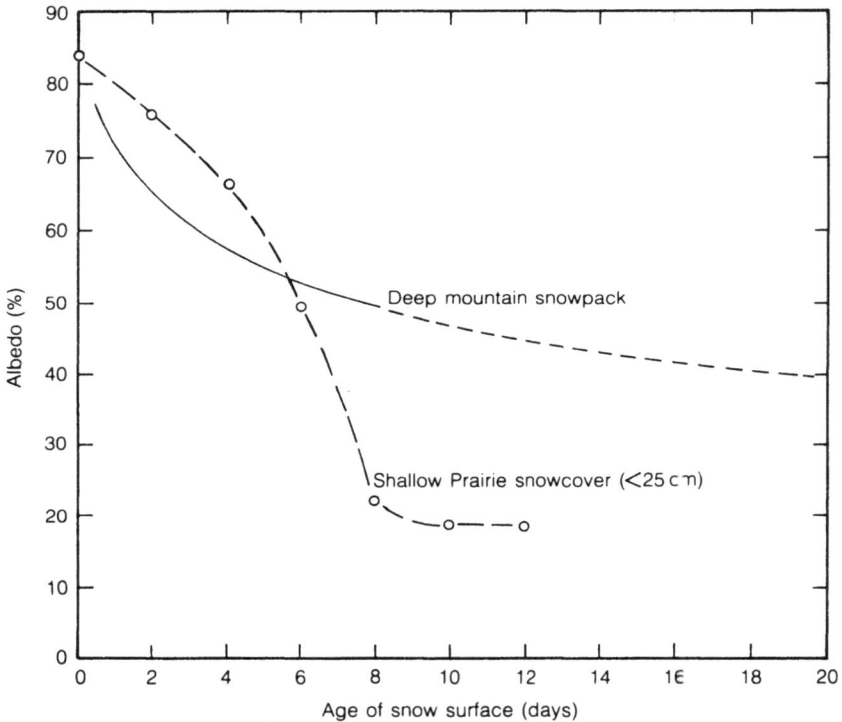

Fig. 9.6 Temporal albedo variations of a melting snowpack (O'Neill and Gray, 1973).

Penetration of solar radiation through snow

Because snow is translucent, radiation is absorbed to some depth within the pack. The penetration of radiation has little practical effect on the snowmelt process of deep packs because the energy reaching the ground is negligible and the amount of melt water produced is independent of the depth over which the radiation is absorbed. However, in very shallow snowcover the solar energy penetration to the underlying ground may affect both melt and infiltration processes: some of the energy absorbed by the ground may be returned to the snow by conduction or by long-wave emission either providing additional energy for melting or for thawing the soil surface to increase the infiltration rate.

Trainor (1947), Dunkle and Bevans (1956), and Giddings and LaChapelle (1961) have found that for a homogeneous snowpack the attenuation of radiation may be approximated by the expression;

$$I_z = I_o \exp(-bz) \qquad\qquad 9.10$$

where I_z = radiation intensity (kW/m^2) at any depth z, (cm),
I_o = radiation intensity at the snow surface (kW/m^2), and
b = extinction coefficient (cm^{-1}).

The magnitude of the extinction coefficient b depends on such factors as wavelength, particle size and snow density. There is considerable uncertainty as to the relative importance of these three factors, so that no reliable functional relationship has as yet been developed relating them. In addition, Manz (1974) has shown that the radiation penetration process is extremely sensitive to the foreign matter content (e.g. dust, organic matter) of the snow, which may often be the major factor affecting the albedo and extinction processes. In Fig. 9.7, taken from Manz (1974), the extinction coefficient is plotted against wavelength for samples of wet and dry snow containing different amounts of foreign material. The figure clearly shows that the magnitude of the extinction coefficient increases directly with foreign matter content independent of wetness or density of the snow. Essentially, the water content of snow has little effect on its optical qualities provided the change in phase of ice to water does not cause extensive changes in the structure of the pack.

Long-wave Radiation

The net long-wave radiation at the snow surface Q_{ln} is composed of the downward radiation Q_{li} and the upward flux emitted by the snow surface Q_{le}. Over snow Q_{le} is normally greater than Q_{li} so that Q_{ln} represents a loss from the snowpack.

Downward radiation is emitted by ozone, carbon dioxide and water vapor from all atmospheric levels; however Geiger (1961) shows that under clear sky conditions the largest portion reaching the earth's surface originates in the lower 100 m of the atmosphere. Ozone emits $\sim$ 2 percent of the total, carbon dioxide $\sim$ 17 percent, and water vapor $\sim$ 81 percent. Variations in Q_{li} are largely due to variations in the amount and temperature of the water vapor. Many investigators have shown that Q_{li} is correlated with air temperature and

	Condition	Density kg/m³	Foreign Matter mg/Litre	Particle Size (cm)
Test 1	Wet	284	51	0.302
Test 2	Wet	338	29.8	0.475
Test 3	Dry	284	50.5	0.259
Test 4	Wet	180	15.5	0.167

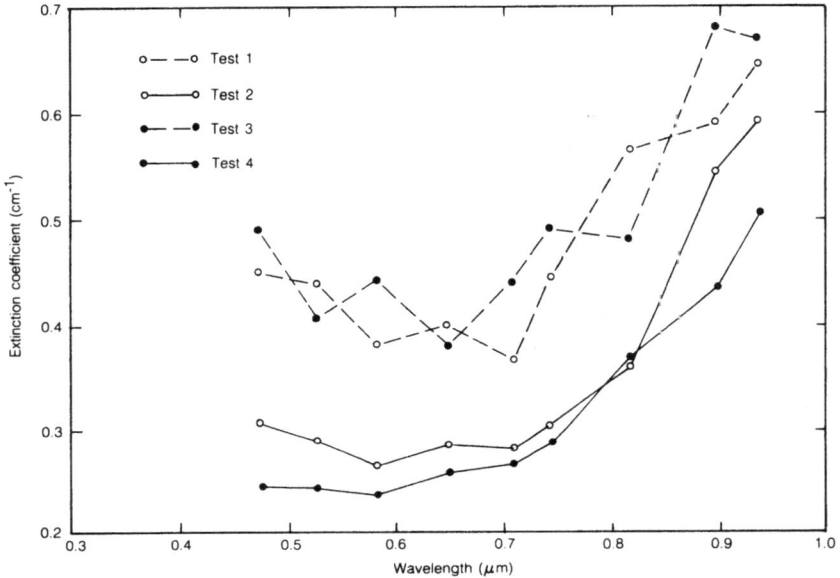

Fig. 9.7 Extinction coefficients of diffuse radiation as a function of wave-length for samples of wet and dry snow (Manz, 1974).

vapor pressure, measured at a height from 1.5 to 2 m above the surface. The most widely quoted correlation was first suggested by Brunt (1952):

$$Q_{li} = \sigma T_a^4 (a + b\sqrt{e}), \qquad\qquad 9.11$$

where σ is the Stefan-Boltzman constant ($5.67 \times 10^{-11} kW/(m^2 K^4)$) T_a and e are the absolute temperature (K) and vapor pressure (Pa) respectively measured at 1.5 or 2m and a and b are empirical coefficients. Based on long-term measurements over snow in the Russian plains and steppe regions, Kuz'min (1961) suggests that Eq. 9.11 can be expressed as

$$Q_{li} = \sigma T_a^4 (0.62 + 0.005\sqrt{e}) \quad (kW/m^2).$$

He emphasizes that this expression is empirical and points out that the coefficients a and b will change with the time of year and location. Recently, Male and Granger (1979) have shown that for a Prairie environment considerable scatter can be expected in the relationship connecting the daily values of Q_{li} calculated by Brunt's equation (with coefficients a = 0.68 and b = 0.005) and values measured over a continuous snow surface.

The approximate nature of Eq. 9.11 is further emphasized by Kondratyev (1969) who provides convincing evidence that this flux is not a unique function of the air temperature and vapor pressure near the earth's surface but is also a function of the vertical distributions of atmospheric temperature and humidity.

A relatively simple means of including the atmospheric temperature and humidity profile data in estimates of Q_{li} has been developed by Brutsaert (1975). Assuming a constant linear temperature decrease with height (- 0.006°C/m) and an atmospheric pressure of 100 kPa he shows that Q_{li} can be calculated by

$$Q_{li} = 0.642 \ (e_a/T_a)^{1/7} \ (T_a^4), \qquad 9.12$$

where T_a is the air temperature near the ground (K), and e_a is the corresponding vapor pressure (Pa). However, Brutsaert shows that the coefficient 0.642 is relatively insensitive to these assumptions. Equation 9.12 has been modified by Marks (1979) for use in alpine areas under the assumptions that the relative humidity is constant with height and the temperature change with height is equal to the standard lapse rate. This gives good agreement with measured values during the snow season.

There is some evidence that Eq. 9.12 departs from the measured values of Q_{li} at air temperatures below 0°C (Aase and Idso, 1978). Satterlund (1979) suggests the following empirical expression gives improved results in this range and exhibits comparable accuracy at higher temperatures:

$$Q_{li} = (\sigma T_a^4) \ 1.08 \ [1 - \exp{(- \ e_a^{T_a/2.016})}]. \qquad 9.13$$

In this equation e_a is in millibars (1 millibar = 10 kPa).

The long-wave radiation emitted by the snow surface is calculated on the assumption that snow is a near-perfect black body in the long-wave portion of the spectrum. Thus,

$$Q_{le} = \epsilon_s \sigma T_s^4, \qquad 9.14$$

where T_s is the absolute temperature of the snow surface (K) and ϵ_s is the emissivity which takes a value between 0.97 and 1.0 (Kondratyev, 1969; Anderson, 1976).

In alpine areas topographical variations have a significant influence on the long-wave radiation received at a point, e.g., in a valley the atmospheric radiation is reduced because a part of the sky is obscured by its walls. However, the valley floor will gain long-wave radiation from the adjacent slopes in accord with Eq. 9.14 in amounts governed by their emissivities and temperatures; the reflected long-wave radiation from snow and most natural surfaces is almost negligible. Thus in areas of high relief the radiation incident at a site includes long-wave emission from the atmosphere and the adjacent terrain. Lee (1962) suggests the fraction of the sky hemisphere which remains unobscured by any obstacle can be calculated by means of the following expression:

$$V_f = \cos^2 (90\text{-}H).$$

V_f is the thermal view factor and H is the average zenith angle of the topographical obstacle. Thus, in areas of high relief the long-wave radiation incident on a point is given by:

$$Q_{li} = (\epsilon_a \sigma T_a^4)V_f + (\epsilon_s \sigma T_s^4) (1 - V_f),   \qquad 9.15$$

where ϵ_a is the effective emissivity of the atmosphere at an ambient temperature of T_a whose value can be calculated using Eq. 9.12 or 9.13 and $\epsilon_s \sigma T_s^4$ is the long-wave radiation emitted by the slopes. Marks (1979) has shown that under clear sky conditions Eq. 9.15 gives values of Q_{li} which are within 10 to 15% of the mean of the measured values over the snow season.

Cloud cover significantly affects the net long-wave exchange process. To a first approximation the radiation emitted by cloud can be obtained by assuming black-body emission at the temperature of the cloud base. Hence, the net long-wave radiation exchange, Q_{ln_1}, between the overcast sky and the snow can be approximated as an exchange between two black bodies having temperatures T_s (snow surface) and T_c (cloud base), i.e.,

$$Q_{ln_1} = \sigma(T_c^4 - T_s^4).$$

Under scattered and patchy cloud cover it is difficult to obtain accurate estimates of the net long-wave flux. Kondratyev (1969) lists several expressions developed for calculating downward radiation under such conditions as a function of cloudless sky radiation. Over snow, the U.S. Army Corps of Engineers (1956) suggest use of the following equation given by Angström to calculate the net flux under partial cloud cover:

$$Q_{ln_2} = Q_{ln_1} (1 - kN)   \qquad 9.16$$

where   k = coefficient which ranges from 0.25 to 0.90 depending on
            cloud type and the height of the cloud base, and
         N = the fraction of the sky covered by clouds.

Kondratyev (1969) provides a detailed discussion of the limitations of the different empirical methods of estimating the net long-wave exchange for partly cloudy conditions and this work should be consulted before these methods are used. For operational purposes it is recommended the net long-wave flux be measured as part of the net all-wave flux by a net pyrradiometer.

A solid canopy of vegetation covering snow as occurring in a coniferous forest, also acts as a long-wave radiation source. Frequently, it is assumed that the effective temperature of the canopy T_s is the same as the ambient air temperature T_a so that the net long-wave flux between the canopy and the snow surface can be calculated by the equation:

$$Q_{ln_3} = \sigma \, (T_a{}^4 - T_s{}^4).$$

9.17

For conditions other than complete cover the situation is more complex, since the type, density, spacing and age of the canopy vegetation affect the exchange process. Also, canopy air currents and the differential heat exchange between the air and vegetation often invalidate using air temperature as an estimate of the effective surface temperature of the canopy.

Convective and Latent Energy Exchanges

The convective and latent energy exchanges, Q_h and Q_e, respectively, are of secondary importance in most snowmelt situations when compared to the radiation exchange. Nevertheless, these fluxes often play an important role in determining the rate of melt. Both Q_h and Q_e are governed by the complex turbulent exchange processes occurring in the 2 or 3 m of the atmosphere immediately above the snow surface. These fluxes can be defined by the expressions:

$$Q_h = - C_p \, \overline{(\rho w)'T'},$$

9.18

$$Q_e = - h_v \, \overline{(\rho w)'q'},$$

9.19

where C_p is the specific heat of the air at constant pressure, h_v is the latent heat of vaporization of water; $(\rho w)'$, T' and q' represent the departure of the quantities ρw, q and T from their respective mean values; ρ is the air density, w is the vertical wind component, T and q are the air temperature and the specific humidity respectively. In the development of Eqs. 9.18 and 9.19 the product $\overline{\rho w}$ is assumed to be zero (Swinbank, 1951), which is true for a horizontal uniform surface if the period of observation is sufficiently long.

An independent determination of these two fluxes is difficult and, to date, no satisfactory operational procedure is available for estimating them. The eddy correlation technique (Dyer, 1961) that is based on Eqs. 9.18 and 9.19 gives a direct determination of Q_h and Q_e. This method has been used over snow by Hicks and Martin (1972) and by McKay and Thurtell (1978). However, the approach is not practical in many applications because complex instrumentation is required and a suitably flat site is often impossible to find in the study area.

A more common method is to compute these fluxes from measurements of the wind, temperature and humidity profiles above the snow surface. It is assumed that at any height Q_h and Q_e are functions of the time-averaged mean potential temperature, θ, and specific humidity q. Schmidt (1925) and Prandtl (1932) gave the following equations for these fluxes:

$$Q_h = - C_p \rho K_c \, \partial\theta/\partial z, \qquad\qquad 9.20$$

$$Q_e = -h_v \rho K_e \, \partial q/\partial z, \qquad\qquad 9.21$$

where: ρ = air density (kg/m^3),
                C_p = specific heat of air (kJ/(kg·°C)),
                h_v = heat of vaporization (kJ/kg),
                z = height above the snow surface (m), and
K_h and K_e = eddy diffusivities for the convective and latent
                energy transfers, respectively (m^2/s).

The potential temperature θ is Eq. 9.20 is the temperature which a volume of air assumes when brought adiabatically from its existing pressure to a standard pressure (generally that at the surface). At a specific height above the ground, θ_z is related to the air temperature T_z by the expression

$$\theta_z = T_z + \Gamma \, z,$$

where Γ is the adiabatic lapse rate (approximately 1°C per 100 m). An atmosphere with this temperature gradient is said to be neutral.

Heat is transferred to the snow by convection if the air temperature increases with height (commonly occurring when the snow is melting); and water vapor is condensed on the snow (accompanied by release of the latent heat of vaporization) if the vapor pressure increases with height. Equations 9.20 and 9.21 are developed on the assumption that the vertical fluxes are constant with height. Panofsky (1974) extends the height of this constant-flux layer to approximately 30 m above the surface, but points out that its thickness is highly variable and is affected by the strength of the wind, the temperature gradients and the "fetch" or length of relatively flat terrain upwind of the measurement site. A fetch to height ratio of 100 to 1 is a commonly-accepted criterion for determining the height of the constant flux layer at a given point, i.e., if the upwind snow surface is free of buildings, trees or other obstructions for 100 m then the constant flux layer will have developed to a height of approximately 1 m.

In general, two procedures are followed in applying Eqs. 9.20 and 9.21. Sverdrup (1936) expresses the heat balance at the snow surface as:

$$Q_h + Q_e + Q_n + Q_m = 0, \qquad\qquad 9.22$$

assuming that the ground heat fluxes, precipitation and internal energy changes may be neglected. If it is assumed that the eddy diffusivities K_h and K_e are equal and $T \simeq \theta$ near the snow surface then Eqs. 9.20 and 9.21 can be combined with Eq. 9.22 to give:

$$Q_h = (Q_n + Q_m) \, C_p \, (\partial T/\partial z)/[C_p \, (\partial T/\partial z) + h_v \, (\partial q/\partial z)], \text{ and} \qquad 9.23$$

$$Q_e = (Q_n + Q_m) h_v \, (\partial q/\partial z)/[C_p \, (\partial T/\partial z) + h_v \, (\partial q/\partial z)]. \qquad 9.24$$

To use this approach requires direct measurements of the net radiation Q_n and snow melt Q_m. Thus, estimates from Eqs. 9.23 and 9.24 are particularly prone to error near sunrise and sunset because Q_n is difficult to measure accurately at these times. Significant errors in the calculated values may also result when the gradients of T and q are small.

The second method commonly used to calculate Q_h and Q_e requires measuring the gradients of T and q and computing the eddy diffusivities from the wind profile. Equations 9.20 and 9.21 can then be used directly. The simplest way of doing this involves calculating the eddy diffusivity for momentum K_m (m^2/s) on the assumption that

$$\tau = \rho K_m \, \partial U / \partial z, \qquad\qquad 9.25$$

where: τ = shear stress (N/m^2), and
                  U = mean horizontal wind speed (m/s).

Usually, it is assumed $K_h = K_e = K_m$ although the validity of this assumption is frequently questioned.

In situations where the change in temperature with height above the snow is equal to the adiabatic lapse rate ($0.01°C/m$) it can be shown (Munn, 1966) that;

$$\partial U / \partial z = U^*/kz, \text{ and} \qquad\qquad 9.26$$

$$K_m = U^* kz, \qquad\qquad 9.27$$

where

        $U^* = (\tau/\rho)^{1/2}$, known as the friction velocity, and
        k = von Kármán's constant (0.4).

If Eqs. 9.26 and 9.27 are substituted into Eqs. 9.20 and 9.21 and the results integrated between the heights z_1 and z_2 in the constant flux layer the following expressions result:

$$Q_h = \rho C_p \, k^2 \, (K_h/K_m) \, (U_2 - U_1) \, (\theta_2 - \theta_1)/(\ln z_2/z_1)^2, \text{ and} \qquad 9.28$$

$$Q_e = \rho h_v k^2 \, (K_e/K_m) \, (U_2 - U_1) \, (q_2 - q_1)/(\ln z_2/z_1)^2. \qquad 9.29$$

Values of U_1, θ_1, q_1 and U_2, θ_2, q_2 are measured at the heights z_1, and z_2, respectively.

The ratios K_h/K_m and K_e/K_m are usually assumed to be unity for a neutral atmosphere; in many snowmelt studies this value is used regardless of the magnitude of the atmospheric temperature gradient. Deviations from a neutral atmosphere are handled by applying the Monin-Obukhov similarity theory (Monin and Obukhov, 1954), which incorporates dimensionless velocity and temperature gradients (ϕ_m, ϕ_h) defined by the following expressions:

$$\phi_m = (kz/U*)\partial U/\partial z, \text{ and} \qquad\qquad 9.30$$

$$\phi_h = (kz/T*)\partial\theta/\partial z, \qquad\qquad 9.31$$

where $T*$ is defined for neutral conditions by

$$Q_h = -\rho C_p U*T*.$$

Monin and Obukhov (1954) show that ϕ_m and ϕ_h are universal functions of the dimensionless stability parameter z/L. L is a length scale first introduced by Obukhov (1946) who defined it as

$$L = -U*^3 C_p \,\rho T/(gkQ_h), \qquad\qquad 9.32$$

where T is the mean absolute temperature of the boundary layer and g is the acceleration of gravity. From Eqs. 9.20, 9.25, 9.30 and 9.31 it can be shown that

$$K_h/K_m = \phi_m/\phi_h. \qquad\qquad 9.33$$

A dimensionless humidity gradient ϕ_e can also be defined (Dyer, 1968):

$$\phi_e = (kz/q*)\,\partial q/\partial z, \qquad\qquad 9.34$$

where $q*$ is a characteristic humidity ratio analogous to $U*$. Using Eq. 9.34 it can be shown that

$$K_e/K_m = \phi_m/\phi_e. \qquad\qquad 9.35$$

The relationship connecting each of the ratios ϕ_m/ϕ_h and ϕ_m/ϕ_e and the stability parameter z/L has been the subject of many investigations; there are significant differences between the various data sets. Under stable conditions ($z/L > 0$), which is common over a melting snow surface, there is general agreement that K_h/K_m is unity (Anderson, 1976). However, for $z/L > 0.1$ (very light winds), Granger and Male (1978) report that for snow the ratio K_h/K_m decreases with increasing stability (Fig. 9.8a) and is better estimated by the expression

$$K_h/K_m = [1 + 7(z/L)]^{-0.1}. \qquad\qquad 9.36$$

To obtain this expression a log-linear profile was fitted to the measured values of wind and temperature (see Webb, 1970).

The largest variations in the relationships connecting K_h/K_m, and K_e/K_m with z/L reported in the literature occur in the unstable range (see Fig. 9.8b). Anderson (1976) recommends the use of the curve based on the work of Dyer and Hicks (1970) for snow conditions. However, Granger (1977) in his work in a Prairie environment found that the resultant relationship between K_h/K_m and z/L corresponded closely with the equation proposed by Swinbank (1968). In fact, the agreement between the measured and calculated values is sufficiently good as to suggest that Swinbank's equation may be extrapolated to the neutral region with a reasonable degree of confidence. The results reported by Businger et al. (1971) shown in Fig. 9.8b, have higher values of K_h/K_m under neutral conditions than those reported by the other investigators. This can be attributed to their choice of 0.35 as a value for the von Kármán constant k, as opposed to the more commonly accepted value of 0.4. It should be noted that both Dyer and Hicks (1970) and Businger et al.

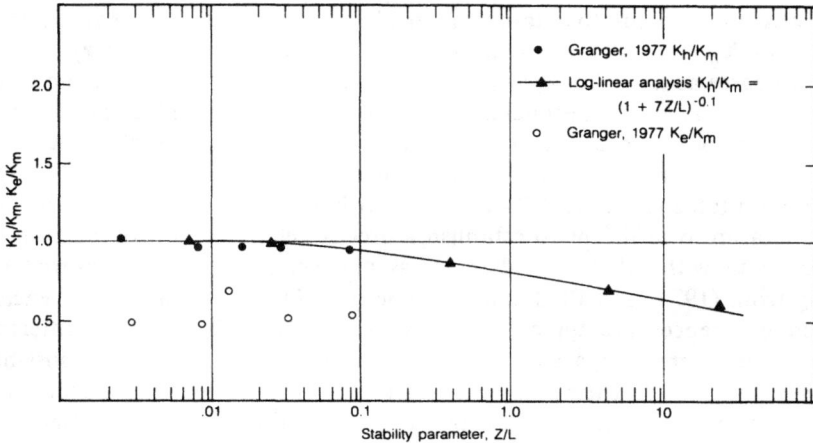

(a) Stable Range

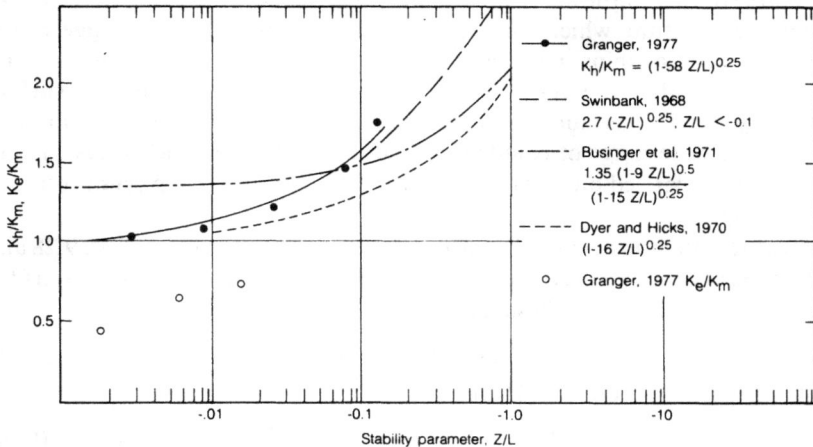

(b) Unstable Range

Fig. 9.8 Comparisons of the relationship connecting the ratios K_h/K_m and K_e/K_m with the stability parameter z/L, as reported by different investigators: (a) stable range, (b) unstable range (Granger and Male, 1978).

(1971) used the eddy correlation technique whereas Granger (1977) and Swinbank (1968) used low level wind velocity and temperature measurements. It is possible that systematic errors are associated with one or both methods of determination which could lead to the different rates of increase of K_h/K_m with increasing values of z/L.

Accurate measurements of humidity profiles near the snow surface have proven more difficult to obtain than profiles of temperature; consequently the ratio K_e/K_m (or ϕ_m/ϕ_e), has not been investigated as a function of z/L to the same degree as the ratio K_h/K_m. The results of Granger (1977) are shown in Fig. 9.8, where each experimental point represents a mean value of individual points grouped according to a range of z/L. These results suggest that K_e/K_m is nearly constant under stable conditions and is approximately equal to 0.5. In the unstable range, K_e/K_m increases with increasing instability although the data are insufficient to establish a precise relationship. Comparison of these data with others for short grass and cereal crops (as reported by Högström (1974) and Blad and Rosenberg (1974)) shows large differences. Such differences are typical when K_e/K_m values obtained by different investigators are compared. McBean and Miyake (1972) suggest a possible reason for this. In their study the fluctuations in velocity, temperature and humidity were analysed in the frequency domain to show the effects of different scales of motion on the turbulent transfer mechanisms near the surface. Their measurements indicate that moisture is a passive scalar, i.e., the moisture content of the air does not affect buoyancy (in contrast to the active scalar temperature which causes convective motion). They suggest that universal relationships pertaining to the transfer of passive scalars are unlikely, and that the transfer mechanisms depend on both the surface boundary conditions and the large scale circulations. This may, in part, explain the wide range of results reported in the literature. Most research to date has been carried out over soils having different covers, e.g., short grass or cereal crops.

Because numerous detailed measurements are required to use Eqs. 9.28 and 9.29, several attempts have been made to develop simplified expressions for estimating Q_h and Q_e. These usually take the form:

$$Q_h = D_h U_z (T_a - T_s), \text{ and} \qquad\qquad 9.37$$

$$Q_e = D_e U_z (e_a - e_s) \qquad\qquad 9.38$$

where: D_h = bulk transfer coefficient for convective heat trans-
                  fer $(kJ/(m^3 \cdot {}^\circ C))$,

            U_z = wind speed at a reference height, z - taken be-
                  tween 1 and 2 m (m/s),

      T_a, T_s = temperatures of the air and the snow surface, re-
                  spectively,

D_e = bulk transfer coefficient for latent heat transfer $(kJ/(m^3 \cdot mb))$, and,

e_a, e_s = vapor pressures of the air and the snow surface, respectively (mb).

In using Eqs. 9.37 and 9.38, e_a and T_a are usually measured at the same height as U_z.

Values for the bulk transfer coefficients reported by numerous investigators are summarized in Table 9.3: the coefficients quoted for the U.S. Army Corps of Engineers (1956), de Quervain (1952) and Sverdrup (1936) have been adjusted to a measurement height of 1 m, assuming the wind velocity, temperature and vapor pressure profiles vary with height according to a 1/6 power law. The U.S. Army Corps of Engineers suggest that this is a reasonable approximation for average conditions. However, since detailed measurements show a wide variation in the profiles during the day as the stability varies the values listed should be considered to be first approximations only.

The table shows that the coefficients are widely different. In part, this may be attributed to instrumental errors and to differences in the heights of measurement used by the investigators. In addition, Yoshida (1962) points out that the values depend on the instrumentation used. Some of the differences can be attributed to differences in the exposure and in weather conditions at the sites where the measurements were made. For example, the coefficients given by the U.S. Army Corps of Engineers were obtained for a mountain region when condensation was much greater than evaporation. The coefficients quoted by Granger (1977) are for Prairie conditions in which the evaporative flux is dominant over a 24-h period. In general, atmospheric stability is different in these two situations. With the exception of the values quoted by Hicks and Martin (1972) the values are averages obtained over relatively long time periods and should not be applied over intervals of less than 24 h. Both the sensible and latent fluxes over snow can show strong diurnal variations (Male and Gray, 1975); consequently the use of the coefficients of Table 9.3 in the morning or evening or during other periods when atmospheric conditions are changing can lead to major errors in the calculated values.

The results given in Table 9.3 illustrate one of the major dilemmas in snow hydrology. On the one hand it is desirable to develop relatively simple expressions for the sensible and latent heat fluxes which can be used on a routine operational basis. On the other hand no simple expression developed to date can adequately describe the complicated interactions which occur continuously in the atmosphere close to the snow surface. A major research effort is needed to develop a suitable theoretical framework for such expressions and to provide carefully instrumented, detailed measurements for

Table 9.3
VALUES OF THE BULK TRANSFER COEFFICIENTS FOR TURBULENT EXCHANGE ABOVE MELTING SNOW.

Author	$D_h \times 10^3$ kJ/m^3·°C	$D_e \times 10^3$ kJ/m^3·mbar	Measurement Height (m)			Comments
			U_z	T_a	e_a	
Hicks and Martin (1972)	1.06	5.15	3.2	2	2	Values based on eddy correlation measurements made in four 1-h periods under highly stable conditions with wind speeds less than 3.5 m/s.
Gold and Williams (1961)	15	25	2	1.2	*	e_a measured 1.6 km from site. D_h and D_e are average values for a two-week period. Assumed $D_h/D_e = 0.6$ (Bowen ratio)
Yoshida (1962)	3.56	6.62	0.7	1.2	1.2	Values apply only for wind speeds greater than 2 m/s.
U.S. Army Corps of Engineers (1956)	1.68	8.0	1	1	1	Coefficients corrected to 1 m using 1/6 power law.
Sverdrup (1936) as quoted by U.S. Army Corps of Engineers (1956)	5.74	10.0	1	1	1	Coefficients corrected to 1 m using 1/6 power law.
de Quervain (1952) as quoted by U.S. Army Corps of Engineers (1956)	--	11.4	1	1	1	Coefficients corrected to 1 m using 1/6 power law.
Granger (1977)	6.69	2.17	1	1	1	Measurements taken above melting prairie snowpack. D_h - average for 6 days; D_e - average for 8 days, based on lysimeter results.

a variety of atmospheric conditions before this problem can be resolved.

Granger and Male (1978) propose a relationship showing some promise in this respect: they related the convective heat transfer with the energy content of the air mass, as indexed by the upper-air temperature at 85 kPa, (see Fig. 9.9). The expression is limited to situations where local advection is negligible (continuous snowcover, no forest cover, or remote from a body of water) and when a frontal zone is not situated above the measurement site.

The outline presented above applies to open areas where vegetation does not protrude above the snow surface. The presence of vegetation further complicates the convective and latent heat exchanges. Well-defined, one-dimensional profiles of wind, temperature, and humidity do not exist above the snow surface under such conditions. Profiles have been measured at the top of the canopy and data analyses similar to those presented above have been made. Such measurements give estimates of the heat flux at the top of the vegetation but not at the snow surface.

Vegetation acts as a source or sink of heat to a varying degree, depending on the type, density and spacing of the canopy, and on atmospheric conditions. Fluxes measured at the top of the vegetation can differ significantly from those at the snow surface, which are generally estimated empirically by applying correction factors to fluxes that have been calculated or measured in nearby open areas.

Internal Energy

The melt release from shallow snowpacks and the upper layers of deep packs often exhibit diurnal cycles, i.e., the snow melts in the late morning and afternoon and refreezes during the night as the pack cools by emitting long-wave radiation. This cyclic process results in changes in the internal energy of the pack such that the night-time energy deficit must be compensated the following day before the pack becomes isothermal at $0°C$ and releases water.

The internal energy consists of components for the solid, liquid and vapor phases of the snow, according to the following expression:

$$U = d(\rho_i C_{pi} + \rho_l C_{pl} + \rho_v C_{pv})T_m, \qquad 9.39$$

where U = internal energy (kJ/m^2),
d = depth of snow (m),
ρ = density, (kg/m^3),
C_p = specific heat, $(kJ/(kg \cdot °C))$,
T_m = mean snow temperature $(°C)$

and i, l, v refer to ice, liquid and vapor.

Assuming the humidity of the air in the snowpack is 100 percent it can be shown that the contribution of the vapor phase term to U is negligible and may be neglected in computing changes in U.

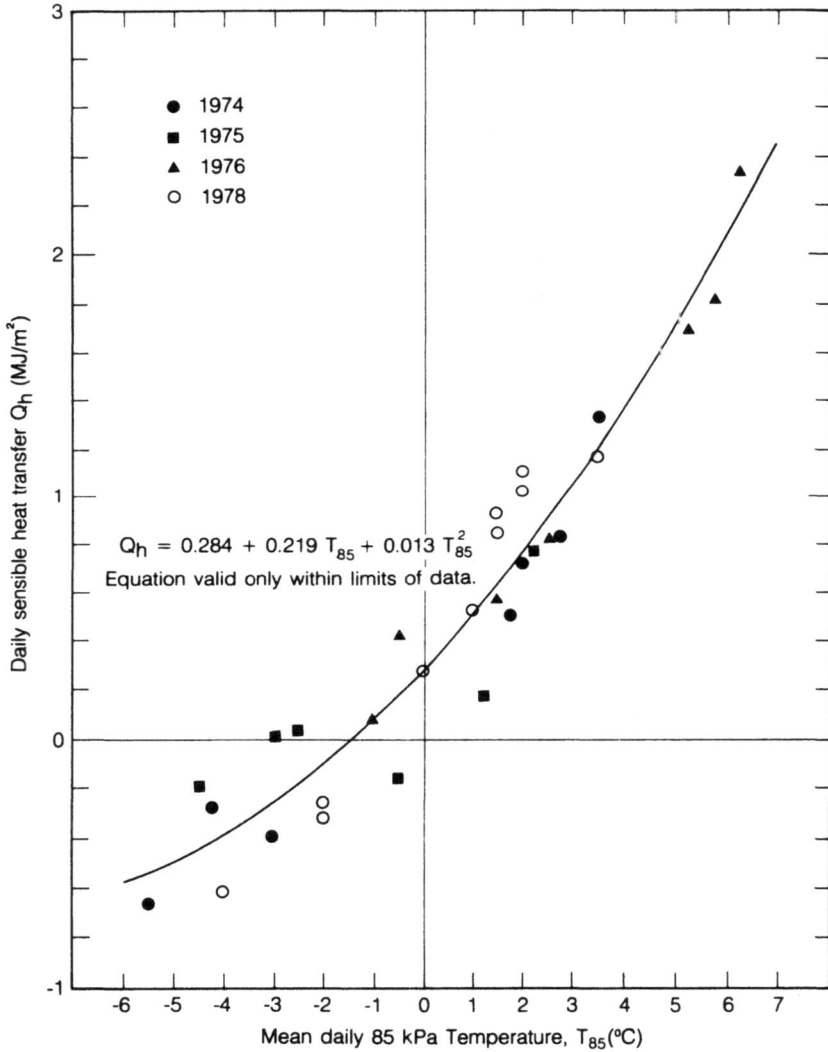

Fig. 9.9 Relation between mean daily totals of sensible heat transfer and the mean daily 85 kPa temperature (Granger and Male, 1978).

Male and Gray (1975) have shown that accurate measurements of internal energy are essential to successfully apply the energy balance principle to shallow snowcovers. In many cases the error involved in energy balance calculations can be associated directly with the error in the internal energy term. During non-melt periods, for all practical purposes, ρ_l can be assumed to be zero, so that evaluation of the U involves measurements of d, ρ_i and T_m. Dybvig (1976) found that during these periods the temperature regime within a snowpack (and hence the internal energy changes) could be predicted using a simple conduction model based on the solution of the non-steady diffusion equation using air and ground temperatures as the boundary conditions.

On days when melt occurs the thermal regime of the snowpack is complex. During the day liquid water exists in the pack and in the evening is often refrozen subjecting the pack to large changes in internal energy. To properly account for these changes with time requires the use of systematic sampling techniques.   The only field methods available to date for measuring the internal energy content under melt/freeze conditions are cumbersome calorimetric techniques unsuitable for operational practice. Of the many remote-sensing methods currently being developed, only the microwave techniques (Meier and Edgerton, 1971; Linlor, et al., 1974) have the capability of measuring both the liquid water and ice contents of the snow.

For deep snowpacks, such as those found in forested, mountainous regions, the magnitude of the internal energy change is relatively insignificant in relation to other components and is therefore frequently neglected in energy balance studies. In such situations the energy balance is calculated for the snow surface rather than for the entire depth of the pack.

Ground Heat Flux

The ground heat flux Q_g is a negligible component in daily energy balances of a snowpack when compared to the radiation, convection and latent heat components (see Table 9.1), so that the total snowmelt produced by Q_g over short periods of time (less than one week) can be safely ignored. However, Q_g does not normally change direction throughout the winter months and consequently its cumulative effect can be significant over a season. This flux also influences the temperature regime of snow near the ground surface and contributes to the conditioning of the snowpack for melt. In areas where snow temperatures remain near the freezing point, melt can be produced as a result of the ground flux. Although the amount of water produced may be small, its resultant effect on the thermal properties and infiltration characteristics of the underlying soil may be significant. Similarly, Q_g affects the state of the soil (i.e., unfrozen, or partially or completely frozen), which in turn influences its infiltration rate at the time of melt. Under certain conditions the soil infiltration rate may be greater when frozen because the size and volume of voids is increased by the structural changes accompanying the freezing process. Conversely, if the soil temperature is below freezing, infiltrating water may be refrozen in the surface layers resulting in the formation of an impervious layer that restricts infiltration and increases the runoff potential. The physics of soil moisture movement under partially or completely frozen conditions is not fully understood, partly because there exist no instruments that will provide accurate *in situ* measurements of the soil physical parameters governing the water transmission.

With respect to snowmelt *per se*, the uni-directional conduction of heat to the pack is given by the equation:

$$Q_g = - k \ \partial T_g / \partial z, \qquad\qquad 9.40$$

where:

T_g = ground temperature, and
k = thermal conductivity.

Thermal conductivity values for silt and clay soils range from 0.4 to 2.1 W/(m · °C) and for sand, from 0.25 to 3 W/(m · °C) depending on density and the moisture content. The latter, in particular, has a strong influence on k. Since the thermal conductivity of ice is approximately four times that of water, frozen soils have a somewhat higher k than unfrozen soils with the same moisture content. A comprehensive review of the factors influencing the thermal conductivity of a soil is given by De Vries (1963). Penner (1970) shows how k may be expected to vary as the soil freezes.

In the absence of significant moisture movement the magnitude of Q_g can be calculated from measurements of the soil temperature profile immediately below the snowcover. However, this procedure is limited by the accuracy with which k can be determined. To estimate k, a soil sample must be obtained which necessarily involves disturbing the snow surface; then its liquid water and ice contents must be estimated.

An alternative way of determining Q_g is direct measurement using a heat flux plate. One form of this instrument particularly suited to measurements in soils has been developed by Fuchs and Tanner (1967).

In snowmelt calculations, average values of Q_g supplied by the soil to the pack over a 24-h period are used, e.g., the U.S. Army Corps of Engineers (1956) suggest a value of 270 kJ/(m^2 · d). Gold (1957) measured an average flux of 860 kJ/(m^2 · d) at Ottawa. Yoshida (1962) observed melt rates equivalent to fluxes of 260 to 360 kJ/(m^2 · d) at the snow-ground interface of a deep pack in Japan. Measurements with heat flux plates, confirmed by corresponding measurements of the temperature profile and thermal conductivity give values in the range 0 to 260 kJ/(m^2 · d) for Prairie conditions (Granger, 1977).

Rain on Snow

Rainfall on a snowcover is a common event in mountainous regions and in eastern parts of North America. The heat transferred to the snow by rain water is the difference between its energy content before falling on the snow and its energy content on reaching thermal equilibrium within the pack. Two cases must be distinguished in this energy exchange:

(1) Rainfall on a melting snowpack where the rain does not freeze,
(2) Rainfall on a pack with a temperature below $0°C$ where the water freezes and releases its latent heat of fusion.

The first case can be described by the expression:

$$Q_p = \rho C_p(T_r - T_s)P_r / 1000 \qquad 9.41$$

where Q_p = energy supplied to the pack by rain, $(kJ/(m^2 \cdot d))$,
ρ = density of water (kg/m^3),
C_p = heat capacity of water $(kJ/(kg \cdot °C))$,
T_r = temperature of the rain $(°C)$,
T_s = snow temperature $(°C)$, and
P_r = depth of rain (mm/day).

Using average values for the parameters: $\rho = 1000$ kg/m^3, $C_p = 4.20$ kJ/ $(kg \cdot °C)$, and $T_s = 0°C$, Eq. 9.41 can be reduced to

$$Q_p = 4.2\, T_r P_r. \qquad 9.42$$

For operational purposes it is usually assumed that T_r is equal to the air temperature.

When rain falls on a snowpack which has a temperature below $0°C$ the situation is more complicated. The pack freezes some of the rain thereby releasing heat by the fusion process. Comparing the latent heat of fusion of water (~ 335 kJ/kg) with the specific heat of snow (2.09 kJ/(kg $\cdot$ °C)), it is obvious that freezing exerts a considerable influence on the thermal regime of the pack. For example, 10 mm of rain at $0°C$ uniformly distributed in a 1-m depth of snowcover having a density of 340 kg/m^3 would, on refreezing, raise the average temperature of the snowpack from -5 to $0°C$. The distribution of heat released by the rain is strongly affected by the way water moves through the pack. In addition, the transmission properties of the pack are usually changed because its internal structure has been changed by the percolating water.

Water Movement Through Snow

Liquid water in the upper surfaces of the snowpack generally percolates very slowly to the ground surface, e.g.,flow velocities ranging from 2-60 cm/min have been observed. The wide range in velocities can be attributed to several factors: the internal structure of the snow, the condition of the snowcover prior to the introduction of water and the amount of water available at the snow surface. On reaching the ground surface the water will infiltrate the soil and, if the infiltration rate is exceeded, some will flow overland in a saturated or "slush" layer. Water movement in this layer normally occurs at speeds from 10 - 60 cm/min although once drainage channels form in the snow the speeds can increase considerably.

Experimental observations of water flow around snow grains of different sizes have shown that movement occurs in a highly irregular manner. At the melting temperature a thin film of water covers the individual grains; de Quervain (1973) showed that in snow having an average grain size of ~ 2 mm, much of the water could flow through this film. Water has also been observed to flow through isolated saturated pores. Wakahama (1968) showed that once the pores have been filled with water, laminar flow of the Hagen-Poiseuille type may occur, which is an extremely efficient mechanism for draining the snowpack. Pressure increases in the pore spaces are unlikely during drainage, since air is free to move through the pores in the snow matrix in response to the movement of liquid water. For fine-grained snow, Colbeck (1974a) suggests that a combination of film flow and droplet movement is possible.

Dye studies conducted in naturally deposited snowcovers which are heterogeneous and anisotropic have shown that the water from film or droplet flow tends to collect in individual capillary pores. On a large scale, Gerdel (1954) describes isolated vertical channels of coarse-grained snow which act as drains or preferred paths for meltwater. The formation of these channels is often associated with rainfall on snow. The coarse grains result from the higher water saturation in the channels, which induces a rapid grain growth; this, in turn, allows more water to drain through the channel.

A second large-scale feature of a naturally-deposited snow is the presence of relatively-impermeable, high density layers, referred to as "ice" layers. Water can accumulate above such layers producing saturation. Dye studies suggest that these layers are not impermeable but rather are characterized by a variable permeability which forces the meltwater to take numerous sideways steps on its route to the ground (Gerdel, 1949; Langham, 1974). An added complication is the fact that ice layers seldom form in horizontal planes but commonly have variable slopes.

Homogeneous snow

Notwithstanding the complex and irregular process which characterizes the percolation of water through snow the net vertical flux or overall movement in the upper layer of a relatively homogeneous snowpack (i.e., in the absence of vertical drainage channels and ice layers) can be described using the concepts of Darcian flow in an unsaturated porous medium (Colbeck, 1972; Colbeck and Davidson, 1973). The theory is applied over an area sufficiently large to enable the influence of spatial inhomogeneities on the percolation rates to be averaged. Evidence from dye tracer studies indicates that the characteristic surface area over which this averaging takes place is small compared to the areal extent of most snowcovers and is of the order of z^2 where z is the depth of snow. The analysis applies to an isothermal snowpack, i.e., a pack which is at $0°C$ throughout.

For an unsaturated snowpack, the air and water fluxes are balanced. Darcy's Law relates the capillary pressure gradient $\partial p_c/\partial z$, and the gravitational acceleration g to q_z, the volume of water flowing per unit area per unit time,

$$q_z = (k_l/\mu_l)[(\partial p_c/\partial z) + \rho_l g], \qquad 9.43$$

where k_l, μ_l and ρ_l are the permeability of the snow, the viscosity and density of the liquid water, respectively, and z is measured in the vertical direction. As a rule, in a snowpack the gravity term is the major driving force and dominates the pressure gradient $\partial p_c/\partial z$. However, local pressure gradients can be significant in some regions. Wankiewicz (1979) suggests there is a pressure gradient in the zone a few centimetres in thickness directly above a saturated layer and at the snow interfaces. Colbeck (1974a) shows that the pressure gradient is significant at low water fluxes, (corresponding to low values of water saturation and high values of water tension) and at the leading edge of the meltwater front as it advances downward through the pack. In the latter case the gradient acts to round-off the otherwise sharp corners of the front but does not influence the rate of movement of the front to any degree (Colbeck, 1974a).

In order to apply Eq. 9.43 to snow it is necessary to develop a relationship between the permeability of snow to the liquid phase k_l and the more easily measured intrinsic permeability of the snow matrix k (m^2). Colbeck and Davidson (1973) relate k_l and k to the effective water saturation S^* (volume of water: volume of voids) by means of the expression:

$$k_l = kS^{*3}, \qquad 9.44$$

where S^* is defined as

$$S^* = (S_w - S_{wi})/(1 - S_{wi}).$$

S_w is the water saturation expressed as a percentage of the pore volume. S_{wi} is termed the "irreducible-water saturation" by Colbeck and is the water permanently retained by capillary forces in the snow (expressed as a fraction of the total pore volume). S^* is termed the "effective water saturation" because it represents that portion of the liquid water free to move under the influence of gravity.

Combining Eqs. 9.43 and 9.44 gives the relationship between the vertical liquid flux through the unsaturated snow q_z and S^* (assuming the pressure gradient is negligible):

$$q_z = \alpha k S^{*3}, \qquad\qquad 9.45$$

$$\text{where } \alpha = \rho_l g / \mu_l.$$

The continuity equation for the liquid water phase applicable to a differential volume of snowcover is

$$\partial q_z / \partial z + \phi_e (\partial S^* / \partial t) = 0, \qquad\qquad 9.46$$

where ϕ_e is the effective porosity i.e., the fraction of the pore volume available for flow, which may be calculated from the expression,

$$\phi_e = \phi \, (1 - S_{wi}),$$

where ϕ is the porosity or ratio of pore volume to total volume. Equations 9.45 and 9.46 can be combined to give the following equation for the flux of water:

$$3\alpha^{1/3} k^{1/3} q_z^{2/3} (\partial q_z / \partial z) + \phi_e (\partial q_z / \partial t) = 0. \qquad\qquad 9.47$$

If an average snow density is assumed this equation can be solved by the method of characteristics to yield the algebraic expression

$$\frac{dz}{dt} \bigg|_{q_z} = 3\alpha^{1/3} k^{1/3} q_z^{2/3} / \phi_e. \qquad\qquad 9.48$$

The important feature of Eq. 9.48 is that the celerity or rate of downward movement, $dz/dt|_{q_z}$, is directly proportional to $q_z^{2/3}$. Thus, large fluxes move more quickly through a snowcover than smaller ones, so that at any given depth the passing wave of meltwater is characterized by a rapid rise in peak value followed by a slow recession. Experimental verification of Eq. 9.48 is reported by Colbeck and Davidson (1973) for isolated columns of repacked, homogeneous snow that were sealed and allowed to drain for several days after a period of intense surface melting.

Equation 9.48 can be applied in hydrological forecasting procedures when values of the intrinsic permeability k and effective porosity ϕ_e of the snow are known. The intrinsic permeability is a function of many physical properties of a snowcover, including its density and grain size, and the distribution, continuity, size, shapes and number of its pores. One of the most useful expressions for estimating the magnitude of k (m^2) is given by Shimizu (1970):

$$k = 0.077d^2 \exp(-7.8\rho_s/\rho_l), \qquad\qquad 9.49$$

where the grain diameter is $d(m)$, and the snow and water densities are ρ_s and ρ_l, respectively. Table 9.4 (Wankiewicz, 1979), based on experimental data reported by Kuroiwa (1968) and Bader et al. (1939) lists some typical values of k for different snowcovers.

Table 9.4
INTRINSIC PERMEABILITY OF SNOW (Wankiewicz, 1979).

Measurement Procedure and Type of Snow	Porosity	Permeability $k \times 10^9 \text{m}^2$
FROM AIR PERMEABILITY (Bader et al., 1954)		
Loose new dry snow	0.86	4
Soft dry wind slab	0.79	3
Hard dry wind slab	0.68	1.4
Fine grain old dry snow	0.59	2
Medium grain old dry snow	0.68	6
Coarse grain old dry snow	0.67	10
Depth hoar	0.74	20
FROM KEROSENE PERMEABILITY (Kuroiwa, 1968[a])		
Newly fallen snow	0.85	4
Fine grain snow	~ 0.7	0.6 - 8
Large grain snow	~ 0.6	3 - 16
Depth hoar	0.59	5

[a] These values may be too small because the flow may not be Darcian.

Since both k and ϕ_e depend on the physical properties of the snowcover the ratio $k^{1/3}/\phi_e$ can be treated as one unknown in solving Eq. 9.48. Granger et al. (1978) found that for a Prairie snowcover values of $k^{1/3}/\phi_e$ may fall in the range from 0.008 to 0.046 $\text{cm}^{2/3}$. During periods when warm air inhibited nighttime refreezing $k^{1/3}/\phi_e$ was observed to increase as melt progressed indicating that k continued to increase and the pack continued to ripen through the melt sequence. However, when nighttime refreezing occurred $k^{1/3}/\phi_e$ decreased as melt progressed indicating that the water refrozen in the snowpack serves to reduce its permeability.

Measured changes in the shape (a function of attenuation and delay) of a diurnal wave of meltwater with depth are shown in Fig. 9.10 (Colbeck and Davidson, 1973). If the flux at the upper surface of the snow were known as a function of time, then a step-by-step application of Eq. 9.48 would show many of the features of Fig. 9.10. In particular the maximum flow rate decreases with depth below the snow surface while the minimum flow rate increases. The leading edge of the wave steepens with depth and eventually forms an almost vertical edge or shock front (curves D and E). The trailing edge of the wave becomes less steep the further it is from the upper surface.

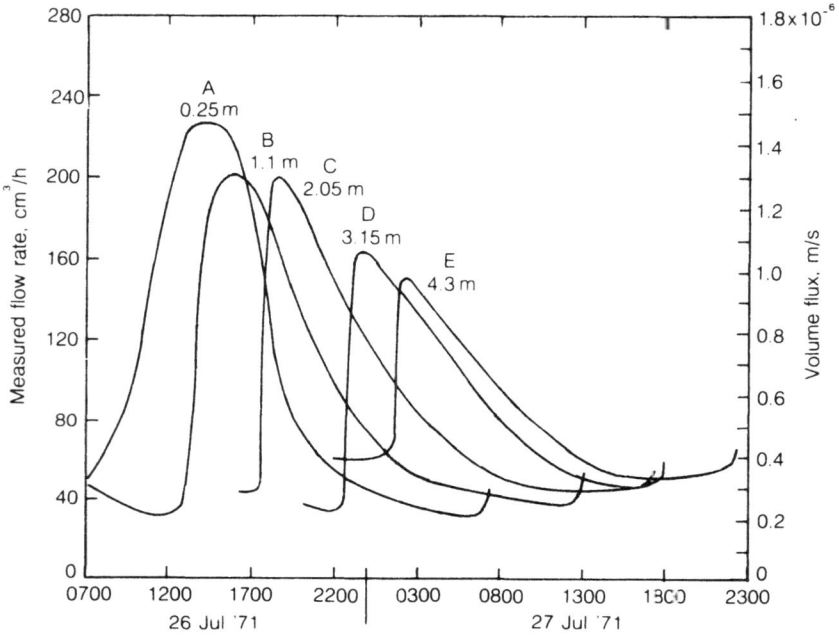

Fig. 9.10 Diurnal melt wave measured at several depths in columns of uniform snow (Colbeck and Davidson, 1973).

According to Eq. 9.48 the celerity or rate of propagation of the volume flux, $dz/dt |_{q_z}$, increases as q_z increases. Slower moving fluxes of q_z produced by early morning melt are overtaken by faster moving fluxes produced later in the day. When two fluxes intersect a shock front is formed. If capillary effects are ignored the water flux is discontinuous across the shock front so that the flow can no longer be described by Eq. 9.48. On imposing on Eq. 9.48 the requirement that the volume of liquid water must be conserved as the front moves downward, the following expression for the rate of propagation of the front is obtained (Colbeck, 1976):

$$dz/dt = \alpha^{1/3}k^{1/3}(q_{z+}^{2/3} + q_{z+}^{1/3} q_{z-}^{1/3} + q_{z-}^{2/3})/\phi_e. \qquad 9.50$$

q_{z+} and q_{z-} are the larger and smaller fluxes, respectively, which have intersected to form the shock front; they change continuously as even larger fluxes overtake the front.

When sufficient melt water reaches the ground surface a form of lateral overland flow is possible in a "slush layer" near the snow-ground interface, for example see Horton (1938). An idealized representation of the shallow saturated region at the ground surface is shown in Fig. 9.11. Colbeck (1974) has developed the basic theory of the flow in this layer. The flow can be described by Darcy's law for small values of slope θ as:

$$q_x = -k/\mu_l \, (\partial p_l/\partial x - \rho_l g\theta) \qquad 9.51$$

where: q_x = volume flow in the x-direction per unit cross-section per unit time,

k = intrinsic permeability,

μ_l = dynamic viscosity of water,

p_l = liquid pressure,

ρ_l = density of water, and

g = acceleration of gravity.

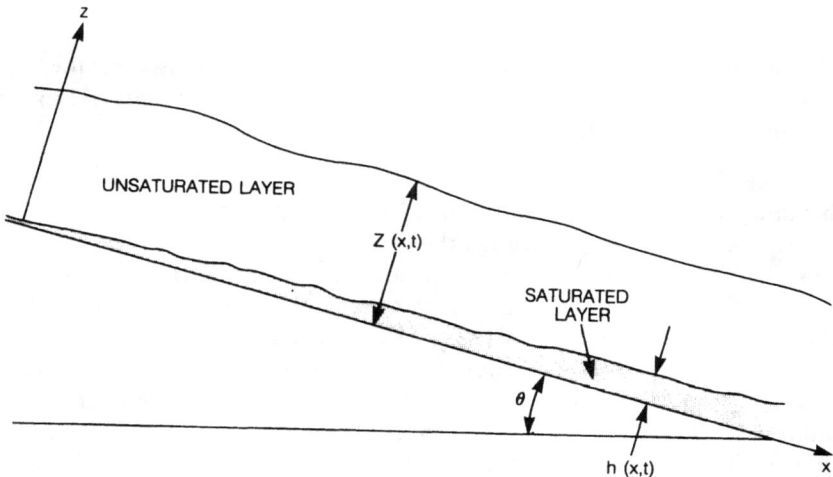

Fig. 9.11 Idealized schematic diagram of meltwater accumulation on the ground surface (Colbeck, 1978).

Metamorphism is rapid in the saturated layer and the average grain size is larger than that in the unsaturated layer. Hence, the intrinsic permeability is also larger and the permeability of snow to the liquid phase in the saturated zone is likely to be several orders of magnitude greater than that in the unsaturated zone. Therefore although the driving potential for flow in the saturated zone is lower than that in the unsaturated zone the velocity of flow in the saturated zone is usually much greater than the velocity of flow in the unsaturated zone.

The continuity equation for the saturated layer of thickness h (Fig. 9.11) is

$$\partial(q_x h)/\partial x + \phi \partial h/\partial t = I(x, t), \qquad\qquad 9.52$$

where $I(x,t)$ is the net flow to the layer, equal to the inflow from the upper saturated zone minus the infiltration to the underlying soil. The governing equation for flow in the saturated layer is obtained by substituting Eq. 9.51 into Eq. 9.52 (Colbeck, 1974b):

$$\alpha k\theta \, (\partial h/\partial x) - \alpha k \, \partial(h\partial h/\partial x)/\partial x + \phi \, (\partial h/\partial t) = I. \qquad 9.53$$

The second term on the left-hand side can be considered to be negligible if it is assumed that the increase in the thickness of the saturated layer with downslope distance, dh/dx, is small compared to the slope θ. With this simplification Eq. 9.53 can be solved using a coordinate system which moves downslope at the velocity of the water particles, c, where:

$$c = \alpha k\theta/\phi. \qquad\qquad 9.54$$

For a saturated layer of length L the solution of Eq. 9.53 shows that the total discharge per unit width can be determined by integrating the inflow I over a time interval Δt given by

$$\Delta t = L/c. \qquad\qquad 9.55$$

This integration does not involve the thickness of the saturated layer directly because the flux of water through the saturated layer varies linearly with h. Thus, the timing of the runoff is independent of the geometrical complexities of the surface. Colbeck (1974b) suggests that if the transit time Δt through a saturated layer of length L is large, any short-term variation in I will be damped as the water moves through it. Conversely, if Δt is small, short-term variations, for example, the diurnal melt cycle of a shallow snowcover, will be maintained. For an isotropic snowcover it may be expected that the travel times in the saturated and unsaturated layers may be approximately equal when the respective path lengths in the layers are in the ratio, 100:1.

Layered snow

A snowpack resulting from a series of individual snowfalls is usually heavily stratified. Single layers accumulating from individual storms can be easily identified by density or textural differences. Frequently, buried crusts or "ice layers" are present, and may originate as an old snow surface which has experienced freezing rain, refreezing of melt water or wind packing. Seligman (1963) also mentions sublimation in a moist wind as a possible mechanism of crust formation.

The stratification of the snowcover, particularly the ice layers, can have an important influence on water movement. However, measurements by Gerdel (1954) suggest that once large amounts of melt water reach the buried ice layers disintegration can occur in a matter of hours through the formation of vertical drainage channels.

A quantitative understanding of water movement through a heterogeneous or layered snowpack has yet to be developed and work on this problem is hampered by a lack of field data. In situ measurements of the distribution of liquid water within the snow are difficult; furthermore, such properties of ice layers or buried crusts as their permeability, thickness and areal extent have not been measured. Developments in the analysis of water flow through layered snow are reviewed by Colbeck (1978) and Wankiewicz (1979).

Wankiewicz suggests that observations on flow through layered snow should be categorized in accord with the general system incorporated in the FINA model. This model postulates that an individual snow horizon or boundary between two snow layers of different textures and densities will impede, have no effect on, or accelerate downward flow (FINA - Flow Impeding, Neutral or Acceleration). The action of the horizon may depend on the flow range, i.e., impeding the flow at low flows but accelerating it at higher flows. This behaviour has been observed in coarse-grained soil but has yet to be confirmed in snow (Wankiewicz, 1979).

Figure 9.12 illustrates, in a general way, the action of impeding and accelerating horizons. Except in the immediate vicinity of the snow horizon gravity flow will occur throughout the snowpack. In the example it is assumed that near the horizon the flux q can be transferred through the differential effect of pressures p_1 in the upper layer and p_2 in the lower layer. If $p_2 > p_1$ the horizon between the two layers will impede flow and water will accumulate immediately above the horizon until $p_2 = p_1$. If the horizon is sloping the stored water will have a component of flow parallel to the horizon. Lateral flow within snowpacks has been observed by Wakahama (1968). In Fig. 9.12a this lateral component is shown to cause a redistribution of the flow such that the horizontal flux is less in section AB but enhanced in section CD.

When $p_2 < p_1$ the horizon will accelerate the passage of the wetting front because the upper layer will be drier adjacent to the horizon. Wankiewicz (1979) assumes the wetting front becomes unstable at this horizon and that fingering occurs in the lower layer (see Fig. 9.12b). Evidence of fingering in snow is mentioned by Wankiewicz and de Vries (1978). Wankiewicz (1979) also examines the effects of ice layers on flow, treating them as compound horizons; he presents the likely flow patterns resulting from various combinations of fine- and coarse-grained snow.

Analytical treatment of the flow of water through snow is still in the initial stages of development. Langham (1973) has analyzed the flow to a drain in a flat impermeable ice layer while Colbeck (1973) has examined a sloping ice layer which has holes or drains at regular intervals. For a snowpack with many discontinuous layers Colbeck (1975) assumes the snow to be an isotropic homogeneous medium and uses a permeability tensor to calculate the components of the flow. At present, results of this analysis have not been compared with field data.

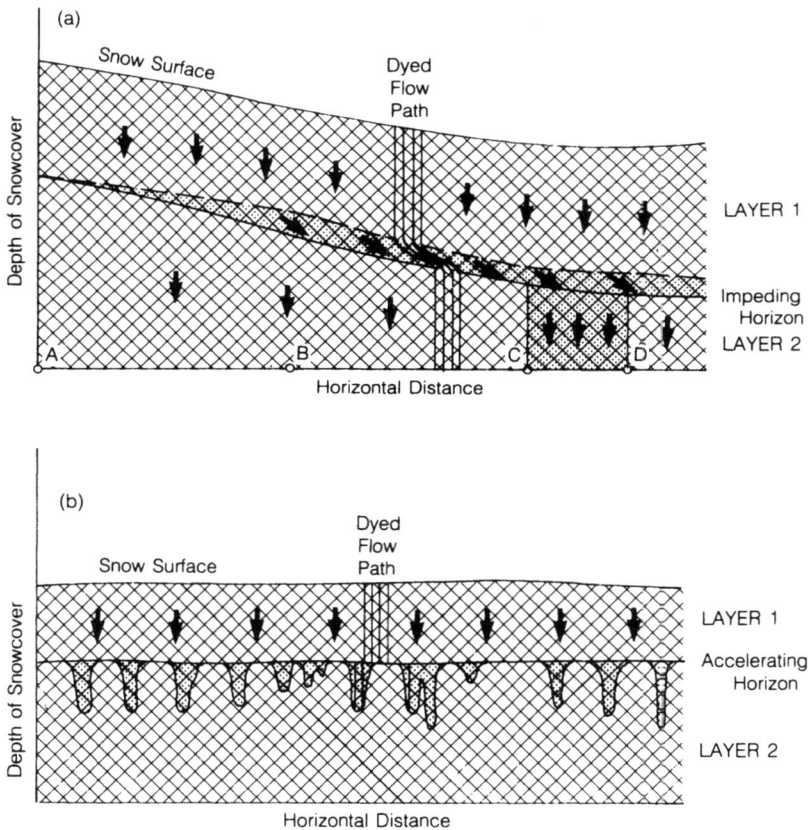

Fig. 9.12 The effects of impeding and accelerating horizons on water movement in a layered snowcover (Wankiewicz, 1979).

SOIL INFILTRATION AND WATER TRANSMISSION

The infiltration process governs, in part, the rate and amount of melt water that enters the ground. Infiltration water derived from snow is an important water resource, since it not only contributes to the replenishment of soil moisture and groundwater reserves that may be used subsequently for plant growth, domestic consumption and other purposes, but also reduces the amount of direct surface runoff thereby reducing peak flows. The infiltration phenomenon is complicated in that it is affected by the surface melt rate, the movement of the melt water through the snowcover, the release of water by the cover, the thermal regime of the soil and the physical properties of the soil that govern water transmission. If the soil is unfrozen, movement of the melt water will occur in a manner similar to water generated by rainfall or ponded on the soil surface. The physics of the infiltration process and water movement in both unsaturated and saturated soils has been intensively studied; since a large number of publications and books are available on this subject, it is not discussed further.

Under many natural conditions the ground is frozen during the snowmelt period; in contrast to unfrozen soils, data on the entry and transmission of water to frozen soils are seriously lacking. As a result, management agencies must resort to the use of bulk empirical relations, (e.g., the ratio of the total volume of streamflow runoff to the volume of snow water equivalent) to estimate the infiltration loss/gain in design calculations. Obviously, the major factors affecting the water transmission characteristics of frozen soil are soil temperatures less than 0° C and ice-filled pores, both of which may impede water movement. For example, snowmelt water may refreeze on entering the soil to form an impermeable layer.

The mechanism of water transport in freezing soil-water systems, accompanied by a change in phase is complicated by the simultaneous transfer of heat and water, the latter occurring in both liquid and vapor phases. Numerous investigators (Dirksen, 1964; Ferguson et al., 1964; Hoekstra, 1966; Jumikis, 1966, Jame, 1977) have studied the movement of water from warm to cold zones during the freezing of unsaturated soils. Water has been found to move from the unfrozen zone to the frozen zone because of the pressure differences imposed by the temperature gradient and because of a local hydraulic gradient caused by the freezing process. Vapor diffusion may also occur because the partial pressure of the water vapor in the unfrozen zone is greater than that in the zone containing ice.

Several models have been proposed to describe coupled heat and mass transfer in unsaturated soil-water systems during freezing and thawing (Harlan, 1973; Kennedy and Lielmezs, 1973; Guymon and Luthin, 1974;

Bresler and Miller, 1975). These require the solution of simultaneous equations for heat and moisture transfer.

The heat transfer or energy equation commonly takes the form:

$$\partial\,[k(z,T,t)\,\partial T/\partial z]/\partial z - C_l\rho_l\,[\partial(V_zT)/\partial z)] = \partial(\overline{C\rho}T)/\partial t, \qquad 9.56$$

where z = position coordinate,
             T = temperature,
             k = thermal conductivity,
             t = time,
             C_l = specific heat of water,
             ρ_l = density of water,
             V_z = flow velocity in z direction, and
             $\overline{C\rho}$ = apparent volumetic specific heat.

Derivation of Eq. 9.56 assumes that the soil matrix does not deform. This equation includes terms for heat transfer by conduction and by the vertical movement of liquid water, but excludes heat transfer by movement of vapor. The apparent volumetric specific heat which incorporates the release of the latent heat of fusion L as the soil water freezes can be estimated from the equation:

$$\overline{C\rho} = \sum_{i=1}^{n}(C\rho)_i\phi_i - L\rho_s\,\partial\theta_s/\partial T \qquad 9.57$$

where $(C\rho)_i$ = volumetric specific heat of i-th constituent,
             ϕ_i = volumetric fraction of i-th constituent,
             ρ_s = density of ice, and
             θ_s = volumetric ice fraction, cm^3/cm^3.

Moisture transfer, which results in changes in the ice and water contents of a volume of soil, is governed by the equation:

$$\partial\,[K(z,T,\psi)\,\partial\phi/\partial z]/\partial z = \partial\theta_u/\partial t + (\rho_s/\rho_l)\,\partial\theta_s/\partial t, \qquad 9.58$$

where K = soil hydraulic conductivity,
             ψ = soil suction potential, and z elevation head,
             ϕ = the total soil water potential, such that $\phi = \psi + z$ and
             θ_u = volumetric liquid water fraction.

In this expression all processes are assumed to be single-valued functions (hysterisis neglected). Jame (1977) presented a numerical solution of Eqs. 9.56, 9.57 and 9.58 that employs the finite difference method using the Crank-Nicholson scheme. A comparison of simulated and measured (laboratory) temperature and moisture profiles showed close agreement. Most of the tests carried out so far to verify the different models of heat and mass transfer have used laboratory measurements on small, homogeneous, isotropic soil cores subjected to freezing temperatures. However, the experimental verification of the models is difficult because reliable estimates of the thermal and hydraulic properties of the soil samples are not easy to obtain.

Analyses and *in situ* measurements of snowmelt infiltration into frozen soils are largely unknown. The few published results clearly show that the infiltration curve (infiltration rate plotted against time) may assume a wide variety of shapes depending primarily on the moisture content and temperature of the frozen soil and the rate of snowmelt. For example, the Russian literature cites four forms of the infiltration curve to frozen soils (see Fig. 9.13). Figures 9.14 through 9.17 are presented to emphasize the wide range of soil water conditions that may be encountered under different snowmelt regimes and land use conditions. The data were derived from point measurements of soil moisture changes obtained with a twin probe gamma system. Calibration of the equipment indicated that the expected errors in the measured infiltration amounts would range from 1 to 4%.

Soil water changes on a wheat stubble site during and after melt are shown in Fig. 9.14; the mass infiltration curve and the change in snow depth with time in Fig. 9.15. These data were obtained in the spring of 1978 at a location where "continuous melt" occurred in the period from March 19 to 30. The soil moisture profiles show orderly development throughout the melt period. After melt, a decrease in soil moisture in the upper layers occurs as a result of evapotranspiration and the downward migration of moisture. It is noteworthy that the wet front 20 days after the snowcover had completely disappeared has advanced only to a depth of 53 cm. At this site the soil moisture changes accounted for approximately 54% of the initial snow water equivalent. Infiltration rates during the ablation period were estimated to range from 4 to 22.6 mm/d, the highest rates occurring near the end of the melt period. It is of interest to note that these high rates (near the end of the melt period), that result from higher rates of melt, of flow in the snowcover and of thaw in the soil, are opposite to entry rates of ponded water to unfrozen soils. Thus the mass infiltration curve for ablation is concave while that for the entry of ponded water to an unfrozen soil is generally convex.

Figures 9.16 and 9.17 show, respectively, the soil moisture profiles and the mass infiltration curve for a summerfallow site in 1979. The snowcover melted discontinuously during the period March 16 through April 18 with rapid ablation occurring from April 15 to 18. Several aspects of the results depicted are of interest:

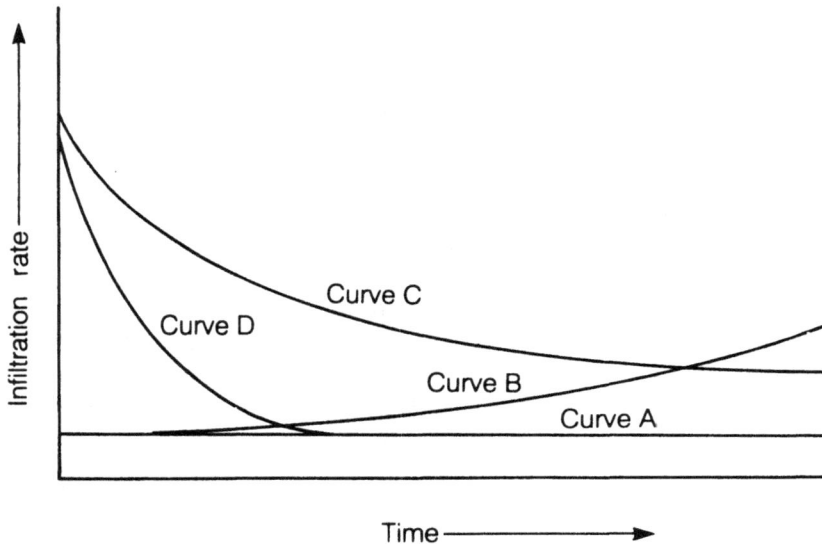

Fig. 9.13 Schematic diagram of the change of infiltration rates of frozen soils with time (Gray et al. 1970).

Curve A: For soil frozen when saturated, or when an impervious ice layer develops on the surface during the melt period. The rate of entry of water is very low, so that the runoff coefficient is high.

Curve B: For soil frozen at a high moisture content (70-80% field capacity). Some of the melt water is able to penetrate the soil, and thus transfer heat to melt the ice-filled pores. Progressively, as the soil warms and the pores melt, the infiltration rate increases. Zavodchikov (1962) cited examples in which the infiltration rate of the soil increased 6-8 times its initial rate during the melting period. Eventually the intake rate decreases because of the high soil moisture content.

Curve C: For soil frozen at a low moisture content, and at a temperature near or above freezing. Only the small pores are filled with ice; these thaw rapidly with the downward movement of water. Infiltration proceeds as under normal unfrozen conditions.

Curve D: For soil frozen at a low moisture content and at a temperature well-below freezing when the snow is melting. Water entering the soil is frozen in the pores and movement is inhibited. Copyright © 1970, Canadian National Committee for the International Hydrological Decade. Reproduced by permission of the National Research Council of Canada.

(1) Snowmelt infiltration during the early stages of the melt sequence, March 16-21, was distributed in the surface layer of soil to a depth of 15 cm. By March 21, 14.5 mm of infiltration had occurred. The infiltration rates in this interval were ~ 1.2 - 2.1 mm/d. Throughout the wetted profile the soil temperatures ranged from -8.6 to -4.5° C (at a depth of 10 cm).

(2) From April 8 to 15 a series of snow and rain showers occurred, resulting in:

    (a) An increase in snowcover depth of ~6 cm.

    (b) An increase in the amount and rate of infiltration. All of the water infiltrated in the period (24.2 mm) was retained in the upper layer of the soil (0-16 cm). The average rate was about 6 mm/d.

    (c) The soil temperature at a depth of 10 cm increased from -5.8° C to +0.4° C.

These findings suggest that appreciable melt was produced by condensation and advection.

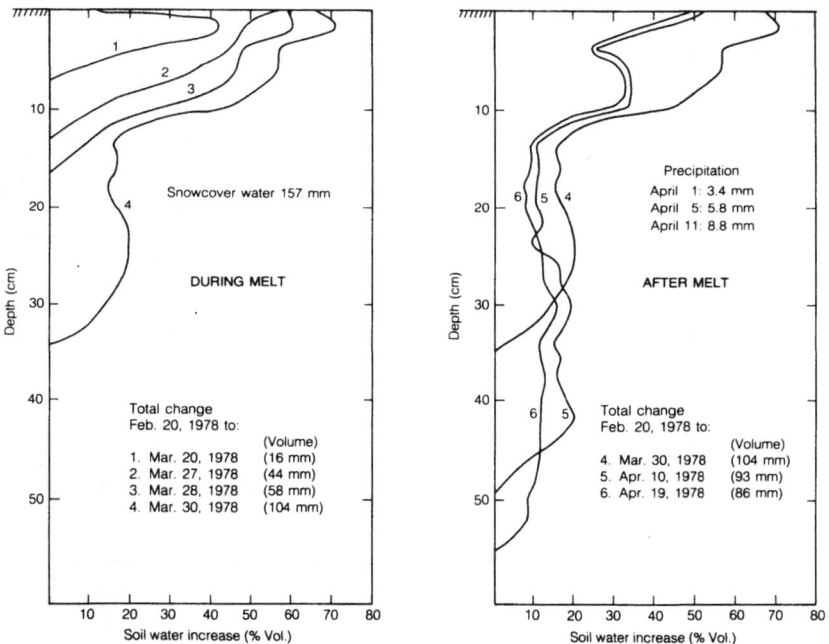

Fig. 9.14 Soil water changes at wheat stubble site, 1978 (during and after snowmelt).

(3) By the time the snowcover had completely disappeared the wet front
     had advanced to a depth of about 16.5 cm, where the soil temperature
     was -0.8° C.

The data presented in the figures are in harmony with those found in other
years. It should be recognized that a major anomaly in the infiltration
characteristics can occur if the soil is highly cracked at the time of melt as a
result of summer soil moisture depletion and freezing. Under these conditions
the entry rate of water is governed primarily by the rate of melt release from
the snowcover since the water will readily penetrate to the bottoms of the
cracks. Entry rates of 33 mm/d to highly-cracked heavy clay soils cropped to
barley have been measured on the Canadian Prairies.

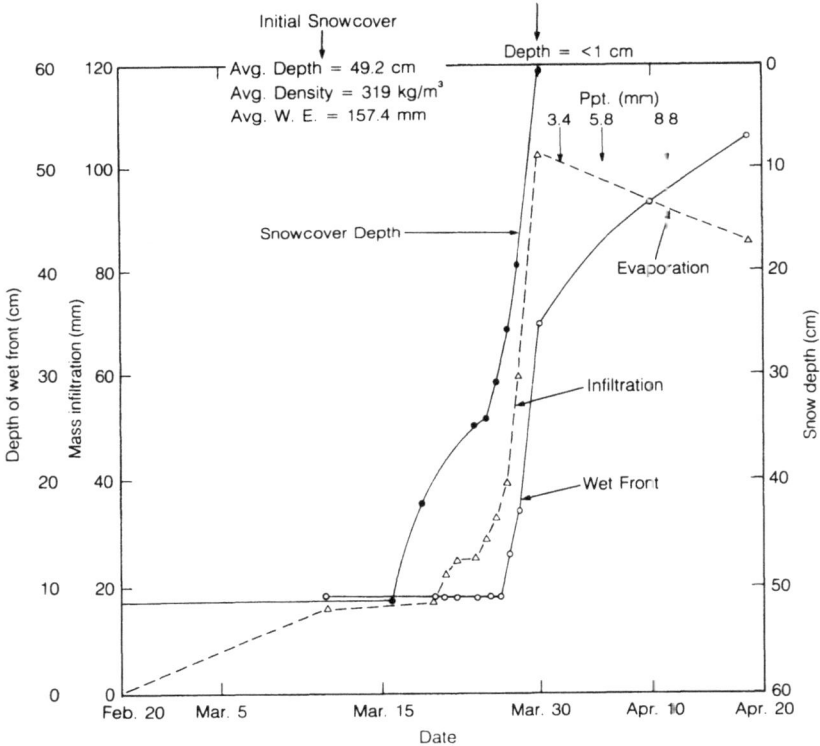

Fig. 9.15 Albation, infiltration and advance of wet front, wheat stubble site: 1978.

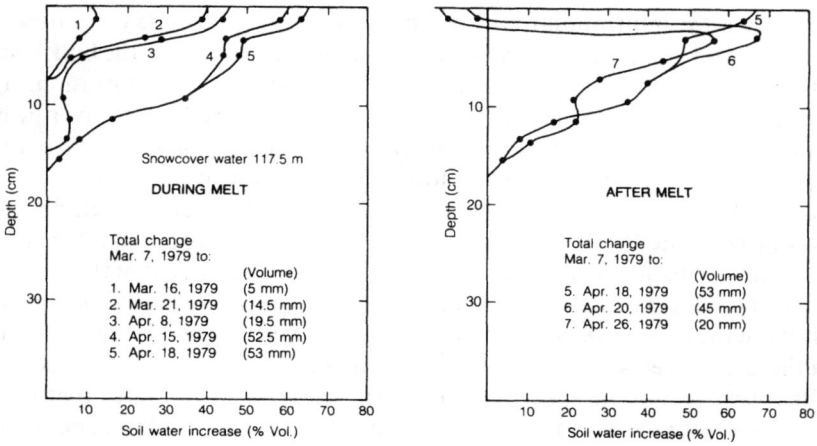

Fig. 9.16 Soil water changes at summerfallow site, 1979 (during and after snowmelt).

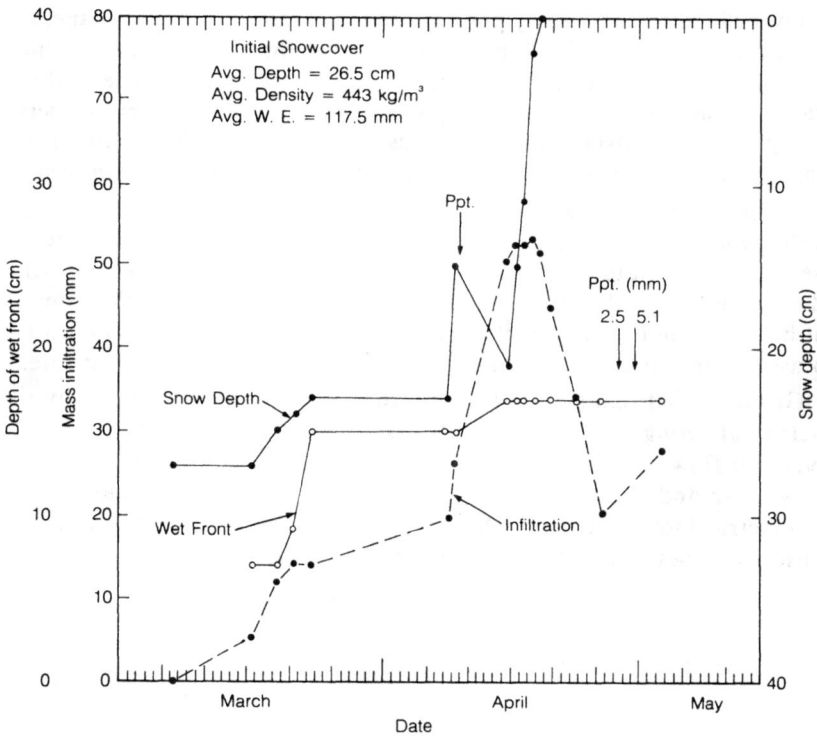

Fig. 9.17 Ablation, infiltration and advance of wet front, summerfallow site: 1979.

Because the infiltration-thawing process is complex and data on infiltration rates are lacking it is common operational practice to consider the bulk loss of the snowcover water equivalent to evaporation, storage and infiltration. The volumetric loss ratio, usually calculated as the initial snowcover water equivalent less measured streamflow runoff divided by the snowcover water equivalent, may be determined for daily, monthly or seasonal periods. Seasonal values are the most common. Use of a constant ratio to determine an average loss rate for short periods of time can lead to substantial errors in the estimates. This arises because the daily loss rate is independent of the melt rate only when it is exceeded by the melt rate. Seasonal values may be adjusted to estimate the losses over shorter time periods by multiplying the daily changes in the snowcover water equivalent by the loss coefficient (snowcover depth is often used because it is easy to measure). Figure 9.15 indicates that in the absence of storage, when melt is continuous and the entry and transmission of water to and within the soil is not impeded by ice lenses, changes in snow depth may usefully serve for estimating the daily infiltration and evaporation quantities. The rate of change in depth approximates the rate of infiltration. Conversely, during the early periods of melt (March 16-26) the shapes of the curves of snowcover depth and mass infiltration are completely different. This can be attributed to the diurnal cycling of melt and storage of the melt water within the snowcover. Similarly, little association exists between changes in snowcover depth and mass infiltration under conditions where impermeable ice lenses form in a soil and advective melt by rain occurs.

Similar problems arise when infiltration is assumed to be a fixed percentage of the heat supply and hence of snowmelt. This assumption is connected with the use of temperature indices for estimating snowmelt (see sections below). The anomaly which occurs is that at low melt rates all melt water may infiltrate with none being available for runoff. Furthermore it is unreasonable to assume that the infiltration rate keeps increasing with the melt rate.

Given the fact that snowmelt losses to streamflow are influenced by all the factors affecting runoff, viz., water transmission through the snowcover, overland flow and flow within the soil and, in addition, by the extent of snowcover and the terrain storage characteristics it is axiomatic that the volumetric loss coefficients show wide variations. Table 9.5 lists typical seasonal values of the coefficient found for different landscapes.

Table 9.5

SEASONAL VOLUMETRIC LOSS COEFFICIENTS: RATIO OF INITIAL SNOWCOVER WATER EQUIVALENT MINUS MEASURED SURFACE RUNOFF TO INITIAL SNOWCOVER WATER EQUIVALENT.

Landscape	Geographical Location	Volumetric Loss Coefficient	Remarks
Undulating terrain, lacustrine and glacial till, loam and silty clay, summerfallow and mixed grains	Prairies	~0.33	Watershed, surface storage included
Undulating terrain, lacustine and glacial till, loam and silty clay	Prairies		
summerfallow		0.60	Small plots, surface storage negligible - normal conditions.
wheat stubble		0.75 - 0.90	
wheat stubble		0.66	Point measurement, infiltration only.
summerfallow		0.40	Point measurement infiltration only, ice lenses present
Varying topography 16° - 49° slopes, sandy soil, pasture.	Vermont, USA	Average: 0.53	Small plots, concrete frost in up-per 3-cm of soil horizon (Dunne and Black, 1971)

OPERATIONAL PROCEDURES FOR ESTIMATING SNOWCOVER RUNOFF — SOME COMMON PRACTICES

Introduction

Over a large portion of the Northern Hemisphere runoff produced by melting snow contributes significantly to the annual water yield and to spring floods. Many federal, provincial, state and private agencies are actively involved in studies concerned with the prediction of particular aspects of the melt sequence including the time of initial melt, the rates and volumes of water released, and the effects of the snowmelt contributions on streamflow discharge rates and volumes. A large number of methods have evolved for forecasting purposes, each of which makes use of empirical formulae and various approximations to describe the melt, transmission and runoff processes. Forecasts are normally required over large areas where there are wide variations in snowcover characteristics, topography, vegetative cover and climate, and where hydrometeorological measurements are scarce.

The direct application of the phenomenological approach to snowmelt as developed in preceding sections of this chapter is not currently feasible for operational purposes. First, methods of describing the melt process on a spatial basis, as would be required for any forecast, are at an elementary stage. Second, forecasters must work with a minimum of data, most commonly air temperature and snowfall or snowcover measurements, obtained on a large spatial scale. Third, most forecasts must be made in a relatively short period of time, e.g., flood forecasts, are usually updated every 6-h or daily if not more frequently, and it is usually impractical to obtain input data at such intervals.

The operational forecast procedures described in the literature show a wide variety of approximations employed, depending on the forecast period (long- or short-term), the data and financial resources available and the characteristics of the watersheds being modelled. Each streamflow forecast model attempts to predict the net effect of different factors on the melt water from the time it is produced at the surface of the snow until it reaches a specific point in a stream or river, for example, a gauging station. Although a given quantity of melt water passes through a series of unique regimes while moving from the snow surface to the ground, along the surface or through the ground, and within a defined stream channel, it is difficult to describe each regime individually by an operational procedure. Hence, it is common practice to group individual processes and to describe them by a single expression or coefficient.

Snowmelt and Transmission Through the Snowcover

The various procedures developed to estimate the contribution of snowmelt to streamflow tend to have some features in common, since experience has shown that for a given set of loss conditions the estimates of the amount and timing of melt release to a channel are sensitive to relatively few factors. The three most important of these are:

(1) The energy available to melt the snow over a given area,
(2) The areal extent of the melting snowcover, and
(3) The effects of storage on the movement of the melt quantities during transit from the snow surface to the steam channel.

Methods of estimating the energy flux to snow

Most operational procedures for snowmelt prediction rely on air temperature as the index of the energy available for melt. There are two reasons for this. First, air temperatures are generally the most readily available data. Second, the temperature-index methods often give melt estimates that are comparable to those determined from a detailed evaluation of the various terms in the energy equation (U.S. Army Corps of Engineers, 1971; Anderson, 1973), particularly when consideration is given to the inaccuracies of weather forecasts.

A review of the expressions presented in the literature show that no single, universally-applicable temperature index of snowmelt exists. Each index is unique to a specific basin and geographical location. Its magnitude varies depending on the prevailing atmospheric conditions (clear, cloudy, rain) and the time of year. Such variability is not surprising, if it is realized that the air temperature is only one factor influencing melt rates and other factors such as wind velocity, atmospheric moisture content and albedo of the snow are not directly related to air temperature.

In general, air temperature is a good index of the energy available for melt in areas covered by dense forest vegetation. In this situation the long-wave radiation exchange between the vegetative canopy and the snow, which is a function of the temperature difference between the two surfaces, is the most important energy flux. Measurements show that the canopy and air temperatures are closely related. In contrast, temperature is not as reliable an index for open areas because short-wave radiation, sensible and latent heat fluxes (none directly related to temperature), can exhibit wide variations depending on weather conditions.

When comparing indices it is necessary to examine the data used for deriving them. Indices used to estimate the energy flux at the snow-air interface, may only be applicable to open or to vegetated areas; some may only apply to periods of clear weather, others to rainy periods. In addition it is important to distinguish between indices applicable to a point from those applicable to a large area.

The simplest and most common expression relating snowmelt to the temperature index is:

$$M = M_f (T_i - T_b), \qquad\qquad 9.59$$

where M = melt produced in cm of water in a unit time,

           M_f = melt factor (cm/(°C · unit time)),

           T_i = index air temperature (most commonly the maximum or the mean daily temperature), and

           T_b = base temperature (usually 0°C).

Yoshida (1962) presents a detailed discussion of one index developed from measurements of the water equivalent of snow at five stations in the Tadami River Basin in Japan, clearly demonstrating the conditions that influence the melt factor, M_f. In this study, T_i was taken as the mean air temperature for that portion of the day when the temperature was above 0°C. Values of M_f were found to range between 4 - 8 mm/(°C · d) depending on location, time of year and meteorological conditions. M_f may change considerably with exposure of the site to solar radiation. A 20 percent decrease in short-wave radiation at a site resulted in a corresponding decrease in M_f of 1 mm/(°C · d). M_f also varies from year to year; often by as much as 1.5 mm/(°C · d). In Yoshida's procedure cloud cover was accounted for by a correction to the index air temperature; T_i was reduced by approximately 0.8°C on overcast days.

In attempting to identify the major variables affecting the melt factor Eggleston et al. (1971) suggest the following relationship:

$$M_f = k_m k_v R_I (1 - A), \qquad\qquad 9.60$$

where k_m = proportionally constant (for mountainous regions $k_m \cong 0.4$),

           k_v = vegetation transmission coefficient for radiation,

           R_I = solar radiation index, and

           A = snow albedo.

The change in albedo with time t (days) is described by

$$A = 0.4 [1 + \exp(-k_e t)], \qquad\qquad 9.61$$

where k_e is a time constant ($\sim 0.2/d$). It is assumed that a fall of new snow increases the albedo to a value of 0.8 while rain reduces it to 0.4. This equation is valid for deep mountainous packs, but, as Fig. 9.6 clearly shows, not for a shallow Prairie pack.

The vegetation transmission coefficient k_v in Eq. 9.60 is assumed to have the form:

$$k_v = \exp(-4C_v), \qquad\qquad 9.62$$

where C_v is the vegetation canopy density (decimal fraction).

R_I, in Eq. 9.60 is the ratio of the radiation received by a surface, with a certain slope and aspect, normalized to that received by a horizontal surface at the same latitude and time of year. In calculating R_I the effects of the atmosphere on the transmission of radiation are ignored so that it is a function of the angle between the normal to the surface and the direct beam radiation component. In Fig. 9.3 R_I is plotted against time of year for different latitudes, slopes and aspects.

For periods of rain the melt factor is calculated by adjusting M_f as follows:

$$M_f (RAIN) = M_f + 0.00126P, \qquad\qquad 9.63$$

where: $M_f (RAIN)$ = melt factor for rain, and
           P = rainfall (mm).

The seasonal variation in the melt factor is well-illustrated by the results obtained from a study reported by Anderson (1973). He developed an index for rain-free periods taking T_i equal to the mean areal temperature computed on a 6-h time interval and T_b equal to $0°C$. Empirical formulae were used to calculate T_i from the daily maximum and minimum temperatures observed at the stations. The melt factor was assumed to vary seasonally following a sine wave function with the maximum and minimum values occurring on June 21 and December 21, respectively. Maximum values of M_f are 2 to 12 times its minimum value, depending largely on the area of the basin forest cover; they were found to fall within the range 1.32 - 3.66 mm/($°C \cdot d$). The procedure adopted by Anderson is particularly well-suited to watersheds where melt is not confined to the spring months but also occurs throughout the winter.

The U.S. Army Corps of Engineers (1956) point out that melt factors derived for one site only and for a single melt season are of limited value for determining basin snowmelt. A more representative estimate of the basin's melt factor can be obtained by analysing data from several points over a period of years. In their approach the index air temperature T_i can be the mean of the daily maximum and minimum temperature or simply the maximum daily temperature. The base temperature T_b is usually taken as $0°C$. Point indices are determined on the basis of changes in the measured water equivalent of the snowpack. The mean value of M_f determined from several years of operation of snow courses at three snow laboratories gave an average point melt factor of 1.74 mm/($°C \cdot d$), using the maximum daily temperature for T_i. However, this value is meaningless for general use, since individual values for each year and location were found to vary greatly, e.g., M_f values for sites with southern exposures could be almost twice those for a sheltered location. Furthermore, the mean all-station factor for each watershed studied gave a higher melt rate than could be accounted for by the measured runoff. This was attributed to measurement sites being situated in open areas that were not representative of the forest on a basin. Indices are often adjusted to account for differences in melt from the forested and open sites by changing the base temperature T_b. For example, at the U.S. Army Corps of Engineers Snow Laboratories respective values of T_b equal to -2.8 and $5.6° C$ were found to be applicable for open and forested sites, if T_i is taken as the maximum daily temperature and the melt factor, M_f is assumed to be 1.83 mm/(°C · d).

Evaluation of a representative basin index should be derived from a complete water balance of the basin in which all gains (precipitation), losses (evaporation, evapotranspiration, infiltration and soil moisture) and components of streamflow (direct runoff, groundwater) can be separated. In addition, the snowcovered area or contributing area should be taken into account in these calculations.

In cases where it can be assumed that the soil moisture deficit and other storage terms are either negligible or satisfied during the initial period of melt, and that the loss to infiltration is small, the value of the melt factor approaches the snowmelt runoff factor. Interestingly, in a study conducted from 1972-1976 on nine small watersheds (1200 to 53560 m^2) in southwestern Saskatchewan each with different land-use and terrain, Erickson et al. (1978) found that the cumulative daily runoff per unit snowcovered area could be linearly related to the product of the fraction of the basin that is snowcovered and the number of degree days (days with $T > 0°C$), i.e.,

$$\sum_{i=1}^{n} Q_i/D = M_Q \sum_{i=1}^{n} (A_i/D) \cdot T_i \qquad\qquad 9.64$$

where: Q_i = mean daily runoff on i-th day (mm),
          D = drainage area of the watershed (m^2),
          M_Q = snowmelt runoff coefficient (mm/(°C · d)),
          A_i = mean area of the watershed snowcovered on the i-th day (m^2),
          T_i = mean daily temperature ($> 0°C$) on the i-th day (°C), and
          n = number of days with runoff.

It was found that the influence of land use, and to a lesser extent terrain, was highly significant. Average values of M_Q for stubble, pasture and summer-fallow were calculated as 0.214, 0.255, and 0.820 mm/(°C · d), respectively. The U.S. Corps of Engineers (1956) report rates (mm/(°C · d) above 0°C) for forested basins with deep snowpacks ranging from 1.70 to 4.07 in April and 1.93 to 4.57 in May. Average runoff rates reported by Erickson et al. (1978) for March and April were 3.90 for stubble, 4.78 for pasture and 10.96 for fallow with corresponding maximum runoff rates (measured in a 5-y period) of 5.04, 6.50 and 18.62.

The results presented in the preceding paragraphs illustrate the very different snowmelt runoff factors that may be expected throughout the year in various climatic and vegetative environments. Notwithstanding these variations they also suggest that under certain conditions (e.g., patchy snowcover) temperature index methods offer potential for forecasting runoff quantities provided some estimate of the snowcovered area is available. Extrapolation of indices within regional soil zones also appears feasible if the procedure includes terrain and land-use parameters.

However, it is not recommended that a temperature index be used to predict the maximum rate of snowmelt or melt release over the short time periods that may be required to derive a maximum design flood. For this purpose it is more appropriate to consider the energy exchange processes in detail so as to discover that combination of factors producing the maximum melt rate.

Areal extent of snowcover

When applying the energy equation to determine the melt from a watershed it is necessary to delineate the snowcovered areas where heat exchange occurs, and each area's contribution to the total melt. The locations of the contributing areas and their characteristics that affect the time of melt release and the transmission of melt to a stream channel influence the shape of the snowmelt runoff hydrograph. For operational purposes, the total area of snowcover has been found to be a good index for improving runoff forecasts. This, together with snow depth and density estimates, defines the volume of water stored on the watershed. Because of the importance of the area of snowcover in hydrologic modelling and streamflow forecasting many studies have been aimed at measuring it. These studies specially emphasize the use of satellites and other remote-sensing devices to delineate the snowcovered areas.

Miller (1953) and Ffolliott and Hansen (1968) report that in mountainous areas snowcover depletion can be expressed as a function of accumulated runoff. Leaf (1969) provides detailed information on the procedure to be followed in establishing this relationship from aerial photographs and runoff data. He points out that the correlations derived from several years of data between streamflow and the area of snowcover also include the effects of the following factors on the annual runoff yield: snowcover depletion, groundwater storage, evaporation and surface storage losses and, to a limited extent, the total water equivalent of the snowpack. Leaf proposes that for a watershed these effects can be studied by comparing control curves obtained by plotting both the observed and generated accumulated runoff values against the area of snowcover. The generated runoff values are obtained from the analysis of the recessions of measured hydrographs; they represent the amounts of melt that result in streamflow or the observed runoff, corrected for transitory storage of water on the surface, in the ground and in the stream channels. The areal extent of snowcover is a weighted average based on the area of snowcover on subunits of the watershed classified according to elevation, orientation, slope and vegetation.   Figure 9.18 schematically represents the curves resulting from this procedure. The difference between the two curves at a given snowcover percentage represents the storage effects. Leaf (1969) found that the storage effects (surface, groundwater and stream channel) of a given watershed did not vary appreciably from year to year, even though the observed and generated runoff values obtained in different years varied about their control curves.   This scatter is attributed to annual differences in the initial snowpack water equivalent, soil moisture recharge and meteorological conditions during melting. Once established, the control curves can be used to synthesize flow or to make simple residual volume forecasts.

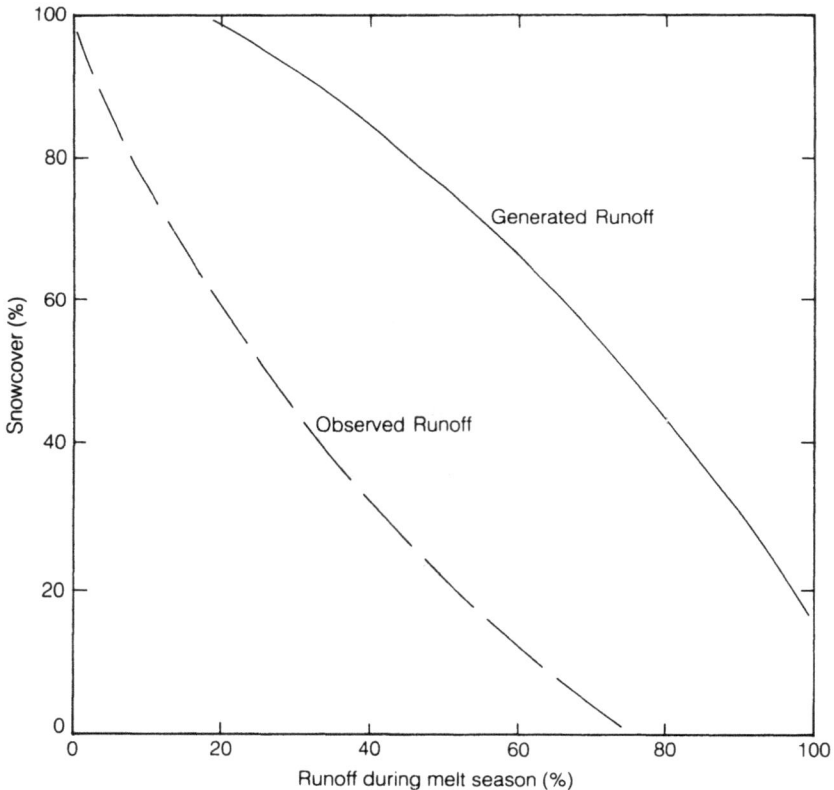

Fig. 9.18 Snowcover depletion as a function of accumulated runoff (Leaf, 1969).

Another procedure for determining the area of snowcover has been suggested by Anderson (1973). It incorporates a plot of the area of snowcover against the ratio of the mean areal water equivalent to an index value A_i, that is the smaller of:

(1)   The maximum water equivalent on the area from the beginning of accumulation, or

(2)   The mean areal water equivalent above which the snowcover is 100 percent.

A typical depletion curve is shown in Fig. 9.19. When snow falls on a partially bare area, the area reverts to 100 percent snowcover for some time before returning to the same percentage snowcover it had before the snowfall. Anderson suggests that the following expression may be used for this situation;

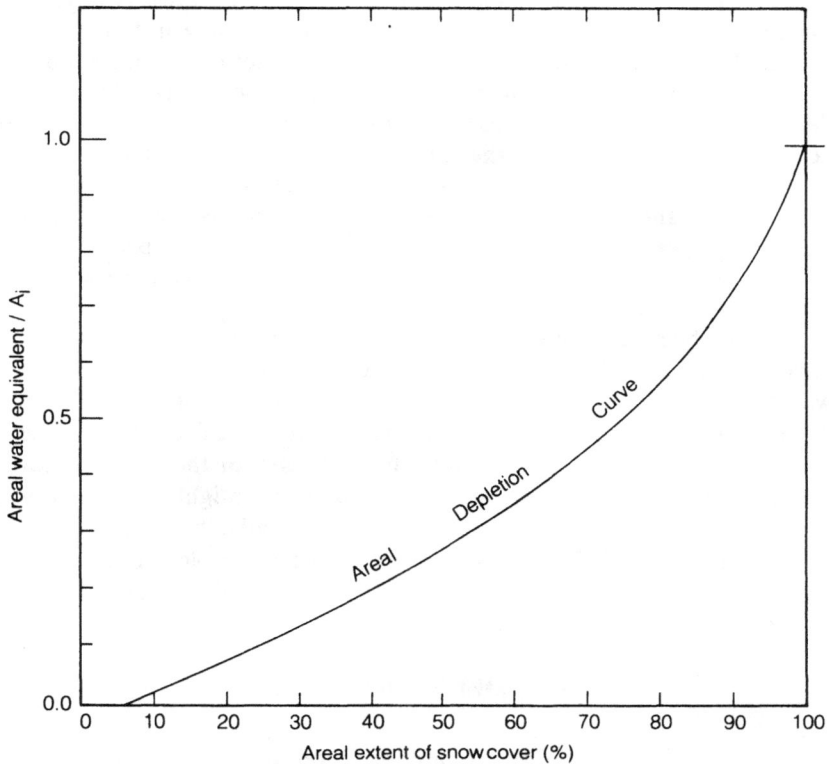

Fig. 9.19 Snowcover areal depletion curve (Anderson, 1973).

$$SBWS = SB + 0.75S, \qquad\qquad 9.65$$

where   SBWS = basin snow water equivalent above which 100 per-
                 cent of the basin is temporarily snowcovered
                 (mm),
          SB = water equivalent of the snowcover on the area
                 just before snow falls (mm), and
           S = water equivalent of new snow (mm).

Equation 9.65 is based on the assumption that the area of snowcover remains
at 100 percent until the snow melted equals 25 percent of the water equivalent
of the new snow. Anderson points out that the value of 25 percent varies from
area to area although the effects of this variation on subsequent runoff are
small so that a more elaborate approach is not warranted.

Another procedure for calculating the area of snowcover is used by Pipes et
al. (1970) in simulation studies of snowmelt runoff in the Fraser River in
British Columbia. In this method the snowcovered area at any time is
assumed to be linearly related to the area covered by snow at the beginning of
melt and the ratio of the generated runoff to the total seasonal runoff.

Yet another approach for determining the snowcovered area has been
devised by the U.S. Army Corps of Engineers (1971). The area of snowcover is
accounted for by dividing the basin into bands of equal elevation and
maintaining an inventory of snow accumulation and melt for each band.
Accumulated snow is assumed to cover an entire band. Any one band must be
either snowfree or completely snowcovered. As the number of bands is
increased the areal extent of the snowcover can be modelled more accurately,
but the computational cost increases.

It should be noted that the operational procedures described in this section
have been developed for mountainous areas, which have high relief, and for
which it has been determined experimentally that a reasonably unique areal
depletion curve exists. In areas of low relief, such as the Prairies, land use and
agricultural practice can have an important influence on the transitory surface
storage of runoff, so that the areal depletion curve might vary significantly
from one year to another. It is possible that the albedo curves reported by
O'Neill and Gray (1972) (see Fig. 9.6) provide a reasonable approximation of
areal snowcover depletion on the Prairies, although direct confirmation of
this has yet to be established.

Time Distribution of Runoff

The shape of the runoff hydrograph produced by snowmelt represents the
integrated effects of the many factors affecting melt and the translation and
storage of melt quantities produced at the surface of the snowpack to the point

of measurement. Delay and attenuation of runoff because of storage in the snow, ground and channels are basic hydrological phenomena. The total time required for some melt water to reach a specific location is the sum of the times required for it to move through the snowpack to the ground surface, to move as overland flow, interflow or groundwater flow to a stream channel and to flow in the channel to the point of measurement. In each flow regime the water is in temporary storage which affects the time distribution. If one is concerned with the time of occurrence and magnitude of the peak discharge rate from a large watershed, storage in the stream channels is generally the most important factor. Conversely, on a small watershed the time for flow through the pack and for overland flow or interflow may be the governing factor. In investigations of long-term sustained water yields the groundwater flow component may be most important.

Snowcover has a marked influence on each of the runoff mechanisms. Wankiewicz (1979) showed that the presence of shallow snowcover on a smooth unbroken 3° slope could be expected to increase the lag time by a factor of 100 compared to that for bare ground. The corresponding factor for a 10° slope is approximately 15. If the slope is rugged it is possible that snow packed into the hollows by wind action could cause a decrease in lag, compared to that on a snowfree slope, by reducing the amount of depressional storage. In basins with a permeable, well-drained soil where much of the runoff is groundwater flow, snowcover can be expected to increase the basin lag only marginally. On the other hand a snowcover with impeding horizons will interrupt runoff near stream channels, increasing the effective overland flow and decreasing the lag time.

Storage and transmission of water in the snowpack

The storage and water transmission qualities of a snowpack depend on the thermal regime of the pack. For operational purposes, it is useful to recognize two snowpack temperature regimes under which melt can occur, namely:

(1)   The snowpack average temperature is less than 0°C, and

(2)   The snowpack is isothermal at 0°C.

When the pack has an average temperature less than 0°C it is assumed that melt water formed at the snow surface or rain falling on the snow will freeze before it reaches the ground. Energy must be added to the snowpack to bring its temperature to 0°C before water will percolate to the ground surface. The most common procedure for calculating the time delay in the movement of water through a pack is to keep a continuous accounting of the heat storage or internal energy changes in the snow. In essence, the snowpack is considered to be an energy reservoir; once the reservoir is full (i.e., snowpack is isothermal at 0°C) melt water moves to the ground.

One of the more widely-accepted procedures for monitoring the energy changes in a snowpack is that proposed by the U.S. Army Corps of Engineers (1956). Their method makes use of the concepts of "cold content" and the "liquid water holding capacity" of snow. Cold content W_c is defined as the heat required per unit area to raise the temperature of the snowpack to $0°C$; it is most conveniently expressed in millimetres of liquid water which, upon freezing within the pack, will warm it to $0°C$ through release of the latent heat of fusion. Thus,

$$W_c = \rho_s dT_s / 160,\qquad\qquad 9.66$$

where ρ_s = average density of the snowpack, (kg/m^3),
           d = depth of the pack (m), and
           T_s = average absolute temperature of the snowpack below $0°C$.

W_c can be increased either by rain or melt water.

The liquid-water-holding capacity is the amount of water held in the pack against gravity. Once this and the cold content of the pack have been satisfied additional melt water will be available for runoff at the ground surface. Amorocho and Espildora (1966) consider the sum of these two quantities to be the permanent storage potential of the pack, since this water is not available for runoff until the pack melts. The delay in transmission through the snowpack is taken to be the total time required to warm the pack to $0°C$ and to satisfy the liquid-water-holding capacity. Anderson (1973) suggests a procedure for determining the time delay, similar to that proposed by the U.S. Army Corps of Engineers, and based on measurements of changes in the snowpack's heat storage or internal energy. The basic assumption of the method is that any negative heat storage (the amount of heat required to bring the snowpack temperature to $0°C$) must be satisfied before melt water or rain can move to the ground or contribute to the storage of liquid water in the pack. When the snowpack has reached $0°C$ throughout its entire depth and the liquid-water-holding capacity has been satisfied water will move downward through the snow. The U.S. Army Corps of Engineers (1956) suggests that the time delay caused by this movement can be ignored for basins that have areas greater than 500 to 700 km^2, and slopes which ensure free flow over the ground surface.

It should be recognized that in shallow snowcovers, such as those encountered on the Prairies, nighttime radiation losses frequently result in a negative heat storage which must be satisfied before snowcover runoff occurs the following day. This gives rise to a diurnal cycle in the runoff hydrograph; a phenomenon, particularly evident during the initial stages of the melt sequence.

For a homogeneous snowcover Colbeck (1978) shows that Eq. 9.48 can be used to calculate the lag time through the unsaturated layer. For periods equal to or longer than the lag between the peak melt rate and the peak snowcover runoff rate, the time delay T_z for vertical movement to the ground surface for a constant flux q_z can be approximated by

$$T_z = \phi_e d / (3\alpha^{1/3} k^{1/3} q_z^{2/3}), \qquad 9.67$$

where d is the snow depth. For a fixed depth and unchanging properties T_z increases as q_z decreases. Equation 9.67 may be applied to stratified or heterogeneous snowpacks provided that variations in their physical properties occur on a scale small compared to their depths (Colbeck, 1972). Under these conditions $k^{1/3}/\phi_e$, which characterizes the important flow properties, will have an average value that could be determined empirically.

In the saturated basal layer all values of flux move with the same speed c:

$$c = \alpha k\theta / \phi. \qquad 9.68$$

Therefore if L is the average distance travelled through the layer the travel time or lag time T_s is

$$T_s = \phi L / (\alpha k\theta). \qquad 9.69$$

If channels form under the snow during the melt period the travel times will be much less. Assuming corresponding values of the porosities and permeabilities of the saturated and unsaturated layers are equal Colbeck (1978) shows that the ratio of the travel times is

$$T_z/T_s = [(1 - S_{wi}) \, d\theta/(3L)](\alpha k/q_z)^{2/3}. \qquad 9.70$$

Values of $(\alpha k/q_z)^{2/3}$ generally lie in the range 10^2 to 10^4, so that the lag in flow through the upper unsaturated zone is likely to dominate the routing except for very shallow snowcovers or snow on flat terrain.

Several empirical expressions have been developed for operational use in routing snowmelt, including those of Anderson (1973) for estimating the transmission time of water through an isothermal pack. These expressions were developed from snow lysimeter data from the Central Sierra Snow Laboratory collected during April and May of 1954. Once the liquid-water-holding capacity has been satisfied the excess water is first lagged and then attenuated. The equation for lag is:

$$\text{LAG} = 5.33 \, [1.0 - \exp(- 0.03 \, \text{WE}/\text{EXCESS})], \qquad 9.71$$

where   LAG = lag time (h),
        WE = water equivalent of the solid portion of the snow-
          pack (mm), and
     EXCESS = excess liquid water available for runoff generated
          in a 6-h period (mm).

The attenuated snowcover outflow is computed from the expression:

$$\text{PACKRO} = (S + I)/[0.5 \exp(-220.4I/\text{WE}^{1.3}) + 1], \qquad 9.72$$

where   PACKRO = snowcover outflow for a 6-h period (mm),
                S = amount of excess liquid water in the pack at the
                    beginning of the period (mm), and
                I = amount of lagged inflow for the 6-h period (mm).

Eggleston, et al. (1971) suggest that the movement of melt water through a pack is analogous to a flood wave moving through reservoir storage, i.e., the basic routing equation is:

$$dS/dt = f(I) - f(O) \qquad\qquad 9.73$$

where   S = amount of water moving through the pack under the
                influence of gravity at any instant (temporary storage)
                (m^3/m^2),
            f(I) = rate of surface melting plus the rainfall intensity at the
                the surface of the snow $(m^3/(s \cdot m^2))$, and
            f(O) = rate at which melt water reaches the ground $(m^3/(s \cdot m^2))$.

If it is assumed that reservoir-type action dominates the flow process and the snowcover runoff is a single-valued linear function of storage, such that $f(O) = k_s S$, then Eq. 9.73 may be written as:

$$dS/dt = f(I) - k_s S \qquad\qquad 9.74$$

where k_s is a storage coefficient (s^{-1}), that can be related to snow depth d (m) by the expression

$$k_s = 1.0 - 0.0984d. \qquad\qquad 9.75$$

Streamflow

Hydraulically, the contribution of snowcover runoff to streamflow represents the case of an unsteady state, spatially-varied flow. Generally, in major streams or large watersheds, routing of these inputs is accomplished by dividing the river into a number of discrete reaches or subdividing a watershed into different subbasins then routing time and space-averaged inputs through the individual reaches or subbasins. The routing procedure involves the solution of the continuity equation:

$$dS/dt = I(x,y,t) - O(x,y,t),$$

where

            dS/dt = rate of change in storage within the reach of the
                    stream channel for which I(x,y,t) and O(x,y,t) are
                    valid,
            I(x,y,t) = inflow rate at a point in the stream as a function of
                    (x,y,t), and
            O(x,y,t) = outflow rate at a point in the stream as a function of
                    (x,y,t).

Many different methods of streamflow routing have been published and their description is beyond the scope of this chapter. Detailed information about these methods, many of which are in routine operational use, may be found in the works of McCarthy (1938), Linsley (1944), Johnstone and Cross (1949), Clark (1945), Chow (1959) and Lawler (1964).

Most routing procedures consider the movement of a flood wave in a relatively unobstructed reach of a channel, i.e., one with no natural obstacles or artificial control barriers. The attenuation of a flood wave by control barriers can be evaluated by reservoir routing techniques. Frequently, on small watersheds during the initial periods of melt the major drainageways are filled with snow which impedes the movement of water and often stores substantial quantities of snowcover runoff. For example, McKay (1970) cites the case on the Canadian Prairies in which 14.5 m^3/m of water equivalent was measured in snow-filled gullies. At the time of the measurement there had been no significant streamflow, although the fields adjacent to the drainageways were free of snow. The factors governing the movement of water through these gullies are largely unknown. However, most field observations show that open channel flow does not occur until channels have formed through the snow. The sudden development of a channel in a drainageway can result in "slug flow" which may produce a higher peak discharge than would result from a normal rate of snowcover runoff. Generally, however, the major effect of snow accumulation in drainageways is to delay the time of occurrence of the peak of a streamflow hydrograph. In addition, snowcover runoff is frequently discharged to rivers that are ice-covered or contain ice floes. The basic equations governing the hydraulics of open-channel flow must be modified to model the flow of water in such channels.

In concluding this chapter it is appropriate to mention the numerous attempts which have been made in North America to develop a snowmelt-streamflow forecast model based to varying degrees on the mechanisms of snowcover ablation and runoff presented above (Anderson and Crawford, 1964; Rockwood, 1964; Crawford and Linsley, 1966; Eggleston et al., 1971; SSARR Model, 1972; Anderson, 1973; Burnash et al., 1973; Anderson, 1976). Most of these models include three identifiable algorithms for: (1) snow accumulation and ablation, (2) soil moisture and groundwater, and (3) streamflow routing. The last algorithm, which is based on accepted open-channel hydraulics procedures, has the most universality. For the most part, the other two subsystem models are based on relatively-weak physical bases. This has been necessary because adequate data about the different processes (on a network scale) are lacking and simple methods are required for operational use. Unfortunately, the gross simplifications of the processes and the empirical relationships used in these models prevent them from being transposed to climatic and physiographic regions that differ from those for which they were developed. Most of the models have been developed for mountainous, forested environments; their reliability for streamflow forecasting in prairie and arctic regions has not been verified.

LITERATURE CITED

Aase, J.K. and S.B. Idso. 1978. *A comparison of two formula types for calculating long-wave radiation from the atmosphere.* Water Resource Res., Vol. 14, pp. 623-625.

Amorocho, J. and B. Espildora. 1966. *Mathematical simulation of the snow melting processes.* WSEP-3001, Dept. Water Sci. Eng., Univ. Calif., Davis.

Anderson, E.A. 1973. *National weather service river forecast system - snow accumulation and ablation model.* NOAA Tech. Memo. NWS HYDRO-17, U.S. Dept. Commer., Washington, D.C.

Anderson, E.A. 1976. *A point energy and mass balance model of a snowcover.* NOAA Tech. Rep. NWS-19, U.S. Dept. Commer., Washington, D.C.

Anderson, E.A. and N.H. Crawford. 1964. *The synthesis of continuous snowmelt runoff hydrographs on a digital computer.* Tech. Rep. 26, Dept. Civ. Eng., Stanford Univ., Palo Alto, Calif.

Bader, H., R. Haefeli, E. Bucher, I. Neher, O. Eckel and C. Thams. 1939. *Der Schnee und seine Metamorphose (Snow and its metamorphism).* Beit Geol. Schweiz., Geotech. Ser., Hydrol., Lief 3, Bern. [English Transl. by U.S. Army Snow, Ice Permafrost Res. Estab., Transl. 14].

Blad, B.L. and N.J. Rosenberg. 1974. *Lysimetric calibration of the Bowen ratio - energy balance method for evapotranspiration estimation in the Central Great Plains.* J. Appl. Meteorol. Vol. 13, pp. 227-236.

Bohren, C.F. and B.R. Barkstrom. 1974. *Theory of the optical properties of snow.* J. Geophy. Res., Vol. 79, No. 30, pp. 4527-35.

Bresler, E. and R.D. Miller. 1975. *Estimation of pore blockage induced by freezing of unsaturated soil.* Proc. First Conf. Soil-Water Problems. Am. Geophys. Union, Calgary, Alta.

Brunt, D. 1952. *Physical and Dynamical Meteorology.* Cambridge Univ. Press, Cambridge, Mass.

Brutsaert, W. 1975. *On a derivable formula for long-wave radiation from clear skies.* Water Resour. Res., Vol. 11, pp. 742-744.

Burnash, R.J.C., R.L. Ferral and A.M. Richard. 1973. *A generalized streamflow system developed by the joint Federal-State river forecast center.* Sacramento, Calif.

Businger, J.A., J.C. Wyngaard, Y. Izumi and E.F. Bradley. 1971. *Flux - profile relationships in the atmospheric surface layer.* J. Atmos. Sci., Vol. 28, pp. 181-189.

Chow, V.T. 1959. *Open-channel Hydraulics.* McGraw-Hill Book Co. Inc., New York.

Clark, C.O. 1945. *Storage and the unit hydrograph*. Proc. Am. Soc. Civil Eng., Vol. 110, pp. 1419-1488.

Colbeck, S.C. 1972. *A theory of water percolation in snow*. J. Glaciol., Vol. 11, pp. 369-385.

Colbeck, S.C. 1973. *Effects of stratigraphic layers on water flow through snow*. Res. Rep. 311, U.S. Army Cold Reg. Res. Eng. Lab., Hanover, N.H.

Colbeck, S.C. 1974a. *The capillary effects on water percolation in homogeneous snow*. J. Glaciol., Vol. 13, pp. 85-97.

Colbeck, S.C. 1974b. *Water flow through snow overlying an impermeable boundary*. Water Resour. Res., Vol. 10, pp. 119-123.

Colbeck, S.C. 1975. *A theory of water flow through a layered snowpack*. Water Resour. Res., Vol. 11, pp. 261-266.

Colbeck, S.C. 1976. *An analysis of water flow in dry snow*. Water Resour. Res., Vol. 12, pp. 523-527.

Colbeck, S.C., 1978. *The physical aspects of water flow through snow*. In Advances in Hydroscience (Ven Te Chow, ed.), Academic Press, New York. Vol. 11, pp. 165-206.

Colbeck, S.C. and G. Davidson. 1973. *Water percolation through homogeneous snow*. The Role of Snow and Ice in Hydrology: Proc. Banff Symp., Sept. 1972, Unesco-WMO-IAHS, Geneva-Budapest-Paris, pp. 242-257.

Crawford, N.H. and R.K. Linsley. 1966. *Digital simulation in hydrology: Stanford Watershed Model IV*. Tech. Rep. 39, Dept. Civ. Eng., Stanford Univ., Palo Alto, Calif.

de Quervain, M.R. 1948. *Ueber den abbau der alpinen schneedecke*. Int. Union Geod. Geophys., Gen. Assem. Oslo. [Vol. 2, Snow and Ice],Int. Assoc. Sci. Hydrol. Publ. 30, pp. 55-68.

de Quervain, M.R. 1952. *Evaporation from the snowpack*. Snow Investigations Research Note No. 8, U.S. Army Corps Eng, North Pacific Div., Portland, Oreg.

de Quervain, M.R. 1973. *Snow structure, heat and mass flux through snow*. The Role of Snow and Ice in Hydrology: Proc. Banff Symp., Sept. 1972, Unesco-WMO-IAHS, Geneva-Budapest-Paris, pp. 203-226.

De Vries, D.A. 1963. *Thermal properties of soils*. In Physics of Plant Environment (W.R. Van Wijk, ed.), North-Holland Publishing Co., Amsterdam, pp. 210-235.

Dirksen, C. 1964. *Water movement and frost heaving in unsaturated soil without an external source of water*. Ph.D. Thesis, Dept. Agron., Cornell Univ., Ithaca, N.Y.

Dozier, J. 1979. *A solar radiation model for a snow surface in mountainous terrain*. Proc. Modeling Snow Cover Runoff (S.C. Colbeck and M. Ray, eds.), U.S. Army Cold Reg. Res. Eng. Lab., Hanover, N.H., pp. 144-153.

Dunkle, R.V. and J.T. Bevans. 1956. *An approximate analysis of the solar reflectance and transmittance of a snowcover*. J. Meteorol., Vol. 13, pp. 212-216.

Dunne, T. and R.D. Black. 1971. *Runoff processes during snowmelt*. Water Resour. Res. Vol. 7, pp. 1160-1172.

Dybvig, W. 1976. *The computer simulation of mass and heat transfer in a one-dimensional snowpack*. M.Sc. Thesis, Dept. Agric. Eng., Univ. Sask., Saskatoon.

Dyer, A.J. 1961. *Measurements of evaporation and heat transfer in the lower atmosphere by an automatic eddy-correlation technique*. Q. J. R. Meteorol. Soc., Vol. 87, pp. 401-412.

Dyer, A.J. 1968. *An evaluation of eddy flux variation in the atmospheric boundary layer.* J. Appl. Meteorol. Vol. 7, pp. 845-850.

Dyer, A.J. and B.B. Hicks. 1970. *Flux-gradient relationships in the constant flux layer.* Q. J. R. Meteorol. Soc., Vol. 96, pp. 715- 721.

Eggleston, K.O., E.K. Israelsen and J.P. Riley. 1971. *Hybrid computer simulation of the accumulation and melt processes in a snowpack.* Rep. PR WG65-1, Utah State Univ., Logan, Utah.

Erickson, D.E.L., W. Lin and H. Steppuhn. 1978. *Indices for estimating prairie runoff from snowmelt.* Pap. presented to Seventh Symp. Water Stud. Inst., Appl. Prairie Hydrol., Saskatoon, Sask.

Ferguson, H., P.L. Brown and D.D. Dickey. 1964. *Water movement and loss under frozen soil conditions.* Proc. Soil Sci. Soc. Am., Vol. 28, pp. 700-703.

Ffolliott, P.F. and E.A. Hansen. 1968. *Observations of snowpack accumulation melt and runoff on a small Arizona watershed.* Res. Note RM-124, U.S. For. Serv., Rocky Mountain For. Range Exp. Stn., Ft. Collins, Colo.

Fuchs, M. and C.B. Tanner. 1967. *Calibration and field test of soil heat flux plates.* Proc. Soil Sci. Soc. Am., Vol. 32, pp. 326-328.

Garnier, B.J. and A. Ohmura. 1970. *The evaluation of surface variations in solar radiation income.* Solar Energy, Vol. 13, pp. 21-34.

Geiger, R. 1961. *Das Klima der bodennahen Luftschicht (The Climate Near the Ground).* [English Transl. by Scripta Technica Inc., Harvard University Press, Cambridge, Mass.]

Gerdel, R.W., 1948. *Physical changes in snowcover leading to runoff, especially to floods.* Int. Union Geod. Geophys. Gen. Assem. Oslo, [Vol. 2, Snow and Ice], Int. Assoc. Hydrol. Sci. Publ. 30 pp. 42-54.

Gerdel, R.W. 1949. *The storage and transmission of liquid water in the snowpack as indicated by dyes.* Proc. 16th Annu. Meet. West. Snow Conf., pp. 81-99.

Gerdel, R.W. 1954. *The transmission of water through snow.* Trans. Am. Geophys. Union., Vol. 35, pp. 475-485.

Giddings, J.C. and E. LaChapelle. 1961. *Diffusion theory applied to radiant energy distribution and albedo of snow.* J. Geophys. Res., Vol. 66, pp. 181-189.

Gold, L.W., 1957. *Influence of snowcover on heat flow from the ground.* Int. Union Geod. Geophys. Gen. Assem. Toronto, [Vol. 4, Snow and Ice], Int. Assoc. Sci. Hydrol. Publ. 46, pp. 13-21.

Gold, L.W. and G.P. Williams. 1961. *Energy balance during the snowmelt period at an Ottawa site.* Int. Union Geod. Geophys. Gen. Assem. Helsinki, [Snow and Ice], Int. Assoc. Sci. Hydrol. Publ. No. 54, pp. 288-294.

Gray, D.M., D.I. Norum and J.M. Wigham. 1970. *Infiltration and the physics of flow of water through porous media.* Handbook on the Principles of Hydrology, (D.M. Gray, ed.), The Secretariat, Can. Nat. Comm. Int. Hydrol. Decade., Ottawa, pp. 5.1-5.58.

Granger, R.J. 1977. *Energy exchange during melt of a prairie snowcover.* M.Sc. Thesis, Dept. Mech. Eng., Univ. Sask., Saskatoon.

Granger, R.J. and D.H. Male. 1978. *Melting of a prairie snowpack.* J. Appl. Meteorol. Vol. 17, No. 12, pp. 1833-1842.

Granger, R.J., D.H. Male and D.M. Gray. 1978. *Prairie snowmelt.* Proc. Appl. Prairie Hydrol. Symp., Seventh Symp. Water Stud. Inst., Saskatoon, Sask.

Guymon, G.L. and J.N. Luthin. 1974. *A coupled heat and moisture transport model for Arctic soils.* Water Resour. Res., Vol. 10, pp. 995-1001.

Harlan, R.L. 1973. *Analysis of coupled heat-fluid transport in partially frozen soil.* Water Resour. Res., Vol. 9, pp. 1314-1323.

Hay, J.E. 1976. *A revised method for determining the direct and diffuse components of the total shortwave radiation.* Atmosphere, Vol. 14, pp. 278-287.

Hicks, B.B. and H.C. Martin. 1972. *Atmospheric turbulent fluxes over snow.* Boundary-Layer Meteorol., Vol. 2, pp. 496-502.

Hoekstra, P. 1966. *Moisture movement in soil under temperature gradients with the cold side temperature below freezing.* Water Resour. Res., Vol. 2, pp. 241-250.

Högström, U. 1974. *A field study of the turbulent fluxes of heat, water vapour and momentum at a 'typical' agricultural site.* Q. J. R. Meteorol. Soc., Vol. 100, pp. 624-639.

Horton, R.E. 1938. *Phenomena of the contact zone between ground surface and a layer of melting snow.* Int. Union Geod. Geophys., [Edinburgh - Rep. and Tech. Notes of the Commision of Snow and Ice], Int. Assoc. Sci. Hydrol. Publ. 23, pp. 545-561.

Jame, Yih-Wu. 1977. *Heat and mass transfer in freezing unsaturated soil.* Ph.D. Thesis, Univ. Sask., Saskatoon.

Johnstone, D., and W.P. Cross. 1949. *Elements of Applied Hydrology.* The Ronald Press Co., New York, N.Y.

Jumikis, A.R. 1966. *Thermal Soil Mechanics.* Rutgers University Press, New Brunswick, N.J.

Kennedy, G.F. and J. Lielmezs. 1973. *Heat and mass transfer of freezing water-soil systems.* Water Resour. Res., Vol. 9, pp. 395-400.

Kondratyev, K., 1954. *Radiant Energy of the Sun.* Gidromet., Leningrad.

Kondratyev, K. Ya. 1969. *Radiation in the Atmosphere.* Int. Geophys. Ser., Vol. 12, Academic Press, New York.

Kuroiwa, D., 1968. *Liquid permeability of snow.* Int. Union Geod. Geophys. Gen. Assem. Bern., [Vol. 5, Snow and Ice], Int. Assoc. Sci. Hydrol., Publ. 79, pp. 380-391.

Kuz'min, P.P., 1961. *Protsess Tayaniya Shezhnogo Pokrova (Melting of Snow Cover),* [English Transl. by Israel Prog. Sci. Transl., Transl. 71].

Langham, E.J. 1973. *Un modèle du manteau nival relié à sa mésostructure.* Bull. Int. Assoc. Hydrol. Sci., Vol. 18, pp. 33-43.

Langham, E.J. 1974. *The occurrence and movement of liquid water in the snowpack.* Adv. Concepts Tech. Study Snow Ice Resour. Interdiscip. Symp., U.S. Nat. Acad. Sci., Washington, D.C., pp. 67-75.

Lawler, E.A., 1964. *Hydrology of flow control. Part II. Flood routing.* In Handbook of Applied Hydrology. (Ven Te Chow, ed.), McGraw-Hill, New York, N.Y., pp. 25-34 to 25-59.

Leaf, C.F. 1969. *Aerial photographs for operational streamflow forecasting in the Colorado Rockies.* Proc. 37th Annu. Meet. West. Snow Conf., pp. 19-28.

Lee, R. 1962. *Theory of the 'Equivalent Slope.'* Mon. Weather Rev., Vol. 90, pp. 165-166.

Linlor, W.I., M.F. Meier, and J.L. Smith. 1974. *Microwave profiling of snowpack free-water content.* Adv. Concepts Tech. Study Snow Ice Resour. Interdiscip. Symp., U.S. Nat. Acad. Sci., Washington, D.C., pp. 729-736.

Linsley, R.K. 1944. *Use of nomograms in solving streamflow routing problems.* Civ. Eng., Vol. 14, pp. 209-210.

List, R.J. 1968. *Smithsonian Meteorological Tables.* 6th Rev. ed. The Smithsonian Institution, Washington, D.C.

Liu, B.Y.H. and R.C. Jordan. 1960. *The interrelationship and characteristic distribution of direct, diffuse and total solar radiation.* Solar Energy, Vol. 4, pp. 1-19.

Male, D.H. and R.J. Granger. 1979. *Energy mass fluxes at the snow surface in a prairie environment.* Proc. Modeling Snow Cover Runoff (S.C. Colbeck and M. Ray, eds.), U.S. Army Cold Reg. Res. Eng. Lab., Hanover, N.H., pp. 101-124.

Male, D.H. and D.M. Gray. 1975. *Problems in developing a physically based snowmelt model.* Can. J. Civ. Eng., Vol. 2, pp. 474-488.

Mantis, H.T. (ed.) 1951. *Review of the properties of snow and ice.* Rep. 4, U.S. Army Snow Ice Permafrost Res. Estab. (Rep. 4), Univ. Minn. Inst. Tech. Eng. Expt. Sta., pp. 52-82.

Manz, D.H. 1974. *Interaction of solar radiation with snow.* M.Sc. Thesis, Dept. Agric. Eng., Univ. Sask., Saskatoon.

Marks, D. 1979. *An atmospheric radiation model for general alpine application.* Proc. Modeling Snow Cover Runoff (S.C. Colbeck and M. Ray, eds.), U.S. Army Cold Reg. Res. Eng. Lab., Hanover, N.H., pp. 167-178.

McBean, G.A. and M. Miyake. 1972. *Turbulent transfer mechanisms in the atmospheric surface layer.* Q. J. R. Meteorol. Soc., Vol. 98, pp. 383-398.

McCarthy, G.T. 1938. *The unit hydrograph and flood routing.* Unpubl. manuscript presented to the North Atlantic Div., Corps of Engineers, War Dept.

McKay, G.A., 1970. *Precipitation.* Handbook on the Principles of Hydrology (D.M. Gray, ed.), The Secretariat, Can. Nat. Comm. Int. Hydrol., Decade, Ottawa, pp. 2.1-2.111.

McKay, D.C. and G.W. Thurtell. 1978. *Measurements of the energy fluxes involved in the energy budget of a snowcover.* J. Appl. Meteorol., Vol. 17, pp. 339-349.

Meier, M.F. and A.T. Edgerton. 1971. *Microwave emission from snow - a progress report.* Proc. 7th Int. Symp. Remote Sensing Environ., Ann Arbor, Mich., pp. 1155-1163.

Mellor, M. 1966. *Some optical properties of snow.* Symposium International sur les Aspects Scientifiques des Avalanches de Neige. 5-10 Avril, Davos, Suisse. Int. Assoc. Sci. Hydrol. Publ. 69, pp. 129-140.

Mellor, M. 1977. *Engineering properties of snow.* J. Glaciol., Vol. 19, pp. 15-66.

Miller, D.M. 1953. *Snow cover depletion and runoff.* Snow Investigation Research Note No. 16, U.S. Army Corps Eng., North Pacific Div.

Monin, A.S. and A.M. Obukhov. 1954. *Basic laws of turbulent mixing in the ground layer of the atmosphere.* Trudy 151, Akad. Nauk. SSSR Geofiz. Inst., pp. 163-187.

Munn, R.E. 1966. *Descriptive Micrometeorology.* Academic Press, New York.

Obled, Ch. and H. Harder. 1979. *A review of snow melt in the mountain environment.* Proc. Modeling Snow Cover Runoff (S.C. Colbeck and M. Ray, eds.), U.S. Army Cold Reg. Res. Eng. Lab., Hanover, N.H., pp. 179-204.

Obukhov, A.M. 1946. *Turbulence in an atmosphere with a non-uniform temperature.* Trudy Institute Teoreticheskio Geofiziki, AN SSSR No. 1 (English Transl. in Boundary-Layer Meteorol. (1971), Vol. 2, pp. 17-29).

O'Neill, A.D.J. and D.M. Gray. 1973. *Spatial and temporal variations of the albedo of prairie snowpack.* The Role of Snow and Ice in Hydrology: Proc. Banff Symp., Sept. 1972, Unesco-WMO-IAHS, Geneva-Budapest-Paris, Vol. 1, pp. 176-186.

Panofsky, H.A. 1974. *The atmospheric boundary layer below 150 m.* Annu. Rev. Fluid Mech., Vol. 6, pp. 147-177.

Penner, E. 1970. *Thermal conductivity of frozen soils.* Can. J. Earth Sci., Vol. 7, pp. 983-987.

Pipes, A., M.C. Quick and S.O. Russell. 1970. *Simulating snowmelt hydrographs for the Fraser River system.* Proc. 38th Annu. Meet. West. Snow Conf., pp. 91-97.

Prandtl, L. 1932, *Meteorologische Anwendung der Strömungslehre.* Beitr. Phys. d. freien Atmos., Vol. 19, pp. 188-202.

Rockwood, D.M., 1964. *Streamflow synthesis and reservoir regulation.* Tech. Bull. No. 22, U.S. Army Corps Eng., North Pacific Div., Portland, Oreg.

Satterlund, D.R. 1979. *An improved equation for estimating long-wave radiation from the atmosphere.* Water Resour. Res. Vol 15, pp. 1643-1650

SSARR Model. 1972. *Program description and user's manual.* Program 724-K5-G0010, U.S. Army Corps Eng., North Pacific Div., Portland, Oreg.

Schmidt, W. 1925. *Der Massenaustausch in freier Luft und verwandte Erscheinungen.* Henri Grand Verlag, Hamburg.

Seligman, G., 1963. *Snow Structure and Ski Fields.* (Reprint of 1936 edit.) Jos. Adam, Brussels.

Shimizu, H. 1970. *Air permeability of deposited snow.* Low Temp. Sci., Ser. A, Vol.22, pp. 1-32.

Stanhill, G. 1966. *Diffuse sky and cloud radiation in Israel.* Solar Energy, Vol. 19, pp. 96-101.

Steven, M.D. 1977. *Standard distribution of clear sky radiance.* Q.J.R. Meteorol. Soc., Vol. 103, pp. 457-465.

Suckling, P.W. and J.E. Hay. 1976. *Modeling direct, diffuse and total radiation for cloudless skies.* Atmosphere, Vol. 14, pp. 298-308.

Suckling, P.W. and J.E. Hay. 1977. *A cloud layer - sunshine model for estimating direct, diffuse and total solar radiation.* Atmosphere, Vol. 15, pp. 194-207.

Sverdrup, H.U. 1936. *The eddy conductivity in the air over a smooth snow field.* Geophysiske Publikasjoner, Vol. XI, Nr. 7.

Swinbank, W.C., 1951. *The measurement of vertical transfer of heat and water vapor by eddies in the lower atmosphere.* J. Meteorol., Vol. 8, No. 3, pp. 135-145.

Swinbank, W.C. 1968. *A comparison between predictions of dimensionless analysis for the constant - flux layer and observations in unstable conditions.* Q.J.R. Meteorol. Soc., Vol. 94, pp. 460-467.

Temps. R.C. and K.L. Coulson. 1977. *Solar radiation incident upon slopes of different orientations.* Solar Energy, Vol. 19, No. 3, pp. 179-184.

Titus, R.L. and E.J. Truhlar., 1969. *A new estimate of average global solar radiation in Canada.* CLA-7-69, Dept. Transport, Meteorol. Branch, Toronto.

Trainor, L.E.H., 1947. *The spectral reflection and absorption of radiation by snow.* M.A. Thesis, Dept. Physics, Univ. Sask., Saskatoon.

U.S. Army Corps of Engineers., 1956. *Snow Hydrology, Summary report of the Snow Investigations.* U.S. Army Corps Eng., North Pacific Div., Portland, Oreg.

U.S. Army Corps of Engineers. 1971. *Runoff evaluation and streamflow simulation by computer. Part II.* U.S. Army Corps Eng., North Pacific Div., Portland, Oreg.

Wakahama, G., 1968. *Infilitration of melt water into snowcover, III Flowing down speed of melt water in a snow cover.* Low Temp. Sci., Ser. A, Vol. 26, pp. 77-86.

Wankiewicz, A. 1979. *A review of water movement in snow.* Proc. Modeling Snow Cover Runoff (S.C. Colbeck and M. Ray, eds.), U.S. Army Cold Reg. Res. Eng. Lab., Hanover, N.H., pp. 222-252.

Wankiewicz, A. and J. de Vries. 1978. *An inexpensive tensiometer for snow-melt research.* J. Glaciol., Vol. 20, pp. 577-584.

Webb, E.K. 1970. *Profile relationships: the log-linear range and extension to strong stability.* Q. J. R. Meteorol. Soc., Vol. 96, pp. 67-90.

Williams, L.D., R.G. Barry and J.T. Andrews. 1972. *Application of computed global radiation for areas of high relief.* J. Appl. Meteorol., Vol. 11, No. 3, pp. 526-533.

Yoshida, S. 1962. *Hydrometeorological study on snowmelt.* J. Meteorol. Res., Vol. 14, pp. 879-899.

Zavodchikov, A.B., 1962. *Snowmelt losses to infiltration and retention on drainage basins during snow melting period in Northern Kazakhstan.* Sov. Hydrol.: Sel. Pap. No. 1, pp. 37-41.

10

SNOW AND ICE ON LAKES

W. P. ADAMS

Graduate Studies, Trent University
Peterborough, Ontario

INTRODUCTION

Approximately eight percent of the surface area of Canada is covered by freshwater and a very considerable proportion of that water is ice- and/or snowcovered for an appreciable period each year. This chapter primarily deals with the various forms of ice and snow which develop on freshwater bodies with particular emphasis on the snow component and its roles in the evolution of the remainder of the cover. One purpose of the chapter is to provide an introduction to the very diverse, dispersed, literature on this topic. References have been selected, in part, to pull together the main threads of the literature, especially those in North America.

For convenience, discussion is focussed on relatively large, deep lakes which experience a fall turnover and are located in cold regions with considerable snowfall. In reality, the factors governing the formation of winter cover vary greatly in response to the freshwater body and to climate and weather conditions. Spectra of cover development situations can be conceived ranging from a highly turbulent river to a very stable lake; from very high snowfall amounts to little or none, and from very cold to relatively warm air temperatures. A focus on relatively large, deep lakes is therefore a very considerable simplification although it is useful for elaborating the role of snow in the evolution of the winter covers of water bodies. Even this "class" of lakes encompasses a range of thermal and circulation regimes which may be greatly modified by local factors of lake morphometry, hydrology, physiography, climate, etc. Variations in freeze-over, ice cover and breakup form one important group of controls of the variability of lake cover. Systematic discussions of the variety of lake types according to their thermal regimes can be found in Welch (1952), Hutchinson (1957), Zumberge and Ayers (1964), Ruttner (1968), MacKay and Löken (1974) and Wetzel (1975). Comprehensive case studies of ice-covered lakes and their thermal regimes, including the perennially-frozen case, have been reported by Brewer

(1958), Barnes (1960), Scott and Ragotskie (1961), Dutton and Bryson (1962), Likens and Ragotskie (1965), Swinzow (1966), Johnson and Likens (1967), Bilello (1968), Parott and Fleming (1970) and Schindler et al. (1974a).

The winter regime of turbulent rivers, which receives little attention below, is comprehensively discussed by Michel (1971). Practical aspects of river ice work are reviewed in Burgi et al. (1974).

An important thesis of this chapter is that the winter cover of most water bodies is more diverse, stratigraphically and spatially, than normally conceived. The ice cover of a lake may consist of a crystallographically columnar component, and various granular components including one or more formed of snow-derived ice. The quantities of the various types of ice can vary greatly in both time and space (see Fig. 10.1). Snowcover of a lake may be similarly varied since the ice surface does not provide a uniform environment for evolution of the snowpack.

COMPONENTS OF THE WINTER COVER OF LAKES

Columnar (Black) Ice

In the late summer and fall, a lake suffers a net loss of energy through radiative, convective and condensation/evaporative exchanges at its surface. Initially this surface cooling results in a turnover of water as the denser, cooler water ($\sim 4°C$) from the surface descends to be replaced by less dense, warmer water from below. Eventually, a situation arises in which less-dense, cooler water ($<4°C$) lies at the surface which can, in the absence of mechanical turbulence by winds or currents, freeze despite the fact that relatively warm water still exists below it. The ice cover which starts to form plays an important role in development of subsequent thermal regimes of the lake. Predictions of freeze-up taking this autumnal decline of air and water surface temperatures into account have been reported by Rodhe (1952), Burbidge and Launder (1957), Bilello (1964a, b), Williams (1965, 1968) and Bilello et al. (1966).

Even in calm water, the initial orientations of the tiny platelets and other crystals of ice which form on the freezing water surface are fairly random, but as the sheet thickens, some crystals grow more vigorously downwards than others, pinching out those which have other orientations, thereby producing a fairly orderly body of crystals. In sea ice, crystals with the initial c-axis (the hexagonal symmetry axis) oriented horizontally are favoured in this selective process (growth is favourable in the basal planes). This growth process is not universal in lake ice where entire covers have the c-axis oriented vertically or horizontally or both ways at once (Perey and Pounder, 1958; Barnes, 1959, 1960; Palosuo, 1961; Knight, 1962; Lyons and Stoiber, 1962; Muguruma and

Kikuchi, 1963; Ragle, 1963; Shumskii, 1964; Pounder, 1965; Hobbs, 1974). However, an ice cover in which the c-axis is vertical seems to be the most common in lakes.

The end result of the freezing process is a sheet of ice composed of columnar crystals, "candles", elongated in the vertical direction, parallel to the heat flux between lake and atmosphere. The organization of the crystals and of the impurities which are concentrated at the crystal boundaries is important to the strength and the final pattern of melt and breakup of the sheet.

After ice has been established, the lake water continues to lose heat at its upper surface, although at a reduced rate, by conduction through the insulating ice sheet. The ice thickens downward in response to this loss. The ice thus formed is not, in fact, uniform. Variations depend on freezing conditions (Shumskii, 1964) and the quantity and type of impurities in the water. Periods of rapid ice growth may result in a larger percentage of embedded air bubbles (Ragle, 1963; Shumskii, 1964; Adams and Jones, 1971) to produce a layering effect which is visible to the naked eye in blocks cut from the ice. The maximum density of "pure" ice at the freezing point and normal atmospheric pressure is 917 kg/m^3; but, Gaitskhoki (1970b) records black ice with bubbles at a density of 895 kg/m^3. There is no reason to believe that lower densities are unusual in this type of ice (Adams, 1976b).

Crystals tend to be larger nearer the base of the sheet (Knight, 1962). An initial spatial variation in thickness, and sometimes in type of ice (Ragle, 1963) results from the spatial variations of conditions at the time of freeze-up, particularly in calmer, shallower parts of a lake which may have a high concentration of impurities acting as nuclei. Further areal variations in black ice, resulting from the differential effects of such factors as currents and snowcover, develop as the season proceeds. Contrasts between the margins and the centres of water bodies are much more marked for rivers.

In very cold regions with low snowfall, growth proceeds more or less unhindered to produce a cover of "black ice" which may be over 3 m thick. Important sources of ice thickness data in North America include *Ice Thickness Data for Canadian Selected Stations* (Canada, AES, 1958-present), *Freeze-up and Break-up Dates of Water Bodies in Canada* (Allen and Cudbird, 1971) and the series *Ice Thickness Observations, North American Arctic and Subarctic* by Bilello et al. (1961-75).

Lake Snowcover

Snow falling onto the initial ice surface affects its evolution. Snowcover exhibits a basic stratigraphy and distribution which depends on the type of precipitation and on weather conditions during deposition (Nakaya, 1954; LaChapelle, 1977), but which is soon altered by wind and internal metamorphism.

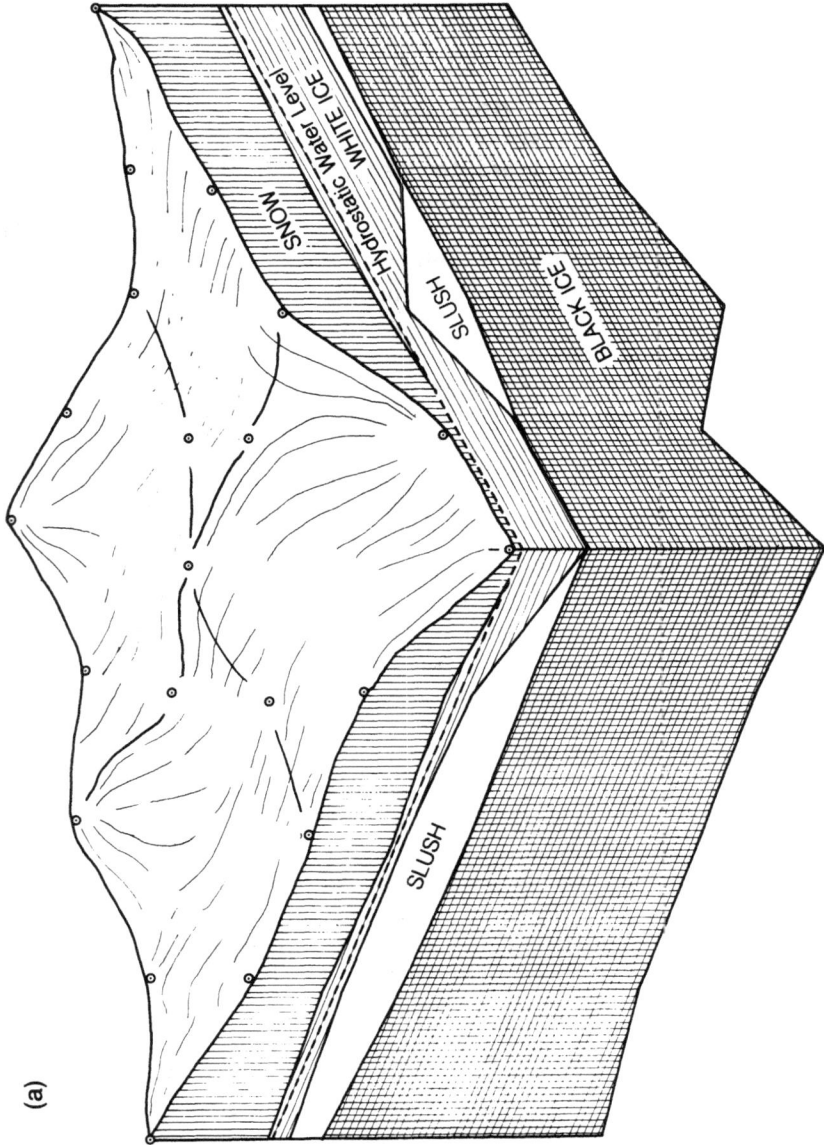

(a)

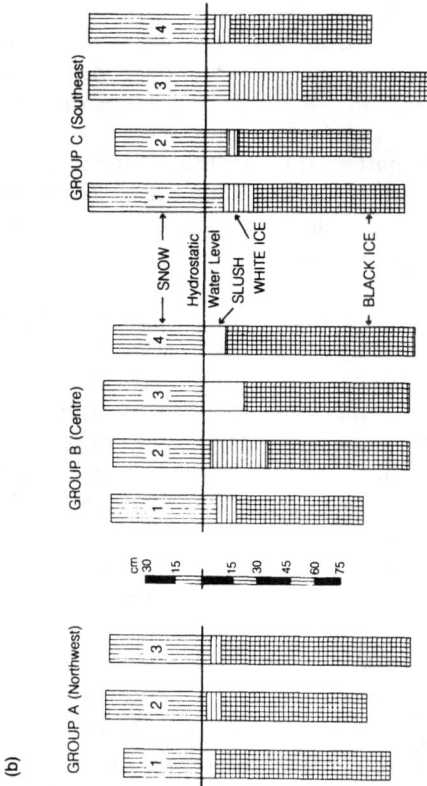

Fig. 10.1 Illustrations of the variability of cover on Knob Lake (187 ha total area) during peak ice season (Adams and Brunger, 1975). (a) The variability within 1,000 m² at a downwind site. Black ice ranged from 71.0 to 132.0 cm, white ice from 0.5 to 45 cm, snow depth from 15.0 to 46.0 cm and hydrostatic water level from –7.5 to +28.0 cm. The hydrostatic water level is the distance between the level of water in a drill hole and the surface of the ice sheet; positive indicates a level *above* the sheet. (b) Sample profiles at groups of sites along an up-wind/down-wind transect of the lake. On the day in question, the extreme ranges (based on a sample of over 300) in the various components were: black ice 33.0 to 109.0 cm (mean 96.5 cm), white ice 2.5 to 61.0 cm (mean 12.5 cm), snow depth 20.5 to 94.0 cm (mean 53.5 cm), hydrostatic water level –7.5 to +30.5 cm (mean +5.0 cm). The range of total ice thickness was 56.0 to 127.0 cm (mean 96.5 cm).

Although snowcover properties vary greatly within almost any given area, the evolution of snowcover at a sheltered site on land, in the absence of melt, follows a general sequence. Each snowfall produces a more or less marked increase in snowcover depth and water equivalent. During the period following snowfall, the water gained tends to persist but, the depth decreases as densification (resulting from the effects of both wind and destructive or equitemperature metamorphism) proceeds. Some types of new-fallen snow are more dense than others but, new snows are generally less dense than old snow already on the ground so that each snowfall lowers the mean density of the cover. The densification process gradually offsets this decrease. Thus, the snowcover can broadly be conceived as gaining in water equivalent during the winter but having marked fluctuations of depth and mean density from individual snowfalls.

Within the snowcover the above pattern involves the gradual densification of each individual layer as the relatively large crystals and air spaces of new-fallen snow are replaced by small, more or less rounded, grains surrounded by small air spaces. This is the essence of the process of "destructive" metamorphism which is normally dominant in isothermal snow (Bader, 1939; Mellor, 1964a; LaChapelle, 1969; Sommerfeld and LaChapelle, 1970). However, if a marked ($> 0.1°$C/cm) temperature gradient exists in the cover a considerable redistribution of mass can occur along the vapor pressure gradient associated with the temperature difference. This transfer of mass in the vapor phase is the characteristic process of "constructive" or temperature-gradient metamorphism which results in the growth of new, markedly-faceted, grains and/or cup crystals within the cover. Generically, this is a hoar-producing (sublimation) process whose most striking product is "depth hoar" (LaChapelle, 1977). Constructive metamorphism is most noticeable at the base of the snowcover at times when a thin cover and/or extremely cold air temperatures produce strong temperature gradients within the snow.

Both types of metamorphism can and often do occur simultaneously at different levels in the same snowcover. Towards the end of winter, a land snowcover which has been allowed to develop without the interference of melt, will characteristically exhibit very light, recognizably atmospheric, snow crystals at the top; rounded, dense products of equitemperature metamorphism in the middle; and less dense, looser products of temperature gradient metamorphism at the base. This top-to-bottom pattern of crystal type and density will be paralleled by related patterns of snowcover properties including grain size (large, small, large) and snow strengths (weak, strong, weak).

The sort of evolutionary sequence and stratigraphic pattern described is apparent in land snowcovers from many diverse regions. It should, however, be stressed that the above description is very general and that there may be very great differences in the development of snowcover at adjacent sheltered and open sites.

Lake snowcover is affected by the same factors and processes as a land snowcover but the lake provides a very distinctive snowcover environment, which cannot be considered simply as an "open" site. The presence of a body of water beneath the lake snowcover produces characteristically high temperature gradients. Thus, on lakes, the effects of constructive metamorphism on snowcover properties, characterized by large grains, relatively low density, and low bearing strength are very marked. Snow on lakes is usually prone to redistribution, densification and compaction by wind. Periodic, abrupt thinnings of the cover resulting from wind action and flooding are other phenomena unique to a lake snowcover. Depth fluctuations affect the metamorphic processes directly through changes in the temperature gradient, generally producing a thinner snowcover with steeper, subnival temperature gradients.

The most striking fluctuations and changes in lake snowcover occur when the ice sheet is flooded, as it may be several times during a winter. Flooding may result in the destruction of the "normal" stratigraphy of the cover and the incorporation of snow into the ice sheet itself. Therefore, at any particular time, the snowcover of a lake is likely to be thinner than expected and stratigraphically "youthful" (despite the more favourable conditions for constructive metamorphism). The mean density of snow of lakes does not, like on land, tend to increase steadily with time after deposition, but rather remains at a fairly low value throughout the winter.

By comparison with equivalent snowcovers on surrounding land areas, a lake snowcover is characterized by a lower depth and water equivalent, very poor stratigraphic development and higher spatial variability (Adams and Rogerson, 1968). This variability extends to such physical properties as depth, density and water equivalent and, presumably, to many others which are essentially depth-density-temperature related (such as hardness and bearing strength). Even though the ice surface on the lake produces a snowcover that is often highly variable (as indexed by the coefficients of variation of depth and water equivalent), nevertheless it is the most plane of the common natural landscape surfaces so that quite systematical spatial patterns of snow can develop upon it. Distinct upwind-downwind, centre-shore trends of snow depth can be an important feature of the winter cover on large lakes. Trends in snow properties are significant in themselves and through their influence on underlying components of the cover which may exhibit reciprocal patterns (Adams and Brunger, 1975). For example, as the distribution in depth controls the light distribution in the lake all phenomena under the snowcover affected by light will exhibit response patterns reciprocal to depth.

A snowcover on an ice surface acts as an insulating blanket because of its lower thermal conductivity. Table 10.1 lists typical values of the thermal conductivity for snow and different types of ice. The low thermal conductivity of snow compared to "ice" (Mellor, 1962, 1964b; Weller and Schwerdtfeger, 1971; Glen, 1974) has the important effect of reducing heat loss from the lake and thereby the downward thickening of the ice. This case of an ice sheet thickening beneath a deep snowcover is the one most frequently treated in methods for predicting ice thickness from atmospheric variables, or as Pounder (1965, p. 134) phrases it, it is the 'freezing exposure' of the ice sheet which is considered. Conversely, a snowcover which reduces heat loss during winter also reduces heat gain during spring.

Table 10.1
THERMAL CONDUCTIVITIES OF ICE AND SNOW (kW/(°C·m)).

Ice/Snow Type	Conductivity	Source
Pure ice at 0°C.	2.24×10^{-3}	Pounder (1965)
White ice		
density: 860 kg/m^3	2.04×10^{-3}	Schwerdtfeger (1963)
density: 780 kg/m^3	1.77×10^{-3}	"
Snow		
density: ~ 250 kg/m^3	0.21×10^{-3}	Mellor (1964b)

The thermal conductivity of ice is considerably affected by embedded air bubbles (reflected by lower density), by ice temperature (Pounder, 1965) and ice structure. The thermal conductivity of snow varies greatly with density and type of snow. Several interesting expressions of snow density-conductivity relationships have been developed, showing that there is a considerable range in conductivity for a given density, presumably resulting from the effects of other snow properties (see Ch. 7, Fig. 7.6).

Snowcover may, however, promote ice growth if it is heavy enough to depress the surface below the hydrostatic water level. In most cases black ice is able to withstand a snow load without cracking. However, should cracking be induced, for example, by rapid changes in air temperature - particularly decreasing temperature - (Wilson et al. 1954; Gold, 1966, 1967; Weeks and Assur, 1969; Jones, 1969, 1970a) while it is depressed, its surface and the overlying snowcover will be flooded initiating a new phase in the evolution of the lake cover. This phase is analogous to some aspects of the development of superimposed ice on glaciers (Adams, 1966). The depression of the ice sheet may also result in the thermal erosion of the base of the ice sheet as it is pushed down into warmer, possibly moving, water.

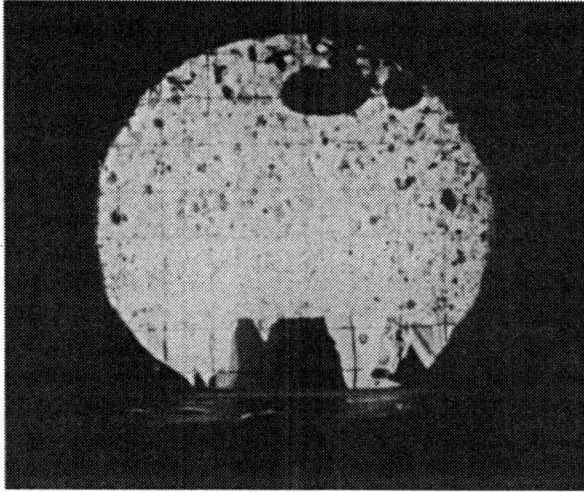

Fig. 10.2 Vertical section of the interface between white ice and black ice photographed under crossed polaroids, with one-centimetre grid. Varying degrees of granularity of the white ice can be seen above, with columnar black ice below.

Water which moves through cracks in the ice rises by capillary action in the snowcover to a depth slightly greater than the depth of depression of the ice surface producing a layer of slush at the snow-ice interface. Given continuous below-freezing air temperatures, the thinly-insulated slush will freeze relatively rapidly from its surface downwards until a new layer of solid ice is formed on top of the original ice sheet. During this freezing process the temperature throughout the underlying ice sheet is for all practical purposes constant, and thus no growth occurs in it. The new layer of ice contains a high concentration of air bubbles and freezing takes place around nuclei provided by the snow. As a result, the ice formed is granular and opaque, appearing white under natural light, quite distinct from the underlying, effectively transparent, columnar, "black" ice.

Granular Types of Lake Ice

The detailed character of "white ice", like that of other ice forms in the lake cover, varies greatly depending, at the time of formation, on the rates of freezing and the stratigraphy of the snowcover (Shaw, 1965; Adams and Jones, 1971; Michel, 1971). The opaqueness, texture and density can vary

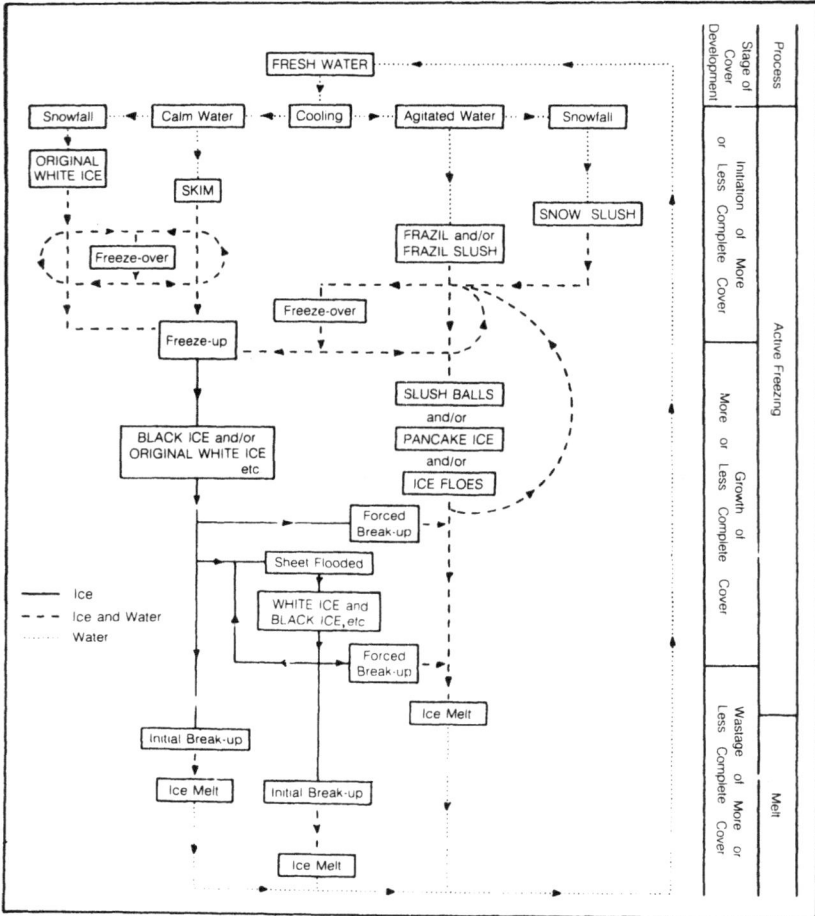

Fig. 10.3 Schematic classification of process and form for freshwater ice. Emphasis is given to the sequences which begin with *Calm Water* (with or without snowfall) and proceed to *Freeze-Over* (temporary complete ice cover)--*Freeze-up--Sheet Flooded*--WHITE and BLACK ICE--*Initial Break-up--Ice Melt*. Adjacent to this 'route', to the right is the fast-flowing river case and, to the left, the normal high arctic, intense cold, low snowfall situation, combinations and permutations of these three broad situations are possible. An example is the evolution of ice cover in an enlargement of a long north-flowing river in a snowy environment, in which *Freeze-up* is delayed by turbulent flow but is eventually achieved as discharge declines but in which *Break-up* is caused by floodwaters from the early-melting headwaters. In such a case, the sequence might be, *Agitated Water*--ICE FLOES etc.--loop

considerably within and between layers. Ager (1962) reports densities ranging from 780 to 910 kg/m^3. It appears likely that widespread sampling of the densities of black ice and white ice would show that white ice generally has a lower mean density but a higher coefficient of variation. The original interface between the white ice and black ice usually remains distinct throughout the winter and can be seen with the naked eye in drill holes. The crystallographic nature of the interface, with columnar ice below and granular above, is illustrated in the vertical section of Fig. 10.2, in the upper part. It should be noted that the newly-formed ice layer contains ice derived from snowfall as distinct from ice formed by the freezing of lake water.

A fairly common variant of the sequence of events establishing a lake ice cover is that where the initial sheet formed is a granular, "white" ice. This occurs when freeze-up takes place during substantial snowfall or when the lake surface is very agitated (Shumskii, 1964; Palosuo, 1965; Weeks and Assur, 1969; Jones, 1969; Michel, 1971). In cold, low-snowfall regions initial white ice may be the only form of white ice present and, although very thin, it plays a significant role in the thermal regime of the lake through its effect on the radiation exchange (Swinzow, 1966). Figure 10.3 schematically illustrates the different processes which can occur during development of a lake cover, and the types of ice produced. Although not central to this discussion the diagram includes the turbulent river case because a lake can be considered as

---

to *Freeze-over* and/or *Freeze-up--Sheet Flooded*-WHITE ICE and BLACK ICE--*Forced Break-up* to an early *Ice Melt* in renewed agitated water. The diagram suggests progressions of events in the initiation phase and in the wastage phase, including a variation in the length of broken ice season, which are associated with the nature of evolution of the sheets concerned. However, there is no attempt to display explicitly the lengths of phases of growth or of the total ice season. The diagram was constructed with reference to the Weeks' classification of sea ice (Mellor, 1964, p. 131) and the schemes of Michel (1971, p. 3) and Wilson et al. (1954). Other interesting sources of information on types and terminology of freshwater ice are Hamelin (1960), Kivisild (1970), Michel and Ramseier (1971), and Pivovarov (1973). The classification used here is elaborated further in Adams (1976c).

An interesting variant of the river "route" are the "naleds" or river icings, a particular type of "White" ice, which can develop after a river has become completely frozen over (Carey, 1973). Another interesting side-light in this area is the fact that records of the dates of the various stages of break-up have been used as indices of long term climatic change (Catchpole et al., 1976). One of the purposes of this diagram is to show that the growth and decay of floating ice sheets are controlled by a variety of factors, some climatic factors and others non-climatic.

"a bulge in a river". There is a continuum of conditions ranging from the large calm lake to the rapidly flowing, narrow stream.

Another example of the various profiles which may be encountered in a lake is given in Fig. 10.4 which also illustrates river-lake relationships. Granular ice has been formed at the base of the cover as well as on top of it. This results from frazil ice (tiny crystals of ice formed in supercooled, turbulent water) being swept under and over the sheet — a common occurrence in rivers. Frazil ice has been extensively studied by Michel (1963, 1967, 1971); the relevant literature was critically reviewed by Williams (1959). Frazil ice and other types of river-borne ice, including those derived from snow, can be an important constituent of lake cover where they are actively brought into the lake. Michel (1971) refers to some lakes as "ice accumulators".

It is interesting to note that "white" ice can form on a bare ice surface when flooded—the "whiteness" persists although the new ice does not contain snow (Ager, 1962; Williams and Gold, 1963; Shaw, 1965). The high albedo of river icings (Harden et al. 1977) is an example of this. Thus "white" ice is not necessarily "snow ice" as implied in some studies. White ice is also produced when a lake snowcover is flooded by rain, springs or river water. In some cases, the entire snowcover can be consumed, but this does not occur where the ice is slushed as a result of being depressed below the lake's hydrostatic water level. Wilson et al. (1954) in their classic study of Lake Mendosa, Michigan classify the types of granular ice which might be present on a lake as "agglomeritic".

The development of white ice, reflecting as it normally does the influence of an appreciable snowcover, appears to be an important factor affecting the temporal and spatial variability of the entire lake cover. Lake snow is subject to redistribution by wind so that both small- and large-scale accumulations of it can produce quite localized environments. The flooding and subsequent white ice formation can take place by small increments in a highly localized fashion (Shaw, 1965) or in a more catastrophic, lake-wide way (Jones, 1970a). The resulting variations in white ice thickness are often accompanied by compensating variations in black ice thickness and associated variations in snowcover. Variations in thickness of the various components of the cover and of the entire ice sheet may be spatially random or may, like the snowcover, exhibit distinct trends related to prevailing wind directions (Adams and Brunger, 1972, 1975), snowcover and white ice tending to be thicker downwind, black ice tending to be thinner.

Roles of Snow and White Ice in Sheet Development

At any particular time, the winter cover of a lake in an area with considerable snowfall is "multililthic". It may consist of a distinctive, spatially- and stratigraphically-varied snowcover underlain by similarly

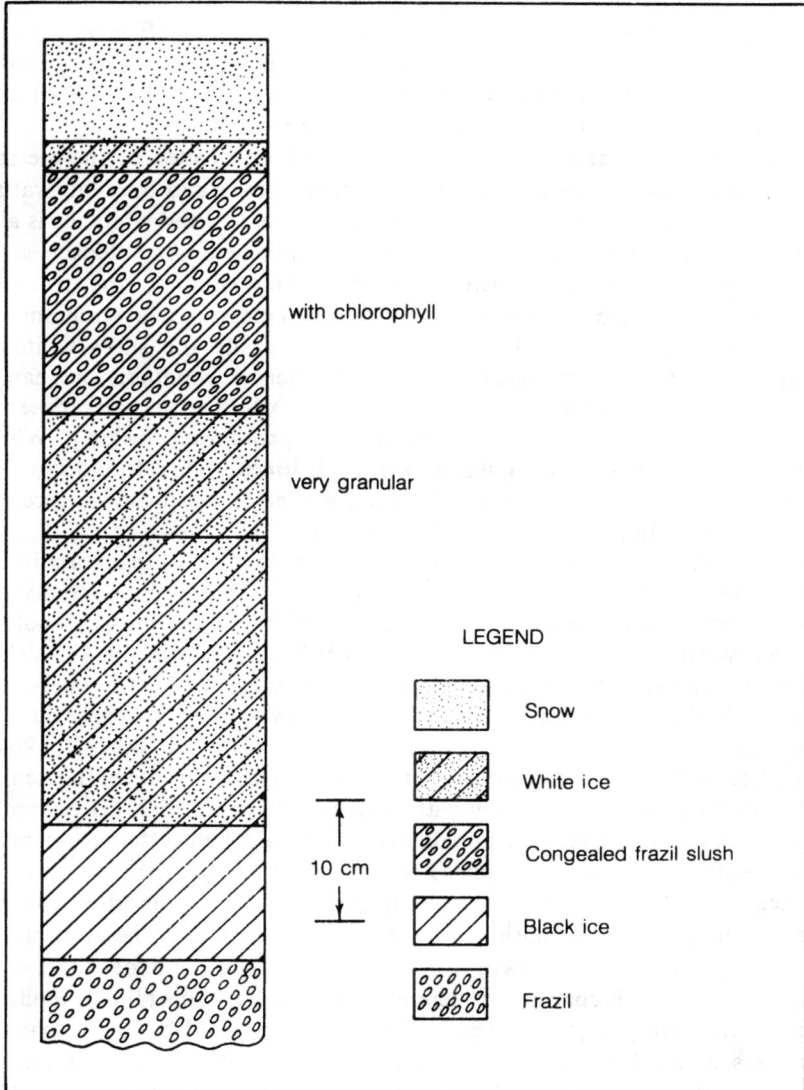

Fig. 10.4 Drill hole profile from Little Lake, Peterborough, Ontario showing evidence of several ice forming events (Adams, 1976a). The "with chlorophyll" stratum was formed by frazil ice* with algae. This lake is essentially a widening of the Otanabee River. The ice symbols are from Michel (1971).

*Tiny crystals of ice formed in supercooled, turbulent water, (Williams, 1969).

varied layers of white ice and/or slush and black ice. The full range of situations between entire profiles of black ice and entire profiles of white ice can be encountered on the same lake. On a lake-wide scale, an entire ice cover may be black ice, as in cases in the high Arctic, or white ice, as in the case of small lakes in subarctic and temperate locations.

The growth rate at the base of the ice sheet is controlled by the thermal conductivities of the snow and ice layers of the cover and the temperature gradients in them. If there is snowcover and white ice, and white ice is also forming, both of these controls are greatly complicated. An extensive slush layer, a phase in white ice growth, can effectively eliminate the heat flux through the ice. Also, the thermal conductivity of individual layers at one site on the lake cover can differ markedly. Generally the snow/white-ice components of the cover, because of their low thermal conductivities, can be assumed to reduce the rate of black ice growth, while at the same time the snow actively assists the growth of white ice by promoting slushing. Where slushing is the major factor of white ice growth Jones (1970a) found snowfall to be the controlling parameter. The effects of early season snow, received while the ice is thin, are particularly important.

The growth process of the lake cover is quite complex so that simplifying assumptions about its temperature gradient and thermal conductivity are normally made in studies of the energy balance of the lake system as a whole or of the cover itself (Scott and Ragotskie, 1961; Schindler et al. 1974a; Welch, 1974). It is remarkable that several methods of predicting ice thickness on a regional basis, using such indices of "freezing exposure" as degree days of frost, are reasonably successful (Shulyakovskii, 1963; Bilello, 1964a; Pounder, 1965). In the shorter term and on a local scale the rate of thickening of an ice sheet can vary greatly with atmospheric conditions. However, it appears that in the longer term and on a regional scale these differences tend to compensate (e.g., a slushing phase retards black ice growth but helps to thicken the whole ice sheet by creating white ice). This allows useful phenomenological "explanations" of ice growth to be made. The similarity of the average rates of ice growth over extensive areas of Canada, for a wide range of ice growth conditions, demonstrated by Williams (1963), tends to support this compensation process. However, it should be noted that the thickness of a lake cover is not a simple function of one or more readily observed atmospheric variables. Michel (1971) presents an excellent discussion of the detailed problem of operational ice thickness forecasting in the presence of white ice.

The existence of qualitative and quantitative differences in an ice cover is often most apparent in the wastage, disintegration, and break-up phases of its evolution. For a sheet of black ice, variations in thickness have little effect on the timing and spatial pattern of the cover's disintegration. In the absence of snow, this disintegration is essentially a result of "candling" produced by radiation melting at the vertical crystal interfaces within the sheet; a process little affected by ice thickness. However, a mixture of snow and white ice components, with their high albedos (ranging from 0.85 for dry, clean snow to less than 0.10 for black ice - Davies, 1962; Polyakova, 1966) and granular textures, can greatly affect the timing of the stages of disintegration and their spatial patterns. The "normal" pattern, involving development of a shore lead followed by the disintegration and melt of ice at the center of the lake, may be completely masked where there are large amounts of snow and white ice.

White ice on the whole or part of a lake delays the initial disintegration of the ice sheet and changes its characteristics. By promoting the formation of ice flows and by prolonging their existence it extends the broken ice season and delays final breakup.

EFFECTS OF SNOW AND ICE COVER ON EXCHANGE PROCESSES TO A LAKE AND ON ITS BIOLOGICAL ACTIVITY

The ice and snowcover of a lake is important in a variety of ways, the most basic being that it cuts the lake off from direct atmospheric influences.

Energy Exchange

The initiation of the cover drastically changes the energy regime of a lake both as to the magnitudes and types of components involved in the balance of energy and the exchange processes. Mass and energy exchange via evaporation/condensation, normally very important during the summer, often cease entirely; the net radiative flux is greatly altered; and sensible exchange, becomes negligible. The change in the energy regime is reflected by its effect on water temperature, which drops before freeze-up, but frequently rises after the lake cover is established (Scott, 1964; Hart and Steinhart, 1965; Bilello, 1968; Parott and Fleming, 1970; Penn, 1970). A decrease in water temperature of a lake following the formation of an ice cover has been reported by Rigler (1974) and Schindler et al. (1974a).

The optical and thermal properties of the different constituents of a lake cover vary greatly (Gaitskhoki, 1970a, 1975). Bubble-free black ice is effectively transparent to short-wave radiation while both snow and white ice reflect a large proportion of it and greatly limit penetration by the remainder. The penetration of snow by short-wave radiation is a function of the density and the size, shape and orientation of the grains (Thomas, 1963; Mellor, 1964b; Weller and Schwerdtfeger, 1971). It is difficult to generalize about solar radiation transmission through snowcover for two equally important reasons: the complex stratification possible, and the pronounced spatial variability in physical properties (Giddings and LaChapelle, 1961; Welch and Kalff, 1974). The poor optical transmissive properties (relative to black ice) of white ice (Lyons and Stoiber, 1959; Pounder, 1965; ; Glen, 1974, Hobbs, 1974) limit the transmission of radiant energy. Williams (1969, 1971a) has plotted graphs of the penetration of short-wave radiation versus thickness for various types of ice (see also Albrecht, 1964, referred to by Wetzel, 1975). The logarithm of the ratio of the radiation intensity received at any depth to that emitted by a source of constant intensity varies indirectly with the magnitude of the extinction coefficient. Williams (1971a) cites extinction coefficients of 0.2 cm^{-1} for snow, 0.07 cm^{-1} for white ice and 0.004 cm^{-1} for black ice. There is, of course, a very large range of values for the extinction coefficients of different covers (Maguire, 1975a, 1975b).

In the case of an ice cover comprised mainly of black ice and snow-free for most of the year, the rise in temperatures below the ice may be the result of direct warming by incoming radiation. Hathersley-Smith et al. (1970) discuss a few examples of this occurring in perennially-frozen lakes of the high Arctic and cite studies of the phenomenon made in Antarctica (Brewer, 1958; Hendy et al., 1972; Wilson and Wellman, 1972). However, where snow and/or white ice are present, a winter rise in temperature is the result of containing heat released from internal sources, notably the bottom sediments (Hutchinson, 1957; Scott, 1964; Johnson and Likens, 1967; Likens and Johnson, 1969; Parott and Fleming, 1970; Penn, 1970). Heat derived from internal sources may contribute a significant portion of the annual energy budget of lakes which freeze over and a dominant portion in at least some small shallow lakes (Schindler, 1971); and may be a disproportionately important part biologically, in the over-wintering of organisms (King, 1973).

Although the lake cover restricts heat losses during winter, it limits heat gains from the atmosphere during spring. As already mentioned, the persistence of the high snow and white ice components of lake cover prolongs break-up and delays the warming of the lake (Allen, 1964; Williams, 1968; Jones, 1970a; Parott and Fleming, 1970).

Snow especially, and white ice (unlike black ice) are very effective in limiting the penetration of light into a lake. Ruttner (1968) describes an interesting experiment which involved making a "window" in a lake snowcover to observe the change in distribution of zooplankton in response to the direct radiation received. He noted that lakes with prolonged snowcovers were relatively low in plankton. Similar findings were reported by others. Wright (1964) found that algae continued to photosynthesize under snowfree ice but the process declined following snowfall. Welch and Kalff (1974) explained the distributions of benthic mosses in Char Lake (in the high Arctic) in terms of corresponding variations in the snowcover. They, and other investigators (Kalff and Welch, 1974) point out that the deep, prolonged, snowcover in some subarctic regimes produces, within a lake, similar conditions of winter darkness as experienced by lakes at high latitudes. In the high Arctic, snowcover is the dominant control of the annual radiation input into a lake (Rigler, 1974) and therefore is critical in determining its productivity.

Gas and Mass Exchange

Black ice, with vertically-oriented optic axes is effectively transparent to light, while the other components in the ice cover are less transparent or opaque and change the radiation distribution in a lake. Although the roles of white ice and black ice are different for radiation, any type of ice cover is relatively impermeable, inhibiting the exchange of gas and other masses between the lake and the atmosphere. For example, the initiation and growth of an ice sheet affects the concentration of oxygen in the unfrozen portion of the lake. The lake is cut off from its atmospheric source of oxygen which still continues to be consumed in the water. The resulting depletion is offset somewhat by the freeze-out of oxygen from that portion of the lake which becomes ice, i.e., the amount of oxygen present in the water in the fall becomes concentrated in a smaller volume of water as the winter progresses. In certain lakes, with low productivity, the net result of these depletion and concentration processes causes the oxygen concentration to increase during the winter (Welch, 1974). On the other hand, in small lakes with a thick snowcover, high productivity and a limited throughput of water (which may also be controlled by the ice) oxygen depletion can be serious (Hutchinson, 1957; Schindler, 1971).

Freeze-out of gases other than oxygen and solids occurs as the black ice sheet thickens producing higher concentrations in the water layer below the freezing interface (Maguire and Watkin, 1975). A double maximum of potassium/sodium/calcium/magnesium etc. concentrations occurs in the vertical lake profiles below the ice. One maximum is just below the ice, the direct result of the freeze-out exclusion process; while the other is the result of sinking of the dense, solute rich water from below the ice (Schindler et al., 1974a). The formation and decay of the ice sheet dominates the annual variation (a later winter peak, open season minimum) of lake chemical concentrations. The exclusion process during freezing gives higher concentrations in the diminishing water body while the melting of the "pure" lake ice sheet in the spring results in pronounced dilution (Hansen, 1967; Rigler, 1974; Schindler et al. 1974a). During ice formation in this annual cycle, the slow freezing of black ice yields a more efficient "freeze-out" of gases and chemicals (Pounder, 1969).

The melting of lake cover in spring does not always have a diluting effect because several month's precipitation, located in and upon the ice sheet, often enters the lake while the cover melts. This could have a considerable impact on the nutrient budget of a lake (Barica and Armstrong, 1971). English (1976) discusses the role of snowcover in the phosphorous loading of a lake in Southern Ontario. Phosphorous originating in snowfall deposited directly on a lake surface (and later possibly incorporated into the ice sheet) has a disproportionate impact on the lake in the spring. Phosphorous in land snowcover within the lake's catchment area, is depleted enroute overland to the lake while that originating in the lake snowcover is not lost.

Extensive literature exists on the relevance of the chemistry of winter precipitation to the pollution of lakes, not exclusively devoted to nutrients, but including studies on "acid" lakes downwind of industrial areas where limnological conditions in the winter season may be very critical (Hagen and Langeland, 1973; Hultberg, 1975).

Other changes affect the biota and productivity of a lake as an ice cover develops, including decrease in water volume, decrease in unfrozen bottom area, and decrease or cessation of water throughput when ice retards flow and restricts outlets and inlets. This last change is one factor contributing to the great reduction in mechanical turbulence which occurs when the lake becomes cut off from the atmosphere. A reduction in turbulence significantly affects the lake's productivity, especially through its effect on the thermal stratification of lake water. Underneath complete ice cover, density currents produced by local differential heating may be significant (Wetzel, 1975).

The topic of the productivity of ice- and snow-covered lakes in relation to oxygen supply and throughput of water (and flushing rates) has important practical implications for the artificial eutrophication of lakes used for sewage disposal. In such lakes, the consumption of oxygen is increased without a compensating increase in its supply. This is particularly serious in the high Arctic because the persistently low temperatures slow the breakdown of organic material in lakes which receive nutrients in a sudden sharp burst during the melt period (Rigler, 1974) and because the season for receiving appreciable radiation is short. Schindler et al. (1974b) discuss the biological processes occurring under these conditions. More practical aspects of sewage problems and ice cover are treated by Dickens (1959) and Osterkamp (1974). The use of variable depth sewage lagoons (deep in winter, shallow in summer), is one practical solution to these problems in permafrost areas.

The detailed individual biological effects of the major components of the lake cover (snow, white ice and black ice) have received relatively little attention. This is quite remarkable since freshwater ice in North America, especially Canada, is very important and extensive work on sea ice has already been accomplished. Welch (1974) notes the limitations of mean snow depth as a measure of the role of snowcover in the light regime of a lake with little or no white ice. He describes the difficulties produced by the stratification and complex spatial distribution of the snowcover. This is one example of the effect of oversimplification on the treatment of one cover component--which seemed to have received the most attention. Because small amounts of solar radiation are received at low angles of incidence during the winter season, it might be assumed that detailed effects of variations in the optical properties of the lake cover would have received greater attention.

Ignoring the presence and/or variability of white ice might influence the interpretation of observations of some biological processes and phenomena. For example, calculation of the ice sheet's share of a lake's energy budget could have a larger error if the maximum density of ice ($917 \, kg/m^3$) were used since the ice sheet has a large spatial variation in density. Also, as mentioned before, the externally-derived (snowcover) portion of white ice significantly affects the heat exchange processes and because of this the spatial variation in white ice will cause different thermal regimes in different parts of a lake.

The significance of white ice and other forms of agglomeritic ice to the freeze-out process appears to be different from that of black ice which has a freezing interface with the main body of lake water. Where freezing takes place within or upon the ice sheet, any gain of gases or solids by the unfrozen body of the lake must occur through cracks in the ice. When a new layer of ice is completely frozen upon an existing ice sheet, gases and chemicals concentrate at the interface between the two ice layers. Such border regions are better situated to receive radiation than the lake beneath and so might have a considerable potential as a habitat for organisms which can live in and on ice; e.g., algae and other life forms are quite commonly observed in lake ice. Dunbar (1968) describes and comments on the standing flora crop which exists on the under-surface of sea ice, specifically noting the "spatial heterogeneity" of this particular habitat. The various combinations of the three principal components of lake cover (snow, columnar ice and granular ice) and their evolution, produce a complex range of biological habitat in and on a lake.

VARIABILITY OF ICE SHEET STRENGTH

The bearing or flexural strength of an ice sheet varies with ice type and thickness across a lake - an important consideration for lake users. Generally applicable values of the strength of ice are not easy to obtain but several sources present methods for measuring and calculating strengths and give experimental results (Mantis, 1951; Butkovitch, 1954; Frankenstein, 1959; 1969; Lliboutry, 1964; Pounder, 1965, 1969; Butiagin, 1966, Glen, 1974, 1975).

The problems of determining ice strength are reviewed by Gold (1968), Lavrov (1969), and Weeks and Assur (1969). Strength values obtained vary because of the density of the ice, temperature, the rate and area of application of stress, size and shape of specimens used in tests, crystal size and its orientation to the direction of stress (Gold, 1960, 1968), the degree of bonding of the crystals, etc. Also, the results cannot easily be transferred from small samples to floating ice sheets (Nevel and Assur, 1968; Stevens and Tizzard, 1969; Nevel, 1970); the strength of small samples appears to be generally high. Furthermore, the loads imparted by stationary and moving objects, even those of the same mass, to an ice surface differ (Gold, 1960; Willmot, 1962; Assur, 1968; Nevel and Assur, 1968; Stevens and Tizzard, 1969; Nevel, 1970). Spacing and speed of those objects relative to the depth of the lake is critical. For stationary objects the duration and spacing of loads (e.g., parked vehicles) is important. The flexural properties of white ice are specifically treated by Stearns (1964).

It is often assumed that white ice is "weaker" than black ice. Jones (1970a) cites the work of Frankenstein (1969) to demonstrate the weakening effect on an ice sheet subject to downbuckling caused by the white ice component. This weakening is critical for over-ice travel. Tabulations of miminum-bearing ice thicknesses often include remarks such as "ice should be clear and blue" or sometimes recommend a safety factor of two for non-black ice (Ontario Ministry of Natural Resources, 1973).

Although several investigators have explicitly studied both white ice and black ice, categorical statements about their differences in strength cannot be made. The various factors affecting the general results of ice strength tests clearly show that the strength characteristics of two types of ice having major crystallographic differences will not be the same and that bearing strength is not a simple function of thickness if there is more than one type of ice. Table 10.2 lists ice thicknesses to support typical loads together with the limitations on the data for safe application. Often the inherent limitations of the data are not sufficiently emphasized.

Ideally, to loosely paraphrase Gold (1968), judgement of the strength of a lake cover requires detailed stratigraphic and spatial observations of the mode of formation and evolution of the layers involved. The properties, areal variations and relative effects of the major cover components must be fully appreciated.

Much work has been done to increase the bearing strength of ice covers for road and air transportation and for timber storage (Ager, 1961; Coble and Kingery, 1963). The simplest methods involve periodic floodings to thicken the ice sheet. Reinforcement with materials such as sawdust, newspaper and fiberglass has been successful. Coble and Kingery present the relevant theory, experimental results and an economic analysis for several methods of increasing the strength of snow and ice roads. Ice and snow roads (roads on lakes are often both ice *and* snow roads) are discussed in Ch. 12.

Table 10.2
**EMPIRICAL SAFE ICE THICKNESSES FOR TYPICAL LOADS. NOTE
THE LIMITATIONS[a] ACCOMPANYING THE DATA (Pounder,
1969). [Granular ice is not differentiated here].**

Load			Safe Thickness (m)	
Object	Mass (kg)	Operation	Freshwater ice	Sea ice
Dakota aircraft	12,800	parking	0.61	1.02
Dakota aircraft	12,800	landing on skis	0.36	0.60
Otter aircraft	3,300	landing on skis	0.18	0.36
Tracked vehicle	9,000	moving slowly	0.43	0.67
Motor toboggan	360	moving slowly	0.1	0.18
Man	90	resting	0.08	0.13

[a] It is assumed that the air and ice surface temperatures are -10°C and that the ice is of uniform thickness, without cracks, so that the ice sheet is effectively infinite. No allowance is made for the possiblity of destructive resonance in the ice cover. This can be dangerous for vehicles moving over an ice sheet with shallow water below.

The values in this table are empirical and provide a safety factor from 1.5 to 2. Numerous theoretrical analyses of the load-bearing capacity of ice exist, and are in good agreement with the empirical results.

The strengths of ice in compression and tension are important influences on the expansion and contraction of an ice sheet in response to temperature fluctuations. The expansion of ice sheets in response to fairly sharp temperature rises has been demonstrated to be important for geomorphological (Zumberge and Wilson, 1952, 1953; Pessl, 1969; Jones, 1970b), biological (Wassen, 1966, 1969; Rigler, 1974) and engineering purposes (Gold and Williams, 1968; Michel, 1970). Ice thickness and lake size and shape are important contributers to ice-shove if there are temperature-induced stresses in the sheet. However, even a very thin snowcover rapidly dampens the air temperature fluctuations which penetrate towards the ice sheet (Montfore and Taylor, 1949; Jones, 1969, 1970a; Michel, 1970). For example a fluctuation of 15°C in air temperature may be reduced to that of 5°C by a 15-cm snowcover having high values of density (400 kg/m^3) and thermal conductivity. The reduction in amplitude of a surface temperature fluctuation with depth by a layered snow/white ice/black ice profile is greater than by a snow/black ice cover because of the low thermal conductivity of white ice compared with black ice (Jones, 1969).

If a sheet of white ice expands, the pushing forces on lake margins may be decreased by the embedded air spaces which absorb some of the compression forces. Consequently, thermally-induced ice-shove effects are less likely to appear where appreciable snow, slush and white ice cover normally occur.

However, the sequence of melt in the spring is snow → white ice → black ice. By prolonging the lifetimes of the complete ice cover and floes, by slowing candling, and by extending the length of the broken season a substantial snow/white ice cover increases the potential for shoreline damage by wind-driven ice (Williams, 1965). This phenomenon is the predominant shore-moulding process during the ice season in much of eastern North America (Tsang, 1974). Delay of break-up until the water levels in the lake are high can affect the lateral scope for ice damage around its margins considerably (Fig. 10.5). The shape and size of a lake and its shoreline characteristics greatly affect the effects of the expansion and wind-induced ice-shove processes.

The exact nature of the ice (some shoreline ice, for example appears "rotted" before melt) and of the ice/rock contact around lake margins has received some attention in the geomorphological and biological literature (Archer and Findlay, 1966; Wassen, 1969; Jones, 1970a; Rigler, 1974; Adams and Mathewson, 1976). With a few notable exceptions the relevant studies emphasize the effects of ice-shove on landforms, vegetation and shore installations but give little attention to the processes concerned. Dionne's annotated bibliographies (1969, 1974) and the Proceedings of the Symposium on the Geological Action of Drift Ice (Montreal, Rèvue de Géographie, 1976) are an excellent *entrée* to the literature.

Thermal stress in the ice cover produces cracks, some of which cause flooding of the ice surface. The importance of cracks in a lake ice sheet has been mentioned above in several contexts including white ice formation and strength of the sheet. Further study of cracking in lakes is desirable not only to understand these and allied processes but also to explain the influence of refreezing of water in cracks on ice sheet expansion and the role of cracks as an exchange link between the "sealed-off" lake and the atmosphere. As the tensile strength of lake ice is lower than its compressive strength, failure (i.e. cracking) can be induced by much smaller changes (in this case decreases) of temperature than those needed for appreciable expansion. Although snowcover limits black ice growth and is instrumental in developing a type of ice which is in some ways weaker than black ice, it may increase the bearing capacity of the ice sheet. This results from a reduction in the brittleness of the ice which, in turn, increases its resistance to thermal cracking. The strength of

Fig. 10.5 Wind induced ice-shove on Astray Lake, Labrador-Ungava. Note the layering which indicated four thicknesses of lake cover in this single mound. A pronounced layer of 'black' ice is visible. The ice passed over alder (*Alnus crispa*) bushes (partly submerged at the existing high water level) which, because of its high resistance to bark damage, its flexibility and its active reproduction through stolons (fast growing new shoots from ice-cut stumps), is often a characteristic feature of ice-shove localities (Wassen, 1969). Ice-shove effects of this type are often not a result of the action of floes but of an essentially complete ice sheet on a lake with narrow leads. This situation provides a large surface area so that even light winds can move ice very effectively. Floes do a great deal of damage where there is a current (see Adams, 1977b).

the ice sheet varies not only with thickness and types of ice but also with the dimensions and distribution of its cracks. Weeks and Assur (1969) point out that; "for large volumes of ice, the basic strength properties become less important than the effective distance between major defects such as cracks. This distance is governed by brittleness which is high for cold fresh (water) ice and low for warm ice". Brittleness and cracking are governed by the depth of snow and, to a lesser degree, the thickness of white ice.

ROLES OF LAKE SNOW AND ICE IN THE HYDROLOGICAL CYCLE

Findlay (1966, 1969), Jones (1969, 1970a), Williams (1971b), Fitzgibbon (1976), and others have discussed the role of lake cover in the hydrological regime of a region. A lake is a reservoir in which water is temporarily stored. Some of the volume in storage in the fall freezes in winter forming part of the ice sheet, e.g., the black ice and perhaps two thirds of the white ice. The snowfall retained on a lake surface is, like snowcover on land, also a temporary reservoir for water, comprising the remaining one third of the white ice as well as the snow on the lake. The ice cover, by partially or completely blocking inlets to and outlets from the lake and by slowing currents, reduces or halts the throughput of lake water.

However, in addition to delaying run-off, the cover may displace a volume of water from the lake. Ideally, the formation of a freely-floating ice sheet would not displace water since the increase in volume because of the change in state is equal to that part of the sheet floating above the water level. Where a snow load is imposed on the sheet, water will be displaced. Jones (1969) assumed that the load of snow perfectly counteracts buoyancy and calculated that over 14% of the winter discharge from the catchment study area (22% lake cover) was "non-precipitation-generated". Since the density of white ice is lower (the volume produced by freezing a given volume of water is greater) and also includes mass (snow) not originally in the lake, it displaces a disproportionately large amount of water. (Jones suggests that this amount is more than three quarters of the 14% of the winter discharge). Thus the pattern of ice growth during a winter and over a lake affects the winter hydrograph. The flexing of the ice sheet by wind and its loading by snow may promote surges of discharge from the lake early in winter (Findlay, 1966, 1969). Increases in winter discharge resulting from melting at the base of a heavily snow-covered ice sheet have also been reported (Williams, 1971b).

Where lakes are a major part of a drainage basin area, precipitation incorporated into the lake ice may be significant to basin-wide precipitation input e.g., an ice sheet flooded to a depth of 25 cm incorporates at least that depth of snow. Furthermore, if the snowcover had a density of 300 kg/m^3, approximately 7.5 cm of the resulting ice thickness is derived from snowfall and not from lake water. In a catchment area which is 23% lake near Knob Lake, Labrador-Ungava, an adjustment of spring snow survey results for this phenomenon resulted in a gain *to the entire catchment* of 2.5 cm water in a winter precipitation total of 30 cm (Adams and Rogerson, 1968).

Since the melting of lake snowcover and white ice and the final dissolution of the ice cover occur after snowmelt on land, lake cover causes further delay in melt runoff from the basin. This is especially true in white-ice environments where slushing promotes a late peak in ice thickness and the high albedo of the ice retards break-up after the snowcover has been removed. At the same time, according to Jones (1969), the volume of water displaced during the winter has to be replaced during melt. The net effect of white ice on the spring hydrograph (providing the volume from its melt is smaller than the volume of water displaced over winter) is essentially the same as the effect of the whole lake cover on runoff. There is a general flattening of the hydrograph, a less steep rising limb and a persistence of fairly high discharge levels after peak melt as the delayed contributions re-enter the system.

Comprehensive information about the lake cover is required for a detailed prediction of the effects.

The use of dust to advance break-up (i.e., to reduce the delay in re-entry) has been shown to be relatively more effective on lakes with a substantial snow and white ice cover (Williams and Gold, 1963; Williams, 1967). However, the objectives of most of these types of experiments were to improve traffic on waterways or to modify the discharge by preventing ice jams. The general effect of lakes and lake snowcover on spring runoff is, in human terms, "beneficial".

MEASUREMENT OF SNOW AND ICE ON LAKES

The diversity of winter cover, both in thickness and types of ice and snow, within and between lakes, presents measurement problems which have been studied at levels of detail which range from remote sensing by satellite to single drill holes in the ice. The four principal questions: "where, when, what and how to measure?", all merit attention. As the design of any system of measurement can only be properly undertaken with reference to a particular objective, e.g., the approach would be different for a detailed study of the energy balance of a lake than for work on the prediction of regional ice thickness—these questions can only be treated in a general way.

Considering "how?" and "when?" first, the standard method of measuring ice cover in North America uses a measuring tape inserted in a drill hole (Canada AES, 1969; Unesco/IAHS, 1972; Bilello, 1974 - see Fig. 10.6). Normally a single weekly measurement on one lake is taken to represent the winter cover on all water bodies in a region. Although frequent monitoring is required to measure the rather complex evolution of ice sheets, regular drilling of white ice can produce anomalous results by promoting flooding. Devices which eliminate the need for drilling appear to be a very desirable alternative for official survey programmes if detailed knowledge of ice conditions is to be improved (Sychev, 1940; Vesso, 1957; Houle, 1961; Adams and Shaw, 1966b; Ramseier and Weaver, 1975).

Even if a better method of measurement is developed, the limitations of using values from a single site as a measure of, for example, the areal distribution are still present (see Fig. 10.7). The problem of the representativeness of one or a few measurements on a lake [part of the "where" portion of the general question posed above; another aspect of the same question arises from the variability between lakes - see Adams and Shaw (1966a) and Jones (1970a)] has been studied by Andrews (1962); Jones (1970a) Adams and Brunger (1972, 1975). Adams and Brunger, using a large, random sample, obtained mean values for cover components which all differed by more than

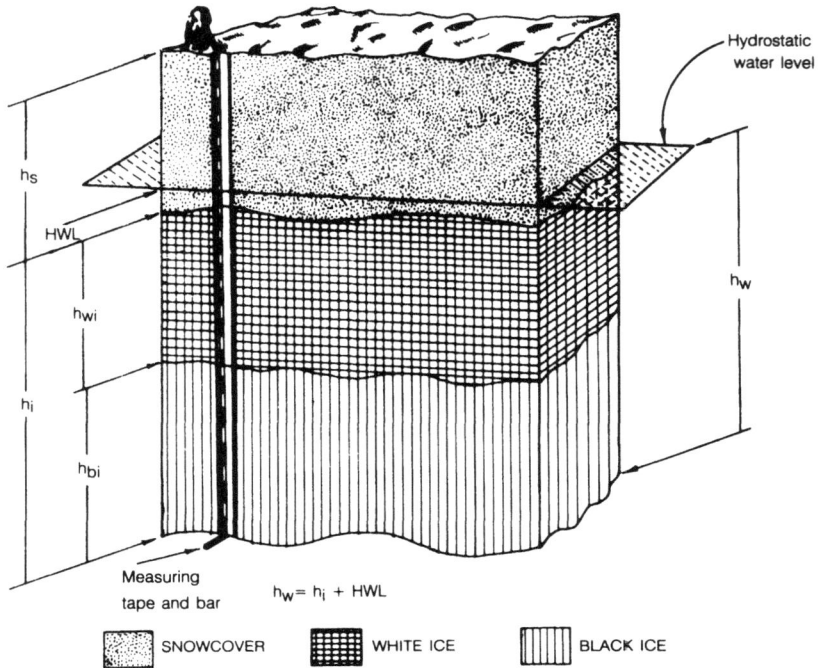

Fig. 10.6 Principal components of lake cover measured by the drill hole method (Adams, 1976a). h denotes thickness and the subscripts s, wi, bi and w refer to snow, white ice, black ice and water respectively. All of these thicknesses are easily obtained in a fairly large diameter drill hole; the density of snow is also easily measured. These values are important for many purposes, e.g., in the buoyancy equation for an ice sheet. Note that only h_s and h_i are measured in the official surveys (see text).

one standard deviation from the officially-reported values for the lake over the same period. They suggest rules of thumb for selecting measuring sites to minimize this problem in areas where only a few measurements are possible and for designing a random survey technique to obtain detailed data. Ideally, a set of measuring devices, areally distributed according to a knowledge of local ice conditions, would provide a firm basis for measurements at official ice data reporting sites.

It is important to realize that any set of observations provide both a measure of central tendency (such as the mean) for the "population" of ice values and indicate the distribution of values around that measure. In many studies of lake ice this distribution (including the extreme range) is a critical statistic more important than the mean.

The use of large random samples allows the range of ice conditions necessary for detailed studies and permits the results of less intensive surveys to be placed in proper perspective.   But this sampling procedure is impractical. Eventually, the same results may be achieved by means of random transects using a mobile instrument for measuring ice thickness (Michel, 1971; Campbell and Orange, 1974) or by a sensitive remote-sensing method still being developed.   These methods are discussed in various articles: Canada DRB (1971); Santeford and Smith (1974); Glaciology (1974).   Since much of the effort in the remote-sensing of ice has been diverted toward the (in some ways) more difficult problems associated with sea ice, the capability of differentiating between freshwater ice types as well as the measurement accuracy are unknown.

This brings us to the "what?" part of the general question posed. It is normal, in routine surveys, to measure the depth of snow and the total thickness of ice.   Reports tend to include notes on the types of ice observed only at freeze-up and break-up.   It is highly desirable that the method of measurement should at least differentiate black ice from white ice; and should ideally provide data on the hydrostatic water level (an indication of the flooding potential on a lake, Fig. 10.6) and the density of snowcover routinely. Depth and density of snow *together* provide a much more useful measure of the properties of ice (such as, strength, insulation properties, water equivalent) than thickness alone (Adams, 1976a).   Easily-measured volumetric properties of the cover are illustrated in Fig. 10.7.

Regarding snow on lakes, it is remarkable how little we know about the snowcover of that increasingly valuable portion of Canada occupied by freshwater bodies (or indeed of that which is under sea water).   Certain difficulties exist in the measurements of the distribution and characteristics of snowcover on land (Adams et al. 1976; Adams, 1977a), however, there are records of hundreds of daily snow-depth measurements across Canada, of numerous weekly and twice-monthly depth, density and water equivalent measurements along snow courses at a great variety of sites, and of occasional studies of snow stratigraphy on land and glaciers.   By contrast, on lakes, measurements of snow depth alone are available from isolated sites mainly for a few lakes chosen principally for convenience.   Since "white" ice is not recorded in these surveys, it is not possible to estimate the proportion of the winter's snow receipts incorporated into the ice sheet.

The work of Bryan (1974), Bryan and Larson (1975), and others using SLAR (Side Looking Airborne Radar) appears to offer considerable promise for differentiating all components of lake cover. However, limitations in the precision of measurement indicate that even if adequate observational coverage of peak ice conditions occurs in the foreseeable future, the monitoring of the evolution of an entire lake cover (very important in many studies of lake ice) would probably continue to be fairly laborious.

BLACK ICE

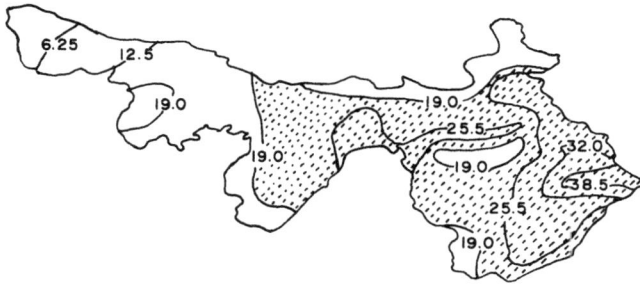

WHITE ICE

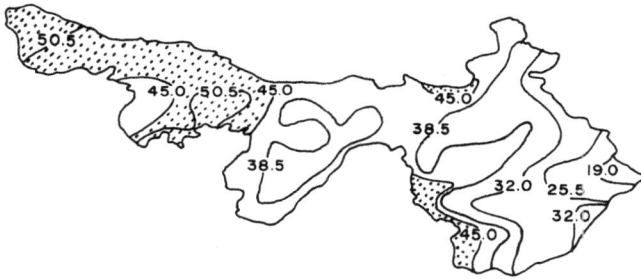

SNOW (water equivalent)

TOTAL ICE

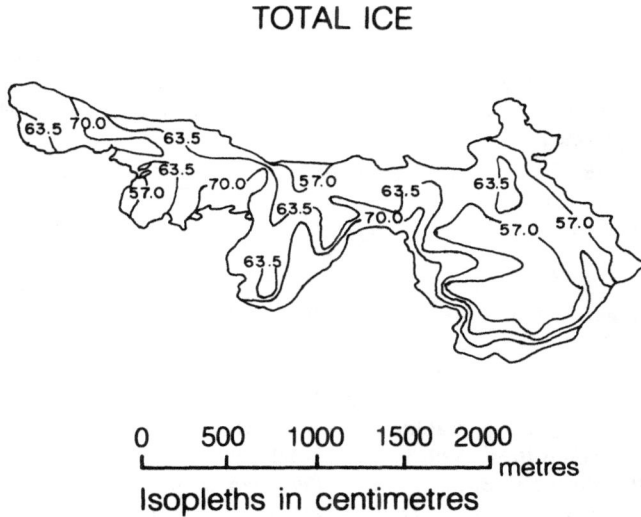

```
0     500   1000   1500  2000
L_____L_____L_____L_____J metres
```

Isopleths in centimetres

Fig. 10.7 An illustration of the spatial variability of lake cover (cm) - Gillies Lake, Bruce Peninsula, Ontario, spring 1971. The isopleths are in cm, the shaded areas are: > 19 cm, black ice; > 45 cm, white ice. The snowcover was thin and relatively uniform with a density < 300 kg/m³ while the white ice was thick and varied. The maps hint at the complementary black ice and white ice distributions. This suggested that a phase of white ice formation had recently been completed, consuming much of the original snowcover. Usually for a lake, the coefficient of variation of snowcover is the highest, followed in order by those of white ice, black ice, and total ice. However, the stage of evolution of the cover is important.

ACKNOWLEDGEMENTS

This chapter was developed from an article which appeared in *MUSK-OX* (Adams, 1976a). The author gratefully acknowledges the assistance of the Ice Climatology Division of the Canadian Atmospheric Environment Services, and the library staffs of the Arctic Institute of North America, Calgary, the Boreal Institute, Edmonton, and the Scott Polar Institute, Cambridge, England for supplying information. The author is also grateful to a number of students and colleagues at Trent University and McGill University, particularly Douglas Barr, and others for their comments while the manuscript was being drafted. Mrs. Grace Dyer and Mrs. Pat Strode typed the manuscript.

LITERATURE CITED

Adams, W.P. 1966. *Ablation and runoff on the White Glacier*. Axel Heiberg Island Res. Rep., Glaciol. No. 1, McGill Univ., Montreal, P.Q.

Adams, W.P. 1976a. *Diversity of lake cover and its implications*. Musk Ox, No. 18, pp. 86-98.

Adams, W.P. 1976b. *Field determination of the densities of lake ice sheets*. Limnol. Oceanogr., Vol. 21, No. 4, pp. 602-608.

Adams, W.P. 1976c. *A classification of freshwater ice*. Musk Ox, No. 18, pp. 99-102.

Adams, W.P. 1977a. *Limitations of the bulk density method of snow course measurement*. J. Soil Water Conserv., No. 32, Vol. 3, pp. 135-137.

Adams, W.P. 1977b. *How spring ice breakups alter our shorelines*. Can. Geogr. J., Vol. 94, No. 2, pp. 62-65.

Adams, W.P. and A.G. Brunger. 1972. *Sampling a subarctic lake cover*. Int. Geography/La Géographie Int. (W.P. Adams and F.M. Helleiner, eds.), Univ. of Toronto Press, Vol. 1, pp. 222-226.

Adams, W.P. and A.G. Brunger. 1975. *Variation in the quality and thickness of the winter cover of Knob Lake, Subarctic Quebec*. Révue de Géographie de Montréal, Vol. XXIX, No. 4, pp. 335-346.

Adams, W.P. and J.A.A.A. Jones. 1971. *Observations in the crystallographic relations of white ice and black ice*. Geophysica, No. 11, Vol. 2, pp. 151-163.

Adams, W.P. and S.A. Mathewson. 1976. *Approaches to the study of ice-push features with reference to Gillies Lake, Ontario*. Révue de Géographie de Montréal, Vol. XXX, Nos. 1-2, pp. 187-196.

Adams, W.P. and R.J. Rogerson. 1968. *Snowfall and snowcover at Knob Lake*. Proc. 25th Annu. Meet., East. Snow Conf., pp. 110-139.

Adams, W.P. and J.B. Shaw. 1966a. *The bathymetry and ice cover of lakes in the Schefferville area*. McGill Sub-Arctic Res. Pap. No. 21, pp. 201-212.

Adams, W.P. and J.B. Shaw. 1966b. *Improvements in the measurement of a lake ice cover*. J. Glaciol., Vol. 6, No. 44, pp. 299-302.

Adams, W.P., B.F. Findlay and B.E. Goodison. 1976. *Improving the data base of snow science*. Proc. 33rd Annu. Meet., East. Snow Conf., pp. 23-40.

Ager, B.H. 1961. *Snow roads and ice landings*. Tech. Pap. 127, Nat. Res. Counc. Can., Div. Build. Res., Ottawa, Ont., pp. 137-146.

Ager, B.H. 1962. *Studies on the density of naturally and artificially formed freshwater ice*. J. Glaciol., Vol. 4, pp. 207-214.

Albrecht, M.L. 1964. *Die Lichtdurchlässigkeit von Eis und Schnee und ihre Bedentung für die Sauerstoffproduktion im Wasser*. Dtsch. Fisch.-Ztg., Vol. 11, pp. 371-376.

Allen, W.T.R. 1964. *Break-up and freeze-up dates in Canada*. CIR 4116, Ice-17, Can. Dept. Transp., Meteorol. Branch, Toronto, Ont.

Allen, W.T.R. and B.S.V. Cudbird. 1971. *Freeze-up and break-up dates of water bodies in Canada*. CLI-1-71., Can. Meteorol. Serv., Downsview, Ont.

Andrews, J.T. 1962. *Variability of lake-ice growth and quality in the Schefferville region, Central Labrador-Ungava*. J. Glaciol., Vol. 4, No. 33, pp. 337-347.

Archer, D.R. and B.F. Findlay. 1966. *Comments on littoral ice conditions at one site on Knob Lake*. McGill Sub-Arctic Res. Pap. No. 21, pp. 189-190.

Assur, A. 1968. *Discussion of bearing capacity of floating ice sheets*. Trans. Am. Soc. Civil Eng., Vol. 127, Part 1, pp. 563-566.

Bader, H. 1939. *Mineralogical and structural characteristics of snow and its metamorphism*. In Snow and its Metamorphism [English transl. by U.S. Army Snow, Ice Permafrost Res. Estab., Tech. Transl. No. 14, pp. 3-56.]

Barnes, D.F. 1959. *Preliminary report on Lake Peters, Alaska ice studies*. Proc. 2nd Annu. Arctic Planning Sessions, Geophys. Res. Dir., U.S. Air Force Cambridge Res. Center, pp. 102-110.

Barnes, D.F. 1960. *An investigation of a perennially frozen lake*. Rept. 129, Geophys. Res. Inst., Air Force Survey in Geophys., U.S. Air Force Cambridge Res. Center, Cambridge, Mass.

Barica, J. and F.A.J. Armstrong. 1971. *Contribution of snow to the nutrient budget of some small northwest Ontario lakes*. Limnol. Oceanogr., Vol. 16, pp. 891-899.

Bilello, M.A. 1964a. *Ice prediction curves for lake and river locations in Canada*. Res. Rep. 129, U.S. Army Cold Reg. Res. Eng. Lab., Hanover, N.H.

Bilello, M.A. 1964b. *Method of prediction of river and lake ice formation*. J. Appl. Meteorol., Vol. 3, No. 38, pp. 38-44.

Bilello, M.A. 1968. *Water temperatures in a shallow lake during ice formation, growth and decay*. Water Resour. Res., Vol. 4, pp. 749-760.

Bilello, M.A., W.P. Adams and J.B. Shaw. 1966. *Prediction of ice formation on Knob and Maryjo Lakes, Schefferville*, McGill Sub-Arctic Res. Pap. No. 21, pp. 213-225.

Bilello, M.A. et al. 1961-75. *Ice thickness observations, North American Arctic and Subarctic*. Special Rep. 43/1, 2, 3, 4, 5, 6, 7, U.S. Army Cold Reg. Res. Eng. Lab., Hanover, N.H.

Bilello, M.A. 1974. *Surface measurements of snow and ice for correlation with data collected by remote systems*. Adv. Concepts Tech. Study Snow Ice Resourc., Interdiscip. Symp., U.S. Nat. Acad. Sci., Washington, D.C., pp. 283-293.

Brewer, M.C. 1958. *The thermal regime of an arctic lake*. Trans. Am. Geophys. Union, Vol. 39, No. 2, pp. 278-284.

Bryan, M.L. 1974. *Ice thickness and variability on Silver Lake, Genesee County Michigan: a radar approach*. Adv. Concepts Tech. Study Snow Ice Resourc., Interdiscip. Symp., U.S. Nat. Acad. Sci., Washington, D.C., pp. 213-222.

Bryan, M.L. and R.W. Larson. 1975. *The study of freshwater lake ice using multiplexed imaging radar*. J. Glaciol., Vol. 14, No. 72, pp. 445-457.

Burbidge, F.E. and J.R. Launder. 1957. *A preliminary investigation into break-up and freeze-up conditions in Canada*. Cir. 2939, Tec-252, Can. Dept. Transp., Meteorol. Branch, Toronto, Ont.

Burgi, P.H., J.M. Childers, G. Frankenstein, J.F. Kennedy, and G.A. Ashton. 1974. *River ice problems: a state of the art survey and assessment of research needs.* J. Hydraul. Div., Proc. Am. Soc. Civil Eng. 100, HYI, pp. 1-15.

Butiagin, I.P. 1966. *Strength of ice and ice cover.* Izdatel'stuo "nauka" Sibirskoe Otdelenie, Novosibirsk, pp. 1-54.

Butkovitch, T.R. 1954. *Ultimate strength of ice.* Res. Pap. 11, U.S. Army Snow, Ice Permafrost Res. Estab.

Campbell, K.J. and A.J. Orange. 1974. *Continuous sea and freshwater ice thickness profiling using an impulse radar system.* Adv. Concepts Tech. Study Snow Ice Resourc., Interdiscip. Symp., U.S. Nat. Acad. Sci., Washington, D.C., pp. 432-442.

Canada, AES. 1958 - present. *Ice thickness data for Canadian selected stations.* Circ. 4372. Continuing series, Atmos. Environ. Serv., Downsview, Ont.

Canada AES. 1969. *Manice, 3rd Provisional Ed.*, Atmos. Environ. Serv., Downsview, Ont.

Canada, DRB. 1971. *Proc. of a seminar on thickness measurement of floating ice by remote sensing.* Def. Res. Board, Ottawa, Ont., Oct. 1970.

Carey, K.L. 1973. *Icings developed from surface water and groundwater.* Mono. III-D3, U.S. Army Cold Reg. Res. Eng. Lab., Hanover, N.H.

Catchpole, A.J.W., D.W. Moodie and D. Milton. 1976. *Freeze-up and break-up of estuaries on Hudson Bay in the eighteenth and nineteenth centuries.* Can. Geogr., Vol. XX, No. 31 pp. 279-297.

Coble, R.L. and W.D. Kingery. 1963. *Ice reinforcement.* In Ice and Snow (W.D. Kingery, ed.), Mass. Inst. Tech. Press, Cambridge, Mass., pp. 130-148.

Davies, J.A. 1962. *Albedo measurements over subarctic surfaces.* McGill Sub-Arctic Res. Pap. No. 13.

Dickens, H.B. 1959. *Water supply and sewage disposal in permafrost areas of Northern Canada.* Polar Rec., Vol. 9, No. 62, pp. 421-432.

Dionne, J.C. 1969. *An annotated bibliography of "glaciel" studies, morphosedimentological aspects.* Révue de Géographie de Montréal, Vol. XXIII, No. 3, pp. 339-349.

Dionne, J.C. 1974. *Bibliographie annotée sur les aspects géologiques du glaciel.* Rapport d'Information LAN-X-9, Centre de Recherches Forestières des Laurentides, Ste-Foy, P.Q.

Dunbar, M.J. 1968. *Ecological Development in Polar Regions.* Prentice-Hall, N.J.

Dutton, J.A. and R.A. Bryson. 1962. *Heat flux in Lake Mendota.* Limnol. Oceanogr., Vol. 7, No. 1, pp. 80-97.

English, M. 1976. *Phosphorus input into a lake from stored winter precipitation.* Trent Stud. Geogr., Trent Univ., Vol. 5, pp. 56-68.

Findlay, B.F. 1966. *The water budget of the Knob Lake area: A hydrological study in Central Labrador-Ungava.* McGill Sub-Arctic Res. Pap. No. 22, pp. 1-95.

Findlay, B.F. 1969. *Precipitation in Northern Quebec and Labrador: an evaluation of measurement techniques.* Arctic, Vol. 22, No. 2, pp. 140-150.

Frankenstein, G.E. 1959. *Strength data on lake ice.* Tech. Rep. 59, U.S. Army Snow, Ice Permafrost Res. Estab.

Frankenstein, G.E. 1969. *Ring tensile strength studies of ice.* Tech. Rep. 172, U.S. Army Cold Reg. Res. Eng. Lab., Hanover, N.H.

Gaitskhoki, B.Y. 1970a. *A photometric model of the snow-ice cover.* The Physics of Ice (V.V. Bogarodskii, ed.) Vol. 295. pp. 48-52 [English transl. by Israel Prog. Sci. Transl., Transl. No. 5849].

Gaitskhoki, B.Y. 1970b. *Spectral transmission of snow and some ice varieties.* The Physics of Ice (V.V. Bogarodskii, ed.) Vol. 295. pp. 44-47 [English Transl. by Israel Prog. Sci.Transl., Transl. No. 5849].

Gaitskhoki, B.Y. 1975. *Optical characteristics of certain varieties of natural ice.* In Physical Methods of Investigation of Ice and Snow (V.V. Bogarodskii and V.P. Gaurilo, eds.), [English Transl. by U.S. Nat. Aeronaut. Space Admin., Transl. II, F-D, 009, 1976, pp. 71-73].

Giddings, J.C. and E. LaChapelle. 1961. *Diffusion theory applied to radiant energy distribution and albedo of snow.* J. Geophys. Res., Vol. 66, pp. 181-189.

Glaciology, J. 1974. *Symposium on remote sensing in Glaciology.* J. Glaciol., Vol. 15, No. 73.

Glen, J.W. 1974. *The physics of ice.* Mono. II-C2a, U.S. Army Cold Reg. Res. Eng. Lab., Hanover, N.H.

Glen, J.W. 1975. *The mechanics of ice.* Mono. II-C2b, U.S. Army Cold Reg. Res. Eng. Lab., Hanover, N.H.

Gold, L.W. 1960. *Field study on load bearing capacity of ice curves.* Pulp Pap. Mag. Can., Woodlands Section.

Gold, L.W. 1966. *Dependence of crack formation on crystallographic orientation for ice.* Can. J. Phys., Vol. 44, pp. 2757-2764.

Gold, L.W. 1967. *Time to formation of first cracks in ice.* In Physics of Snow and Ice, (H. Oura, ed.), Inst. Low Temp. Sci., Hokkaido Univ., Sapporo, Vol. 1, No. 1, pp. 360-370.

Gold, L.W. 1968. *Elastic and strength properties of freshwater ice.* Ice pressures against structures, NRC No. 9851, Nat. Res. Counc. Can., Assoc. Comm. Geotech. Res., Ottawa, Ont., pp. 13-23.

Gold, L.W. and G.P. Williams, eds. 1968. *Ice pressures against structures.* Tech. Memo. 92, Nat. Res. Counc. Can., Assoc. Comm. Geotech. Res., Ottawa, Ont.

Hagen, A. and A. Langeland. 1973. *Polluted snow in southern Norway and the effect of the meltwater on freshwater and aquatic agencies.* Environ. Pollut. Vol. 5, No. 1, pp. 45-57.

Hamelin, L.E. 1960. *Classification générale des glaces flottantes.* Naturaliste Canadien, Vol. LXXXVII, No. 10, pp. 209-227.

Hansen, K. 1967. *The general limnology of arctic lakes as illustrated by examples from Greenland.* Medd. om Gronland, Vol. 178, No. 3.

Harden, D., P. Barnes and E. Reimnitz. 1977. *Distribution and character of naleds in northeastern Alaska.* Arctic, Vol. 30, No. 1, pp. 28-40.

Hart, S.R. and J.S. Steinhart. 1965. *Terrestrial heat flow: measurements in lake bottoms.* Science, Vol. 149, pp. 1499-1501.

Hathersley-Smith, G., J.E. Keys and H. Serson, 1970. *Density stratified lakes in northern Ellesmere Island.* Nature, Vol. 225, No. 5527, pp. 55-56.

Hendy, C.H., M.J. Selby and A.T. Wilson. 1972. *Deep Lake, Cape Barnes, Antarctica.* Limnol. Oceanogr., Vol. 17, No. 3, pp. 356-362.

Hobbs, P.V. 1974. *Physics.* Clarendon Press, Oxford.

Houle, J.L. 1961. *Two instruments for the measurement of ice thickness.* DRNL Tech. Note S/61, Can. Def. Res. Board, Ottawa, Ont.

Hultberg, H. 1976. *Thermally stratified acid water in later winter - a key factor inducing self-accelerating processes which increase acidification.* Water, Air, Soil Pollut. Vol. 7, pp. 279-294.

Hutchinson, G.E. 1957. *A Treatise on Limnology.* John Wiley and Sons, New York.

Johnson, N.M. and G.E. Likens. 1967. *Steady-state thermal gradient in the sediments of a micromictic lake.* J. Geophys. Res., Vol. 72, pp. 3049-3052.

Jones, J.A.A.A. 1969. *The growth and significance of white ice at Knob Lake, Quebec.* Can. Geogr., Vol. 13, pp. 354-372.

Jones, J.A.A.A. 1970a. *The growth and significance of white ice at Knob Lake, Quebec.* McGill Sub-Arctic Res. Pap. No. 25, pp. 6-160.

Jones, J.A.A.A. 1970b. *Ice-shove - a review with particular reference to the Knob Lake area.* McGill Sub-Arctic Res. Paper No. 25, pp. 223-231.

Kalff, J. and H.E. Welch. 1974. *Phytoplankton production in Char Lake, a natural polar lake, and in Maretta Lake, a polluted polar lake, Cornwallis Island, N.W.T.* J. Fish. Res. Board Can., Vol. 31, No. 5, pp. 621-636.

King. R. 1973. *Winter heating lakes.* Trent Stud. Geogr., Trent Univ., Peterborough, Ont., Vol. 2, pp. 11-19.

Kivisild, H.R. 1970. *River and lake ice terminology.* Int. Assoc. Hydraul. Res., Symp. Ice and its Hydraulic Action, Reykjavik, Iceland, Vol. 1, pp. 1-14.

Knight, C.A. 1962. *Studies of arctic lake ice.* J. Glaciol. Vol. 4, pp. 319-335.

LaChapelle, E.R. 1977. *Field guide to Snow Crystals.* Univ. of Washington Press, Seattle.

Lavrov, V.V. 1969. *Deformation and strength of ice.* [English Transl. by Israel Prog. Sci. Trans., Transl. No. 5824].

Likens, G.E. and N.M. Johnson. 1969. *Measurement and analysis of the annual heat budget for the sediments in two Wisconsin lakes.* Limnol. Oceanogr., Vol. 14, No. 1, pp. 115-135.

Likens, G.E. and R.A. Ragotskie. 1965. *Vertical winter motions in a small ice -covered lake.* J. Geophys. Res., Vol. 70, pp. 2333-2344.

Lliboutry, L. 1964. *Traité de Glaciologie*, Tome I, Masson et Cie, Paris.

Lyons, J.B. and R.E. Stoiber. 1959. *Orientation fabrics in lake ice.* J. Glaciol., Vol. 4, No. 32, pp. 367-370.

MacKay, D.K. and O.H. Löken. 1974. *Arctic hydrology.* In Arctic and Alpine Environment (J.R. Ives and R.G. Barry, eds.), Methuen, London, pp. 111-132.

Maguire, R.J. 1975a. *Effect of ice and snowcover on transmission of light in lakes.* Sci. Ser. No. 54, Environ. Can., Inland Waters Dir., Ottawa, Ont.

Maguire, R.J. 1975b. *Light transmission through snow and ice.* Tech. Bull. No. 91, Environ. Can., Inland Waters Dir., Ottawa, Ont.

Maguire, R.J. and N. Watkin. 1975. *Effects of ice cover on dissolved oxygen in Silver Lake, Ontario.* Tech. Bull. No. 89, Environ. Can., Inland Waters Dir., Ottawa, Ont.

Mantis, H.T., ed. 1951. *Review of the properties of snow and ice.* Rep. No. 4, U.S. Army Snow, Ice Permafrost Res. Estab.

Mellor, M. 1962. *The physics and mechanics of snow as a material.* Mono. II-B, U.S. Army Cold Reg. Res. Eng. Lab., Hanover, N.H.

Mellor, M. 1964a. *Snow and ice on the earth's surface.* Mono. II-C1, U.S. Army Cold Reg. Res. Eng. Lab., Hanover, N.H.

Mellor, M. 1964b. *Properties of snow.* Mono. III-A1, U.S. Army Cold Reg. Res. Eng. Lab., Hanover, N.H.

Michel, B. 1963. *Theory of formation and deposition of frazil ice.* Proc. 20th Annu. Meet., East. Snow Conf., pp. 130-148.

Michel, B. 1967. *Morphology of frazil ice.* Physics of Snow and Ice (H. Oura, ed.), Inst. Low Temp. Sci., Hokkaido Univ., Sapporo, Vol. 1, No. 1, pp. 119-128.

Michel, B. 1970. *Ice pressure on engineering structures.* Mono. III-B1b, U.S. Army Cold Reg. Res. Eng. Lab., Hanover, N.H.

Michel, B. 1971. *Winter regime of rivers and lakes.* Mono. III - B1a, U.S. Army Cold Reg. Res. Eng. Lab., Hanover, N.H.

Michel, B. and R.O. Ramseier, 1971. *Classification of river and lake ice.* Can. Geotech. J., Vol. 8, pp. 36-45.

Montfore, G.E. and F.W. Taylor. 1949. *The problem of an expanding ice sheet.* Proc. 19th Annu. Meet., West. Snow Conf., Vol. 11, pp. 30-46.

Montréal, Révue de Géographie. 1976. *Proc. 1st Int. Symp. Geological Action of Drift Ice,* Vol. XXX, Nos. 1-2.

Muguruma, J. and K. Kikuchi, 1963. *Lake ice investigation at Peters Lake, Alaska.* J. Glaciol., Vol. 4, No. 36, pp. 689-708.

Nakaya, U. 1954. *Snow Crystals, Natural and Artificial.* Harvard University Press, Cambridge, Mass.

Nevel, D.E. 1970. *Moving loads on a floating ice sheet.* Res. Rep. 261, U.S. Army Cold Reg. Res. Eng. Lab., Hanover, N.H.

Nevel, D.E. and A. Assur. 1968. *Crowds on ice.* Tech. Rep. 204, U.S. Army Cold Reg. Res. Eng. Lab., Hanover, N.H.

Ontario Ministry of Natural Resources, 1973. *Ice safety.* In Safety Bits and Pieces Ser., Toronto.

Osterkamp, T.E. 1974. *Waste-water sludge ice.* J. Glaciol., Vol. 13, pp. 155-157.

Palosuo, E. 1961. *Crystal structure of brackish and fresh-water ice.* Int. Union Geod. Geophy., Gen. Assem. Helsinki, [Snow and Ice], Int. Assoc. Sci. Hydrol. Publ. 54, pp. 9-14.

Palosuo, E. 1965. *Frozen slush on lake ice.* Geophysica, Vol. 9, No. 2, pp. 131-147.

Parott, W.H. and W.M. Fleming. 1970. *The temperature structure of a mid-latitude dimictic lake during freezing, ice cover and thawing.* Res. Rep. 291, U.S. Army Cold Reg. Res. Eng. Lab., Hanover, N.H.

Penn, A.F. 1970. *Lake water temperatures in relation to ice cover near Schefferville, P.Q.* McGill Sub-Arctic Res. Pap. No. 25, pp. 204-222.

Perey, F.G.J. and E.R. Pounder. 1958. *Crystal orientation in ice sheets.* Can. J. Phys., Vol. 36, pp. 494-502.

Pessl, F. 1969. *Formation of a modern ice-push-ridge by thermal expansion of lake ice in southeastern Connecticut.* Res. Rep. 259, U.S. Army Cold Reg. Res. Eng. Lab., Hanover, N.H.

Pivovarov, A.A. 1973. *Thermal conditions in freezing lakes and rivers.* [English Transl. by Israel Prog. Sci. Transl., Publ. by Halstead Press, John Wiley, N.Y.].

Polyakova, K.N. 1966. *Characteristics of the melting of the ice cover and the opening of the middle Lena River,* Sov. Hydrol., No. 3, pp. 276 - 292.

Pounder, E.R. 1965. *Physics of Ice.* Pergamon Press, Toronto, Ont.

Pounder, E.R. 1969. *Strength and growth rates of sea ice.* Ice Seminar, Can. Inst. Mining and Metallurgy, Sp. Vol. 10, pp. 73-76.

Ragle, R.H. 1963. *Formation of lake ice in a temperate climate.* Res. Rep. 107, U.S. Army Cold Reg. Res. Eng. Lab., Hanover, N.H.

Ramseier, R.O. and R.J. Weaver. 1975. *Floating ice thickness and structure determination—heated wire technique.* Tech. Bull. 88, Environ. Can., Ottawa.

Rigler, F.H. 1974. *Char Lake Project Final Report.* Can. Comm. Int. Biol. Prog., Toronto, Ont.

Rodhe, B. 1952. *On the relation between air temperature and ice formation in Baltic.* Geogr. Annaler, Vol. 34, Nos. 3-4, pp. 175-202.

Ruttner, F. 1968. *Fundamentals of Limnology.* 3rd Ed., Univ. of Toronto Press, Toronto, Ont.

Santeford, H.S. and J.L. Smith. (compilers) 1974. *Advanced Concepts and Techniques in the Study of Snow and Ice Resources.* Nat. Acad. Sci., Washington, D.C.

Schindler, D.W. 1971. *Light, temperature and oxygen regimes in the experimental lakes area, N.W. Ontario.* J. Fish. Res. Board Can., Vol. 28, pp. 157-169.

Schindler, D.W., H.E. Welch, J. Kalff, G.J. Brunskill and N. Kritsch. 1974a. *Physical and chemical limnology of Char Lake, Cornwallis Island (75° N lat.)* J. Fish Res. Board Can., Vol. 31, No. 5, pp. 585-607.

Schindler, D.W., J. Kalff, H.E. Welch, G.J. Brunskill and N. Kritsch. 1974b. *Eutrophication in the high arctic - Meretta lake, Cornwallis Island (75° N lat.).* J. Fish. Res. Board Can., Vol. 31, No. 5, pp. 647-662.

Schwerdtfeger, P. 1963. *The thermal properties of sea ice.* J. Glaciol., Vol. 4, No. 36, pp. 789-807.

Scott, J.T. 1964. *A comparison of the heat balance of lakes in winter.* Tech. Rep. 13, Univ. Wisc., Dept. Meteorol., Madison, Wisc.

Scott, J.T. and R.A. Ragotskie. 1961. *Heat budget of an ice covered inland lake.* Tech. Rep. 6, Univ. Wisc., Dept. Meteorol., Madison, Wisc.

Shaw, J.B. 1965. *Growth and decay of lake ice in the vicinity of Schefferville, Quebec.* Arctic, Vol. 18, No. 2, pp. 123-132.

Shulyakovskii, L.G., ed. 1963. *Manual of forecasting ice formation for rivers and inland lakes*. [English Transl. by Israel Prog. Sci. Transl., Transl. No. 1553].

Shumskii, P.A. 1964. *Principles of Structural Glaciology*. [Transl. by D. Kraus], Dover Press, New York, N.Y.

Sommerfeld, R.A. and E. LaChapelle. 1970. *The classification of snow metamorphism*. J. Glaciol., Vol. 9, No. 55, pp. 3-17.

Stearns, S.R. 1964. *Flexural properties of snow and snow ice*. Sp. Rep. 59, U.S. Army Cold Reg. Res. Eng. Lab., Hanover, N.H.

Stevens, H.W. and W.J. Tizzard. 1969. *Traffic tests on Portage Lake ice*. Tech. Rep. 99, U.S. Army Cold Reg. Res. Eng. Lab., Hanover, N.H.

Swinzow, G.K. 1966. *Ice cover of an arctic proglacial lake*. Res. Rep. 155, U.S. Army Cold Reg. Res. Eng. Lab., Hanover, N.H.

Sychev, K. 1940. *Register of ice thickness*. Sov. Arktika, Vol. 5, pp. 1-4 (in Russian).

Thomas, C.W. 1963. *On the transfer of visible radiation through sea ice and snow*. J. Glaciol., Vol. 4, No. 34, pp. 481-484.

Tsang, G. 1974. *Ice pilings on lake shores*. Sci. Ser. 35, Environ. Can., Canada Centre for Inland Waters, Burlington, Ont.

Vesso, J.J. 1957. *An instrument for the measurement of ice thickness in the field*. NL Tech. Note 8/75, Can. Def. Res. Board, Ottawa, Ont.

Unesco/IAHS. 1972. *Guide to world inventory of sea, lake and river ice cover*. Tech. Papers in Hydrol., 9, Paris.

Wassen, G. 1966. *Gardiken. Vegetation und flora eines lappländischen Seeufers*. Kungl. Ventenskaps-akademlens avhandlinger i naturskyddsarenden, Vol. 22.

Wassen, G. 1969. *Some aspects of lakeshore vegetation in central Labrador-Ungava*. McGill Sub-Arctic Res. Pap. No. 24, pp. 7-32.

Welch, P.S. 1952. *Limnology*. (2nd ed.), McGraw-Hill, New York, N.Y.

Welch, H.E. 1974. *Metabolic rates of arctic lakes*. Limnol. Oceanogr., Vol. 19, No. 1, pp. 65-73.

Welch, H.E. and J. Kalff. 1974. *Benthic photosynthesis and respiration in Char Lake*. J. Fish. Res. Board Can., Vol. 31, No. 5, pp. 609-620.

Weeks, W.F. and A. Assur. 1969. *Fracture of lake and sea ice*. Res. Rep. 269, U.S. Army Cold Reg. Res. Eng. Lab., Hanover, N.H.

Weller, G.E. and P. Schwerdtfeger. 1971. *New data on the thermal conductivity of natural snow*. J. Glaciol., Vol. 10, No. 59, pp. 309-311.

Wetzel, R.G. 1975. *Limnology*. Saunders, Toronto, Ont.

Williams, G.P. 1959. *Frazil ice, a review of its properties with a selected bibliography*. Eng. J., Vol. 42, No. 11, pp. 55-60.

Williams, G.P. 1963. *Probability charts for predicting ice thickness*. Eng. J., Vol. 46, No. 6, pp. 31-35.

Williams, G.P. 1965. *Correlating freeze-up and break-up with weather conditions*. Can. Geotech. J., Vol. II, No. 4, pp. 313-326.

Williams, G.P. 1967. *Ice-dusting experiments to increase the rate of melting of ice*. Tech. Pap. 239, NRC 9349, Nat. Res. Counc. Can., Div. Build. Res., Ottawa, Ont.

Williams, G.P. 1968. *Freeze-up and break-up of freshwater lakes.* Tech. Memo. 92, Nat. Res. Counc. Can., Div. Build. Res., Ottawa, Ont., pp. 217-229.

Williams, G.P. 1969. *Water temperature during the melting of lake ice.* Water Resour. Res., Vol. 5, No. 5, pp. 1134-1138.

Williams, G.P. 1971a. *Predicting the data of lake ice break-up.* Water Resour. Res., Vol. 7, No. 1, pp. 323-333.

Williams, G.P. 1971b. *The effect of lake and river ice on snowmelt runoff.* Hydrol. Symp. No. 8, Inland Waters Branch, Can. Energy Mines Resour., Ottawa, Ont., pp. 61-80.

Williams, G.P. and L.W. Gold. 1963. *The use of dust to advance the break-up of ice on lakes and rivers in Canada.* Proc. 20th Annu. Meet.. East. Snow Conf., pp. 31-53.

Willmot, J.L. 1962. *A safe ice load meter and a motor-driven coring tool for estimating bearing capacity of an ice sheet for transportation purposes.* Proc. 19th Annu. Meet., East. Snow Conf., pp. 149-162.

Wilson, A.T. and H.W. Wellman. 1972. *Lake Vanda: an Antarctic lake.* Nature, Vol. 196, pp. 1171-1173.

Wilson, J.T., J.H. Zumberge and E.W. Marshall. 1954. *A study of ice on an island lake.* Rep. 5, Part I, U.S. Army Snow, Ice Permafrost Res. Estab.

Wright, R.T. 1964. *Dynamics of a phytoplankton community in an ice-covered lake.* Limnol. Oceanogr., Vol. 9, pp. 163-178.

Zumberge, J.H. and J.C. Ayers. 1964. *Hydrology of lakes and swamps.* In Handbook of Applied Hydrology (V.T. Chow, ed.), Mc-Graw Hill, New York, pp. 23-1 - 23-33.

Zumberge, J.H. and J.T. Wilson. 1952. *Ice push studies on Wampler's Lake, Michigan.* Geol. Soc. Am. Bull., Vol. 63, p. 1318.

Zumberge, J.H. and J.T. Wilson. 1953. *Quantitative studies on thermal expansion and contraction of lake ice.* J. Geol., Vol. 61, No. 4, pp. 374-383.

11

AVALANCHES

P. A. SCHAERER

Division of Building Research, National Research Council of Canada, Vancouver, British Columbia.

INTRODUCTION

An avalanche is a rapid downslope movement of a large mass of snow. Snow-slide is a term used to describe the same phenomenon, although the term avalanche is becoming more common. The essential elements for avalanches are deep snow and steep slopes, but owing to a great variation of snow and terrain conditions, avalanches appear in various forms and sizes.

Publications dealing with specific technical aspects of avalanches are referenced within this chapter. Books treating all aspects of avalanches comprehensively include those of Perla and Martinelli (1976) and Fraser (1978).

The start of avalanches

The start of an avalanche is the outcome of a contest between the stress and strength within the snowpack on an incline - the strength being the loser. When snow lies on a slope, the force parallel to the slope caused by its weight produces shear stresses while the force perpendicular to the slope produces compressive stresses (see Fig. 11.1). Because the snow deforms readily under its load resulting in settlement, creep, and glide (see Fig. 11.2), additional tensile and compressive stresses appear at anchor points such as rocks, trees, and flat parts of the terrain (see Fig. 11.3). Failure occurs when the stress exceeds the strength at some point. Stresses in a snowpack are increased by the weight of additional snowfall or the accumulation of drifting snow, but a decrease of strength most frequently results from a rise in temperature. After the snow has failed at one point the stresses are concentrated in immediately adjacent areas, which in turn become overloaded and fail, causing a rupture to propagate over a wide area.

The stress-strength relationship of an inclined snowpack is very complicated because the snow is layered and viscous, its mechanical properties depend on temperature, the slope is variegated, and the snow depth is not uniform. Because of these factors, it is impractical to attempt a rigorous,

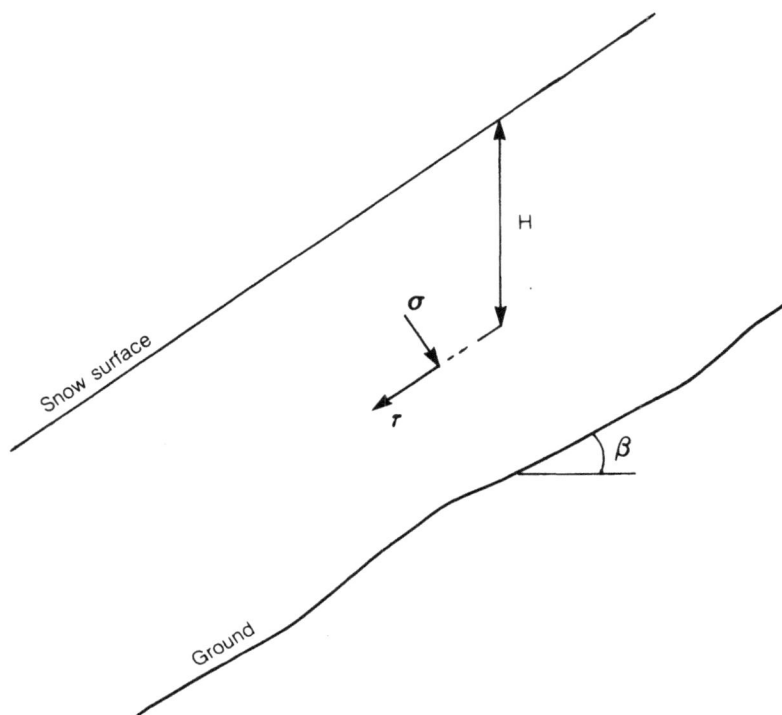

Fig. 11.1 Stresses in a snowpack due to weight.

τ = a shear stress parallel to the ground at any depth H in the snowpack ($\tau = \rho gH \sin\beta$),

σ = normal stress (to slope) at any depth H in the snowpack, ($\sigma = \rho gH \cos\beta$),

ρ = average density of snow at depth H,

g = acceleration due to gravity, and

β = slope angle.

theoretical analysis of the stability of a given snowpack. Predictions of the start of avalanches are made, therefore, on an intuitive and empirical basis.

Motion

The dislodged snow usually accelerates rapidly on a steep slope and, as it moves downhill, breaks delicately balanced snow in its path. The character of the motion depends on the type of snow and the terrain. During the initial stage of movement the snow has a gliding and rolling motion which becomes turbulent as the speed increases. Large blocks break up into masses of rounded and pulverized material. If the snow is dry, the fine particles mix with the air to form a powder. The component that follows the ground is known as a *flowing avalanche*; that carried by the turbulent air motion, a *powder avalanche*. Often both forms are present. A drop in the terrain may cause all the snow to mix with air producing a pure powder avalanche.

On a steep slope an avalanche may attain a high speed and exert great pressures on obstacles in its path, thereby becoming destructive. As the slope of the terrain decreases the avalanche decelerates and finally stops.

Terrain

An *avalanche path* is the specific locality in which a snow mass moves, and is generally divided into the *starting zone* at the top where the snow initially breaks away, the *runout zone* at the bottom where the snow decelerates and stops, and the *track* that connects the starting zone with the runout zone. On the track, the speed of the avalanche may increase, remain steady, or decrease, however, its mass remains more or less constant. Often there is no clear separation between each part.

Minimum inclines of around 25° are required to initiate avalanches and maintain their motion. The runout zone begins where the angle of the slope drops below this minimum value and can usually be recognized by a break in the terrain.

Avalanche hazard

Deep snow and steep slopes, the essential elements for initiating and propagating avalanches occur in mountain regions throughout the world. Although thousands occur every winter, the majority have no effect on human activities and are unnoticed. Hazards develop whenever man and his constructions are exposed to the sliding snow. They have existed ever since the mountains became inhabited, and in recent years have increased with the growth in recreational activities and traffic.

Protection

The most effective method of reducing the avalanche damage is to locate structures and outdoor activities properly. If avalanche paths cannot be avoided, either temporary safety measures must be instituted or the avalanche itself must be controlled. Avalanche controls influence its start or course and can be divided into two categories: those that modify the terrain and those that modify and stabilize snow.

Information concerning avalanches and ways of minimizing their destructive effects have been developed in several countries. The oldest and foremost avalanche research centre is the Federal Institute for Snow and Avalanche Research at Davos, Switzerland. In Canada, the Division of Building Research (National Research Council) and the Glaciology Division, Inland Waters Directorate (Fisheries and Environment Canada) are collecting information about the formation and dynamics of avalanches and are studying the feasibility of protective measures under Canadian conditions. In British Columbia avalanche control and safety programs are carried out at highways and ski areas of the National Parks, by the Department of Highways, at numerous ski developments, and by some mining companies.

THE START OF AVALANCHES

The properties of snow determining its strength are described in detail in Ch. 7. Those significant to avalanche formation are the layered structure of the snowpack and the strong dependence of layer strength on local temperature. Therefore, measurements of the characteristics of the snow layers, known as snow profile observations, including temperature are important for evaluating the stability of the pack and the formation of avalanches.

Effect of Metamorphism on Snowpack Properties

Under temperature conditions usually found in mountain areas, the snow on the ground goes through a change of structure known as metamorphism (see Ch. 7).

The newly deposited snow crystals - usually finely branched, star-shaped dendrites - first undergo equitemperature metamorphism which tends to produce more rounded crystals. With time, the average grain size decreases while the snowpack density increases resulting in an observable settlement of the snow. The process occurs at temperatures considerably below the melting point, but is more rapid at temperatures close to 0°C. Also, during metamorphism water moves as a vapor to the corners between the individual snow grains building ice necks between them. This process, known as sintering, produces strong bonds between the grains, so that the end product of equitemperature metamorphism is a dense, high-strength snow.

If a temperature difference exists over a very small depth of the snowpack, the snow undergoes temperature-gradient metamorphism, characterized by growth of selected snow crystals into angular, facetted shapes. The end product is depth hoar, sometimes called sugar snow. Temperature gradients produce large loosely-spaced crystals, very little sintering, and result in weakening of the snow. A strong temperature gradient is found usually in a shallow snowpack during cold weather, but also at the surface of deep snow, when - because of radiation heat loss - the surface is colder than the underlying snow. In the latter case, the surface snow changes into facetted grains having little cohesion.

Another important snow type, although not resulting from metamorphism, is surface hoar formed by deposition of water vapor from the atmosphere onto the cold snow. Surface hoar, which develops mainly during cold clear nights, consists of sparkling, angular crystals in layers a few millimeters thick, but sometimes reaches a thickness of 50 mm. Surface hoar has characteristics similar to those of depth hoar and has little strength.

If the temperature gradient necessary for metamorphism is not maintained because of a rise in temperature or increase in snow depth, depth hoar, facetted grains at the surface and surface hoar experience equitemperature metamorphism. In this process the rounding and sintering of the crystals, in contrast to dendritic new snow, is very slow, so that the snow remains in a loose, weak (strength) state for a long time.

Melt-freeze metamorphism occurs whenever fluctuations in air temperature cause repeated melting and refreezing. The result is the formation of large, irregular grains, known as corn snow when loose and wet. In its frozen state this snow forms a solid crust.

Usually layers of snow within a deep snowpack can be distinguished, reflecting differences in the stages of metamorphic development. Each layer tends to have its own mechanical and physical properties.

Snow Strength

Like other materials, snow has strength properties which can be tested by applying and measuring the forces required to cause a failure. Significant to the start of avalanches are the shear and tensile strengths. One difficulty with testing snow is its inherent fragile nature and low density which require that observations be carried out in situ with considerable care and skill; another is its viscoelastic behaviour, i.e. the strength depends on the rate of loading.

Because shear and tensile tests are time consuming and difficult to make, hardness is often used as an index of strength. Hardness is defined as the resistance of the snow to penetration by a specified object, and is related to the tensile, shear and compressive strengths.

Field methods have been developed for measuring the shear strength and hardness of snow. To determine shear strength a frame of area 0.01 m^2 or larger is inserted into the snow and rapidly loaded until failure occurs, the load being measured by a spring balance (Perla, 1977). Hardness observations are made either by means of ram penetrometers and spring-loaded plates to give quantitative indexes (Perla and Martinelli, 1976) or by means of simple hand held equipment to yield qualitative data (Commission on Snow and Ice, 1954).

Strength observations are generally not made within the starting zone of an avalanche because it is inaccessible and hazardous. Furthermore, since the strength varies widely within the starting zone it is difficult to obtain the value at the weakest point. Therefore, only index strength properties are monitored on selected safe test sites.

The strength of the snow is related to its temperature and density, and the type and bonding of the grains. Of primary concern is the strong dependence of strength on temperature, e.g., shear strength decreases as the temperature approaches 0°C. When the snow begins to melt and contains liquid water there is a further reduction of strength.

Conversely an increase in temperature accelerates equitemperature metamorphism which may increase the shear strength. However, if the temperature rise is rapid the increase in strength from metamorphism lags the decrease from warming so that the snow becomes weaker. If the temperature rises slowly, metamorphism produces a high strength snow which probably would not fail even at 0°C. These counteracting tendencies illustrate the complex nature of snow and the difficulty of making predictions about the time of failure.

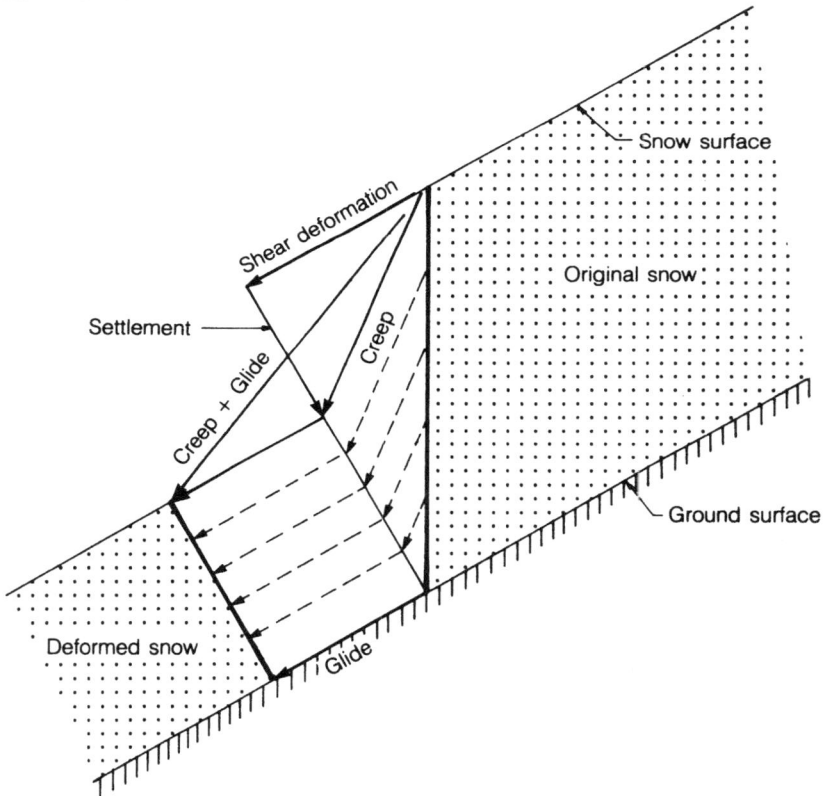

Fig. 11.2 Schematic diagram of creep and glide deformations.

Snow Creep

The normal and shear stresses in a sloping snowpack produce permanent deformations because of its viscous nature. Figure 11.2 shows the motion of the snow particles. In pure creep the snow adheres to the ground, while in gliding it moves along the ground. Gliding and creep occur together; often no distinction is made between the two motions.

The microrelief of the ground surface, vegetative cover, and slope have the most influence on creep and glide. High rates of glide have been observed on smooth, steep ground, e.g., steep, grass-covered slopes (In der Gand and Zupancic, 1966). The creep and glide velocities also increase with an increase of snow temperature, internal liquid water, and depth. Thus, the velocities change at places where features of the ground surface change, e.g., along a transition between smooth rock and boulders; a break of terrain; a change in snow depth (say between shallow snowcover and deep drifts). The variation in creep velocity along a slope produces different tensile and compressive stresses in the snow (see Fig. 11.3). It is important to identify zones of high stress because snow fractures and starts an avalanche at places of high tensile stress (see Fig. 11.4).

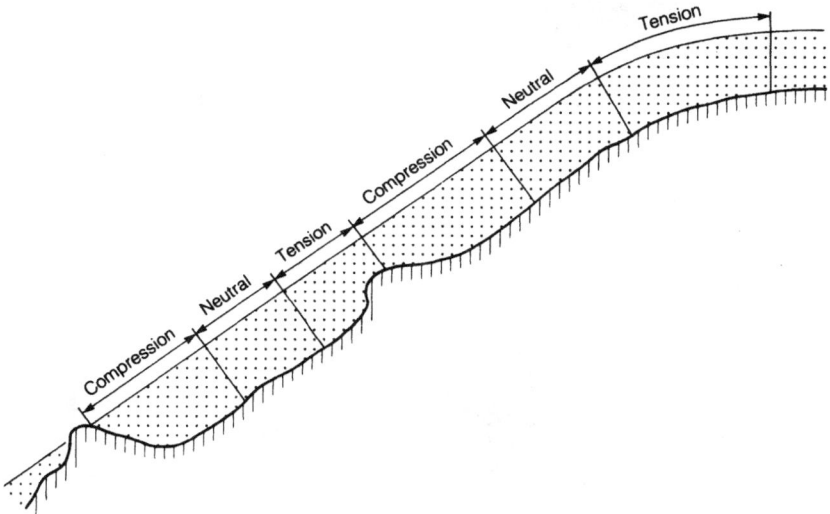

Fig. 11.3 Stress distributions developed under creep and glide deformations.

Failure of Snow Slopes

Two different types of failure are recognized, leading to two types of avalanches. Slab avalanches occur in cohesive snow; loose snow avalanches in cohesionless snow where intergranular bonding is negligible so that it behaves like dry sand.

It is not always possible to classify an avalanche as slab or loose snow. An avalanche may start as a loose snow avalanche on the surface, then the moving snow overloads weak layers deep inside the snowpack initiating a slab avalanche farther down the slope.

Slab avalanches develop in snow where a relatively strong layer overrides a weaker one, sometimes called the lubricating layer, which may be only a few millimeters thick. A weak layer may form at the surface of a snowpack as a result of metamorphism (loose facetted grains), the development of surface

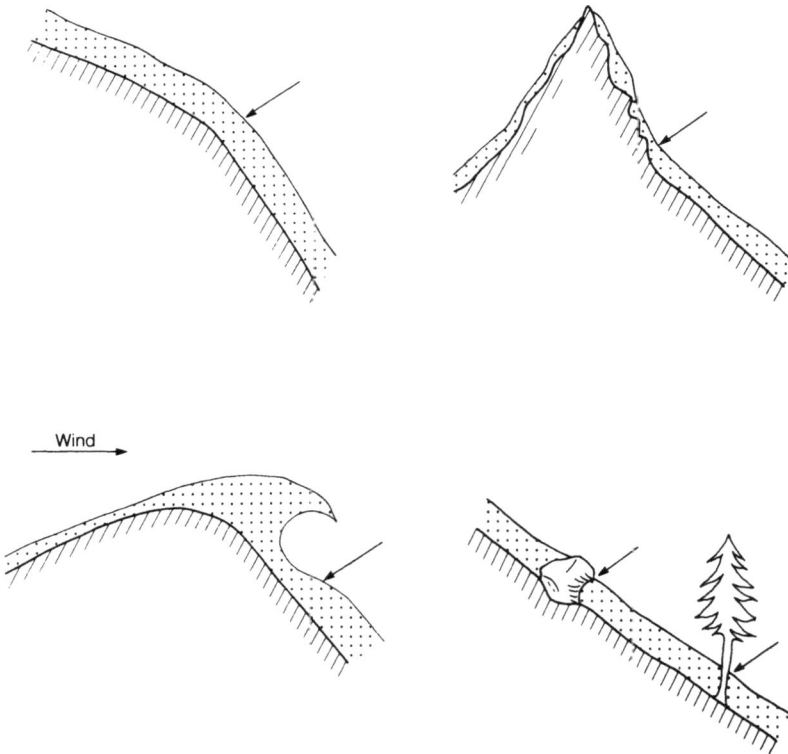

Fig. 11.4 Typical starting places for avalanches.

hoar or the deposition of new snow having a crystalline structure other than dendritic and then buried by subsequent snowfalls. Otherwise it may develop within the snowpack, e.g., when depth hoar or waterlogged snow overlies a crust. A sharp, jagged fracture line is the most significant and visible characteristic of the slab avalanche (see Fig. 11.5). The relatively strong slab layer that breaks away may contain snow deposited by one or several snowfalls and may range in depth from about 10 cm to 3 m.

The exact mode of failure depends on the mechanical properties of the snow and its stress distribution. In the simplest model, failure begins when the shear stress τ at one point exceeds the shear strength of the weak layer.

$$\tau = \rho gH \sin \alpha, \qquad\qquad 11.1$$

where ρ is the density of the snowcover, g is the acceleration of gravity, H is the depth of snow above the weak layer and α is the slope angle of the weak layer (see Fig. 11.1). This may result from an increase in shear stress, a decrease in

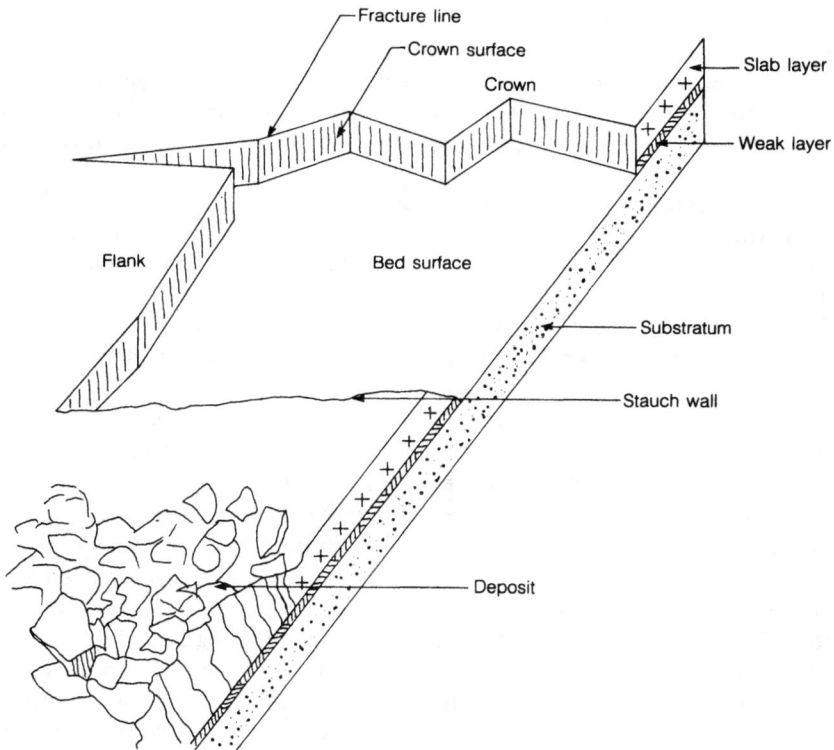

Fig. 11.5 Characteristics of a slab avalanche.

strength, or their combination. The most common cause for an increase in stress leading to failure is the added weight imposed on the system by accumulating snowfalls or drifting snow. Hence most avalanches occur during snowfalls and/or periods of high winds. Other factors producing high stresses and failure are the weight of skiers, explosive shocks, and traffic vibrations. Because skiers often initiate the failure, they are frequent victims of avalanches. An increase in snow temperature caused by warm air or intense solar radiation usually causes a decrease in strength of the weak layer. Temperature gradient metamorphism can also lead to weakening of the snow and its subsequent failure.

After the snow has failed at one location the loads are quickly transferred to adjacent areas which in turn become overstressed and fail, thus propagating a rupture by shear over a wide area. Finally, when the shear strength is exceeded, the tensile stress at the crown increases and reaches the breaking point. For a given slope it is not possible to predict where the initial failure will occur, i.e., at the crown or at a lower part of the slope. Many cases have been documented where skiers, while crossing the bottom of a slope, have caused a slab avalanche to be released above them.

Another possible rupture sequence starts with failure by tension at points of stress concentration, such as rocks, trees, breaks of terrain (see Fig. 11.4), or a ski track, and continues by spreading into the weak shear plane at the bed surface.

Another possibility for slab fracture can be attributed to the initial collapse of the fragile layer underlying the slab layer. After losing its support perpendicular to the slope, the slab would initially fail as a result of high bending stresses.

Slab avalanches are probably started by a combination of the different failure mechanisms; however, regardless of the mechanism, the shear strength of the weak layer at the bed surface must be exceeded.

Loose snow avalanches contain cohesionless snow and are analogous to slides in dry, sandy soil. The motion develops from a point on the surface when a small clump of snow is moved, e.g., by a snowball falling from a tree, or by skis. The disturbance propagates downward and sideways setting into motion increasing quantities of snow. Loose snow avalanches may occur in either wet or dry snow. Newly-fallen "dry" snow is cohesive and remains stable on steep slopes because the fine, sharp branches of the crystals interlock. However, it may become cohesionless through metamorphism. During initial periods of equitemperature metamorphism the crystals lose their sharp ends; sintering between the grains does not take place as rapidly as the rounding process. Thus, dry snow can temporarily lose its cohesive strength. The angle of repose for cohesion lies between 30° and 40°, so that the snow is unstable on steeper inclines. Generally, "dry" loose avalanches only contain snow from the snowcover surface and therefore are minor. They occur most frequently a few days after a snowfall.

A snowpack is wet and cohesionless when the bonds between the coarse crystals are destroyed by melting. This occurs most often in corn snow and depth hoar in the spring (sometimes called "rotten snow"). In contrast to dry snow avalanches, wet loose snow avalanches can be very large and contain deep snow layers.

Starting Zone

The two characteristics of terrain most important to avalanche formation are its slope and its orientation to the wind direction. Avalanches start most frequently on slopes with average inclinations between 30° and 45°; but generally not on slopes with inclinations $\geqslant 60°$ since small sluffs occur during snowfalls unloading the slopes continuously. Similarly there are infrequent avalanches on inclinations less than 25° and only if the snow is highly unstable or if fractures propagate from steeper slopes above.

Avalanches are more common on the leeward than on the windward side of slopes because they favour the deposition of wind-transported snow, which increases the snow load resulting in high stresses in the snowpack. Lee slopes lie not only along the high crests of mountains, but along any ridge at the side of a mountain.

The exposure of the slope to the sun has little influence on the average frequency of avalanches, but must be considered in the day-to-day evaluation of the avalanche hazard. In the Northern Hemisphere slopes facing north and east are more hazardous between December and March when they become colder because of the low amount of incident incoming radiation and therefore are more likely to have unstable surface-hoar and depth-hoar layers than other slopes. In the spring, however, the sun's radiation is much stronger and causes weakening and melting of the snow, particularly on the south- and west-facing slopes.

Ground surface obstructions, e.g., boulders, stumps, logs, shrubs, must be covered with some minimum snow depth before avalanches can slide: 30 cm on smooth ground, such as rock slabs, grass or fine scree; 50 cm on average mountain terrain above the treeline; 120 cm on very rough ground with boulders.

Summary of Conditions and Their Effects on Avalanche Formation

The complex system of environmental conditions that cause avalanches was summarized by the Working Group on Avalanche Classification of the

International Association of Scientific Hydrology (1973) in a scheme of conditions and effects (see Table 11.1). This scheme illustrates the difficulties associated with the prediction of the movement of avalanches because under a given set of terrain, weather and snow conditions, some factors produce instability in the snowpack, while others lead to stability.

Attempts have been made to analyze the stability of snow on a slope and to predict its failure (Jaccard, 1966; Haefeli, 1967; Mellor, 1968; Perla, 1975; Smith and Curtis, 1975). Theoretical models have led to an understanding of the mechanics of failure and a qualitative evaluation of the causes of avalanche triggering. These models generally assume certain values of a number of parameters such as snow density, viscosity and strength. In practice, it is very difficult to obtain continuous measurements of changes in the snow parameters and to predict the stability of a snowpack in real-time.

MOVEMENT OF AVALANCHES

Nature of Motion

Gliding motion. In order to start moving after failure has occurred a snow slab must overcome the static friction at its bed surface. This frictional resistance is often large enough to prevent development of an avalanche; the only sign of snow failure is cracks at the snow surface.

Once static friction is overcome and motion occurs, the slab breaks into blocks which accelerate rapidly. The equation of motion of a snow block can be written as

$$m \, (dV/dt) = mg \, (\sin \beta - \mu \cos \beta), \qquad\qquad 11.2$$

where m is the mass of the block, V is its velocity, g the acceleration of gravity, β is the inclination of the slope, and μ is the coefficient of kinetic friction between the block and its bed surface.

Little is known about the coefficients of friction: the static coefficient ranges from 0.3 to 0.6; the kinetic, from 0.2 to 0.5. The latter decreases with increasing speed; however, no conclusive data are available on this relationship.

The original slab quickly breaks into smaller fragments through friction, and collision. At low speeds there is little mixing, so that persons caught in the avalanche may remain at the surface by making swimming motions.

Turbulent motion. Avalanches accelerate rapidly on slopes with inclinations steeper than 30° and after attaining speeds greater than about 10 m/s their motion becomes turbulent. The original blocks disintegrate, and the smallest particles mix with the air at the front and along the upper surface to

Table 11.1
SCHEME OF CONDITIONS AND EFFECTS
[abbreviation: av = avalanche(s)] (Working Group on Avalanche Classification, 1973)

Condition	Effect on avalanche activity

A. Fixed framework

(1) Terrain Conditions

(1.1) Relative altitude

General topographic situation:	Effect depending on latitude and level of surrounding mountains.
— zone of crests and high plateaux	Strong wind influence, cornices, local slab av.
— zone above timberline and below crests	Extended areas of slab av. formation.
— zone below timberline	Reduced wind influence. Reduced slab av. soft type prevailing.

(1.2) Inclination of slope

>35°C	Formation of loose snow av. possible.
>25°C	Formation of slab av. possible.
>15°C	Stationary or accelerated flow.
<20°C	Retarded flow or deposition. (Slush av. at very low angles).

(1.3) Orientation of slope

— relative to sun	On shady slopes enhanced slab av. formation. On sunny slopes enhanced wet av. formation.
— relative to wind	On lee slopes increased drift accumulation; enhanced slab av. formation. On luff slopes vice versa.

(1.4) Configuration of terrain

— open, even slopes	Unconfined av.
— channels, funnels, ridges	Confined, concentrated, channelled av.
— changes in gradient	Slab or loose snow fracture at convex gradients.
— steps	Powder av., cascade formation

(1.5) Roughness

— smooth ground	Snow glide (on wet ground); full depth av. favoured.
— protruding obstacles (rocks, cross ridges)	Surface layer av. above level of roughness.
— vegetation	Grass: promoting snow glide, and full depth av. shrubs: reduction of av. formation if not snow covered. forests: prevent av. formation if dense.

Table 11.1 (cont'd.)
SCHEME OF CONDITIONS AND EFFECTS
[abbreviation: av = avalanche(s)] (Working Group on Avalanche Classification, 1973)

Condition	Effect on avalanche activity
B. Genetic variables	
(2) Recent Weather (period ~ 5 days back)	
(2.1) Snow fall	Increasing load. Increasing mass of low stability. *Most important factor of av. formation.*
— type of new snow	Fluffy snow: loose snow av. Cohesive snow: slab av.
— depth of daily increment of new snow	Increasing instability with snow depth. New or old snow fracture.
— intensity of snow fall	Progressive instabililty with higher intensity; promoting new snow fracture; expanding danger to low inclination.
(2.2) Rain	Promotion of wet loose snow av. or soft slab. av. Mixed snow and land slides.
(2.3) Wind	Two effects: enhanced local snow deposit (see 1.3) and increased brittleness of snow.
— direction	Increased slab av. formation on leeward slopes. Formation of cornices.
— velocity and duration	Local slab av. formation increased with increasing velocity and duration.
(2.4) Thermal conditions	
Significant factors: Temperature and free water content of snow.	Ambivalent effect on strength and stress, i.e. on av. formation: Rise of snow temperature causes crisis, but ultimately stability. Rise of free water content promotes av. formation.
— air temperature	Similar effect to all exposures.
— sun radiation	Dominant effect on sun exposed slopes.
— temperature radiation	Cooling of snow surface at night and in shadow; important with cloudless sky. Promotion of surface and depth-hoar formation (see 3.2)
(3) Old Snow Conditions	
Integrated past weather influences of the whole winter season	
(3.1) Total snow depth	Not dominant factor for av. danger. Influences mass of full depth av. Important to compaction and metamorphism of snow cover. Surface layer av. see (1.5).

Table 11.1 (cont'd.)
SCHEME OF CONDITIONS AND EFFECTS
[abbreviation: av = avalanche(s)] (Working Group on Avalanche Classification, 1973)

Condition	Effect on avalanche activity
B. Genetic variables (cont'd.)	
(3.2) Stratification	
Sequence of strength	Stability governed by weakest layer with respect to state of stress.
— surface layer	Looseness (surface hoar), brittleness, roughness important to subsequent snow fall.
— interior of snow cover	Old snow fractures caused by weak intermediate layers (old surfaces) and depth hoar.
(4) Triggering Conditions	
(4.1) Natural release	Natural av.
— internal influences	Spontaneous av.
— external (non human) influences	Naturally triggered av.
(4.2) Human release	
— accidental triggering	— Accidental av. (triggering).
— intended release	— Artificial av. (triggering).

form a cloud of powder snow. Unstable snow at the base may also be incorporated and debris may be deposited. Therefore the mass may not remain constant.

Persons caught in turbulent avalanches are tossed about, can do little to stay at the surface, and must ride along as best as possible.

Powder, Dry-flowing, Wet-flowing and Mixed Avalanches

For engineering purposes it is convenient to distinguish between powder, dry-flowing and wet-flowing avalanches, even though combinations of these types are frequently observed. Each type consists of different material and has different flow characteristics.

Powder avalanches, also called snow dust avalanches, are an aerosol of fine, diffused snow which behaves as a sharply bounded body of dense gas. The avalanche may flow in deep channels, and generally is not influenced by obstacles in its path. After leaving a deep gulley this type can move in a straight line across irregular terrain (see Fig. 11.6). The speed of a powder avalanche is approximately equal to the wind speed. Its mass density generally lies between 3 and 15 kg/m³ (cf. air density of 1.3 kg/m³). Because of their greater densities powder avalanches are more destructive than wind storms.

Dry-flowing avalanches are avalanches of dry snow that have travelled over steep and irregular terrain, usually consisting of particles that range between powder and balls with an average diameter of about 0.2 m. These avalanches move along the surface, follow well-defined channels and are relatively uninfluenced by small irregularities in the terrain. On open slopes the depth of

Fig. 11.6 Powder avalanche.

the flowing snow is slightly greater than the depth of the original slab layer and usually is 0.3 to 3 m. However, when concentrated in a channel the depth can greatly exceed these values. The mass density of the flowing material ranges between 50 and 150 kg/m^3.

Wet-flowing avalanches are composed of wet snow either as rounded particles (0.1 to several metres in diameter) or as a mushy mass. They have an average depth between 0.2 and 2 m on open slopes, but are much deeper in channels. The wet snow tends to flow in channels and is easily deflected by small irregularities of the terrain. After leaving a channel the avalanche often retains the shape of the channel. Characteristics of a wet avalanche are the grooves and flow marks it makes in the snow left in the track and, as well, the channels in the deposited snow. The mass density of the flowing wet snow is between 300 and 400 kg/m^3.

Powder and dry avalanches frequently occur together as a mixed avalanche, however, for studies of their dynamics they are treated as separate entities. On steep terrain the powder and the flowing snow move together (see Fig. 11.7), while on gentle slopes the powder often moves ahead of the flowing snow and travels farther. Although powder avalanches usually have densities less than 15 kg/m^3, a transition type has intermediate densities.

Because some of the avalanche energy is converted into heat which melts its snow, many avalanches originating in dry snow are moist on arrival in the

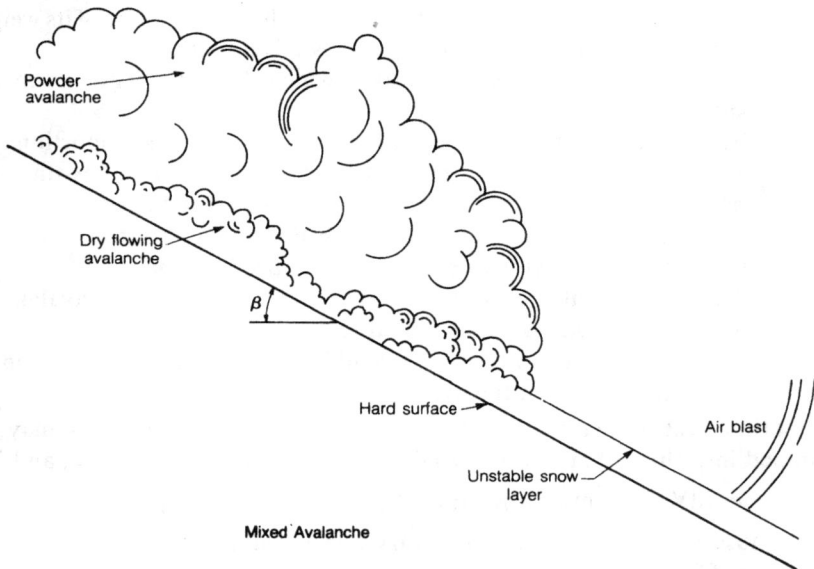

Fig. 11.7 Mixed avalanche.

valley. Moist snow avalanches have flow characteristics and densities intermediate between those of dry and wet-flowing avalanches.

An air pressure wave may precede an avalanche, as it has been observed that
objects have been moved and structures damaged before the avalanche front
reached them. There is no firm and quantitative information about this
phenomenon and the conditions required for its formation. Most investigators have envisaged a displacement of the air caused by the descending
avalanche acting like a piston to produce a pressure wave ahead of itself,
analogous to a bow wave ahead of a moving vessel. Destructive airblast is the
exception, rather than the rule, even though the powder component of mixed
avalanches is often mistaken as airblast. It can also be surmised that
irregularities of the terrain create strong eddies in the avalanche boundary
layer which act like miniature tornadoes.

Speed

The speed of the front varies with avalanche type. Some typical speeds
observed are: powder avalanches, 20 to 70 m/s; dry-flowing avalanches, 15 to
60 m/s; and wet-flowing avalanches, 5 to 30 m/s. It must be assumed that
somewhat higher speeds prevail in the cores of powder and dry-flowing
avalanches. Avalanches have been observed to attain speeds up to 100 m/s
while descending through free air.

The speed V of an avalanche is determined by the driving force of its weight
parallel to the slope and the following forces of resistance:

a) Kinetic friction at the bed surface which decreases with increasing
speed.

b) Viscous shear in the moving snow, whose magnitude is proportional to
V. Rapid viscous shear has not been studied, but its resistance is
probably negligible compared with those of other factors.

c) Turbulent resistance against the bed (similar to the turbulent
resistance of fluids) whose magnitude is proportional to V^2.

d) Air resistance at the front and at the upper surface of the avalanche,
whose magnitude is proportional to V^2.

e) Resistance of the snowcover, after additional snow has been broken up
and incorporated in the avalanche, independent of V.

Equation 11.2 is therefore modified so that the resistance forces may be
lumped into those that are independent of V, and dependent on V, and V^2:

$$m \, (dV/dt) = mg \sin \beta - [(a_{11}/V) + a_0 + a_1 V + a_2 V^2]. \qquad 11.3$$

Because m, β, a_{11}, a_0, a_1, and a_2 vary during the avalanche an analytical
solution of Eq. 11.3 is not possible. Therefore, the equation must be simplified
to obtain practical solutions. For example, if an avalanche of constant mass is

assumed to move down a uniform incline such that the coefficient of kinetic friction remains constant and the resistance forces proportional to V can be neglected then the terminal speed of the avalanche is:

$$V_T = [(mg \sin \beta - a_0)/a_2]^{1/2}. \qquad 11.4$$

In practice Eq. 11.4 can be written more conveniently in a form similar to that of the equations used to model open channel flow of a fluid, i.e.,

$$V_T = [\zeta \cdot R(\sin \beta - \mu \cos \beta)]^{1/2} \qquad 11.5$$

where: ζ = coefficient of turbulent friction, including the influences of all resistance forces proportional to V^2,

μ = coefficient of kinetic friction, and

R = hydraulic radius, equal to the height of flowing snow on open slopes.

When μ is negligible (in high speed flowing avalanches and powder avalanches) Eq. 11.5 becomes:

$$V_T = [\zeta R \sin \beta]^{1/2}, \qquad 11.6$$

which is similar to the Chezy equation for the motion of fluids in a channel where $\sqrt{\zeta}$ = C is the Chezy discharge coefficient.

Speeds calculated from Eq. 11.6 using tabulations of C (to obtain ζ) in hydraulic handbooks agree satisfactorily with the observed speeds of flowing dry and wet avalanches (Schaerer, 1975). Recommended values of ζ for various terrain features are listed in Table 11.2.

The influence of various factors on μ is not well known but its magnitude decreases with increasing speed and depends on the type of snow. The effect of kinetic friction on powder avalanches and on avalanche flow at greater than

Table 11.2
RECOMMENDED VALUES OF THE COEFFICIENT
OF TURBULENT FRICTION, ζ.

Terrain Features	ζ (m/s^2)
Smooth, hard snowpack with uniform incline, no trees, no visible rocks	1200-1600
Average open mountain slope, no trees	750
Open slope, brush, rocks	500
Average gully	400-600
Rough terrain with boulders, winding gully	300
Forests	150

50 m/s can be neglected. Realistic values of μ for flowing avalanches are 0.1 to 0.15 for speeds between 30 and 50 m/s, and 0.2 to 0.3 for speeds between 10 and 30 m/s.

Determination of the flow height and the hydraulic radius requires experience. The thickness of the slab layer that breaks away, the surface area of the starting zone, and the cross section of the track must be taken into account.

There is no reliable method for calculating the speed of powder avalanches. Equations have been suggested but have not been confirmed by enough observations.

Impact Pressures

Powder avalanches

The impact of a powder or light, dry-flowing avalanche on an obstacle, oriented perpendicular to the direction of flow, can be likened to the impingement of a jet of water on a fixed surface. Thus, the pressure P_p or drag force per unit surface area on the obstacle can be expressed by the relationship

$$P_p = C\rho_a V^2/2, \qquad\qquad 11.7$$

where C is a drag coefficient whose magnitude depends on the size and shape of the obstacle, ρ_a and V are the density and the speed of the avalanche respectively. Values of C may be found in tables of wind pressure on structures, such as contained in Building Codes. A value of C = 2 seems to be adequate for large objects such as walls.

Flowing avalanches

Dense, flowing snow hitting a rigid obstacle is first compacted and then flows around the obstruction. The initial plastic deformation results in a short-time peak pressure P_i which drops to the stagnation pressure P_a, where;

$$P_a = \rho_a V^2. \qquad\qquad 11.8$$

The initial peak pressure P_i is a function of the deformation properties of the snow, which in turn, depend on the density, the temperature, the amount of free water and the grain size. Observed impact pressures and theoretical studies suggest that the peak pressure P_i is 2 to 3 times P_a but the data are insufficient for a detailed analysis (Salm, 1964; Mellor, 1968; Schaerer, 1973; Kotlyakov et al., 1976).

The impact pressure P_a has been observed to fall within different ranges, depending on avalanche type, e.g. 2 to 30 kPa for powder avalanches and 20 to 300 kPa for flow avalanches.

On striking an object an avalanche may be compressed or deflected so as to produce forces in directions other than in the direction of flow. The deflection of snow in the vertical direction is particularly important for designing buildings, since the uplift forces exerted by the flowing snow on the soffit, (that part of the roof extending out from the wall), may cause failure. Most buildings are not designed to resist this type of loading. Further, if a structure is struck by a large avalanche composed of several waves of snow the series of collisions may produce dangerous vibrations which can lead to destruction. Observations of the variations in impact loads with time produced by this type of avalanche are not available.

Leaf and Martinelli (1977) provide considerable information about calculating pressures exerted by avalanches. It must be stressed, however, that the accuracies of impact pressures calculated by numerical expressions depend largely on realistic estimates of the speed and density of the moving snow.

Runout Distance

In the runoff zone avalanches decelerate and stop on terrain with a low inclination. However, they can advance a considerable distance into a valley and even climb the opposite side. The equation commonly used to estimate the runout distance D is

$$D = V^2 / \{2g[\mu \cos \psi + \tan \psi + (V^2/2\zeta h_m)]\}, \qquad 11.9$$

where: ψ = inclination of the slope in the runout zone,

         μ, ζ = friction coefficients as defined in Eq. 11.5 based on the surface conditions in the runout zone. (Due to the low average speed of an avalanche in the runout zone μ is generally 0.25 to 0.3) and

         h_m = average depth of the avalanche in the runout zone.

One of the main problems affecting the application of Eq. 11.9 and other similar equations is the lack of information about the values of the friction coefficients. Another problem is the selection of the beginning of the runout zone and the point from which D should be measured. This is particularly difficult in terrain having a gentle change of slope. Therefore much experience is required to make a realistic estimate of D.

CLASSIFICATION OF AVALANCHES

The primary purpose of an avalanche classification scheme is to establish uniform descriptive terms that can be used to exchange information about accidents, safety measures, and control. Another purpose is the grouping of avalanche events for statistical analysis, e.g., to find relationships between avalanches and the factors responsible for their formation, such as terrain, weather, the characteristics of the snowpack, or to make decisions for planning and implementing protective measures.

Morphological

A widely accepted classification is based on observations of the immediate phenomenon, e.g., the properties and type of motion of the snow. Table 11.3 presents such a classification developed by the Working Group on Avalanche Classification (1973). Essentially qualitative observations are used, supplemented on occasion by such quantitative properties as the depth of fracture and the depth, width and length of the deposit.

Table 11.3
MORPHOLOGICAL AVALANCHE CLASSIFICATION (Working Group on Avalanche Classification, 1973.)

Zone	Criterion	Alternative characteristics and symbol designation	
Origin (Starting)	A. Manner of starting	A1 starting from a point (loose snow avalanche)	A2 starting from a line (slab avalanche) A3 soft A4 hard
	B. Position of sliding surface	B1 within snow cover (surface layer avalanche) B2 (new snow B3 (old snow fracture) fracture)	B4 on the ground (full-depth avalanche)
	C. Liquid water in snow	C1 absent (dry snow avalanche)	C2 present (wet snow avalanche)
Transition (Track)	D. Form of path	D1 path on open slope (unconfined avalanche)	D2 path in gulley or channel (channelled avalanche)
	E. Form of movement	E1 snow dust cloud (powder avalanche)	E2 flowing along the ground (flow avalanche)
Deposit (Runout)	F. Surface roughness of deposit	F1 coarse (coarse deposit) F2 angular F3 rounded blocks clods	F4 fine (fine deposit)
	G. Liquid water in snow debris at time of deposition	G1 absent (dry avalanche deposit)	G2 present (wet avalanche deposit)
	H. Contamination of deposit	H1 no apparent contamination (clean avalanche)	H2 contamination present H3 rock debris, H4 branches, soil trees

Genetic

A genetic classification relates avalanche events to the conditions that are responsible for their formation; for example, the shape of terrain, weather, and properties of the snowcover. Genetic classifications have been suggested but are not satisfactory because the process of formation is often so complex that it does not permit the avalanche to be related to one or two causes. Table 11.1 lists the criteria that could be used for a genetic classification.

Classification by Magnitude

When avalanches are classified by magnitude they are grouped according to their size (for example, the mass or volume of snow moved) or destructive force. The following nominal classification scheme, according to the destructive effect of the avalanche, is widely used in Western Canada:

- size (1) - sluff; any small amount of snow which would not injure a human,
- size (2) - could injure a human,
- size (3) - could damage buildings, automobiles, break a few trees;
- size (4) - could destroy large vehicles, or forests with areas up to 4 ha, and
- size (5) - unusual, catastrophic events, which could damage villages or destroy large areas of forest.

AVALANCHE HAZARDS

The various ways in which avalanches interfere with the life of man and his works may be grouped as follows:

(1) Catastrophic accidents - people either remaining inside or outside of dwellings or travelling are buried in an avalanche that starts independently of their actions.

(2) Tourist accidents - people travelling in the mountains start an avalanche and are caught in it. Skiers and mountaineers are the most frequent victims of this type of accident.

(3) Property damage - avalanches damage or destroy buildings, bridges, powerlines, and other structures, forests, vehicles on roads, trains.

(4) Communications interruptions - the avalanche blocks traffic on roads and railways or breaks power and telephone lines. The cost arising from delays in traffic and interruption of services are indirect but can be high.

The basic conditions for avalanche formation (steep slopes and snow depths sufficient to cover irregularities of the ground surface) are generally found in mountain ranges, e.g., those of western Canada, western United States, and the Alps of western Europe. However, people have been buried in avalanches that have occurred in other parts of Canada; in Manitoba, at Toronto and near Quebec city.

Avalanche hazards exist where transportation routes, settlements and recreation areas are developed very close to avalanche paths, e.g., in the West Kootenay, in the transportation corridors through the Pacific Coast Mountains, in Banff National Park and at mining developments. Rogers Pass, the crossing of the Trans Canada Highway over the Selkirk Mountains, has the highest avalanche hazard in Canada because a major highway and railway line both cross numerous avalanche paths where avalanches are frequent. At Rogers Pass the greatest single avalanche disaster in Canada occurred in 1910, when 58 railway workers perished. The hazard is particularly high in the densely populated Alps of Switzerland, France and Austria where the pressure of population has forced buildings and communication lines into avalanche paths. Furthermore, these mountains are visited by a large number of skiers who expose themselves to avalanches, which kill between 20 to 30 people in Switzerland every year (Bündnerwald, 1972).

Survival Chances

The causes of death of people buried in avalanches are:

(1) Suffocation caused by snow dust in the lungs, compression of the chest, or a lack of oxygen,

(2) Injuries resulting from collapsing buildings, collisions with trees and rocks, and falls over cliffs,

(3) Shock (terror, fright), and

(4) Exposure (hypothermia).

Avalanche accidents in the United States between 1910 and 1974 have been analyzed for the length of burial of the victims (Williams, 1975). Such statistics must be treated with caution, however, because many avalanches that cause minor damage are not reported. Nevertheless, studies indicate that the chances of survival of an avalanche victim depend on how deep and how long he has been buried. In the USA approximately one third of the people who were buried completely away from the protection of buildings or vehicles were rescued alive. The survival chances decrease rapidly with time. After 30 minutes of burial only 50 per cent of the victims can be expected to survive. The rapid decrease of chance of survival with increasing time of burial stresses the importance of a speedy rescue.

People buried in buildings and in vehicles on roads have the best chance of survival, because of the airspaces which exist under roofs. People trapped in buildings have been rescued alive after 10 days of burial (Fraser, 1978).

Identification of Avalanche Hazards

The information used in making decisions on the safe location of facilities, such as roads, structures, ski runs, and the selection of an avalanche control method is developed by identifying the avalanche paths, estimating the frequency of occurrence and the sizes of avalanche paths, and evaluating the potential damage.

Avalanche paths can be recognized from features of the terrain (incline, channels, characteristic starting points), from vegetation, and from deposited avalanche snow (Martinelli, 1974). In the heavily forested mountains of southern British Columbia and Alberta, avalanche paths can be identified by examining the age and species of trees at different locations on the slope and the sharp trimline. Terrain and vegetation features of avalanche paths can best be recognized from airphotos, but should be verified by ground inspection. Tree growth must be evaluated carefully and compared with possible flow patterns of the avalanches. One must be aware that not only avalanches but also fires, debris flows, logging operations, soil, and exposure to sun and wind influence the growth of trees.

It is a difficult task to estimate the frequency of occurrence, type and size of large avalanches. The most reliable procedure is to observe avalanche occurrences over many years. Data show that on the average once every 12 to 20 years there is a winter or a series of winters with large, unusual avalanches. Often the observation period may not be long enough to contain a winter with maximum snowfall and hence the historical records must be supplemented by examining trees for their ages and damage and by analyzing climatic data.

The most important factor for planning the location of facilities outside of avalanche prone areas is the maximum runout distance. In forested areas the extent of the runout zone of very large avalanches that have occurred is often made visible by the sharp boundary between trees of different age or different species. This boundary can best be recognized by comparing old and new airphotos of the same area.

The observations of tree growth must be supplemented by calculations of the runout distance according to the procedures outlined above. Obviously a calculation of the runout distance of the largest avalanche possible is the only method applicable to avalanches that advance into treeless terrain which have no historical observations.

Ives and others (1976) demonstrate through a case history approach the techniques for estimating the locations, frequency, and maximum runout distance of avalanches.

The probability that an object lies in the path of an avalanche together with the estimated frequency of avalanche occurrence determines the hazard. The hazard is the probability that certain consequences will occur, e.g., a car being buried on a highway, a building being destroyed, or a skier being caught in an avalanche.

Evaluating the Daily Hazard

Evaluating avalanche hazards from day to day during the period of operation of facilities such as ski resorts is called avalanche hazard forecasting. This procedure involves three steps:

(1) Conducting a snow stability analysis to obtain an estimate of the likelihood that avalanches will occur within a given time,

(2) Estimating the size of the avalanches that could develop. This depends on the amount of unstable snow in the starting zone and the track, and

(3) Estimating the extent and amount of damage that could occur.

Avalanche hazard forecasting is largely an empirical art based on field experience. Quantitative analysis is difficult because the snow, weather and terrain includes many interrelated variables that cannot be quantified. It is usually impossible to predict the exact time of occurrence of an avalanche, but it is possible to estimate the degree of instability of the snow and the chance that a given size of avalanche will occur within a few hours.

Analysis of snow stability

The technique usually applied to snow stability analysis involves observing several factors and determining whether each has reached a critical level affecting the stress-strength relationship of the snowpack (see Table 11.4). The critical factors are weighted according to their importance and then combined to indicate the probability of avalanche release and to identify potential avalanche slopes. The critical levels given in the Table are based partly on experience and partly on the results from analyzing available data. They are applicable in most avalanche areas. When sufficient observations of the snowpack, the weather and avalanche frequency are available for a specific area, the critical values of the factors can be modified to provide a more reliable prediction of the snow stability for that area.

The amount of information used in any given study depends on the accessibility of the starting zone, observations of local climate, and data obtained from avalanche control methods. In all forecasts, observation of the current weather is essential. Highway operations rely heavily on additional observations of avalanche occurrences. Slope tests and snowcover observations are standard practice in ski areas and for ski touring.

A few comments about the factors applied in a snow stability analysis (Table 11.4) are given below. Additional information can be found in LaChapelle (1978) and Perla and Martinelli (1976).

Snow stratification and snow temperature: the characteristics of the snow layers are examined (from pits dug at representative locations near starting zones) according to standard methods as set forth by the Commission on Snow and Ice (1954). The most important information used to evaluate whether a slab avalanche condition exists is obtained by identifying and testing the relative strengths of weak, intermediate layers in the snowpack, for example, buried surface hoar, depth hoar, interfaces of crusts and very wet snow. The strength of individual layers is usually observed subjectively with a hand test. In certain cases, measurements of shear strength and hardness are made. Skiers obtain information about the stratification of the snow by pushing a ski pole deep into it and feeling the changes of hardness. This method has the advantage of being quick, thus allowing for more frequent

Table 11.4
FACTORS USED IN SNOW STABILITY ANALYSIS

Factor	Generally Critical Condition
Terrain in Starting Zone	Slope 25° and greater, abrupt change of incline smooth surface, lee side of prevailing wind
Snow Depth	Greater than 30 cm on smooth ground, greater than 60 cm on average ground
Snow Stratification	Weak intermediate layers
Snow Temperature	0°C
Depth of Snowfall	30 cm new snow and greater
Precipitation	20 mm and greater
Rate of Precipitation	2 mm/h and greater
Settlement of New Snow	Less than 15% per day
Wind Speed	Greater than 4 m/s
Wind Direction	On lee slopes
Air Temperature	0°C and higher, -10°C and lower
Solar Radiation	Slopes exposed to the sun
Avalanche Occurrences	No avalanches in last snowstorm, avalanches running presently
Slope Tests	Release of avalanche by a ski; easy breaking with a shovel
Weather Forecast	Prediction of snowfall, wind or high temperature

tests on slopes with variable exposure. However, it fails to detect weak layers imbedded deeper than the length of a ski pole.

Snow temperature is usually measured by inserting a thermometer into an exposed but shaded face of a snowpit; these data are essential for estimating the strength of the snow, particularly during the melt period.

Weather: precipitation, wind speed, wind direction, and temperature have the most direct influence on the stability of snow. Catastrophic avalanches are usually the result of heavy snowfalls, high winds and high temperatures. Ski resorts, highway departments, and industrial operations that maintain avalanche safety and control programs must monitor the precipitation (snowfall), wind and temperature at carefully selected sites located as close as possible to the avalanche starting zones.

The relative stability of a snowcover can be predicted only with the aid of a weather forecast; the avalanche hazard forecast is only as good as the weather forecast. Therefore, close cooperation usually exists between the avalanche forecasting service and the weather office.

Avalanche occurrences: to assess the avalanche hazard of an area, the amount of snow that has been removed from the location by previous avalanches must be considered. Obviously, snow removed from the slopes will not be available for the development of subsequent avalanches. For this reason, and to evaluate the success of the control measures, records must be maintained of the avalanches occurring in each path. In most avalanche areas the snow on certain slopes becomes unstable first; small avalanches occur on these slopes before large avalanches become a hazard. Activity on these "indicator" slopes is an essential consideration in the stability analysis.

Slope tests: tests that attempt to induce sliding of the snow are carried out on short, steep, accessible paths with exposures equal to that of large avalanche paths. The means for testing the stability are skiing, cutting a section of snow with a shovel, and explosives.

Test skiing is an attempt to release avalanches by skiing in a calculated manner, if necessary with jumps and turns, along the normal fracture line. An avalanche released by this procedure signifies unstable snow conditions, and furthermore permits the depth and nature of the sliding layer to be examined. Test skiing is practical during snowfalls when the new snow only slides, but is dangerous on deep and hard slab layers that result from several snowfalls or snow drifting.

The "shovel test" is carried out by cutting a block of snow with vertical faces 0.5 to 1 m long. When the snow is highly unstable the weak layer fractures as soon as all four sides of the block are cut. If no fracture occurs a failure may be induced by pushing the block with the shovel in the direction of the slope. The force required to produce a fracture is a subjective index for the stability of the snow.

The stability of slopes can be tested also by exploding hand-placed charges or artillery shells. Fracturing of the snow and release of an avalanche would indicate instability, but no result (i.e., no avalanche release) does not necessarily indicate safe conditions. In this event it is necessary to evaluate the slope test together with other factors.

Estimation of avalanche size

The size of an avalanche that develops after snow has failed is a function of the depth and the width of the slab released, the length of the starting zone, and the depth of unstable snow in the track. Very often a shallow slab layer is released, and after being set into motion, also releases a deeper avalanche on a lower bed surface. The possibility of this happening must be recognized from observations of the snowcover and other avalanche occurrences.

The width of the slab depends on the width of the slope and the distance that the fracture propagates. The fracture propagates more easily in hard, brittle snow formed under the action of wind and moist air than in soft, new snow.

METHODS OF PROTECTION AND CONTROL

Objectives

The primary objectives of avalanche protection and control are:

(1) To minimize the loss of life and injury,
(2) To minimize the damage to structures such as buildings, power lines, ski lifts, highways and railways,
(3) To minimize the interruption of traffic, and
(4) To maintain safe, open ski areas for recreational purposes.

Other benefits which accrue from these measures include the reduction of snow removal and the prevention of damage to forests.

The choice of a particular avalanche protection measure depends on the level of protection required, the terrain, the type of avalanche prevalent in the area and cost. In some instances avalanche control, or the use of temporary safety measures can be effective. Avalanche control measures influence the start or course of avalanches and can be divided into two categories: those that modify the terrain and those that modify and stabilize the snowpack. The implementation of control measures which would completely eliminate the avalanche hazard and damage is impractical and uneconomical. Rather, it is common practice to accept those measures that reduce the hazard to some acceptable level.

The measures applied to different types of structures for different levels of protection in Canada are listed in Table 11.5. The measures adopted by different countries depend on population density and cost. For example, in the densely-populated mountainous areas of Switzerland, supporting structures and galleries are frequently used for control, whereas in many parts of Canada, because of their sparse populations, explosives are utilized.

Table 11.5
PROTECTIVE MEASURES IN CANADA

Application	Primary Protection	Secondary Protection
Buildings	Avoidance by location	Deflecting dykes Splitting wedges
Major highways	Avoidance by location Closures Warning signs Artillery	Deflecting dykes Galleries Retarding structures
Secondary roads	Closures Warning signs	Avoidance by location
Railways	Detection Galleries	Deflecting dykes Artillery
Electric Transmission Lines	Splitting wedges of earth	Avoidance by location Reinforced structure
Ski areas	Avoidance by location Hand explosive charges Ski stabilization	Closures Artillery
Touring, Helicopter Skiing	Avoidance by route selection	No travel

Location of Structures

One of the primary requisites for avalanche protection is location of facilities outside "avalanche prone" areas. Many accidents in the past might have been prevented if this general rule had been followed during the early planning stages of buildings, roads, power lines, or ski runs. Also, simple control devices could have been installed to prevent avalanche damage to the structures.

Buildings

People who settled in the mountainous areas of Europe quickly learned to build their dwellings in places safe from avalanches. However, as the population increased, marginally-safe and unsafe areas were occupied and many buildings and complete villages were destroyed by avalanches (Bündnerwald, 1972). In Canada and the United States the mountains are not nearly as densely populated as those in Europe, but many disasters have still occurred. In 1971, avalanches damaged seven cabins, killed four people, and injured another four people in a new development at Stevens Pass, Washington (Williams, 1975). In 1974, near Terrace, B.C. a restaurant was destroyed in which seven lives were lost. Atwater (1968) reports that mine buildings, in particular, have suffered avalanche damage with heavy losses.

Construction of a facility in an avalanche area presupposes that the builder is prepared to assume a level of "risk". The process of restricting buildings in avalanche areas is known as "Avalanche Zoning", and involves consideration of technical, political and legal problems. The technical work required to identify dangerous zones can be carried out with reasonable accuracy using the procedures outlined above. Determination of the acceptable risk and the enforcement of building restrictions are political and legal matters. Frutiger (1970) and Oppliger (Bündnerwald, 1972) describe the technical and legal work in Switzerland. Zoning laws for constructing buildings in hazardous areas are in force in Switzerland and in the USA at several places (Baker and McPhee, 1975). However, in Canada no generally accepted avalanche zoning legislation is in existence although provincial, regional, and municipal authorities have the power to restrict the location of buildings through the normal process of issuing building permits.

Total avoidance of potential avalanche sites through zoning procedures is probably too restrictive. In some areas a more practical zoning alternative has been used which distinguishes between a zone of high hazard and a fringe zone of moderate hazard. Buildings may be permitted in the fringe zone if they meet certain criteria based on their use and structural design (Bündnerwald, 1972).

Roads, conveyance systems and ski runs

In the past it has been common practice to construct roads and railways without considering their potential damage by avalanches. Today, however, the selection and final location of these routes in mountains usually is based on identification of potential avalanche sites and evaluation of their hazard to traffic. Obviously, the benefits to be gained in locating transportation routes so as to avoid avalanche sites must be weighed against satisfying design requirements concerned with soil stability, grades, curves, maintenance and environmental damage. Where roads already exist, the relocation of exposed sections is often chosen as the most appropriate method of avalanche protection.

The proper design and location of the facility (e.g., near the end of the runout zone), and elimination of through-cuts, high embankments and extra-wide ditches can often mitigate the effect of avalanches if the right-of-way cannot be relocated.

Although most avalanches would normally pass harmlessly underneath electric transmission lines, aerial tramways, and chairlifts, the location of these structures within potential avalanche sites should be avoided. Large avalanches, although infrequent, may occur and their forces are often unpredictable. The supporting towers of conveyance systems are their most vulnerable structural component and should be placed outside the influence of maximum avalanches, or at least outside the principal flow channels. Trees which have been uprooted by avalanches and retained within the flow may damage cables and towers. Hence, it is common practice to remove those trees from the upslope area that could be broken by avalanches.

In planning a ski area it is important for public safety to identify, map and evaluate the hazard of its potential avalanche sites. At least one ski run, which would be open during storms, should be located in safe terrain, free of avalanches. The other ski trails must be located in areas that are readily accessible for avalanche control. Care should be exercised in establishing ski runs so that they do not form avalanche paths.

Modification of Terrain

In avalanche control and protection work, terrain modifications include those structures and earth works that are constructed either to prevent the release of avalanches or deflect the sliding snow away from the facilities to be protected.

Supporting Structures

Supporting structures or retaining barriers are used in the starting zones of avalanches (Bündnerwald, 1972); their functions are:

(1) To provide external support to the snowcover, thereby reducing the internal stresses within the snow,

(2) To produce a discontinuity in the snowcover, thereby limiting the propagation of a fracture and the resultant size of the avalanche, and

(3) To stop small avalanches before they gain sufficient momentum to cause major damage.

The first supporting structures used in avalanche control were built in the 19th century in European alpine countries. These were earth terraces, posts and masonary walls. Today the structures are made out of wood, steel, aluminum, concrete, and various combinations of these materials (see Fig. 11.8).

Supporting structures are expensive because of their large physical size. They must be at least as high as the deepest snow, usually between 3 and 5 m, capable of resisting the forces produced by creeping snow and small avalanches, and of covering the full width and length of the starting zone. Because of their high cost, supporting structures can be justified only for the protection of inhabited areas or for installation at sites where the starting zone is small. The Swiss Federal Forest Department, through the Institute of Snow and Avalanche Research has developed guidelines for the design and location of supporting structures (Oberforstinspektorat, 1968).

Temporary low cost structures of wood are often used on reforestation projects to protect young trees from snow creep and avalanches. These temporary structures weather and decay. However, they generally last sufficiently long for the trees to grow large enough to provide natural avalanche protection (Bündnerwald, 1972).

Fig. 11.8 Slatted fence retaining structure.

Snow Fences and Wind Baffles[1]

Collector fences and baffles control drifting snow that would contribute to avalanche formation. These structures have proven effective in reducing avalanches but do not eliminate them. They are generally used in combination with supporting structures for control of avalanches as well as cornices. Cornices can be a hazard when they break and roll on a ski run or a road; occasionally they may start an avalanche. The function of collector fences, which are usually 4 to 6 m high and located on the windward side of ridges is to decelerate the wind velocity and to retain blowing snow. Conversely, wind baffles are either vertical walls about 4 m high and 2 m wide or jet roofs (blower fences) made of inclined boards 4 m long (Mellor, 1968; Bündnerwald, 1972).

Deflectors

Deflecting structures are used as protection devices in the track and runout zones of avalanches. Three major types are commonly used: dykes and walls, splitting wedges and galleries. .

Deflecting dams or walls intercept avalanches and direct the flow to an area where they can run out harmlessly (see Fig. 11.9). Guiding dams or walls are constructed parallel to the direction of the avalanche and confine it in a narrow channel; they are often used in combination with galleries. Most are built as earth banks but may be concrete or steel walls, gabion walls and cribs.

Dykes and walls are effective against flowing avalanches, but do not control powder avalanches. Deflecting dykes for avalanche control can essentially only be used in areas having enough space for the deflected avalanche to run out harmlessly, but this often limits their application. The deflection of the slug of snow must occur gradually otherwise the avalanches will overflow the structure. It has been found that the angle α between the direction of the avalanche and the deflector (see Fig. 11.9) should not exceed 20 degrees for an earth bank, and 30 degress for a wall.

The minimum height H_d of the deflector can be calculated by the relationship:

$$H_d = H_s + H_a + H_v,$$
(11.10)

where: H_s = depth of snow deposited either by snowfall or previous
                 avalanches,

---

[1]   For additional material on this subject the reader should consult Ch. 16.

H_a = flow depth of the "design" avalanche,

H_v = $(V_a \sin \alpha)^2/2g$, and

V_a = average velocity of the "design" avalanche (see Fig. 11.9).

When avalanches impinge on a deflecting wall, forces are exerted on it because of the impact and friction. The impact pressure P_v acting perpendicular to the wall can be calculated by considering linear momentum:

$$P_v = \rho_a(V_a \sin \alpha)^2, \qquad\qquad 11.11$$

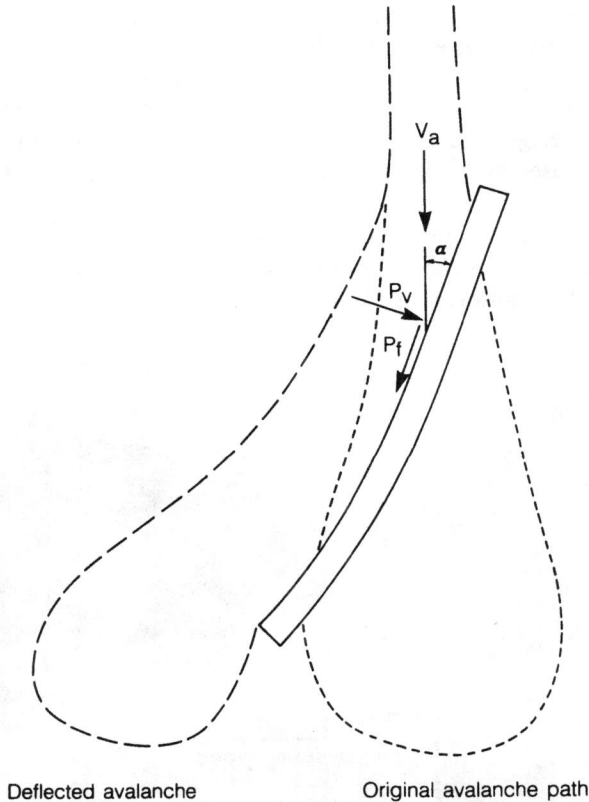

Deflected avalanche   Original avalanche path

Fig. 11.9 Action of a deflecting wall.

P_f = friction stress which acts parallel to the wall,

P_v = pressure normal to the deflecting wall produced by impact of the avalanche

V_a = average "design" velocity of the avalanche, and

α = angle of incidence of the avalanche to the wall.

and the friction force per unit surface area P_f acting parallel to the wall by the relationship:

$$P_f = \mu P_v, \qquad\qquad 11.12$$

where: ρ_a = density of the avalanche, and

μ = coefficient of friction (sliding) whose values are estimated to range between 0.3-0.5.

Wedges have been proven effective for protecting single objects, e.g. a building, power line or chairlift tower. They can be made from earth, concrete, steel, wood, or other materials. They are placed directly in front of a structure and serve to direct the flow around it. Wedges can also be incorporated into the structure, e.g. a building may be built with a strong wedge-shaped wall facing the avalanche zone. Similarly the bases of towers may be encased in concrete or earth. The height of the wedge and the forces which they must be designed to withstand can be determined in a manner similar to the procedure used in the design of dykes and walls (see Eqs. 11.10, 11.11 and 11.12).

Galleries, which are now called snow sheds, are roofs designed to allow the avalanche to pass over the object to be protected (see Fig. 11.10). They are generally used on railways and highways and are sometimes also incorporated

Fig. 11.10 Snow shed.

into the design of buildings. Schaerer (1966) reports that galleries are very expensive and their proper design requires reliable estimates of both the width of the avalanche and the forces which will be exerted on the structure by the moving snow. These forces are (see Fig. 11.11):

(1) Friction force due to moving snow, F_f,
(2) Vertical force due to the weight of moving snow, F_v,
(3) Vertical force due to the weight of the deposited snow, F_d, (for galleries located in runout zones the pressures caused by this load may approach 60 kPa),
(4) Force exerted on the side by deposited snow, F_s, and perhaps impact loads caused by avalanches originating on the opposite side of the valley, and
(5) Dynamic force caused by deflecting the moving snow in the vertical direction, R.

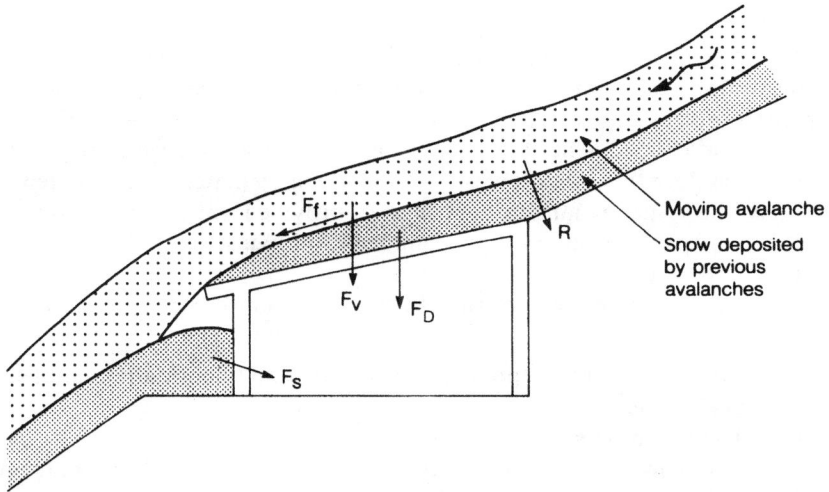

Fig. 11.11 Forces on a gallery.

F_f = friction force produced by moving snow,
F_v = vertical force due to the weight of moving snow,
F_D = vertical force due to the weight of deposited snow,
F_s = force exerted on the side of the gallery by deposited snow or impact loads,
R   = dynamic force caused by deflecting avalanche.

Retarding Works

Retarding works, also called breakers or arresters, are obstacles located in the paths of avalanches, whose function is to slow down or stop the avalanche. These works are effective in controlling wet flow avalanches but they are ineffective against powder avalanches. The most common type, which has proven both economical and efficient, is massive earth mounds (Bünd-nerwald, 1972), which are usually from 4 to 10 m high arranged in two or more rows. Other types are walls and earth and snow dykes, which are placed perpendicular to the avalanche flow direction. Dykes stop very slow avalanches but are easily flooded by rapidly moving large avalanches.

The best location of a retarding structure is on flat terrain near the end of the runout zone. As a general rule, retarding structures should not be built on slopes steeper than 20 degrees.

Protection by Forests

The cause of many early avalanche disasters can be traced to the extensive deforestation accompanying habitation. In Europe, as the population in the mountainous regions increased, forests were often removed to obtain more grazing land (LaChapelle, 1967). In North America, extensive logging has taken place close to mining towns to supply building materials; also extensive cuts have been made along railway lines to provide right-of-way clearances. Avalanches began to run on these denuded mountains where there had been no history of occurrence.

A forest with high trees in the starting zone inhibits the formation of avalanches because:

(1) The tree trunks support the snowcover and anchor a potential slab avalanche,
(2) There is little snow drifting,
(3) The crowns of the trees retain snow and release it gradually to form a stable cover on the ground,
(4) The forest canopy moderates variability in the net energy exchange with the snow surface which tends to produce a uniform snow temperature distribution and stable snow.

To be effective for avalanche control, the forest must be dense, i.e. the spacing between young trees should not exceed 3 m. An open forest offers no protection against avalanches (Martinelli, 1974).

A forest in the track and runout zone would probably stop small and slow avalanches but would not inhibit the progress of large avalanches (de Quervain, 1968). When avalanches break trees, they are carried in the flow, thereby increasing its mass and its destructive power.

In potential avalanche zones the forest must be protected from fire. Also logging projects on steep terrain should be investigated as to their potential impact on the avalanche hazard. The most important consideration is the preservation of trees in potential avalanche starting zones and at ridgetops.

A large avalanche denudes the slopes of trees. These areas can only be reforested with great difficulty; meanwhile temporary supporting structures must be constructed to inhibit avalanches until the newly planted trees are large enough to provide natural control.

Use of Explosives

The prevention and control of avalanches by snowcover modification is a more versatile and usually a much cheaper procedure than terrain modification. However, it is only a temporary measure that must be undertaken every winter. The advantages of creating avalanches artifically are:

(1) Avalanches can be released at a predetermined time when the paths are unoccupied, that is, when roads are blocked to traffic, ski runs are closed, etc. After an avalanche has run, the affected area is considered safe until changes in weather and snowcover produce another potentially dangerous condition.

(2) Snow can be released as several small avalanches rather than as a single large one. Ideally, to protect highways, numerous small avalanches should be produced that would run short of the roads.

The artificial release of avalanches by explosives is the most widespread protective method. Explosives are most effective in inducing avalanches if they are placed in the starting zone at the time when the stress-strength relationship of the snow is critical, yet *before the unstable snow is deep enough to produce large avalanches.*

A primary criterion for effective control by explosives is detection of unstable snow through evaluation of the daily hazard. Since explosives are most successful in releasing dry slab avalanches they are best applied during or immediately after snowfalls. Because of human judgement, delays caused by bad weather, site inaccessibility, and equipment breakdowns, explosives are sometimes applied late so that large avalanches result. Therefore, explosives are not always recommended for protecting buildings and other structures.

Several methods are used to place explosives in an avalanche.

Hand thrown charges: This method is used in ski areas where the starting zones are easily accessible. The charge, commonly consisting of about 1 kg of explosive with a high detonation velocity, a cap with an attached fuse and a pull-wire igniter, is tossed by hand into the area.

Case bombs: The explosives are placed at the base of a starting zone, e.g. on short slopes adjacent to a road. A charge of 20 kg of dynamite may be required to produce a large shock wave that will propagate through the snow or the ground causing snow to be released from areas higher up the slope.

Artillery: If the avalanche starting zone is inaccessible, yet visible, the explosive charge is best delivered by artillery. Military weapons such as mortars, howitzers, recoiless rifles, and bazookas have been used successfully. The avalauncher, a gun that uses compressed gas as the propellant, was developed specifically for avalanche control. Mellor (1968) suggests that the choice of the weapon depends on the range, the accuracy, the cost and availability of ammunition, and safety and manpower requirements. A hazard occurs when shells do not explode and remain in the avalanche path.

Helicopter bombing: Charges containing about 5 kg of explosive are dropped into the avalanche site. This method has proven to be fast and inexpensive where several avalanche sites can be controlled during the same flight. Because good flying weather is mandatory, helicopter bombing can not be used when avalanches must be released during snowfall.

Preplaced charges: The method is suited for sites that are inaccessible in winter and cannot be reached by artillery. Explosive charges, which may be detonated individually by a radio or cable-transmitted signal, are placed in the avalanche starting zone before winter.

Other Types of Snowcover Modification

Protective skiing

Frequent skiing on avalanche slopes either releases the snow in small amounts or increases the strength of the snow by compaction. Compaction brings the snow grains into close contact and in this manner accelerates the sintering or age hardening process. The resulting high-strength snow is less prone to avalanches than natural snow, even when the snow begins to melt.

Protective skiing as an avalanche control method is successful only when it is carried out frequently because the compaction effect is confined mostly to the surface layers. Weak snow at the base of a deep snowcover cannot be stabilized by skiing. For safety reasons protective skiing must be done only on short slopes and with extreme precaution.

Foot packing

Packing snow by foot is a more effective method of avalanche control than protective skiing because a greater depth of snow is compacted. However, foot packing is used only on small areas because considerable human effort is required.

Chemicals

LaChapelle and Stillman (1966) report that chemicals may be used as a control measure in cases where layers of depth hoar are the frequent cause of avalanches. The purpose of the chemicals is to inhibit the recrystallization of snow and the formation of depth hoar. Experiments have shown that if benzaldehyde and N-heptaldehyde are spread on the ground surface or the surface of the first snow, effective control of depth hoar formation resulted. However, at the present time there is no chemical commercially available that is both effective and economical for controlling avalanches under field conditions yet does not damage vegetation or contaminate water supplies.

Safety Measures

Several safety measures are employed in an attempt to reduce the possibility of disaster from avalanches, the most common including: closures, avalanche detection and warning systems, and warning signs. The simplest means of preventing disaster is to impose restrictions on the use of roads, ski runs, buildings and work areas during high hazard periods. The effectiveness of these measures depends on reliable evaluations of the daily hazard by a person capable of recognizing when a dangerous condition may begin and end. Preventive closures of roads with low traffic volume are frequent but are accepted for major traffic routes only when they are short and infrequent.

A major problem in effecting closures as a safety measure is enforcement. Traffic can usually be controlled by warning notices posted on low volume private roads, such as forest and mine roads, but only by strong, physical barriers and police patrols on public highways.

Ski areas are generally too large to be closed by physical means. Signs placed where skiers might enter hazardous areas are usually obeyed for short periods as long as the hazard is obvious to the layman, e.g. during a snow storm.

Avalanche tracks may be instrumented with trip wire, pressure plates, light beams, geophones, or other devices indicating avalanche occurrence. The signal is transmitted to a control center which in turn activates closure signals. Such systems are used on railway lines where a series of trip wires beside the track are connected to the block signals. This method has also been tried on roads, (e.g. on the access road to a ski area at Banff, Alberta) but the difficulties of maintaining the sensors and traffic signals were extensive. Furthermore the road users did not readily accept and obey a red light on a stretch of road with no intersection.

Warning systems consist of a network of field stations that make regular snowcover, weather and avalanche observations and report to a control center. After consulting with the weather office, the center issues an avalanche hazard forecast for the area, indicating, for example, that large avalanches could reach the valley and block highways, or that skiers could release local slab avalanches on slopes of certain aspect. Avalanche warning systems are in operation in some National Parks of Canada, Switzerland, Austria, and for mountain passes in the States of Colorado and Washington (Judson, 1976).

In avalanche-prone areas it is common practice to erect permanent signs on highways to warn travellers of the potential danger. These signs take two forms, "Avalanche area, do not stop" followed by signs which indicate "End of avalanche area". Travellers should not stop in these areas and should watch for avalanche snow deposited on the highway.

Skiers and mountain climbers are the most frequent victims of avalanches. While enjoying these activities skiers can greatly reduce the probability of accidents originating from avalanches by following simple safety rules. Books on mountaineering (e.g. Brower, 1969) include chapters on the recognition of unstable snow, the selection of safe routes, and safety measures.

The preferred routes to travel are located on ridges, on the windward sides of ridges, in the center of wide valleys, and in dense timber. Terrain that should be avoided include steep slopes, gullies, and the leesides of ridges. A further precaution is not to stop or to camp in potential avalanche paths.

The safety rules to follow in crossing dangerous slopes include:

(1)   Traverse the slope at the highest elevation possible,
(2)   Allow only one person in the avalanche path at a time; the others should remain outside,
(3)   Remove hands from wrist loops of poles and open the ski boot safety straps,
(4)   Carry a rescue beacon or wear an avalanche cord,
(5)   Wear jackets, mitts, hat and
(6)   Think about what to do if an avalanche starts; plan an escape route.

Methods of search and rescue for persons buried in avalanches are well developed and are improved continuously by mountain rescue groups. Summaries of these procedures are given in the works of LaChapelle (1978) and Perla and Martinelli (1976).

LITERATURE CITED

Atwater, M.M. 1968. *The Avalanche Hunters*. MacRae Smith Co., Philadelphia, Pa.

Baker, E.J. and J.G. McPhee. 1973. *Land use management and regulation in hazardous areas*. Mono. NSF-RA-E-75-008, Inst. Behav. Sci., Univ. Colo., Boulder.

Brower, D. 1969. *Manual of Ski Mountaineering*. Sierra Club.

Bündnerwald, 1972. *Lawinenschutz in der Schweiz.* (Avalanche Protection in Switzerland)Beiheft No. 9, Selva, Chur, Switzerland, [English Transl. by USDA For. Serv., Rocky Mtn. For. Range Exp. Stn., Fort Collins, Colo. General Tech. Rep. RM-9, 1975].

Commission on Snow and Ice (IASH). 1954. *The International Classification for Snow.* Tech. Memo. 31, Assoc. Comm. Geotech. Res., Nat. Res. Counc. Can., Ottawa. Abstract published in Seasonal Snow Cover; Tech. papers in Hydrol. No. 2, Unesco-IASH-WMO, Paris, 1970.

de Quervain, M.R. 1968. *Die Rolle des Waldes beim Lawinenschutz.* Schweiz. Z. Forstwes. Vol. 119, No. 4/5, pp. 393-399.

Fraser, C. 1978. *Avalanches and Snow Safety.* Scribner's Son, New York, N.Y.

Frutiger, H. 1970. *Der Lawinenzonenplan (The Avalanche Zoning Plan)* Schweiz. Fortwes. Vol. 121, No. 1, pp. 246-276. [English Transl. by Alta Avalanche Study Centre, USDA For. Serv., Rocky Mtn. For. Range Exp. Stn., Fort Collins, Colo. Transl. 11].

Haefeli, R. 1967. *Some mechanical aspects on the formation of avalanches.* Physics of Snow and Ice, (H. Oura, ed.) Proc. Int. Conf. on Low Temp. Sci., Hokkaido Univ., Sapporo, pp. 1199-1213.

In der Gand, H.R. and M. Zupancic. 1966. *Snow gliding and avalanches.* Avalanches and Physics of Snow, Symp. of Davos 1965., Int. Assoc. Sci. Hydrol., Publ. No. 69, pp. 230-242.

Ives, J.D. and others. 1976. *National hazards in Mountain Colorado.* Ann. Assoc. Am. Geogr., Vol. 66, No. 1, pp. 129-144.

Jaccard, C. 1966. *Stabilitè des plaques de neige.* Avalanches and Physics of Snow, Symp. of Davos, 1965. Int, Assoc. Sci. Hydrol., Publ. No. 69, pp. 170-181.

Judson, A. 1976. *Colorado's avalanche warning program.* Weatherwise, Vol. 29, No. 6, pp. 268-277.

Kotlyakov, V.M., B.N. Rzhevskiy and V.A. Samoylov. 1976. *The dymanics of avalanching in the Khibins.* J. Glaciol., Vol. 19, No. 81, pp. 431-439.

LaChapelle, E.R. 1967. *Timberline reforestation in the Austrian Alps.* J. For., Vol. 65, No 12, pp. 868-872.

LaChapelle, 1978. *The ABC of Avalanche Safety.* The Mountaineers, Seattle, Wash.

LaChapelle, E.R. and R.M. Stillman. 1966. *The control of snow metamorphism by chemical agents.* Avalanches and Physics of Snow, Symp. of Davos 1965, Int. Assoc. Sci. Hydrol., Publ. No. 69, pp. 261-266.

Leaf, C.F. and M. Martinelli, Jr. 1977. *Avalanche dynamics: Engineering applications for land use planning.* Res. Pap. RM-183., USDA For. Serv., Rocky Mtn. For. Range Exp. Stn., Fort Collins, Colo.

Martinelli, M. 1974. *Snow avalanche sites, their identification and evaluation.* Agric. Inf. Bull. 360, U.S. Govt. Printing Off., Washington, D.C.

Mellor, M. 1968. *Avalanches.* Mono. III-A3d, U.S. Army Cold Reg. Res. Eng. Lab., Hanover, N.H.

Oberforstinspektorat. 1968. *Lawinenverbau im Anbruchgebiet. (Avalanche Control in the Starting Zone)* Mitteilungen des Eidg. Institutes für Schnee-und Lawinenforschung, Nr. 29, Davos, Switzerland, [English Transl. by USDA For. Serv., Rocky Mtn. For. Range Exp. Stn., Fort Collins, Colo., 1962, Pap. 71].

Perla, R.I. 1975. *Stress and fracture of snow slabs.* Int. Symp. on Snow Mechanics, Grindelwald, Int. Assoc. Sci. Hydrol., Publ. No. 114, pp. 208-221.

Perla, R.I. 1977. *Slab avalanche measurements.* Can. Geotech. J., Vol. 14, No. 2, pp. 206-213.

Perla, R.I. and M. Martinelli, Jr. 1976. *Avalanche Handbook.* Agric. Handb. No. 489, U.S. Govt. Printing Off., Washington, D.C.

Salm, B. 1964. *Anlage zur Untersuchung dynamischer Wirkungen von bewegtem Schnee.* Math. Phys., Vol. 15, No. 4, pp. 357-375.

Schaerer, P.A. 1966. *Snow shed location and design.* J. Highw. Div., Am. Soc. Civil Eng., Vol. 92, pp. 21-33.

Schaerer, P. 1973. *Observations of avalanche impact pressures.* Adv. North Am. Aval. Tech. Gen. Tech. Rept. RM-3, USDA For. Serv., Rocky Mtn. For. Range Exp. Stn., Fort Collins, Colo., pp. 51-54.

Schaerer, P. 1975. *Friction coefficients and speed of flowing avalanches.* Proc. Int. Symp. on Snow Mechanics, Grindelwald, Int. Assoc. Hydrol. Sci., Publ. No. 114, pp. 425-432.

Smith, F.W. and J.O. Curtis. 1975. *Stress analysis and failure prediction in avalanche snowpacks.* Proc. Int. Symp. on Snow Mechanics, Grindelwald, Int. Assoc. Sci. Hydrol., Publ. No. 114, pp. 332-340.

Williams, K. 1975. *The snowy torrents: avalanche accidents in the United States 1967-71.* Gen. Tech. Rep. RM-8, USDA For. Serv., Rocky Mtn. For. Range Exp. Stn., Fort Collins, Colo.

Working Group on Avalanche Classification. 1973. *Avalanche classification.* Bull. Int. Assoc. Sci. Hydrol., Vol. 18, No. 4, pp. 391-402.

PART III

SNOW AND ENGINEERING

519

12

TRAVEL OVER SNOW

K. M. ADAM

Interdisciplinary Engineering Company, Winnipeg, Manitoba.

INTRODUCTION

The difficulty of travel over snowcovered terrain as compared to movement over bare ground has been a source of frustration to man for centuries. The advantage of travel on compacted rather than loose snow was sensed by animals possibly even before the existence of man. The great herds of caribou migrate in single file across the arctic and sub-arctic landscape in winter. The advantage of single file movement through snow was exemplified in early Canadian history by the men of the 104th Regiment from New Brunswick sent to defend Upper Canada against an anticipated American attack (MacNutt, 1963). In the bitter winter of 1813 the unit set out from Fredericton for Quebec. By a kind of leap frog action, each officer and man took a turn at breaking trail, then stepped aside until the others had passed, took off his snowshoes and walked in the rear on the hard beaten path until his turn came up again. Once the unit reached St. Andre, the remainder of the march to Quebec was made easily over an existing snow road.

More recently, compacted snow has been utilized as a road material to good advantage by the pulp and paper industry. When horses were used for hauling logs, compacted snow was formed into guide ruts for the sleigh runners and water added to strengthen the road and to maintain the desired form. Even now, logs are hauled by truck over similar winter roads.

Most provinces of Canada have a winter road program to facilitate transport of supplies to their northern communities. Often construction materials for remote hydroelectric sites, mines and other developments are also hauled over winter roads.

Since the 1960's oil and gas exploration in arctic regions was restricted to the winter months, so that winter roads have become of great interest to these industrial activities. Besides providing a level surface over hummocky terrain, they offer protection to vegetation and the peat layer, thereby maintaining the integrity of the permafrost. Their construction techniques are also applied to constructing level pads for pipeline assembly work in arctic and subarctic regions (Canadian Arctic Gas Pipeline Limited, 1974).

The most significant contribution to oversnow travel in the twentieth century has been the development of a wide range of mechanized oversnow vehicles which can operate away from roads or trails. In Canada, the development of these machines can be largely attributed to the effort of two persons: Joseph Armand Bombardier of Valcourt, Quebec, and W. Bruce Nodwell of Calgary, Alberta.

In 1922, at the age of 15, Armand Bombardier built his first oversnow vehicle from a Model T Ford and parts of a sleigh. By the mid 1930's he had designed a unique drive sprocket and track, which was patented in 1937, and formed the basis for the tracked vehicles which he developed later. In 1959 Bombardier realized his lifelong dream - to perfect a small, personal snowmobile, the Ski-Doo that started a new winter sport and a new industry.

During the winter of 1959-60 Bombardier sold 225 Ski-Doo machines; by the end of the 1964 season, when others were producing similar machines, the total sales of snowmobiles reached 15,000. One decade after the first Ski-Doo was made, about 60 companies were manufacturing snowmobiles, so that by 1971 the annual sales numbered more than half a million.

While Bombardier was perfecting the lightweight snowmobile, Bruce Nodwell was directing his efforts to producing large, heavy duty transportation equipment for the oil industry. In 1953 he adapted a Ford tractor for use over snow and in 1954 he developed a unit that could handle 4,500 kg of seismic drilling equipment in heavy muskeg. This proved to be a break-through; by 1957, 9,000- and 10,000-kg oil well rigs could be moved through severe muskeg conditions with Nodwell vehicles. In 1965, a vehicle called the "Spider" was developed for use in deep snow and was found to be particularly useful in grooming ski slopes. Another version has been used in the far north.

During 1965 Nodwell and five other track-vehicle associates left Robin-Nodwell Manufacturing Limited, an outgrowth of the original company, and together with Jack Nodwell formed Foremost Development Limited. By 1967 they had developed and produced the "Husky", a vehicle with a carrying capacity of 36,000 kg, featuring 142-cm wide rubber and steel tracks, which produce a ground bearing pressure of only 30 kPa, and a two-point steering system which allows sharp turns to be made.

The evolution of this wide range of off-road vehicles as well as the use of various types of winter roads has reduced considerably the difficulties of travel over snowcovered terrain and made exploration and development activities possible in areas previously inaccessible to the great numbers of men and heavy machinery required for these twentieth century activities.

In this chapter the development of oversnow vehicles in North America is briefly outlined. The characteristics of each type of winter road are described along with the techniques for their construction. Finally the environmental impact of oversnow vehicles and snow roads are mentioned.

OVERSNOW VEHICLES

Development of Oversnow Vehicles

Much of the material contained in this section is derived from the paper of Wilson and Nelson (1968) who outlined the development of oversnow vehicles in the United States.

In 1913 Virgil D. Wright of New Hampshire invented what was possibly the earliest oversnow vehicle (see Fig. 12.1). It was a modified Ford with skis attached to the front. Both sides of the machine had cleated traction belts which were powered by the rear wheels and looped around them and an additional set of wheels mounted on the back which served as idlers. Keels were fitted to the skis for steering and to prevent side-slipping. Wright worked on these modifications for several years perfecting the details before the machine was marketed in 1922.

Fig. 12.1 "Snowmobile Attachment" Model T Ford (Courtesy of Western Snow Conference).

In 1922, Armand Bombardier built the first propeller-driven snow vehicle; at about the same time, E.M. Tucker invented a unique spiral or auger-type traction system in which two spiral augers were mounted on cantilevered springs parallel with the direction of travel (Fig. 12.2). An air-cooled engine powered the augers with most of the weight being carried on three skis. In 1926 the drive was reduced to one spiral.

Fig. 12.2 Double Spiral Vehicle (Courtesy of Western Snow Conference).

The Eliason Motor Toboggan (Fig. 12.3) was developed in the 1930's, consisting of a cleated track suspended through the bottom of a toboggan, powered by a motorcycle engine. Front twin runners were used to steer the toboggan which had good speed but little ability to climb steep slopes.

By 1937, Armand Bombardier had patented the sprocket-track design that allowed him to develop larger oversnow vehicles used primarily by the logging industry.

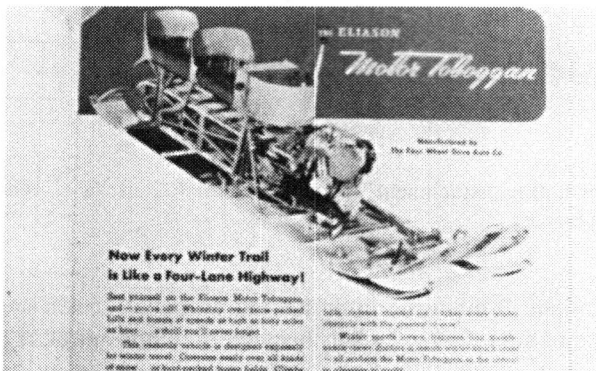

Fig. 12.3 Eliason Motor Toboggan (Courtesy of Western Snow Conference).

During the 1940's several oversnow vehicles were developed, many undoubtedly inspired by the war effort, e.g., in 1941, the "Iron Fireman", for use by the U.S. Forest Service (Fig. 12.4). This machine was powered by a 71-kW engine mounted inside a large pontoon. The track encircled the entire pontoon. A sled behind the power unit was used for steering; lateral stability was achieved by supports placed on both sides of the sled which rode on the snow. This machine was designed for heavy loads, and being heavy itself, moved very slowly.

Fig. 12.4 Iron Fireman (Courtesy of Western Snow Conference).

A vehicle powered by a large, cleated drum was built by Robert Allway at Montana State College in 1942 (Fig. 12.5). The power unit was mounted inside the drum while steering was accomplished by a sled located in front. This unit probably had little success in deep snow.

The need for oversnow vehicles by the military resulted in the development of the first metal-track oversnow vehicle, the M-7, early in World War II (Fig. 12.6). This vehicle had a two-passenger cab mounted between two tracks consisting of metal pads fastened together with rubber-clad cable. The tracks were driven by large, rubber-coated sprockets. The power unit was a conventional 4-cylinder water-cooled jeep engine with power being transferred to the sprockets through a conventional transmission and modified differential. Various grousers for the M-7 were tested under different snow conditions (Diamond, 1956). The "Weasel" having a power supply drive and track assembly similar to that of the M-7, was also developed during World War II (Fig. 12.7). Larger than the M-7, it had a seating capacity of four to six passengers, and had a different steering mechanism incorporating a clutch-brake assembly for each track. The "Weasel" was found to be very useful for light hauling when coupled to a light cargo sled (Lanyon, 1961). Detailed trafficability tests in deep snow have been performed with this vehicle, (for the results see Knight, 1965).

Fig. 12.5 Cleated Drum Vehicle (Courtesy of Western Snow Conference).

Fig. 12.6 M-7 (U.S. Army) (Courtesy of Western Snow Conference).

Fig. 12.7 Weasel (U.S. Army) (Courtesy of Western Snow Conference).

The Tucker Sno-Cycle (Fig. 12.8), developed during the mid 1930's became one of the better-known American oversnow vehicles. An automotive engine, transmission and differential was used to bring power to an open link endless track on roller bearings. The machine was supported on the snow by oblong steel pontoons around which the track was driven by metal sprockets. A later adaption called the Tucker Kitten was developed in the early 1950's, but did not perform as well as the M-7 (Diamond, 1956). A large, four-pontoon Tucker Sno-Cycle using a four-wheel drive was also developed (Fig. 12.9). Independent tracks on each wheel allowed the vehicle to be steered by the front tracks.

Fig. 12.8 Tucker Sno Cycle (Courtesy of Western Snow Conference).

Fig. 12.9 Tucker 4-Pontoon (Courtesy of Western Snow Conference).

Another vehicle developed in the early 1940's was the Montana Snow Bug, which consisted of a toboggan powered by a washing machine motor (Fig. 12.10). Later the toboggan was replaced by a pontoon while the unit was equipped with a 15-kW air-cooled engine and a transmission that gave nine forward and two reverse speeds. By 1960, a two-pontoon "Bug" had been produced but it was clumsy and steering was a problem.

Fig. 12.10 Original Montana Snow Bug (Courtesy of Western Snow Conference).

The Hobson Snow Traveller (Fig. 12.11) was developed in Idaho in the mid-1940's. It resembled a large toboggan with a protective cab, and was steered by turning the front toboggan-like unit and was powered by a Wisconsin combine motor driving a wooden-cleated track. This machine was used for snow surveys in southeastern Idaho until 1956.

Fig. 12.11 Hobson Snow Traveller (Courtesy of Western Snow Conference).

Also during this period the Nelson Sno-Poke was developed by Ivan Nelson at Boise, Idaho. It was one of the first vehicles to use a single-track, but later its design was changed to a double-belt track. Hardwood cleats were attached to the belts for traction, and steering was executed by turning two front-mounted skis. The early model had a remarkable resemblance to the modern-day snowmobile.

Between 1940 and 1945, the Agricultural Engineering Department of Montana State College modified an old Chevrolet into an oversnow vehicle. The power unit was constructed by shortening the chassis of the car, removing the steering mechanism on the front wheels, and placing an endless belt track containing wooden cleats around the front and rear wheels on each side (Fig. 12.12). It was steered from a sled mounted behind the power unit using reins to apply individual braking against the differential. In 1945 the belt track was replaced by the Ford Motor Company's "Blue Bird Snow Chains" which were then in common use on the western plains for coverting Model A Fords to oversnow vehicles. In 1943 a larger two-passenger unit with longer chassis, was built, having a belt track with wooden cleats and equipped with a conventional auto-steering mechanism with skis attached to the front wheels.

Fig. 12.12 Modified Overland Automobile (Courtesy of Western Snow Conference).

Fig. 12.13 Early Frandee (Courtesy of Western Snow Conference).

In the mid-1940's Ross Eskelson and the Utah Scientific Foundation developed the Eskelsled which was powered by a motorcycle engine which drove a single-cleated track mounted around the frame and steered by a front ski. However, it had little use until it was modified into a double track unit. By the late 1940's the Foundation achieved some success with several models of a vehicle, called the Frandee (Figs. 12.13, 12.14), which was equipped with a tandem, cleated, endless belt track around idler and drive wheels, and used either brake or ski steering.

Fig. 12.14 Frandee Snoshu Cab-over (Courtesy of Western Snow Conference).

In 1956 the Utah Scientific Foundation produced its largest oversnow vehicle called the Trackmaster (Fig. 12.15), which was essentially an enlarged Frandee but with individual transmissions for each track. Turning was accomplished by speeding up one transmission and slowing the other. The tracks were steel-cleated, endless belts driven by rubber-coated sprockets.

Kam Industries of Boise, Idaho produced their first oversnow vehicle, called the Sno Ball, in 1957 (Fig. 12.16). This had a double belt assembly with dual drive wheels and idlers, and a planetary differential for steering. A unique feature of this vehicle allowed most automobile cabs to be fitted over the power unit by removing the running gear and differential, mounting the chassis over the tracks, fastening the springs to the axles of the track assembly and connecting the drive line to the planetary differential. The final models of the Sno Ball used aluminium custom cabs and framing.

Fig. 12.15 Trackmaster (Courtesy of Western Snow Conference).

Fig. 12.16 Kam Sno-Ball (Courtesy of Western Snow Conference).

In the late 1950's the Kristi oversnow vehicles were manufactured using airplane-type construction with tubular framework and were equipped with a cab encased by a thin metal skin (Fig. 12.17). A Volkswagen engine powered the metal-cleated, double-belt track through a standard transmission and differential. A unique feature was a hydraulically-operated cab leveller that allowed the cab to remain level while negotiating hillsides. The first models of this machine had the motor in the rear but later versions had the power unit in front.

Fig. 12.17 Kristi - Motor in Front (Courtesy of Western Snow Conference).

In 1957, Armand Bombardier produced the first commercially-successful, small, single-track vehicle (Fig. 12.18), the "Ski-Doo". It was followed by many similar single-track vehicles, all of which had the same appearance and basic construction as the "Ski-Doo". As many as 60 independent companies have produced and marketed this type of snowmobile in recent years. Usually they are powered by two- or four-cycle engines which drive a single track from 32.5 to 75 cm wide fitted with metal, hardwood or rubber cleats. The vehicle is steered by either one or two front skis attached to handle bars which also provide hand control of the throttle and braking systems. The average machine can climb slopes of 45%; and racing models can travel at speeds well over 100 km/h.

Fig. 12.18 Skidoo - Alpine (Courtesy of Western Snow Conference).

The snow plane vehicle was modified further in the 1960's (Fig. 12.19) and included an airplane type construction. It was pushed forward by a propeller powered by an airplane engine, fastened behind the cab, which was mounted on three skis, the one in front being used for steering. Braking was effected by a drag system also mounted on the skis.

Fig. 12.19 Snow Plane (Courtesy of Western Snow Conference).

Large industrial-type oversnow vehicles have been improved continuously since 1937, when the Bombardier Company patented the sprocket track design. Currently, Bombardier offers a wide range of medium to lightweight double tracked vehicles including the Muskeg Carrier, Muskeg Tractor, Qua-Trac, J-5 Tractor and the Skidozer. The Skidozer (Fig. 12.20) is a wide-track, lightweight vehicle having a low ground pressure which allows it to traverse deep, soft snow with ease. It has been very useful as a grooming vehicle for ski trails. The Bombadier Snowmobile Bus (Fig. 12.21) can travel at speeds up to 80 km/h and carry 12 passengers, and is widely used for transporting people in snowbound areas.

Since Nodwell built his first tracked vehicle in 1953 his companies have evolved a full range of tracked or wheeled vehicles in the medium-to-heavy payload range which are now marketed under the name of Foremost International Industries Ltd., located in Calgary, Alberta. Tracked vehicles are produced in a number of series: TVS, Yukon, Dawson and Husky, which are capable of carrying loads of 8000 to 36,400 kg. The Husky 8 (Fig. 12.22), recognized as the world's largest off-road tracked vehicle, only exerts a

Fig. 12.20 Bombardier Skidozer 252.

Fig. 12.21 Bombardier Snowmobile Bus.

Fig. 12.22 Foremost Husky 8 (Courtesy of Foremost International Industries Ltd.).

ground bearing pressure of about 31 kPa under full load. It features a two-point steering system, and is powered on the four-track principle (the same as vehicles in the Dawson series). The TVS and Yukon Series are two track carriers. Foremost's wheeled vehicles are the Delta 2, Delta 3, Delta Commander and HDRV (Heavy Duty Recovery Vehicle). The Delta 2 (Fig. 12.23) has two axles, a four-wheel drive and a capacity of 4,500 kg; it can negotiate 1.2-m snowbanks with its large tires, which work effectively at pressures from 34 to 2000 kPa (ground pressures exerted by a vehicle are normally equivalent to tire pressures) and can be air freighted to remote areas by Hercules Aircraft. The Delta 3 has three axles, can handle up to 13,600 kg, and can operate on a 60% forward grade and a 50% side slope. Available models are equipped with a four- or eight-man cab (Fig. 12.24). It is useful in oil fields and pipeline operations, moving rigs, hauling freight and fuel and transporting materials. The Commander (Fig. 12.25), with a payload of about 27,000 kg, is the largest all-terrain vehicle. A unique feature of this vehicle allows tire pressures to be adjusted to values suitable for the terrain by a self-contained air pressure system. It is competitive with trucks even on well-constructed roads. The HDRV (Fig. 12.26), Foremosts's wrecking vehicle, has a six-wheel drive and a telescopic boom which can swing and load up to 10,000 kg. The machine is suitable for commerical, civil and military applications wherever there are off-highway recovery problems.

Fig. 12.23 Foremost Delta 2 (Courtesy of Foremost International Industries Ltd.).

Fig. 12.24 Foremost Delta 3 (Courtesy of Foremost International Industries Ltd.).

Fig. 12.25 Foremost Commander (Courtesy of Foremost International Industries Ltd.).

Fig. 12.26 Foremost Heavy Duty Recovery Vehicle (Courtesy of Foremost International Industries Ltd.).

The Utah Scientific Foundation turned over production of its machines in the early 1960's to the Thiokol Chemical Corporation, which modified the existing models and developed new ones. Many of these are modifications of the earlier Frandee and Trackmaster models. The most significant change in the original models is in the steering mechanism, which is now a planetary controlled differential system allowing braking and power transfer to individual tracks. This company produced a large quad-track model called the Thiokol Juggernaut in 1963.

Canadian Flextrac, a company formed in 1969, whose assets were acquired by Foremost International Industries in 1976, built a full range of all-terrain tracked vehicles. Their CF 23 carries a driver and co-driver and has a payload capacity of about 900 kg, exerting a ground pressure of 8.25 kPa. The CF-10/5 Axle Floater is a lighter vehicle but like the CF 23 is amphibious, and exerts a low ground pressure. The CF 20 model has about the same capacity as the CF 23, but is slightly faster and has a fording depth of 80 cm. The FN20W/B has a payload of 1600 kg and slightly higher ground bearing pressure (10 kPa) when fully loaded. In all, the company produced vehicles which will carry payloads in the range 900 to 27,000 kg.

Canadair Flextrac also offered three models of wheeled vehicles with low ground pressure tires. The CF 100-TT can carry a payload of 4,500 kg at a ground pressure of 28 kPa; the FN Norcan 200, 9,000 kg at 31 kPa, the FN Norcan 300, 13,650 kg at 38 kPa.

Environmental Effects of Oversnow Vehicles

The use of oversnow vehicles affects the environment in several ways. The effects of noise, particularly on wildlife, is a major concern. Terrain degradation is potentially serious in permafrost areas, since the vegetative cover and the integrity of the peat layer can be damaged.

Industrial oversnow vehicles

Little terrain damage will result from using industrial oversnow vehicles provided sufficient snowcover (depth and extent) exists. However, if there is insufficient snowcover or if vehicles are used too early in the fall or late in the spring, severe terrain damage will result even on properly constructed roads, particularly in permafrost areas. To maintain the integrity of the peat layer, winter roads should be constructed so that vehicles do not significantly compact or wear the layer, preserving its insulative capacity and preventing permafrost degradation. In permafrost areas where winter roads have been inadequately constructed or improperly used, or where summer travel by low ground pressure vehicles has taken place, damage to the terrain is manifested by rutting, thermal erosion, gullying on slopes, and formation of thermokarsts. The importance of maintaining the peat layer and vegetative cover in those areas cannot be overemphasized - a conclusion following from many observations in such areas (Porkhaev and Sakovskii, 1959; Brown et al., 1969; Murrmann and Reed, 1972; Hernandez, 1973).

The terrain disturbance caused by low ground pressure vehicles is well-documented by Hok (1969) and Burt (1971). In general, off-road travel in summer in permafrost regions should be avoided even with low, ground-pressure vehicles. In non-permafrost terrain these vehicles can be used with little permanent damage to vegetation.

Snowmobiles

Most of the environmental damage resulting from snowmobiles can be avoided if snow trails are used. These trails limit the impact of machine noise on wildlife or humans. They can be located along existing paths or tracks thereby limiting the number of trees to be cut and, where necessary, can be routed to avoid ecologically sensitive areas.

The snowmobiler can be enticed to use snow trails if their routes are planned through scenic areas and if they are properly constructed and maintained. Each trail should be at least 3-m wide with traffic restricted to one direction. A trail can safely handle about 12 snowmobiles per kilometre per day. Those at least 100 km long are the most attractive, provided interconnections exist. They should be prepared in summer to be free of brush, trees, rocks and stumps; with overhanging branches clipped back to 2.5 m above the average snow depth. Slopes greater than 2 to 1 and curves on steep grades should be avoided. Curves should be gradual and have a radius of curvature greater than 8 m. Standards and methods of trail maintenance and grooming are available (Bombardier Limited, 1972). Environmental damage by snowmobilers could be limited if snowmobilers practised the "Snowmobilers Code of Ethics" (Bombardier Limited, 1971). This document sets conduct guidelines which, if followed, would improve the public image of the sport.

Injuries and fatalities in snowmobiling mainly result from collisions with fixed objects, motor vehicles, trains and other snowmobiles or from falling off or being thrown from a snowmobile. Most accidents could be avoided if snowmobile trails were used more extensively and if the Code of Ethics were followed by all participants in the sport.

A 1970-survey of snowmobile owners showed that about 47% of all snowmobile use is for recreational trail riding, 40% for other recreational use, 8% for hunting and fishing, 3% for required transportation and 2% for racing. With well over 2,000,000 snowmobiles in North America, the potential for environmental damage from recreational use alone is large. Noise is the greatest single objection to snowmobiles (Committee on Environmental Quality, 1970). Other adverse effects from snowmobiles and snowmobilers include damage to trails and vegetation, destruction of fish and wildlife, conflicts with other forms of recreation, vandalism, theft, trespassing and liability (Baldwin and Stoddard, 1973). Strict regulations for off-road vehicles, better law enforcement and zoning of areas for vehicle use are recommended to minimize these effects. A bibliography of off-road vehicles compiled by Lodico (1973) includes topics on snowmobiles. The environmental impact has been considered extensively as it relates to vegetation, mortality of small mammals, and noise (Chubb, 1971; Masyk, 1973). Mechanical compaction of the snow reduces the snow depth and destroys the subnivean air spaces while increasing the density and thermal conductivity. These effects in turn inhibit the movement of small mammals beneath the snow and produce lower temperatures which subject the mammals to greater temperature stresses.

The beneficial uses of snowmobiles are many. They are used to advantage by law enforcement agencies, in the feeding of wildlife during severe winters, for hauling cargo, trapping, power and pipeline surveillance, and for farm and ranch use. In the Arctic and Subarctic they have been used extensively by Indians and Eskimos, and also by the Skott Lapps of North-eastern Finland for herding reindeer. However, circumstantial evidence indicates that the use of snowmobiles in herding operations has adversely affected the number of calves born in the spring (Pelto, 1973). Some Eskimos have abandoned snowmobiles because of their unreliability in very severe weather conditions; the warmth that can be provided by sled dogs cannot be replaced as a source of heat for survival.

WINTER ROADS

Types

The term winter road refers to any of the following four types of road, built of snow or ice, that remain functional only during the winter season:

(1) Winter trail - established by a single pass of a wheeled vehicle using a "blade" if necessary to gain access. It is not usually acceptable in permafrost regions since surface smoothness is often obtained by blading off the tops of hummocks.

(2) Snow road - which can be either a compacted, or a processed snow road; i.e., a road in which the snow is "agitated" or "processed" before compaction.

(3) Ice-capped snow road - a snow road to which water has been added to produce a bond between snow particles and thus give added stability to the roadway.

(4) Ice road - constructed by sprinkling water on the surface to produce an ice surface of suitable thickness to support traffic (Adam, 1974); for areas with a scarcity of snow. Ice aggregate roads constructed from ice chunks chipped from the frozen surface of lakes and rivers are a more recent form of ice road, particularly on the north slope of Alaska.

Another type of winter road, also often referred to as an ice road, is a road across frozen lakes or rivers. This type is very common because its surfaces are naturally smooth, no clearing of trees is required, and overall road preparation is minimal. To avoid confusion this type is referred to as a "winter road on ice".

Required Properties for Trafficability

The season of winter road use refers to the period when the road remains functional for its intended purposes. In permafrost regions, this period is related to those structural and physical properties of the road which offer protection to the underlying surface.

The time to begin construction of any type of winter road occurs when frost has penetrated deeply enough into the ground so that it can support a light vehicle. If snowcover inhibits frost penetration, low ground-pressure vehicles can be used to compact the snow. The resultant increase in thermal conductivity and decrease in snow depth would then considerably reduce the insulating capability of the snowcover. Hence the rate of frost penetration would be increased so that heavier construction vehicles could be used to complete the road construction at an earlier date.

The winter road season ends when the snow or ice surface starts to deteriorate naturally. Runoff from the roadway itself generally occurs before the adjacent snowpack starts to melt. This is partly due to the lower albedo of a well-travelled winter road because of its darker surface.

The required properties for winter road trafficability are critical whenever the temperature of the surface approaches $0°C$. At high latitudes normally this occurs only twice a year; early in the fall and late in the spring. At these times the bearing strength of the surface is low so that traffic penetrates and destroys the road surface. The winter road thus becomes nonfunctional due to rutting and roughness, and in permafrost regions its ability to protect the terrain surface is lost.

The ability of a winter road to withstand traffic can be expressed in terms of the density and hardness of the snowcover. In general, the density of dry newly-fallen snow ranges from 100 to 250 kg/m^3. Changes in snow density and hardness resulting from environmental conditions are discussed in detail in Ch. 7 but are mentioned below if relevant to the construction of winter roads. Newly-deposited snow is initially subjected to a process of isothermal or destructive metamorphism characterized by a rapid decrease in porosity and an increase in density. During this process the dendritic snow crystals decompose into single fragments. Then, the smallest particles dissipate in favour of larger ones until an aggregate of rounded, isometric, oblong or irregular grains is formed.

While the snowcover is undergoing destructive metamorphism, bonds develop between the grains which are in contact, resulting in an increase of the strength of the snow. This process, known as sintering or age hardening, commences at the time of snowfall and becomes significant once the initial density increase is nearly ended. Ramseier and Keeler (1966) suggest that the major mechanism responsible for the growth of bonds is evaporation-condensation, i.e., evaporation occurs on the convex parts of the aggregate where the higher vapor pressure promotes mass transport. The water vapor then diffuses through the local voids and condenses where the grains are in contact because of the lower vapor pressure there.

Sintering, or age hardening, is the most important process when planning construction of snow roads. For snow to withstand wheeled traffic, studies show that it must be compacted to an average density of at least 500 kg/m^3. This compaction process breaks up the existing bonds and the snow grains resulting in a more uniform distribution of grain size and an increase in the number of contacts between grains. Sintering begins immediately after compaction is completed with the most rapid increase in strength occurring in the first two or three days.

A third metamorphic process, constructive metamorphism, operates in a deposited snow layer as soon as there is a temperature gradient. This process is characterized by the growth of selected crystals and a decrease of crystal number per unit volume. A brittle aggregate of reduced density and strength called depth hoar results. Normally, the best conditions for depth hoar formation occur in the lower layers of the snowcover. The basic mechanism involved is diffusion. Water evaporates from a warmer particle into the adjacent pore space and then condenses on a neighboring colder particle. It is assumed that the vapor pressure of each particle is approximately at saturation with respect to the local temperature. Once constructive metamorphism has started it is very difficult to stop. For this reason compacted snow roads are more easily constructed with new snow rather than older snow which has been subjected to a temperature gradient for some time.

The rammsonde hardness number R is the generally-accepted index of snow hardness. Although an arbitrary index, it indicates the resistance offered by snow to vertical penetration of a standard metal cone rammed into the snow surface. The hardness reading for the depth reached by the tip of the cone represents the mean hardness through the depth of penetration, where hardness is defined as the ratio of the pressure to the depth of penetration

(United States Navy, 1955). Hardness of compacted snow depends on geographical location, season of the year, type and condition of the snow, ambient and snow temperatures, and the length of the age hardening period (Camm, 1960). Strength is also related to hardness and, to a lesser extent, temperature and crystal size (Gold, 1956).

The Rammsonde hardness instrument is a cone penetrometer consisting of a hollow, 2-cm diameter stainless steel shaft with a 60° conical tip, a guide rod and a drop hammer. The standard cone has a diameter of 4 cm and a height of 3.5 cm; the total length of the penetrometer cone element (to the beginning of the shaft) is 10 cm. The guide rod, inserted into the top of the shaft, guides the drop hammer. Complete details of the testing procedures and equipment are given by Clark et al. (1973, Appendix B).

The ram hardness number R is computed as

$$R = (Whn/X) + W + Q, \qquad\qquad 12.1$$

where: W, Q = mass of the drop hammer,
                  and penetrometer, respectively (kg),
          h = height of drop (cm), and
          X = penetration after n blows (cm).

Because the loading end of the instrument is conical, wide ranges of hardness numbers may be obtained near the surface. To compensate for this effect the R reading for the 0 to 5 cm depth must be multiplied by 4.7, for the 5 to 10 cm depth by 1.6, and for the 0 to 10 cm depth by 3.0 to obtain the correct vlaue of R (Wuori 1963; Clark et al., 1973).

It has also been suggested that hardness values are not reliable when R > 800. For winter roads this poses no problem since surfaces with R > 800 will certainly handle most, if not all, wheeled vehicles. However, this problem might limit the instrument for indexing the hardness of snow runways used for heavy aircraft.

Studies revealed that the relationship between R and density ρ (kg/m³) for processed snow is of the form:

$$R = 3.826 \, \rho - 1618 \qquad\qquad 12.2$$

(Joint Snow Compaction Program, 1954). They also showed that R > 350 is necessary to carry repeated passes of military traffic. Tests conducted at Norman Wells, N.W.T. indicated that a value of R > 350 is necessary to carry wheeled vehicle traffic used for pipeline construction (Adam and Hernandez, 1977).

Attempts have been made to use several existing tests for soils to evaluate snow as a construction material, e.g., the California Bearing Ratio (CBR) test and the modified Standard Proctor test. However, their application to snow has met with only marginal success. The CBR values fall within a narrow, limited range; the Proctor mould requires a cover plate to retain snow within it and more compactive energy to obtain reproducible densities. According to Weber and Zyla (1975) freshly-fallen snow subjected to loads exhibits a large change in volume and compacts easily but will compact only to densities of 320 to 390 kg/m^3. Higher densities in the range 490 to 520 kg/m^3 were obtained with older, granular snow and a mixture of old and new snow which compacted to densities of 520 to 540 kg/m^3 under a given compactive force. These data support the results of field tests which show that higher densities may be obtained by mixing or processing snow.

To date, the rammsonde instrument has been found to be the most useful instrument for producing a reliable index of snow strength and for monitoring the age hardening process. However, even a large number of hardness measurements made during the construction of a road may not provide a reliable indication of its capacity for supporting traffic-loads, since undetected soft spots may be present (Clark et al., 1973). Adam (1973) cites an example of this problem in tests conducted on snow roads at Norman Wells, N.W.T.

Construction Techniques

Each type of winter road entails a different construction method which is frequently selected according to intended road use. Wheel loads, tire pressures, and number of passes of the traffic expected must be considered in planning. A few passes of tracked equipment would necessitate less rigid standards. Data on future volume of traffic is needed before construction to avoid over- or under-designed roads and to eliminate unnecessary expense.

Winter trail

Seismic operations often use winter trails, which are usually constructed by a single pass of a wheeled or tracked vehicle using a blade, if necessary, to gain access. Initial access is frequently made with a caterpillar tractor capable of moving heavy snow and levelling ground features that would impede less maneuverable vehicles. Winter trails through forested regions are not particularly suited to rubber-tired vehicles because of the danger of tire puncture by sharp stumps or debris.

Winter trails are not normally used in permafrost terrain because of the likelihood of severe surface disturbances that may lead to ground subsidence of the ice-rich soils; blading of the ground surface should be avoided. On non-permafrost terrain they should be used only when a few passes of tracked vehicles are anticipated.

Snow roads

Both the compacted snow road and the processed snow road have received considerable attention in the literature. The compacted road is normally constructed by repeated passes of a caterpillar tractor which pulls a snow drag or roller. Compacted snow roads are used either to obtain a smooth surface over hummocky terrain or to give surface protection in permafrost regions.

The construction of a processed snow road is carried out in two stages: (1) breaking up and mixing the snow particles by means of tooth or disc harrows, rotary snow plows or other specifically designed equipment and, (2) compacting the snow. Processing the snow enables a higher density surface to be achieved.

Compacted snow roads

Construction usually begins using a light-tracked oversnow vehicle to compact the early snowfalls. This initial compaction destroys the insulating effect of the snowcover and allows muskeg and bogs to freeze to sufficient depths to carry heavier vehicles. After a solid working base is established the final preparation of the snow road commences.

The compacted snow road requires snow to be physically moved onto or off the roadway to establish the desired snow thickness, or to be compacted in situ. Usually this is accomplished by blading the snow into position with a caterpillar tractor, then compacting it by several passes of the same vehicle, often pulling a snow drag or roller. Frequently, a heavy snow drag is used to level the road by cutting off mounds and filling depressions. However, the surface of the resulting road is usually not hard enough to withstand repeated loadings of a heavy-wheeled vehicle (Camm, 1960). Drags and rollers tend to harden snow roads more at the top than at the bottom. Figure 12.27 is a schematic representation of the change in the vertical hardness distribution of a compacted snowcover after the use of rollers.

Ager (1955) reports that in Sweden compacted snow roads constructed without depth processing will not stand up to wheeled vehicles. But, like the United States Navy (1955), he found that hardness increases with time or with weather conditions (e.g. decreasing temperature). This phenomenon of age hardening (or sintering), is widely recognized and used to advantage by winter road builders. The increase in hardness of snow with time due to sintering is illustrated in Fig. 12.28 (Adam, 1973) which also indicates a decreasing trend of hardness with increasing temperature. If the road is allowed to "age" a couple of days after final compaction it can withstand greater loads than those which would have destroyed it immediately after final compaction.

Other Swedish field tests (Leijonhufvud, 1955) showed that surface compaction by drags or rollers results in about the same hardness and can provide a load-bearing capacity sufficient to withstand wheeled tractors for

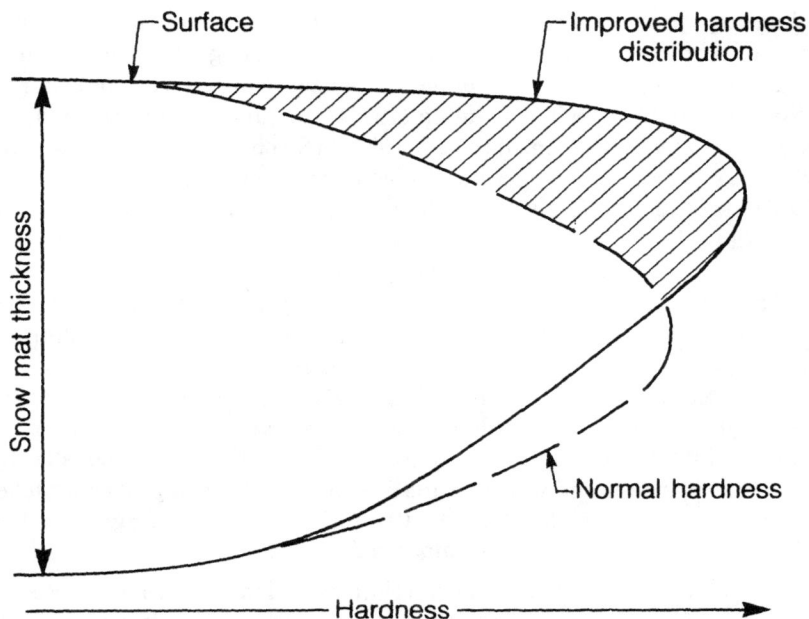

Fig. 12.27 Improved hardness distribution in compacted snow after special rolling (U.S. Naval Civil Engineering Laboratory, 1962).

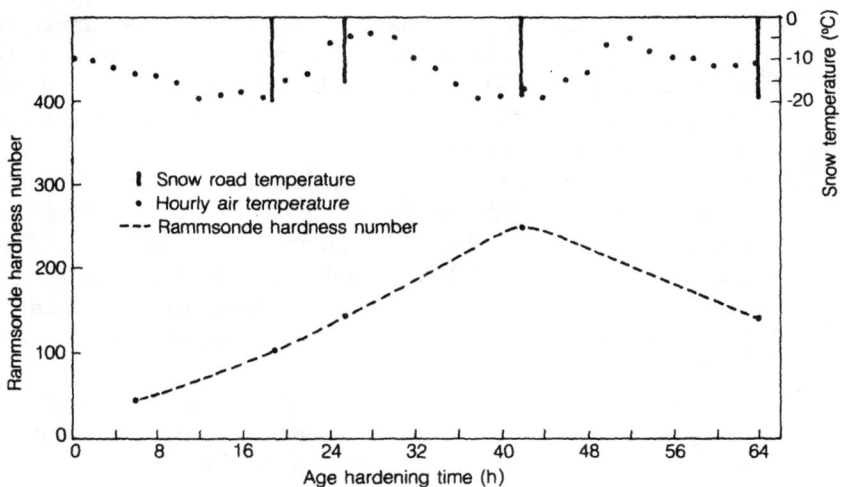

Fig. 12.28 Rammsonde hardness number as a function of age hardening time and snow temperature (Adam, 1973).

hauling logs provided the road gradient does not exceed four percent; in addition surface hardness increased with decreasing snow temperature though with a slight lag. This phenomenon was also observed in the subarctic at Norman Wells, N.W.T. by Adam (1973). Leijonhufvud concluded that compacted snow roads seemed to be feasible in Sweden where the snow depth is 40 cm and the mean temperature is -6°C or lower.

Rollers are more effective than drags in compacting deep snow (United States Navy, 1955). Tests indicated that one hour after several passes of a roller, the snow surface was twice as hard as after a single pass. After 12 hours, the hardness was 8 to 10 times that obtained by a single pass. If no age-hardening period is allowed, working hard dry snow only results in churning of the snow with no increase in compacted hardness.

Several extensive snow compaction trials were carried out in North America in the early 1950's using drags and rollers of various designs and weights (Joint Snow Compaction Program, 1954). All drags were constructed on the same principle and were essentially frames with cutting edges attached to them. The rollers weighed from 1450 to 10,000 kg. The findings from these trials and from other tests are summarized below.

(1) Tests on vehicles at Fort Churchill, Man. indicated that their speeds and the tire pressures are important factors in determining the compacted snow road trafficability. If either speed or tire pressure must be reduced below normal recommended levels the road either is inadequately compacted or an adequate density cannot be achieved.

(2) Tests at Kapuskasing, Ont., showed that heavy rollers compacted snow in two or more passes to 540 kg/m³, regardless of type. To increase the density to 600 kg/m³, (which ensures the snow will not fail under heavy traffic), other procedures had to be used such as adding water to decrease the porosity,

(3) In tests at Norman Wells, N.W.T. light drags (less than 1000 kg) produced snow densities of approximately 540 kg/m³. However, expected rammsonde hardness numbers, (greater than 300) as calculated by Eq. 12.2, were not obtained. The largest value was 167. The compacted snow road failed with less than 25 passes of a 15-passenger bus having a tire pressure of 200 kPa (Adam, 1973).

(4) Camm (1961) reports that a snow compacting roller, 4,650 kg in weight and 2.4 m in diameter, produced effective compaction at speeds up to 9 km/h. The equipment could be maneuvered in all types of snow except extremely soft, deep, new snow and could be operated at temperatures to -45°C. A 13-wheeled, pneumatic-tired roller was most effective as a surface compactor when the tire penetration was 1.27 cm or less.

(5) Camm (1960) tested different snow-levelling and snow-finishing drags. He found that a 420 kg, Douglas fir drag was needed to maintain a level surface for other compaction equipment but that a 1,300 kg, steel finishing drag was needed to produce hard, smooth finishes on compacted snow areas.

(6) Tests with a relatively lightweight 465 kg, 2.4-m diameter roller, developed by the U.S. Navy for high speed compressive compaction of large snow areas, showed that the effective depth of compaction of natural snow by rolling alone is 20 to 25 cm (Moser, 1963).

A compacted snow road can be constructed by using heavy rollers or drags but heavy rollers and processing are generally needed for snow roads that are expected to carry wheeled vehicles. Drags are required mainly for the production of a level surface and for general maintenance.

Processed snow roads

Processed snow roads are used where roads capable of carrying the traffic cannot be constructed by compaction alone. Processing of snow before compaction enables a denser road surface to be constructed. The purpose of processing is twofold: to break up snow particles, thereby increasing the number of contact points; and to mix layers of snow of different temperatures and consistencies. Processing at low temperatures increases the possibility of bonding between particles and not only increases the snow density but also promotes and accelerates age hardening.

Processing of snow before compaction entails using tooth or disc harrows, pulvimixers, pulvimixers supplemented by heat, or rotary snow plows. Since tooth harrows have no moving parts, they are easy to maintain in a cold climate. They are usually constructed of structural steel sections with steel bar teeth having various spacings and projections. Tooth harrows are of limited use in deep snow because of their tendency to ride over the snow. This difficulty can be overcome by adding weights. However, tooth harrows do not produce vertical intermixing of snow layers and they do not leave a smooth surface. Disc harrows are not as successful for processing as tooth harrows. Usually the draw bar and axle of these harrows are set too low so that under deep snow conditions they tend to plow up the snow ahead of the discs. To be effective the harrows should be mounted on large runners or skis so that the depth of penetration can be controlled (United States Navy, 1955). Even in fairly shallow snow, the discs may break at cold working temperatures (Adam, 1973). However, they are useful in chopping up a hardened snow surface.

A pulvimixer is a machine that scoops snow into a closed compartment where it is pulverized into fine particles by a chain or hammer flail before being redistributed to the surface. For any type of pulvimixer to be effective in deep snow it should be powered by tracked equipment. Hot air can be forced into the pulvimixer chamber to induce melting thereby adding water to the agitated snow particles. During construction of snow roads, if pulvimixers are both preceded and followed by compaction equipment, good trafficable surfaces can be constructed.

Moser (1963) reports that the peripheral speed of the pulverizer rotor is the most important factor affecting the performance of this equipment for processing snow. For example, increasing the maximum peripheral speed from 12 to 29 m/s resulted in more thorough pulverizing and also increased the processing speed from 1.4 to 3.1 km/h. A full-width ski placed behind the rotor was found useful for extruding and compacting the snow immediately after processing.

If many miles of snow road must be processed modified rotary snow plows are recommended because of their high capacity and ability to direct the disaggregated snow where required. The Peter Junior Snow Miller has been extensively tested as a snow processing machine (Wuori, 1963). This machine has a horizontally-mounted, closed drum with cutting blades spiraling around it, and can cut a 3-m wide swath in snow up to 1.6-m deep. At higher drum speeds, the machine is more effective in milling the snow and expelling it through the overhead rear ejection chutes. Other standard rotary snow plows gave similar results. A small model Rolba performed slightly better in regard to the resultant grain size and density of the processed snow.

Two disadvantages of rotary plows as snow processors are: (1) the large amounts of energy required to blow the snow from the front of the machine up and over to the rear, and (2) the need for levelling the snow after processing because it is unevenly distributed from the rear ejection chutes. Both problems may be overcome by using a front extended mount allowing the processed snow to be returned to the surface before the prime mover passes over the completed section.

Snow roads constructed over deep snow in Antarctica using a ski-mounted snowblower were comparable to conventional processed snow roads. According to Thomas and Vaudvey (1973) this method reduced construction time by 40 percent and produced roads that stood up to wheeled traffic for two months.

Snow should be compacted immediately after processing and levelled so that the energy of compaction is not wasted in breaking bonds produced by age hardening. Equipment for compacting snow roads also serves for compacting processed snow. Vibratory methods for compacting processed snow have been successful since the snow initially acts like a cohesionless, granular material similar to sand. Moving a heavy, tracked vehicle such as a D-8 tractor back and forth to compact processed snow gives near-adequate densities for wheeled vehicles, but it is a laborious method because a fairly small area is compacted per unit of time.

Processors introducing hot air to dry snow at a compacted density of 540 kg/m³ must melt about 60 kg of snow in each cubic metre to raise the density to 600 kg/m³; a density which assures that the snow road is adequate to carry wheeled vehicles. At Kapuskasing, Ont., under average climatic conditions, five passes of the heater travelling at 3.0 km/h and producing 0.76 MW were required to provide sufficient water to increase the snow density by 60 kg/m³ (Joint Snow Compaction Program, 1954). This operation would require 975 l/h of fuel oil, or 300 l/km for a roadway 2.4 m wide.

Similar tests by the U.S. Army Cold Regions Research and Engineering Laboratory showed that a very hard snow-ice layer could be obtained by using a 5.85-MW burner; however fuel oil consumption at the rate of 606 l/h was a real disadvantage (Wuori, 1963). As with other processing methods, even with the application of heat, compaction is necessary after processing.

In 1950, tests at Camp Hale, Colorado (United States Navy, 1955) evaluated the capacities of different types and combinations of equipment to produce compacted snow, including a 24-m diameter, variable-weight roller, drag, pulvimixer, disc harrow, and 72.7-kW heater. The most effective combination of equipment for compaction was the drag, the pulvimixer, and the roller, when operated in that order.

The wide variations in trafficability of processed snow roads suggest that the physical properties of the snow and atmospheric conditions play an important role in determining the success or failure of a road. The lack of documentation of these factors combined with the questionable results of most field tests and their poor experimental procedures make it extremely difficult to evaluate the different procedures used in constructing processed snow roads. The most promising method of building trafficable roads appears to be dragging to reduce the bulk of snow, followed by rotary blade processing, then compaction by heavy rollers. Wuori (1963) comments on using additives (when readily available) to increase snow strength when dry processing and other compaction techniques prove inadequate. Tests with several types of additives applied to the surface (direct heat, water and sawdust) showed that mixing sawdust into the surface snow layer, made it very strong after normal age-hardening; the sawdust also served as an effective insulator and retarded the softening of the surface during periods of thaw so that traction on the compacted surface increased appreciably. Sawdust as an additive must be used cautiously since it may be environmentally unacceptable, particularly wherever it may enter bodies of water and be hazardous to fish.

Ice-capped snow roads

Ice-capped snow roads are generally used where compacted snow cannot carry the traffic. They first require construction of a compacted snow road to establish a firm sub-base. Water is then applied to the surface to produce a hard, trafficable ice-snow pavement.

Although not described by this name, ice-capped snow roads were first tested by the U.S. Army Cold Regions Research and Engineering Laboratory (Wuori, 1963) in an attempt to reduce fuel consumption when using a heat processor. In the tests, reprocessing using heat to melt snow was confined to the top 20 to 30 cm of the surface layer, and required much less heat than that for processing the entire 86-cm layer.

It is stressed that heat control is important if only the surface layer is processed. Only enough heat should be used to produce as much moisture as the snow layer can retain before freezing. Excess water encourages drainage into the lower dry processed layer which could result in destruction of the ice-capped zone.

Another method tested in this program was spraying the snow surface with water from snow melted by large heaters. From 5 to 25 mm of water was sprayed on different test lanes to produce snow-ice pavements ranging in thickness from 12 to 25 cm. Disadvantages included the elaborate methods needed to keep the water tanks, pipes, and nozzles from freezing.

Adam (1973) reported that at Norman Wells, N.W.T. reconstruction of a compacted snow road that had been destroyed under very light traffic, followed by applying less than 25 kg/m^2 (2.5 cm) of water transformed the road into an adequate ice-capped snow road, whose average density was 630 kg/m^3 (over the full depth of 28 cm of compacted snow). Average densities over the top 10 and 5 cm were 790 kg/m^3 and up to 950 kg/m^3, respectively. This road withstood nearly 36,000 vehicle passes of simulated pipeline traffic with only minor maintenance, much of which could have been avoided if water had been more uniformly applied.

A laboratory study of the ice-capping process (Wilson and Adam, 1975) showed that the infiltration of water at 0°C into a packed snow bed results in two distinct zones of penetration. In the top zone, penetration is uniform to the front penetration depth; this is the ice-capped zone. With an excess application of water a zone develops below the front penetration depth in which the depths of penetration are highly variable because of the flow that takes place through discrete channels or fingers which develop. Over-application of water causes desaturation of the ice-capped zone which may accelerate wear of the surface when it is subjected to traffic. Also, it is a wasteful practice. Wilson and Adam found that the front penetration depth required to store a given amount of water was slightly greater than the depth of the snow bed saturated to 95% containing an equivalent volume of water. In a snow with an initial density of 500 kg/m^3, the front penetration was approximately 2.2 cm for each 10 kg/m^2 of water applied.

Channels formed below the front penetration depth are not well defined. The deeper the channels or fingers, the more extensive the desaturation in the ice-capped zone. High water and snow temperatures, large amounts of water, and low snow densities favour the formation of channels and hence desaturation. Figure 12.29 shows the maximum depth of water that should be applied to snow of given temperature and density to minimize desaturation effects.

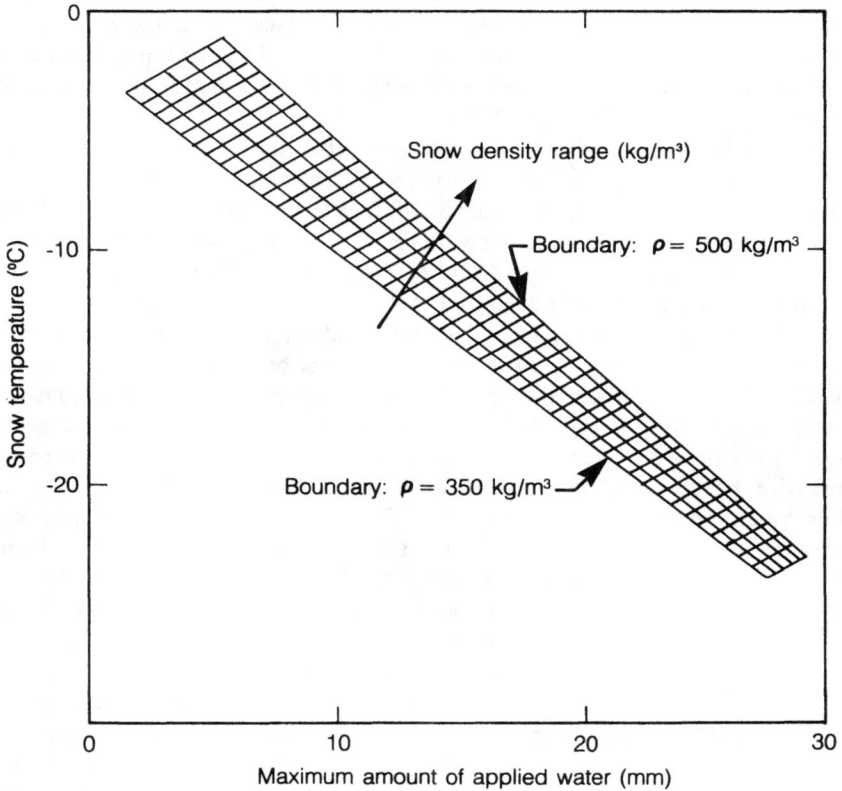

Fig. 12.29 Maximum depth of water application to minimize desaturation of a compacted snow surface (Wilson and Adam, 1975).

Ice roads

Ice roads are normally constructed where a lack of snow prohibits constructing another type of winter road, and is different from the ice-capped snow road in not requiring a base course of snow. Water is sprinkled or hosed onto the surface when the air and ground temperatures are low enough so that freezing occurs almost immediately. Once a layer of ice has developed and the surface is fairly impervious, larger amounts of water can be applied to fill the depressions. Water should be applied until a few centimetres of ice is formed over the highest undulations or hummocks of the ground surface.

The newest concept in winter road construction is the ice aggregate road. It is formed from chunks of ice chipped from the frozen surface of lakes or rivers. Once chipped, the ice is loaded on trucks by light-weight front-end loaders, hauled to the right-of-way, and end-dumped in place. A tractor with a blade levels the aggregate, which is then bonded together by sprinkling with water. Ice aggregate roads have been used extensively by major oil companies operating in the north slope of Alaska.

Ice roads are used mainly in arctic and subarctic regions where, in some years, no snow occurs until after December 31. For example, at Tuktoyaktuk, N.W.T. the accumulated depth of snowfall to December 31 is less than 20 cm once every three years, while the probability that the mean daily temperature is $< 0°C$ becomes 100% after October 10 (Burns, 1974).

Although ice roads have been built by at least one oil company operating in the Mackenzie Delta region, little information is available about them (Kerfoot, 1972). Adam (1973) reports on tests on an ice road at Norman Wells, N.W.T. constructed as part of a test loop on a right-of-way that was essentially cleared of snow to simulate low-snow conditions. Two passes of a Euclid tire drag produced a smooth surface except for the ridges left between the tire passes; the depressions were filled with packed snow and the hummocks were fairly clear of snow. The ice section of the test loop was sprayed with water, initially with about 2.5 mm, to provide a good seal over the moss. Over a one-week period the road was gradually built up, first by sprinkling water from spray bars, and later by hosing water to fill the depressions. An equivalent depth of 13.5 cm of ice at an average density of 890 kg/m^3 was formed over the 10-m wide roadway.

Under wheeled vehicle traffic, this ice road performed better than an ice-capped snow road and withstood nearly 36,000 wheeled vehicle passes with only one easily-repairable failure due to traffic. Over 30% of the equipment had tires with pressures between 550 and 620 kPa; many single axle loads exceeded 8,150 kg. However, under tracked vehicles the ice road did not perform as well as the ice-capped snow road. The latter supported one hundred passes of a D7E caterpillar tractor travelling over the same path without the tractor cutting through to the ground surface, whereas the ice road was shattered leaving the ground surface exposed after 22 passes. The durability of the ice-capped snow road was attributed to the protection offered by the underlying compacted snow.

Rammsonde hardness numbers for the ice road exceeded 600 from the time the road was partially completed and were ~ 1200 at temperatures of -20°C. An ice road is almost always trafficable if its rammsonde number is greater than 450. Since the hardness test provides only an index of the strength characteristics at a single point, careful sampling procedures must be employed to apply results as a measure of road trafficability. This is particularly true for ice roads because in the spring they can be almost completely melted while the remaining ice patches often have hardness values greater than 450. Therefore, the instrument has only a limited use for evaluating ice roads and has no value for determining an appropriate date to close the road. Either the minimum protective ice thickness or the percentage of exposed surface are the best criteria for determining this data.

Although this method of construction is very laborious, the study (Adam, 1973) proved that ice roads are practicable if constructed with proper hauling equipment using large amounts of water. The speed of construction of long ice roads largely depends on the speed at which water can be hauled over rough ground. Obviously, it would be advantageous and, sometimes necessary, to start construction of the road at the source of water. Nevertheless, to optimize construction it is essential to have enough road under construction so that the ice formation rate equals the water haul rate. Otherwise, runoff will occur from the road resulting in a lowering of construction efficiency.

Winter roads on ice

Winter roads across frozen lakes and rivers are common in remote "cold regions" with their alignments often following the waterways. Their ice surfaces are usually reasonably flat and smooth, and are therefore ideal for winter roads. The major construction problems encountered result from insufficient thickness or low bearing capacity of the ice, large cracks or pressure ridges, and snow clearing requirements.

Although ice thickness is an important criterion for determining whether ice on a lake or river can support a particular vehicle, estimates based on this alone can be fatal (Willmot, 1962). Other factors affect the carrying capacity of ice include ice quality, crack spacing and pattern, vehicle speed, and the water depth.

Following an ice accident at Minaki in 1957, when five lives were lost, Ontario Hydro developed a "safe" ice-load meter, whose readings are expressed in terms of the thickness of clear and white ice, crack conditions, and the number of days of thawing weather. The thickness data are determined by coring and serve as input for the meter, which is an analogue computer for solving the equation:

$$P = 5.9 \ (h_c + 0.5 \ h_w)^2 \ x \ C_1 \ x \ C_2, \qquad\qquad 12.3$$

where

P = safe load (kg),

h_c, h_w = thicknesses of clear and white ice
(each 0 - 90 cm), respectively, and

C_1, C_2 = crack and thaw factor, respectively.

To date this is the most practical method of evaluating the safe load for travel on ice. Both C_1 and C_2 include safety factors.

Equation 12.3 is mainly based on the work of Korunov (1960) and Gold (1960). However, thermal stresses and fatigue have been found to play a role in ice cover failures (Gold, 1971) so that the equation must not be the sole criterion for calculating the safe load for ice travel. Because of the indeterminate effects of these two factors it is common practice to apply safety factors to the 'calculated' loads. When repeated loadings are expected, the use of higher safety factors is justified. The speed of vehicle travel, the water depth and the spacing of vehicles in convoy also affect safety because of the resonant hydrodynamic wave formed in the ice cover (Gold, 1971).

Cost of Winter Road Construction[1]

The cost of most winter roads depends mainly on the effort required to produce a density sufficient to withstand the anticipated wheel loads. Winter trails cost about an average of $275/km (Engineering News Record, 1969; Mauro, 1969). But, a maintenance cost of $27/km plus an inflation factor must also be considered to estimate the total costs.

If the costs of machine time, man-hours and fuel are considered, a high quality snow road can be constructed for about $1600/km (Joint Snow Compaction Program, 1954). Ice-capping requires more work so that the cost is increased by about $200/km. Trafficable winter roads in remote areas may cost up to $3,250/km.

If a snow road is to be built every winter at a cost of $3,250/km or more consideration should be given to building a permanent road instead. The two roads cost about the same over a couple of decades assuming the permanent road has an initial cost of $32,350/km, an annual maintenance charge of 1.5% and annual interest charges at 8.5% to discharge the debt.

ICE BRIDGES

Ice bridges are artificially-thickened sections of ice constructed at river crossings, usually perpendicular to the river alignment. In choosing their locations river sections with fast water should be avoided since localized turbulence may erode the ice sheet. Both the natural thickness of ice and its frequency of formation are less in highly turbulent reaches than in more tranquil flows.

---

[1] The reader is cautioned against accepting the cost values listed in this section as current. They are presented solely to demonstrate the comparative costs of different types of winter roads.

In the past, corduroy construction was used for ice bridges, i.e., trees were trimmed and placed along the traffic lane and then snow and water were compacted over the timber to obtain a smooth surface. In arctic and subarctic regions insufficient supplies of suitable timber usually limit application of this technique.   Furthermore, this method is often subject to environmental constraints, for example, in Northern Canada it is only permitted provided all excess timber and debris are removed from the site after use (Government of Canada, 1971).

Ice bridges are often constructed by building a low dyke on both sides of the right-of-way using light vehicles to windrow the snow. Removal of snow from the surface of the roadway increases the natural ice thickness. The dykes are iced or, if snow is lacking, logs are used to confine the area into which water is pumped in 3- to 5-cm lifts and then allowed to freeze. Ice bridges ~ 30 to 60 cm thick can be built in this manner.

The construction methods are well documented (Carson, 1955, 1958; U.S. Department of the Army, 1962). All methods depend on a critical depth of ice cover, which can be estimated by Stefan's equation:

$$h = \alpha\sqrt{\theta}, \qquad\qquad 12.4$$

where

  h = estimated ice thickness (m),
  α = a coefficient (0.58 - 0.95) representing the combined effects
        of local conditions, e.g., snowcover, river conditions, and
        water properties (Rhoad, 1973), and
  θ = number of degree-days below freezing (°C).

By predicting ice thicknesses from meteorological data by means of Eq. 12.4 costly early mobilization of the field crews can be avoided. Even though the calculated thickness may indicate a safe limit, it should still be checked 'in situ' by core sampling before heavy equipment is allowed on the ice cover.

The construction techniques and the principles of winter roads on ice apply to ice bridges when the river crossing is relatively wide. Ice bridges have been analysed theoretically in terms of bridge dimensions, ice properties, and ice quality for both concentrated and distributed loads (Nevel, 1965, 1970). Lofquist (1951) estimated the bearing capacities of ice sheets for specific static-load conditions.

In constructing ice bridges in permafrost terrain cutting of the banks should be avoided and compacted snow should be used as fill to achieve acceptable grades. If structural failures occur at the contacts of the ice with the shorelines of rivers and lakes the approaches must be reinforced.

In general, ice bridges require minimal maintenance. Snow clearing and dragging of the surface to remove ruts may be required periodically. Also, the ice thickness should be checked periodically to ensure that the ice is not eroding from the bottom. Erosion may occur if the velocity of the water is substantially increased by the sagging of an ice bridge under its own weight.

ENVIRONMENTAL EFFECTS OF SNOW ROADS

Winter roads are constructed to provide a smooth trafficable surface across all types of terrain and to protect the surface vegetation and peat layer in permafrost areas. Traffic tests have confirmed that properly constructed winter roads can withstand thousands of passes of wheeled vehicles without suffering adverse wear. Undoubtedly tracked vehicles with proper grousers could make hundreds of passes without causing adverse wear. However, tracked D-7 caterpillar tractors can wear an ice road down to the ground in as few as 22 passes but a snow road surface can withstand up to 100 passes without being adversely affected (Adam, 1973).

In the boreal forest regions, another more indirect form of terrain damage can occur. Clearing of the forest exposes the feather mosses to direct sunlight which kills certain plant species and exposes the underlying peat layer. Normally, tamarack and black spruce seedings are not removed by clearing, but their numbers are noticeably reduced on that area where the winter road was constructed. Most shrub species are taller and more abundant on cleared areas than on trafficked areas.

At the winter road test site at Norman Wells during the summer immediately following heavy winter road use, live plant cover was reduced to 12% of its original level, rising to 31 and 44% after one and two years, respectively (Adam and Hernandez, 1977). Similar observations of seismic lines in the Mackenzie Delta showed an increase from 1.4% plant cover in the first summer after disturbance to 21% in the second and 67% by the third summer (Hernandez, 1973).

In the tundra, peat depth is the major factor controlling thaw depth and ground temperatures (Haag and Bliss, 1974a) while in the boreal forest removal of the forest canopy alone increases soil temperatures and thaw depth (Haag and Bliss, 1974b). However, the intensity of an environmental disturbance and its time of occurrence during the year determines the degree of compaction of the peat and its thermal conductivity. Winter roads should be built wherever conventional wheeled vehicles must be used, especially if more than a few passes of either tracked or wheeled low ground-pressure vehicles are necessary. A properly-constructed winter road will protect the peat layer and vegetation so that recovery of the terrain to a stable condition is possible.

LITERATURE CITED

Adam, K.M. 1973. *Report on Norman Wells winter road research study, 1973.* Winnipeg Interdiscip. Syst. Ltd., Winnipeg, Man.

Adam, K.M. 1974. *Impact from winter road use and misuse.* Res. Reps., Environmental impact assessment of the portion of the Mackenzie gas pipeline from Alaska to Alberta. Winnipeg Environ. Prot. Board, Winnipeg, Man., Vol. IV, Ch. 2, pp. 29-36.

Adam, K.M. and H. Hernandez. 1977. *Snow and ice roads: ability to support traffic and effects on vegetation.* Arctic J., Arct. Inst. North Am., Vol. 30, No. 1, pp. 13-27.

Ager, B.H. 1955. *Den snöpackade vägen. II Väderlekens betydelse (The compacted snow road. II Climatic considerations).* Svenska Skogsvardsföreningens Tidskrift, Vol. 2, pp. 201-227 [English Transl. by Nat. Res. Counc. Can., Div. Build. Res., Ottawa, Tech. Transl. 816].

Baldwin, M.F. and D.H. Stoddard. 1973. *The off-road vehicle and environmental quality.* The Conserv. Found., Washington, D.C.

Bombardier Limited. 1971. *Play safe with snowmobiles for more winter fun.* Bombardier Ltd., Valcourt, P.Q.

Bombardier Limited. 1972. *A guide to the development and maintenance of good snowmobile trails.* Bombardier Ltd., Valcourt, P.Q.

Brown, J., W. Richard and D. Victor. 1969. *The effect of disturbance on permafrost terrain.* Sp. Rep. 138, U.S. Army Cold Reg. Res. Eng. Lab., Hanover, N.H.

Burns, B.M. 1974. *The climate of the Mackenzie Valley - Beaufort Sea.* Clim. Stud. No. 24, Vol. II, Environ. Can., Atmos. Environ. Serv., Downsview, Ont.

Burt, G.R. 1971. *Travel on thawed tundra.* Symp. on Cold Regions Eng., Vol. 1, Univ. Alaska, College, Alaska.

Camm, J.B. 1960. *Snow-compaction equipment - snow drags.* Tech. Rep. 109, U.S. Nav. Civ. Eng. Lab., Port Hueneme, Calif.

Camm, J.B. 1961. *Snow-compaction equipment - snow rollers.* Tech. Rep. 107, U.S. Nav. Civ. Eng. Lab., Port Hueneme, Calif.

Canadian Arctic Gas Pipeline Limited. 1974. *Applications and supporting documents to the National Energy Board and to the Department of Indian Affairs and Northern Development.* Can. Arct. Gas Pipeline Ltd., Toronto, Section 13a.

Carson, P.J. 1955. *Ice bridges.* R. Eng. J. Vol. 49, pp. 166-176.

Carson, P.J. 1958. *Ice bridges.* Polar Rec., Vol. 9, pp. 18-19.

Chubb, J. 1971. Proc. 1971 Snowmobile and Off the Road Vehicle Res. Symp. Mich. State Univ., East Lansing, Mich., pp. 55-71.

Clark, E.F., G. Abel and A.F. Wuori. 1973. *Expedient snow airstrip construction technique.* Sp. Rep. 198, U.S. Army Cold Reg. Res. Eng. Lab., Hanover, N.H.

Committee on Environmental Quality. 1970. *Rep. of the Committee on Environmental Quality.* Int. Snowmobile Cong., Duluth, Minn.

Diamond, M. 1956. *Studies on vehicular trafficability of snow (Part 1).* Rep. 35, U.S. Army Snow Ice Permafrost Res. Estab.

Engineering New Record. 1969. *Arctic roadbuilders open up North Slope.* Eng. News Rec., Vol. 182, No. 7, pp. 36-37.

Gold, L.W. 1956. *The strength of snow in compression.* J. GlacioL, Vol. 2, pp. 719-725.

Gold, L.W. 1960. *Field study on the load bearing capacity of ice cover.* Tech. Pap. 98, Nat. Res. Counc. Can., Div. Build. Res., Ottawa, Ont.

Gold, L.W. 1971. *Use of ice covers for transportation.* Can. Geotech. J., Vol. 8, pp. 170-181.

Government of Canada. 1971. *Territorial land use regulations.* Territorial Lands Act, Government of Canada, Ottawa, Ont.

Haag, R.W. and L.C. Bliss. 1974a. *Energy budget changes following surface disturbance to upland tundra.* J. Appl. Ecol., Vol. 11, pp. 355-374.

Haag, R.W. and L.C. Bliss. 1974b. *Functional effects of vegetation on the radiant energy budget of boreal forest.* Can. Geotech. J., Vol. 11, pp. 374-379.

Hernandez, H. 1973. *Natural plant recolonization of surficial disturbances, Tuktoyaktuk Peninsula Region, N.W.T.* Can. J. Bot., Vol. 51, pp. 2177-2196.

Hok, J.R. 1969. *A reconnaissance of tractor trails and related phenomena on the North Slope of Alaska.* U.S. Dept. Land Manage., U.S. Dept. Interior.

Joint Snow Compaction Program. 1954. *Report on snow compaction trails 1952-53.* DSIS Doc. L7167, Can. Def. Res. Board, Ottawa, Ont.

Kerfoot, D.E. 1972. *Tundra disturbance studies of the Western Canadian Arctic.* Arct. Land Use Res. Rep. 71/72, Dept. Indian Affairs North. Dev., Ottawa, Ont.

Korunov, N.M. 1956. *O gruzopodyemnosti ledianogo pokrova pri transportirovke lesa (Load carrying capacity of ice for timber transport).* Lesnaia Prom. Vol. 11, pp. 18-19. [English Transl. Nat. Res. Counc. Can., Div. Build. Res. Tech. Transl. 863].

Lanyon, J.J. 1961. *Conservation of M29C Weasel tracks.* Sp. Rep. 42, U.S. Army Cold Reg. Res. Eng. Lab., Hanover, N.H.

Leijonhufvud, A.C. 1955. *Den snöpackade vägen III Fältförsök och praktiskt tillämping (The compacted snow road: III Field tests and practical applications).* Svenska Skogsvardsföreningens Tidskrift, Vol. 4, pp. 337-394 [English Transl. by Nat. Res. Counc. Can., Div. Build. Res., Tech. Transl. 909].

Lodico, N.J. 1973. *Environmental effects of off-road vehicles.* Bibliography Ser. 29, U.S. Dept. Interior, Office Libr. Serv.

Lofquist, B. 1944. *Lyftkraft och Bärförmaga hos ett Istäcke (Lifting force and bearing capacity of an ice sheet).* Teknisk Tidskr, ft, No. 25. Stockholm. [English Transl. by Nat. Res. Counc. Can., Div. Build. Res., Tech. Transl. 164].

MacNutt, W.S. 1963. *New Brunswick, a History 1784-1867.* MacMillan of Canada, Toronto, Ont.

Masyk, W.J. 1973. *The snowmobile, a recreational technology in Banff National Park: Environmental impact and decision making.* Nat. Park Ser. 5, Univ. Calgary, Calgary, Alta.

Mauro, A.V. 1969. *The Royal Commission Inquiry into Northern Transportation,* Province of Manitoba, Winnipeg, Man., pp. 185, 268-271, 371.

Moser, E.H. (Jr.). 1963. *Navy cold-processing snow-compaction techniques.* In Ice and Snow: properties, processes and applications (W.D. Kingery, ed.), The M.I.T. Press, Cambridge, Mass. pp. 459-484.

Murrman, R.P. and S. Reed. 1972. *Military facilities and environmental stress in cold regions*. Sp. Rep. 173, U.S. Army Cold Reg. Res. Eng. Lab., Hanover, N.H.

Nevel, D.E. 1965. *Ice bridge analysis*. Res. Rep. 148, U.S. Army Cold Regions Res. Eng. Lab., Hanover, N.H.

Nevel, D.E. 1970. *Moving loads on a floating ice sheet*. Res. Pap. 261, U.S. Army Cold Reg. Res. Eng. Lab., Hanover, N.H.

Pelto, P.J. 1973. *The snowmobile revolution: technology and social change in the Arctic*. Cummings Publishing Co., Menlo Park, Calif.

Porkhaev, G.V. and A.V. Sadovski. 1959. *Zemlyanoe polotno dorog i aerodromov (Beds for roads and airfields)*. Osnovy geokriologii (merzlotovedeniya), Chast'vtoraya, Inzhenernaya geokriologiya, Glava VIII. Akademiya Nauk, SSSR, Moskva. pp. 231-254. [English Transl. by Nat. Res. Counc. Can., Div. Build. Res. Tech. Transl. 1220].

Ramseier, R.O. and C.M. Keeler. 1966. *The sintering process in snow*. J. Glaciol., Vol. 6, No. 45, pp. 421-424.

Rhoad, E.M. 1973. *Ice crossings*. North. Eng., Vol. 5, No. 1, pp. 19-24.

Thomas, M.W. and K.D. Vaudvey. 1973. *Snow road construction technique by layered compaction of snowblower processed snow*. Tech. Note 1305, U.S. Nav. Civ. Eng. Lab., Port Hueneme, Calif.

U.S. Department of the Army. 1962. *Arctic construction*. Tech. Manual TM5-349, Dept. of the Army, Washington, D.C.

U.S. Naval Civil Engineering Laboratory. 1962. *Snow compaction equipment, sprayers and dusters*. Tech. Rep. R111, Port Hueneme, Calif.

United States Navy. 1955. *Arctic engineering*. Tech. Publ. Navy Docks TP-PW -11, Bureau of Yards and Docks, Washington, D.C.

Weber, B. and J. Zyla. 1975. *Compaction of snow*. M.Sc. Thesis, Dept. Civ. Eng., Univ. Man., Winnipeg, Man.

Willmot, J.G. 1962. *A safe ice load meter and a motor-driven coring tool for estimating the bearing capacity of ice*. Proc. 7th Annu. Meet. East. Snow Conf., pp. 149-160.

Wilson, J.A. and M.W. Nelson. 1968. *A history of the development of oversnow vehicles*. Proc. 36th Annu. Meet. West. Snow Conf., pp. 9-18.

Wilson, T.M. and K.M. Adam. 1975. *The interrelationships of water application rate, ambient temperature, snow road properties and ice-cap during construction of winter roads*. Can. Arct. Gas Study Ltd., Dept. Civ. Eng., Univ. Man., Winnipeg, Man.

Wuori, A.F. 1963. *Snow stabilization studies*. In Ice and Snow: properties, processes and application (W.D. Kingery, ed.), The M.I.T. Press, Cambridge, Mass., pp. 438-458.

13

SNOW AND BUILDINGS

D. W. BOYD (retired)
Atmospheric Environment Service, Environment Canada, Ottawa, Ontario.

W. R. SCHRIEVER
Division of Building Research, National Research Council of Canada, Ottawa, Ontario.

D. A. TAYLOR
Division of Building Research, National Research Council of Canada, Ottawa, Ontario.

INTRODUCTION

Snow can adversely affect different structural, safety and access features of buildings; the snow accumulation on the roof may result in overloading and possible collapse, the formation of ice and ice damming may cause water leakage under shingles and over flashings, wetting caused by the infiltration of windblown snow into attics and other spaces may cause interior damage, snow slides from sloped roofs may endanger pedestrians, and drifting around buildings may impede access by people and vehicles. All these aspects must be considered in designing buildings. This chapter is mainly concerned with the problem of predicting roof snow loads for the structural design of buildings in Canada.

Snow loads are extremely variable, not only from place to place and from winter to winter, but also from building to building depending on the shape of the roof and its exposure to wind and sun. Such variations will be discussed in two parts: first, the geographical variations under the heading *Snow Load on the Ground*, and second, the variations peculiar to the buildings themselves under the heading *Snow on Roofs*.

SNOW LOAD ON THE GROUND

Development of Loads for the National Building Code

The load resulting from the accumulation of snow on a roof in a severe winter must be taken into account in designing buildings in most parts of Canada. Until a few decades ago the strength of the roof was determined simply by experience. Later, a few of the larger cities included minimum roof loads in their building by-laws, which were apparently used throughout large regions without considering differences in normal snow conditions (Thomas, 1955).

Two events in 1941 led to a gradual change towards more rational snow loads. One was the publication of the first edition of the National Building Code (1941). The other was the beginning of routine measurements of the accumulated depth of snow on the ground at many weather stations.

In 1941 the only data on which snow load estimates could be based were the snowfalls for each month averaged over a number of years. For roofs having a slope of 20 degrees or less, the National Building Code (1941, p. 162) required that buildings be designed for a snow load of 20, 30 or 40 psf (0.96, 1.44, 1.92 kPa). As a guide to selecting the load, the following formula was to be used:

$$L = S + R, \qquad\qquad 13.1$$

where:

S = the sum of the average snowfalls in January, February and March, (inches), and

R = the sum of the average rainfalls in January, February and March, (inches).

The parameter, L, is related to the design snow load as follows:

For $L < 20$ (51 cm)   load = 20 psf (0.96 kPa),

For $20 \leqslant L \leqslant 30$ (51 $\leqslant L \leqslant$ 76 cm)   load = 30 psf (1.44 kPa),

For $L > 30$ (> 76 cm)   load = 40 psf (1.92 kPa).

Equation 13.1 is purely empirical, and involves the dimensionally incorrect addition of inches of snowfall (with an average density of about 100 kg/m^3) and inches of rainfall (with a density of 1000 kg/m^3), and the expression of their sum in pounds per square foot (psf). For many parts of Canada, however, it gave remarkably good results. Its worst fault probably lay in limiting the design snow load to 40 psf, or 1.92 kPa.

A revised edition of the National Building Code (National Building Code of Canada, 1953) gave a more rational approach to estimating snow loads.

Actual measurements of the accumulated depth of snow on the ground were available for each winter since 1941 for a number of weather stations. These records were searched for the largest of the seasonal maxima at each station (Thomas, 1955); the values were converted to loads by assuming a snow density of 200 kg/m^3 and adding the load resulting from the estimated maximum one-day rainfall during the season when the maximum snowcover might be expected. The resulting snow loads averaged about the same as the values of L calculated by Eq. 13.1 but they are probably more accurate because the effects of topography and proximity to open water on snow accumulation and melt are automatically taken into account. The removal of the upper limit of 40 psf (1.92 kPa) led to higher design loads in areas of deep snow accumulation, particularly near the Gulf of St. Lawrence, in some areas east of Lake Huron, and in some parts of British Columbia.

The same basic method of computing snow loads (an extreme snow depth with an assumed density of about 200 kg/m^3, increased by a one-day rainfall) has been used in Canada since 1953. There have been a few refinements in the calculation procedures, however, with some revisions based on measurements in later winters.

By 1960 up to 18 years of snow-depth records were available for several stations. It became feasible to use a statistical analysis of the seasonal maximum depths of snow and to select a design snow load with a given probability of occurrence instead of merely selecting the one extreme value as had been done earlier (Boyd, 1961a, b). The extreme value distribution was chosen and fitted to the observed seasonal maximum depths of snow using the method described by Gumbel (1954). The 30-year return period value, i.e., the value that has one chance in 30 of being exceeded in any particular year, was selected as a basis for design loads.

For convenience the snow density was assumed to be 12 lb/ft^3 (192 kg/m^3) so that one inch of snow corresponded to a load of one pound per square foot (1 cm equivalent to 18.85 Pa). The fact that the density of old snow ranges from 200 to 400 kg/m^3 or even more would suggest that a value of 192 kg/m^3 is rather low for computing loads. However, the value required is the average snow density at the time of maximum snowcover in a 30-year period. This maximum will almost certainly occur immediately after an unusually heavy snowfall and hence a large portion of the snowcover will have a low density.

Roof failures in Canada are often associated with early spring rains, and, therefore, it is advisable to add to the snow load the load of rain water that might be retained in the snow. Because rainfall is measured daily at all precipitation stations, it was convenient to use the maximum one-day rainfall that might occur at the time of year when snow depths were greatest.

Revisions

Records of snow depths from over 200 weather stations were used to prepare a snow load map for Canada in 1961 (Boyd, 1961a, b). However, even 200 stations are not enough to define adequately the snow load distribution in all areas. In mountainous terrain and near large bodies of water snow loads may differ significantly over distances of 10 to 20 km; the average distance between weather stations with good snow-depth records is more than 200 km. In several areas where snow loads were uncertain, careful analyses were carried out using all short-term, snow-depth records as well as the records from stations where data were missing in many months. The latter records may be misleading, but when compared winter by winter with those obtained from more reliable stations they still can provide some idea of the differences in snow conditions. Such analyses have been used to estimate new design snow loads and modify old ones for some of the interior valleys in British Columbia, the Huron Slopes in Ontario, and a few other small areas.

Effect of Elevation

In selecting a design load for a location at an elevation different from that of the station where snow depths were measured, the change in load with elevation must be considered. This is particularly important for buildings such as those used in mining and skiing operations that are often built far above the valley floor. In remote locations no weather stations exist and hence no continuous snow-depth records are available. During 1935, the British Columbia Water Resources Service instituted a snow data collection program known as snow surveying to obtain information that would be useful to them in predicting spring run-off. By 1970 there were over 200 snow courses in British Columbia operated by many different companies and government departments (Hunter, 1971).

From these data it was possible to estimate the increase in snow load with elevation in the Rocky Mountain Trench. At several locations where weather stations and snow courses were situated at nearly the same elevations and only a few kilometres apart it was discovered that the snow loads obtained from the snow survey measurements were considerably higher than those obtained from weather station observations. It has been suggested that the reason for the lack of agreement is because snow courses are often deliberately located in areas where the snow is particularly deep.

Although the absolute values of snow loads computed from snow-survey data are often too high (for use in design) they at least indicate the increases to be expected in the loads with elevation; these vary from about 0.6 to 1.9 kPa per 100 m in British Columbia. Since 1967 some observations of snow loads have been made at several elevations on each of several mountains (Schaerer, 1970), confirming the increase with elevation; but the rate of increase is slightly less, 0.5 to 1.4 kPa per 100 m.

Urban Effect and Climatic Change

Much has been written in recent years about the effects of urban areas on local climate. There is no doubt that quite marked differences in air temperature between urban and rural areas do exist under certain meteorological conditions. Long-term averages also show a difference, but usually only a few degrees. Differences in precipitation amount have been reported, but the effects of urban development on the accumulation of snow on the ground and on snow loads are still in doubt.

Some climatologists have warned that the increasing use of fossil fuels (including uranium in atomic plants) may increase the average world temperature significantly. On the other hand, the trend in the last two or three decades seems to be toward lower temperatures, so that some climatologists are warning that this downward trend may continue. Any changes in temperature will probably be accompanied by changes in precipitation and in snow loads. Such changes should be reflected in the design snow load, but revisions every ten or twenty years should be frequent enough to allow for this slow trend.

No changes are anticipated in the basic approach for determining design snow loads on the ground, but values will be revised as more data become available. Ground snow loads, based on the most recent analysis of the available snow-depth measurements, are listed in Table 13.1 for several cities in Canada (National Building Code of Canada, 1980a).

SNOW ON ROOFS

As late as 1960 the problem of designing roofs for snow loads was not well understood or well developed in building codes and standards. The fact that every year throughout Canada many roof failures occurred attested to the fact that roofs had not been designed, or constructed to support the amount or distribution of snow that could have been expected to occur in the lifetime of the structure. It is true, however, that some of these failures happened in buildings and houses which were constructed without reference to building codes, and therefore are not significant in assessing current design requirements. Other failures, particularly those of larger roofs, are important in showing that some of the assumptions made for determining snow loads in recent codes were unsatisfactory for predicting the amount and distribution of

Table 13.1
RECOMMENDED DESIGN SNOW LOADS ON THE GROUND
(National Building Code of Canada, 1980a). Reproduced by permission of the
National Research Council of Canada.

Province	City	kPa
British Columbia	Vancouver	1.9
	Victoria	1.5
Alberta	Calgary	0.9
	Edmonton	1.5
Saskatchewan	Regina	1.7
	Saskatoon	1.5
Manitoba	Winnipeg	2.1
Ontario	Ottawa	2.9
	Toronto	1.8
	Windsor	1.1
Quebec	Montreal	2.7
	Quebec City	3.8
New Brunswick	Saint John	3.0
Nova Scotia	Halifax	2.2
Prince Edward Island	Charlottetown	3.3
Newfoundland	St. John's	3.5

the snow accumulation. In particular, the design snow load was frequently assumed to be uniformly-distributed over the entire roof and was assigned a nominal value such as 30 or 40 psf (1.44 or 1.92 kPa).

Snow loads on roofs depend on two factors: the weather and the physical properties of the roof. The weather varies with geographical location, elevation and exposure to or shelter from the wind and sun and, of course, it may also vary considerably from one winter to another. In addition the shape, size, colour and heat loss properties of the roof are very important. Therefore determination of the basic ground load is only a first step in calculating roof loads. Factors which modify this load, namely, local topography and building properties, must also be considered before a realistic and reasonably accurate prediction of roof snow loads can be made. Field observations are essential in development of design requirements.

Snow Load Surveys

A survey of snow loads on roofs was started by the Division of Building Research, National Research Council in 1956 and carried out with the help of volunteer observers in many parts of Canada (Peter et al., 1963). The survey involved the measurement of depth and density of the snowcover on different types of roofs and also observations of shelter conditions in the vicinity of each roof, unusual snow load distributions and roof collapses (Schriever et al., 1967; Lutes and Schriever, 1971). This survey continued over a period of 11 years and yielded results which permitted development of a set of "snow load coefficients" relating the snow load on the roof to the ground load. Special surveys, initiated more recently to examine the problem of snow loads on large multi-level flat roofs and on curved roofs, are continuing.

The Structure and Density of Snow on Roofs

Some of the problems in predicting snow conditions on roofs in a country as large as Canada arise because the condition of the snow varies with geographical location. Variations in the properties of newly-deposited snow are mainly caused by differences in air temperature and humidity during snowfall. On the east and west coasts and in the Great Lakes Regions, wet and sticky snow is more frequent than in the Prairies and northern areas where very cold temperatures usually result in a dry snow that tends to drift. The type and character of snow found in Quebec and Ontario usually lies between these two extremes.

The density of snow on roofs, obtained from measurements at a number of stations across Canada was found to vary from 200 to 400 kg/m^3. A value of 240 kg/m^3 is adequate for use in building design (National Building Code of Canada, 1980a).

Effects of Wind on Snow Accumulation

Falling snow is deposited on roofs in uniform layers only if the wind speed is low. Schneider (1959) found that with wind speeds in the range from 4 to 4.5 m/s, much of the falling snow is carried past roof areas whose shapes cause the wind to accelerate to areas of "aerodynamic shade" where it is deposited in drifts. Drifts accumulate at sudden changes in roof level, behind buildings or in other locations where the wind speed is reduced. The loads created by drifts may be significantly larger than the existing roof load. If the wind speed increases above 4 to 4.5 m/s snow particles can be picked up from the snow-cover and carried along in the flow, leading to depletion of the snow-cover in areas of high wind speed and re-deposition of the snow in drifts in wake areas such as valleys, the lee sides of peaked or arched roofs, lower roofs in the lee of higher roofs, or behind obstructions on roofs. In the redistribution process a considerable quantity of snow may be blown off the roof and fall to the ground. Sometimes freezing rain coats the existing snowcover with hard crust, making it essentially invulnerable to scour or drifting.

Obviously, in many cases redistribution of snow on a roof by drifting is an important factor governing the snow load; in certain sheltered conditions it is insignificant. Two classes of shelter may be considered: complete and partial. In completely-sheltered locations snow is deposited in relatively uniform layers. These exposure conditions are found most often in valleys of the mountains of Western Canada and occasionally at individual sites surrounded on all sides by high trees serving as windbreaks. Prediction of roof snow loads for these sites is relatively easy if field data on ground loads are available. Partial or local shelter occurs more frequently and prediction of drift loads is more difficult. Snow may blow from almost any direction; the shelter provided by the surroundings, other buildings and trees, may change during the lifetime of the structure. Hence major accumulations may occur in these areas from storms coming from different directions.

Except for special cases, it is usually too costly to carry out individual experimental studies, such as wind-tunnel tests, to determine the effect of exposure conditions on snow loads. It has therefore become accepted practice to design roofs to resist three loading conditions: (a) an uniformly-distributed load of some fraction of the ground load, depending on the exposure conditions; (b) drift loads in selected wake areas, assuming that the wind can blow from any direction; and (c) a certain degree of unevenness in the deposition and depletion of the snowcover. An adequate design for this latter condition can be obtained by assuming the snowcover is distributed with the specified design load on any one portion of the area and half of this load on the remainder, whichever arrangement produces the greatest effect on the structural members of concern.

A general assessment of the results of the roof load survey conducted by the National Research Council of Canada indicated that the average roof loads are considerably lower than the ground loads in the vast majority of cases. On the average 50, 20 and 6% of the observed flat roofs had an average load that reached 30, 60 and 80% respectively, of the ground load. For the peaked roofs observed it was found that on average 50, 6 and 1% of the roofs had average loads that reached 15, 60 and 80% respectively, of the ground load (Schriever and Otstavnov, 1967). These values were obtained mainly from measurements taken on single residential buildings located in exposed areas and therefore may not be representative of all types of roofs. Based on these data, the recommended design for a uniformly-distributed load on flat and sloped roofs was lowered to 80% of the ground load for average exposure and to 60% of the ground load for completely exposed roofs.

Solar Radiation and Heat Loss

Other factors than wind modify snow loads on roofs, although some of these, such as solar radiation and heat loss, are important only under special weather conditions. For example, solar radiation has little effect in reducing loads provided the ambient air temperature remains well below freezing, and the snowcover is essentially continuous over the roof. In regions with mild winters, and in colder regions towards the end of winter, the effects of solar radiation in reducing the snow load can be considerable, particularly if part of the roof is bare. In spite of this no allowance should be made for a reduction in load by solar radiation when determining the design load for structural calculations.

Heat loss through the roof may also cause a significant reduction in the possible maximum roof loads, especially in regions where the maxima result from the accumulations of several snowstorms. Many roofs have been saved from collapse because heat loss has reduced the load by either melting or initiating slides by melting. However, it is not prudent to count on such reductions when choosing a design load, particularly with the present trend towards better insulated roofs. In special cases roofs have been designated for reduced loads, for example, where a heating system has been incorporated into the roof as a means of clearing the snow periodically during the winter.

Roof Slope and Texture

On unheated roofs sheltered from the wind the depth of a uniformly-distributed snow load decreases as the roof slope increases. Falling snow particles which strike the sloped surface of the snowcover on the roof may rebound over the edge because of their momentum or they may be carried from the roof by light air currents which occur even in the sheltered zones. Furthermore, snow will slide off an inclined roof if the tensile strength of the snowcover or the coefficient of friction at the roof surface are exceeded. The texture of the surfacing material largely determines the coefficient of sliding friction. Smooth surfaces such as sheet metal or glass have lower coefficients than asphalt or wood shingles. In addition, heat loss through the roof will sometimes produce a thin film of water at the roof-snow interface, which reduces the coefficient of friction.

The angle of repose of dry snow grains also limits the maximum slope of the snowcover and hence the maximum load which may accumulate on a sloping roof. However, when the snow is wet and cohesive it can stand vertically or even overhang the edge of a roof. Under these conditions the depth of snow on sloped, unheated roofs located in a completely sheltered exposure might be almost as deep as the ground snowcover (see Fig. 13.1). The movement or creep of snow down the roof slope may reduce the snow load or in some cases cause a load to shift to a lower part of the roof.

In Canada, at the present time, a simple linear reduction from the applicable load at roof slopes of up to 30° to zero load at slopes of 70° or more is widely used for design purposes (National Building Code of Canada, 1980b). A less conservative relationship might apply to slippery roof surfaces, but in any case care must be taken to use a slope-reduction relationship intelligently, especially if ice loading is known to be significant.

Fig. 13.1 Glacier, B.C., January 9, 1959. Effect of shelter and roof slope. The load on the Alpine Club house roof (slope 38°) was 90% of the ground load and exceeded that indicated by the slope reduction formula.

Redistribution Due to Sliding and Melting Snow

Major readjustments of the roof snow load may occur because of sliding, or melting and subsequent refreezing. Obviously, the load will be reduced considerably when snow slides off the roof and falls to the ground; in these cases, the safety of humans and equipment is a major factor in roof design. Also, the shifting of the snow load to a lower or adjacent roof greatly increases the load there. In another situation snow may slide off a heated portion of a building and collect on an unheated roof area or it may be retained by a valley, a parapet, ice dam (Baker, 1967) or other obstructions.

At air temperatures below freezing, melting of snow on a roof may be caused by heat from solar radiation, losses through the roof, and heating cables, steam lines, or extractor fans. On flat roofs (of zero slope), the melt water collects in areas of greatest deflection, usually at the mid-span of the roof joists. If this water is not drained it will cause the roof to sag and accumulate more water or ice until either a stable situation is established or the roof collapses. Ice may also build up on gently-sloping roofs when melting occurs on one part and the melt water freezes or is absorbed by snow at another.

As snow is generally a poor conductor, snow on the roof of a heated building tends to increase the resistance to heat loss. As the depth of snowcover increases the $0°C$ isotherm will move up into the snow layer causing melting at the interface between the upper surface of the roof and the snow. This melting reduces friction at the snow-roof interface and if the weight of the snowpack is sufficient to overcome its tensile strength a part of the pack will break away from the better-anchored parts and slide off.

Snow Removal

Some roofs require regular snow removal to keep them safe. This need can arise if roofs have been poorly designed, the snow and ice loads badly underestimated, or an addition to a building or changes in exposure conditions have resulted in increased drift accumulation onto an existing roof. Some roofs, such as roof parking decks must be cleared to remain useable. Clearing of roofs may be done by snowblower, shovel, heating systems or, in the case of parking roofs, even by front end loaders and trucks.

During mechanical snow removal operations, depending on the pattern of removal, severe load imbalances may arise which could jeopardize the safety of the structure. For example, loads may be significantly increased in some areas of the roof by piling the snow removed from other areas. This may occur on parking garages or on large roofs cleared by a snowblower. Thus, sometimes load conditions more severe than those given by building codes may have to be considered by the designer.

Basic Roof Load

The "snow load coefficient" C_S, defined as the ratio of the roof snow load to the ground load, is used for calculating both the average roof loads and the drift loads. Using the results of the roof load surveys started by the Division of Building Research of the National Research Council in 1956, and judgement and experience, the "basic" snow load coefficient for uniformly-distributed snow on roofs (except for completely sheltered mountain valleys) has been set at 0.8, and for roofs that are, and will remain entirely exposed to the wind, at 0.6.

There are, however, cases when the uniformly-distributed roof design load might be reduced further. Whenever the roof of a building consists mainly of glass or other materials having a heat loss sufficient to melt the snow between storms it may be reasonable and safe to design for a "reduced" load, possibly for the worst one-, two- or three-day snowstorm. In the future adequate heat may not be available for melting the snow because of possible energy shortages. For a roof designed for even less snow than this "reduced" load, and which relies on heating systems to melt the snow, a further hazard could arise if a power failure occurs during a snow storm.

Drift Loads

Major modifications to the basic snow load coefficients are required to account for the effects of wind on the distribution and the amount of accumulation. Because wind at most sites can be highly variable in both speed and direction, drifting must be considered likely for all directions.

The manner in which snow is deposited in drifts on and around buildings is determined by wind speed and direction, air temperature and humidity, snow deposition rate, external geometry of the structure and its surroundings, and upstream terrain roughness (Isyumov, 1971; Isyumov and Davenport, 1974).

As discussed by Boyd (1961a, b) the basic ground loads used to obtain roof design loads are those reached or exceeded, on the average, once in 30 years. When these loads are adjusted to obtain drift loads on a roof, the statistical association between the calculated drift loads and the ground loads becomes tenuous because of the influence of local exposure, orientation, and the shape of the roof as they affect the accumulation patterns. The snow load coefficients for drifts are based on general observations, judgement and detailed field measurements. Unfortunately, insufficient field data are available from enough stations across Canada to derive reliable estimates of the coefficients for different conditions on a purely statistical basis.

To provide a better understanding of the complexity of designing structures to support drift loads some basic roof shapes and their possible drift accumulations are considered. For additional references on this subject the reader should consult Schriever et al. (1967), Lutes and Schriever (1971) Taylor (1980), and the supplement to the National Building Code of Canada (1980b).

Simple flat and shed roofs

On a simple flat roof the snow may be fairly uniformly distributed; then depending on the local shelter, the design load is assumed to be either 60 or

80% of the ground load. It is common in Canada to apply the slope reduction relation (p. 571) to the 60 or 80% load to obtain the design load for sloped roofs.

Gable or hip roofs

If the wind direction is perpendicular to the ridge of a gable roof the windward side of the roof may be scoured clear and some of the snow removed and redeposited in a fairly even layer on the opposite (lee) side (see Fig. 13.2). To design these roofs it is recommended that the windward side of the roof be assumed to be free of snow, and the lee side loaded with 100% of the ground load modified according to slope (National Building Code of Canada, 1980b).

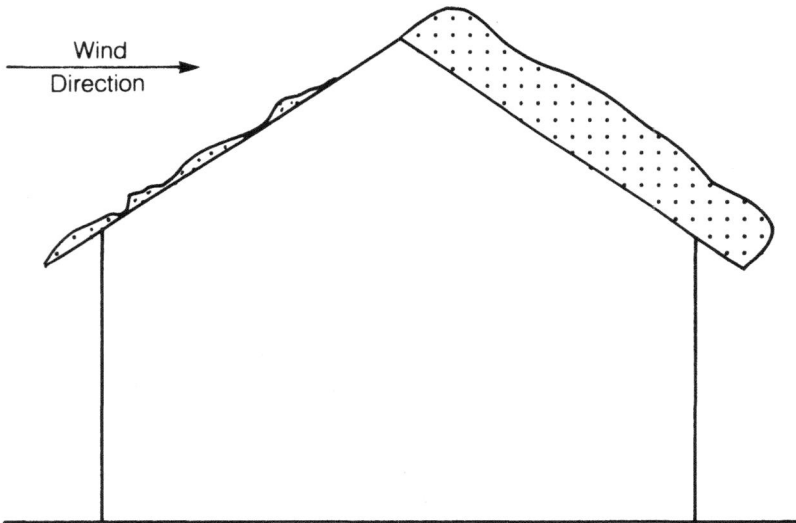

Wind
Direction

Fig. 13.2 Unbalanced load on gable roof.

Arched or curved roofs

Under calm conditions the pattern of snow deposited on an arch is symmetric, with the deepest accumulation occurring at the crown and the depth tapering to zero at about 70° slope. If windy conditions prevail during or following deposition, so as to cause drifting, the drift patterns formed on the roof are quite unsymmetrical. Although some load may accumulate on the windward side, a large drift with a horizontal upper surface starting at the crown normally forms on the leeward side (see Fig. 13.3). The heaviest part of this drift will seldom be more than twice the 30-year ground load. The slope-reduction relationship can be applied to arched and curved roofs.

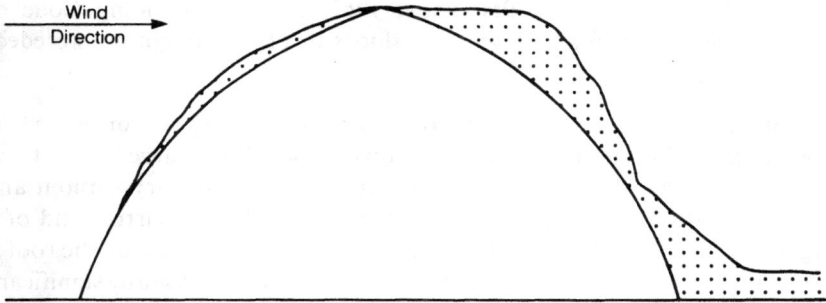

Fig. 13.3 Drift load on curved roof.

A further consideration in the design of these roofs is the load acting on the sides near the base caused by snow which slides off the curved surface. This load should be added to that produced by wind drifting.

Valley areas of two-span and multi-span curved or sloped roofs

For a valley roof, such as shown in Fig. 13.4, there appears to be two mechanisms of snow accumulation which result in an increase in the weight of snow in the valley. The most important is the deposition, in the aerodynamically-sheltered valley, of falling snow entrained in the wind and snow eroded from the windward side of the roof. Of secondary importance is the creep of snow down the slopes towards the bottom of the valley.

Observations in Russia indicated that the maximum weight of snow accumulation in the valley is approximately 1½ times the 30-year ground

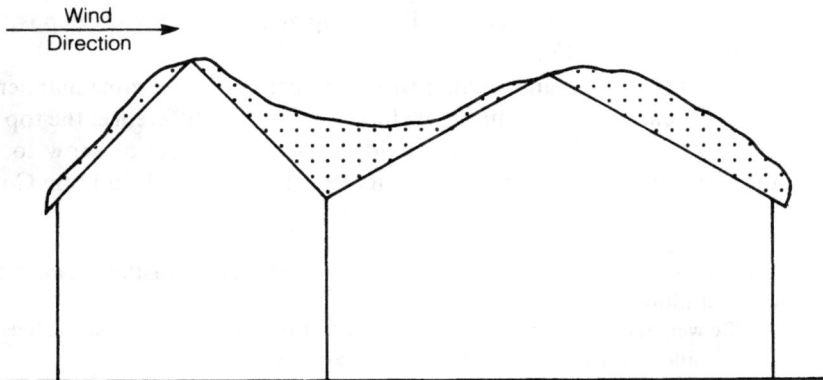

Fig. 13.4 Roof showing accumulation in valley due to drifting and creep.

snow load (Schriever and Otstavnov, 1967; National Building Code of Canada, 1980b), but for long or steep slopes this figure might be exceeded.

Multi-level roofs

Multi-level roofs, obstructions, and parapets can all be considered as geometrical variations of a rectangular "box" situated on a large flat roof. As such, the drifting of snow around and over this box will vary without any sharp delineation from one case to another. If the box is narrow and of a height less than the design depth of uniformly-distributed snow on the roof it may be regarded as "non-obstructing" (that is, *it does not* cause any significant drifts); if it is higher, it "obstructs". In the latter case, if it is wide enough to have a significant deposit of snow on its upper surface, it becomes an "upper level roof".

Wind blowing across multi-level roofs deposits snow in drifts below the lee edge of the upper level (see Fig. 13.5)[1]. These drifts are usually approximately triangular in cross-section with a base width of from 3 to 10 m. The snow load surveys conducted by the National Research Council of Canada showed that although a drift sometimes reaches the height of an upper roof, for the most common situation of a lower roof which is about one story below the upper roof the maximum drift load will seldom exceed three times the ground load (that is, the snow load coefficient, C_s = 3.0).

Snow deposited in the drifts originates from two sources: falling snow, and snow scoured by the wind from the upper and lower roofs. It is important to remember, in estimating the locations of these drifts, that it is not uncommon for the wind direction to vary by 180° from the principal direction during a storm. The only safe assumption therefore is that the wind may blow from any direction.

In the case of sloping multi-level roofs, snow sliding off an upper slope will accumulate, usually in a triangular drift, on top of the snow already deposited on the lower roof (see Fig. 13.6).

Snow accumulates behind smaller roof obstructions in the same manner it accumulates behind steps in multi-level roofs, with one difference: the top of the obstruction is too narrow to provide a large reservoir of snow to be entrained by the wind and deposited in its lee. The National Building Code

---

[1] In Figs. 13.5-13.7 inclusive and in subsequent discussions in this chapter the following definitions of symbols apply:

γ = specific weight of the snowcover (kN/m³) equal to the product of the snow density (kg/m³) and the acceleration of gravity (9.81 m/s²),

G = weight of the ground snowcover with a 30-year return period (kN/m²), values for Canada are given in the National Building Code (1980a), and

h = height of the step (m).

recommends that for design purposes the depth of the drift behind an obstruction should be taken as 2/3 of that behind a step of the same height h on a large multi-level roof, i.e. the snow load coefficient $C_s = 2\gamma h/3G$.

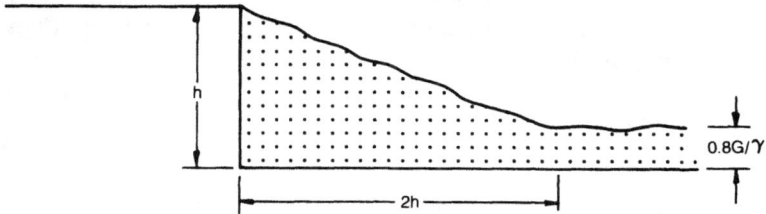

Fig. 13.5 Drift below upper level roof.

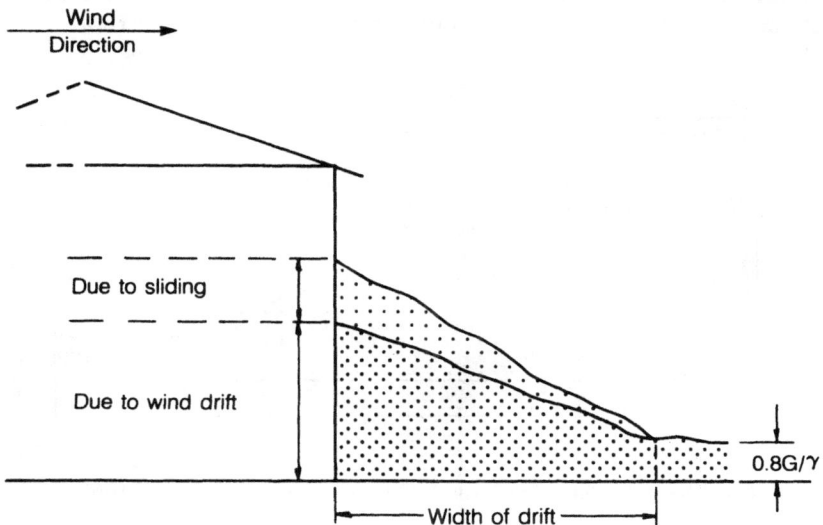

Fig. 13.6 Drift below sloping upper level roof.

It is sometimes difficult to establish when a structure should be considered as an "obstruction" or as an "upper level roof". An empirical method of deciding is by assuming that some portion of the snow, say 50% of the design ground load, blows from the upper roof and is deposited behind the structure in a triangular-shaped drift having a base width equal to 2h, with the restriction that 3 m < 2h < 10 m (see Fig. 13.7). Then the minimum width of an upper level roof which gives a maximum snow load coefficient equal to $2\gamma h/3G$, due to the deposition of drifted snow, can be computed by a volume calculation. If the calculated width is less than the actual width of the structure then it may be considered to be an obstruction.

Parapets are a special case of roof obstruction which affect airflow and can cause drift formation. For design purposes, if a parapet is lower than the depth of the uniformly-distributed design snowcover on the roof (i.e., less than $0.6\,G/\gamma$ or $0.8\,G/\gamma$), its effect on snow load may be ignored; otherwise it should be treated as an obstruction.

Domes

There is very little field or experimental data on the snow loading of domes (Negoita and Mateescu, 1969), however, snow does accumulate on them. Uniformly-distributed loads, reduced accordingly as the slope increases, and unbalanced loads, which may be due to drifting or sliding of snow from upper areas, can occur. The possibility of instability of the structure and of localized buckling must also be examined in designing domes as arched roofs.

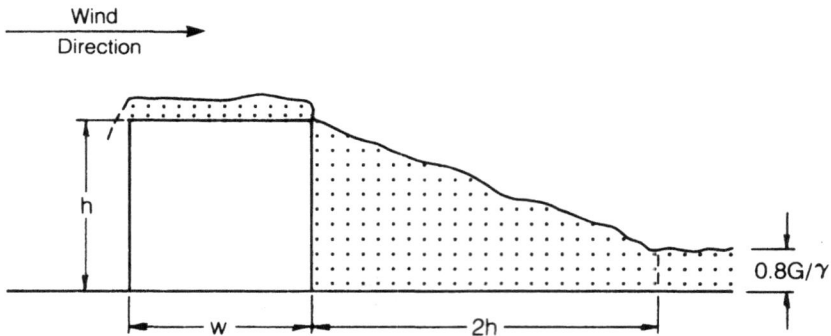

Fig. 13.7 Drifting against a limiting obstruction/upper level roof.

LITERATURE CITED

Baker, M.C. 1967. *Ice on roofs.* Canadian Building Digest, CBD 89, Div. Building Res. Nat. Res. Counc. Can., Ottawa, Ont.pp. 1-4.

Boyd, D.W. 1961a. *Maximum snow depths and snow loads on roofs in Canada.* Proc. 29th Annu. Meet. West. Snow Conf. pp. 6-16. (Res. Pap. 142, Div. Build. Res., NRC 6312, Nat. Res. Counc. Can., Ottawa, Ont., pp. 6-16.)

Boyd, D.W. 1961b. *Climatic information for building design in Canada.* (Supplement No. 1 to the National Building of Canada). NRC No. 6483. Assoc. Comm. Nat. Build. Code, Nat. Res. Counc. Can., Ottawa, Ont.

Gumbel, E.J. 1954. *Statistical theory of extreme values and some practical applications.* U.S. Dept. of Commerce, NBS Appl. Math. Ser. 33, Washington, D.C.

Hunter, H.I. (compiler). 1971. *Summary of snow survey measurements in British Columbia 1935 to 1970.* Water Investigations Branch, Dept. Lands, For. Water Resour., Victoria, B.C.

Isyumov, N. 1971. *An approach to the prediction of snow loads.* Ph.D. Thesis, Res. Rep. BLWT-9-71, Faculty Eng. Sci., Univ. Western Ont., London.

Isyumov, N. and A.G. Davenport. 1971. *A probabilistic approach to the prediction of snow loads.* Can. J. Civil Eng., Vol. 1, No. 1, Sept., pp. 28-49.

Lutes, D.A. and W.R. Schriever. 1971. *Snow accumulations in Canada: Case Histories: II.* Tech. Pap. 339, Div. Build. Res., NRCC No. 11915, Nat. Res. Counc. Can., Ottawa, Ont., pp. 1-17.

National Building Code. 1941. Canada, Dept. of Finance and Nat. Res. Counc. Can., Ottawa, Ont., NRC No. 1068.

National Building Code. 1953. *Part 2, Climate.* NRC No. 3188, Assoc. Comm. Nat. Build. Code, Ottawa, Ont.

National Building Code. 1980a. *The Supplement, Climatic Information for Building Design in Canada,* NRCC No. 17724, Assoc. Comm. Nat. Build. Code, Ottawa, Ont.

National Building Code. 1980b. *The Supplement, Commentaries on Part 4 of the NBC of Canada 1980.* NRCC No. 17724, Assoc. Comm. Nat. Build. Code, Ottawa, Ont., pp. 139-280.

Negoita, A. and C. Mateescu, 1969. *Recherches Roumaines concernant la Surcharge de neige.* Annales de l'institut technique du bâtiment et des travaux publics, No. 259-260, pp. 1172-1194.

Peter, B.G.W., W.A. Dalgliesh, and W.R. Schriever. 1963. *Variations of snow loads on roofs.* Trans. Eng. Inst. Can., Vol. 6, No. A-1, pp. 1-11. (Res. Pap. 189, Div. Building Res., NRC 7418, Nat. Res. Counc. Can., Ottawa, Ont.).

Schaerer, P.A. 1970. *Variation of ground snow loads in British Columbia.* Proc. 38th Annu. Meet. West. Snow Conf., pp. 44-48. (Res. Pap. 479, Div. Build. Res., NRCC 11910, Nat. Res. Counc. Can., Ottawa, Ont., pp. 44-48).

Schneider, T.R., 1959. *Schneeverwehungen und Winterglätte (Snow drifts and winter ice on roads).* Eidgenössisches Institut für Schnee-und Lawinenforschung. Interner Bericht Nr 302, p. 141. [English Transl. by Nat. Res. Counc. Can., Div. Build. Res., 1962, Tech. Transl. 1038].

Schriever, W.R., Y. Faucher and D.A. Lutes. 1967. *Snow accumulations in Canada: Case Histories: I.* Tech. Pap. 237, Div. Build. Res., NRC No. 9287. Nat. Res. Counc. Can., Ottawa, Ont., pp. 1-29.

Schriever, W.R. and V.A. Otstavnov, 1967. *Snow loads - Preparation of standards for snow loads on roofs in various countries, with particular reference to the USSR and Canada.* Res. Rep. No. 9, Int. Counc. Build. Res., Studies and Documentation (CIB), pp. 13-33.

Taylor, D.A. 1980. *Roof snow loads in Canada.* Can. J. Civil Eng. Vol. 7, No. 1, pp. 1-18.

Thomas, M.K. 1955. *A method of computing maximum snow loads.* Eng. J., Vol. 38, No. 2, pp. 120-123.

14

CHEMICALS AND ABRASIVES FOR SNOW AND ICE CONTROL

J. HODE KEYSER

Section Génie Urbain et Transport, Faculté d'Aménagement et Ecole Polytechnique, Université de Montréal, Montréal, Quebec.

INTRODUCTION

The purpose of applying de-icing chemicals and abrasives for winter road maintenance is to ensure that a roadway is safe, that is, to eliminate slippery and hazardous winter driving conditions and to allow an acceptable flow of uninterrupted traffic under inclement weather conditions. Figure 14.1 shows the increase with time in the coefficient of sliding friction of a pavement, initially covered with 0.6 cm of compacted snow, following the application of sodium chloride and a 3:1 mixture of sodium chloride and calcium chloride (Caird and Young, 1971). The coefficient increases with time because of the increase in the amounts of snow melted. As the coefficient of sliding friction increases, the traction increases and the stopping distances decrease. Table 14.1 given by Smith et al. (1971) illustrates the decrease in stopping distances of an automobile if an icy pavement is sanded or the ice or snow surface is melted by salt to provide a wet pavement.

Table 14.1
THE EFFECT OF SALTING AND SANDING ON THE STOPPING DISTANCE OF A VEHICLE (Smith et al., 1971)

Pavement condition	Stopping distance	
	m	% of icy road distance
Icy road (at -1°C)	143	100
Sanded (at -1°C)	55	38
Bare wet pavement	20	14

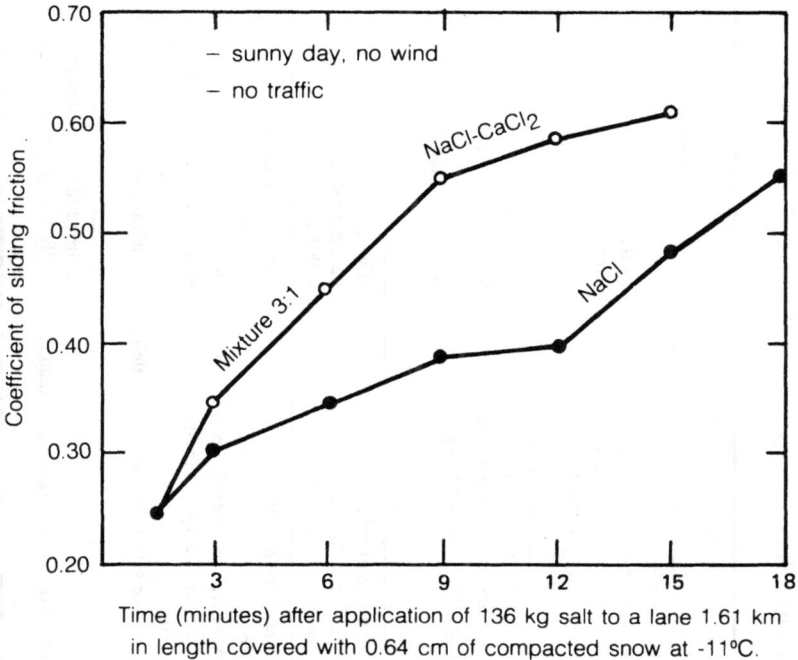

Fig. 14.1 The effect of de-icing salts on the improvement of the coefficient of sliding friction with time (Caird and Young, 1971).

A summary of the principal advantages and disadvantages of sodium chloride and calcium chloride for controlling snow and ice conditions and different abrasives to increase the traction and skid resistances on road surfaces is given in Table 14.2.

PROPERTIES OF DE-ICING CHEMICALS

Chemicals act to control snow and ice conditions by preventing the formation of ice films, by weakening the bond between the snow and the road surface, by melting fresh snow as it falls and by melting the compacted snow that remains after plowing.   Several investigators and agencies (Allied Chemical, 1958; Brohm and Edwards, 1960; Schneider, 1960; Salt Institute, 1962b; National Research Council, 1967) state that a chemical suitable for snow and ice control on roads must have the following characteristics:

(1) It must lower the freezing point of water to normal winter temperatures.

Table 14.2
QUALITATIVE EVALUATION OF CHEMICALS AND ABRASIVES USED IN SNOW AND ICE CONTROL

Material	Main Purpose	Suitability for Use	Principal Advantages	Principal Disadvantages
Sodium Chloride (rock salt) NaCl	To melt snow and ice	Very effective when the temperature is above -3.8°C; effective between -3.8 and -9.5°C; marginal between -9.5 and -12.3°C; and not effective below -12.3°C.	Provides immediate traction. Salt particles bore, penetrate and undercut the ice layer. Freezes dry on pavement surface. Low cost.	Low rate of solution. Ineffective at very low temperatures.
Calcium Chloride CaCl₂		Normally used when the temperature is below -12.3°C; effective down to -29.1°C; marginal between -29.1 and -34.7°C; and not effective below -34.7°C.	High rate of solution. Liberates heat on going into solution. Effective at low temperatures	High cost. Melting action could take place at the ice surface. Pavements remain wet
Mixtures of NaCl and CaCl₂		In cold weather down to -17.9°C when snow and ice must be melted in a short time	High rate of solution. Effective at low temperatures. Chemical more stable on the road.	High cost. Pavement stays wet longer than with CaCl₂
Mixtures of abrasives and salt	To increase the sliding friction immediately	In very cold weather, when salt is not effective or where clean ploughing is impossible if immediate protection is necessary	Free flowing material. No freezing of stockpiles. Abrasives more stable on the road. Quick anchoring of abrasive to the road. Improve skid resistance immediately.	Creates spring clean up problems. Does not remove ice or snow which causes slipperiness May damage vehicles travelling at high speed.
Abrasives treated with salt				
Abrasives			Improve skid resistance immediately.	As listed for other abrasives. Easily brushed off the road by tires.

(2)   It must melt the snow or ice within a reasonable time.

(3)   It must penetrate the snow and ice layers and break any bonds between the snow and ice and the pavement.

(4)   It must act as a non-lubricant on the road surface when spread on a dry pavement or when it is in solution.

(5)   It must be available in bulk quantities at low cost.

(6)   It must be easy to store, handle and apply.

(7)   It must not be toxic to people, animals or plants.

(8)   It must not cause serious damage to metallic structures, pavement or clothing.

Of the several chemicals and mixtures of chemicals which have been used in practice, calcium chloride, sodium chloride and their mixtures satisfy the above requirements to the highest degree (Allied Chemical, 1958; Boies and Bortz, 1965; Monsanto Research Corporation, 1965; OECD, 1969). Other more expensive chemicals have been found suitable for de-icing under special conditions.   For example, the Monsanto Research Corporation (1965) proposes the use of a mixture of 75% formamide and 25% tri-potassium phosphate for runways to reduce corrosion to aircraft. Also, it has been suggested that a composite de-icing agent of urea and calcium formate mixture would reduce the rate of corrosion of highway structures from that produced by other de-icing chemicals (Boies and Bortz, 1965).

Sodium Chloride and Calcium Chloride

Sodium chloride ($NaCl$) is usually supplied either as rock salt (salt obtained by evaporation from solution mining) or solar salt (salt obtained from evaporation of sea water) (National Research Council, 1967). The physical characteristics of sodium chloride, that is, the sodium chloride content and grading, which are important when the chemical is used as a de-icing salt, are described in the ASTM Standard Specification D 632 (American Society for Testing Materials, 1972). Sodium chloride with a moisture content of less than 2% can be considered as being dry (Lang and Dickinson, 1960); with a moisture content less than 1% by weight, there is never a moisture problem (Kelley, 1969).

Calcium chloride ($CaCl_2$) is supplied in the form of flakes or pellets containing, respectively, 77 to 80% and approximately 95% $CaCl_2$ by weight. Specifications covering calcium chloride as a de-icing chemical for snow and ice control on roads are given in the ASTM Standards Specification D 98 (American Society for Testing Materials, 1977).

The phase diagrams in Fig. 14.2 for sodium chloride and calcium chloride show the following important characteristics:

(1)   The solubility of each salt, i.e. the amount of salt that can be dissolved
       in the liquid phase to form a solution, increases very slowly with
       temperature,
(2)   The freezing point of each solution decreases with concentration to
       the eutectic composition, i.e. the concentration of the solution that
       possesses the lowest freezing point,

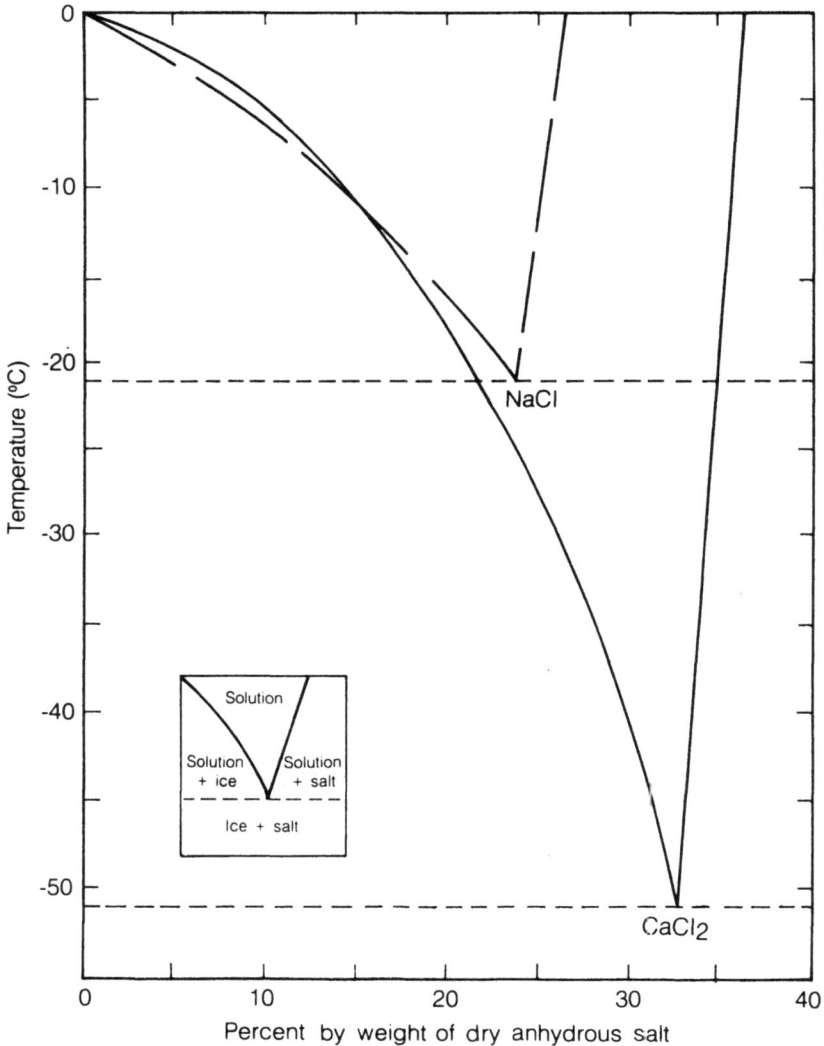

Fig. 14.2 Phase diagrams of NaCl and CaCl₂ solutions (Schneider, 1960).

(3)   Solutions having a concentration less than the eutectic composition have a freezing point less than the melting temperature of pure ice, and

(4)   The eutectic temperature, i.e. the temperature at the eutectic point, of a calcium chloride - water system is lower than the eutectic temperature of a sodium chloride - water system.

In Fig. 14.2 it can be seen that the eutectic composition for the calcium chloride-water system is approximately 68% H_2O and 32% $CaCl_2$ (by weight), which remains liquid at temperatures as low as -51°C. The eutectic composition for the sodium chloride-water system is 23.3% NaCl and 76.7% H_2O which freezes at about -21°C. These differences explain why calcium chloride rather than sodium chloride is used to remove ice from highways in very cold climates.

Table 14.3 compares those properties of sodium chloride and calcium

Table 14.3

COMPARISON OF THE PROPERTIES OF SODIUM CHLORIDE AND CALCIUM CHLORIDE

Property	Calcium Chloride	Sodium Chloride
Eutectic temperature (lowest temperature at which ice melts)	-51°C	-21°C
Rate of solution (Rate of ice melting)	High	Low
Ice melting capability given unlimited time	Less	More down to -12°C
Moisture attraction for solution	Attracts moisture	Does not attract moisture
Heat of solution	Releases heat when going into solution	Absorbs heat when going into solution
Characteristics of the salt in solution	Remains in solution	Recrystallizes
Cost	2 to 3 times the cost of NaCl	Lower

chloride which are important for snow and ice protection. In general, sodium chloride is used at temperatures above -12° C because the amount needed for de-icing at lower temperatures becomes prohibitively large. Similarly, the use of calcium chloride is generally limited to temperatures above -35° C. Calcium chloride costs about 2 to 3 times more than sodium chloride.

Sodium chloride has a much lower rate of solution or rate of melting ice than calcium chloride. Table 14.4 shows a comparison of the amounts of ice melted per unit application of these salts as a function of time after application (Calcium Chloride Association, 1949). It is evident that calcium chloride has a much more rapid rate of solution than sodium chloride. Nevertheless, over unlimited time, and provided it is used above its eutectic temperature, sodium chloride is superior to calcium chloride as to the amount of chemical required to melt a given quantity of snow (see Table 14.5). Because of its relatively slow rate of solution individual coarse particles of sodium chloride will bore and penetrate ice and weaken any bonds which may have developed between the road surface and the layer of ice. This enables traffic or plows to break up the compacted snow or ice crust and cast it to the side of the road (Salt Institute, 1962b). One disadvantage is that its slow rate of solution may result in large amounts of the salt being wasted with frequent plowing operations (Commonwealth of Massachusetts, 1965a). Also, when pavement dries, sodium chloride recrystallizes leaving a dry, salt-covered surface which may be blown from the pavement.

In comparison, lesser amounts of calcium chloride are lost during frequent plowing because of its faster rate of solution. Also, it is not easily blown from bare pavement and wasted because it is deliquescent; which however, requires that it must be stored in dry or low moisture areas.

Table 14.4
AMOUNTS OF ICE MELTED BY CaCl₂ AND NaCl WITH TIME (Calcium Chloride Association, 1949).

Temperature	kg of ice melted per kg of chemical applied							
	15 min		30 min		1 h		6 h	
°C	$CaCl_2$	NaCl	$CaCl_2$	NaCl	$CaCl_2$	NaCl	$CaCl_2$	NaCl
-18	1.3	0.1	1.7	0.1	2.0	0.3	2.3	1.9
-12	1.7	0.5	2.0	0.9	2.4	1.6	2.8	3.8
- 7	2.5	1.6	3.1	2.6	3.8	4.1	5.3	7.5
- 3	2.6	1.8	3.4	2.9	4.4	4.3	7.1	9.5

Table 14.5
AMOUNT OF SALT REQUIRED TO MELT 100 GRAMS
OF ICE (UNLIMITED TIME) (Schneider, 1960).

Temperature °C	g to melt 100 g of ice	
	Calcium chloride[a]	Sodium chloride
- 1	3.1	1.5
- 3	8.2	5.2
- 6	14.7	10.3
- 9	20.2	14.7
-15	29.3	23.2

[a] $CaCl_2 \cdot 2H_2O$

Sodium Chloride and Calcium Chloride Mixtures

When calcium chloride and sodium chloride are combined they complement each other as de-icing agents. The reaction between calcium chloride and moisture is exothermic (that is heat is released) whereas that between sodium chloride and moisture is endothermic (heat must be supplied or absorbed for the salt to dissolve). Thus, when combined, the deliquescent calcium chloride absorbs moisture releasing heat and thereby increasing the rate of solution of the sodium chloride. These reactions produce rapid melting over a large temperature range, and produce a brine which sustains the melting action of the sodium chloride over a long period thereby providing prolonged protection (Coulter, 1965; Calcium Chloride Institute, 1968). Because the mixture has a much faster melting action than that of sodium chloride alone, bare pavement conditions are obtained in less time and with less equipment. As the temperature drops below -1°C, the difference in melting time between the mixture and sodium chloride becomes greater (Dickinson, 1959; Brohm and Edwards, 1960; Lang and Dickinson, 1960).

Tests conducted by several different organizations (National Research Council, 1967; Calcium Chloride Institute, 1968; Kelley, 1969) have indicated that the relative proportions of NaCl and $CaCl_2$ in the mixture should be varied according to air temperature and depending on whether the temperature is rising or falling (see Fig. 14.3). For example, under falling temperatures in the range from approximately -6.7 to -12.5°C, a mixture of 3 parts NaCl to 1 part of $CaCl_2$ may be used with a trend to a ratio of 1:1 as the air temperature falls below -17.8°C. Following a storm, if the temperatures are about -7° C a pavement can be cleaned within 15 to 20 min after spreading of a 3:1 mixture of the two salts (Lang and Dickinson, 1960).

The National Research Council (1967) suggests that satisfactory mixing can be achieved by feeding both salts to a conveyor belt through adjustable hopper gates pre-set to obtain the desired proportion. Usually, the cost of mixing is of the order of 10 to 30% of the material cost (Carr, 1969).

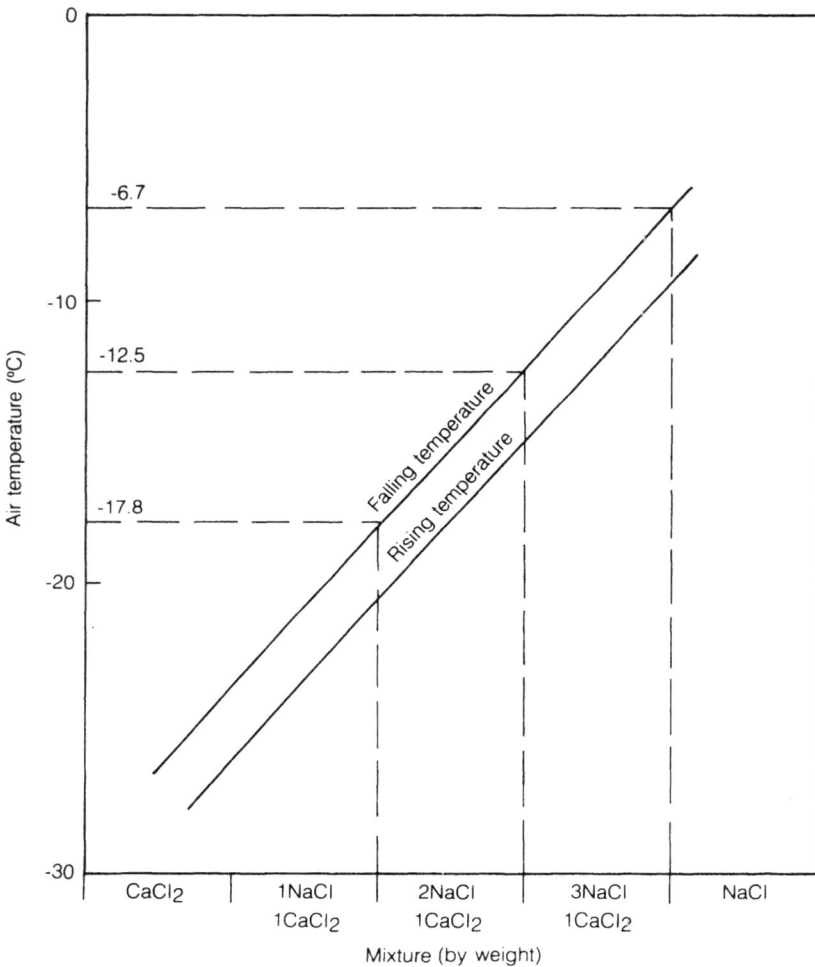

Fig. 14.3 Use of mixtures of sodium chloride and calcium chloride at different temperatures.

ABRASIVES

Natural sand, manufactured sand and crushed slag or cinder are abrasives usually applied to a road surface covered with snow or ice to provide an immediate increase in skid resistance. Abrasives are especially useful at very low temperatures where de-icing chemicals are not active (Nichols and Price, 1956; Road Research Laboratory, 1968). Table 14.6 lists a few of the desirable characteristics of abrasives used for snow and ice control on highways.

Table 14.6
DESIRABLE CHARACTERISTICS OF ABRASIVES. (Schneider, 1959)

Characteristic	Reason
Great resistance to compression, crushing, impact, and grinding	Resist degradation under the action of traffic Could be recovered in spring
Angular shape	Greater stability Prevent its being blown away
Darkish color	Absorbs heat to melt itself into the surface of ice
Uniform grain size	Uniform spreading pattern Less likely to damage equipment

The maximum aggregate size is generally limited to 1.3 cm because there is a risk that the larger particles may damage vehicles and injure pedestrians (Road Research Laboratory, 1968). Particles which pass through a number 50 mesh sieve (less than 300-μm diameter)should not be used as abrasives since they do not significantly increase the skid resistance of the surface (Hegmon and Meyer, 1968). Generally, to obtain good skid resistance large quantities of abrasives must be applied. Spreading rates varying from 11.5 to 23 kg/km on a 2-lane highway have proven effective whereas rates of less than 5.6 kg/km have not (Schneider, 1959).

Experience has shown that on heavily-travelled highways, untreated abrasives can be whipped off the road by traffic and even by winds of moderate speed (Highway Research Board, 1954). Schneider (1959) found that under high-speed traffic the initial slippery conditions of a highway were re-established after only 10 to 15 passes of an automobile following application.

Treatment of an abrasive with an amount of salt equal to approximately 1/30th of its weight prevents the particles from forming frozen lumps and enables the abrasives to penetrate and to become anchored to the surface of the hard snow or ice; hence, they are not easily swept off the roadway by wind or traffic. Treatment also facilitates the loading of the abrasive from stockpiles and improves the performance of mechanical spreaders enabling them to apply a uniform spread (Calcium Chloride Association, 1949). Additional details of the methods of preparation and the storage of treated abrasives are described by Highway Research Board (1954, 1962) and the National Research Council (1967).

Several organizations have used mixtures of either calcium chloride or sodium chloride and abrasives to decrease the slipperiness of a road and to melt the snow during periods when the air temperature increases. Experiments on roads have shown that, for mixtures to be effective, a coarse-graded salt should be used so that individual salt particles can penetrate into the snow and ice layer allowing most of the abrasives to remain on the surface and maintain traction (Brohm and Edwards, 1960). If fine-graded salt is used, the salt softens the snow and ice surface and the abrasive moves into the ice layer under the action of traffic, thereby losing its effectiveness as a skid preventative.

Brohm and Edwards (1960) and Himmelman (1963) found that sodium chloride is less effective as a melting agent when used in a mixture with sand. Figure 14.4 shows that fine-crushed sodium chloride and sand mixtures are much more effective in melting ice than coarse-crushed mixtures.

Some investigators suggest the use of coarse salt alone as an abrasive for increasing traction at low temperatures: when the air temperature increases the salt increases the melting rate; also there is no need to clean streets and catch basins in the spring (National Research Council, 1967).

In a study conducted by the New York Throughway Authority (Lang and Dickinson, 1960) it was found that a satisfactory mixture of an abrasive and de-icing salt, useful under almost all storm conditions, contained one part by weight of mixed chemical salt (proportion 1:3 calcium chloride to sodium chloride by weight) combined with one part of abrasives. This mixture was found to be nearly as effective in its melting action as an equivalent mixture of salts and in addition provided the abrasive necessary for skid protection.

FACTORS INFLUENCING THE MELTING RATE OF SNOW AND ICE

Physical Characteristics of Snow and Ice

The rate of removal of solid ice and snow from pavements is greatly influenced by the density, thickness and uniformity of the ice or snow layer.

The density of loose, newly-fallen snow generally varies from 40 to 112 kg/m³ (Amberg and Williams, 1948; Coulter, 1965; Minsk, 1965). The density of ice is about 918 kg/m³ under average conditions. Hence, the density of a snow layer on a pavement can vary widely depending upon the amount of compaction.

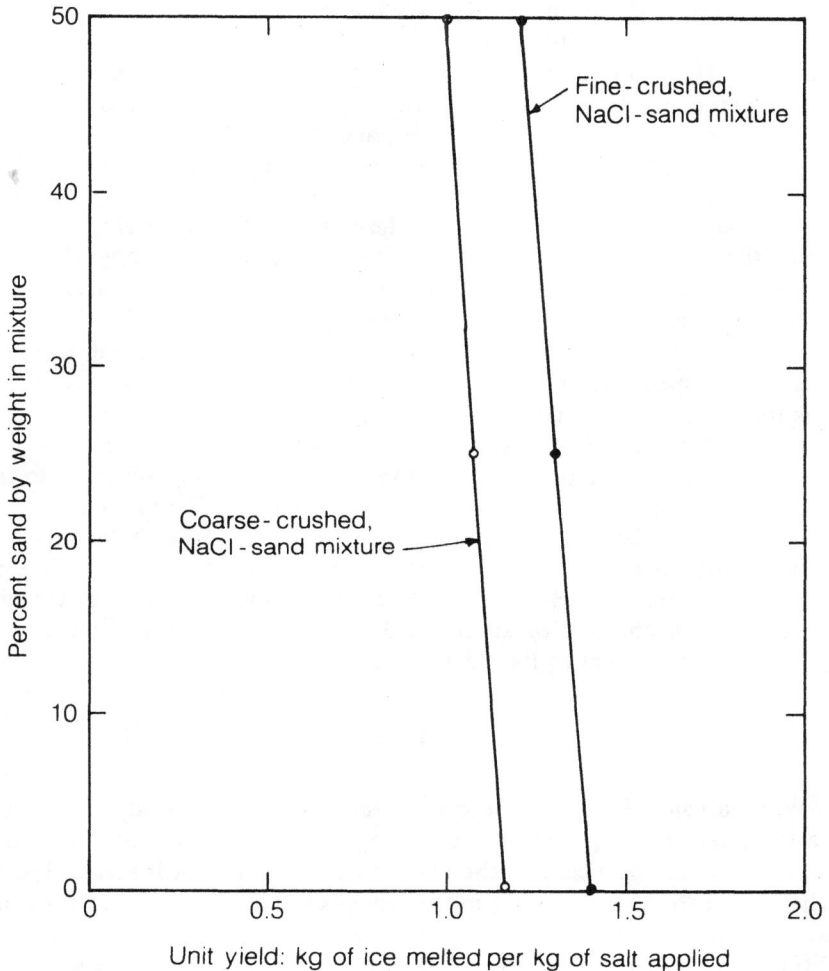

Fig. 14.4 The mean effects of sand on the melting capacity of salt-sand mixture. Data is for the following conditions: temperatures -1.1, - 6.7, -12.2 and -17.8°C; concentrations 0.39, 0.98, 1.95 and 7.2 kg/m²; reaction times 15, 30, 45 and 60 min (Brohm and Edwards, 1960).

Coulter (1965) reports that the density of new snow depends mainly on the air temperature at the time of the storm. He suggests that the density increases as an exponential function of air temperature defined by the expression:

$$\rho = 16.018 \exp(1.907 + 0.0835\ T), \qquad\qquad 14.1$$

where:

ρ = average density of snowfall (kg/m^3), and
T = temperature ($^\circ$C).

Snow which is not removed from a roadway varies widely in its physical properties from day to day depending upon weather conditions, type of precipitation and traffic density (Caird and Young, 1971).

The thickness of an ice layer on a pavement may also vary greatly. Very thin, scarcely-visible films may be formed by direct precipitation of supercooled atmospheric moisture on the surface or by the freezing of the thin liquid film produced by the pressure of a tire on the snow or ice crystals. Thick ice crusts may result from freezing rain, the freezing of wet snow accumulated on the surface or the freezing of melt water.

To prevent the compaction and bonding of snow to the pavement surface by traffic, an application of salt to the road surface should be made as early as possible during the storm, but preferably, before the storm (Amberg and Williams, 1948; Salt Institute, 1962b; National Research Council, 1967). A weak brine solution left from a previous salting is usually adequate to free compacted snow from a road surface (Amberg and Williams, 1948).

When a thick layer of ice or compacted snow is bonded to the pavement surface large quantites of chemicals applied over a long period of time are required to remove it. Himmelman (1963) has shown that the rate at which either calcium chloride or sodium chloride will effectively remove ice is inversely proportional to the thickness of the layer.

Type of Chemical

When a solid chemical is spread on ice, during the period the chemical particles are entering solution the melting rate depends on both the rate of solution of the chemical and the rate of ice solution (Salt Institute, 1962a). The rate of solution of the chemical depends on its solubility and grain size while the rate at which ice or snow is melted by a brine solution depends on the diffusion of ions from the highly-concentrated parts of the brine to the less concentrated parts of the solution at the ice or snow contact surfaces. The rate of diffusion depends upon the concentration gradient, the mobility of the ions and the temperature.

At moderate temperatures ($> -7^\circ$C), given sufficient time, a unit volume of sodium chloride solution of given concentration will melt about twice as much ice as a unit volume of calcium chloride solution with the same concentration. This difference is attributed to the larger number of ions in the sodium chloride solution and their greater mobility (Highway Research Board, 1962).

Pellets of calcium chloride melt ice more rapidly than flakes, and because they have a higher concentration of calcium chloride, melt about 20% more ice than an equal weight of flakes (National Research Council, 1967).

Field and laboratory studies have shown that a mixture of three parts of sodium chloride and one part of calcium chloride will melt a greater quantity of ice than equivalent amounts of either chemical (Dickinson, 1959; Brohm and Edwards, 1960; Lang and Dickinson, 1960; Himmelman, 1963; Coulter, 1965). Figure 14.5 shows the differences in the depths of ice melted in two hours by equal weights of the two salts and a 3:1 mixture of NaCl:CaCl₂ as a function of temperature; the mixture will melt a greater quantity of ice than either chemical at all temperatures above about -27° C. The colour of the chemical and the amount of impurities in it may also influence its melting action, however their effects have not been quantitatively evaluated (National Research Council, 1967).

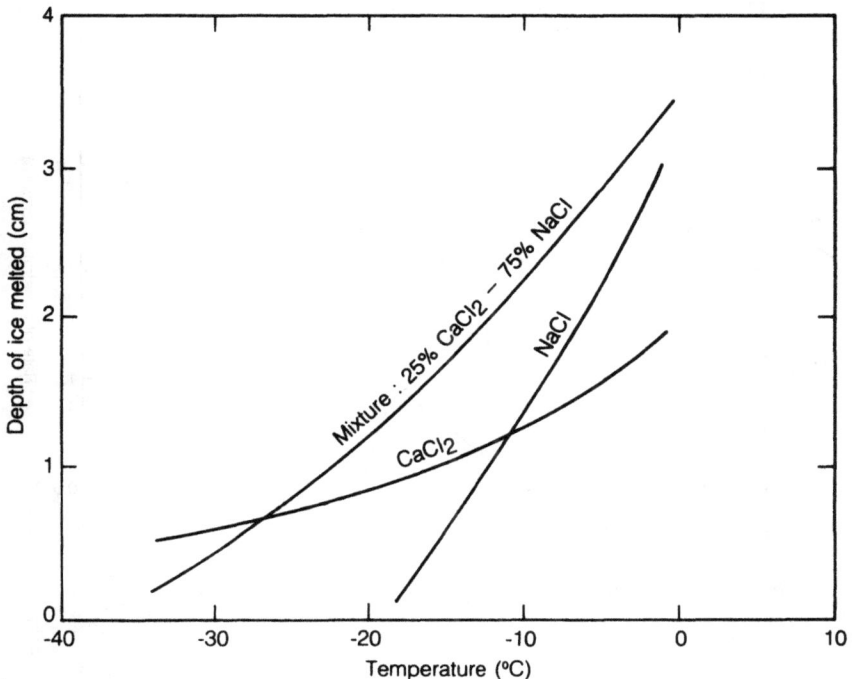

Fig. 14.5 Depth of ice melted in 2 h by equal weights of salt (Dickinson, 1959).

State of the De-Icing Chemical

Chemical de-icing salts are supplied either as aqueous solutions or granulated solids. Aqueous solutions can be used very effectively for preventative and maintenance purposes to control thin layers of ice or snow. They may be applied either before or during a snow storm and are particularly effective during cold, dry periods (Scotto, 1970; Keyser, 1971). To obtain the maximum efficiency aqueous solutions should be applied close to the pavement and spread as uniformly as possible. They will not melt thick layers of ice or hardened, packed snow.

The effectiveness of granulated de-icing chemicals depends on the grading of particle sizes. In general, the melting action of fine-graded salt is greater than that of coarse salt, a fact which is attributed to the larger specific surface area of the fine salt (Amberg and Williams, 1948). To be efficient for snow and

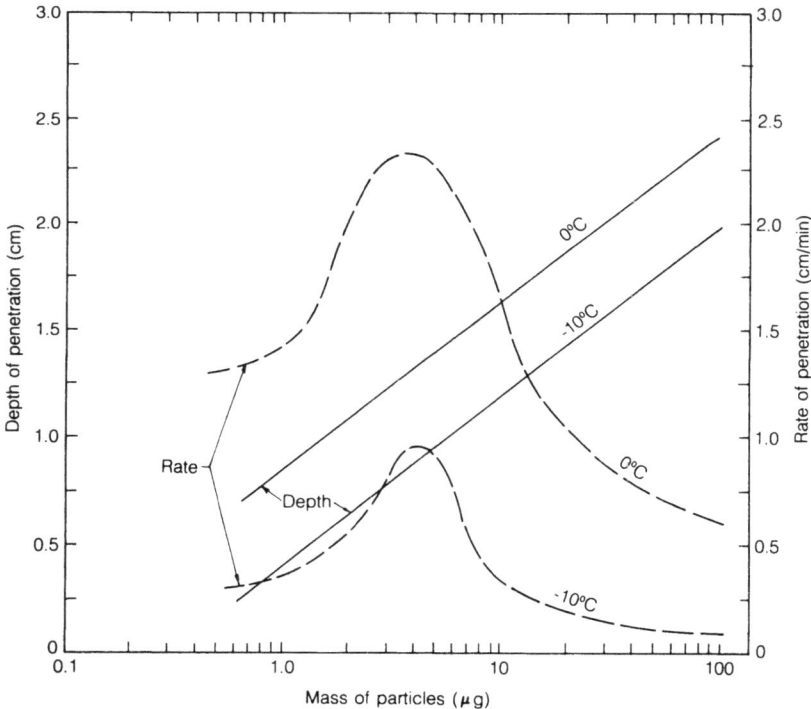

Fig. 14.6 The influence of the mass of salt particles on the depth and rate of penetration into a layer of ice (Amberg and Williams, 1948).

ice removal granulated salts should contain a sufficient amount both of fine particles to initiate rapid melting at the surface and of coarse particles to penetrate the pavement to break the bond between the pavement and ice. Once these bonds are broken the movement of traffic breaks up the ice sheet and casts the slush and smaller pieces from the roadway. The coarse particles also act as an abrasive for a limited period of time immediately following application (National Research Council, 1967).

Amberg and Williams (1948) have studied the influence of the mass of salt particles on depth and rate of penetration into a layer of ice (see Fig. 14.6). The depth of penetration increases with the mass of particles; the maximum rate of penetration occurs at approximately 4 μg, corresponding to a particle diameter of approximately 16 mm.

Concentration of Salt

The effect of concentration (the amount of salt applied per unit area, kg/m^2) on the amount of ice melted varies with different chemicals. Brohm and Edwards (1960) compared the effects of different concentrations of sodium chloride and calcium chloride (pellets) on the mean total melt produced (see Fig. 14.7). The mean total melt by calcium chloride increased almost directly with concentration whereas that by sodium chloride was almost independent of concentration at levels greater than about 1.5 kg/m^2. The difference is attributed to the fact that calcium chloride dissolves at a relatively uniform rate whereas sodium chloride dissolves more slowly with increasing concentration. The unit yield (melting efficiency) or the amount of ice melted per kilogram of NaCl applied increases with decreasing concentration (see Fig. 14.8). At low concentrations, ($< \sim$ 0.3 to 0.4 kg/m^2) the total melt produced by each salt is approximately the same, and very low.

The relationship between the amount of ice melted and the concentration of the chemical is time dependent. Brohm and Edwards (1960) reported on the interaction between the concentration of NaCl, the amount of ice melted and the time of application (see Fig. 14.9). Each concentration of salt requires a certain reaction time to produce the maximum total melt; the concentration must be increased with time to sustain the maximum rates. Thus, the most efficient and economical application of de-icing salt is at concentrations sufficient to sustain maximum melt. At more than these amounts some salt does not undergo reaction and is wasted, while at less the ice may not be completely melted. Schneider (1960) suggests the application rate be based on the eutectic properties of the salt so that the prevailing temperatures need not be considered during application and melting will take place rapidly.

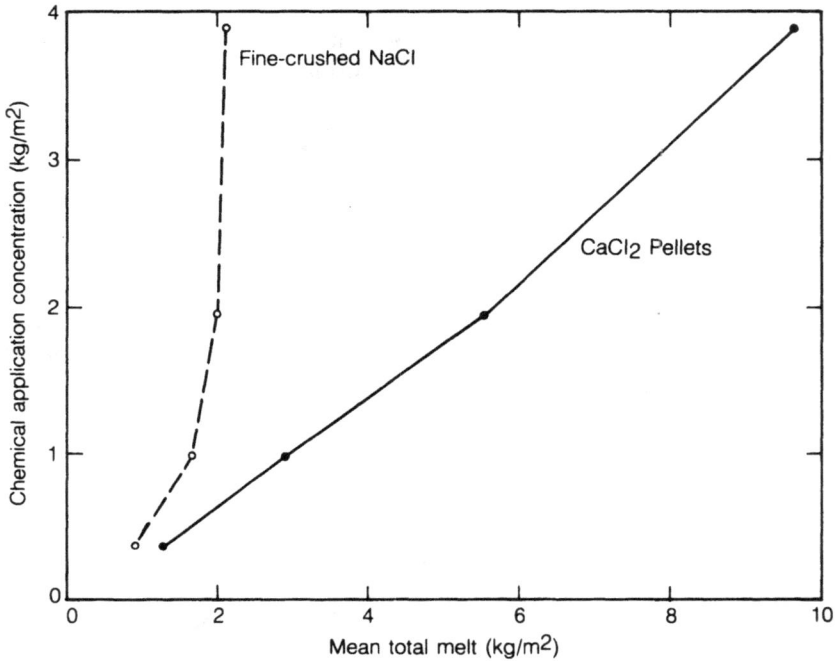

Fig. 14.7 Mean concentration effects on the mean total melt: data for temperatures of -17.8, -12.2, - 6.7 and -1.1°C and times of 15, 30, 45 and 60 min (Brohm and Edwards, 1960).

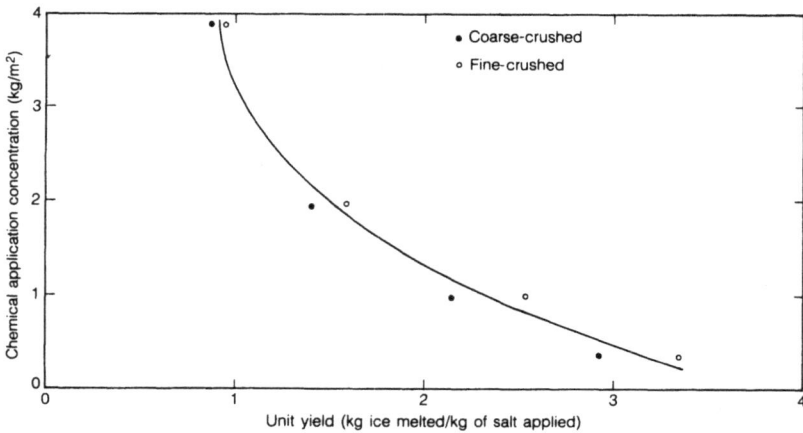

Fig. 14.8 The effect of concentration of NaCl on unit yield: data for coarse-crushed and fine-crushed NaCl at - 6.7°C and melting time 1 h (Brohm and Edwards, 1960).

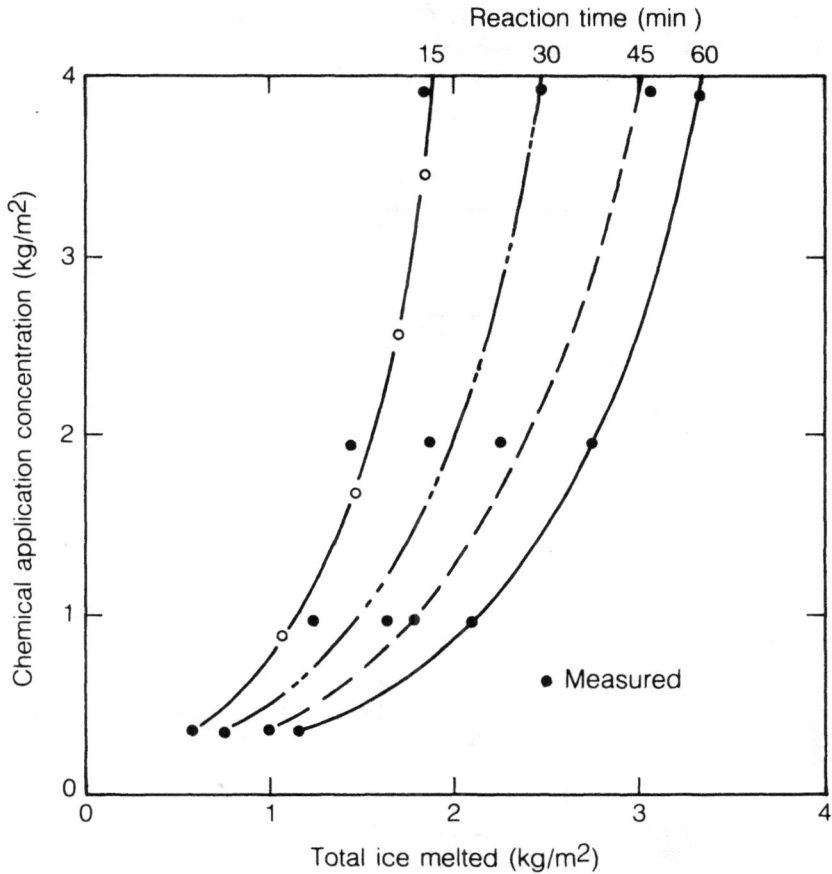

Fig. 14.9 The effect of concentration on total melt: data for coarse-crushed NaCl at - 6.7° C (Brohm and Edwards, 1960).

Time of Solution

In general, the longer the de-icing chemical is permitted to act, the larger the amount of melt. The rate of melting by calcium chloride is very rapid immediately following application of the chemical and decreases with time; but the rate for sodium chloride is slow following application and increases with time to a magnitude which equals or exceeds that produced by calcium chloride (Kersten et al., 1959; Dickinson, 1959; Himmelman, 1963). As shown

in Fig. 14.10 the rates of melt by CaCl₂ or NaCl are highly temperature
dependent and increase progressively as the temperature approaches 0° C. In
practice, when it is possible to allow a relatively long reaction time, a saving in
the cost of chemicals may be realized.

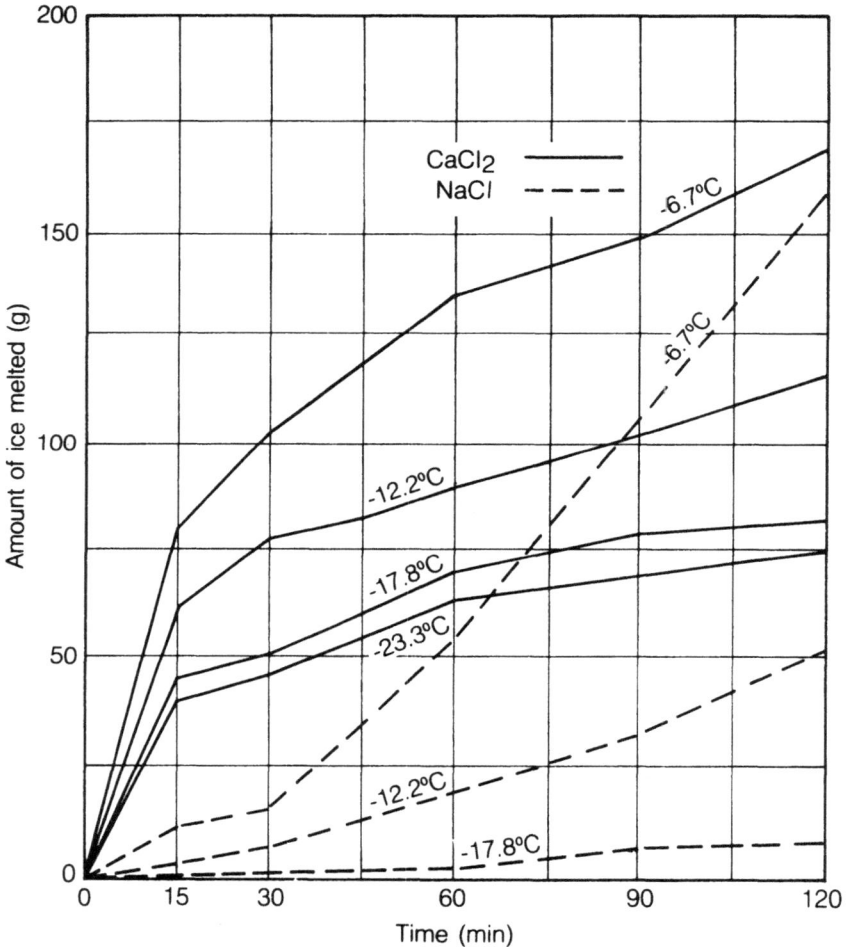

Fig. 14.10 Amount of ice melted in 15 to 120 min; 350 g of ice in 17.8 cm by 27.9 cm
pan; 40 g of 100% NaCl or CaCl₂. No NaCl reaction at -23.3° C (Dickinson,
1959).

Temperature

The temperature of the pavement, which may be different from that of the ambient air, is an important factor influencing the amount of chemical required and the rate of melting (Salt Institute, 1962a; National Research Council, 1967). The solubility diagram (Fig. 14.2) shows that the amount of chemical required to maintain a liquid solution increases almost linearly with decreasing temperature to the eutectic point (see also Table 14.5). Figure 14.11 shows that the mean total melt produced by different types and gradings of salts increases with reaction temperature.

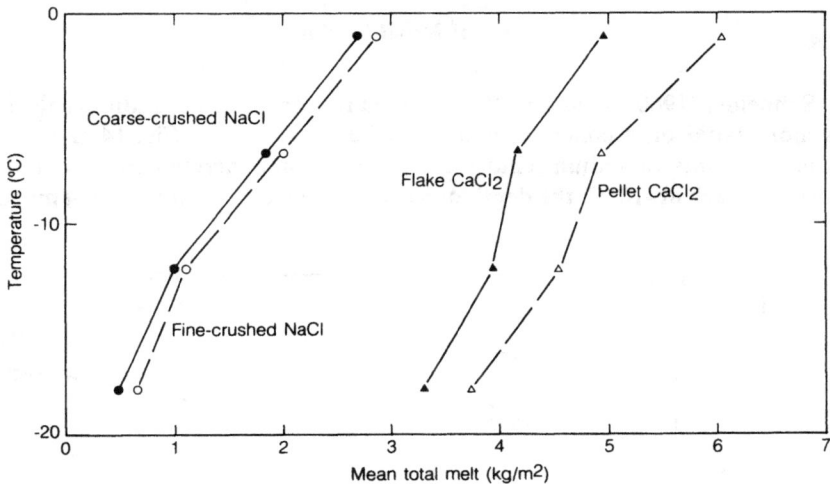

Fig. 14.11 The effect of mean reaction temperature on mean total melt: data for concentrations of 0.39, 0.98, 1.95 and 3.90 kg/m² and melt times of 15, 30, 45 and 60 min (Brohm and Edwards, 1960).

Radiation, Wind and Humidity

Much of the solar radiation incident on an exposed base or partially snowcovered pavement is absorbed, causing its temperature to be higher than the ambient air temperature by as much as 6°C (National Research Council, 1967). As the temperature of the pavement surface increases the amount of chemicals required for melting a given mass of snow is reduced. On the other hand, in non-exposed, shaded areas the amount of chemicals required is increased. Therefore the effects of exposure leads to variable application rates along a highway. It should also be recognized that under clear-sky and low humidity conditions appreciable cooling of the bare, exposed pavement may occur owing to night-time long-wave radiation loss.

The melting action of calcium chloride is enhanced by high humidities since, being deliquescent, it readily absorbs water from the atmosphere at relative humidities from 46 to 60% and air temperatures from 0 to -9°C (Eberhard, 1959).

High winds may reduce the efficiency of chemicals. If chemicals are applied to roads during high winds either they may be swept from the road or they may cause blowing snow to stick to the pavement; an untreated surface is kept bare by the wind.

Type of Road Surface

Schneider (1960) calculated that the melt rate produced by sodium chloride is more rapid on a concrete than an asphalt surface (see Fig. 14.12); when equal amounts of sodium chloride are spread on concrete and bituminous surfaces, each at -10°C, the depth of ice crust melted in 30 minutes is 4 mm on

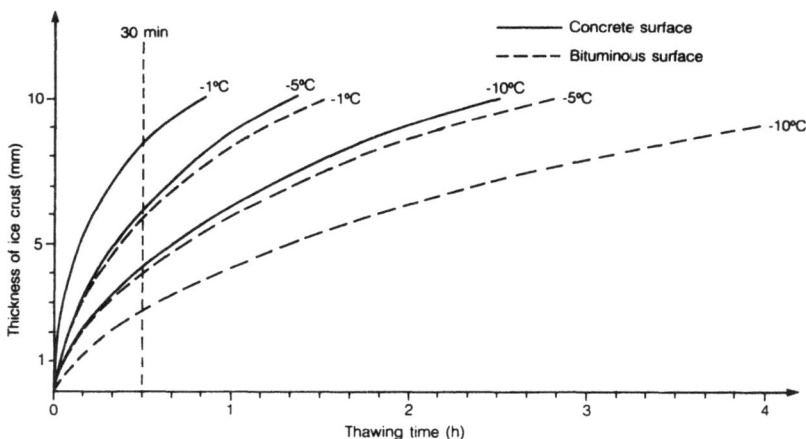

Fig. 14.12 Effect of type of surface on the time needed for NaCl to thaw ice. The curves are calculated on the assumptions: (a) that the conductive capacities of concrete, asphalt and ice are 1.71, 0.865 and 0.649 $kW/°C \cdot s^{1/2}$ respectively; the air, ice and pavement surface have a temperature of -5°C; the wind speed is low, ~0.5 m/s; the heat transfer coefficient is 5.8 $W/m^{2 \cdot °}C$; and the interface boundaries between ice and pavement and ice and air are cooled rapidly to -20°C and remain at this temperature until the ice is completely melted (Schneider, 1960). Reproduced by permission of the National Research Council of Canada.

the concrete and 2.5 mm on the bituminous surface. This difference is attributed to the fact that the concrete pavement releases heat more rapidly than the bituminous surface. These calculations neglect the effect of solar radiation and assume little or no wind over the road surface. In the field, an asphalt pavement absorbs more global radiation which increases melt and the increased advection of heat from exposed asphalt surfaces accounts for the earlier disappearance of snow from adjacent areas.

Topography

The topographic conditions favouring ice formation include features which either directly or indirectly affect the energy exchange at the pavement. The most important are those which reduce the exposure of the pavement to direct beam solar radiation. The effects of slope and aspect on the radiation received by a surface are summarized in Ch. 9 In the construction of new roads attempts should be made to minimize "shaded" areas by using relatively flat side slopes (Milloy and Humphreys, 1969).

Traffic

Traffic assists in removing snow and ice from a road surface in several ways:

(1)   The pressure on the surface caused by the weight of vehicles lowers the melting point of ice and increases the rate of heat transfer to the melting layer (Amberg and Williams, 1948). This effect is significant when equilibrium conditions exist near the melting point.

(2)   Tire friction supplies heat to the surface (Coulter, 1965).

(3)   Traffic mixes chemicals with the snow and breaks up ice layers that have been weakened by salt action (National Research Council, 1967).

(4)   Vehicular activity removes slush (Amberg and Williams, 1948).

Schaerer (1970) found that the effects of traffic on the physical properties of snow and its removal depends on the free water content (see Table 14.7). Other investigators (Nichols and Price, 1956; Kersten et al., 1959; Brohm and Edwards, 1960; Himmelman, 1963) have found that only about 30 to 50% of the ice on a pavement needs to be melted so that traffic, having a density of more than 30 vehicles per hour, can cast the slush away from the surface.

Width and Time of Application

In general, for a given amount of salt the melt rate can be increased by applying it in narrow strips on the road surface. However, the amount of

Table 14.7
THE EFFECT OF TRAFFIC ON SNOW AT DIFFERENT FREE WATER
CONTENTS (Schaerer, 1970)

Free Water Content (by weight)	Effect of Traffic
15%	Compaction of snow into ice crust
15-30%	Snow stays in a soft loose state
30%	Adhesion of snow to tire
30% +	Slush is easily removed by traffic

snow melted is the same, independent of the strip width. One advantage of narrow strips is that the snow or ice underneath melts relatively quickly, exposing bare pavement which absorbs solar energy and increases the rate of melt on the entire roadway (Amberg and Williams, 1948). Chemicals applied before snowfall or freezing rain occurs should be spread uniformly on 4.5-m strips of the road (Quebec, 1967). After plowing, chemicals are usually applied on the middle one-third of the pavement in narrow strips of about 0.3 to 1 m (Brohm and Edwards, 1960; Quebec, 1967; Ontario, 1967).

The most important factor affecting the success of chemical treatment for snow clearing is the time of application (Nichols and Price, 1956). Chemicals used for de-icing have their effectiveness significantly increased if applied within 30 min of the beginning of snowfall (Salt Institute, 1962b). A small amount of chemical mixed with loose snow causes some melting and produces granular snow that is not packed by traffic and that can be easily removed by plows. Approximately 15% of the newly-fallen snow must be melted to obtain this condition.

Combined Effects

As indicated above, many factors influence the melting rate of snow and ice by salts. However, the effects of any material on the rate of melt are valid only when other factors remain constant. In practice, the interaction of all the variables, whose magnitudes are governed by prevailing conditions, determine the melting rate. According to Himmelman (1963) the volume of ice that could be melted by sodium chloride and calcium chloride after 90 min at temperatures from -12 to -6°C can be estimated from the following equations:

sodium chloride

$$V = A\sqrt{RT^3}/I, \qquad\qquad 14.2$$

calcium chloride

$$V = BR^2T/I^{0.8}, \qquad\qquad 14.3$$

where   V = volume of ice melted on a lane of unit length after 90 min,
              R = application rate (mass per unit lane length),
              T = temperature ($^\circ$C),
              I = thickness of the ice layer (mm), and
    A,B = empirically-derived coefficients.

Comparison of Eq. 14.2 and 14.3 shows that the volume of ice melted by calcium chloride is more sensitive to the application rate while the volume melted by sodium chloride is more sensitive to temperature.

Brohm and Edwards (1960) found from laboratory studies that the depth X of newly-fallen snow melted by a given chemical could be expressed by the equation:

$$\log X = \log D + C_1 D + C_2 T + C_3, \qquad\qquad 14.4$$

where   D = concentration of chemical mixture,
            T = temperature, and
   C_1, C_2, C_3 = constants whose magnitudes depend on the type of chemical and the reaction time.

EFFICIENT SNOW AND ICE CONTROL USING CHEMICALS

In establishing a snow and ice control system consideration must be given to the level or quality of service which can be economically justified to obtain effective control and to the type of organizational structure and management procedures that will produce the most effective and efficient results. The choice of the system, based on economic considerations relative to the level of service, involves the analysis of the component costs; capital investment, operation and maintenance and usage in terms of losses arising from accidents, delays and immobilization. Figure 14.13, given by Tighe and Webber (1971) illustrates a procedure for choosing a system which provides a maximum level of service at a minimum total cost. The total system cost must be calculated for different levels of service.

To obtain the most efficient, effective results, information should also be provided on purchasing methods, storage specifications for the types and number of machines to be used in the system, and the selection and use of the de-icing compounds. The design of the system should include an evaluation of the qualifications of the personnel involved, recommendations related to training programs, methods and procedures of chemical application, the

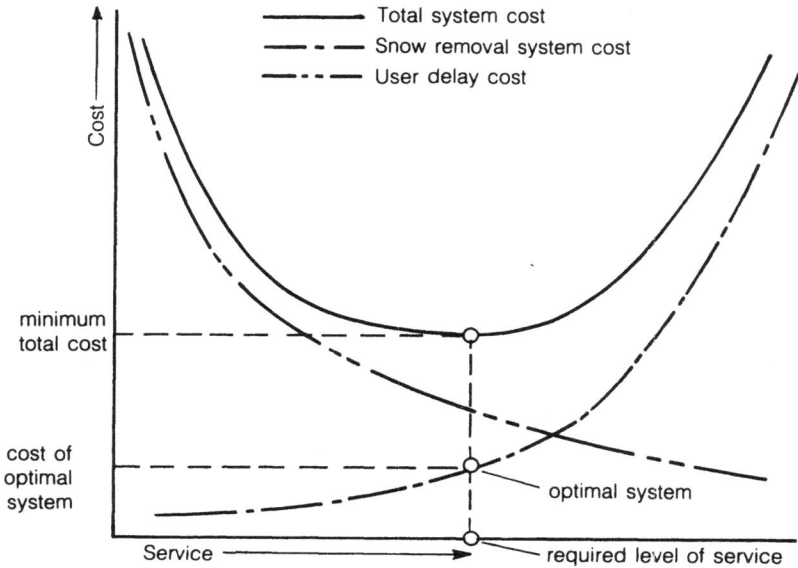

Fig. 14.13 The choice of snow and ice control system based on minimum total cost (Tighe and Webber, 1971).

standards under which only minimum performance would be realized and the procedures and methods for continuing and updating control and cost analyses.

Rate of Application of De-Icing Chemicals

Since no two storms are alike, no single set of standards can be written establishing the spreading rates of de-icing chemicals. As a general rule, for light traffic conditions, chemicals should be applied only in quantities sufficient to produce 30% melt within a maximum of 30 min from the beginning of snow accumulation (Schaerer, 1970; Caird and Young, 1971). With increasing traffic the time of solution of the chemicals, and therefore the time for applying them, decreases accordingly. A spreader capable of applying chemicals at rates from 90 to 1400 kg/lane-km with a precision of 23 kg is recommended by the American Society for Testing Materials (1974).

Main roads and arteries carrying a daily traffic of 1000 or more vehicles generally require bare pavement conditions, and the control treatment includes snow plowing, spreading of different combinations of chemicals and abrasives and removing partially melted snow and ice (Highway Research Board, 1962).

The quantities of chemicals recommended for ice control and snow removal under different conditions are given in Table 14.8, which summarizes the findings reported by the National Research Council (1967) and Caird and Young (1971) as well as the experience of road departments and the results of laboratory studies (Amberg and Williams, 1948; Salt Institute, 1962b; Schneider, 1959; Highway Research Board, 1962; Himmelman, 1963; Coulter, 1965; Ontario, 1967; Quebec, 1967; Road Research Laboratory, 1968; Kelley, 1969; Schaerer, 1970; and American Society for Testing Materials, 1974). The rates of application for chemicals listed in Table 14.8 are for a solution time of less than 30 min.

SIDE EFFECTS OF DE-ICING CHEMICALS

Environmental

Salt moved from a roadway and deposited at the sides may damage certain species of plants growing there. Plant biologists do not know precisely how salt injures plants, how much salt is needed to harm them, what the exposure time must be, or what the symptoms of salt-caused injury are (Hanes et al., 1970a; American Public Works Association, 1971). However, quantitative information about the salt tolerance of many important plants has been obtained and the proximate causes of salt injury have, in some cases, been carefully analyzed. For example, Zelazny and Blaser (1970) have tabulated the relative salt tolerances of trees and ornamental shrubs (see Table 14.9). Landscape architects have also identified salt-resistant plants and developed methods of improving the salt resistance of biota (Hayward and Bernstein, 1958; Hanes et al., 1970a, 1970b; American Public Works Association, 1971). The use of salt tolerant plants is a realistic approach to achieving growth and maintaining vegetation where deposition of salt from roadways produces problems.

Salt used in ice and snow control on highways is also a potential pollutant of water supplies, but is mostly limited to shallow wells, small ponds and streams which receive runoff directly from roadways. Based on the research in Maine, it is recommended that wells be located at least 12 m from a highway and preferably 25 m away.

Some salt may drain to adjacent areas and infiltrate the ground, thereby reducing soil fertility and damaging plants (Highway Research Board, 1967). In an attempt to determine the contribution of highway salting practices to the contamination of soils, Hutchinson and Olson (1967) measured the sodium and chloride concentrations at 27 sites in soils adjacent to salted highways over different periods up to 18 years. They found that the concentrations of

Table 14.8

RECOMMENDATION FOR THE RATE OF APPLICATION OF CHEMICALS ON PAVED ROAD WITH AN AVERAGE DAILY TRAFFIC OF 500 VEHICLES OR MORE.

Air Temperature	Rate of application - kg/(km of 2-lane road)			
	Application before snowfall or freezing rain	Melting per centimetre of loose snow	Clean-up removal of thin crusts after ploughing	Thick crust of hard-packed snow and ice
a) - 4°C in the shade and higher	55 to 115 kg of NaCl or 55 to 115 kg CaCl₂-NaCl mixture if NaCl is removed by wind or traffic	110 kg of NaCl	85 kg of NaCl	170 kg of NaCl or 85 kg of 3:1 NaCl - CaCl₂ mixture
b) -7 to - 4°C in the sun or on warm pavement				
a) - 4°C and higher if temperature is falling		170 to 225 kg of NaCl or 135 kg of 3:1 mixture	130 kg of NaCl	170 to 280 kg of NaCl or 170 kg of 3:1 NaCl-CaCl₂ mixture
b) -7 to - 4°C in the shade				
c) -12 to -7°C in the sun or on warmer pavement				
a) -7 to - 4°C with falling temperature	70 to 140 kg of NaCl or 1:3 CaCl₂-NaCl mixture	165 kg of 3:1 NaCl-CaCl₂ mixture	170 kg of 3:1 NaCl-CaCl₂ mixture	280 kg of 3:1 NaCl-CaCl₂ mixture
b) -12 to -7°C in the shade				
c) -18 to -12°C in the sun or on warmer pavement				
-18 to -12°C in the shade	No application of chemical	No application of chemical	210 kg of 3:1 NaCl-CaCl₂ mixture	340 kg of 3:1 NaCl-CaCl₂ mixture or 1700 kg of treated abrasive
Below -18°C			Abrasive mixed with salt	

both ions had increased with the greatest increases occurring at the edges of the road embankment and where salt had been applied longest. Hutchinson (1969) and Zelazny and Blaser (1970), report that the increases were more pronounced at the 15-cm depth than at 45 cm but the sodium concentration increased more than chloride with distance perpendicular to a major highway (see Fig. 14.14).

Table 14.9
THE RELATIVE SALT TOLERANCE ON PLANT TISSUE FOR TREES AND ORNAMENTALS (Zelazny and Blaser, 1970).

Low (0-2000 ppm chloride)	Moderate (2000-5000 ppm chloride)	Good (5000-6000 ppm chloride)
Sugar maple	Birch	Mulberry
Red maple	Aspen	White oak
Lombardy poplar	Cottonwood	Red oak
Larch	Hard maple	Hawthorne
Rose	Beech	Scotch elm
Black walnut	White spruce	White poplar
	Blue spruce	Grad poplar
	Balsam fir	Silver poplar
	Green ash	

Effect of Salt on Concrete, Bridge Decks and Automobiles

The brine resulting from the melting of snow and ice by the salt penetrates into concrete affecting its durability. The primary mechanism causing the deterioration is the osmotic pressure gradients arising from differences in salt concentrations in adjacent areas of the slab. The greater its permeability and the lower its compressive strength, the more the concrete is subject to deterioration. In general, concrete that contains sufficient entrained air (not to be confused with air voids) and is impervious and well-hardened will be resistant to salt action.

When de-icing salts come into contact with reinforcing steel an electrolytic action begins that causes corrosion. As the corrosive particles build up in the presence of water they expand against the concrete surrounding the steel. The only way to avoid steel corrosion in highway decks exposed to salt is to prevent contact of salt and steel, usually by sealing the decks with an impervious membrane; other methods of control involve the use of a non-corrosive de-icing agent or artificial heat sources rather than salt (Spellman and Stratfull, 1970; Stewart, 1971).

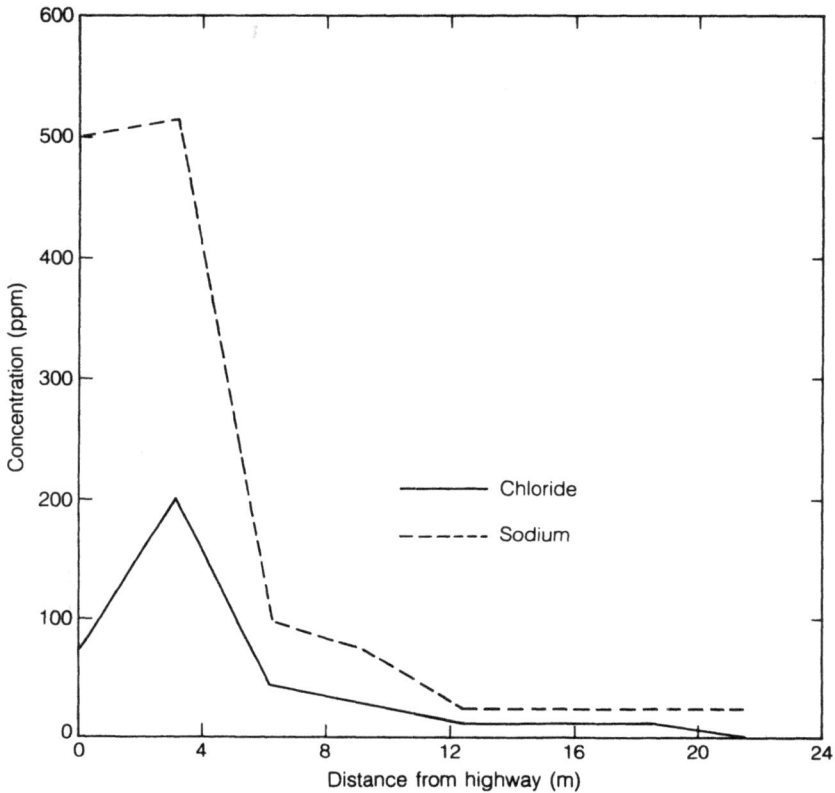

Fig. 14.14 Variation in sodium and chloride concentrations with distance from the highway (Hutchinson, 1969).

Corrosion is not caused by salt and brine but rather by oxygen in the air uniting with bare steel in the presence of moisture to produce oxide or rust. Salt, by itself, is not harmful to the lacquers and enamels coating modern automobiles. However, corrosion results after this protective coat has been pitted or damaged, so as to expose the underlying metal surface. Once started, corrosion will continue to spread under the paint film, lifting it and exposing the rusted metal. Obviously, this problem is much more severe in cities than in open areas because of the large quantities of salt used (Wirshing, 1957; Commonwealth of Massachusetts, 1965a).

The OECD (1969) gives details about various types of protective coatings for use on automobiles. Webster (1961) suggests that corrosion control can best be achieved by incorporating a high level of protection in vehicle construction.

INHIBITORS

In the mid-Sixties, significant developments were reported on the chemical inhibition of the corrosive effects of salt on vehicles. Inhibitors, in the form of chemical additions to salt either reduced the electrolytic effect of melted ice and the effect of the concentration cell produced or increased the passivity of a steel surface. Corrosion protection obtained by using inhibitors diminishes as they become diluted; they are only effective in quantities of the order of 7% by weight of salt. However, this much inhibitor increases the cost of de-icing salt substantially. Such an increase in public expenditure could probably be justified only if complete protection from corrosion can be obtained (Bishop, 1972). Inhibitors have been found to be both uneconomical and ineffective (Salmins, 1966; Fromm, 1967; Ontario, 1968; OECD, 1969).

LITERATURE CITED

Allied Chemical. 1958. *Calcium Chloride*. Bull. No. 16, Tech. and Eng. Ser., Ind. Chem. Div., Morristown, N.J.

Amberg, C.R. and L.E. Williams. 1948. *Rock salt for ice and snow control*. Bull. 3, Ceramic Res. Dept., New York State Coll. Ceramics, Alfred, N.Y.

American Public Works Association. 1971. *Facts you should know about the effects of de-icing salts*. Published as a supplement to the January 1971 issue of the Reporter of the APWA.

American Society for Testing Materials. 1972. *Standard specification for sodium chloride*. ASTM D632-72.

American Society for Testing Materials. 1974. *Standard methods of sampling and testing calcium chloride for roads and structural application*. ASTM D345-74.

American Society for Testing Materials. 1977. *Standard specification for calcium chloride*, ASTM D98-77.

Bishop, R.R. 1972. *The development of a corrosion inhibitor for addition to road de-icing salt*. Rep. LR489, Transport and Road Res. Lab., Her Majesty's Stationery Off., London.

Boies, D.B. and S. Bortz. 1965. *Economical and effective deicing agents for use on highway structures*. NCHRP Rep. 19, Highw. Res. Board. NAS-NRC, Washington, D.C.

Brohm, D.R. and H.R. Edwards. 1960. *Use of chemicals and abrasives in snow and ice removal from highways*. Res. Bull. 252, Highw. Res. Board, NAS-NRC Publ. 761, Washington, D.C.

Caird, J.C. and R. Young. 1971. *Field test report on Calsel 31 road deicer/anti-icer*. Allied Chemical (Canada) Ltd.

Calcium Chloride Association. 1949. *Calcium chloride for ice control*. Washington, D.C.

Calcium Chloride Institute. 1968. *The mixer's manual*. Washington, D.C.

Carr, F.H., Jr. 1969. *Massachusetts finds premixed chemicals boon in winter mainte-nance.* Better Roads, Vol. 39, No. 7, pp. 20-23.

Commonwealth of Massachusetts. 1965a. *Highway salt and automobiles.* Mass. Highw. Salt Study, Mass. Legis. Res. Counc. Rep.

Commonwealth of Massachusetts, 1965b. *The use of highway salt and its effect of pavements.* Mass. Highw. Salt Study, Mass. Legis. Res. Counc. Rep., Senate No. 2.

Coulter, R.G. 1965. *Understanding the action of salt.* Public Works, Vol. 96, No. 9, pp. 78-80.

Dickinson, W.E. 1959. *Ice-melting properties and storage characteristics of chemical mixtures for winter maintenance.* Res. Bull. 220, Highw. Res. Board, NAS-NRC, Washington, D.C. pp. 14-24.

Eberhard, J. 1959. *An evaluation of the relative merits of sodium and calcium chloride for highway ice treatment.* Eng. Exp. Stn., Ohio State Univ., Columbus.

Fromm, H.J. 1967. *The corrosion of auto-body steel and the effects on inhibited de-icing salts.* Rep. No. RR135, Dept. Highw., Toronto, Ont.

Hanes, R.E., L.W. Zelazny and R.E. Blaser. 1970a. *Effects of deicing salts on water quality and biota - Literature review and recommended research.* NCHRP Rep. 91. Highw. Res. Board, NAS-NRC, Washington, D.C.

Hanes, R.E., L.W. Zelazny and R.E. Blaser. 1970b. *Salt tolerance of trees and shrubs to de-icing salts.* Highw. Res. Rec. 335, NAS-NRC, Washington, D.C., pp. 16-18.

Hayward, H.E. and L. Bernstein. 1958. *Plant growth relationships on salt affected soils.* Botanical Rev., Vol. 24, pp. 584-635.

Hegmon, R.R. and W.E. Meyer. 1968. *The effectiveness of antiskid materials.* Highw. Res. Rec. 227, Highw. Res. Board, NAS-NRC, Washington, D.C., pp. 50-56.

Highway Research Board. 1954. *Recommended practice for snow removal and treat-ment of icy pavements.* Current Road Problems, No. 9, 3rd Rev., Dept. Maint., Comm. Snow and Ice Control, Highw. Res. Board, NAS-NRC, Washington, D.C.

Highway Research Board. 1962. *Current practices for highway snow and ice control.* Current Road Problems. No. 9, 4th Rev., Dept. of Maint., Comm. on Snow and Ice Control., Highw. Res. Board, NAS-NRC, Washington, D.C.

Highway Research Board. 1967. *Environmental considerations in use of deicing chem-icals.* Highw. Res. Rec. No. 193, Highw. Res. Board, NAS-NRC, Washington, D.C.

Himmelman, B.F. 1963. *Ice removal on highways and outdoor storage of chloride salts.* Highw. Res. Rec. 11, Highw. Res. Board, NAS-NRC, Washington, D.C.

Hutchinson, F.E. and B.E. Olson. 1967. *The relationship of road salt applications to sodium and chloride levels in the soil bordering major highways.* Highw. Res. Rec. 193, Highw. Res. Board, NAS-NRC, Washington, D.C. pp. 1-7.

Hutchinson, F.E. 1969. *The influence of salts applied to highways on the levels of sodium chloride ions present in water and soil samples.* Office of Water Resour. Res., U.S. Dept. Interior, Washington, D.C.

Kelley, J.F. 1969. *Industry pre-mix of salt and calcium chloride.* New England Con-struction, Jan., Mass. Dept. Public Works.

Kersten, M.S., L.P. Pederson and A.J. Toddie Jr. 1959. *A laboratory study of ice removal by various chloride salt mixtures.* Highw. Res. Bull. 220, Highw. Res. Board, NAS-NRC, Washington, D.C. pp. 1-13.

Kyser, J.O. 1971. *Brine solution removes stubborn ice.* Public Works, Jan., pp. 67 and 89.

Lang, C.H. and W.E. Dickinson. 1960. *Chemical mixture test program in snow and ice control.* Highw. Res. Bull. 252, Highw. Res. Board, NAS-NRC, Washington, D.C., pp. 1-8.

Lang, C.H. and W.E. Dickinson. 1964. *Snow and ice control with chemical mixtures and abrasives.* Highw. Res. Rec. 61, Highw. Res. Board, NAS-NRC, Washington, D.C., pp. 14-18.

Milloy, M.H. and J.S. Humphries. 1969. *The influence of topography on the duration of ice-forming conditions on a road surface.* Rep. LR 274, Road Res. Lab., Her Majesty's Stationery Off., London.

Minsk, L.D. 1965. *Snow and ice properties pertinent to winter highway maintenance.* Highw. Res. Rec. 94, Highw. Res. Board, NAS-NRC, Washington, D.C.

Monsanto Research Corporation. 1965. *Chemical means for prevention of accumulation of ice, snow and slush on runways.* Rep. AD615420D, prepared for the Federal Aviation Agency, U.S. Dept. Commer., Springfield, Va.

National Research Council. 1967. *Manual on snow removal and ice control in urban areas.* Tech. Memo. No. 93, NRC 9904, Nat. Res. Counc. Can., Ottawa, Ont.

Nichols, R.J. and W.I.J. Price. 1956. *Salt treatment for clearing snow and ice.* The Surveyor, London, Vol. 115, pp. 886-888.

OECD. 1969. *Corrosion des véhicules automobiles de influences des fondants chemiques.* Rapport préparé par un groupe de recherche routière de OECD, Organization for Economic Cooperation and Development.

Ontario. 1967. *Use of salt for snow and ice control.* Dept. Transp. and Communications, Maint. Div., Toronto, Ont.

Ontario. 1968. *Corrosion inhibitor study (Summary).* Dept. Highw., Toronto, Ont.

Quebec. 1967. *Manuel de l'entretien.* Ministère de la Voirie du Quèbec, Quebec City, P.Q.

Road Research Laboratory. 1968. *Salt treatment of snow and ice on roads.* Road Note No. 18, Dept. Sci. and Indust. Res., 2nd Ed., Her Majesty's Stationery Off., London.

Salt Institute. 1962. *Technical analysis of salt (sodium chloride) for ice and snow removal.* Chicago, Ill.

Salt Institute. 1962b. *Salt for ice and snow removal.* Chicago, Ill.

Salmins, G. 1966. *Does salt corrode cars?* Eng. and Contract Rec., Vol. 79, pp. 52-53.

Schaerer, P.A. 1970. *Compaction or removal of wet snow by traffic.* Sp. Rep. 115, Highw. Res. Board, NAS-NRC, Washington, D.C., pp. 97-103.

Schneider, T.R. 1960. *Die berechnung der zur auflösung von schnee-und eiskrusten notwendigen salzstreumengen (The calculation of the amount of salt required to melt ice and snow on highways)* Eidgenössisches Institut für Schnee-und Lawinenforschung, Weissfluhjoch - Davos, Interner Bericht Nr. 328. [English Transl. by Nat. Res. Counc. Can., Div. Build. Res., Tech. Transl. 1004].

Schneider, T.R. 1959. *Schneeverwehungen und Winterglätte. (Snowdrifts and winter ice on roads).* Interner Bericht Nr. 302, Eidgenössisches Institut für Schnee-und Lawinenforschung. [English Transl. by Nat. Res. Counc. Can., Div. Build. Res. Tech. Tranl. 1038].

Scotto, G.E. 1970. *Liquid treatments of commerical $CaCl_2$ in winter road maintenance.* Sp. Rep. 115, Highw. Res. Board, NAS-NRC, Washington, D.C., pp. 156-171.

Smith, R.W., W.E. Ewens and D.J. Clough. 1971. *Effectiveness of studded tires.* Highw. Res. Rec. 352, Highw. Res. Board, NAS-NRC, Washington, D.C., pp. 39-49.

Spellman, D.L. and R.F. Stratfull. 1970. *Chlorides and bridge deck deterioration.* Highw. Res. Rec. 328, Highw. Res. Board, NAS-NRC, Washington, D.C., pp. 38-49.

Stewart, C. 1971. *Deterioration in salted bridge decks.* Sp. Rep. 116, Highw. Res. Board, NAS-NRC, Washington, D.C., pp. 23-28.

Tighe, D. and D.B. Webber. 1971. *Planning airport snow and ice removal using operations research techniques.* Highw. Res. Rec. 359, Highw. Res. Board, NAS-NRC, Washington, D.C., pp. 9-15.

Webster, H.A. 1961. *Automobile body corrosion problems.* Corrosion, Vol. 17, No. 2, pp. 9-12.

Wirshing, R.J. 1957. *Effect of de-icing salts on corrosion of automobiles.* Bull. 150, Highw. Res. Board, NAS-NRC, Washington, D.C., pp. 14-17.

Zelazny, L.W. and R.E. Blaser. 1970. *Effects of de-icing salts on roadside soils and vegetation.* Highw. Res. Rec. 335, Highw. Res. Board, NAS-NRC, Washington, D.C., pp. 9-12.

15

THERMAL METHODS OF CONTROL

G. P. WILLIAMS

Division of Building Research, National Research Council of Canada, Ottawa, Ontario.

INTRODUCTION

Thermal methods of snow control can be divided into two general categories: pavement heating systems where the snow is melted in place on the pavement surface and snow melters where the snow is transported to a tank or container for melting.

Pavement heating systems, sometimes referred to as "in situ melting systems" are usually incorporated into a road or structure and are used to keep facilities such as building entrances, sidewalks, parking areas, toll plazas, ramps, roofs and bridge approaches free of snow. Common examples of these systems are embedded pipes, embedded electrical cables, and overhead infrared lamp installations. Because of their large capital investment and high operating cost, pavement heating systems are too expensive for snow control on a large scale. They are recommended only for sites where mechanical snow removal would be difficult, the use of chemicals would damage structures, traffic delays could not be tolerated or safety is an important consideration.

Two basic types of snow melters are available: stationary or pit melters to which snow is transported, dumped and melted, and mobile melters which move along streets, accumulate the snow and melt it in place. Melters come in several different sizes and designs and are attractive if their annual operating cost is equal to or less than the cost of removing snow by other methods (for example the hauling to and disposal of snow at snow dumps).

HEATED PAVEMENTS

Heat Requirements

Pavement heating systems must provide sufficient heat to melt snow and to offset heat losses both to the atmosphere and to the ground underneath. The total heat requirements Q_T of a slab of pavement for a given time interval can be expressed by the heat-balance equation:

$$Q_T = Q_m + Q_u + Q_{ep} + Q_{cp} + Q_{lp} + Q_{gp} + Q_p \ [kJ/m^2],\qquad 15.1$$

where

Q_m = heat required to melt snow at the fusion temperature,

Q_u = heat required to raise the temperature of the snow to the melting point,

Q_{ep} = heat loss by evaporation from the base portion of pavement,

Q_{cp} = (sensible) heat loss by convection from bare pavement,

Q_{lp} = heat loss by long-wave radiation from bare pavement,

Q_{gp} = heat loss to the ground (including the heat required to bring that part of the pavement affected to its operating temperature), and,

Q_p = edge heat loss around the perimeter (including heat required to bring that part of the pavement affected to its operating temperature).

The designer needs to calculate all components of Eq. 15.1 in order to obtain a proper estimate of the total heat required at a particular site. Q_m, the heat required to melt snow (depending on the desired rate of melting), is a large part of the total requirement. The sensible heat Q_{cp} is a small proportion and can usually be ignored in the calculations. Surface heat loss $(Q_{ep} + Q_{cp} + Q_{lp})$ depends on weather conditions and the percentage of heated area to be maintained free of snow; the designer must specify the standard of snow removal to be achieved and take into account weather conditions expected at a site. Ground and edge heat loss $(Q_{gp} + Q_p)$ will depend on the type of installation (i.e., with or without insulation) and on the method of operation (steady or intermittent). With so many variables it is important that designers understand the principles governing heat loss and the limitations of calculating the different components of the heat-balance equation.

Heat required to melt snow

The calculation of the heat required to melt snow Q_m is difficult because it depends on the required rate of melting. If a completely bare pavement is required, the system is designed to melt the maximum hourly rate of snowfall anticipated at a site. The hourly rate of snowfall is not measured at Canadian meteorological stations and were such measurements available they would be uncertain because the accurate monitoring of snowfall during high winds is difficult. The only information usually available from meteorological records is the 6-h average snowfall or the total snowfall deposited by a storm.

The power requirement of most heaters is based on the amount of heat needed to melt the snow at a rate equal to the average hourly snowfall for the storm (usually calculated from 6-h or daily values). This power can be calculated by the expression:

$$Q_m = 2.78 \times 10^{-4} \; h_f \; WE, \qquad\qquad 15.2$$

where

Q_m = power required, kW/m^2,

h_f = heat of fusion, kJ/kg, and

WE = water equivalent of the snowfall rate, mm/h.

Using Eq. 15.2 and assuming h_f = 333.5 kJ/kg and WE = 25 mm/h the power needed to melt the snow is approximately 2.32 kW/m^2. Obviously, much higher rates of heat input are needed to melt the maximum hourly snowfall rates as these may be two or three times the average rate. Likewise, the power requirements of heaters are increased if they must also melt the snow which drifts on the site.

Atmospheric heat losses

Many formulae are available for calculating the atmospheric heat losses from a heated pavement by long-wave radiation, convection and evaporation (Williams, 1975a). Most of these relationships do not consider the heat gained at the surface by solar radiation (short-wave radiation) because systems are designed for periods of maximum heat losses which usually occur during the night.

Figure 15.1 shows the effects of air temperature and wind speed on the combined heat loss by radiation, convection and evaporation from a bare, exposed wet concrete surface at 1.1°C. The formulae used for the calculations were based partly on field measurements of surface heat loss at an experimental heated pad and partly on other formulae reported in the literature (Penman, 1956; Williams, 1975a). Wind speeds were converted to a height of 15.2 m above the ground. It is assumed that there is complete cloud cover and that the dew-point temperature equals the air temperature during a snow storm.

The relative contribution of long-wave radiation, convection, and evaporation to the total heat requirement varies with weather conditions. For high wind speeds, heat loss by convection is the largest component; for low wind speeds and large air/surface temperature differences, heat loss by radiation becomes dominant. Because of the wind effects, heat loss will depend on the degree of exposure to wind. The rate of evaporation is limited by the vapor pressure difference between the surface and the air. It is a significant component of the total heat balance at high wind speeds, particularly from wet pavements after snow has melted. Thus, it is important to provide adequate drainage for snow melting systems so that a pavement can dry quickly, thereby reducing the heat loss by evaporation.

Small heated areas such as steps, narrow sidewalks or heated wheel tracks have high rates of heat loss by evaporation and convection. The results of

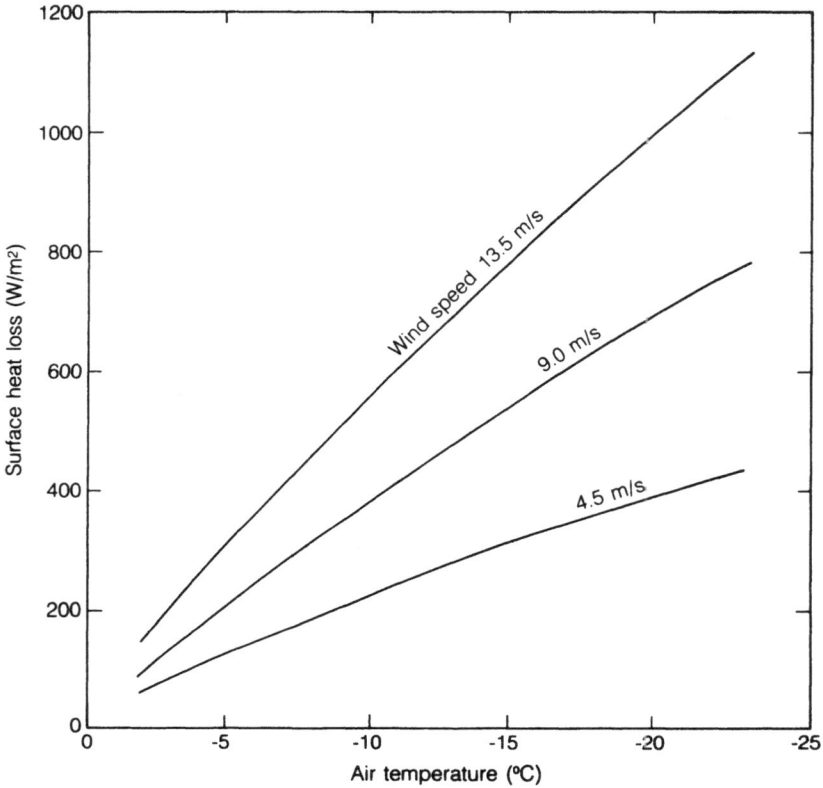

Fig. 15.1 Combined surface heat loss (radiation, evaporation, convection) from bare, heated pavement at 1.1°C.

experiments conducted by Williams (1976) indicated that the convective heat loss from a heated pavement, 0.9 m x 0.9 m, was about double the loss from a 4.9 m x 4.9 m slab under similar weather conditions. Calculations by Coulter and Herman (1964) indicate that the size of the heated area is not a significant factor affecting the convective heat loss from slabs having lengths and widths greater than 3 m.

Surface heat losses by evaporation, convection, and radiation are reduced in direct proportion to the percentage of heated pavement covered by snow. If the area is half-covered with snow, surface heat losses are reduced to about one half of the value of a completely bare area; if the pavement is completely covered by even a thin layer of snow, the only surface heat loss beyond that for melting occurs through the snowcover by conduction. Losses of heat by conduction and exchange at the snow/air interface may be significant in wet or high-density snows, for example, when slush or an ice layer forms on the surface, or when the snow is compacted.

Ground and edge heat losses

Allowances must be made in design for the heat loss to the ground under embedded pipes or electrical heating coils. Recommended values for this loss can be as high as 30 to 50% of the total surface heat loss from a bare pavement (Coulter and Herman, 1964). Losses from the underside of a heated ramp have been measured at from 10 to 15% of the total heat loss (George and Wiffen, 1965). The ground heat loss is a substantial part of the "total design heat load" for systems operated intermittently, e.g., when the slab is heated only during the period snow is falling or expected to fall. With these systems the slab loses heat to the ground between storms and cools. Hence, large amounts of energy may be required to warm the slab for each individual storm. Conversely, for a slab which is heated continuously the heat stored in the slab may provide a substantial part of the total amount of heat required to melt the snow which falls on it, reducing the "design" heat requirements by embedded pipes or heating coils. It is important, however, to recognize that the total loss of heat to the ground from systems operated continuously throughout the winter is much greater than the loss by systems operated intermittently over the same period of time. A continuously-operated system maintains the slab and the ground surface at warm temperatures. Over the relatively short time snow falls this stored heat contributes to the melting of snow.

A substantial amount of heat can be lost from the edges of a heated area to the adjacent soil or pavement surfaces. In determining a design heat load this loss is most important for a system that is operated intermittently. For example, on narrow heated areas such as sidewalks or vehicle tracks, in which the heating system is operated intermittently, the edge heat loss can be as much as 20 to 30% of the total design heat-load requirement. Both edge and ground heat losses can be reduced substantially by the use of insulation placed under and at the edges of the slab.

Design Heat Loads

Standards of removal

Only general guidelines can be given as standards for snow removal by snow melting systems. Research and experience indicate that it is neither economical nor necessary to design systems to always maintain bare pavement. An assumed average snowcover of 50 percent of the heated area for the more severe storms expected in any locality appears to be reasonable for most traffic conditions (Jorgenson and Associates, 1964). This standard may well be reduced to allow up to 75 percent snowcover for situations where heavy traffic will assist in melting and clearing. A further reduction to allow complete snowcover of the heated area may be satisfactory for residential requirements.

The ASHRAE Handbook (1973) recommends that sites be classified as follows according to the melt rate required:

Class I (minimum) - Residential walks, roadways,
Class II (moderate) - Commerical sidewalks and driveways,
Class III (maximum) - Toll plazas, highways and bridges, aprons
and loading areas of airports.

Heating systems designed for Class I sites assume that the pavement will be snowcovered during a storm, but that sufficient heat will be supplied to prevent excessive snow accumulations; for Class II, that 50 percent of the pavement will be snow covered during storms; for Class III, that the pavement will be bare under practically all weather conditions.

In regions such as the Canadian Prairies, which are subject to low temperatures and high winds, the weather may be such that the heat required to maintain an ice-free surface after a storm may be greater than the heat to melt snow during the storm. In such situations, the calculation of heat requirements can be based on the rate of surface heat loss from bare, wet pavements (Williams, 1975a).

Design heat calculations

Regardless of the snow melting standard selected, the system must be adequate for some combination of weather factors. Two approaches have been suggested to obtain information on the design weather conditions. In one method outlined in the ASHRAE Handbook (1973), a frequency analysis is performed on all snowfalls recorded at the site. Using these data a storm is chosen, based on its probability of occurrence, to design the heating system. A system designed in this manner has sufficient capacity to handle all storms less than the design storm.

In the other approach a design storm is chosen that is reasonably representative of weather conditions expected at the site. Williams (1974a) suggests the use of data from five snow storms occurring over a ten-year period having the highest rates of snowfall and highest wind speeds. The average values of air temperature, wind speed and snowfall observed during the five storms are then taken as the basis for design. This method is simpler and probably produces results sufficiently accurate for most practical situations.

The procedure recommended by ASHRAE has often proved to be unduly conservative. The capacities of installed systems are frequently much lower than the values calculated using ASHRAE procedures particularly for Class III installations (see Fig. 15.2).

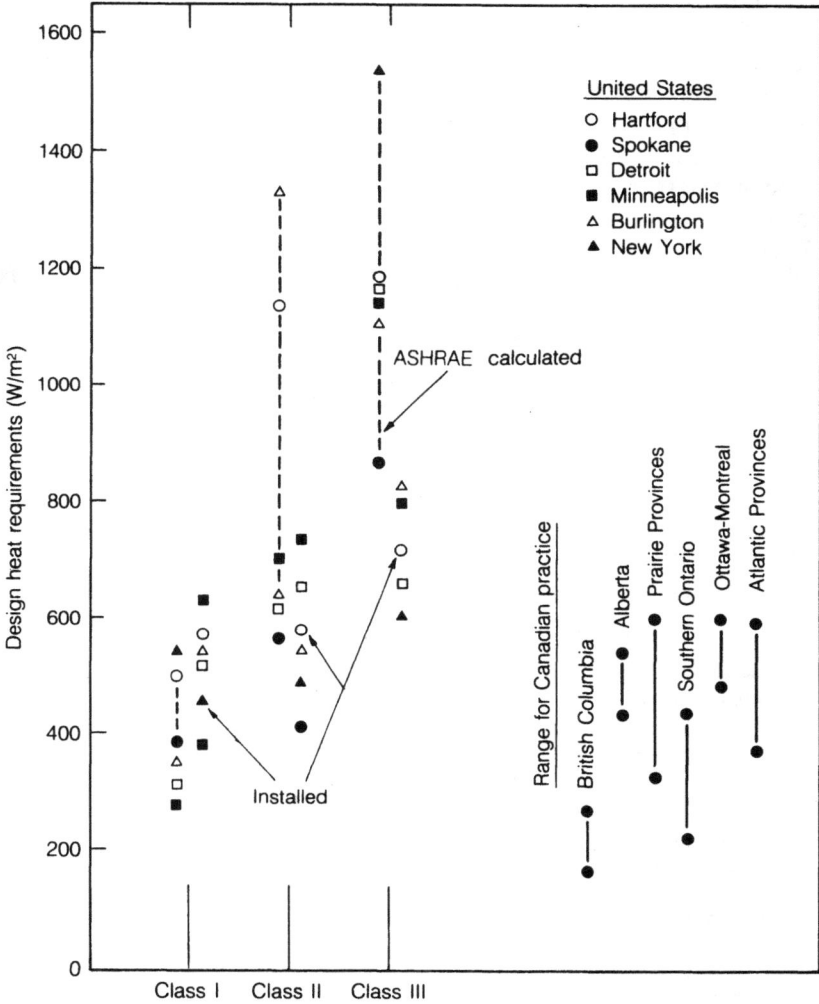

Fig. 15.2 Comparison of design heat requirements for snow melting systems: ASHRAE calculated, installed (USA), Canadian practice (Williams, 1974a).

Design heat guidelines

The design heat requirements for snow melting systems vary widely depending on the system's purpose and climatic conditions. Hence, it is possible to establish only broad ranges of values for different installations. Table 15.1 compiled by Williams (1974b) list suggested values of heat losses from bare pavement following a storm and heating rates for snow melting systems during storms for six Canadian urban centers. Adjustments for ground and edge heat losses should be made to these values if the system is to be operated intermittently, and adjustment is specially important for narrow heated areas such as sidewalks, since ice will form on the edges of an underdesigned system (particularly if drainage is poor).

Design capacities used in Europe are generally much lower than those in North America, reflecting the milder climatic conditions (Felix, 1971). The maximum requirements for exposed and cold conditions seldom exceed 320 W/m^2; the lowest design value is 100 W/m^2 for areas requiring just enough heat to prevent ice from forming under mild temperatures.

Table 15.1
DESIGN GUIDELINES FOR HEATED PAVEMENTS AT
SIX URBAN CENTRES IN CANADA (Williams, 1974b).

Surface heat losses (W/m^2) for maintaining a bare wet pavement after design storm conditions

Toronto	Halifax	Quebec City	Ottawa	Winnipeg	Edmonton
193	310	353	535	705	696

Heating rates of snow melting systems during a storm (W/m^2).

	Toronto	Halifax	Quebec City	Ottawa	Winnipeg	Edmonton
Average Exposure	268-321	375-428	436-449	428-482	589-642	589-642
Sheltered	268	321	375	375	428	428
Extremely Exposed	375-428	482-535	535-589	535-589	642-696	642-696

Electrical Systems

Types of electrical systems

Electrical systems for heating pavements can be divided into two groups:

(1)   Systems which use insulated heating cables in the medium to high voltage range of 60-240V. Systems with very high voltage (1000V) have also been tried (Schneider, 1962) but are not in general use because of their potential hazard.

(2)   Systems which employ uninsulated heating cables, wire-mesh grids or electrically-conductive mats in the low voltage range of 10-30V.

Systems using insulated heating cables have been in use for the past 25 years. Kobold and West (1960) noted that the first electrical system for pavement heating in Canada was installed about 1960 by Ontario Hydro. Within the last decade heating cables in pavements have become commonplace in Canada and other countries. Wiffin and Williamson (1963) have reported on extensive research and development of medium voltage systems using heating cables to prevent the formation of ice and frost on roads in Great Britain. Some of the earlier systems failed because the heating cables were corroded by salt or the sulphates in the concrete. The corrosion problem has been eliminated by using special, mineral-insulated heating cables with nylon jackets. Potter (1967) discusses the special cables embedded in materials such as concrete and asphalt. Such precautions minimize the possibility of failure of the system and subsequent high replacement costs.

Low-voltage systems which have been used range from special electrically-conductive mats for heating sidewalks or steps to the extensive wire-mesh grids for a test installation on the Gardiner Expressway in Toronto (George and Wiffen, 1965). Their main advantages are ease of installation and safety; their disadvantages are the extra cost of transformers, the high voltage losses in feeder cables and the problem of placing transformers nearby. In Britain, such systems have limited use because of these disadvantages. Electrically-conductive materials such as graphite have also been added to road materials to make the surface layer electrically-conductive without using metal heating grids.

Operation, cost

Electrical heating systems can be operated in three modes: continuous, intermittent, and night-time only.

Systems designed for continuous operation whenever the air temperature is below 0°C and no snow is falling would run almost continually in cold climates. Since the annual atmospheric heat loss from a bare heated pavement surface during such periods might need an annual power requirement up to 20 times (or more) of that required to melt the snow alone, most systems in Canada are designed for intermittent operation.

Systems designed for intermittent operation require heat both for warming the system before a snowfall, and for melting snow and drying the pavement afterwards. If heating cables are installed well below the surface, several hours may be needed to warm the surface to 0°C. The use of insulation under the heating cables will reduce the warm-up time by reducing ground heat losses.

Systems designed for night-time operation only are favoured by electrical generating companies since the snow-melting load generally occurs during "off-peak" hours. The heat stored in the pavement during the night could melt snow that falls during daylight hours.

Systems can also be designed and operated to make efficient use of the heat stored in concrete or asphalt. For example, they may be operated to take advantage of low off-peak hourly rates by either reducing or cutting power for limited periods thereby reducing costs. A two-level heat system with a base heating load and a higher snow melting load is another system which is efficient and economical. The base load is usually determined by the requirement for immediate melting when the snow starts to fall.

Various methods of manual and automatic control of intermittently-operated systems have been tried. Snow detectors are useful for controlling melting systems, particularly when the air temperature is close to 0°C, and almost a necessity for the most economical operation. Jorgensen and Associates (1964) report there is a lack of information on the reliability of snow detectors. Some systems have simple air or surface temperature controls that automatically start the system when temperatures reach preset values.

The cost of installing and operating electrical snow-melting systems is a major factor in determining their feasibility. Costs depend on power rates, placement and availability of transformers, method of operation, climatic conditions, anticipated life of the system and repairs and maintenance. The high annual cost (including the amortized installation cost) is estimated to be 30 to 40 times that of snow control by mechanical or chemical methods. However, cost information quickly becomes out of date because of the rapid increases in the cost of energy. Felix (1971) summarizes the costs of electrically heating pavements in various countries.

Pipe Systems

General characteristics

Embedded pipe snow-melting systems became relatively commonplace in the 1940's partly because of the work of Adlam (1950), whose book on the subject is a source of practical information for designing such systems. In 1957, the National Research Council of Canada prepared an extensive bibliography on snow-melting systems, most of which have embedded pipes (Division of Building Research, 1957).

Embedded pipe pavement heating systems are similar to electrical systems except that heat is transferred to the pavement by a network of pipes. The piping material may be steel, wrought iron, copper or plastic. The most widely-used fluid is water which is mixed with ethylene glycol to prevent freezing. Oil has also been used in iron and steel pipes. Heat is usually supplied by steam from conventional gas or oil-fired boilers or building steam-heating plants. Unfortunately steam is corrosive and may freeze if the system must be shut down for any length of time; also its high operating temperature imposes significant thermal stresses on the system and therefore may cause damage. Hence, heat exchangers are commonly used with water or oil as the circulating fluid to transfer heat from the steam plant.

Design, operation

A major consideration in the design of pipe heating systems is the determination of its hydraulic characteristics. The procedure to be followed for proper hydraulic design is completely discussed in the ASHRAE Handbook (1973). The final design must allow for changes in viscosity of the fluid with temperature since the fluid friction in the piping circuit increases with viscosity and decreases the flow rate. A smaller flow rate requires a longer time for the system to reach its operating temperature.

Precautions must be taken during installation to prevent corrosion and operational and thermal stress problems. Manufacturers of pipe snow-melting systems usually issue information booklets giving advice on installing and operating procedures. These systems are usually controlled manually but may be automatically controlled as in the case of electrical systems.

Even though the installation costs of pipe and electrical systems are comparable, the annual operating costs of embedded-pipe systems are usually much lower, particularly if they are connected to a building heating system. However, the latter cost is about 15 times more than for chemical or mechanical removal. A pipe system is usually not as flexible as an electrical system, is not as well-suited for intermittent operation and is usually restricted to new construction since it cannot be installed in existing pavement.

Infrared Systems

Infrared lamps provide heat directly to snow or pavement surfaces. Both electrical and gas-fired infrared units have been used successfully to prevent snow accumulation, but the requirement that the lamps should be mounted above the heated area has confined their use to limited areas including short sidewalks.

The principal advantages of infrared systems over embedded systems are their short warm-up times and their provision of heat for pedestrians. The energy consumed may be from 20 to 40% higher than that used by embedded systems, resulting in higher costs. Also infrared lamps and reflectors are prone to deterioration. Frier (1964) and the ASHRAE Handbook (1973) give guidelines for the design of these systems.

MOBILE SNOW MELTERS

General Characteristics

As early as 1854, a Canadian patent was issued for a type of mobile snow melter for clearing snow from railway tracks. It was designed to run on a track and consisted of a type of furnace, into which the snow was placed and melted. Many similar patents have been issued, mostly in the United States.

In contrast to embedded heating systems, the design heat requirements of these snow melters are easy to calculate since most of the heat is used for melting. For a well-designed system, up to 90 percent or more of the heat supplied is used for melting snow and for raising the temperature of the melt water to prevent ice problems during disposal.

Various methods have been developed for heating the melt-producing medium, usually water, and transferring the heat to the snow that is to be melted. Usually, hot combustion gases are passed through tubes immersed in the water. Unfortunately, the amount of technical information about the design specifications of these exchangers and the advantages and disadvantages of each system is very limited.

The combustion units on most of the snow melters, both stationary and mobile, use a light fuel oil rather than heavy oil because the associated combustion equipment is cheaper and maintenance costs are less. Heating by natural gas would simplify the installation of stationary melters by eliminating fuel storage tanks and pumps. However, natural gas has not been used because demand for snowmelting may occur at peak heating periods when the highest commercial rate would apply. Other fuel sources such as coal and electricity have not been used to any extent.

Types

Many types of mobile snow melters are commercially available in several sizes. The size of the melter is usually quoted on its ability to melt a given quantity of snow in one hour. These ratings can be exceeded for short periods of time particularly if slush, with a high snow content is discharged.

A small melter with a rated capacity from about 22.5 to 67 5 Mg/h is towed to the melting site, usually by equipment such as a front end loader which also loads snow into it. Such melters are designed for removing snow from parking lots and shopping centers, where high mobility is required and where space is limited.

Large mobile melters having rated capacities from 91 to 137 Mg/h are usually mounted on a large trailer chassis. However, some of the larger melters may require separate vehicles for hauling; others are completely self-loading and self-propelled, and have a conveyor belt located at the front of the machine to move the snow to a large melting tank. Large fuel tanks are needed for continuous operation over several hours.

Operation

Portable melters are most frequently used to remove snow that has been pushed to the sides of streets by conventional snow clearing equipment. Some of the larger melters are designed to move along streets, simultaneously melting the accumulated surface snow. Where possible the melt water is drained into the storm drain system. Under good drainage conditions the melt water is allowed to flow in gutters, thereby flushing debris and salt from the streets. Under poor drainage conditions and cold weather, hoses are used to carry the water to the sewers which must be large enough to allow the melting tank to be emptied quickly so as to minimize delays in the removal operation and traffic interference. Catch basins must be cleaned and thawed before the snow-melting operation is started.

Mobile melters are economical only if they cost less than other conventional snow removal methods. In making cost comparisons, the total cost of removal must include the equipment depreciation cost plus costs of plowing, placing the snow in windrows and maintaining and operating the melter. For some areas, such as the downtown cores of large cities, the convenience of using snow melters, which help to reduce traffic delays, must also be considered.

STATIONARY SNOW MELTERS

General Characteristics

Stationary melters range in size from small melters, with rated capacities from 22.5 to 91 Mg/h, designed for removing snow from areas such as parking lots, to large units with capacities from 225 to 730 Mg/h. The large melters match the capacity of several large snowblowers, each loading a fleet of six or more large trucks. Their snow melting pits must be large enough to handle these large loads.

The water in the pits is usually heated by direct-fired oil burners, which fire downward, directing the hot combustion gases through stainless steel tubes immersed in the water. Water or air under pressure may be added by jets to cause turbulence in the water to aid in circulating the snow-water mixture. The jets also help to break up the snow, thus increasing the rates of heat transfer and melting. Dirt, sand and debris associated with the snow settles to the bottom of the pit, and is periodically removed with a clamshell or backhoe. Mechanical conveyors for automatic removal of sediment are no longer incorporated into current designs because debris dumped into the pit with the snow tends to jam the conveyors.

The water and slush from the pit flows over a weir at one end of the melter, where a paddle wheel or slush raker breaks large clumps of snow. These rakers assist in pushing slush over the weir, which can then be flushed into the sewer system by the melt water. The slush in the overflow often contains a large amount of suspended solids and trash which can create problems if dumped into an undersized collector.

In cold weather large amounts of steam originating from the pits may produce low visibilities on adjacent roads or streets. The amount of steam produced can be kept to a minimum by constantly feeding snow into the melter and by controlling its water temperature.

Cost Factors

Pit melters are justified only when the cost of snow disposal is equal to or less than the cost of collecting snow in snow dumps. The cost of hauling to the melter, and installing, maintaining, and operating the pit must be less than the combined cost of hauling the snow to the dump and maintaining and operating the dump. The total cost of a pit melter must include the costs of plowing, placing snow in windrows, loading, hauling to the melter, and installing, maintaining and operating the melting plant, as well as the cost of the land where the melter is situated. A similar cost analysis of snow dumps should include the availability of a suitable site and any attendant problems such as pollution and appearance. The debris remaining after the snow has been melted is not only unsightly but is a contributor of pollutants, such as salts and lead, to groundwater supplies.

No general cost guildlines for stationary melters can be given since they vary from region to region and are subject to rapid change in periods of inflation. Richards and Associates et al. (1973) published one of the few comprehensive evaluations on snow disposal for the National Capital area in Ottawa, Ont. Table 15.2, taken from their report, lists the capital costs and rated snow melting capacities of various installations throughout Eastern Canada.

Table 15.2
LIST OF EXISTING STATIONARY SNOW MELTING
INSTALLATIONS IN EASTERN CANADA (Richards and Associates et al.,
1973).

Date Installed	City	Capital Costs	No. of Burners	Basic Rated Capacity Mg/h
1962	Westmount	$ 252,250	10	364
1963	Montreal	---	10	364
1964	Outremont	---	10	364
1964	Mount Royal	---	5	182
1966	Montreal	---	11	400
1966	Montreal	---	11	400
1967	Quebec	---	10	364
1968	Montreal	726,000	20	728
1969	Montreal	389,000	10	364
1969	Quebec	515,000	20	728
1970	Montreal	505,671	10	364
1971	Quebec	540,000	20	728
1972	Montreal North	700,000	20	728
1972	Montreal	---	6	218
1972	Ottawa	94,000	3	109

SPECIAL SNOW MELTING SYSTEMS

Alternative Sources of Heat

Most snow melting systems use conventional sources of heat such as gas, oil, electricity and coal. However, as the cost of these fuels increases other sources must be considered. One obvious alternative source is the heat surplus discharged by thermal power plants if practical methods can be developed for using it.

Several existing snow-melting systems now use novel sources of heat. Jorgensen and Associates (1964) report that one of the earlier snow melting systems used water from natural hot springs. Winters (1970) reports than an experimental heated pavement designed to evaluate the use of geothermal energy or heat stored in the ground for snow melting has been installed in New Jersey. This system extracts heat from the earth by circulating an ethylene glycol solution through a buried pipe heat exchanger linked to the pipes embedded in the pavement. During the summer it is used to transport solar heat absorbed at the surface for storage in the ground. Such a system is probably most valuable in mild climates. Worden (1972) suggests that the heat normally dissipated in a cooling tower would be another potential source for heating and melting snow on sidewalks. Similarly, Mason (1971) proposes that the waste heat from a diesel generator's jacket water cooling system be used for snow melting.

Special Applications

Several snow melting applications are special and do not fall into the categories discussed earlier. These include the removal of snow by heated compressed air or by direct flame or heated attachments with modified turbo-jet engines (Jorgensen and Associates, 1964), compressed air, or by direct flame or heated attachments with modified turbo-jet. For example, with modifications, weed-burning and chemical-spraying equipment may be used for snow removal. Other heating devices have been developed to remove snow and ice from accumulating on roof overhangs, mainly by preventing ice dams from forming.

There has been much interest in developing special heating devices for removing snow and ice from highways and airport runways. These utilize methods of transferring heat to concrete or asphalt surfaces to free compacted snow and ice which is difficult to remove by conventional snow plows. The British Columbia Research Council has initiated a pilot research project on the use of radiation to loosen ice on pavements. A device attached to a vehicle emits radiation of a certain wavelength, which penetrates the thin layers of surface ice providing enough heat to the pavement below so as to loosen the ice, which can then be removed easily by a snow plow blade.

LITERATURE CITED

Adlam, T.N. 1950. *Snow Melting*. The Industrial Press, New York, N.Y.
ASHRAE. 1973. *Handbook and Product Directory. Systems*. Chapter 38: Snow Melting. Am. Soc. Heat., Refrig. Air Cond. Eng., New York, N.Y.
Coulter, R.G. and S. Herman. 1964. *Control of snow and ice by induced snow melting*. The City College Res. Found., New York, N.Y.
Division of Building Research. 1957. *Literature survey of papers dealing with the use of heat for keeping roads sidewalks and parking areas free from snow and ice*. Bibliography No. 8, Nat. Res. Counc. Can., Ottawa, Ont.
Felix, R. 1971. *Le Chauffage Electrique des Chaussées*. Dunod, Paris.

Frier, J.P. 1964. *Design requirements for infrared snow-melting systems.* Illum. Eng., Vol. 59, pp. 686-694.

George, J.D. and C.S. Wiffen. 1965. *Snow and ice removal from road surfaces by electrical heating.* Highw. Res. Rec. 94, Highw. Res. Board, NAS-NRC, pp. 45-60.

Jorgensen, R. and Associates. 1964. *Non-chemical methods of snow and ice control on highway structures.* Nat. Coop. Highw. Res. Prog. Rep. 4, Highw. Res. Board., NAS-NRC, Washington, D.C.

Kobold, A.E. and G.H. West. 1960. *Snow melting by electrical means.* Ont. Hydro. Res. News, Vol. 12, No. 37, pp. 18-24.

Mason, H.G. 1971. *Heat pipe: Economical Snow Melting Device?* Heat.-Piping-Air Cond., Nov., pp. 75-79.

Potter, W.G. 1967. *Electric snow melting systems.* ASHRAE J., Oct., pp. 35-44.

Penman, H.L. 1956. *Evaporation; An introductory survey.* Neth. J. Agric. Sci., Vol. 4, No. 1, pp. 9-29.

Richards, J.L. and Associates; Labrecque, Vezina and Associates. 1973. Snow disposal study for the national capital area, Ottawa, Ont.

Schneider, T.R. 1962. *Schneeverwehungen und Winterglätte (Snowdrifts and winter ice on roads).* Interner Bericht Nr. 302, 1959. Eidgenössisches Institute für Schnee-und Lawinenforschung. (English Transl. by Div. Building Res., Nat. Res. Counc. Can., Tech Transl. 1038).

Wiffin, A.C. and P.J. Williamson. 1963. *Electrical heating of roads to prevent the formation of ice and frost.* Heating (London), Vol. 24, pp. 41-47.

Williams, G.P. 1974a. *Heat requirements of snow melting systems in Canada.* Tech. Pap. 418, Div. Building. Res., Nat. Res. Counc. Can., Ottawa, Ont.

Williams, G.P. 1974b. *Design heat requirements for snow melting systems.* Can. Build. Digest 160, Div. Building Res., Nat. Res. Counc. Can, Ottawa, Ont.

Williams, G.P. 1975a. *Surface heat losses from heated pavements during snow melting tests.* Tech. Pap. 427, Div. Build. Res., Nat. Res. Counc. Can., Ottawa, Ont.

Williams, G.P. 1976. *Design heat requirements for embedded snow melting systems in cold climates.* Transp. Res. Rec. 576, Transp. Res. Board, NAS-NRC, pp. 20-32.

Winters, F. 1970. *Pavement heating, snow removal and ice control research CRREL,* Sp. Rep. 115, Highw. Res. Board, NAS-NRC, pp. 129-145.

Worden, A. 1972. *Heat reclaimed in office air conditioning melts snow full time.* Heat.-Piping-Air Cond., June, pp. 59-63.

16

DRIFT CONTROL

R.W. VERGE (retired)

Meteorological Applications Branch, Atmospheric Environment Service, Downsview, Ontario

G.P. WILLIAMS

Division of Building Research, National Research Council of Canada, Ottawa, Ontario

INTRODUCTION

The literature on snow drifting is extensive and is a measure of the practical experience which has accumulated on drift control methods for highways, railways and buildings. This chapter summarizes some of the available information and experience and provides references to additional details on the subject. The information presented is likely to prove most useful to engineers who have a limited experience with snow drifting problems or to those who are responsible for controlling drifts at an unfamiliar location.

Drifting snow creates well-known problems in the prairie and northern regions of North America. Severe drifting can also occur in southern Ontario, in the Atlantic Provinces and even in the southern United States. In regions where blowing snow is a common occurrence, buildings, highways, and railways are often designed to minimize drifting and frequently, control structures, such as snow fences, are installed each winter season. As a result, problems with drifts in these areas are often less serious than in regions where severe drifting of snow is relatively rare.

BASIC PRINCIPLES

A detailed discussion of the basic mechanics of the movement of snow by wind and the simulation criteria used to study the process via models is given in Ch. 8. In this chapter the basic principles governing the process are briefly reviewed.

The three major factors governing the formation of drifts are the amount and type of snow, the speed and direction of the wind, and the terrain (including obstacles) over which the windblown snow is carried. As each factor is extremely variable, it is difficult to predict the amount of drifting that will occur at a given site without the aid of field observations taken over a number of years.

The physical properties of the snow, particularly the cohesion of its surface, determines the ease of drifting. If a hard crust is formed by thawing and subsequent refreezing of the surface, the individual snow grains will be firmly cemented. Wind action itself is a compacting mechanism, creating, after some hours, a surface capable of withstanding velocities that would ordinarily drift loose, fresh snow. This implies that most drifting occurs when snow is falling or during the first day or so immediately following the end of the snowfall. Once snow accumulates into drifts it tends to remain until it melts. However, high wind speeds can erode layers of compacted snow, recommencing the drift process. This is especially true in the Arctic and the Prairies where the air is dry and cold and the snow surface does not normally form the crust associated with diurnal thawing and freezing, except perhaps during the period immediately before the spring melt.

The wind speed controls the amount of snow that can be transported and it's direction determines the place it accumulates. Snow transported by wind moves parallel to the ground with the largest quantities moving at heights less than 1.5 m above the surface. Most of the movement takes place by hopping or rolling of the particles (saltation) along the snow or ground surface. The minimum wind speed required to move fresh, dry snow is assumed to fall in the range 3.5 to 4.5 m/s, although this depends on the snow condition (Schneider, 1959). Experiments have shown that the quantity transported by wind varies with the cube of the wind speed, so that even slight changes in wind speed may greatly influence the amount of snow that is transported.

Snow is deposited because some obstacle reduces the wind speed; the shape and size of a drift depends on the shape and size of the obstacle and its orientation to the wind direction. Rikhter (1945) classified drifts into two general groups: crescent-shaped drifts formed by strong gusty winds, and wave-like drifts formed by medium-speed winds which move large amounts of loose snow. The form and shape of drifts not falling into these two groups vary widely. Typical shapes created by a few basic types of obstacles are shown in Fig. 16.1. At the base of a vertical wall facing the wind (Fig. 16.1a) a trough will form, out of which the snow whirls to form a drift downwind. A vertical wall facing away from the wind will normally have a snow cornice and a drift downwind (Fig. 16.1b). The accumulation pattern caused by a narrow ridge with steep slope facing downwind is similar (Fig. 16.1c); a reverse effect occurs when the steep slope faces upwind (Fig. 16.1d).

Snow accumulation depends on the shape of the obstructions, their number and the wind speed and direction so that the general patterns shown in Fig. 16.1 are at best approximate guides to what can be expected in nature. Gerdel (1960) states that some of these patterns will be considerably modified under Arctic conditions.

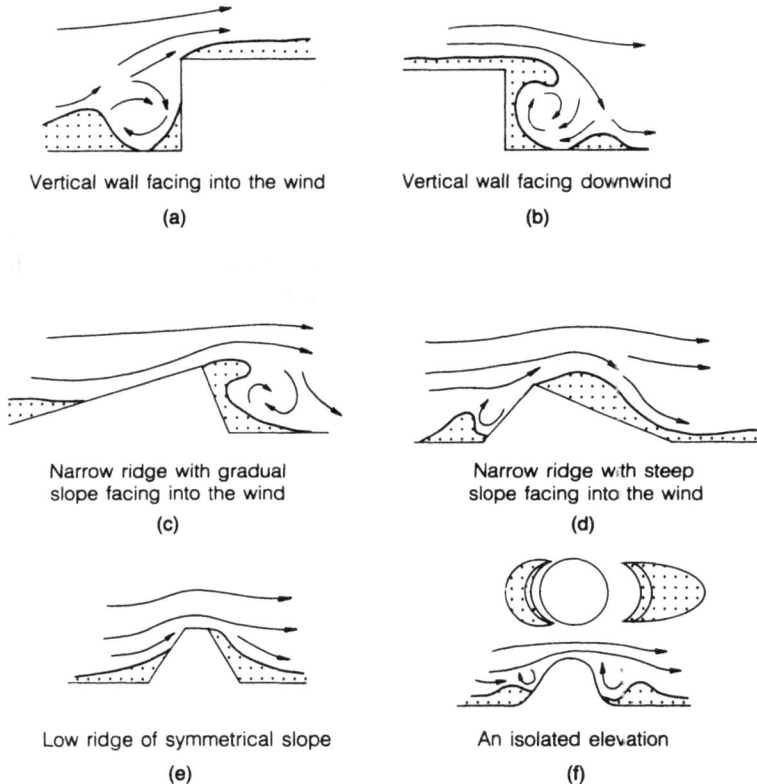

Vertical wall facing into the wind

(a)

Vertical wall facing downwind

(b)

Narrow ridge with gradual
slope facing into the wind

(c)

Narrow ridge with steep
slope facing into the wind

(d)

Low ridge of symmetrical slope

(e)

An isolated elevation

(f)

Fig. 16.1 Accumulations of drifting snow caused by some typical obstacles (Rikhter, 1945).

DRIFT CONTROL BY DESIGN AND LAYOUT OF STRUCTURES

The dependence of snow drifting on variable wind speed, snow and site conditions has prevented the development of specific design criteria for structures. However, the findings of several wind-tunnel studies and field tests can be formulated into general guidelines.

Roots and Swithinbank (1955) recommend the following practices:

(1) In Arctic regions, it may be advantageous to elevate structures above the surface, allowing the wind to accelerate beneath them to carry the suspended snow through the spaces and to deposit it to the leeward.

(2) A rectangular object should be placed with its long axis in the direction of the prevailing drift-producing wind.

(3) The upper surface of objects should be as streamlined as possible. In some cases, e.g. when equipment or material is stored in a drift area, it may be worthwhile to erect a flat, smooth roof.

(4) The downwind distance between structures should be at least 30 times their height if coalescence of drifts is to be avoided. Where sufficient space exists, objects should be placed along a line normal to the prevailing wind to avoid the possibility of over-lapping drifts.

Gerdel (1960) elaborates on some of these general rules. He notes that drift accumulations to the leeward of a low obstruction may extend horizontally more than 10 times the height of the structure. Roots and Swithinbank (1955) state that in the Antarctic, snowdrifts are frequently many hundred times longer than they are wide and recommend that successive objects be spaced downwind at distances never less than thirty times their heights. For obstructions up to about 2-m high, the size of the drift increases roughly in linear proportion with height. For higher obstructions, which project above the heavily snowladen layer of air, the proportionate increase in the size of the drift with the height of the obstruction decreases considerably. A limit is reached when the object is too high to be submerged in a wind-formed drift during a storm; obstacles higher than this limit do not produce larger drifts. Gerdel (1960) also confirms that high buildings that extend well above the zone of blowing snow tend to deflect the moving snow and create different drift patterns than those created by low objects. When the air temperature is above -10° C overhanging snow cornices may develop above the leeward wall of a structure producing heavy snow loads on the edge of the roof.

Drifting problems can be minimized if the long side of a building is placed parallel to the direction of the prevailing drift-producing wind (Schaerer, 1972). The drift formed behind the short side may be longer, but is usually less bothersome than the one formed behind the long side. A building having a simple rectangular exterior shape and walls without internal corners or projections is best for minimizing undesirable accumulations. The most serious drifting may occur if the wind strikes the sides of the building obliquely. Where several buildings are located close together individual drifts usually overlap to form a large one. Basically, buildings should be placed close together in rows with their long axes parallel to the direction of the drift-causing wind. The most suitable orientation for roads is usually parallel to the wind direction, between the rows of buildings. Although terrain often restricts the layout of roads and buildings this rule should be adhered to as closely as possible.

The underlying principle in the design or layout of structures is to maintain the speed of the wind and hence prevent snow accumulations. Finney (1939) found that by streamlining slopes, drift accumulations were reduced (Fig. 16.2). Obstacles having a steep slope (1:1) had large drift accumulations; those having a gradual slope (1:6) had much smaller accumulations. Natural snow drift deposits have a slope of approximately 1:6. In flat country, roads should be constructed at a grade above the expected snow depth in order that the

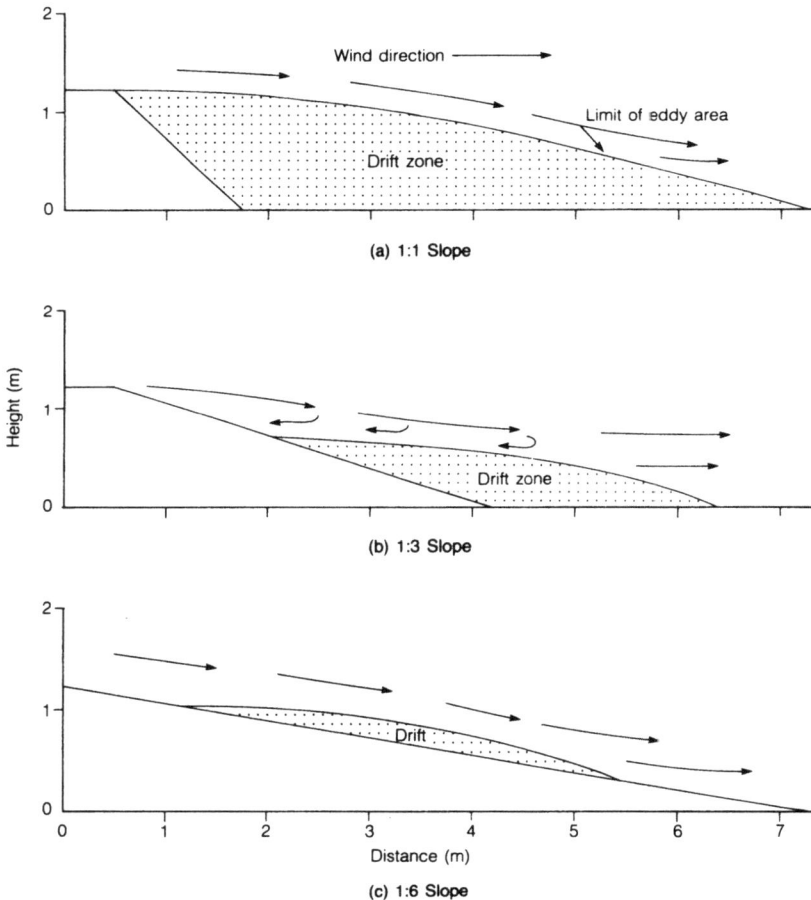

Fig. 16.2 Effect of embankment slope on the amount of drifting (Finney, 1939).

wind speed will be high over the road surface. When highways or roads must be placed perpendicular to the direction of drift-producing winds, the embankments should be about 20 percent higher than those for routes parallel to the wind direction (Schneider, 1959). Highway cuts are most susceptible to drifting if they are oblique to the direction of the drift-causing winds. Transitions from cuts to embankments cause a decrease in wind speed and thus are also susceptible to drifting and in mountainous areas, cuts along steep slopes are particularly prone to severe drifting. In complex situations where irregularly-shaped structures are to be constructed in hilly or vegetated areas, the optimum arrangement may be determined by model tests conducted in a wind tunnel or water flume (Schaerer, 1972).

SNOW DRIFT CONTROL WITH SHELTER BELTS

The objective in planting trees or shelterbelts to control snow drifting is to cause the snow to be deposited before reaching the area to be protected. Hedges and rows of trees provide effective protection and help to beautify the landscape. The major disadvantage of vegetative control methods is the large amount of space they require.

Finney (1937) investigated snow control by tree planting and concluded that:

(1) In determining the proper spacing distance for a shelterbelt both the spread of the tree in relation to its age and its shape must be considered.

(2) It is not necessary to plant more than three rows of trees for effective snow control. Two rows of trees can be very effective if properly placed.

(3) The spacing of trees should be close for a single row and not more than twice the diameter of the crown for three or more rows.

(4) On the average, the effective length of the area that will be protected is about fifteen times the height of the barrier, but with considerable variations.

(5) Wind speed has no effect on the size of the eddy area produced by a barrier, but it does affect the position of the drift as high winds tend to move snow back toward the barrier.

The selection and spacing of trees for a shelterbelt will depend not only on their effectiveness as snow barriers but also on available space, ownership of land, soil and climatic conditions, their susceptibility to disease, their life expectancy, and their rate of growth. It is often assumed that trees or shrubs can be left alone after planting but this is not true. Without continuous care they will never provide optimum protection. It has been estimated that this method of snowdrift control costs only one-third to one-half as much as the annual cost of erecting wooden fences; after seven years trees are more effective than fences (Schneider, 1959).

Where soil and climatic conditions are suitable for their growth, conifers are usually planted rather than deciduous trees since the latter have their lowest canopy density at the time when they are most needed for snow protection. However, Schneider (1959) reports that the lower branches of conifers tend to die off as the trees mature so that their canopy density becomes inadequate in the snow drifting layers. Williams (1949) cites a case where extensive snow drifting took place in the space between the ground and the tree branches. It was recommended a row of shrubs be planted to plug this gap to control drifting.

Some shrubs and trees suitable for planting as snow fences for the prairie and plain areas of the northern United States are listed by Bates and Stoeckler (1942) (see Table 16.1, Fig. 16.3). They recommend that species should be low-growing, dense near the ground, frost resistant and adapted to a wide range of soil and climatic conditions. By the end of the winter the hedges may be covered with deep snow, therefore the individual trees and shrubs must be able to carry the load.

The main disadvantages in using trees and hedges for shelterbelts are that they require several years to grow and once planted are not easily moved to correct for any errors made in the initial site selection and orientation. Thus the optimum arrangement should be determined using moveable fences before planting. Hedges need not be higher that 2.5 m. The trees and shrubs need not be planted in long rows; clumps of trees are equally effective in controlling drifting, even though greater space is required.

SNOW FENCES

General Considerations

The most common method of snow drift control is with snow fences. Because the snow-carrying capacity of the wind is approximately proportional to the cube of its speed even small reductions in speed will produce substantial deposits of snow. A snow fence is designed to reduce the wind speed, therefore causing snow to deposit. Normally, places where drifting occurs are known from experience, or in the case of new installations, have been observed during one winter. Without this information the effectiveness of a snow fence will be uncertain. It is placed upwind of the area to be protected and oriented perpendicular to the direction of the snow-carrying wind so that snow is deposited in front of and behind it (Schaerer, 1972).

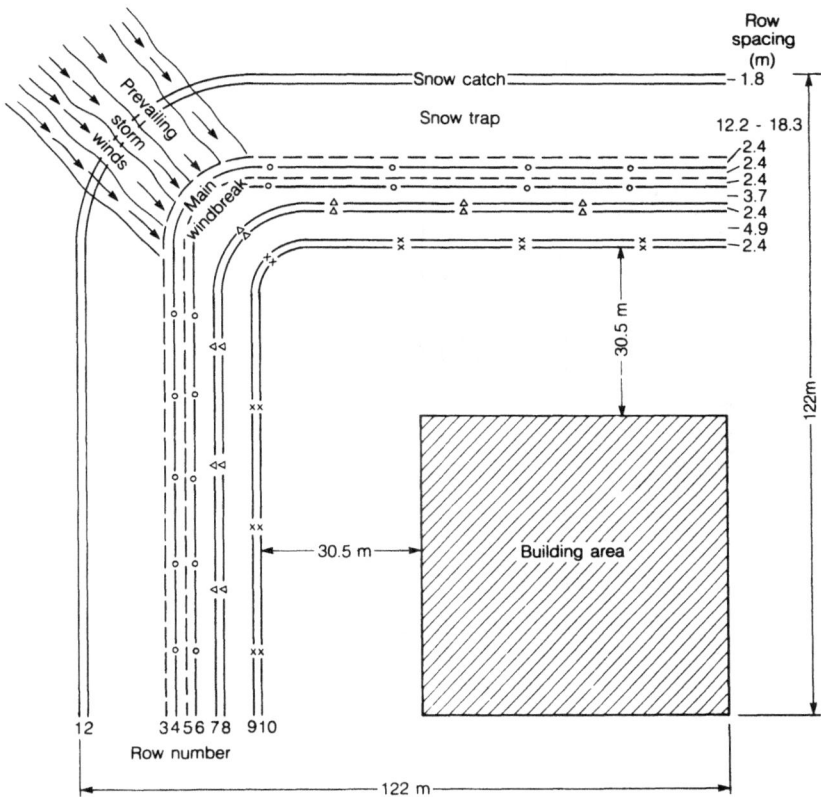

Fig. 16.3 Planting plan for snow protection in prairie and plains areas (Bates and Stoeckler, 1942).

When snow fences are being considered as a control measure for a new building, a careful assessment should be made of the result which could be obtained by improved building design and layout. This may increase the construction cost substantially but a single expenditure of this kind may produce a permanent improvement and a solution that in the long run will pay for itself.

Rikhter (1945) has indicated the types of deposit that can be expected for the solid fence, the sloping solid fence, a wind-deflecting shield and the slatted (open) fence (Fig. 16.4). Because sloping a fence reduces its effective height these types are rarely used.

Economics, availability and space limitations determine the type of materials and arrangement used for fences. Because the wind pressure is less

Table 16.1
SHRUBS AND TREES SUITABLE FOR PLANTING AS SNOW FENCES (Bates and Stoeckler, 1942)

Number of rows in belt	Type of wind-break[b]	Windward Row 1	Choice of species by rows[a] Row 2	Row 3	Leeward Row 4
1	Shrub	Caragana Chokecherry Honeysuckle American plum			
2	Shrub Shrub	Caragana Chokecherry Lilac Honeysuckle American plum	American plum Chokecherry Russian olive Buffaloberry		
2	Shrub Evergreen	Caragana Chokecherry Lilac Honeysuckle American plum	Rocky Mountain red cedar Eastern red cedar Colorado blue spruce Ponderosa pine		

3	Shrub	Cargana	Green Ash	Rock Mountain red cedar
	Low tree	Chokecherry	Boxelder	Eastern red cedar
	Evergreen	Lilac	Hackberry	Colorado blue spruce
		Honeysuckle	Chinese elm	Ponderosa pine
		American plum	American elm	
4	Shrub	Caragana	Green ash	American plum
	Low tree	Chokecherry	Boxelder	Chokecherry
	Low tree	Lilac	Hackberry	Russian olive
	Shrub	Honeysuckle	Chinese elm	Buffaloberry
		American plum	American elm	Lilac
				Honeysuckle
				Caragana

[a] Any one of the three or more species can be selected for any well-drained, non-alkaline soil, ranging in texture from loamy sand to clay loam. Soils with coarse gravel subsoils and shallow topsoil should be avoided. Where it is necessary to cross low spots, plant 1-year-old rooted cuttings of Russian golden willow. Evergreen, especially spruce, could well be limited to a line east of the 100th meridian of longitude.

[b] Spacing between rows can range from 1.2 to 2.5 m. Spacing between plants in the rows should be as follows: All shrubs in Row 1 should be 0.6 to 1 m apart; all shrubs in Rows 2, 3 or 4 can be from 0.9 to 1.2 m apart; all low broadleaf trees in Rows 2 or 3 should be 1.2, 1.5 or 1.8 m apart. All evergreens should be 1.2, 1.5 or 2 m apart in the rows, and 3.7 to 4.9 m away from tall-growing trees like boxelder or elm. These spacing recommendations are contingent upon the assumption that any dead plants will be replaced.

on the slatted fence, it can be fabricated from lighter material. Vertically slatted fences are the most common. Since a solid fence or wall produces a shorter deposit on its leeward side than on its windward side. it is most suitable when space is limited. The major disadvantages of solid fences are the high cost of materials and the need for a strong foundation.

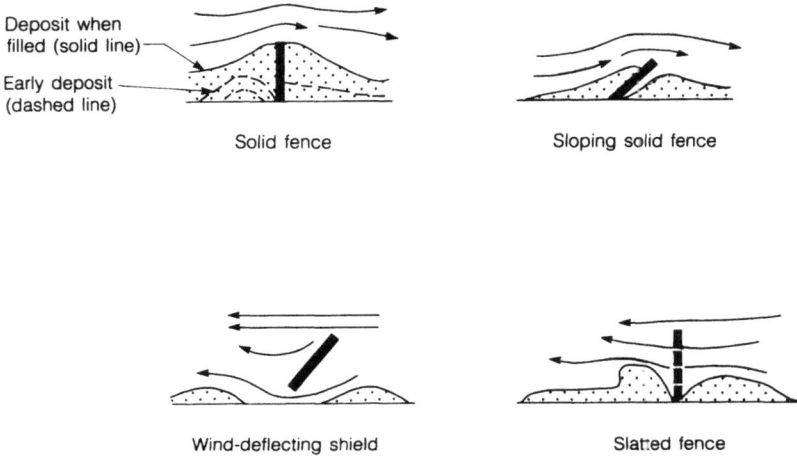

Fig. 16.4 Accumulation of snow at various types of snow fence (Rikhter, 1945).

Fence Height

Tabler (1974) states that tall fences are more efficient than short ones in trapping snow. The reduction in wind speed behind a fence increases with its height. However, since most of the blown snow is transported in a shallow layer adjacent to the ground or snow surface the reduction of wind speed in this layer is probably responsible for the increase in catch rather than the capture of additional snow in the higher layers. Tabler, who used 3.8-m high fences for a highway snow protection project in Wyoming, considers that such fences are the most economical on the basis of cost per cubic metre of water-equivalent storage. He estimates that these fences cost about one-third of that for a system of four, 1.8-m fences of equal storage capacity. The construction cost per given volume of water-equivalent storage is plotted as a function of fence height in Fig. 16.5, which suggests that a 3.8-m fence may be near the maximum height to give the lowest cost per unit of water-equivalent.

When moveable fences are used, their heights may be reduced. However, considerable time and expense are required to relocate them in snow-bound conditions.

The expected depth of snowfall and the quantity of drifting snow associated with the snowfall are decisive factors in selecting fence height. Observations of the conditions under which drifting occurs must be made. If drifting occurs when the snow is shallow the required fence height will be substantially less than that if drifting only occurs during and after heavy snowfalls (Schneider, 1959).

For areas with light to moderate snowfalls fence heights between 1.2 and 1.8 m are usually sufficient. Even in places of heavy drifting two parallel rows of relatively low, inexpensive slatted snow fence may be more economical than one high fence which is more expensive to construct.

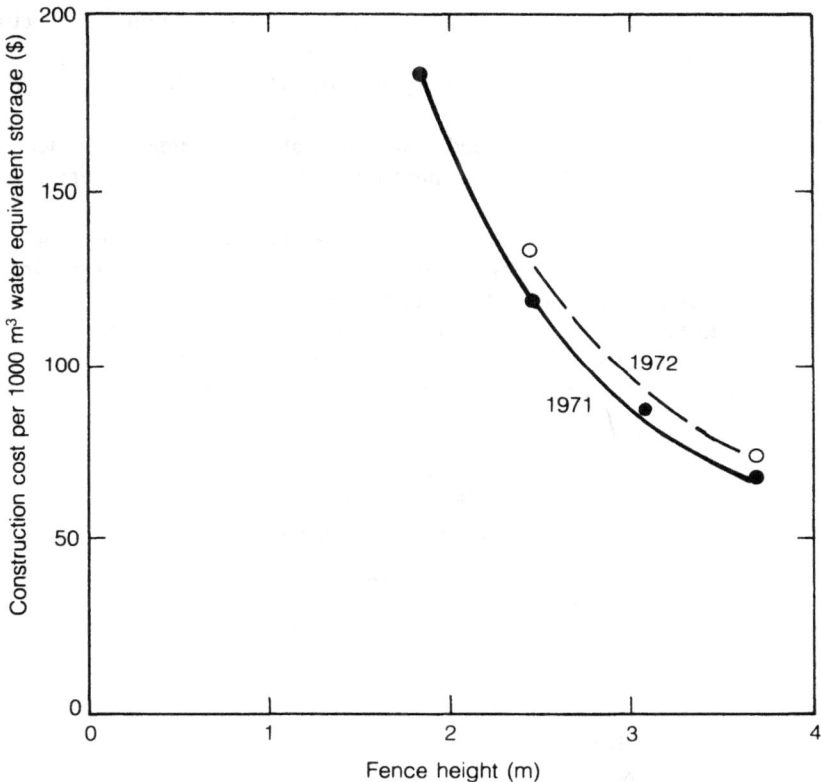

Fig. 16.5 Construction costs C, ($/ 1000 m³) of water-equivalent storage as a function of fence height H, (m) (Tabler, 1974).

Design and Location

Several criteria are generally applicable to the design and location of most types of snow fence. The base of the fence should lie above ground level. A ground gap tends to produce vorticies immediately in front and behind the fence preventing filling in or choking at least until the deposit reaches the bottom of the fence; the gap also reduces the snow pressure on the fence; also, there is less tendency for the lower parts of the fence to rot due to moisture. The size of the gap will vary with the type of fence and amount of snow. Some authors recommend a gap about one-seventh the height of the fence, i.e., $\sim$ 17 cm for a 1.2 m fence (Schneider, 1959).

The smaller the density ratio (the ratio between the solid area and the total area of the fence), the longer and shallower the drift. The maximum collecting capacity of an open fence occurs for a density ratio between 40 and 60 percent. Price (1961) considered that the density of an open fence is the most important factor in determining the volume of snow deposited. Tests in the field and in wind tunnels have demonstrated that the slat arrangement (vertical or horizontal, slightly inclined) or material (wood, metal or other) is not important (Schaerer, 1972).

The optimum distance of the fence from the object or area to be protected will vary with wind and snow conditions. In practice, unless better information is available, open fences should be placed upwind at a distance fifteen to twenty times their height. Using various sources Schneider (1959) compiled a list of recommended distances from the object to be protected. Some of this data is reproduced in Table 16.2 where the distances have been normalized to

Table 16.2
LENGTH OF SNOW DEPOSIT CAUSED BY A FENCE 1 m
HIGH, 50% DENSITY (Schneider, 1962).

Authors	Recommended distances
Croce (1943)	18
Croce (1956)	25
Finney (1934)	14
Hjort (1928)	10-18
Kreutz (1954)	22
Lawrence (1947)	15
Petersen (1942)	15-20
Price (1954)	22
Sund (1929)	10-20

a fence 1 m high having a 50% density. The variability is typical of what can be expected when this type of data is compared.

Figure 16.6 shows the relationship between fence density and the distance of the fence from the protected area as given by Pugh (1950); e.g., a fence with a height of 2 m and a density of 50% should be placed approximately 26 m from the protected area. In practice, other factors such as land ownership or building location often determine the siting of a fence.

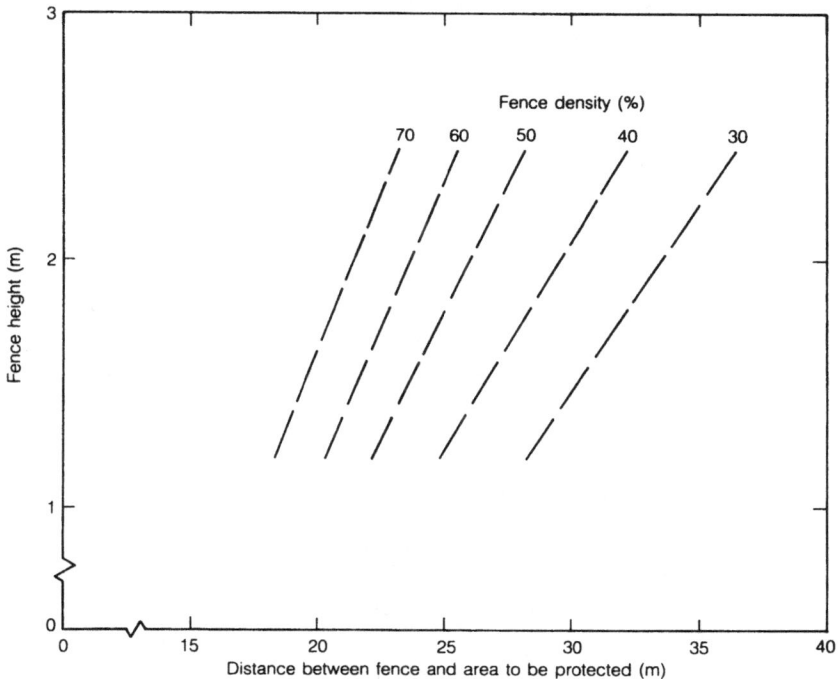

Fig. 16.6 Relation between fence height and the distance between the fence and the area to be protected (Pugh, 1950).

Since the length of the deposit formed behind a solid fence is about ten times the height of the fence, this distance should be maintained between the fence and a building or road. When a solid fence is longer than the width of the building to be protected and closer than five times its height, a strong eddy occurs between the fence and the building so that this space is usually snow free. It could be easily filled, however, if the wind direction was oblique rather than perpendicular (Schaerer, 1972).

As a general rule if the wind-blown snow originates from more than one direction, which is frequently true, several protective fences may have to be installed. In regions where large quantities of drifting snow can be expected, snow fences are arranged in rows, at a recommended separation of about 10 times their height. Pugh (1950) presented some basic row arrangements (Fig. 16.7). The arrangement selected depends on the prevailing wind direction and its expected variation. Other factors such as available space, soil conditions, and the depth of snow also must be considered. The first arrangement will rarely give satisfactory results; field observations of the shapes of the deposits are necessary before the best locations can be made.

Special Types

Blower fences (see Fig. 16.8) also called jet roofs, can be likened to inclined table tops mounted on posts. Wind passing underneath is accelerated, eroding the snow behind the fence to a distance of about 6 m. Blowers are best used for preventing local snow accumulations behind ridges or depressions. The incline of the roof should approximately be equal to the slope of the land on the lee side, but not less than 30 degrees (Schaerer, 1972).

Deflector fences are vertical baffles or walls 2.4- to 3-m high that deflect the wind in a horizontal direction to produce acceleration and local erosion of the snow. Schneider (1959) cites an example of the protection of a railroad line in Switzerland by the erection of fence-like walls several metres high which are used to deflect the wind; several years of experimenting was necessary before satisfactory results were obtained. Such a protective wall is especially useful in mountainous regions where the very large quantities of drifting snow quickly saturate ordinary snow fences.

Deflector fences can also be used effectively near buildings to keep entrances free of snow (see Fig. 16.8); the one principal objection is their lack of aesthetics. Schaerer (1972) suggests their use and positioning must be carefully considered because they can act as solid collection fences at some wind directions.

Other Considerations

Basic requirements for all temporary snow fences are economy, ease of installation in the fall and removal in the spring for storage. Durability of material is also important. In recent years the trend has been toward mass-produced, factory-built fences. For an extensive program of snow control by fences, economy and the space available for installation are often the dominant factors governing the type used and its location rather than any aerodynamic considerations.

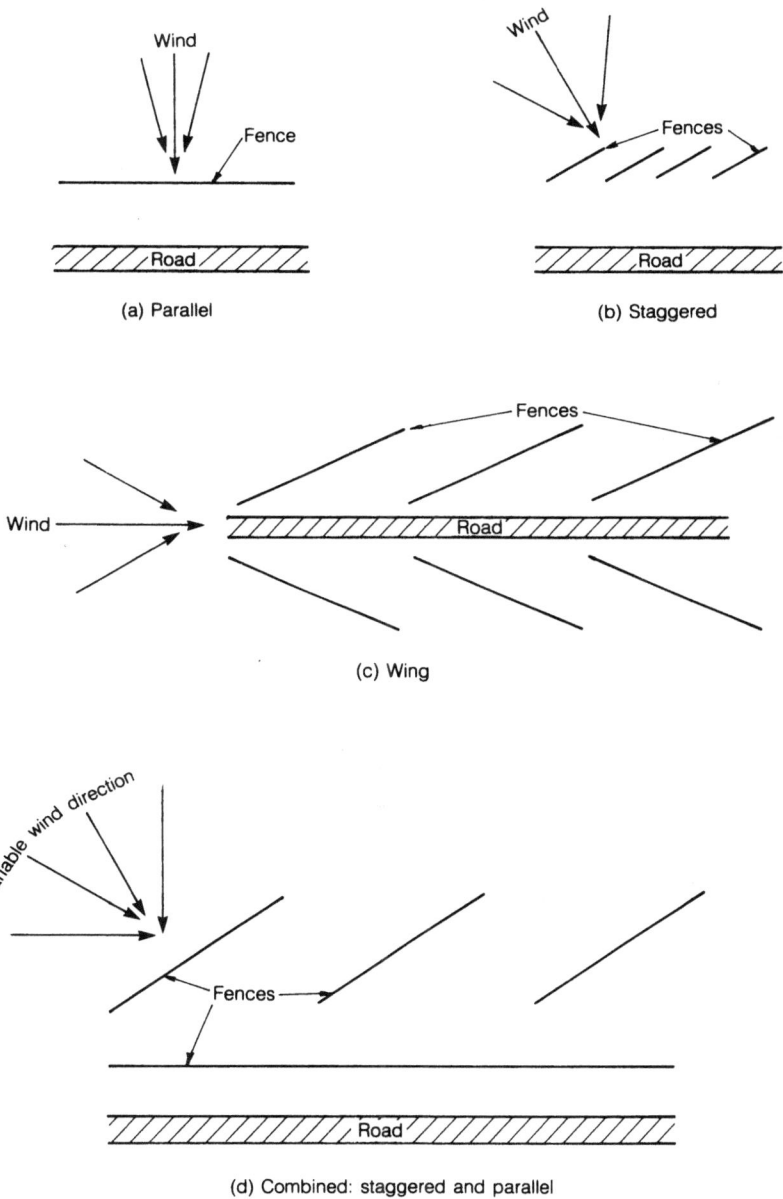

Fig. 16.7 Basic arrangements of snow fences (Pugh, 1950).

Guide rails, property fences, and other structures can act as snow fences and cause drifts in undesirable areas. Some drifting can be prevented by properly planning the location of guide rails and restricting erection of property fences and structures close to roadways and buildings. Banks and piles formed by snow plowing can also cause undesirable drifts but these can be used to advantage if properly located. If snow removal equipment and adequate space are available, windrows of snow about 1.2 m high may be plowed about 15 m to the windward side of the object or area to be protected. This can only be done after sufficient snow has fallen and must be repeated when the rows become covered (Schaerer, 1972).

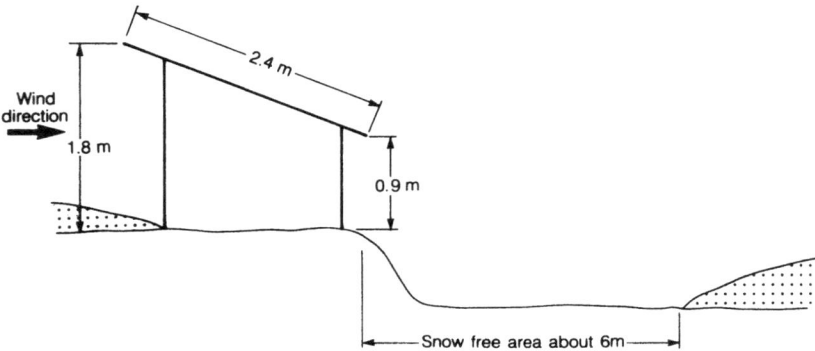

(a) Blower fence

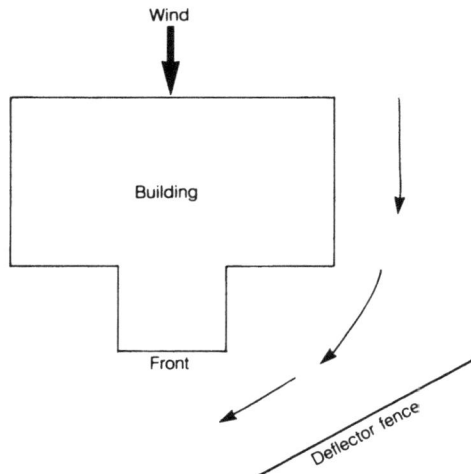

(b) Deflector fence to keep front of a building free of snow

Fig. 16.8 Types of blower and deflector fences (Schaerer, 1972).

LITERATURE CITED

Bates, C.G., and J.H. Stoeckler, 1942. *Planting to control drifts on highways.* Eng. News Rec., Vol. 129, p. 92.

Croce, K. 1943. *Messversuche an Schneezäunen in Wintern 1940/41 und 1941/42.* SLF Fremdber. Nr 2034.

Croce, K. 1956. *Messversuche an Schneezäunen in Winter 1943/44.* Arbeitsbericht C 3 des Bauhofes für den Winterdienst in Inzell.

Finney, E.A. 1934. *Snow control on the highways.* Bull. 57, Mich. Eng. Exp. Stn., Ann Arbor, Mich.

Finney, E.A. 1937. *Snow control by tree planting.* Bull. 75, Mich. Eng. Exp. Stn., Ann Arbor, Mich.

Finney, E.A. 1939. *Snowdrift control by highway design.* Bull 86, Mich. Eng. Exp. Stn., Ann Arbor, Mich.

Gerdel, R.W. 1960. *Snow drifting and engineering design.* Meteorol. Mono. 22, Am. Meteorol. Soc., Vol. 4, pp. 57-64.

Hjort, I. 1928. *Ona snösskydd, narga erfarenheter fran den gangna vintern.* Swenska Vägförening, Tid. 15.

Kreutz, W. 1954. *Untersuchung von Schneeablagerungen an Schutzgittern.* Strasse und Autobahn 5, Heft 2.

Lawrence, N.G. 1947. *Drift control by snow fences.* The Surveyor and Municipal and County Eng., Dec. 12.

Petersen, F. 1942. *Schutzanlagen gegen Schneeverwehungen an Strasses.* Bautechnik 20, Heft 50/51.

Price, W.I.J. 1954. *How to use fences to prevent roads being blocked by snow.* Roads and Road Constr., No. 32., pp. 7-10.

Price, W.I.J. 1961. *The effect of the characteristics of snow fences on quantity and shape of the deposited snow.* Int. Union. Geod. Geophys., Gen. Assem. Helsinki, [Snow and Ice], Int. Assoc. Sci. Hydrol., Publ. 54 pp. 89-98.

Pugh, H. 1950. *Snow fences.* Road Res. Pap. 19, Road Res. Lab., H.M. Stationary Office., London.

Rikhter, G.D. 1945. *Snezhnyi Pokrov, Ego Gormirovanie i Svoista (Snow Cover, its Formation and Properties).* Izv. Akad. Nauk. SSSR Moscow. [English Transl. by U.S. Army Snow, Ice Permafrost Res. Est., Transl. 6].

Roots, E.F. and C.W.M. Swithinbank. 1955. *Snowdrifts around buildings and stores.* Polar Rec., Vol. 7, No. 50, pp. 380-387.

Schaerer, P.A. 1972. *Control of snow drifting about buildings.* Build. Digest, CBD 146, Div. Build. Res., Nat. Res. Counc. Can., Ottawa, Ont.

Schneider, R. 1959. *Schneeverwehungen und Winterglätte (Snowdrifts and winter ice on roads).* Interner Bericht Nr. 302. Eidgenössisches Institute für Schnee-und Lawinenforschung, (English Transl. by Nat. Res. Counc. Can., Tech. Transl. 1038).

Sund. J. 1929. *Sneskjermet langs veiene i Vestfold:* Medd. Vegdirektren No. 4.

Tabler, R.D. 1974. *New engineering criteria for snow fence systems.* Transp. Res. Rec. 506, Transp. Res. Board, NAS-NRC, pp. 65-78.

Williams, R. 1949. *Trees beat the blizzard.* Am. For., Vol. 55, pp. 26-27, 42-43.

17

SNOW REMOVAL EQUIPMENT

L.D. MINSK

Applied Research Branch, U. S. Army Cold Regions Research and Engineering Laboratory, Hanover, New Hampshire.

INTRODUCTION

Wheeled vehicles normally cannot travel on level snowcovered pavements through depths exceeding half the wheel diameter. This fact, considered together with the increased probability of collision and the economic dislocations which would be caused by blocked roads dictates the removal of snow from the travelled way. Snow clearance is also required from airport runways, taxiways, aprons, and ramps, where limitations to visibility and the potential development of slush are of equal concern as vehicle movement. Railroad tracks must be cleared to reduce the frontal resistance to locomotives and to prevent the wheel flanges from riding up on the compacted snow, causing derailment. All major transportation systems have a common need to reduce snow accumulation to a level which allows safe passage. This chapter emphasizes equipment for clearing airfields and road networks.

Snow falling on a paved surface may be removed by chemical, thermal or mechanical means. Chemical and thermal methods are discussed in Chs. 14 and 15. Mobile mechanical equipment is discussed in this chapter, including devices for removing ice. Frequently the same equipment is used to remove both snow and ice, however, because of the high-strength adhesive bonds which may form between ice or compacted snow and pavement specialized equipment is frequently required for satisfactory results.

TYPES OF EQUIPMENT

The relationships among the various equipment designs are depicted in Figs. 17.1a, b and c. The majority of snow removal devices fall into two categories, displacement or blade plows and rotary plows. Displacement plows are classified primarily according to the location of the plow mounting on the carrier vehicle. Rotary plows are classified according to the number and types of rotating elements.

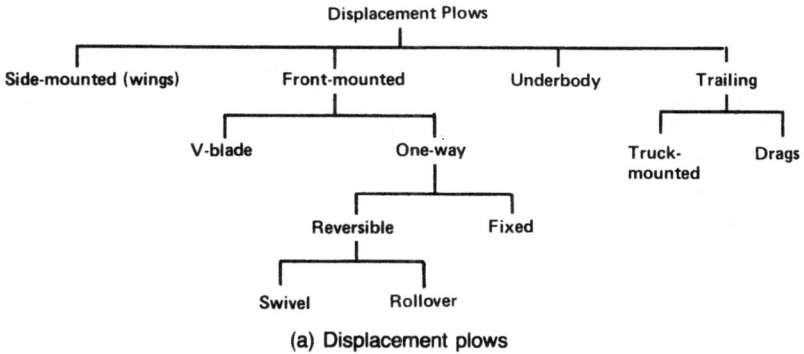

(a) Displacement plows

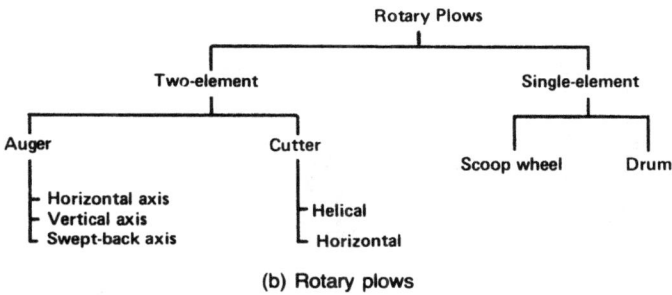

(b) Rotary plows

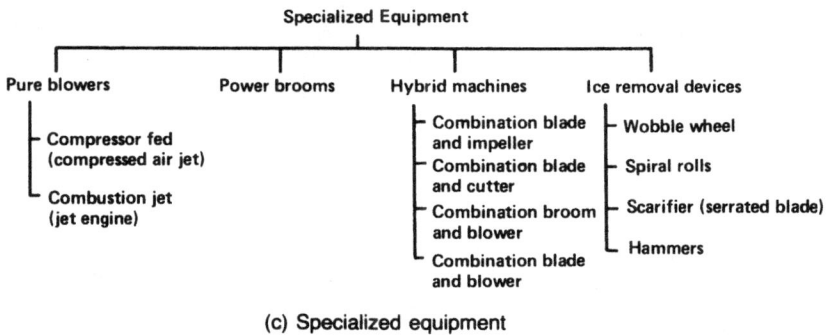

(c) Specialized equipment

Fig. 17.1 Family tree of snow removal equipment. (a) Displacement plows. (b) Rotary plows. (c) Specialized equipment.

Blade or Displacement Plows

Blade or displacement plows are the most common type of snow removal device. These are generally mounted on the front of a vehicle and displace snow in a direction approximately at a right angle to the vehicle movement. A one-way blade plow (Fig. 17.2) casts snow to one side, usually to the right shoulder or curb in those countries with right-hand traffic. Reversible blades can be adjusted to cast to either the right or left. Two types of mechanisms are employed to change the cast direction:

(1) The rollover, which rotates about a horizontal axis parallel to the direction of vehicle movement and whose blade is symmetrical about a plane parallel to the pavement, (see Fig. 17.3a), and

(2) The swivel, whose blade is symmetrical about a vertical plane normal to the blade and which is shifted from one cast direction to the other by a crank, pivot, or eccentric linkage rotating about a vertical axis (see Fig. 17.3b).

Fig. 17.2 One-way blade plow mounted on front of a 15,350 kg GVW truck; a wing plow is also mounted on the right side.

A V-blade can cast snow to the right and left simultaneously. They are usually mounted on trucks, from 230 kg utility vehicles to 24,500 kg GVW units. Front-end loaders and graders (Fig. 17.4) are also used as plow carriers.

Common variants of displacement plows include:

(1) Underbody blades mounted underneath trucks (Fig. 17.5) which are used for removing either uncompacted snow or compacted snow or ice when a downward pressure is applied to the blades,

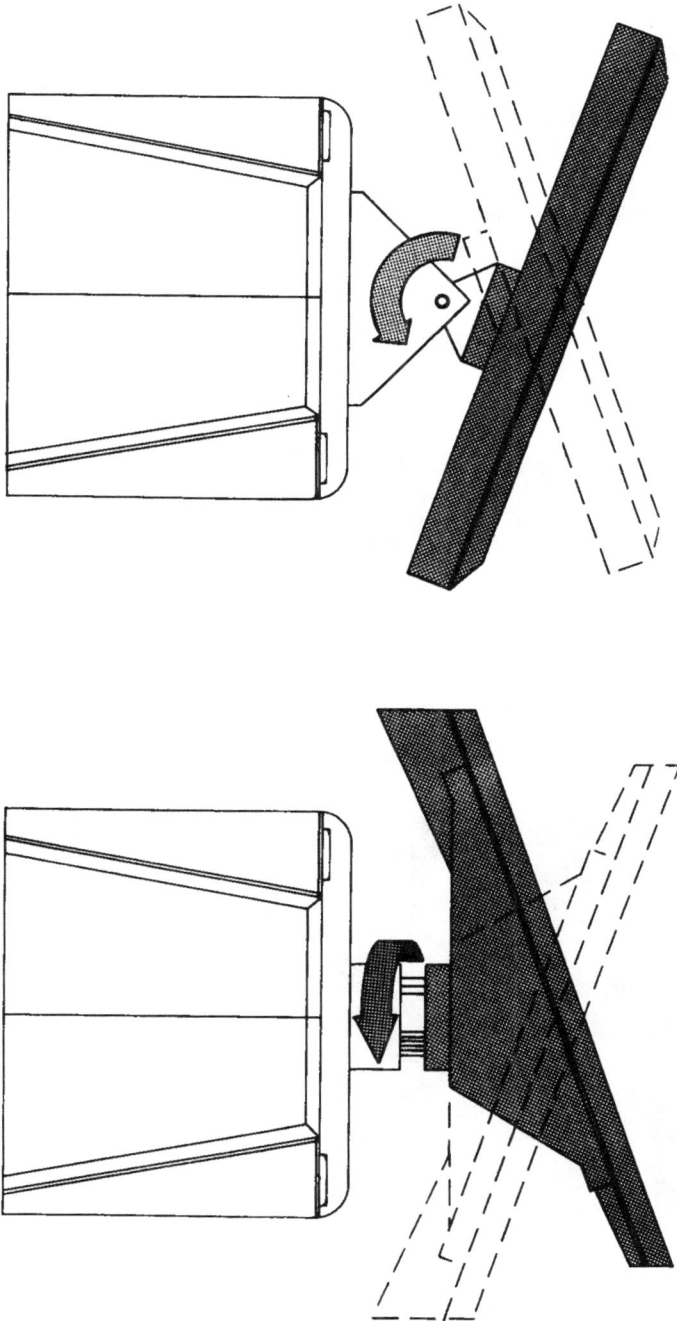

Fig. 17.3 Mechanisms for changing blade cast direction: (a) the rollover reversible plow which changes direction of cast by rotating around a horizontal axis parallel to the direction of vehicle travel. (b) the swivel reversible plow, which rotates around a vertical axis to change cast direction.

(2)   Trailing or tailgate blades, and
(3)   Wing or side-mounted plows (Fig. 17.2).

The principal advantage of underbody, trailing or tailgate blades over the front-mounted blade is the improvement in the driver's range of vision. Wing or side-mounted plows may produce a cloud of snow which can impair vision. Such plows are generally used to extend the width of cut or to cut down high banks or windrows at the side of a road ("high winging").

Fig. 17.4 Grader (motor patrol) plowing snow with a front-mounted blade.

Fig. 17.5 Segmented underbody blade (Finland).

Rotary Plows ("Snowblowers")

The limited range of cast achieved by blade plows and their inability to displace deep or very hard snow has led to rotating cutting devices, with one or more rotating elements. The snow is cut into pieces of varying sizes which are then cast to the side through a directional chute.

Single-element rotary plows use the same rotating device to both disaggregate the snow and cast it aside. Two designs are illustrated in Fig. 17.6: the scoop wheel whose axis of rotation is parallel to the direction of vehicle movement (axial rotation, Fig. 17.6a); and the milling drum in which the axis of rotation is horizontal and normal to the direction of vehicle movement (transverse rotation, Fig. 17.6b).

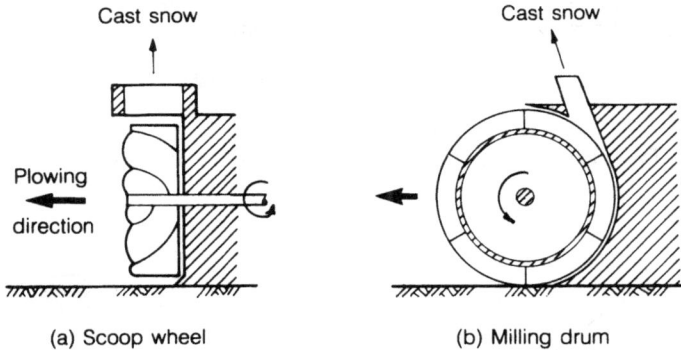

Cast snow

Cast snow

Plowing

direction

(a) Scoop wheel

(b) Milling drum

Fig. 17.6 Single-element rotary plow design: (a) scoop wheel, (b) milling drum.

Two-element plows have separate components for disaggregating and casting (see Fig. 17.7). Disaggregating elements rotate in the transverse direction while impellers rotate axially in the direction of motion. Some disaggregators utilize a pair of augers (Fig. 17.7a) or an open-helix cutter (Fig. 17.7b). Impellers consist of a number of flat or contoured blades attached to a web or disk (see Fig. 17.8a and b, respectively).

Power Brooms

Brooms (Fig. 17.9) are generally used for snow removal on airfield runways. Since little directional control is possible, the snow swept up is usually cast in the direction of the prevailing wind. The bristles are made of steel or fiber (natural or synthetic) and are incapable of removing ice or well-bonded compacted snow from the pavement; however, a jet of compressed air directed at grazing incidence can assist in removing both residual and bonded snow which cannot be dislodged by the bristles. Feasibility studies have been conducted on the use of a high velocity stream of air from nozzles, high

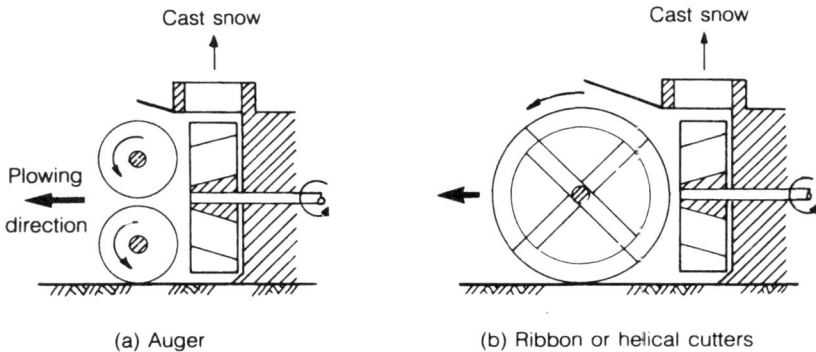

(a) Auger   (b) Ribbon or helical cutters

Fig. 17.7 Two-element designs of rotary plows: (a) auger, (b) ribbon or helical cutters.

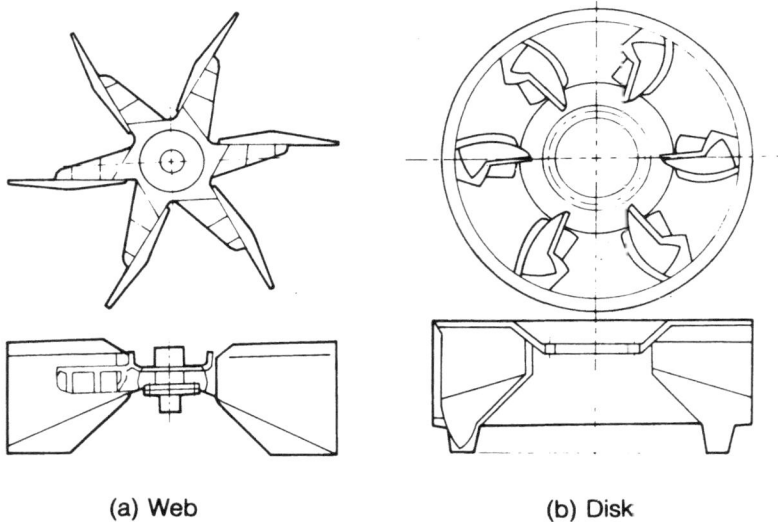

(a) Web   (b) Disk

Fig. 17.8 Typical impeller designs: (a) web, (b) disk.

volume blowers, and jet engines (similar to those in aircraft) for snow removal (Kinker, 1952; Caird, 1964; Hawkins, 1964). In general they found that the thermal content of the air stream was insufficient to melt significant snow or ice unless the travel speed was very low. None of these moving-air devices have achieved any significant success.

Fig. 17.9 Power broom.

Hybrid Machines

Machines which feature various combinations of blades, blowers, and rotating elements have been developed but have not proven sufficiently successful to be used in operating practice. The most common hybrid consists of a blade which displaces snow into a rotating casting element (Fig. 17.10).

Ice or compacted snow which is bonded to the pavement is extremely difficult to remove by any mechanical method. Usually the cutting edge of a blade is pressed downward to assist snow removal under these conditions. Similarly, blades with serrated edges are used to increase the force of penetration at the points of contact. Wobble-wheel rollers, spiral rolls, and hammers are also used to help the blade penetrate the layer by imparting either a shear force or an impact load in addition to the compressive force. However, the risk of damaging the pavement is high with all force-penetration equipment.

Fig. 17.10 Example of hybrid machine; a displacement plow feeds snow into a rotary element.

OPERATING PRINCIPLES

Displacement (Blade) Plows

Snow removal is fundamentally a material-handling problem. The material being handled (snow) is, however, very unusual because of its wide range of physical properties and bonding characteristics. Energy must be supplied both to break the intergranular bonds of settled snow and to compress the assemblage. The task of a displacement plow is to transport the snow lying in its path to a selected location at the edge. Compaction of the snow will not occur in the passage over the plow if its shape is that of a developable surface (that is, a surface which can be flattened into a plane without stretching it) and if the speed of the plow does not exceed a critical value. Croce (1941) analyzed the movement of a layer of snow across a non-distorting surface. The simple case of a wedge lifting a snow layer considered in two dimensions is illustrated in Fig. 17.11 and for three dimensions in Fig. 17.12. Model studies and mathematical analyses based on Croce's work have also been made for conical plows (Kihlgren, 1961) and cylindrical plows (Kihlgren, 1961; Vinnicombe, 1968).

The simplest analysis assumes a plow moving on a horizontal road at a constant speed V such that the snow layer acts as a coherent mass with no change in density or shape. For this case the principle of mass conservation

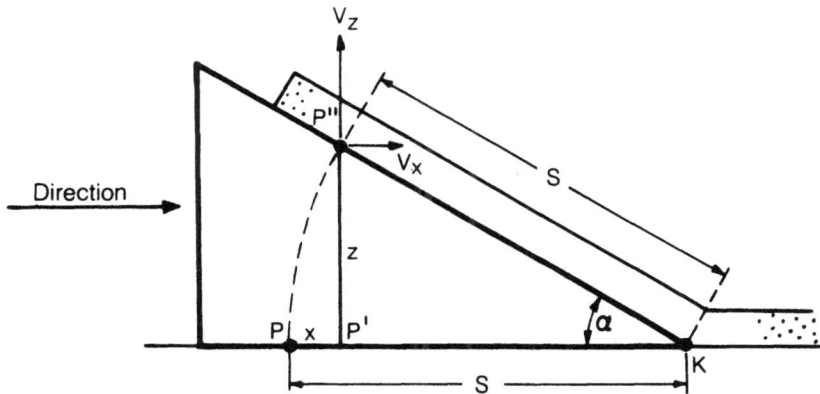

Fig. 17.11 A simple wedge lifts a snow layer without distorting it. The point P on the ground will be lifted to P″ on the surface of the wedge. With respect to the ground, however, point P has been translated to P′. By simple trigonometry, $z = s \sin \theta$ and $x = s (1 - \cos \theta)$. If the wedge is moving at a velocity $V_p = ds/dt$ then $V_x = V_p(1 - \cos \theta)$ and $V_z = V_p \sin \theta$ (Croce, 1941).

requires that the ejection of the snow relative to the blade be equal in magnitude to the entrance velocity V. Since a snow layer probably does not behave in this ideal manner a correction factor C is introduced, making the exit velocity equal to CV with respect to the blade (Mellor, 1964).

Assume the plow is moving in the x-direction of an orthogonal system x,y,z (Fig. 17.12b). The relative velocity of a random snow particle with respect to the plow surface is independent of the particle's position, and is equal to or greater than the absolute velocity of the plow. The components of the exit velocity relative to the ground are

$$V_x = V(1 - C \cos \alpha \cos \beta), \qquad 17.1$$
$$V_y = CV \cos \alpha \sin \beta, \text{ and} \qquad 17.2$$
$$V_z = CV \sin \alpha. \qquad 17.3$$

Forces on the blade are found by considering the changes in momentum which the snow undergoes in its travel over the blade. If it is assumed the strength of the snow can be neglected the force component F in any direction is equal to the rate of change of momentum in that direction. Thus

$$F_x = BD\rho V^2 (1 - C \cos \alpha \cos \beta), \qquad 17.4$$
$$F_y = BD\rho CV^2 \cos \alpha \sin \beta, \text{ and} \qquad 17.5$$
$$F_z = BD\rho CV^2 \sin \alpha,$$

where:

B = width of the plow cut,
D = depth of the snow in the cut, and
ρ = density of the snow.

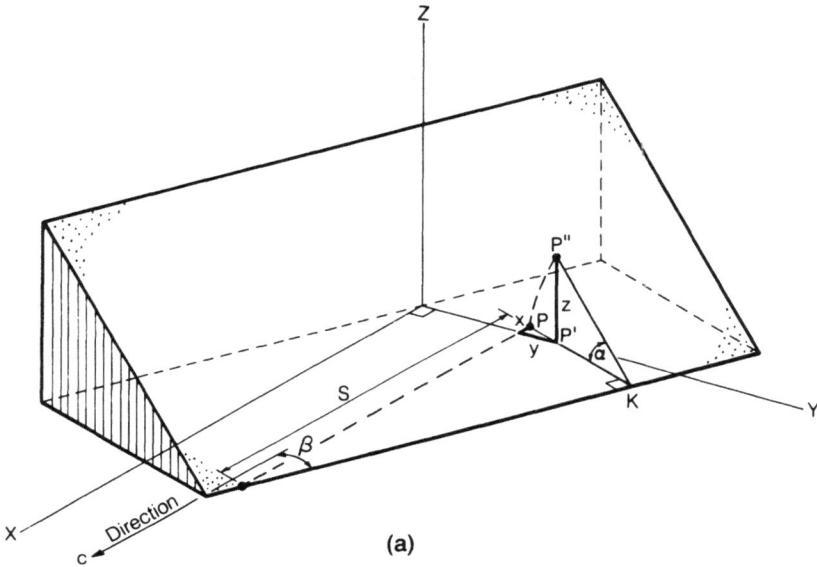

Fig. 17.12a The simple wedge considered in three dimensions. It is moving parallel to
the x axis and casting in the y direction. A snow particle lying on the
ground at P will be lifted to P″ on the surface of the wedge, having undergone the translations x, y and z. The wedge intersects the xz plane in the
angle ϕ, and its face is inclined to the xy plane at an angle θ. Noting that PK
= P″K, it can readily be shown that the particle velocity at P″ can be re-
solved into components $V_x = V_p \sin^2\phi \, (1 - \cos \theta)$, $V_y = V_p \sin \phi \cos \phi$
$(1 - \cos \theta)$ and $V_z = V_p \sin \theta \sin \phi$, where $V_p = ds/dt$ (Croce, 1941).

The force offered by the snow in the direction of F_x must be overcome by a
tractive force from the driving wheels. If the engine power is adequate, the
limit to plowing capability is governed by the maximum traction developed
before slip occurs. The lateral force F_y tends to thrust the front of the truck to
the side opposing the direction of cast. This turning moment is countered by
either the side friction developed by the tires, or by the driver "crabbing" the
wheels, that is, steering in the direction of cast while maintaining a straight
course. If the side friction developed by the rear tires drops below that of the
front tires, then the rear of the truck will rotate in the direction of cast. F_z, the
downward force produced on the plow, is accommodated in a number of
ways: the plow may be operated in "floating" position, not contacting the
pavement but entirely supported by the vehicle, shoes or caster wheels may be
used to support the plow so that the blade does not contact the pavement, or
the full plow weight and the snow forces may be applied to the pavement along
the cutting edge of the blade. The frictional forces of shoes, casters, and blades

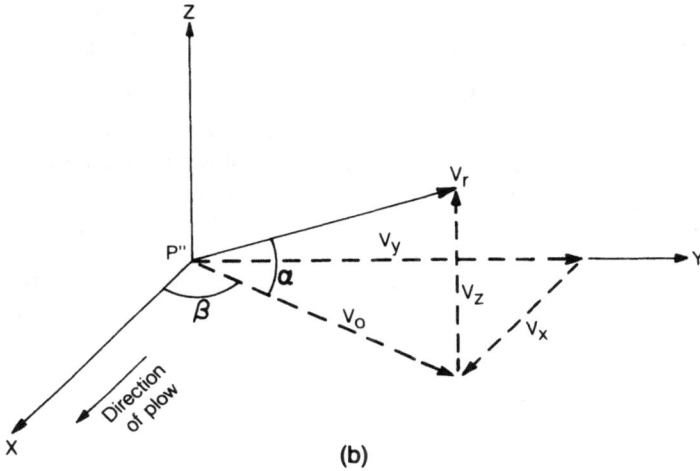

Fig. 17.12b Trajectory V_r of a snow particle cast from a curved blade at an angle β from the direction of plow movement and at an elevation angle α from the horizontal plane. V_r is resolved into components V_x, V_y, V_z in the orthogonal x, y, z system.

are directly proportional to F_z and must be added to the snow force F_x to determine the maximum tractive effort required. It is important to note that the forces on the blade increase as the square of the vehicle speed.

On ejection the snow has an absolute velocity V_r which is the resultant of the components V_x, V_y and V_z. Summing the three components vectorially gives

$$V_r = V(1 + C^2 - 2C \cos \alpha \cos \beta)^{1/2}. \qquad 17.7$$

Neglecting air resistance, particle interaction, and any differences in elevation between the point of departure of a snow particle from the plow surface and its point of impact with the adjacent snowcover, the distance the snow will be cast in the y direction is

$$L_y = 2V_yV_z/g = 2(C^2V^2\sin \alpha \cos \alpha \sin \beta)/g, \qquad 17.8$$

where   g = acceleration of gravity.

Equation 17.7 gives the absolute velocity with which a snow particle leaves the plow surface. The minimum speed that gives the same cast occurs when the x-component of velocity is zero and the initial direction of the snow particle makes an angle of 45° with the horizontal. Under these conditions the minimum speed is $\sqrt{gL_y}$ so that Eq. 17.8 yields the expression:

$$\sqrt{gL_y} = CV \sqrt{2 \sin \alpha \cos \alpha \sin \beta}. \qquad 17.9$$

As the kinetic energy is proportional to the square of the velocity the ratio of gL_y to V_r^2 can be used to define a dynamic efficiency η_d, given by

$$\eta_d = gL_y/V_r^2 = 2C^2 \sin \alpha \cos \alpha \sin \beta/(1 + C^2 - 2C \cos \alpha \cos \beta) \qquad 17.10$$

in which η_d is a measure of the efficiency of the blade in casting snow a given distance. This applies only to that part of the plow surface over which the particle has moved. η_d varies in magnitude along the blade for those designs which change the angle of departure of the snow; the maximum value occurs at $\alpha = 45°$, $\beta = 90°$.

During its passage over the blade a snow particle with mass m gains kinetic energy KE in an amount given by

$$KE = (1/2)mV_r^2 = (1/2)mV^2(1 + C^2 - 2C \cos \alpha \cos \beta). \qquad 17.11$$

The mechanical efficiency η_m of the plow may be defined as

$$\eta_m = [(1/2)mV^2(1 + C^2 - 2C \cos \alpha \cos \beta)]/W, \qquad 17.12$$

 where W = work done by the plow in producing the kinetic energy of the particle.

The casting work H of a plow is defined as the product of the mass of snow displaced by the plow per unit distance of movement in the forward direction and the distance from the centre of cut to the centre of mass of the windrowed snow y', that is

$$H = \rho_o Ay', \qquad 17.13$$

 where ρ_o = average density of the snow in the plow path, and

    A = face area of plow cut.

The theoretical casting work H' for a single particle of mass m is

$$H' = L_y m = (2C^2V^2 m/g) \sin \alpha \cos \alpha \sin \beta \qquad 17.14$$

and, the casting efficiency is

$$\eta_c = H/H'. \qquad 17.15$$

Therefore

$$H = (2mC^2V^2 \eta_c/g) \sin \alpha \cos \alpha \sin \beta, \qquad 17.16$$

and from Eq. 17.12:

$$H/W = 2 \eta_d\eta_c\eta_m/g, \qquad 17.17$$

where η_d, the dynamic efficiency, is defined by Eq. 17.10. η_m and η_c cannot be separated experimentally, hence an overall efficiency η_o defined as

$$\eta_o = \eta_m\eta_c \qquad 17.18$$

is used. Combining Eqs. 17.17 and 17.18, it follows that

$$\eta_o = (H/W) (g/2\eta_d). \qquad 17.19$$

The preceding equations have been developed from analysis of the velocities and forces on a single snow particle of a given mass. To arrive at practical values for the different efficiencies bulk or integrated average values of the different terms must be used in the expressions. In tests conducted on V-plows, Vinnicombe (1968) found the values of η_o to range from 7.0 to 9.7% and η_d from 53 to 68%.

Rotary Plows

Whether a rotary plow is single- or two-element, transverse or axial rotational, it must cut through a portion of the snowcover into which it is being propelled and accelerate the snow through a directional chute for casting. Ballistic analysis for the cast distance L, neglecting air resistance, shows that

$$L = (V^2/g) \sin 2\alpha, \qquad \text{17.20}$$

where   V = snow exit velocity, and

   α = angle of exit stream from horizontal.

The maximum drag exerted on the particles in the snow stream reduces L. Experiments reported by Croce (1950) have established that the effect of aerodynamic drag on L for $\alpha = 45°$ can be accounted for by the expression:

$$L_c = (V^2/g) - 0.0023V^3. \qquad \text{17.21}$$

L_c, the corrected cast distance is in metres when V is in m/s and g = 9.81 m/s^2. The theoretical energy E required to cast a mass of snow M, at $\alpha = 45°$ is

$$E = MV^2/2 = MgL/2. \qquad \text{17.22}$$

If this mass is cast in t seconds, the power requirement is

$$P = (0.5MgL/t) \times 10^{-3}, \qquad \text{17.23}$$

in which P is in kW, M in kg, g in m/s^2, and L in m.

Davis (1952) has reported on tests of two-element auger-type rotary plows operated at an elevation of about 2450 m a.m.s.l. in the Sierra Nevada mountains of California. The plows were operated at an average rate of 203 kg/s and produced an average cast distance of 23 m. Under these conditions the theoretical power, calculated from Eq. 17.23 is 23 kW. The unit was equipped with an engine having a sea-level rating of 224 kW and assuming the engine was operating near its rated capacity the efficiency was only about 10%, power losses being attributed to aerodynamic, mechanical and frictional effects.

Impellers

The impeller in a two-stage machine receives the snow collected from the cutters, and then imparts momentum to it by a series of radial blades mounted on a rotating plate or web (see Fig. 17.8 and 17.13). Snow rotating with the blades is confined within a casing up to the point where the casing opens to a directional chute which allows the snow to be flung out. The capacity of a two-stage rotary plow depends on the speed of the tip of the blades, the radial thickness of the snow in the impeller housing, and the blade dimensions (Croce, 1950). The thickness d of the layer of snow in the impeller depends

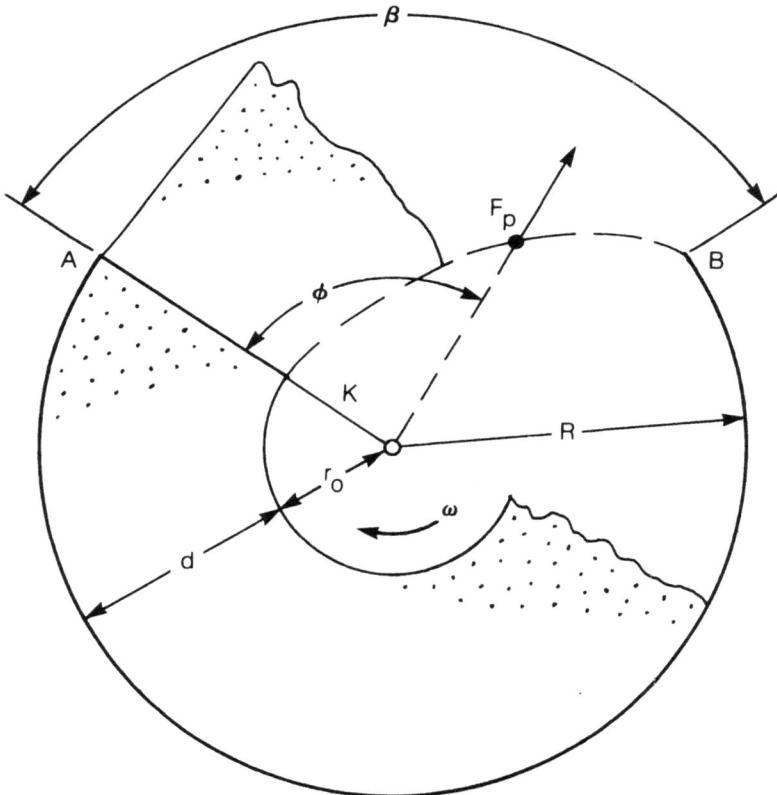

Fig. 17.13 Section view of an impeller housing which is perpendicular to the impeller shaft. Sector angle β of the opening in the impeller housing is a function of snow thickness $d = R - r_0$. Only one blade is shown (at K).

upon the size of the opening in the casing. In Fig. 17.13 the opening AB encloses an angle β and the line AK represents a plane radial blade, revolving around the centre point with a constant angular velocity ω, which has reached the beginning of the opening AB. Snow of mass M at the point K (radius r_0) in contact with the blade will slide to its outer edge (radius R) by the time the blade tip rotates to point B. The centrifugal force F_p propelling the snow outward along the blade surface is

$$F_p = Mr\omega^2, \qquad\qquad 17.24$$

where   r = radius of the snow mass with respect to the centre point.

If other forces, such as internal friction and sliding friction along the blade surface, are neglected then,

$$F_p = Mr\omega^2 = Ma_r = M(d^2r/dt^2). \qquad\qquad 17.25$$

Integration of Eq. 17.25 with the boundary conditions; $r=r_0$ at $t=0$ and $dr/dt = 0$ gives

$$\phi = \omega t = \ln (r + \sqrt{r^2-r_0^2}/r_0), \qquad\qquad 17.26$$

where ϕ = angle between an arbitrary blade position and the initial casing opening (radians).

For $\phi = \beta$, the angle occurring where the snow is at radius R,

$$\beta = \ln(R + \sqrt{R^2-r_0^2}/r_0) = \cosh^{-1} R/r_0. \qquad\qquad 17.27$$

Equation 17.27 defines the sector angle β for complete discharge of the snow layer $d = R - r_0$ in the impeller. $\beta = 70°$ is the angle often used for impeller design. If the sector angle is less than the value calculated by Eq. 17.27 some of the snow will not be discharged, but will be retained in the housing, eventually plugging it. As β is only a function of the radii (Eq. 17.27) its magnitude is unaffected by the speed of rotation (Croce, 1970).

Cutters

The function of the cutter is to remove a layer of snow from the cutting face and feed it to the discharge chute at a rate not exceeding its delivery capacity. The cutting depth varies as the cutter progresses into the snow when the vehicle moves forward. For the condition illustrated in Fig. 17.14 Mellor (1975) calculated the depth of cut to be

$$l = (V \sin \theta)/n\omega, \qquad\qquad 17.28$$

where l = depth of cut at the angle, θ,
          V = forward speed of the vehicle,
          n = number of cutting elements,
          ω = speed of rotation, and
          θ = position angle of the cutting element measured from the
               vertical.

According to Eq. 17.28 the cutting depth increases as the forward speed of the vehicle increases and decreases as the rotational speed of the cutter and the number of cutting elements increases. Single-flight, helical augers have only one cutting element, therefore their cutting depth depends only on the forward and rotational speeds.

Capacity

The volume which an impeller with flat radial blades can handle is called the absorption capacity, S; assuming negligible friction loss it is given by the expression

$$S = \pi \; nT \; (R^2 - r_0^2), \hspace{3cm} 17.29$$

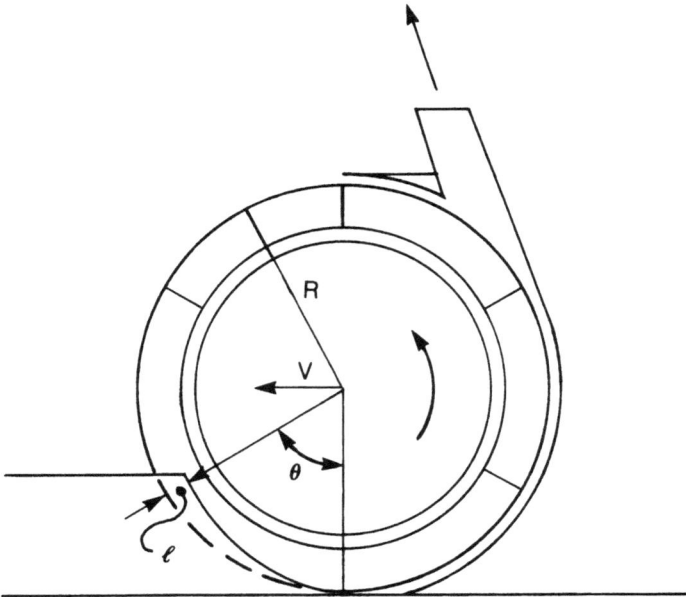

Fig. 17.14 Depth of cut l of a transverse rotational device.

where   n = revolutions per unit time,
          T = width of the blade (parallel to rotational axis),
          R = radius of impeller housing, and
          r_0 = inner radius of the snow layer.

If friction losses are assumed to be 25%, the absorption capacity can be approximated by (Croce, 1950)

$$S \simeq 140nT\ (R^2 - r_0^2)\ [m^3/h]. \qquad 17.30$$

The absorption capacity can also be approximated by the simple relationship

$$S = VdT, \qquad 17.31$$

since the exit speed V of the snow differs little from the circumferential speed of the impeller tip. The absorption capacity varies directly with exit speed and therefore with cast distance. However, power is wasted if the cast distance is greater than required.

PLOW CARRIERS

Motorized vehicles are required to force both the displacement and rotary plows into a snowcover. Wheeled vehicles are preferred for operation on prepared or frozen surfaces with a shallow snowcover; continuous track-laying vehicles are necessary when deep snowcover must be removed, the ground is unprepared, or low ground pressures are required. The goal in efficient utilization is selection of equipment having year-round usage.

The dump truck is the most versatile maintenance vehicle for all-season use: plowing in winter, hauling in winter or summer. The specifications of trucks commonly used in snow removal are listed in Table 17.1.

Wheeled bucket loaders (Fig. 17.15) are frequently used for clearing snow in small areas or intersections; they can also be used to remove ice or compacted snow by setting the bucket edge at a small angle with the pavement. In addition one-way, reversible and V-plows can be mounted on wheeled

Table 17.1
SPECIFICATIONS OF TRUCKS USED IN SNOW REMOVAL SERVICE.

Condition	Snowcover Depth cm	Density kg/m^3	Class (kg x 1000)	Truck Recommended Engine Power kw	Wheel Configuration
Light	1-15	50-100	10.5-12.5	97-101	4 x 2
Moderate	15-45	50-200	12.5-15.5	108-172	4 x 4, 6 x 4, 6 x 6
Heavy	45-90	100-300	15.5-28	120-172	6 x 4, 6 x 6

loaders. The short wheelbase wheeled loaders are highly maneuverable, but this capability may result in "porpoising" or bouncing of the plow on the pavement, leaving unplowed strips.

Fig. 17.15 Bucket of a front-end or wheeled loader scraping ice from pavement.

APPLICATIONS

Blade plows are most commonly used for snow removal because they are less expensive to operate. A rotary plow should be considered only for deep or hard snows.

In urban areas snow is generally windrowed before hauling to disposal sites. Snow can be loaded onto trucks by various equipment. Rotary plows are very efficient because they are fast and maneuverable. Front-end loaders equipped with either standard front-tip buckets or side dump scoops (Fig. 17.16) can be used, but they require more area to maneuver than the rotary plow. Conveyor snow loaders (Fig. 17.17) are specialized devices. All of this equipment except the rotary plow can be used for operations other than snow removal.

Since sidewalks are narrower and built on weaker subgrades than road pavements, heavy plows and vehicles cannot be used to clear snow from them. Small wheeled farm tractors with front-mounted displacement blades are sometimes used. However, continuous track-laying tractors (Fig. 17.18) having narrow widths are most popular because they are faster, more maneuverable, and less likely to damage the sidewalk or adjoining grounds. For areas with light snowfall V-plows are preferred but in areas of heavy snowfall a rotary plow may be necessary. The width of way cleared by a V-plow is less than that by a rotary plow if snow slides back into the cut behind the plow.

Fig. 17.16 Side-dump bucket mounted on wheeled loader used for truck loading.

Fig. 17.17 Truck being loaded with windrowed snow by a self-propelled loader.

Fig. 17.18 Small track-laying tractor with V-plow for sidewalk plowing. A hopper sand distributor is being towed.

EQUIPMENT EVALUATION

In selecting and evaluating different equipment for snow and ice removal several factors other than power requirements, efficiencies and costs must be considered, namely safety, speed of operation and maintenance.

A snowplow driver must be able to see the road and traffic clearly; likewise other vehicles must be able to see him. As trite as this may appear, failure of one vehicle operator to see the other leads to many accidents. Plowing speed is often limited, not by inadequate vehicular power, but by poor visibility conditions created by a cloud of snow whipping over the top of a blade plow (Fig. 17.19). Therefore it is recommended that the blade design should be evaluated for overspill at the speeds likely to be used during clearing operations. Snow removal equipment can be made more visible to other vehicles by mounting either flashing lamps (rotary or strobe) on the cab or light-bars across the cab or above the tailgate.

Snow removal equipment must be selected according to its most frequently intended use, e.g., high-speed plowing on rural routes or expressways, or low-speed, limited-cast plowing in built-up areas. City operation involves speeds from 8 to 30 km/h; expressway operation sometimes exceeds 80 km/h. The most useful criterion of plow performance is the removal rate per unit of input energy (kilograms per kilowatt-hour).

Fig. 17.19 A snow cloud created by a displacement plow can limit speed of clearance.

The maintenance policy for a road dictates the most appropriate type of cutting edge to use. If a thin layer of snow on a pavement is tolerable adjustable shoes or caster wheels provide clearance between the blade and pavement; a blade with a mild steel cutting edge has adequate wear and strength characteristics. If bare pavement is desired, the cutting edge must contact the pavement, and high rates of wear occur with either mild or hardened steel edges. When snow becomes compacted and tightly bonded to the pavement, steel blades with tungsten carbide inserts are effective. Their higher initial cost is offset by their longer life (5 to 6 times that of a hardened steel edge), the longer distances travelled (5000 to 10,000 km), and the savings in labour resulting from the longer intervals between blade changes. However, tungsten carbide is brittle, and is more likely to chip upon striking a solid obstruction. Spring-loaded trip mechanisms can reduce the amount of this damage.

In mild climates where the snow may have a high water content, rubber blades which have a squeegee action can be used effectively. Their life is comparatively long, particularly when the rubber edge is backed with a steel plate to within 5 to 6 cm of the face edge; over 4800 km can be plowed before the rubber blade must be changed.

LITERATURE CITED

Caird, J.C. 1964. *RCAF snow removal and ice control procedures: the development of equipment and techniques.* Snow Removal and Ice Control, Tech. Memo. No. 83, Assoc. Comm. Soil Snow Mech., Nat. Res. Counc. Can., Ottawa, Ont., pp. 93-96.

Croce, K. 1941. *Der heutige stand der Schneepflugtechnik (The present status of snow-plow technology).* In Winterdienst auf Strassen und Reichsaatobahnen. Forshungsarbeiten aus dem Strassenwesen, Band 31. Volk und Reich Verlag., Berlin, pp. 62-76.

Croce, K. 1950. *Messversuche an Schneeräummaschinen für Landstrassen, Entwurfs-grundlagen (Measurements relative to performance and efficiency of snow removal machines for highways)*. Fortschrittsberichte aus dem Strassen-und Teifbau. Band 4, (English Transl. by U.S. Army Engineers Waterways Expt. Station, Vicksburg, Miss. for U.S. Army Cold Reg. Res. Eng. Lab., Hanover, N.H. Transl. 51-5.]

Croce, K. 1970. *Principles of snow removal and snow-removal machines*. Snow Removal and Ice Control Research, Sp. Rep. 115, Highw. Res. Board, NAS-NRC, pp. 231-240.

Davis, C.W., Jr. 1952. *Snow removal equipment tests at Sierra test site during winter season of 1951-52*. Tech. Memo M-073, U.S. Naval Civil Eng. Lab., Port Hueneme, Calif.

Hawkins, L.M.E. 1964. *Runway snow removal and ice control methods at airports maintained by Department of Transport*. Snow Removal and Ice Control, Tech. Memo. No. 83, Assoc. Comm. Soil Snow Mech., Nat. Res. Counc. Can., Ottawa, Ont., pp. 101-105.

Kihlgren, B. 1961. *Undersökningar rörande snöplogar (Snow plow investigations)*. Rep. 38 Statens Väginstitut, Stockholm.

Kinker, E.C. 1952. *Snow removal*. Symp. on Snow Removal and Compaction Procedures for Airfields. Rep. GG-ES 200/1, Res. Develop. Board, Comm. Geophys. Geogr., U.S. Dept. of Defense, Washington, D.C., pp. 30-37.

Mellor, M. 1964. *Design of high-speed snow plows*. Unpubl. Note, U.S. Army Cold Reg. Res. Eng. Lab., Hanover, N.H.

Mellor, M. 1975. *Mechanics of cutting and boring. Part 1: Kinematics of transverse rotation machines*. Sp. Rep. 226, U.S. Army Cold Reg. Res. Eng. Lab., Hanover, N.H.

Vinnicombe, G.A. 1968. *Comparative tests on model vee snow ploughs*. Rept. LR 180, Road Res. Lab., Crowthorne, U.K.

18

APPLICATIONS OF REMOVAL AND CONTROL METHODS

Section 1: RAILWAYS

L.D. MINSK

Applied Research Branch, U. S. Army Cold Regions Research and Engineering Laboratory, Hanover, New Hampshire.

Section 2: HIGHWAYS

D.R. BROHM
S. COHEN

Maintenance Operations Office, Maintenance Branch, Ontario Ministry of Transportation and Communications, Downsview, Ontario.

Section 3: AIRPORTS

L.M.E. HAWKINS

Airport Facilities Branch, Transport Canada, Ottawa, Ontario.

RAILWAYS

HISTORICAL DEVELOPMENT

Precedence must be given to the railroads for developing mechanical methods of snow removal. Not until the advent of the motor car in the late 19th century was there a great need for clearing highways in North America. Before that time, compacting the snow with large rollers gave a road surface for carts and sledges far superior to the frozen ruts that formed the roads before the first heavy snowfall. Snowcover on railroad tracks, however, can either impede the movement of trains or lead to derailment if sufficient amounts build up around the tracks to lift the wheel flanges above the rails. In addition, snow and ice can interfere with switch operation, so that their accumulations must be removed or prevented to permit complete movement of the switch point or frog.

In low to moderate snowfall areas frequent passages of trains can keep tracks clear of snow by the scouring action of the wind produced on each passage. In these regions, at the most, a small pilot plow added to the front of the locomotive is adequate for snow removal. This equipment proved satisfactory during the early extension of rail lines in the United Kingdom, Europe and eastern North America. However, the westward expansion of railways in North America led across the Sierra Nevada and Rocky Mountains where heavy and recurring snowfalls, avalanches, and heavy wind-drifting frequently immobilized trains equipped with the lightweight equipment. This led to the development of wedge-type plows in 1865, and later to the design of large, rotary plows which could cut through deep snow.

J.W. Elliot, a dentist in Toronto, Ontario, designed a "Rotary Snow Shovel" which he patented in 1869, but he built only a small model. In the same year, C.W. Tierney of Altoona, Pennsylvania, was granted a patent for a two-stage rotary plow that consisted of a revolving screw to feed snow to a fan behind it; but he also only built a model (Best, 1966). The first successful full-sized rotary plow was designed by Orange Jull, a mechanically proficient owner of an Orangeville, Ontario, flour mill, and built by the Leslie Machine Shop of that city in 1884 (Winterrowd, 1920). This "Rotary Steam Snow Shovel", (Fig. 18.1) with a number of refinements, is still used, though the early units have been converted to diesel-electric drive. The "shovel" is basically a rotating set of knives which cuts the snow and feeds it to an impeller wheel located immediately behind the knives.

REQUIREMENTS AND CONTROL METHODS

Mainline

The linear arrangement of a mainline track facilitates handling snow at high speeds and thereby simplifies removal except in deeply-drifted sections. Terrain features play a major role in governing the locations of major accumulations. Small pilot plows can prevent small snowfalls from building up along level, unobstructed sections, whereas larger, rigid nose plows (snow dozers), wedge plows or flangers may be necessary to handle deep snow (Fisher and Hurley, 1964). Cuts through which the roadbed passes may accumulate large deposits of snow if not protected upwind by natural obstacles or snow fences. Off-track equipment, usually tracked bulldozers, are most frequently used to clear accumulations from cuts, pushing the spoil to the downwind side to reduce the redrifting potential (Anonymous, 1973).

Fig. 18.1 Leslie type rotary plow, designed by Orange Jull in 1884. Unit shown was built for the Southern Pacific; picture was taken in 1890 in the Sierra Nevada Mountains (Southern Pacific Company).

Yards and Terminals

Terminals contain many areas that accumulate snow that must be cleared: platforms, mail and cargo areas, walkways, industrial areas, roads, and switching leads. The congestion in most yards makes snow clearance difficult, while the lack of storage space for snow removed from tracks and other facilities frequently makes hauling to disposal areas necessary (Anonymous, 1964). Experiments have been conducted in Japan on the problem of reducing hauling by dumping the cleared snow in flowing streams or by sprinkling water on the road beds (Kikuchi, 1975). The Japanese National Railway has also developed a snow loader which uses a rotary plow to feed a longitudinal chute extending over three or four open-top gondola cars pulled behind the plow. The hauling cars are fitted with four, 9-m^3 capacity air-bag snow dumping containers; air blown into the containers lifts the snow and dumps it automatically (Ishibashi and Takahashi, 1971).

Small walk-behind snowblowers, rotary power brooms, and manual shovelling are common methods of clearing station platforms. Melting of thin accumulations of snow and ice by chemicals is also an effective method, sodium chloride being the most commonly used chemical because it is inexpensive.

Switches

Remote operation of switches, particularly those automatically controlled on centralized traffic control (CTC) sections, requires that they should be snowfree. Conventional moving-point switches cannot function when snow falls between the moving and stationary parts and becomes compacted, usually by switch operation. The point rails swing transversely on slide plates and are machined to fit closely to the head and base of the stock rails. As little as 0.6 cm of snow, when sufficiently compacted between the fixed and moving rails, may produce a gap that could cause derailment (Ringer, 1979). Cleaning snow and ice from switches is one of the costliest winter maintenance programs for a railroad. The early method of track workers cleaning switches with brooms and shovels is still practiced, but the large number of switches in yards and the need for the remote control of switches far removed from normal work areas have led to the development of mechanical or thermal switch-clearing devices.

Early methods of snow and ice control of switches used some type of heating equipment. Many variants of switch heaters have been developed which work quite satisfactorily. Electric resistance heaters placed against the rails along the switch are preferred by some railroads because of their reliability and adaptability to automatic operation. Fuel-fired heaters which use propane, fuel oil, kerosene, or natural or manufactured gas, placed so as to direct open flames on the rails are common. Manual ignition of fuel-fired heaters is feasible in yards, but automatic igniters are required at remote or difficult-to-reach locations. The kerosene pot burner was one of the earliest types of switch heater. Though largely replaced now by other control measures, because it is quickly installed and put into operation, it is still used today at outlying locations not normally protected by a permanent heater. Fire damage to ties is not unusual with pot burners but can be lessened by placing asbestos sheets on the sides of ties. Portable or backpack weed burners as well as weed-burner rail cars have been used to melt the snow with an open flame. Another control method involves the use of steam coils permanently installed between the ties in and around an interlocking plant.

This method has not gained wide acceptance because the large quantities of steam required render it uneconomical.

Other attempts have made use of heating pads installed in cribs at switches. Fluid at 200° C circulated through the pad provides a heat source to melt snow or ice. In certain installations the use of an infrared heater mounted over the switches has proven to be adequate. However, the use of these heaters is limited to locations having a power supply.

Portable or permanently-installed compressed air jets have also been tested as control devices. The portable type generally consists of a number of compressed air lances or small jets which are located around a yard, and are powered by a centrally-located compressor. Jet engines and gas or diesel-powered fans have been used to blow snow from switches.

A number of switch heaters have been designed and tested by the National Research Council of Canada (Ringer, 1979). Coldroom tests of a full-scale, forced convection system indicated that a thermal requirement of 73 kW was needed to maintain a 6.6-m long switch snow free under test conditions of 2.5 cm/h snowfall, a windspeed of 32 km/h and an ambient temperature of -18°C. One system used a cyclone burner combustion chamber to provide hot air which was forced by an electric fan through a conical re-entrant outlet to a crossduct and distribution outlet (Fig. 18.2). The power for driving the fan (1-2 kW) requires that a generator be supplied to operate the system at a remote

Fig. 18.2 Horizontal air curtain type of switch protector. High velocity air from a blower in the shelter at left is ducted along the rails.

location thereby rendering it uneconomical. A pulse-jet combustion heater was adapted to the switch duct system to eliminate the need for electric power.

A non-thermal switch protection device which uses a horizontal air curtain to prevent snow from falling on the switch was also developed by the National Research Council. This system requires only a 5 kW electric motor to drive the centrifugal fan so that the power requirements are far less than the 73 kW (thermal) and 1-2 kW (electric) required by the thermal design. The air curtain device, however, cannot remove accumulated snow and therefore must be put into operation before or at the start of a snowfall. In remote locations this requires using sensors to detect snowfall and automatically activate the system. Thermal switch heaters, in contrast, can clear a switch that has been covered by snow.

Certain problems and precautions are inherent in the use of thermal systems to control snow and ice around switches, including:

(1)   Drainage for the melt water away from the switch must be provided so that it will not accumulate and refreeze to create a situation which could cause derailment.

(2)   Hot air and open-flame heaters may result in the burning-out of ties that support the switch.

(3)   Sustained operation of thermal systems may induce thawing of the ballast supporting the switch. This may lead to differential settlement between the heated and unheated sections of the track.

Grade Crossings

Flangeways must be kept free of compacted snow at highway grade crossings. Manual clearing with pick, shovel, and broom is still practiced, but mobile power brooms, compressed air jets, and the application of chemicals have replaced much of the hand work. A protection method recently developed involves placing elastomeric material alongside the rails and at rail height; these deflect under the flanges and rebound to prevent snow buildup (Evenmo, 1974).

Tunnels and Snow Sheds

Tunnels are not generally a winter maintenance problem. If icicles form on the roof they are usually knocked down manually from a work train. Drainage must be maintained within the tunnel to prevent accumulation and freezing of seepage. Snow sheds are artificial tunnels constructed in mountainous areas where snowslides are common and alternative means of preventing the snow from reaching the track are unavailable or more costly than sheds. They are usually open on the downslope side to reduce construction costs, to allow snow clearance and other maintenance and to enable passengers to view the scenery.   In early times they were constructed of wood; today they are frequently made of reinforced concrete.

MECHANIZED EQUIPMENT

On-Track

Small accumulations of snow can be removed by pilot plows mounted on the front of locomotives. These remove snow only to track height, making it necessary to use flangers to remove the snow between the rails. A flanger is a steel blade that is dropped below rail level between the rails. A flanger car clears snow from between the rails and for a short distance away from the outer sides (Fig. 18.3).   An operator controls the position of the flangers, raising them when approaching switches and crossovers.

Deep snow accumulations, which cannot be removed by flangers, are removed using either a wedge ("bucker") plow or a rotary plow. Because of their lower cost of operation wedge plows are used whenever it is possible to

Fig. 18.3 A flanger is a steel blade which clears snow between the rails below rail level to pemit unimpeded movement of wheel flanges. Here the blade is mounted underneath a flanger car.

operate them at speeds of 50-65 km/h. At these speeds snow is cast a sufficient distance outside the track area so that it will not spill back into the cleared path. Usually two diesel locomotives, instead of a single unit, are used to power the plow because of the high frequency with which a single locomotive becomes wet and disabled. A cardinal principle followed is that the clearance operation should be kept going continuously to keep the line from being plugged. A stalled train creates its own drift which becomes a major factor in preventing restoration of the line to service (Jordan, personal communication).

A wedge plow is designed to cut into the snow in a plane making a relatively flat angle to the horizontal, and to lift and deflect it to the side. Single-track plows are symmetrical and cast the snow in both left and right directions (Fig. 18.4). A double-track plow corresponds to a one-way highway plow and casts the snow to the right only (in North America) to avoid displacing the snow onto the adjacent track (Fig. 18.5). Wings mounted on the side of the plow car widen the cuts. A type of wedge plow fitted with wings is the Russell plow, named after its manufacturer (Parkes, 1961).

Rotary plows are used to remove very deep snow or snow too hard to allow a wedge plow to operate at its optimum speed. These plows are now powered by electric motors and require two diesel locomotives for operation, one to

Fig. 18.4 Single-track wedge plow casts snow on both sides. Low horizontal angle of blade front exerts downward force to reduce possibility of derailing.

Fig. 18.5 Double-track wedge plow for casting snow in one direction only to avoid covering adjacent track. The wing on the side on the plow car is partially extended as it would be for cut widening.

provide electric power for the plow, the other for propulsion. Clearance of deep drifts is slow, the plow generally moving at about 8 km/h. The snow is cast well beyond the cut, but a straight-walled trench the width of the plow is formed; if left in this form the trench would rapidly trap and accumulate snow under blowing conditions. Therefore, it is necessary to cut back the high walls, a job frequently accomplished with a cut widener teamed with a wedge plow or a spreader or spreader-ditcher. The cut widener consists of wing blades shaped as a "Vee" with the wide end in the front. It is pulled behind a locomotive and shaves the sides of the cut, depositing the snow in the center of the track where it is picked up by the ditcher or wedge plow and cast or shoved aside. Since the cut widener moves at 8-16 km/h, and the plow at 50-65 km/h, careful scheduling is required to prevent the plow from overtaking the cut widener. Whenever possible, the cut widener is operated in only one direction, cutting only what can be handled in one pass. When repeated passes are necessary to cut back a high drift, extreme care is needed to ascertain the locations of the two units at all times (Jordan, personal communication).

The spreader was originally designed for grading and ditching the ballast and the right-of-way. It is a highly versatile piece of equipment, however, and is used frequently during winter. Its side blades, analagous to the wings of highway plows, can be extended to clear a width up to 13.7 m. These are frequently referred to as "Jordan spreaders" or "McCann spreaders" after their manufacturers.

Off-Track

Tracked bulldozers, wheeled bucket or front-end loaders, motor graders, clamshell buckets on cranes, and hydraulic boom clamshells ("Gradalls") are used to clear snow from the right-of-way. Their main asset is their economy as general purpose equipment, since they can be used for maintenance tasks throughout the entire year. Great care must be exercised when operating off-track equipment close to the track to avoid turning over a rail or damaging ties. Some railroads use off-track equipment only as a last resort because of the damage they may cause and the high cost of operating them with extra care to avoid damage; instead they rely on the use of on-track equipment. Tracked bulldozers are commonly used in snowslide or avalanche areas because large rocks and trees are generally entrained in the snow.

Deposits of blowing snow can reach depths, particularly in cuts, such that they can only be removed by specialized equipment (heavy, track-mounted plows or blowers). At these locations preventative measures are employed: (a) using motorized tractors and scrapers to widen the snow cuts and flatten the backslopes, (b) removing trees and brush that cause the formation of drifts within the right-of-way, (c) plowing a number of windrows of snow upwind of the right-of-way; these may be cleared periodically to maintain their trapping effectiveness and (d) placing snow fences upwind of cuts. Snow fences are generally erected only at locations where control can be proven to be an economical alternative to snow removal. When located within the right-of-way, usually a 2.4-m high fence is placed 60 m upwind of the top of the cut. In other cases, where the fence must be located outside the right-of-way, either an easement to erect the fence from the property owner may be required or title to an extra width of right-of-way must be obtained. Where seasonal fences are erected, the standard 1.2-m wood-slat fence is used.

HIGHWAYS

HISTORICAL CHALLENGES

Snow clearing is a necessity to permit efficient and economical mass transportation of people and goods in winter. Widespread introduction of motor cars and trucks in the early 1900's was followed closely by an increasing demand for highway snow clearing to permit full utilization of roads by automobiles through the winter months.

Harrison (1920) reported that the first recorded efforts to maintain passable winter highways in North America occurred in 1918 and 1919. At first, winter maintenance consisted of plowing snow from the road surface and treating icy locations with abrasives, usually sand or cinders, to reduce skidding. The principal disadvantages of this procedure were:

(1) The high cost of the abrasive treatment,
(2) Sand and cinders, the principal abrasives, were whipped off the road surface by traffic, and
(3) The abrasive treatment must be repeated daily on road sections with high traffic volumes.

Piles of stored abrasives wetted by fall rains can freeze to a considerable depth during the winter. Generally, either calcium chloride or sodium chloride is mixed with the abrasive to prevent freezing and also to "cap" the pile. Experience gained with chemically-treated abrasives has indicated that they assist snow and ice removal (National Research Council, 1967); untreated abrasives increase only the skid-proofing effect of the surface. An increase in the chemical content of the mixture increases the rate of snow and ice removal roughly in proportion to the amount of chemical added (Brohm and Edwards, 1960).

The use of sodium chloride, without an abrasive, has proven to be an effective method for snow and ice removal (see Ch. 14). It has been found that this chemical produces extremely rapid melting of snow and ice. The costs of treatments with the chemical alone are generally less than those with chemically-treated abrasives. In addition, since the chemical is not whipped off the road, repeated daily treatments are not required during periods of clear cold calm weather.

Snowplowing activities have varied depending upon terrain and climatic conditions. In windswept terrain, crawler tractors, heavy motor graders and all-wheel drive heavy duty trucks equipped with plows came into common use by 1930. These methods of snow removal are still popular. The plow used with these units normally has a V-configuration, so that snow is directed to both sides of the road (see Fig. 18.6). Where the roadway is protected and

Fig. 18.6 All wheel drive truck with V-plow. (Photo credit: - Ministry of Transportation and Communications, Ontario).

Fig. 18.7 Truck with one way plow. (Photo credit: - Ministry of Transportation and Communications, Ontario).

drifting of snow is not a major problem, the prime snow-plowing vehicle is a truck of the common cartage type, normally in the size range from 10,000 to 18,000 kg gross vehicle mass[1], equipped with a plow which directs snow to one side (see Fig. 18.7).

---

[1]   In this section vehicle size is expressed in units of mass with the kilogram as the basic unit.

Snow clearing patterns vary by region and are determined by the amount and frequency of blowing snow, and to a lesser extent, by the ambient air temperature.

Where temperatures are favourable, i.e. -18°C or above, and drifting snow is not a problem, an integrated program of abrasive treatment, chemical treatment and snowplowing can provide bare pavements during most of the winter. Drivers have become accustomed to bare pavements in winter and expect that they will be provided. Snow plows characteristically operate throughout a storm with bare pavement as the clearance goal; chemicals are applied judiciously so that traffic displaces the slushy water-snow mixture over the road.

Over flat and slightly undulating terrain drifting snow may control the snow removal operation. In these regions, if the snowfall is accompanied by high winds so that significant drifting occurs, many rural and urban roads become impassable and remain closed until equipment can reopen them.

Snow clearing (and roadway ice control) practices also vary widely according to geographical location. Mechanical clearing (snow plowing) is the primary method used in North America. Chemical treatment is the most effective in the British Isles, where the main problems arise from very light snowfalls or from thin layers of ice which form after water condenses on the roads and freezes.

The safety of the winter road user must be given first priority in a winter maintenance program. The road must be maintained in such a condition that it can be used safely; otherwise it must be closed to traffic until it becomes safe. For highways experiencing high volumes of high speed vehicles, safe travel requires that the pavement be maintained in a bare condition. Snow-packed, abrasive-treated surfaces may be utilized by low volumes of traffic which should be restricted to moderate speeds.

In Canada, the removal of snow and the treatment of ice-covered surfaces are neither common law requirements nor statutory requirements. Nevertheless a program of well-planned and effective snow removal and ice control is imperative. The service to be provided must, from a legal standpoint, be reasonable in consideration of the character and volume of traffic which will use the facility; the program must also have a consistent level of service from year to year.

TRACTION CONSIDERATIONS

Motorists have used many methods to obtain improved traction and rolling resistance on a snow or ice surface. Those which have found general acceptance include tires with special tread patterns for snow, tire studs, steel tire chains and the application of chemical sprays. Development of new methods to improve traction has been complicated by the requirement to provide smooth quiet operation of the automobile at highway speeds on bare pavement.

Snow Tires and Studded Tires

Tires with special tread patterns have been designed to provide improved traction. It is estimated that during the winter months, 60 percent of the passenger cars in Canada are equipped with snow tires on the driving axle. The treads on these tires have an open design with prominent lugs on the outside edges which greatly improve traction in loosely-packed snow when the tire penetrates deep enough to grip. Because of the open tread pattern, snow tires generate noise on bare pavement when the vehicle is operated at high speeds. Since tires with heavy lugs tend to run hotter at high speeds than smooth tires they are usually removed from the automobile during the summer.

Tests conducted in Europe (National Rubber Producers Research Association, 1967) have indicated that using natural rubber instead of synthetic rubber for the tread produces a tire with better skid resistance at low temperatures. The use of an oil-extended natural rubber tire fabricated by mixing natural rubber, carbon black and oil, is claimed to extend the temperature range over which this improvement applies. Tests in Canada (Damas and Smith, Ltd., 1971) however, have not shown that there is any consistent decrease in stopping distance on clear ice or sanded ice nor is there any improvement in the starting traction of tires that have natural rubber in their treads.

Tungsten studs protruding from the tire tread were first used in the Scandinavian countries during the late 1950's before coming into general use in Canada during the mid 1960's. The percentage of passenger cars equipped with studded tires increased in the Province of Ontario from 2% in the winter of 1966-67 to 32% in 1969-70, when the province announced that studded tires would be banned after the 1970-71 winter because of the increased highway wear they produced. By 1970, 50% of the automobiles in some areas of Quebec and the Maritime provinces were equipped with studded tires. Tests have shown that studded tires do result in shorter stopping distances and improved starting traction on glare ice and hard-packed snow (Smith and Clough, 1972). They are most effective at surface temperatures just below the freezing point and less so at lower temperatures (see Figs. 18.8 and 18.9). Although studded tires on all four wheels lead to much shorter stopping distances than can be obtained with studded tires on only two wheels of one axle, most car owners install these tires only on the drive axle.

The increase in traction with studded tires is proportional to the number of studs and to the number of circumferential rows of studs in the tread surface.

The studs in the tire tend to lose their effectiveness after approximately 16,000 km. Smith and Schonfeld (1969), found over the 10-year period, 1960-1969, that studded tires on heavy passenger vehicles resulted in increased wear in the wheel tracks of highways, particularly those with large volumes of traffic. This damage was particularly noticeable in the passing lanes, at traffic lights and on expressway ramps.

The nature of the highway wear depends on the hardness of the aggregate in the pavement. If the pavement contains soft aggregate the wear tends to be uniform over the life of the pavement. If either concrete or bituminous pavements contain hard aggregate the studs tend to wear away the softer matrix and to dislodge the coarser aggregate. The wear resistance of a pavement to studded tires can be increased by increasing the stone content and hardness of its aggregate. Needless-to-say, the cost of replacing or repairing a pavement that has been damaged by studded tires is extremely high. Also, highway damage caused by studded tires causes unsafe driving conditions when the roads are wet or icy. In summer, the deep ruts worn in the wheel tracks trap water and promote hydroplaning; in winter they contain icy patches and promote skidding.

Extensive use of studs on bare pavements may cause rapid wear of the centre lane lines, particularly on winding roads where automobiles frequently cross the centre line. This presents a traffic hazard when the lines become unrecognizable and is particularly serious during the winter, since the lines cannot be readily repainted. Because of the extensive roadway damage that these tires may cause they have been banned in many areas of Canada and the northern United States.

Tire Chains and Other Removeable Devices

Tire chains, which can be secured to the tire, have been used for many years to improve the starting and stopping traction on road surfaces covered with snow or ice. Chains usually enable shorter stopping distances to be obtained on ice and packed snow than can be obtained with snow tires; they also provide the best starting traction on packed snow.

Reinforced steel tire chains are very noisy and therefore must be removed when the vehicle is operated on bare pavement. Several types of rubber tire chains have been marketed; these can be attached more easily to the tires than steel chains and provide additional traction because of the abrasive surfaces mounted on the cross bars. Some designs appear to provide good performance, although some problems have been encountered because of their low durability (Ontario Department of Transportation and Communications, 1971).

Other Methods

With the banning of studded tires, a number of alternative methods have been developed to improve traction on snow- and ice-covered roads. These include using tires equipped with studs that can be retracted into the tire tread by air pressure, a chemical spray that may be applied to the tire to give increased starting traction, and different types of attachments that clamp to the tire and can be retracted manually.

Radial Ply Tires

Today, radial ply tires are used extensively throughout most of the world including North America. These tires have superior traction properties, compared with bias ply tires, on most icy and snowcovered roads because they have a higher contact patch (Browne, 1973) and the ground pressure distribution across the tread is relatively uniform. However, on packed snowcover bias ply tires may perform better because the higher pressures under the shoulder enable these tires to penetrate and interlock with the snowcover.

Radial ply tires can be built with larger grooves than bias ply tires without sacrificing tread life. Tires with large grooves give better traction in heavy, deep snow. Also, the radial tire construction results in a tire with low tread squirm and therefore superior traction; squirm causes snow ridges to be sheared off laterally thus reducing traction.

Summary

Snow tires provide a definite improvement in traction under some conditions. However, even with their use high traffic volumes will eventually pack the snow leaving a slippery surface which greatly reduces the traffic capacity of the highway. The only realistic solution to improving traction and reducing the driving hazard created by snow on roadways is to clear most or all of it from the road surface. Figures 18.8 and 18.9 summarize the stopping distances for different tire combinations on clear ice at two temperatures. Note that in the figures the stopping distances of all snow tires, except tires that are studded or fitted with chains, are significantly longer on clear ice at -1°C than on clear ice at -18°C.

TIRE COMBINATION

TIRE COMBINATION		
HIGHWAY TIRES, FOUR WHEELS		
SYNTHETIC SNOW TIRES, REAR ONLY		
SYNTHETIC SNOW TIRES, FOUR WHEELS		
STUDDED SYNTHETIC SNOW TIRES, REAR ONLY		MEAN STOPPING DISTANCE
STUDDED SYNTHETIC SNOW TIRES, FOUR WHEELS		
NATURAL RUBBER SNOW TIRES, REAR ONLY	95% CONFIDENCE LIMIT OF MEAN	
NATURAL RUBBER SNOW TIRES, FOUR WHEELS		
STUDDED NATURAL RUBBER SNOW TIRES, REAR ONLY		
STUDDED NATURAL RUBBER SNOW TIRES, FOUR WHEELS		
CONTROLLED PROTRUSION STUDDED SNOW TIRES, REAR ONLY		
CONTROLLED PROTRUSION STUDDED SNOW TIRES, FOUR WHEELS		
ELASTOMERIC TIRE a/ ATTACHMENT, REAR ONLY		
REINFORCED STEEL TIRE CHAINS, REAR ONLY		

a/ Rubber Tire Chains

75   80   90   100   110   120

Average Stopping Distance (m)

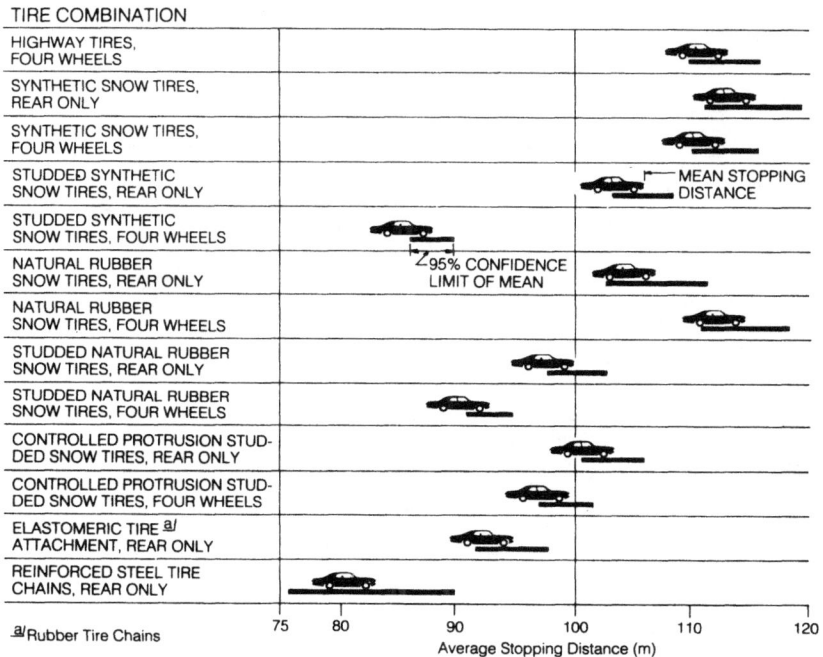

Fig. 18.8 Stopping distances on clear ice at -1°C for different tire combinations. (Ontario Department of Transportation and Communications, 1971).

MANAGEMENT OF SNOW REMOVAL AND ICE CONTROL[1]

To the motorist, highway snow removal and ice control are the most important aspects of winter highway maintenance. As an example of this concern in Ontario, more than half of the annual road maintenance resources of the Ministry of Transportation and Communications are allocated to snow removal and ice control. The primary criteria governing the assignment of this level of resources are the safety of the drivers and passengers of vehicles and their demand for an uninterrupted flow of traffic.

---

[1] This portion of the text draws heavily on management practices followed in the Province of Ontario. The highway snow removal and ice control program of this Province was taken as a model to illustrate the standards, procedures and other factors to be considered in the design and discharge of an effective, efficient winter maintenance program. In most cases the material presented has been simplified and reduced in content from material given in "Standards, Specifications and Instructions" established by the Province. For additional information on this subject the reader is referred to the appropriate references.

TIRE COMBINATION

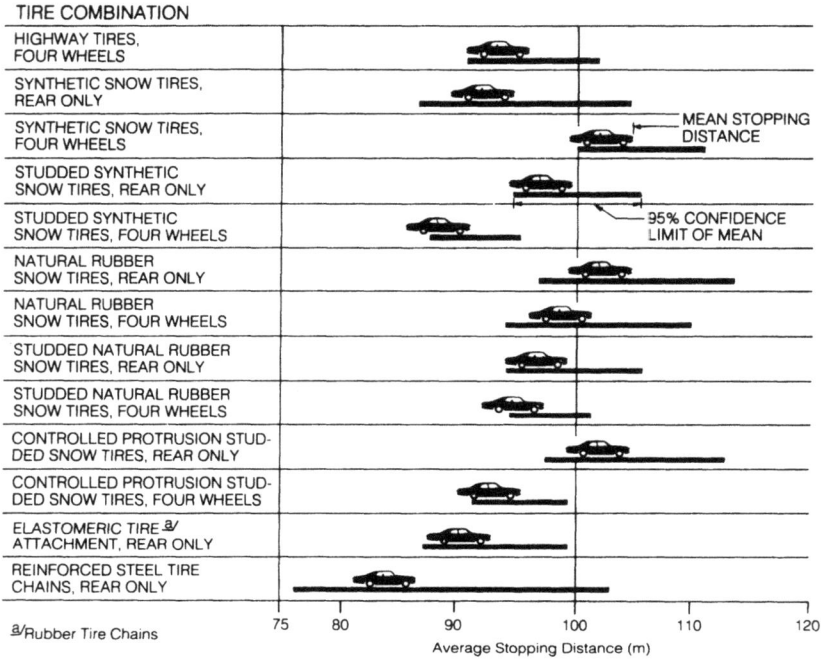

HIGHWAY TIRES, FOUR WHEELS	
SYNTHETIC SNOW TIRES, REAR ONLY	
SYNTHETIC SNOW TIRES, FOUR WHEELS	MEAN STOPPING DISTANCE
STUDDED SYNTHETIC SNOW TIRES, REAR ONLY	
STUDDED SYNTHETIC SNOW TIRES, FOUR WHEELS	95% CONFIDENCE LIMIT OF MEAN
NATURAL RUBBER SNOW TIRES, REAR ONLY	
NATURAL RUBBER SNOW TIRES, FOUR WHEELS	
STUDDED NATURAL RUBBER SNOW TIRES, REAR ONLY	
STUDDED NATURAL RUBBER SNOW TIRES, FOUR WHEELS	
CONTROLLED PROTRUSION STUD-DED SNOW TIRES, REAR ONLY	
CONTROLLED PROTRUSION STUD-DED SNOW TIRES, FOUR WHEELS	
ELASTOMERIC TIRE a/ ATTACHMENT, REAR ONLY	
REINFORCED STEEL TIRE CHAINS, REAR ONLY	

a/Rubber Tire Chains   75   80   90   100   110   120

Average Stopping Distance (m)

Fig. 18.9 Stopping distances on clear ice at -18°C for different tire combinations. (Ontario Department of Transportation and Communications, 1971).

Considerable effort is required by any agency involved with the maintenance of highways to determine and implement control procedures that are consistent with environmental protection requirements. In Ontario, such procedures, specifying levels of service applicable in specific circumstances, have been established and issued to Ministry personnel in the form of *Quality Standards and Operating Instructions* (Brohm and Cohen, 1973, Appendix A). The standards were developed by a panel of experienced engineers and field supervisors from District and Head Office units of the Ministry. From their deliberations a document resulted which defines maintenance procedures relevant to winter operations for all highways in Ontario. Based on available resources, current acceptable practice, and traffic needs, each segment of the highway network is assigned to one of four basic service levels, each having a maximum allowable accumulation of snowfall. The levels of service are:

Bare pavement: *Class I* (Fig. 18.10a). This service applies to all hard-surfaced highways having an average daily traffic level of 1500 vehicles or more in southern Ontario. In northern Ontario this number is reduced to 1200 vehicles. For this service snowplowing is continuous and snow accumulation on the road is not to exceed 2.5 cm. Salt and sand are used where necessary.

Bare Pavement: *Class II.* This service is applied to all hard-surfaced highways having an average daily traffic level in the range from 500 to 1500 vehicles in southern Ontario (400 to 1200 vehicles in northern Ontario). For this service snow plowing is continuous such that the accumulation of snow on the road does not exceed 4 cm. Salt and sand are used where necessary.

Centre Bare Pavement: (Fig. 18.10b). This service is applied to all hard-surfaced highways having an average daily traffic level in the range 250 to 500 vehicles in southern Ontario (200 to 400 vehicles in northern Ontario). For this service an attempt is made to remove all snow and ice from the centre 2 to 3 m of pavement within 24 h of the end of any storm. During a storm the accumulation of snow is not to exceed 5 cm. When favourable weather prevails the pavement should be bare for its full width. Sand and salt are used as necessary.

Snowpacked: (Fig. 18.10c). This level is applied to all gravel roads and all highways having an average daily traffic of less than 250 vehicles in Southern Ontario (200 vehicles or less in northern Ontario). This level requires that the road be maintained in a snowpacked condition and that plowing be conducted so that the accumulation of new snow does not exceed 6.5 cm at any time. Sand is used as required, salt is never used.

Performance Return from Highway Service Level

The test of any Service Level Classification system is its usefulness in everyday application. Experience has proven that the level of service can be related to such factors as the type of road, the winter average daily traffic and the snowfall rate and depth of accumulation. Table 18.1 shows a winter service level classification for different types of roads and winter average daily traffic based on:

(1) The maximum snowfall rate serviced,
(2) The maximum depth of snow allowed to accumulate on a pavement, and
(3) The percentage of storms for which no more than the maximum depth of snow (item 2) is allowed to accumulate.

Using this information, the number of snowplows needed to service the road can be adequately calculated from the expression:

Fig. 18.10 Levels of service for snow removal - Ontario highways. (a) Bare pavement. (b) Centre bare. (c) Snowpacked. (Photo credit: - Ministry of Transportation and Communications, Ontario).

$$\frac{\text{Number of Plow}}{\text{Vehicles}} = \frac{\text{Single Lane (km)} \times \text{Snowfall rate (cm/h) for Percent of Storms Serviced}}{\text{Plowing Speed (km/h)} \times \text{Maximum Allowable Snow Accumulation (cm)}}$$

In Table 18.1 the number of plows per two-lane kilometre are listed for different service levels, roads and daily traffic conditions. As an example of

Table 18.1

CLASSIFICATION OF LEVELS OF WINTER MAINTENANCE (Brohm and Cohen, 1973).

Highway Level of Service Code	Type of Road and Winter Average Daily Traffic (WADT)	Snow Plowing Activity			
		Level of Service			
		Snowfall Rate cm/h	Maximum[a] Allowable Snow Accum. cm	Snow[b] Storms Serviced %	Calculated Snowplow Req/2-lane kilometre
Urban Expressway	Multi-lane divided over 50,000 WADT	1.9	2.5	98.0	0.047
1 A₄	Multi-lane divided over 10,000 WADT	1.9	2.5	98.0	0.047
1 A	Undivided	1.9	2.5	98.0	0.047
1 B₄	Multi-lane divided	1.9	2.5	98.0	0.047
1 B	1200-9999 WADT.N[c] 1500-9999 WADT.S[d]	1.4	2.5	95.0	0.036
2	400-1199 WADT.N 500-1499 WADT.S	1.4	3.8	95.0	0.024
3	200-399 WADT.N 250-499 WADT.S	1.0	5.1	90.0	0.013
4	under 200.N under 250.S	0.5	5.1	75.0	0.0069

[a] Maximum allowable snow accumulation (MAA) is the maximum depth of snow (cm) permitted to accumulate on the road.

[b] Snow storms serviced are the storms in which no more than the maximum allowable snow accumulation is permitted on the road.

[c] N = Northern Ontario.

[d] S = Southern Ontario.

the use of these data, consider the case of 60 km of 4-lane divided highway, class 1_A; the equivalent to 120 km of 2-lane undivided pavement of the same class. From the Table it can be seen that 98 percent of all snow storms are serviced to a maximum accumulation of 2.5 cm of snow on the pavement, at a snowfall rate of 1.9 cm/h or less using 0.047 snowplows/2 lane-km. The number of snowplows required for the highway is therefore the product 120 x 0.047 = 5.64 or 6 plows. A similar calculation provides an estimate of the number of sander units required.

Daily traffic parameters, such as those listed in Table 18.1, employed for southern Ontario highways differ from those applied to highways in northern Ontario. Practical observation indicates that winter trips in these two areas differ considerably because in the north; (a) the typical trip is longer, (b) the average temperature is considerably lower, and (c) the vehicle speeds tend to be higher. In addition to these factors the sparser settlements along northern highways make the results of an accident more severe in the north than in the south. For these reasons, northern highways that would ordinarily be assigned to Class 2 and Class 3 level of service are often given preferential maintenance treatment.

Snow Removal and Ice Control Guide

A review of field work practices in Ontario revealed that field managers use various methods to clear snow, and to remove or control ice under specific road type and traffic conditions. For example, with freezing rain some managers believed that sand should be applied in sufficient quantities to provide traction over the ice; others preferred to apply enough chemicals to permit traffic to remove the ice. In order to provide similar levels of service to all roads having similar traffic patterns and physical conditions Operating Instruction M-700-C, *Winter Maintenance Operations During Winter Season* (Brohm and Cohen, 1973, Appendix B) was prepared. This document provides a comprehensive list of instructions and directions for operating and maintaining sanding, salting and plowing equipment and for inspecting roads during and after storms. Inspection includes checking road surfaces, railway crossings, mail boxes and electrical installations and the drainage of melt water from the sides of the roads. The operating instructions also contain procedures for reporting type of precipitation, visibility, condition of the pavement, wind direction and speed, and the temperature trend (rising or falling). Essentially the documents provide all supervisors and field managers with a set of recommendations for treating each storm condition ensuring a uniform operational approach to highway snow removal and ice control.

Table 18.2 illustrates the information used to guide plowing, salting and sanding operations to maintain a bare pavement level of service. The major activities are categorized according to:

(1) Temperature,
(2) Type of precipitation, i.e. dry snow, rain, etc.,
(3) Road condition, snow packing on pavement, etc.,
(4) Temperature trend (rising or falling), and
(5) Storm stage with different treatments listed for the beginning-of-storm, during-the-storm, and after-storm periods.

Practical experience with the programmed multiple-choice wall summary chart (Table 18.2) has indicated that it is a useful guide for selecting treatment.

Pre-Season Preparation

As well as establishing the levels to which all highways should be maintained it is necessary to ensure that the staff, materials and equipment are adequate to render the desired level of service.

In Ontario, a Standards Panel was assigned the task of preparing instructions for pre-season preparations. These were summarized in the document, Operation Instructions M-700-C, *Pre-season Preparations - Winter Maintenance Operations* (Brohm and Cohen, 1973, Appendix C) which was issued to all field managers. It details the procedures concerning men, equipment and materials to be completed before the onset of winter weather. For example, the standards require that:

(1) The names, addresses and telephone numbers of all employees qualified to perform winter duties be compiled and posted in each work area.
(2) All staff have a thorough understanding of their duties and responsibilities regarding shift operations, call-back regulations, and plowing speeds and methods, and be familiar with the plowing, sanding and salting routes in their assigned areas.
(3) Equipment be allocated, inspected and overhauled. Stocks of replacement parts and small tools at patrol headquarters must be checked and replacements ordered. Communications equipment should also be inspected and all employees instructed in their use.
(4) Sand and salt be stockpiled in sufficient quantities, and snow fences installed to control abnormal drifting.

Thus, by the time the first snow falls, all staff in supervisory positions would be ready to manage the maintenance operations.

Table 18.2

BARE PAVEMENT LEVEL OF SERVICE (Brohm and Cohen, 1973)

Temperature Range °C	Type of Precipitation	Road Condition	Temp.	Activity	Beginning of Storm	RECOMMENDED TREATMENT[a]	
						During Storm[b]	After Storm[b]
1. Below -18°	Dry Snow	No packing Dry Pavement	Rising	Plowing	After 1 cm snow accumulates	Continuously to bare pavement	Wing back shoulders/clean-up
				Sanding	Follow after plowing	As necessary after plowing	Icy spots only
				Salting	No	No	Icy spots only
			Falling	Plowing	After 1 cm snow accumulates	Continuously to bare pavement	Wing back shoulders/clean-up
				Sanding	Follow after plowing	As necessary after plowing	Icy spots only
				Salting	No	No	No
2. -18° to -12°	Dry Snow	No packing Dry pavement	Rising	Plowing	After 1 cm snow accumulates	Continuously to bare pavement	Wing back shoulders/clean-up
				Sanding	Follow after plowing	No	Icy spots only
				Salting	No	No	Icy spots only
			Falling	Plowing	After 1 cm snow accumulates	Continuously to bare pavement	Wing back shoulders/clean-up
				Sanding	Follow after plowing	No	Icy spots only
				Salting	No	No	No
3. -18° to -12°	Dry snow	Packing	Rising	Plowing	½ hour after salting	Continuously to bare pavement	Wing back shoulders/clean-up
				Sanding	No	Follow after plowing	Icy spots only
				Salting	Before 0.5 cm snow accumulates	As necessary after plowing	Icy spots only
			Falling	Plowing	After 1 cm snow accumulates	Continuously to bare pavement	Wing back shoulders/clean-up
				Sanding	Follow after plowing	Follow after plowing	Icy spots only
				Salting	No	No	No
4. -17° to -7°	Dry Snow	No packing Dry Pavement	Rising	Plowing	After 1 cm snow accumulates	Continuously to bare pavement	Wing back shoulders/clean-up
				Sanding	No	No	Icy spots only
				Salting	No	No	Icy spots only
			Falling	Plowing	After 1 cm snow accumulates	Continuously to bare pavement	Wing back shoulders/clean-up
				Sanding	Follow after plowing	Icy spots only	Icy spots only
				Salting	No	No	No

No.	Temperature	Condition	Trend	Operation			
5.	-12° to -7°	Dry Snow	Packing				
			Rising	Plowing	½ hour after salting	Continuously to bare pavement	Wing back shoulders/clean-up
				Sanding	No	As necessary after plowing	No
				Salting	Before 0.5 cm snow accumulates	As necessary after plowing	Icy spots only
			Falling	Plowing	After 1 cm snow accumulates	Continuously to bare pavement	Wing back shoulders/clean-up
				Sanding	No	As necessary after plowing	Icy spots only
				Salting	No	No	No
6.	Above -7°	Wet Snow	Packing, Wet Pavement				
			Rising	Plowing	½ hour after salting	Continuously to bare pavement	Wing back shoulders/clean-up
				Sanding	No	Icy spots only	No
				Salting	Before 0.5 cm snow accumulates	As necessary after plowing	Icy spots only
			Falling	Plowing	½ hour after salting	Continuously to bare pavement	Wing back shoulders/clean-up
				Sanding	No	As necessary after plowing	Icy spots only
				Salting	Before 0.5 cm snow accumulates	As necessary after plowing	No
7.	Above -7°	Sleet or Freezing Rain	Possible Icing, Wet Pavement				
			Rising	Plowing	No	No	Remove slush
				Sanding	No	Yes	Icy spots only
				Salting	When icing starts	Yes	Icy spots only
			Falling	Plowing	No	No	Remove slush
				Sanding	No	Yes	Icy spots only
				Salting	When icing starts	Yes	No
8.	After storm Any temperature	No Precipitation	Road snow-packed or icy				
			Rising	Plowing			Continuously to bare pavement
				Sanding			As necessary
				Salting			When above -18°
			Falling	Plowing			Continuously to bare pavement
				Sanding			As necessary
				Salting			Icy spots only
9.	After storm Any temperature	No Precipitation	Drifting				
			Rising	Plowing			Continuously to bare pavement
				Sanding			No
				Salting			Icy spots only
			Falling	Plowing			Continuously to bare pavement
				Sanding			Icy spots only
				Salting			No

[a] NOTE: Recommended treatment for various conditions shown on this chart should be used in MOST cases. However, unusual circumstances may necessitate departure from the recommended treatment.

[b] NOTE: Salt should not be applied in "during storm" or "after storm" situation during night hours (to 5 a.m.)

SPECIAL CONSIDERATIONS

Snow Removal and Ice Control on Large Bridges

Lengthy bridge spans, either level with the surrounding terrain or elevated, pose unusual snow removal and ice control problems. Elevated bridges can be subject to considerable drifting owing to their high exposure to wind, the geometry of their structural elements and the presence of illumination standards. An observation and relay system giving the condition (snow depth, drifts, ice, etc) of the road surface on these bridges is very important for safety and traffic flow. Information about bridge conditions is normally obtained from the public, police or maintenance staff, and is relayed to the supervisor who decides on the corrective action.

The first treatment of snow on bridges, occurring with the start of precipitation, consists of spreading an abrasive, such as sand, containing only sufficient chemical additive to prevent freezing of the storage pile. Since sodium chloride and calcium chloride can cause steel structures to corrode and cause concrete to deteriorate, they are undesirable for treating snow on bridges. When the accumulation is sufficient to warrant snowplowing, the snow is plowed into a windrow to the side or centre of the roadway, and then loaded on trucks for removal. The waste is not pushed over the sides of the bridge because it might fall upon a roadway at a lower elevation or onto private land, inviting injury to pedestrians and vehicles or damage to property. Snow containing solid wastes or chemicals may also produce environmental damage if dumped directly into water.

Snow removed from large bridges is hauled to a prepared land site, located away from bodies of surface water, providing a place for the snow to melt and the meltwater to run off slowly, leaving the solid waste for disposal.

Snow Control at Complex Interchanges

Where two roadways cross at different levels (grade separation) unusual snow removal problems may develop. For example, the grade separation may have one road which is partially elevated, while the other is depressed at a level below normal terrain. Under drifting conditions extensive drift formation can be expected on the lower road. Similarly, solid structures, such as interchange bridges, may cause heavy drifts to form on the downwind side of each bridge; the drifts may exceed 1-m depth, depending upon storm conditions and the physical layout of the interchange. Theakston (1973) found that the following construction practices, devices and design procedures assist in alleviating snow problems at complex interchanges.

(1) Embankment and structure orientation: where some choice of orientation exists, the axis of the structure and principal embankments should be aligned parallel to the direction of the prevailing wind. This will minimize drifting caused by the embankment.

(2) Embankment slopes, elevated or depressed roadways: snow accumulation due to embankments can be minimized by flattening their slopes, to a gradient as low as 1 in 20. With the flattened slopes, wind-borne snow is not deposited as drifts on the travelled surface.

(3) Snow collection devices: snow hedges and snow fences are very useful, collecting snow before it can reach the interchange area.

(4) Interchange design analysis: complex interchange designs proposed for snow-prone areas with good wind exposure should be analyzed for snow drifting and the necessary protection treatment built into the design. Such analyses usually involve testing models of the interchange in a wind tunnel or a water flume.

White-Out: Highway Hazard

When a road becomes unsafe because of poor visibility its continued use by the public cannot be allowed. Blowing snow reduces visibility to a level depending on the concentration of particles in the snow-laden wind.

When dense clouds of wind-borne powder snow are blown around obstructions, such as a tree, or an isolated building, the large-scale turbulence may produce an opaque, white cloud (Fig. 18.11) which reduces the visibility

Fig. 18.11 White-out. (Photo credit: - Ministry of Transportation and Communications, Ontario).

to zero (white-out). In a white-out, automobile drivers travelling at normal separations cannot see a vehicle ahead so that the chance of collision is greatly increased. On routes carrying heavy volumes of traffic a white-out often causes multiple car pile-ups resulting in great toll of life and property. During a white-out, motorists can only be protected by stopping all traffic, and permitting individual automobiles to travel through the white-out area one at a time.

The long-term protective measures to be applied to white-out prone areas include installing snow fences or planting hedges to reduce the amount of blowing snow.

ACKNOWLEDGEMENTS

The authors gratefully acknowledge the assistance of J. Hugh Blaine, P. Eng. in preparing this manuscript.

AIRPORTS

REQUIREMENTS, PRIORITIES, STANDARDS

In airport management, "Snow and Ice Control" includes all actions taken to reduce and/or eliminate the effects of snow and ice on airport operation. These actions include plowing, blowing, sweeping, compacting or melting snow, erecting snow fences, applying chemicals to reduce ice and/or snow buildup, or applying abrasives to improve traction on aircraft movement surfaces.

The requirements for snow and ice control on airports are determined by the type and number of aircraft movements and the area's prevailing weather. Canadian winter conditions range from extreme cold with heavy snowfalls to moderate temperatures with little or no snow. The standard for active runways as published in the IFR (Instrument Flight Rules) Supplement of Canada Air Pilot reads as follows: "Insofar as practicable, snow removal and ice control are carried out to provide airport surfaces which will permit safe operational use at all times." The active runway, adjoining taxiways, access to the terminal apron and aircraft parking areas are maintained in a serviceable condition on a "priority one" basis. The remaining manoeuvering and movement areas are cleared on a lesser priority basis in accordance with operational requirements. Except for the extraordinary storm condition, the staff and equipment resources available at international airports are such that a 23-m wide centre area of the priority runway can be cleared within 30 min after the snowfall ends. The snow control operation commences as soon as the first flakes fall. The factors considered while assigning snow removal and ice control priorities to airport areas are:

(1)   The type and volume of aircraft movements,
(2)   The anticipated weather conditions,
(3)   The anticipated type and quantity of precipitation,
(4)   The prevailing wind direction and velocity,
(5)   The type of snow removal and ice control equipment available, and
(6)   Fire routes.

The following list delineates priority areas recommended for international airports.   However, changes in area priorities may be made to meet operational requirements:

Airside (Area restricted to the operation of aircraft or related support facilities).

The airside priority areas are established as follows:

(1)   *Priority I Area* - This is the minimum area that must be cleared on a continuous basis throughout the storm to maintain the minimum airside operational capability of the airport.

This area is usually composed of the following surfaces:

   (a)   One runway,
   (b)   One taxiway,
   (c)   Sufficient apron area (accounting for at least 20% of the total apron area) to accommodate aircraft and passenger terminal and cargo requirements,
   (d)   The entrance and exit access associated with the runway, taxiway and apron areas, and
   (e)   Access roads from the firehall to all of the above areas.

(2)   *Priority II Area* - This is the area that will be cleared during the storms so that, in the event of a change in the prevailing wind direction the other runway can be made operational on short notice. This area is composed of the following surfaces:

   (a)   Secondary runway,
   (b)   Secondary taxiway,
   (c)   The entrance and exit access routes associated with the above runway and taxiway,
   (d)   Associated apron access to taxiways if different from those cleared under Priority I clearance operations, and
   (e)   An area of apron required to gain access to the minimum 20% cleared under Priority I clearance operations.

(3) *Priority III Area* - This area is composed of surfaces which are cleared after a snow storm. Such areas include:

    (a) Airside service roads,
    (b) Runway, taxiway, shoulder areas,
    (c) Pre-threshold areas,
    (d) Glide path sites, and
    (e) Remaining airside areas required to permit full operational use of the airport.

Groundside

The groundside priority areas are established as follows:

(1) *Priority I* - This is the minimum area that must be cleared in order to allow access of employees and passengers to the airport, and permit manning of essential airport services (e.g. air traffic control, snow removal, etc.). These areas are:

    (a) Employees' parking areas,
    (b) Public parking area(s) as required,
    (c) Primary terminal access routes, and
    (d) Air traffic control tower access route(s).

(2) *Priority II* - This is the remainder of the groundside area not cleared under groundside Priority I clearance operations. This area includes:

    (a) All remaining airport access routes to service roads,
    (b) Remaining public parking areas, and
    (c) Remaining employee parking areas.

(3) *Priority III* - This is the area of the groundside which is cleared after the storm:

    (a) Fire training area,
    (b) Remote Navigational-Aid sites, and
    (c) Drainage systems/areas.

Degree of Cleanliness

All paved runways, taxiways and aprons shall be cleared to bare pavement surface using a combination of plowing, sweeping and/or blowing.

Maximum Allowable Clearing Times on Priority Areas

The maximum allowable clearing times for the different priority areas, as defined above are set as follows:

(1)	Airside Priority Areas - International Airports	Clearing Time (h)
	Priority I	2
	Priority II	2
	Priority III	48-60

(2)	Groundside Priority Areas - International Airports	Clearing Time (h)
	Priority I	2
	Priority II	2
	Priority III	48

USE AND DESCRIPTION OF SNOW AND ICE CONTROL MEASURES

Snow fences are erected along roads where drifting snow may hinder snow removal operations. They help to prevent windblown snow from accumulating on roadways, and usually are erected 30 to 90 m to the windward side.

The common snow and ice control chemicals, sodium chloride and calcium chloride, are only used sparingly and with care in groundside areas because of their corrosive action on aircraft, their toxicity to grass and their adverse chemical effects on concrete. Urea, a commercial fertilizer with a nitrogen content of approximately 46%, applied in pellet or granular form is frequently used as a chemical control measure for airports. Its density is about one-half that of sand and each pellet or grain of urea is coated with clay or mineral oil which prevents the chemical from absorbing moisture which can cause caking and bridging. This material is free-flowing and therefore can be applied with conventional sanding equipment or agricultural fertilizer spreaders. It is effective as an ice preventative for pavement temperatures down to -9.4°C. Because it can form a bond with the surface it is not readily displaced by the wind. Urea is most effective when used before ice formation. Its action in lowering the freezing point of water allows time for the water to be removed before it freezes, usually by sweeping. Normal applications are 7 g/ m^2, with approximately 15 g/ m^2 being required for de-icing.

In some cases, following application, sanding may still be necessary. Application of urea before sanding increases the bonding of the sand to the ice and provides better traction, thereby facilitating ice removal by mechanical means. Urea has a prolonged residual effect lasting several days and is both non-toxic and non-corrosive. Since it is manufactured under strict quality control procedures it is currently available from only six approved Canadian suppliers.

Depending on the surface temperature the application of urea to the runways will:

(1) At temperatures between -4.4 and 0° C, lower the freezing point of water thereby increasing the time available for removing water and slush by sweeping, and

(2) At temperatures between -9.4 and -4.4° C, effectively prevent snow or ice from bonding to the runway surface and permit removal by plow, grader blade or sweeping when precipitation stops.

The effectiveness of urea in snow and ice control depends also on exposure (radiation) conditions at the time of application. On a cloudy day, de-icing reaction may not be obtained with temperatures of up to -1° C while on a bright sunny day de-icing can be expected with temperatures as low as -17.8° C.

As the use of urea increases, the use of abrasives decreases. By reducing the amount of abrasives on runways the potential for damage by particles impinging on critical aircraft components is reduced. The use of sand has been discontinued at coastal airports where temperatures are high enough for urea to control ice formation and removal completely.

The search for more effective ice-control chemicals is continuing with the objective of developing a material that will be effective in any icing situation. Every new chemical investigated is judged on the following ice control performance criteria:

(1) For anti-icing, the ice control chemical, when applied before or during freezing rain, or decreasing temperature, shall prevent the coefficient of friction from falling below one half the value for clean dry pavement and will render possible the removal of the ice/slush contaminant by physical means, and

(2) For de-icing, the control chemical application shall raise the coefficient of friction of the ice-covered surface to at least one half the value for the clean dry pavement within one hour after application and will render possible the removal of the ice/slush contaminant by physical means.

All chemicals are subject to exhaustive testing to ensure that they do not damage aircraft structural components and engines.

Abrasives

Sand has been the long-time standard material used to improve traction on runways during icing conditions, especially during the pre-jet years when both traffic volumes and the aircraft landing speeds were lower. However, sand has its limitations.

(1) Its usefulness is strictly "after the fact" as it is of no value whatsoever in preventing icing.

(2) In some areas it is difficult to obtain suitable grades.

(3) Because of its bulk, it is expensive to transport over long distances and heavy equipment is required for handling.

(4) Heated, dry storage is required to keep it in an usable condition.

(5) Since it is an insoluble material, successive applications may be required, often within a short period of time.

The grading of particle size must be such that the material provides optimum traction and minimum aircraft engine ingestion and impingement damage. Hence, strict quality control of the material is exercised during delivery and storage so as to prevent contamination by foreign matter, e.g., stones and oversized particles.

The standard operational practice is to remove the sand from the runway with sweepers as soon as the ice has been removed.

APPLICATION OF SNOW AND ICE CONTROL EQUIPMENT

During activities to remove snow and ice from airports three main types of equipment are used: high-speed sweepers, truck-mounted plows and snow blowers. The number of different pieces of equipment operated at any specific time depends on the area to be cleared, the snowfall, icing and drifting conditions, and the operational commitment or level of service required.

Canada airport runways are 61 m wide and from 1.5 to 1.7 km long. The edges of these pavements are lined with specially designed electric lights 46 cm in height, spaced 61 m apart. These units delineate the landing surface for the aircraft pilot. The immediate and continuing objective in snow removal is to provide a clean pavement surface 23 m in width for the full runway length, to allow safe aircraft landing or takeoff. The snow removal operation begins with the high-speed sweeper which uses a 4.3-m broom angled at 30° to the lateral axis to clear a 3.7-m path. These sweepers operate at speeds of 6.5 to 13.5 m/s. Their use alone is limited to snowfalls less than 7 or 8 cm. Once several widths have been broomed from the centre of the runway, plow trucks are operated together with the sweepers to pick up and cast the snow into flat windrows 7- to 15-cm deep. The plow-sweeper combination clears a 4.9-m wide path at speeds up to 13.5 m/s. Blowers, operating at speed of 7-11 m/s follow and blow the windrowed snow from the runway. Snow is removed to a minimum of 8 m beyond the edge of the runway to allow unobstructed lighting visibility, and snow banks are tapered or sloped to minimize any obstruction hazard they might present to aircraft and to reduce drift formations.

Runway and taxiway edge lights are kept free of snow to provide unobstructed visibility. Off-pavement snow clearance along taxiways is carried out as necessary to prevent ingestion of snow by aircraft engines. A recently established aircraft landing aid designed for poor visibility conditions includes, among other things, runway center-line lighting. This system involves a 30.5 cm diameter metal fixture embedded in the runway. The top of the unit is 1.3-cm higher than the pavement and contains an optic lens and a 200-W bulb. A runway may have over 400 of these fixtures mounted along its centerline. This facility is part of what is known as a Category II instrument landing system (ILS) and has complicated procedures for snow removal and ice control. Plows and blowers used on runways equipped with the Category II ILS facilities must be fitted with rubber contact areas or the steel edges must be carried above the pavement on casters so as not to damage the fixtures. Also, the steel shoes or casters normally used to support plow frames, blades or blower bankhead assemblies must be replaced with pneumatic or semi-pneumatic rubber tires. On these runways greater use must be made of the sweepers for snow removal and of urea for ice control.

Ice tends to build up around the fixtures of in-pavement lighting systems. When the lights are on, some of the drifting snow touches them, melts and depending on the pavement temperature, forms deposits of water or ice, which accumulate on the surrounding pavement. If the pavement temperature is above -9.4°C urea can be used effectively to minimize the buildup of ice rings around the lights.

Another Category II facility that requires strict snow control is the large infield area ahead of the glide path array. To prevent glide path signal wave distortion and reflection, snow accumulations in this area should not be allowed to exceed 30 cm. Also, there should be no abrupt snow banks or drifts. Conventional snow removal equipment with pneumatic tires cannot be operated on the infield. However, a tracked vehicle equipped with a 3.7-m snowplow can be used there to maintain the operational integrity of the Category II system without requiring the complete infield area to be stabilized and paved (Transport Canada, 1976a).

On aircraft apron areas graders with wing blades are used to clear and windrow the snow before it is loaded by blowers and loaders into trucks and hauled to a dump site. Because of escalating trucking costs and the concern expressed by environmentalists about the impact of snow dumps, other methods of snow removal have been evaluated, but not yet implemented. One of these is the use of mobile melters. Tests of a 1400 kg/h capacity mobile melter has established that this method is highly competitive in cost with present disposal methods where long truck hauls are becoming more common (Transport Canada, 1976b).

The introduction of Category II operations as a standard requirement and the increasing frequency of aircraft movements demands a new approach to runway snow removal. This involves using high speed sweepers and blowers, operating at 11 m/s or higher; the practice of using low speed plows to windrow snow will be terminated. The conventional sweeper has been modified by replacing its mechanical drive by a hydraulic drive. This allows an effective sweeping speed in excess of 11 m/s by reducing the vibration or "bounce" of the broom experienced with the mechanical drive.

A system has been developed that indicates when frost, ice or snow forms on a runway. The system comprises remote, runway-mounted sensors connected to a display unit which is located in an appropriate place where it can be monitored 24 h a day. The advance warning provided by this system allows field maintenance personnel to utilize their snow and ice control equipment and materials more efficiently by making it possible to operate an "anti-icing", as opposed to a "de-icing", runway maintenance program. This system is not presently in use at any Canadian airports; however, evaluation programs being conducted by other agencies are being closely monitored to assess its potential application at Canadian International Airports.

LITERATURE CITED

Anonymous. 1964. *Tampers, track liners and snow equipment: Improvements in snow and ice control and removal.* Committee reports. Proc. 76th Annu. Convent. Roadmasters Maint. Way Assoc. Am., pp. 52-61.

Anonymous. 1973. *Prairie snowfighting: the trials and techniques.* Railw. Track Struct., Vol. 69, No. 8, pp. 14-17.

Best, Gerald M. 1966. *Snowplow Clearing Mountain Rails.* Howell-North Books, Berkeley, Calif.

Brohm, D.R. and S. Cohen. 1973. *Management and control of winter operations.* Minist. Transp. Commun., Toronto, Ont.

Brohm, D.R. and H.M. Edwards. 1960. *Use of chemicals and abrasives in snow and ice removal from highways.* Bull. 252, Highw. Res. Board, NAS-NRC Publ. 761, Washington, D.C., pp. 9-30.

Browne, A.L. 1973. *Traction of pneumatic tires on snow.* Res. Pub. GMR-1346, Gen. Motors Res. Lab., Warren, Mich.

Browne, A.L. and D.F. Hays (eds.). 1974. *The Physics of Tire Traction: Theory and Experiment.* Plenum Press, New York, N.Y.

Damas and Smith, Ltd. 1971. *Winter testing of tires.* Can. Safety Counc., Ottawa, Ont.

Evenmo, O. 1974. *Mechanized snow clearance keeps Norway's trains on the move.* Railw. Gaz. Int., Vol. 130, No. 11, pp. 423-426.

Fisher, E.H. and E.T. Hurley. 1964. *CN's fight with snow.* Snow Removal and Ice Control, 1964, Tech. Memo. No. 83, Nat. Res. Counc. Can., Ottawa, Ont., pp. 58-63.

Harrison, J.L. 1920. *Eastern states plan their snow removal works for coming winter.* Public Roads, Vol. 3, Sept.

Ishibashi, T. and O. Takahashi. 1971. *Development of a snow loader with longitudinal troughs.* Quart. Reps. Railway Tech. Res. Inst., Tokyo, Vol. 12, No. 3, pp. 141-143.

Kikuchi, I. 1975. *Measures against snow damage.* Jpn. Railw. Eng., Vol. 16, No. 1, pp. 7-10.

National Research Council. 1967. *Manual of snow removal and ice control in urban areas.* Tech. Memo. 93, Assoc. Comm. Geotech. Res., Nat. Res. Counc. Can., Ottawa, Ont.

Natural Rubber Producers Research Association. 1967. *Oil-extended natural rubber in winter tyre treads.* N.R. Tech. Bull., Great Britain.

Ontario Department of Transportation and Communications, 1971. *Studded tire fact sheets, Oct. 1971 edition.* Res. Rep., Res. Br., Dept. Transp. Commun. Downsview, Ont., pp 7-12, 7-13.

Parkes, G.R. 1961. *Railway snowfighting equipment and methods.* Published by the author. Hadfield, Hyde, Cheshire, England.

Ringer, T.R. 1979. *Protection methods for railway switches in snow conditions.* Snow Removal and Ice Control Res., Proc. 2nd Int. Symp. Snow Removal Ice Control Res., Hanover, N.H., Sp. Rep. 185, Transp. Res. Board, NAS-NRC, Washington, D.C., pp. 308-313.

Smith, P. and R. Schonfeld. 1970. *Pavement wear due to studded tires and the economic consequences in Ontario.* Rep. R.R. 152, Ont. Dept. Highw., Toronto, Ont.

Smith, R.W. and D.J. Clough. 1972. *Effectiveness of tires under winder driving conditions.* Highw. Res. Rec. 418, Highw. Res. Board, NAS-NRC, pp. 1-10.

Theakston, F.G. 1973. *Snow control on highways.* Proc. First Nat. Conf. Snow Ice Control, Roads and Transp. Assoc. Can., Ottawa, Ont.

Transport Canada. 1976a. *Report on evaluation of Bombardier muskeg tractor equipped with 12 ft. plow for snow control in I.L.S. glide path areas, 1973 to 1976.* Rep. AK-71-09-121, Airp. Const. Serv. Dir., Transp. Can., Ottawa, Ont.

Transport Canada. 1976b. *Report on the evaluation of jet melt snow melter.* Rep. AK-71-09-008, Airp. Const. Serv., Airp. Fac. Br., Transp. Can., Ottawa, Ont.

Winterrowd, W.H. 1920. *The development of snow fighting equipment.* Railw. Maint. Eng., Vol. 16, pp. 458-462.

PART IV

SNOW AND RECREATION

707

19

SKIING

R. PERLA

National Hydrology Research Institute, Environment Canada, Canmore, Alberta.

B. GLENNE

Department of Civil Engineering, University of Utah, Salt Lake City, Utah.

INTRODUCTION

Without skis or snowshoes, travel on foot through deep snow is tedious or impossible. In the earliest account of a snow disaster, Xenophon describes his men dying of exhaustion while trying to march through deep snow in Armenia about 400 B.C. (Anabasis IV.V, 4-9). Today skiing is generally viewed as a form of recreation, but for some people it was a means of survival.

Skiing is at least 4000 to 5000 years old, and perhaps older. These figures are corroborated by a Stone-Age carving (Norwegian Ski Museum, Oslo; see Fig. 19.1), and by ancient ski fragments discovered in Scandinavian bogs (Berg et al., 1941). The first proficient skiers were probably tribes having an historical link with the present inhabitants of Northern Europe and Soviet Asia. Early Southern Europeans were apparently unfamiliar with skiing, although tribes in the Caucasus developed snow-shoe mountaineering over 2000 years ago (see Strabo, Geography, Book XI.5, 5-7). Evidence of skiing in ancient times has not yet been found in the Western Hemisphere where snow-shoe travel also developed independently.

Snowshoes are practical under conditions requiring a short, wide surface (deep soft snow, unconsolidated granular snow), however, skis are faster and more efficient under most conditions. On prepared ski tracks, a top cross-country skier can cover a marathon distance (50 km) averaging better than one kilometre in three minutes (which is slightly faster than the best times in a running marathon). However, skiing requires more skill and practice than snowshoeing, and can be quite frustrating at first to the uninitiated. Thus, credit should be given to the early skiers for patiently experimenting with a

Fig. 19.1 Pictograph from Rödöy, Norway circa 2000 B.C. (courtesy Norwegian Ski Museum).

method of snow travel that had no immediately obvious utility. The history of skiing is further discussed by Richardson (1909), Dudley, (1935), Firsoff (1943), Lid (1949), Lunn (1952), Flower (1976) and Vaage (1979), among others.

The rather recent history of skiing in Canada is documented in the Canadian National Ski Museum, Ottawa, and is recounted by Lund (1971) and Marsh (1975) who estimate that nearly ten percent of the Canadian population ski during some period in their lives.   Recreational skiing throughout North America has increased astronomically since World War II. Herrington (1976, personal communications) estimates that there are approximately ten million skiers in the United States alone while Greenberg, (1978, personal communications) places the number closer to 14 million. Skiing is the major recreational activity at Canadian Mountain Parks where skier visits increased 23% annually from 1958 to 1972 (Parks Canada, 1973). There are over 1000 ski areas in North America (Enzel, 1979).

In the mid-1970's, downhill skiers enjoyed, on the average, over 3000 vertical metres of skiing per day (Parks Canada, 1973); a distance made possible by advances in ski lift technology (Schneigert, 1966; Dwyer, 1975; Judge, 1979) snow making equipment (Ericksen, 1980a, b) and ski area planning (USDA Forest Service, 1973a,b).   The ski industry has made a significant impact on the North American economy, especially in those mountain communities that were once declining but are now developing rapidly in response to the demand for recreational facilities (Goeldner et al. 1973; Goeldner and Farwell, 1975). In fact, this rapid development has raised important social and environmental issues by outstripping existing sanitation and housing facilities, and sometimes causing irreversible losses of soil, wildlife and vegetation in these mountainous areas. Environmental issues reached a climax in Colorado when the electorate voted against sponsoring the 1976 Winter Olympics; further, in Southern California a major new development was blocked in the Supreme Court (McCloskey and Hill, 1971). As an alternative to downhill lift-skiing there is an increasing interest in cross-country skiing.

Ski equipment has also evolved rapidly in the latter part of the twentieth century with notable improvements in the construction of skis, poles, boots, safety bindings, clothing, goggles, and many speciality items to match the rich variety of skiing styles (Scharff, 1975; Brady, 1979).

The physical and mental pleasures derived from skiing are great, but are accompanied by risks of injury which are discussed in sports medicine publications, e.g., FIMS (1972). Injury prevention through proper training and conditioning is emphasized by Foss and Garrick (1978).

Thus, skiing is a diverse activity that can be approached from many viewpoints, some quite removed from the subject of snow. In keeping with the theme of this book, the scope of this chapter is restricted to the scientific study of skiing with emphasis on the interaction of snow and ski, and is further limited to the study of skis worn by humans, with only a few references to aircraft skis (Klein, 1947; McConica 1951, 1952) and to general trafficability on snow (see Knight, 1965 and Ch. 12 for further information on travel on snow).

Earlier studies on the science of skiing with special emphasis on snow were made by Seligman (1936), Lliboutry (1964), and the Society of Ski Science (1972), whose comprehensive treatment of the subject includes an extensive bibliography.

SNOW PARAMETERS

Skiing performance can be related to the following snow parameters: density, temperature, liquid water, hardness and texture.

Density is a fundamental and unambiguous parameter that correlates directly with snow hardness, other parameters remaining constant. The density distribution in a snowcover, especially near the surface, crucially determines ski maneuverability, safety and the overall enjoyment of the sport. Density is easily measured by weighing known volumes, to an error ranging from less than 1% for large sample volumes ($\sim 10^{-3}$ m^3) to 10% or more for small volumes ($\sim 10^{-4}$ m^3), e.g., thin layers and crusts.

Snow temperature is also a fundamental and unambiguous parameter that is easily measured with standard thermometers, although the influence of solar and terrestrial radiation complicates measurements in the surface layer. Snow hardness correlates inversely with temperature, all other factors held constant. Thus, for a specific snow texture, hardness decreases as the temperature approaches 0°C. Snow temperatures are routinely measured by competitive skiers to match ski wax to snow hardness, which is difficult to measure.

At 0°C, the other important parameter affecting skiing performance is the amount of liquid water in the snow. Ski drag increases markedly as liquid water increases. Liquid water (sometimes called free water) can be measured using calorimetric, centrifugal and dielectric techniques, or by determining the dissolving rates of chemicals. However, these measurements are either time consuming or difficult to calibrate, and hence have not yet been made systematically in ski studies. For these reasons, data reported below are based on the qualitative nomenclature of the International Classification of Snow (Unesco, 1970): *dry, moist, wet, very wet and slush.*

The objective measurement of snow hardness (or more properly, an index of snow hardness) is also relatively time consuming and complex. First, it is important to distinguish between hardness at the microscale dimension of individual crystals (~ 1 mm) and at the large scale (samples >10 mm). Microscale hardness is of great practical importance for ski friction and waxing, whereas large scale hardness is relevant to the penetration of the ski. At present writing, only the large-scale hardness measurements have been made systematically in evaluating ski performance (Kinosita, 1960; Kuroiwa and LaChapelle, 1972). Microscale hardness measurements are bypassed and temperature is used as an index instead.

The indices of large-scale hardness reported in the following sections were measured with the Swiss Rammsonde instrument (see Fig. 19.2) which is driven down into the snow by dropping a hammer with mass H from a height h. If the apparatus (total mass M) sinks a distance z, the hardness index N (Ram number - see also Ch. 12) is:

$$N = K(h/z) + M, \qquad\qquad 19.1$$

where K is a function of M, H and η the coefficient (restitution) for the efficiency of collision between hammer and penetrometer, given by the equation (Bader et al., 1939):

$$K = H\{H(M - H)(1 + \eta)^2 + [H - \eta(M - H)]^2\}/M^2. \qquad 19.2$$

Assuming a semi-elastic collision, η is about 0.5 which is midrange between the purely elastic value (1.0) and the purely inelastic value (0.0). The Ram number has the dimension of mass (usually kilogramme) but is often reported as a non-dimensional index. For hard snow (N > 1), a heavy-weight Rammsonde (M ≃ 1.0 kg) must be used; for soft snow (N < 1), a light-weight Rammsonde (M ≃ 0.1 kg) is used.

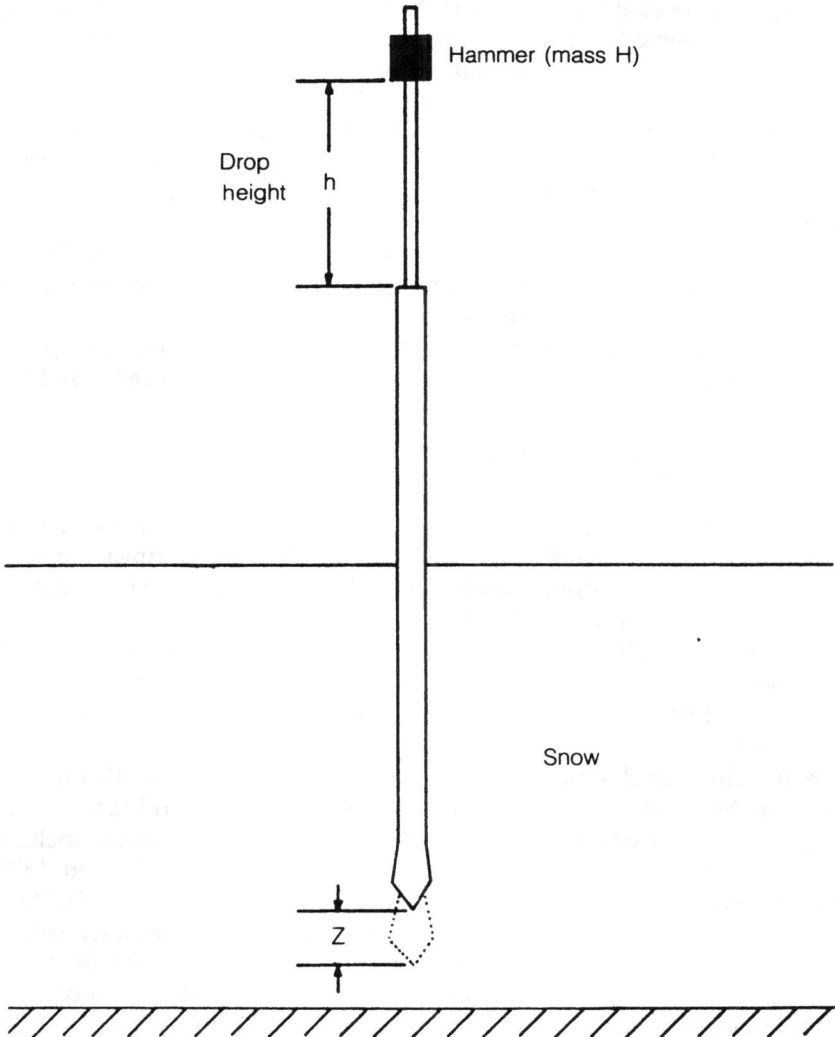

Fig. 19.2 Swiss Rammsonde for measuring a snow hardness index.

The term texture is meant to describe the shape, size, and bond inter-connections of snow grains, and is the most difficult parameter to describe numerically. The objective measure of texture remains, at least for the present, a matter for research (Kuroiwa and LaChapelle, 1972; Yosida, 1972; Good, 1975; Kry, 1975). However, there are many ways that snow texture can be described qualitatively, provided it is understood that English language descriptions are somewhat awkward, since English lacks the rich vocabulary possessed by some Northern people for expressing the wide variety of snow types. For example, Norwegians have over 20 special terms for snow conditions and the Inuit people even more. Irrespective of their native tongues, competitive skiers learn, through experience, to judge differences in texture qualitatively or subjectively. This is especially true for selecting wax based on texture and temperature, since both factors correlate with the elusive parameter, microscale hardness.

The nomenclature for texture used below is composed of English terminology adapted from the International System for Classification of Snow (Unesco, 1970). Emphasis is placed on observing the shapes of individual grains. In the field, snow samples are disaggregated into separate grains, and examined by means of a hand-held magnifier or the naked eye.

SKIING AND SNOW METAMORPHISM

The charm of skiing is that it works on an endless variety of snow conditions. Seligman (1936) provides an encyclopedic treatment of snow conditions including a discussion of newly fallen snow, metamorphosed snow, crystallization at the surface, wind deposition, wind erosion, cornice formation, avalanches, and many other related topics. As explained by Seligman, one of the most important factors determining skiing conditions is the type and degree of snow metamorphism, which can be divided into the three categories described below.

Newly fallen and partially metamorphosed snow. (see Fig. 19.4a). The original shapes of crystals formed in the atmosphere are still recognizable despite modifications due to wind fragmentation, sublimation, melting, sintering, and recrystallization. In the International System (Unesco, 1970) for classification of snow grain shape, this category is sometimes called "new and partially settled snow". Skiers often use the term "powder snow". Enjoyable downhill "powder" conditions occur when the snow is dry or slightly moist, its Ram number is relatively low ($\leqslant 1$), and its density is less than about 200 kg/m^3, steadily increasing with depth so that skis are supported by underlying strata. Typical excellent powder conditions occur when the surface layer has a density of 50 to 100 kg/m^3 which increases to about 150 kg/m^3 at about 30 cm depth. This increase in density with depth is a normal result of freshly fallen snow settling under its own weight, but can be off-set by wind action, which may compact the surface layer to a higher density, or by rising temperature during snowfall. Two examples of different density distributions measured at Whistler Mountain, B.C. are shown in Fig. 19.3. Ski maneuverability is inhibited by a density inversion because skis tend to dive under the denser surface layer. Assuming the density criterion is met, almost all forms of snow crystals (prisms, plates, dendrites) allow satisfactory downhill powder skiing. Even a layer of hail or graupel deposited on a firm base makes excellent downhill skiing possible because its "ball-bearing" texture offers good compressive support and little shear resistance to ski turns (see Fig. 19.4f).

For dry new snow, the cross-country skier usually applies a relatively hard wax to obtain optimum grip and glide. The traditional explanation is that the unmetamorphosed crystals are hard and sharp and easily penetrate ski wax (Palosuo et al., 1979). In general, the hardness of the ski wax must be increased as the temperature of the powder snow decreases.

Moist and wet new snow is notoriously difficult for skiing. The liquid water adds viscous resistance and surface tension, which seriously slow down both cross-country and downhill skiers. To match the "softness" of the wet snow grains, the cross-country skier typically changes to a softer wax. In practice, the most difficult waxing problems arise when newly fallen snow begins to

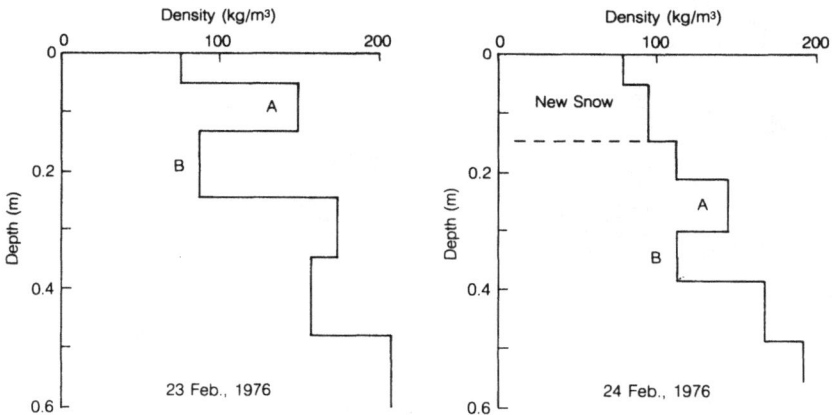

Fig. 19.3 Density versus depth measured at a Whistler Mountain, B.C. ski run (elev. 1800 m). On 23 February, powder skiing was difficult because skis penetrated under wind stiffened layer A into softer layer B. On 24 February, powder skiing was excellent after a new snowfall moved layer A to a ski support level. Ski penetration was 0.3 m on 23 February and 0.2 m on 24 February.

melt at 0° C because then the microscale hardness is changing too abruptly for a single wax to give optimum grip and glide. If increased melting converts the snow to a "very wet" state or to a "slush", the cross-country skier is forced to use a very soft and tacky coating known as klister.

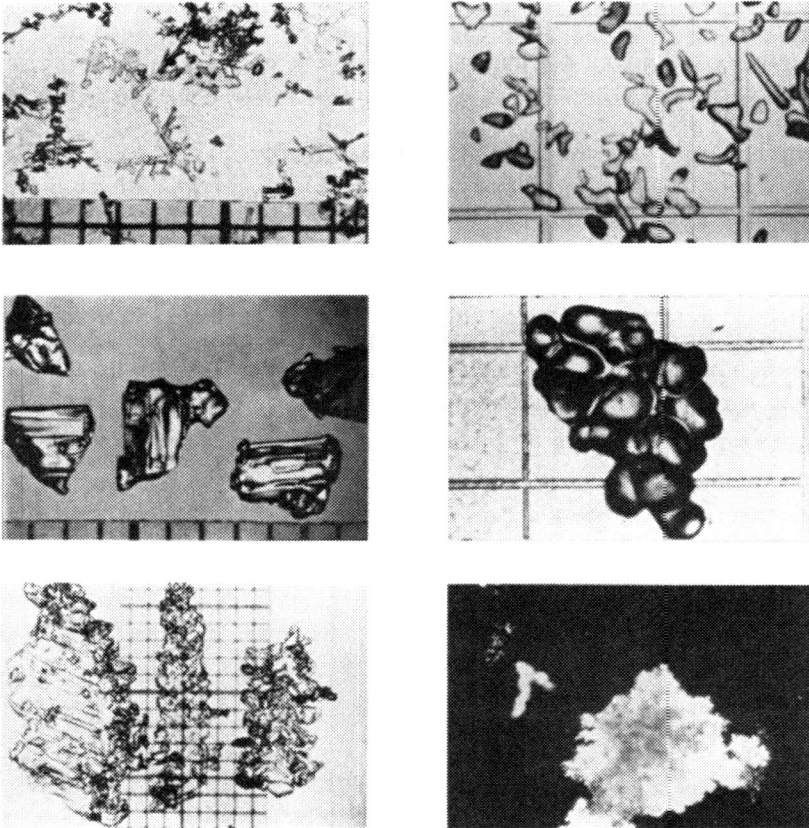

Fig. 19.4 Snow Crystals from natural ski fields
    (a) Newly-fallen snow (1-mm grid).
    (b) Equitemperature metamorphosed snow (2-mm grid).
    (c) Temperature gradient metamorphosed snow (1-mm grid).
    (d) Melt-refreeze aggregate (2-mm grid).
    (e) Surface hoar (1-mm grid).
    (f) Graupel (about 2-mm diameter).

Dry metamorphosed snow. Dry snow metamorphism includes two competing processes: the tendency for crystals to evolve into sizes and shapes with minimum free energy, and the tendency for crystals to evolve in response to significant snow temperature gradients ($> 10°C/m$). The first process is sometimes called equitemperature (ET) metamorphism; the second process is sometimes called temperature-gradient (TG) metamorphism (Sommerfeld and LaChapelle, 1970; Perla, 1978).

ET metamorphism tends to produce small, rounded grains ($\leqslant$ 1-mm diameter), with relatively strong interconnections due to sintering. The snow is generally firm ($N > 0.5$, density > 150 kg/m^3), and typically supports the skier at the surface. Ski waxes seem to work very well on ET snow, and can be selected for high performance in terms of grip and glide according to hardness (temperature).

TG metamorphism produces coarse large grains (1 mm to 10 mm size), with relatively weak interconnections. Weak TG layers may suddenly collapse under the skier's weight. This is especially troublesome in the spring when melting further weakens the snow. Large TG crystals that form at the bottom of the pack are known as depth hoar (see Figs. 19.4b, c).

The rate of change of newly fallen snow into metamorphosed snow is mainly a function of air temperature, radiation, snowpack thickness, and initial density and shape of the crystals. This snow may convert into TG snow in a few days if the snowpack is thin, porous, and shaded so that its surface temperature drops well below the ground temperature, producing a strong temperature gradient across it. Conditions are conducive for TG snow in the drier continental ranges (eastern British Columbia and the Cordillera); however in the coastal ranges (western British Columbia, Washington, California), the deeper snowpack normally metamorphoses into smaller more rounded grains (like ET snow).

Melt-freeze metamorphosed snow. Metamorphism is greatly accelerated when liquid water is present. Large grains (> 1 mm) grow rapidly at the expense of small grains since diffusion is accelerated by the presence of the liquid phase. Surface energy effects are more important compared to the dry case where there is only the vapor phase (Colbeck, 1978). In the spring, ski fields typically undergo alternating melt-freeze cycles, with a hard, icy surface usually forming at night. The most enjoyable downhill skiing regularly occurs as the surface begins to soften in the later morning but before the texture melts completely to slush. Repeated melt-freeze cycles form large, coarse grained aggregates known as corn snow, which often provide easy maneuverability and delightful downhill skiing conditions (see Fig. 19.4d).

Waxing cross-country skis for melt-freeze conditions usually involves various combinations of klister. While it is difficult to choose a wax for unmetamorphosed, newly fallen snow just beginning to melt, it is fairly easy to select a klister that gives efficient grip and glide on melt-freeze metamorphosed snow at 0°C, and during the transition from dry to moist grains.

In a perennial snowpack the net result of repeated melt-freeze cycles and the pressure exerted by overlying layers is to convert the snow to a high density (over 500 kg/m³) and coarse-grained sintered texture known as *firn* that is evolving toward ice as the end product (non-communicating pores, density about 800 kg/m³). Ski maneuverability on firn largely depends on the amount of surface softening.

SPECIAL SURFACE CONDITIONS

There are several categories of surface conditions of special interest to the skier:

Wind slab. Wind action can fragment and sublimate newly fallen snow crystals into extremely small particles that pack to relatively high densities, sometimes as high as 450 kg/m³. For wind slab, N typically ranges from 1 to 10, and occasionally as high as 50. Ski penetration is generally erratic; the skis may ride along smoothly on the surface, but then suddenly dive under. Ski glide is reduced because of the small particle size and the increased number of particle contacts. Waxing for optimum grip and glide may be difficult.

Surface irregularities. Seligman (1936) and Oura (1967) give examples of various surface irregularities resulting from wind erosion and deposition. Ranked by increasing size, these irregularities are known as ripples (~ 10 mm high), wind ridges, barchans, and sastrugi (up to about 1 metre high). The larger features are especially troublesome to the skier.

Crusts. Crusts are thin (5 to 50 mm thick), hard layers formed by strong winds (wind crust), or by a melt-freeze process (sun crust, rain crust), having a wide range of density from about 200 to 800 kg/m³. With respect to ski support, crusts are further subdivided into "breakable" and "unbreakable" types, both of which may pose difficult and dangerous skiing conditions. Icy, unbreakable crusts can be hard enough (N ≃ 1000) so that a steel ski-edge may not bite into the surface, allowing little possibility for edge control and maneuverability. On the other hand, breakable crusts may be strong for a while but suddenly let a skier crash through into a weaker sublayer to suffer complete loss of control. To obtain grip and glide on crusts, the cross-country skier usually applies a klister with sufficient toughness to resist abrasion. Ordinary wax (non-klisters) will wear off quickly, especially if crust temperatures are below 0°C.

Surface hoar. Surface hoar consists of large (> 1 mm), flat crystals, which grow by vapor deposition from the atmosphere onto a radiation-cooled snow surface. The thicknesses of surface hoar layers are usually less than about 10 mm. A thick layer (10 to 50 mm) on a firm base can provide excellent ski conditions because its large crystals are poorly bonded, as in depth hoar. Initial density of surface hoar is about 100 kg/m^3 (see Fig. 19.4e).

Surface recrystallization. A loose, coarse-grained snow, similar to depth hoar but generally involving smaller grains ($\sim$ 1 mm), which is often observed at, and just below the surface of snowpacks in the Cordillera and other continental mountain ranges may result from surface recrystallization. Its' loose texture, which offers very good powder-like skiing, presumably is due to temperature gradients localized near the surface, although the exact mechanisms for its formation have not yet been identified. Favourable conditions for surface recrystallization are believed to be a dry, cold atmosphere, which allows radiation to escape easily from the snow surface. Another special case of surface recrystallization, called "radiation recrystallization" (LaChapelle, 1970), occurs when an intense temperature gradient at the surface, produced by solar radiation absorbed at $\sim$10 to $\sim$100 mm depth, causes increased emission of terrestrial radiation from the surface.

Glaze and rime. Thin, breakable surface crusts (glaze) are caused by supercooled droplets freezing onto the snow surface. A glazed cross-country track is usually quite fast unless the supercooled droplets freeze in the atmosphere before impact, and fall as tiny rime particles, which then reduce glide considerably. The amount of deposited glaze and rime increases with increasing wind.

COMPACTION

Heavy ski traffic greatly alters the natural ski field, producing what is generally known as "hardpack", a layer with a higher density and Ram number than the undisturbed snow (see Fig. 19.5). In ski areas hardpack slopes are further compacted by the movement of snowcats and large rollers, usually immediately after a new snowfall, and by the general smoothing operations on the slope with racks drawn by snowcats. Systematic compaction improves snow durability; the snow is less subject not only to uneven scraping by skiers, but also to wind erosion. Smoothing operations include removing troublesome ruts and moguls (snow mounds). In some ski areas icy slopes are chopped up to produce a corn-like surface. Compaction, smoothing, ice-chopping, etc., all augment the average skier's enjoyment and safety.

There is also a trend to groom competitive slopes to optimum conditions. Studies by Yosida (1972) and Kuroiwa and LaChapelle (1972) point out that for competitor safety and maneuverability, slalom and downhill courses should be compacted to densities of about 500 kg/m^3, and to a hardness which

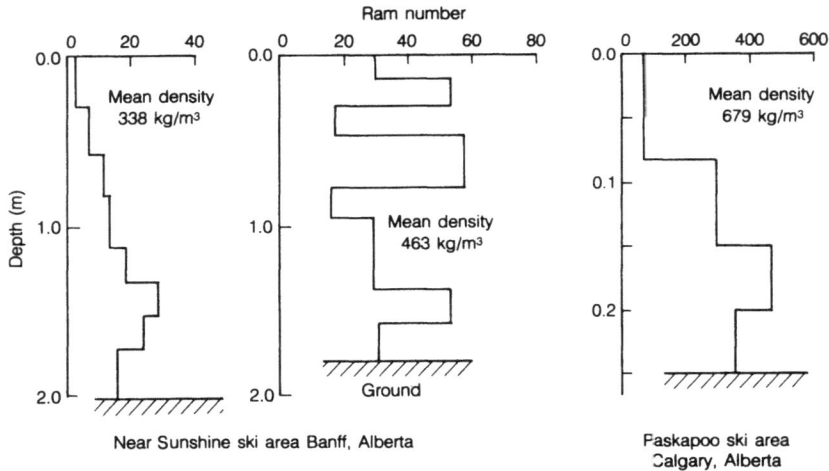

Fig. 19.5 Ram penetrometer profiles of: left, a natural ski field; center, a hardpack from natural snow; and right, a hardpack from mixed natural and artificial snow.

resists penetration by a steel shovel ($N \simeq 100$). This condition can usually be achieved by boot-packing (trampling snow with ski boots) but not by ski or snowcat compaction, since pressures under a ski or snowcat are about one-sixth that under a boot. Boot pressure of a person standing on one leg is about 4 kPa. To obtain maximum effect boot-packing should begin with the first snow before the snowpack thickness exceeds about 30 cm, and before sintering and recrystallization affect its texture. After boot-packing, the slopes should be smoothed with ski and snowcat, and time allowed for sintering, which can take from about one to two weeks, depending on temperature. Sintering becomes more rapid as the temperature approaches 0° C. Taking into account such factors as manpower, recrystallization, ground heat, and skier safety, Yosida (1972) suggests that an optimum pack thickness is about 40 cm.

Ski slopes can also be chemically treated with salts, including sodium chloride, ammonium nitrate, and calcium chloride, to soften ice, refreeze slush, or promote durability depending on snow temperature, (Stillman, 1962; MacConnell et al., 1974). Despite environmental objections and problems of cost and corrosion, chemical treatment appears to be gaining popularity, especially in ski areas where snow is manufactured artificially, since the chemicals can be added directly into the snow-making equipment.

A current problem is how to standardize the reporting of ski conditions to the public. In the eastern U.S.A., such reporting of snow conditions has essentially stopped because of the lack of agreement regarding terminology. The Ontario Ski Resorts Association uses guidelines for groomed slopes as follows:

> *Good.* Packed natural, man-made or machine-groomed snow on a packed base with no bare spots and less than 10% icy spots on developed runs.
>
> *Fair.* Machine-groomed snow on a packed base with more than 10% icy spots and less than 10% bare spots or wet snow on normal runs.
>
> *Poor.* Icy slopes with more than 10% bare spots on normal runs.

The public expects well-groomed slopes with major hazards identified. Some questions remain unsettled concerning the risks that a person assumes while skiing in a developed area. In one controversial legal suit, a severely injured skier was awarded $1.5 million because he hit a "hidden" bush on a supposedly groomed ski trail. Some lawyers now advise ski area operators that they are in a sounder legal position if they do not identify (i.e. flag) rocks, bushes, and other hazards.

LOW FRICTION OF ICE AND SNOW

The ease and speed of ski travel results from the relatively low coefficient of friction (< 0.1) at the snow-ski interface, which largely depends on the low friction of the individual ice crystals. Seligman and Debenhan (1943) speculated that the "ball-bearing" action of snow grains could also be a friction-reducing factor; however, snow grains rarely occur as unsintered spheres, so that any lessening of friction because of this "ball-bearing" action is probably of secondary importance compared to the intrinsic low friction of ice.

Even though the low friction of ice is important to such diverse activities as skating and highway transportation, and to physical phenomena involving ice forces against structures, the fundamental explanation for this low friction remains controversial. The traditional explanation of Bowden and Hughes (1939), that low friction is *primarily* due to a thin film of melt-water induced by heat at the ice-slider contact, continues to receive support (Evans et al., 1976). In principle, there is sufficient energy derived from sliding to melt a thin film, even at relatively cold temperatures; water films have been observed by many investigators. However, many investigators remain unconvinced that a frictionally-melted water film is the primary explanation, e.g., Huzioka (1963) found that melt water increased friction, presumably because of viscous resistance.

McConica (1950), University of Minnesota (1955) and Niven (1959) argued that the low friction of ice and snow also derives from some unique characteristics of the ice surface, and not only from a film of melted water. They considered ice as a solid lubricant, analogous to graphite, which has relatively low friction since adjacent crystallographic planes can slide easily over one another, and can reorient themselves to reduce friction when rubbed. Surface films, such as oxides, sometimes play a role in solid lubrication, but they are not necessarily produced by melting. In perhaps the most comprehensive study to date on ice and snow friction, the University of Minnesota (1955) modeled the asperities (surface roughness features) of an ice surface as elastic-plastic microscopic protrusions. From their experimental and theoretical work they concluded that ice asperities have an uniquely low shear strength. They also speculated that the ice surface has an intrinsic "liquid-like structure", irrespective of frictional melting, which could be considered a separate effect and cited Weyl's (1951) conjecture that the "liquid-like" surface layer could have a thickness of several hundred molecular layers, which could be significant compared to the height of the asperities.

Following the study by the University of Minnesota, extensive theoretical arguments and experimental results indicate that mobile protons, disordered atoms, and larger scale defects are relatively abundant on the surface of ice; see Golecki and Jaccard (1978) for a summary of experimental results. At -1°C various measurements indicate that: the defect layer is $\sim 10^4$ nm thick (electrical conductivity); the layer of atomic disorder is $\sim 10^2$ nm thick (proton scattering) and the layer of proton mobility is ~ 10 nm thick (nuclear magnetic resonance). In a sense, the combination of defects, disorder, and mobility could imply a "liquid-like" surface; however, Golecki and Jaccard surmise that the structure of the surface is sufficiently different from a liquid that the term "liquid-like" is misleading, and can therefore be abandoned. Regardless of terminology, the surface of ice is unique compared with other materials, and according to Fletcher (1973), even compared with other hydrogen-bonded substances.

Defects, disorder, and mobility should significantly contribute to the relatively low shear strength of ice asperities, and hence to a relatively low friction. As the temperature decreases from 0°C, disorder and mobility decrease, and defects become more difficult to activate. Consequently, asperities are more difficult to shear (or reorient) so that ice friction increases as temperature decreases--a well-known fact to inhabitants of cold climates. The number of mobile molecules on the surface should also be affected by humidity. It is interesting to note that competitive skiers pay close attention to humidity, and generally soften their waxes when it is relatively high.

The above remarks are not meant to imply that frictional melting is unimportant for reducing friction, but rather that other effects should also be considered. The complex nature of the "friction" phenomenon is summarized by the University of Minnesota (1955) and by Kuroiwa (1977), who discuss the combined effects of plastic deformation, asperity fracture, and lubricating films.

RESISTANCE TO SKI MOVEMENT

The resistive force exerted by the snow can be separated into two components. One is due to the ploughing, shearing and compressing action of a ski which penetrates into a macroscopically soft snowpack and disturbs it to the depths greater than the snow grain dimensions, sometimes to about one metre. The second component is the frictional interaction at the ski-snow interface, and is due to the microscopic effects discussed in the previous section.

The component due to ploughing, shear and compression has been discussed by only a few investigators (Huzioka, 1957; Nakaya et al., 1972; and Kuroiwa, 1977), and mostly in terms of qualitative observations of snow disturbance. In principle, its magnitude may be approximated by assuming that the skis cause a perfectly plastic disturbance of the snow. If a ski advances at a speed V through snow with initial density ρ, then the resistive pressure P_r can be approximated by the expression;

$$P_r \cong \rho \, (\rho + \Delta\rho) \, V^2/\Delta\rho, \qquad 19.3$$

where $\Delta\rho$ is the snow density increase due to compression (Dolov, 1967; Wakahama and Sato, 1977). For example, consider Huzioka's (1957) measurements of a ski compressing snow from an initial density of 65 kg/m³ to a final density of 165 kg/m³. If the cross-sectional area of each ski track is 70 mm x 70 mm, then the skier would be overcoming a resistive force of 10 N, which is about half the value expected for the friction component. According to Eq. 19.3, P_r should increase with speed and density; in addition, as the speed increases, the resistive force vector rotates somewhat toward a direction perpendicular to the ski thus providing a flotation force which forces the fast-moving ski to the surface. At slow speeds, say < 1 m/s, the approximation for P_r (Eq. 19.3) probably becomes invalid and the resistive forces are controlled by the material properties of the snow rather than by inertial effects (i.e., the forces may no longer be proportional to ρV^2).

At present writing, the vast majority of ski studies have dealt with the frictional component and have tacitly assumed that the skis are supported sufficiently close to the surface so that the other component could be neglected. These studies showed that friction depends on many variables; some examples are given below, not listed in order of importance:

Speed. Friction is a complex function of speed V. One simplification is to define a static coefficient of friction μ_s for the special case, V = 0, and a kinetic coefficient μ_k, for the case $V \neq 0$. The values of μ_k for $0.1 \leqslant V \leqslant 10$ m/s are less (often considerably) than those for μ_s (Palosuo et al., 1977; Keinonen et al., 1978). As V increases ($\geqslant 10$ m/s), μ_k increases exponentially and may eventually exceed μ_s, presumably because of viscous and inertial effects (Kuroiwa et al., 1969; Shimbo, 1972; and Kuroiwa, 1977). For $V \leqslant 0.1$ m/s, Shimbo (1972) found that μ_k increases toward the values of μ_s; however, the University of Minnesota (1955) emphasized that a finite time of stationary contact must elapse before the full adhesion effects of static friction are established.

Surface temperature. Most investigators have confirmed the well-known experience that ski friction is a minimum close to the melt point, and increases both with decreasing temperatures or increasing liquid water. McConica (1950) found that minimum friction occurs at a few degrees below 0°C rather than at 0°C. He claimed that the solid lubrication mechanism, was more important than the friction-melting mechanism and speculated that even the smallest amount of melt water increased rather than decreased the friction.

The rate of increase of friction with decreasing temperature is controversial, probably depending on variables such as speed and slider material. Examples of a strong dependence are shown in Fig. 19.6, from data by Bowden and Tabor (1964) for μ_k at very slow, but unspecified speeds. Note that at -25°C, the friction of lacquered and paraffinned skies approach that of

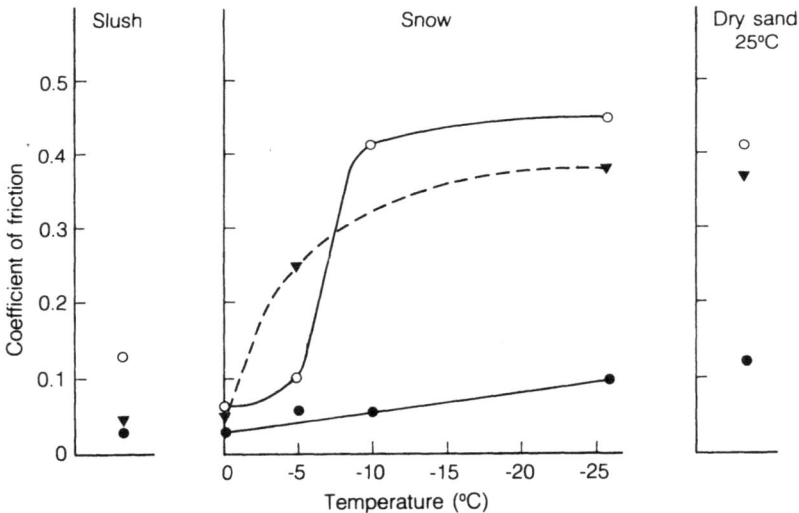

Fig. 19.6 Influence of temperature on the friction of a ski sliding at very low speeds. The ski was covered with: o, ski lacquer; ▼, paraffin wax; •, Teflon. (Bowden and Tabor, 1964). Copyright © Oxford University Press 1964. Reproduced by permission of Oxford University Press.

a ski on sand. The increased friction of slush for lacquered skis is apparent. Friction is shown to increase more slowly for Teflon than for treated wood. Palosuo et al. (1977) found similar results for polyethylene skis; but at $V \approx 2.5$ m/s, some of their data even indicate a decrease in μ_k with decreasing temperature — μ_s was more sensitive than μ_k to temperature variations. However, they recognized the influence of the air temperature gradient, since higher values of μ_s and μ_k were observed when the air temperature above the snow was colder than that of the snow surface. An opposite gradient produced little effect.

The reduction of friction near $0°C$ suggests that friction on cold, dry arctic snow could be minimized by using electrically heated runners. This was tested in at least one experiment (Pfalzner, 1947), with negative results that are somewhat puzzling since, irrespective of its method of production, heat should lower friction.

Contact pressure. Eriksson (1949) found that friction decreased with increasing contact pressures greater than 10 kPa. This phenomenon is supposedly exploited by sled drivers (Wakefield, 1938), who believe that after a sled has been heavily loaded the frictional drag increases only slowly with increasing load. However, for average pressures exerted by typical skiers (see Table 19.1), the frictional decrease with increasing load appears to be negligible. For example, holding other variables constant, Shimbo (1972) found that both μ_s and μ_k remained quite constant for the pressure range of Table 19.1 (see Fig. 19.7); Palosuo et al. (1977) obtained similar results. Moreover, competitive downhill skiers reach their fastest times on relatively long skis (typically 2.23 m long); their lower average pressures at the ski-snow contact does not apparently result in any significant increase in friction which could affect the stability advantage of the longer ski.

Table 19.1
TYPICAL SNOW PRESSURE EXERTED BY
RECREATIONAL SKIERS

Skier	Weight kg	Recommended Ski Dimensions length m	width mm	Snow-Ski Contact Length m	Average Pressure kPa
Alpine	80	2.05	74	1.80	3.0
Cross-country	80	2.10	44	1.80	5.0
Alpine	55	1.90	73	1.65	2.2
Alpine	30	1.50	70	1.28	1.7

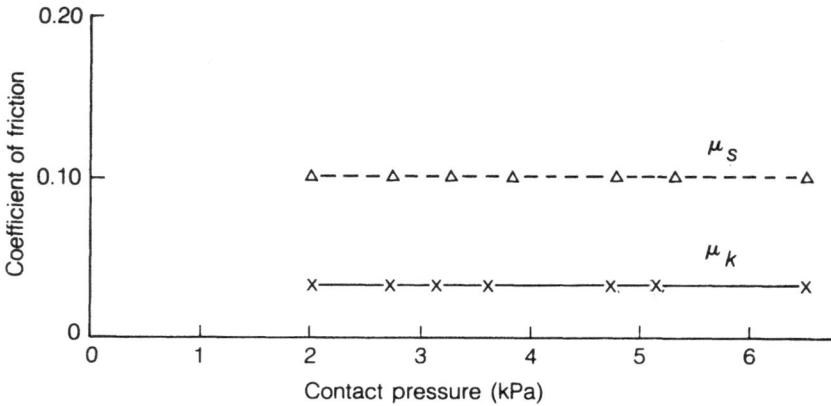

Fig. 19.7 Coefficient of friction versus contact pressure. Flourine resin slider. Air 7°C; snow 0°C, wet and compact (Shimbo, 1972).

Contact area and grain size. Very little is known about the area of ice actually in contact with the ski, yet this parameter is probably crucial to the effects produced by melt water films and solid lubrication mechanisms. According to Torgersen (personal communication), the ice contact area for dry new snow at -7°C is less than 1% of the area of the ski, but for new snow melting at 0°C increases to nearly 80%. Kuroiwa (1977) summarizes the work of Huzioka (1963) and other investigators that confirms the occurrence of small ice contact areas for initially dry snow and the production of melt water in the contact areas, as photographed through glass plates.

A snow surface with larger, less numerous grains is considered to have a lower μ_k; in the limit, the fastest competition times are set on icy tracks. Eriksson (1949) and Palosuo et al. (1979) found that ski resistance increases rather strongly with decreasing grain size from the ice-limit to small grains characteristic of newly fallen snow, but it is not clear how much of this increase can be attributed to friction as opposed to ploughing.

Slider material. Both μ_k and μ_s strongly depend on the adhesion between the snow and the slider material. One convenient measure of adhesion is the contact angle, i.e., the angle formed between a droplet of water and a flat surface of material on which it is resting (Jellinek, 1957; Palosuo et al., 1979). Contact angles range from about 0° for most metals to about 110° for several polymeric materials; in general, the higher the contact angle, the lower the coefficient of friction. Jellinek (1957) measured the contact angles of an assortment of substances including resins, waxes and plastic coatings and found, as expected, that high contact angle was correlated with symmetry or non-polarity of the chemical structure. An example of a well-known material with a high contact angle (about 104°) is "Teflon" (polytetrafluoroethylene or PTFE) which has the highly symmetrical, non-polar structure shown in the following diagram where F, C represent flourine and carbon, respectively.

```
        F    F    F    F    F    F
        |    |    |    |    |    |
  ~     C    C    C    C    C    C   ~
        |    |    |    |    |    |
        F    F    F    F    F    F
```

Bowden and Tabor (1969) compared descent times on gentle slopes of skis coated with PTFE, ski wax, and protective lacquer: PTFE-coated skis were 100% faster than the lacquered skis and 25% faster than paraffinned skis. PTFE has not been found to be practical as a ski base due to its poor mechanical properties; however, ski manufacturers are experimenting with various ways of incorporating PTFE into the more durable bases. Almost all modern downhill skis and an increasing number of cross-country skies are manufactured with a high density polyethylene base (contact angle $\sim 90°$) which has a low coefficient of friction and is lightweight, inexpensive, and easy to patch.   Manufacturers are experimenting with various polyethylene densities to optimize friction and other properties. It appears the higher the density, the lower the μ_k.   Manufacturers are also experimenting with chemical and mechanical base treatments to maximize downhill speed.

Roughness. At first it seems reasonable that the smoother the slide surface, the lower the μ_k, but this is not always true. Competitive skiers know that if the snow is wet, a highly polished wax surface will in fact be slower than a deliberately-roughened wax surface.   Shimbo (1972) found that μ_k may increase to $3\mu_s$ for a very smooth slider (bumps < 0.01 mm) on wet snow. He calls this phenomenon "inversion of the coefficient of friction". Perhaps surface tension and viscous resistance become increasingly important as the snow becomes wet, in which case some useful expressions for these effects relevant to snow friction are derived by Moore (1975).

SKI WAX

Five categories of wax-like compounds are presently manufactured to help skiers optimize glide and grip: downhill wax, glider wax, cross-country wax, klister, and base wax.

Downhill Wax

Downhill wax is intended to give downhill or alpine skiers the fastest possible glide for specific snow conditions. Competitive skiers believe that a properly waxed ski is faster than any known polymeric base, including PTFE, and will spend much time impregnating downhill wax into the polyethylene. For competition, the wax is applied hot, scraped off, and then polished; the process is repeated 10 or more times. The final thin coat may be scratched (parallel to the length of the ski) if the snow is wet. Some competitive skiers try to break the "inversion phenomenon" by implanting a shingle-like pattern into the wax. It would seem that the harder the wax, the faster the glide, but this does not agree with experience. Skiers often find that a softer wax provides faster glide; the physics is virtually unknown. Torgersen (personal communications) considers that the hardness of the wax should be matched to the *microhardness* of the snow crystal to optimize glide, and that if the hardness is too great, energy is lost as the wax scratches the snow. Similarly, if the wax is too soft, energy is lost as the snow asperities scratch the wax. A variable which further complicates the physics of gliding is the skiers speed. In anticipation of speeds faster than 80 km/h, downhill skiers sometimes wax softer than slalom and giant slalom racers (Torgersen, personal communications). Another complicating factor is the thickness of the wax layer. Shimbo (1972) found that the static coefficient increases greatly with wax thickness, but the sliding coefficient μ_k increases only slightly (see Fig. 19.8). Competitive skiers sometimes increase the thickness of the layers as the liquid water content of the snow increases, but for dry snow the wax is kept as thin as possible.

Glider Wax

Glider wax is fairly new and was originally designed for the competitive cross-country ski. It is applied to the front- and rear-thirds of the skis to optimize glide (as opposed to wax applied to the middle-third for both grip and glide). Today, recreational as well as competitive skiers are learning to use glider wax. Manufacturers claim that glider and downhill waxes are essentially the same, except the "gliders" contain a more durable texture that resists abrasion, and therefore can remain on the ski for track lengths characteristic of cross-country events (30 km, 50 km). However, during the 1980 Winter Olympics at Lake Placid, at least one top downhill contender used glider wax; and several cross-country contenders used downhill wax on the glide portions of their skis, so that differences between the waxes are probably small. In any case, the remarks concerning the physics and the application of downhill wax also pertain to glider wax.

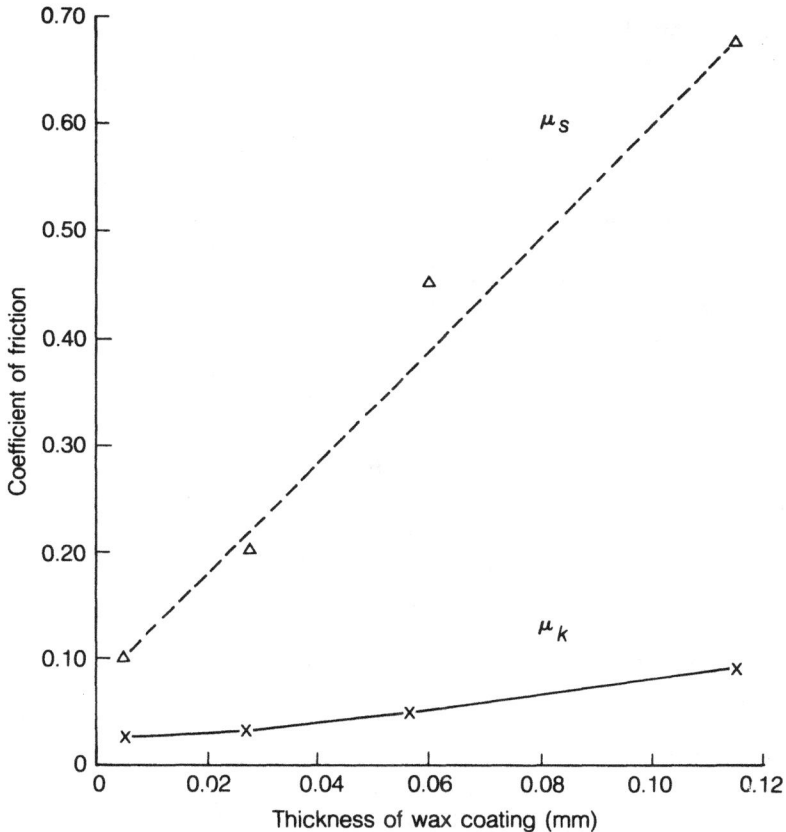

Fig. 19.8 Coefficient of friction versus thickness (mm) of cross-country wax coating. Air 8 to 9° C; snow 0° C, wet and compact. (Shimbo, 1972)

Cross-country Wax

Cross-country wax is the general purpose wax used by the cross-country skier for both "grip" and "glide". The thickest amount is usually applied to the middle of the ski beneath the foot to cover about one third of the ski towards the tip and tail. However, in some cases the wax may cover the entire ski, depending on snow conditions, the skier's weight, the ski camber and base, and whether the tips and tails have been waxed with "glider".

Proper cross-country skiing technique demands that the skier shift his weight from ski to ski. As the skier transfers weight onto a ski, the middle of the ski flexes down onto the snow, and snow crystals are pushed into the wax.

Depending on ski camber, the peak contact pressure may be an order of magnitude higher than the contact pressures listed in Table 19.1. This enables the ski to "grip" the snow, so that the skier can push along on flat terrain, or climb uphill. Ski manufacturers have measured binding forces as high as twice the skier's weight in the brief interval of the grip phase of motion (Stafner, 1979). As weight is released, the centre of the ski springs off the snow; optimally during the forward motion of the ski only the tips and tails, which are waxed more for glide than grip, remain in contact with the snow. Obviously during the weight transfer process there is an interval during which the mid-portion of the ski is in contact with the snow during the glide action. This requires that the wax have reasonable glide characteristics in addition to its gripping ability. Snow crystals which stick in the wax during the grip phase must come out of the wax during the glide phase.

Klister

Klisters are paste-like blends of synthetic resins, mineral oils, and synthetic rubber (patent claimed in 1913), used to provide grip and glide on crusts and slushy snow, conditions that are notoriously troublesome for the cross-country skier. They are remarkably resistant to abrasion, even on cold, icy tracks. In the same way as cross-country wax, they are applied to the middle portion of the cross-country ski. A competitive cross-country skier may own several pairs of skis, differing according to camber, and each intended for a different combination of wax or klister. Normally, the camber of a "klister" ski is "high", that is, its mid-section (klistered) rides with ample clearance above the snow during the glide part of the movement.

Base Wax

Base wax is used to bond cross-country wax onto the middle portion of the ski, and is not intended to contact snow, but must be covered with the cross-country wax. It is not applied to the front or rear of the ski, where the glider wax is "worked" directly into the polyethylene. Bonding with base wax is especially necessary to resist the abrasion of cold, coarse-grained crystals and crusts. It is believed that waxes wear off layer-by-layer; therefore cross-country skiers anticipating a long tour will wax with many thin layers, usually applying the hardest layer immediately above the base wax. Owing to their relatively high synthetic resin and synthetic rubber contents, some of the more viscous klisters bond quite well onto polyethylene without the intermediate base wax.

Early recipes for ski wax included tar, bacon rind, "calves" blood, old bicycle tires, gramphone records, and a high proportion of beeswax (Bo,

1968). Today, beeswax is replaced by mineral waxes because of their resistance to the high temperatures which occur during production and during application with a hot iron (300°C). Also mineral waxes contain less uncontrollable impurities than animal and vegetable waxes, and are available in a wide range of textures (Torgersen, personal communication). The two categories of mineral waxes used by the ski wax industry are paraffin and micro-crystalline wax, whose chemical structures are illustrated in the diagrams where H, C represent hydrogen and carbon, respectively.

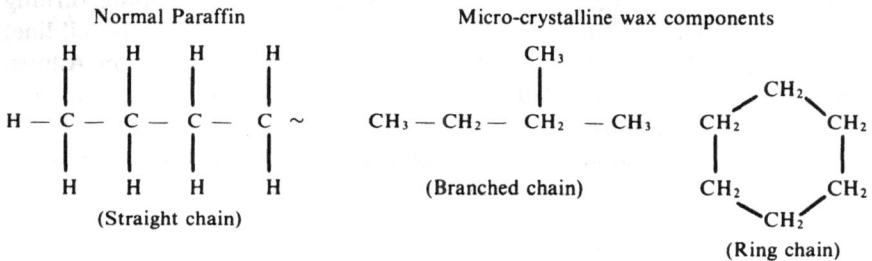

Other substances such as synthetic resins, plasticizers, rubbers, mineral oils, and aluminum powder are added to control hardness and durability; also colouring, for identification. Some waxes, sold as pastes or in spray cans, are softened by solvents that are intended to evaporate after application.

Pioneer studies on the science of ski wax were made by Hultberg (1947) who emphasized the importance of the contact angle and the rheology of the wax, i.e., the balance between its elastic, plastic, and viscous characteristics. Normal paraffin has a high contact angle (105°) and gives a good glide on wet snow, but a poor glide on dry, cold snow. Micro-crystalline wax has a tougher texture than paraffin, and provides a good glide on dry snow, but a poor glide on wet snow. Optimum glide therefore requires selecting a wax with the proper proportions of paraffin and micro-crystalline wax to match the temperature, texture, and water content of the snow.

The grip of cross-country wax or klister is largely a function of its viscosity which is controlled by adding mineral oil to the blend. The wax industry evaluates the kinematic viscosity by measuring the time for a calibrated weight to penetrate a wax sample. Values range over a factor of 300 from 3 MNs/m^2 for the hardest waxes to 10 kNs/m^2 for the softest klisters (Torgersen, personal communication). In general, high viscosity is required to achieve the proper balance of glide and grip on new and cold snow; low viscosity, on warm or wet snow.

Many skiers are interested neither in waxing for high performance nor cleaning and rewaxing the ski base for each new condition. Several non-wax bases are available to provide both grip and glide, e.g., mohair strips, fish-scale patterns, and mica impregnation. These non-wax bases perform as well as conventionally waxed skis only under certain conditions (for example, new snow at $0°$ C), but are significantly inferior for most conditions. Future development could make non-wax bases a more attractive option.

SKIING DYNAMICS

Examples of high speed skiing maneuvers are schussing, jumping, turning and braking. The dynamics of schussing (straight running down the fall-line) have been studied by Sprague (1970), Fukuoka (1971), and Von Allmen (1975). As shown in Fig. 19.9, the schussing skier is pulled by gravity and resisted by air and snow friction. At relatively high speeds (~ 28 m/s), the aerodynamic lift force also becomes significant. Summing these forces, the equation of motion for the schussing skier is:

$$m(dV/dt) = mg \sin \theta - \mu_k(mg \cos \theta - L) - (C_D\rho_aV^2A \cos \theta)/2, \qquad 19.4$$

where:
- m = skier's mass,
- V = skier's velocity at a given time, t.
- dV/dt = skier's acceleration,
- g = acceleration of gravity,
- θ = fall-line slope angle,
- μ_k = kinetic friction between ski and snow (assuming negligible penetration on hardpack),
- L = aerodynamic lift force,
- C_D = drag coefficient,
- ρ_a = air density, and
- $A \cos \theta$ = projected area of the skier in the direction of motion.

In the case where L is small, the terminal velocity from Eq. 19.4 is

$$V_{max} = [D(\tan \theta - \mu_k)]^{1/2}, \qquad 19.5$$

where

$$D = 2mg/C_D\rho_aA.$$

In general a heavy adult will have a higher D-value than a light child, and therefore could develop significantly higher schuss speeds.

For a skier with total mass of about 80 kg (skier, ski, boots, etc.) the smallest value of D, determined by modeling a slow schussing, erect skier as a cylinder in a moving fluid at a Reynold's Number of 6.4 x 10^4 and $C_D = 1.2$, is about 1000 m^2/s^2. The largest value of D similarly determined by modeling a fast, tucked skier as a sphere at a Reynold's Number of 1.7 x 10^6 and $C_D = 0.2$ is about 8000 m^2/s^2 (Olson, 1980). The kinetic friction μ_k on hardpack probably lies in the range from 0.02 to 0.20, depending on temperature, snow texture and related factors (Shimbo, 1972). A graphical solution of Eq. 19.5 is given in Fig. 19.10 for various values of μ_k and air drag.

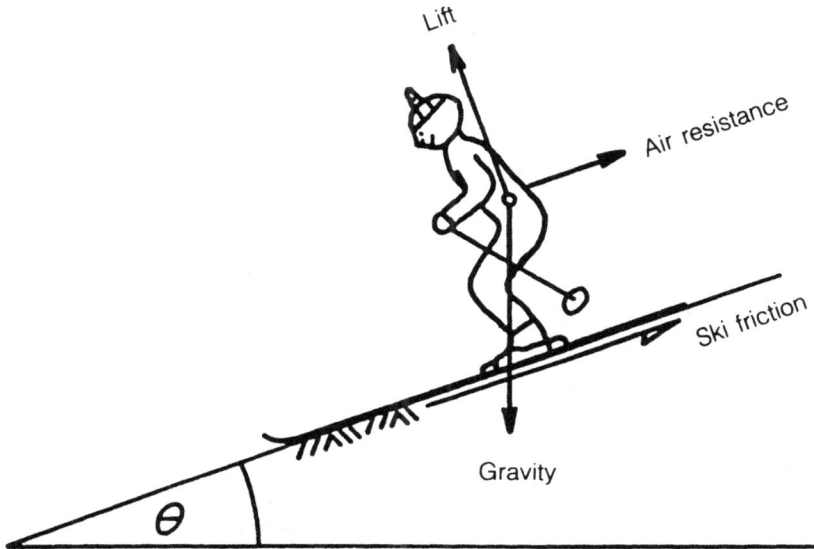

Fig. 19.9 Forces on a schussing skier.

From wind-tunnel experiments, it appears that the air resistance force exerted on a skier moving at about 28 m/s is about 90 to 180 N, depending on his projected area, clothing and body position (Eidgenössiches Flugzeugwerk, 1967). Taking into account aerodynamic lift, the forces against an 80-kg skier moving at about 28 m/s on a 10° slope with $D = 5700$ m^2/s^2 and $\mu_k = 0.04$ are approximately as shown in Fig. 19.11.

Wind-tunnels have also been used to study the mechanics of ski-jumping (Tani and Iuchi, 1972; Iguro and Yamaki, 1972). The dynamical equation for jumping is similar to Eq. 19.4 excluding the friction term, μ_k, and contains terms which can model a skier leaning considerably farther forward than during schussing and otherwise changing his body position to minimize air drag. Tani and Iuchi (1972) give detailed guidelines for optimum body position for ski jumping. Iguro and Yamaki (1972) specify optimum shapes for ski jumping hills.

The dynamics of turning have been examined in many published and unpublished reports. Waller (1946) showed how the skier's turning radius

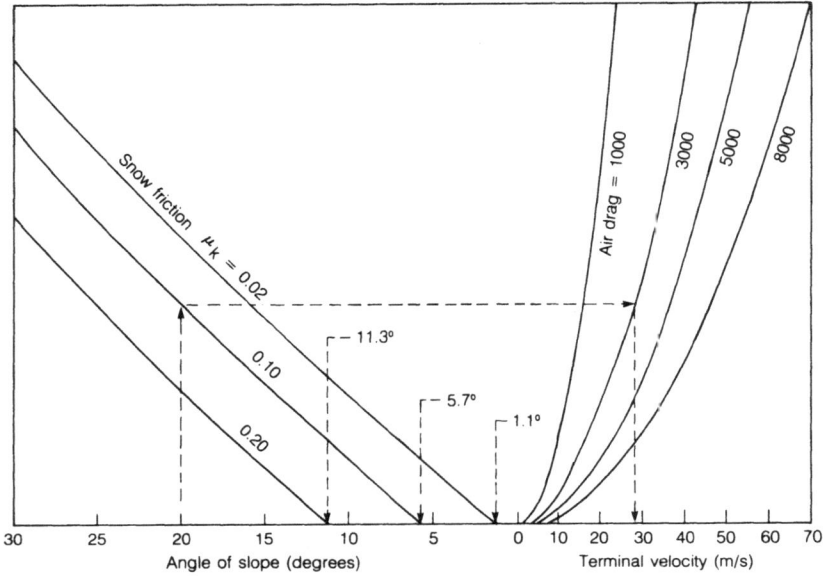

Fig. 19.10 Nomograph for estimating schuss velocities. Air drag factors (m²/s²): 1000—slow speed, skier erect; 3000—medium speeds; 5000—high speeds, high tuck; 8000—high speeds, egg shape.

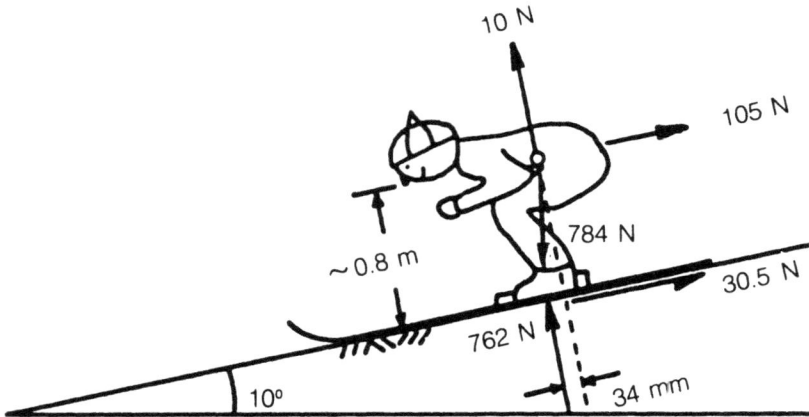

Fig. 19.11 Forces on an 80-kg skier schussing at 28 m/s on a 10° slope.

tends to increase after the skier crosses the fall-line of the slope. Fukuoka (1971) and Morawski (1973) studied the wedeln style of turning; Fukuoka also measured knee and boot forces.   Brandenberger (1974) incorporated mechanics into Swiss ski instruction. Glenne and Von Allmen (1979) modeled contemporary turning styles. Other studies include those by Moser (1958), Vagners (1972), Mote et al. (1976), Howe (1977), Karlsson et al. (1978), Larsson and Major (1978), Schultes (1978). A well-executed turn involves carving an arc (radius = r) with minimum skidding, when the force on the skier is increased by a centrifugal term mV^2/r. Values of the centrifugal force, calculated by Kopta (1974) from measurements of skier mass, velocity, and turning radius were found to range as high as 1010 N for expert skiers on 19 to 27° slopes.

The mechanics of braking involve skidding or sideslipping the skis with gradual indentation of the edges. Theoretically, braking forces should equal the energy gradient during the maneuver. If a skier brakes from a speed V in a distance s on a slope with angle θ the average braking force F would approximately be given by:

$$F = mV^2/2s + mg \sin \theta. \qquad\qquad 19.6$$

For example, the average force exerted against an 80-kg skier braking from a speed of 10 m/s in a distance of 5 m on a 15° slope would be $\sim$ 1 kN.

Although forces of turning and braking are not exceptionally high, nevertheless they can produce dangerous torques to the skier's legs. Several studies have specifically treated the safety aspects of ski maneuvers: Outwater and Woodward (1967), and Outwater (1969) studied the stress transmitted from ski to leg; Mote and Hull (1976) measured the release characteristics of safety bindings; Von Allmen (1974, 1975) studied the kinetics of fallen skiers.

It seems that an analytical treatment of skiing dynamics should be a prerequisite for ski design, but as discussed in Ch. 20 the reverse often occurs: advances in ski construction determine the style and standards of skiing.

FUTURE DEVELOPMENTS

Skis and skiing styles will continue to evolve. Five thousand years from now a very strange skier will unearth a very obsolete pair of skis. Meanwhile, there will be improved techniques for monitoring snow parameters which are directly relevant to skiing. Grooming standards for ski areas and for cross-country tracks will improve. Chemical treatment of artificially-produced snow will become widely accepted.

Research will produce more definitive conclusions about the fundamental physics of friction on snow and ice, and about the relative importance of melt water films and solid lubrication mechanisms. The physics of waxed ski surfaces will be better explained toward the practical objective of optimizing glide for various combinations of snow parameters. Such curious phenomenon as why glide surfaces have to be softened or roughened for certain conditions will also be explained. Waxes and klisters will not become obsolete; instead wax companies will constantly improve their products. At the same time, major breakthroughs in developing wax-free bases will occur and provide a relief to many cross-country skiers.

ACKNOWLEDGEMENTS

Background material was supplied by R. Frederking, National Research Council of Canada, Ottawa; S. Jones, Environment Canada, Ottawa; A. Lenes, Canadian Mountain Holidays, Banff; R. Miller, USDA Forest Service, Denver; D. Slotfeldt-Elligsen, Central Institute of Industrial Research, Oslo; and L. Torgersen, Astro-Wallco Corporation, Skarer, Norway.

LITERATURE CITED

Bader, H., R. Haefeli, E. Bucher, J. Neher, O. Eckel, C. Thams and P. Niggli. 1939. *Der Schnee und seine Metamophose (Snow and its Metamorphism)*. Beit. z. Geol. Schweiz. Geotech. Ser. Hydrol., Lief. 3. Bern. [English Transl. 14 by U.S. Army Snow, Ice Permafrost Res. Estab.]

Berg, G., G. Lundqvist, A. Zetterstern, and E. Granlund. 1941. *Finds of skis from pre-historic time in Swedish bogs and marshes*. Esselte, Stockholm.

Bowden, F.P. and T.P. Hughes. 1939. *The mechanism of sliding on ice and snow*. Proc. R. Soc., Ser. A, Vol. 172, pp. 280-298.

Bowden, F.P. and D. Tabor. 1964. *The Friction and Lubrication of Solids, Part II*. Oxford Univ. Press, London.

Brady, M. 1979. *Cross Country Ski Gear*. Mountaineers, Seattle, Wash.

Brandenberger, H. 1974. *Skimechanic: Skimethodik*. Derendingen-Solothurn, Switzerland.

Bo, O. 1968. *Skiing Traditions in Norway*. Det Norske Samlaget, Oslo.

Colbeck, S.C. 1978. *The physical aspects of water flow through snow*. In Advances in Hydroscience Vol. 11, (Ven Te Chow, ed.), Academic Press, New York, pp. 165-206.

Dolov, M.A. 1967. *Deformatsiia snega pri bol-shikh impul-snykh nagruzkakh (Snow deformation in the case of large load impacts)*. Nal-chik, USSR Vysokogornyi Geofizicheskii Inst., Trudy, No. 9, pp. 3-12.

Dudley, C.M. 1935. *Sixty Centuries of Skiing*. Stephen Daye Press, Brattleboro, Vermont.

Dwyer, C.F. 1975. *Aerial tramways, ski lifts and tows*. Rep. EM-7320-1, USDA For. Serv., U.S. Gov. Printing Off., Washington, D.C.

Eidgenössischen Flugzeugwerk. 1967. *Luftwiderstandmessung an Skifahrer*. Bericht FO 882, Emmen, Switzerland. (Unpubl.)

Enzel, R.A. 1979. *The White Book of Ski Areas*. Rand McNally, Chicago, Ill.

Ericksen, N. 1980a. *A short history of snowmaking*. Ski Area Management, May, pp. 70-71.

Ericksen, N. 1980b. *Snowmaking: the state of the art*. Ski Area Management, July, pp. 32-38.

Eriksson, R. 1949. *Medens friktion mot snö och is (Friction of runners on snow and ice)*. Foren skogsarbet., Foren Skogsarbet., Kg., Domänstyrelsens arbetsstud, Med. 34-35. Stockholm. [English Transl. 14 by U.S. Army Snow, Ice Permafrost Res. Estab.].

Evans, D.C.D., J.F. Nyre and K.J. Cheesman. 1976. *The kinetic friction of ice*. Proc. R. Soc. London, Ser. A, Vol. 347, pp. 493-512.

FIMS. 1972. *First International Congress of Winter Sports Medicine, Sapporo, Japan*. Fédération Internationale de Médecine Sportive, Torino.

Firsoff, V. 1943. *Ski Tracks on the Battlefield*. A.S. Barne and Co., New York, N.Y.

Fletcher, N.H. 1973. *The surface of ice*. In Physics and Chemistry of Ice (E. Whalley, S.J. Jones and L.W. Gold, eds.), R. Soc. Can., pp. 132-136.

Flower, R. 1976. *The History of Skiing and Other Winter Sports*. Methuen, Toronto, Ont.

Foss, M.L. and Garrick, J.G. 1978. *Ski Conditioning*. John Wiley & Sons, New York, N.Y.

Fukuoka, T. 1971. *Zur Biomechanik und Kybernetik des Alpinen Schilaufs*, Limpert Verlag, Frankfurt/M, Germany.

Glenne, B. and B. von Allmen. 1979. *Basic mechanics of alpine skiing*. Proc. 3rd Int. Conf. Ski Trauma Skiing Safety, Queenstown, New Zealand.

Goeldner, C.R., K. Dicke, and Y. Sletta. 1973. *Bibliography of skiing studies*. Res. Div., Grad. School Bus. Admin., Univ. Colo., Boulder.

Goeldner, C.R. and T. Farwell. 1975. *Economic analysis of North American Ski areas*. Res. Div., GRAD. Sch. Bus. Admin., Univ. Colo., Boulder.

Golecki, I. and C. Jaccard. 1978. *Intrinsic surface disorder in ice near the melting point*. J. Phys. C: Solid State Phys., Vol. 11, pp. 4229-4237.

Good, W. 1975. *Numerical parameters to identify snow structure*. Proc. Int. Symp. on Snow Mechanics, Grindelwald, Int. Assoc. Hydrol. Sci. Publ. 114, pp. 91-102.

Howe, J. 1977. *The physics of skiing*. National Coaches Clinic, Boulder, Colo., Sept. 1977. (Unpubl.)

Hultberg, S.O. 1947. *Skidvallor (Skiwax)*. Teknisk Tidskrift 79, Sweden, pp. 935-942.

Huzioka, T. 1957. *Skii no kenkyy (3) (Studies on ski (3))*. Low Temp. Sci., Ser. A, No. 16, pp. 31-46.

Huzioka, T. 1963. *Studies on the resistance of snow sledge VI. Friction between snow and a plate of glass or plastic*. Low Temp. Sci., Ser. A, No. 21, pp. 31-43.

Iguro, M. and A. Yamaki. 1972. *Design of ski jumping hill for 1972 Winter Olympic Games*. Scientific study of skiing in Japan, Soc. Ski Sci. Hitachi, Tokyo, pp. 53-62.

Jellenik, H.H.G. 1957. *Contact angles between water and some polymeric materials*. Res. Rep. 36, U.S. Army Snow, Ice Permafrost Res. Estab.

Judge, P. 1979. *"State-of-the-art" Gondola.* Ski Area Management. Nov., pp. 48-49, 66.

Karlsson, J., A. Ericksson, A. Forsberg, L. Kallberg and P. Tesch. 1978. *The physiology of alpine skiing.* [English Transl. by W. Michael, U.S. Ski Coaches Assoc., Park City, Utah, 1978.]

Keinonen, J., E. Palosuo, P. Korhonen and H. Suominen. 1978. *Lumen ja suksenpohjamuovien välisen kitkan mittauksia (Measurements of friction between snow and sliding materials of ski).* Rep. 10, Univ. Helsinki, Rep. Ser. in Geophys.

Kinosita, S. 1960. *The hardness of snow.* Low Temp. Sci., Ser. A, No. 19, pp. 119-134.

Klein, G. 1947. *The snow characteristics of aircraft skis.* Aeronaut. Rep. AR-2, Nat. Res. Counc. Can., Ottawa, Ont.

Knight, S.J. 1965. *Trafficability of snow and muskeg.* Proc. 15th Alaskan Sci. Conf., Univ. Alaska, College, pp. 355-374.

Kopta, D. 1974. *The centripetal force produced by skiers.* Civil Eng. Dept., Univ. Utah, Salt Lake City. (Unpubl.)

Kry, P.R. 1975. *Quantitative stereological analysis of grain bonds in snow.* J. Glaciol., Vol. 14, No. 12, pp. 467-477.

Kuroiwa, D. 1977. *The kinetic friction on snow and ice.* J. Glaciol., Vol. 19, No. 81, pp. 141-152.

Kuroiwa, D. and E.R. LaChapelle. 1972. *Preparation of artificial snow and ice surface for XI Olympic Winter Games, Sapporo.* The Role of Snow and Ice in Hydrology, Proc. Banff Symp., Sept. 1972, Unesco-WMO-IAHS, Geneva-Budapest-Paris, Vol. 2, pp. 1350-1361.

Kuroiwa, D., G. Wakahama, K. Fujino and R. Tanahashi. 1969. *Coefficient of sliding friction between skis and chemically treated or mechanically compressed snow surfaces.* Low Temp. Sci., Series A, No. 27, pp. 229-245.

LaChapelle, E.R. 1970. *Principles of avalanche forecasting.* Ice Engineering and Avalanche Forecasting and Control, Tech. Memo. No. 98, Nat. Res. Counc. Can., Ottawa, Ont. pp. 106-111.

Larsson, O. and J. Major. 1978. *Passive, accelerated and instant inclination of the skier's general axis in slalom and giant slalom turns.* J. U.S. Ski Coaches Assoc., Vol. 2, No. 2, Park City, Utah, pp. 18-25.

Lid, N. 1949. *Ski tracks across time and space.* In World Ski Book (F. Elkins and F. Harper eds.), Longmans, Green, and Co., New York.

Lliboutry, L. 1964. *Traite de Glaciologie, Tome I.* Masson and Cie, Ed., Paris.

Lund, R.T. 1971. *A history of skiing in Canada prior to 1940.* Masters Thesis, Dept. Phys. Educ., Univ. Alta., Edmonton.

Lunn, A. 1952. *The Story of Skiing.* Eyre and Spottiswoode, London.

MacConnell, W.P., D.L. Mader and L.F. Whitney. 1974. *Fertilize your snow for a faster harvest.* Ski Area Management, Vol. 13, No. 1, pp. 42-43

Marsh, J. 1975. *The changing ski scene in Canada.* Can. Geograph. J., Vol. 90, No. 2, pp. 4-13.

McCloskey, M. and A. Hill. 1971. *Mineral King: Wilderness versus mass recreation in the Sierra.* In Patient Earth (J. Harte and R.H. Socolow. eds.), Holt, Rinehart and Winston Inc., New York, N.Y.

McConica, T.H. 1950. *Sliding on ice and snow*. Contract W 44-109 qm-2113, U.S. Army. Off. of the Quartermaster General, U.S Army, Res. Devel. Branch.

McConica, T.H. 1951. *Aircraft ski performance*. WADC Tech. Rep. No. 52-19, U.S. Airforce, Wright Air. Dev. Centre, Ohio.

McConica, T.H. 1952. *Aircraft ski performance*. WADC Tech. Rep. No. 52-318, U.S. Airforce, Wright Air Dev. Centre, Ohio.

Moore, D.F. 1975. *Principles and Applications of Tribology*. Pergamon Press, London, pp. 321-325.

Morawski, J.M. 1973. *Control systems approach to a ski-turn analysis*. J. Biomech., Vol. 6, pp. 267-279.

Moser, G. 1958. *Untersuchung der Belastungsverhaltnisse bei der alpinen Schitechnik*. Theori und Praxis der Korperkultur, H.6, S.551.

Mote, C.D. Jr. and M.L. Hull. 1976. *Fundamental considerations in ski binding analysis*. Orthopedic Clinics North Amer., Vol. 7, No. 1.

Mote, C.D. Jr., M.L. Hull, R.L. Piziali and D.A. Nagel. 1976. *Experimental measurements of forces and moments during downhill skiing*. In Skiing and Safety II, Forum, Davos, Switzerland, Sept., pp. 161-176.

Nakaya, U., M. Tada, Y. Sekido and T. Takano. 1972. *The physics of skiing: Preliminary and general survey*. Scientific Study of Skiing in Japan, Soc. Ski. Sci., Hitachi, Tokyo. pp. 1-32.

Niven, C.D. 1959. *A proposed mechanism for ice friction*. Can. J. Phys. Vol. 37, pp. 247-255.

Olson, R.M. 1980. *Essentials of Engineering Fluid Mechanics*. Intext Publ., New York, pp. 384-401.

Oura, H. 1967. *Studies on blowing snow*. Physics of snow and ice, (H. Oura, ed.), Int. Conf. on Low Temp. Sci., Hokkaido Univ., Sapporo. Vol. 1, Pt. 2, pp. 1085-1117.

Outwater, J.O. and M.S. Woodward. 1967. *Skiing forces and fractures*. Mech. Eng., Vol. 189, No. 12, pp. 27-30.

Outwater, J.O. 1969. *On the friction of skis*. Dept. Mech. Eng., Univ. Vermont, Burlington.

Palosuo, E., T. Hiltunen, J. Jokinen and M. Teinonen. 1977. *Lumen kitkan vaikutus suksen luistoon (The effect of friction between snow and skis)*. Rep. 6, Univ. of Helsinki, Ser. in Geophys.

Palosuo, E., J. Keinonen, H. Suominen and R. Jokitalo. 1979. *Lumen ja suksenpohjamuovien välisen kitkan mittauksia, oas 2. (Measurement of friction between snow and ski running surfaces, Part 2.)*. Rep. 13, Univ. Helsinki, Ser. in Geophys.

Parks Canada. 1973. *Downhill ski areas: utilization and capacity*. Rep. PR-73-3, Policy and Planning Res. Div., Western Region, Calgary.

Perla, R.I. 1978. *Temperature-gradient and equi-temperature metamorphism*. Proc. 2nd Int. Conf. Snow Avalanches, Grenoble.

Pfalzner, P.M. 1947. *The friction of heated sleigh runners on ice*. Can. J. Res., Vol. F 25, pp. 192-195.

Richardson, E.C. 1909. *The Ski Runner*. Published by the author. London, (In the collection of the National Ski Museum, Ottawa).

Scharff, R. 1975. *Ski Magazine's Encyclopedia of Skiing*. Harper and Row, New York.

Schneigert, Z. 1966. *Aerial Tramways and Funicular Railways*. Pergamon Press, New York.

Schultes, H. 1978. *Der Alpinski*. Haller and Jenzer A.G., Switzerland.

Seligman, G. 1936. *Snow Structure and Ski Fields*. MacMillan and Co. Ltd., London.

Seligman, G. and F. Debenham. 1943. *Friction on snow surfaces*. Polar Rec., Vol. 4, No. 25, pp. 2-11.

Shimbo, M. 1972. *Friction on snow of ski soles unwaxed and waxed*. Scientific Study of Skiing in Japan. Soc. of Ski Sci., Hitachi, Tokyo. pp. 99-112.

Society of Ski Science. 1972. *Scientific study of skiing in Japan: Papers in European Languages*. Hitachi, Tokyo.

Sommerfeld, R.A. and E.R. LaChapelle. 1970. *The classification of snow metamorphism*. J. Glaciology, Vol. 9, No. 55, pp. 3-17.

Sprague, R.C. 1970. *Parallel Skiing for Weekend Skiers*. Excelsior Printing Co., North Adams, Mass.

Stafner, A. 1979. *Kick and glide mechanics of cross-country skiing*. Unpublished seminar presented at College Eng., Univ. Utah, Salt Lake City.

Stillman, R.M. 1962. *Effects of methods of slope packing as determined by the ram penetrometer*. Misc. Rep. 5, Alta Avalanche Study Centre, USDA For. Serv., Salt Lake City, Utah.

Tani, I. and M. Iuchi. 1972. *Flight mechanical investigations of ski jumping*. Scientific Study of Skiing in Japan, Soc. Ski Sci., Hitachi, Tokyo, pp. 33-52.

University of Minnesota. 1955. *Friction on snow and ice*. Tech. Rep. 17, U.S. Army Snow, Ice Permafrost Res. Estab., Wilmette, Ill.

USDA Forest Service. 1973a. *Planning considerations for winter sports resort development*. U.S. Gov. Printing Off. 1974-781-126, Washington, D.C.

USDA Forest Service. 1973b. *Winter Sports Symposium 1972*. USDA For. Serv., Rocky Mountain Reg. Denver, Colo.

Unesco. 1970. *Seasonal snow cover*. United Nations Educ. Sci. Cult. Organization, Paris.

Vaage, J. 1979. *Skienes Verden*. Hjemmenes Forlag, Oslo.

Vagners, J. 1972. *Biomechanics Manual*. Prof. Ski Inst. Am.

Von Allmen, B. 1974. *Unzeckmässige Bekleidung - eing Unfallgefahr beim Ski-fahren*. Neue Zürcher Zeitung.

Von Allmen, B. 1975. *On sliding and collision dangers of fallen skiers*. Proc. Am. Soc. for Testing and Materials, Las Vegas, Nev.

Wakhama, G. and A. Sato. 1977. *Propagation of a plastic wave in snow*. J. Glaciol., Vol. 19, pp. 175-183.

Wakefield, W.E. 1938. *Efficiency of logging sleighs for pulpwood operations in different types of terrain*. Project 107, For. Product Lab., Dept. North. Affairs and Nat. Resour., Ottawa, Ont.

Waller, I. 1946. *The Dynamics of Skiing*. The British Ski Year Book for 1946, pp. 103-122.

Weyl, W.A. 1951. *Surface structure of water and some of its physical and chemical manifestations*. J. Colloid, Sci., Vol. 6, pp. 389-405.

Yosida, Z. 1972. *Investigations on snow conditions for the XI Olympic Winter Games, Sapporo, February 3-13*. Scientific Study of Skiing in Japan, Soc. Ski Sci., Hitachi, Tokyo, pp. 113-126.

20

MECHANICS OF SKIS

B. GLENNE

*Department of Civil Engineering, University of Utah,
Salt Lake City, Utah.*

INTRODUCTION

The discussion of the mechanics of snow skis would rightfully begin with an analysis of the way one skis in order to quantify the forcing functions to which skis are subjected. However, since we do not yet understand very well how we ski, and since skis are still designed and constructed empirically, we are obliged to examine skis and skiing from symptoms and causes.

Successful ski design (which precedes successful skiing) is a compromise of the design variables (construction modes, geometry, elastic properties, dynamic behavior, economics) to match the skiers' weight, technique and use. Since there exists a large array of ski models and ski types on the market, it is important that the factors affecting ski-skier-use compatability should be well understood.

The emphasis in this chapter is on alpine skis and skiing (slalom and downhill). However, cross-country skiing (in pre-made tracks and virgin snow) is rapidly growing; our quantitative information about this activity and the skis used is improving.

SKI HISTORY

Traditionally, ski technique has developed as ski equipment has improved. Sondre Norheim from Telemark in Norway is often termed the father of modern skiing. About 1850 he began using a stiff binding which was fastened around the heel of the boot for better ski control. He also promoted ski sidecamber[1], an improvement which reduced sidewall friction and eased turning (Vaage, 1967). Until that time the sidewalls of a ski added liveliness or

---

[1] The reader is referred to Table 20.1 for explanations and definitions of the ski terminology used in the chapter.

Table 20.1
SKI CHARACTERISTICS (see also Fig. 20.2).

1. **Ski Elements**

Tip of Ski:	The extreme forward end of the ski.
Tail of Ski:	The extreme rear end of the ski.
Shovel of Ski:	The forward turned-up section of the ski.
Heel of Ski:	The rear turned-up section of the ski.
Shoulder Contact:	The forward contact line where a 0.5 mm thick feeler gauge impinges when sliding it from the tip between the ski and a flat surface against which the ski body is pressed.
Heel Contact:	The rearward contact line where a 0.5 mm thick feeler gauge impinges when sliding from the tail between the ski and a flat surface against which the ski body is pressed.
Body of Ski:	The part of the ski between the Shoulder Contact Line and Heel Contact Line.
Forebody of Ski:	The half of the ski body toward the tip (or toward the shovel).
Afterbody of Ski:	The half of the ski body toward the ski tail (or toward the heel).
Load Carrying Layers (fibers):	The longitudinal layers in the ski which carry the major tensile, compressive or shear stresses when the ski is bent or twisted.
Core of Ski:	The internal part of the ski which normally does not carry high stresses when the ski is bent or twisted.
Neutral Axis (surface):	The longitudinal axis (surface) within the ski which is not strained when the ski is bent.
Sandwich Construction:	Composite construction with high-strength materials in layer(s) near the running surface and the top of the ski.
Box Construction:	Composite construction with high-strength material arranged in box or tube configuration

2. **Geometric Terms**

Nominal Length (Material Length):	The length of the ski as measured between the tip and tail along the ski bottom (often used to denote ski size).
Chord Length:	The straight-line distance from the tail to the tip.
Projected Length:	The projected distance between the tip and heel of the ski on a flat surface when it is pressed against the surface.
Contact Length:	The distance between the shoulder contact and the heel contact.
Support Length:	The projected length minus 25 cm.
Ski Body Centre:	Located at half the distance between the shoulder contact and the heel contact.
Ski Support Centre:	Located at the middle or centre of the support length.
Heel Width:	The width of the ski in the widest part of the heel.
Waist of Ski:	The width of the ski in the narrowest part of the heel.
Shoulder of Ski:	The width of the ski in the widest part of the shovel.
Weighted Bottom Camber:	The maximum height of the ski bottom in the binding mounting area above a flat and horizontal surface on which the ski is placed.
Free Bottom Camber:	The maximum distance measured between the ski bottom in the binding mounting area and a flat surface when the ski is mounted in a weightless position with its shoulder and heel in contact with the surface.
Sidecut:	The departure of the ski contour from a straight line measured by the distance between the bottom edge and a flat surface when the ski is placed on its side in contact (at the shovel and heel) with the surface.
Sidecamber:	The maximum distance between a line connecting the shoulder contact to heel contact (along the outer edge of the ski) and the sidewall of the ski.

Table 20.1
SKI CHARACTERISTICS (see also Fig. 20.2) (cont'd).

3. **Physical Properties**

Polar Moment of Inertia:	A measure of a ski's resistance to rotate laterally about an axis drawn perpendicular to the running surface which passes through the ski's centre of gravity.
Bending - Centre Spring Constant:	The load applied at the support centre necessary to deflect the support centre one centimetre when the ski is simply supported at the heel and shoulder support lines.
Bending - Forebody Spring Constant:	The load applied at the shoulder necessary to deflect the shoulder one centimetre when the ski is clamped at the support centre (cantilevered).
Bending - Afterbody Spring Constant:	The load applied at the heel necessary to deflect the heel one centimetre when the ski is clamped at the support centre (cantilevered).
Bending - Spring Constant Balance:	The ratio of the afterbody spring constant to the forebody spring constant (in bending).
Torsional - Forebody Spring Constant:	The moment applied at the shoulder necessary to twist the bottom surface at the heel one degree when the ski is clamped at the support centre.
Torsional - Afterbody Spring Constant:	The moment applied at the heel necessary to twist the bottom surface at the heel one degree when the ski is clamped at the support centre.
Torsional - Spring Constant Balance:	The ratio of the afterbody spring constant to forebody spring constant (in torsion).
Sideflex - Centre Spring Constant:	The load, applied at the support centre, necessary to deflect the support centre one centimetre when the ski is simply supported at the sidewall and at the heel and shoulder support lines with its camber depressed.
Natural Frequency:	The number of vibrational cycles per second (Hz) that the ski or part thereof will exhibit when allowed to vibrate freely.
Damping - Half-Life:	The time required to reduce the ski's vibrational amplitude from 2 to 1 mm.

springiness (high stiffness) while hickory gave high strength. The art of laminating two or more wood layers to increase strength and reduce weight and warpage developed at about the turn of the century (Vaage, 1967).

By the 1930's ski design included the different types of skis we know today, that is, the cross-country racing ski and the touring, jumping and alpine skis. Even then, the durability and friction of the bottom of a ski were important elements of ski design. In 1944, Cellolix, the forerunner of modern polyethylene ski bases was developed by Michal in France (Lund, 1975).

Shortly after World War II in the United States Howard Head produced a ski of aluminum sandwich construction (see Fig. 20.1) which had high-strength materials on the exterior of the ski where they are needed. Beginning in 1960, neoprene rubber was laminated into the more expensive "Head" skis to provide shear dissipation and to dampen the vibration of the ski on snow. Before 1950 little or no data about the physical properties of skis appeared in the technical literature, the differences in the dynamic properties of wood and metal skis prompted several investigations (Glenne, 1957; Schultes, 1962; Ohnishi, 1964).

In the early 1960's development of the more responsive fiberglass-wood ski with its excellent damping properties meant almost immediate success in the alpine racing field. Initially, its ski-camber durability was poor, however, improvements in fiberglass technology overcame this problem leading to a reduced ski weight as well. Several investigators have documented the physical properties of fiberglass skis (Piziali and Mote, 1970; Deak et al., 1975; Glenne et al., 1975).

During the 1970's foam or hollow cores, preimpregnated fiberglass, carbon and kelvar fibers, and viscoelastics were introduced into ski construction resulting in light, responsive, well-damped and highly durable skis. Fiberglass materials have proven their applicability in cross-country skis as well as alpine skis.

SKI PERFORMANCE[1] AND SKI PROPERTIES

Ski design usually starts with the consideration of characteristics connected with performance or use: for alpine skis these are associated with learning, recreation, and competitive activities. For cross-country skis, these are associated with mountaineering, touring, light-touring and racing activities.

In alpine skiing the use or performance characteristics mainly depend on the speed of skiing, the type of turning, and the snow surfaces encountered. That is, a sharp ski edge is important when skiing on steep slopes with hard snow while ease of turning initiation is appreciated on soft snow. Unfortunately, some of the performance characteristics become antonyms in ski design and tradeoffs must be made.

The two main types of turning are the carved turn, made without much sideslipping where the main controlling force is imparted at the ski shovel and heel, and the skidded turn, made with sideslipping (skidding), where the controlling force is imparted under the foot. The main properties of alpine skis that are determined by the performance priorities are durability, geometry (lengths and widths), bending (flex and torsion) and vibrational characteristics.

In cross-country skiing the use or performance characteristics depend mainly on the snow or track conditions (virgin snow, premade track, loose snow, hard snow, etc.). The main cross-country ski properties determined by the performance priorities are durability, weight, geometry (widths), flex pattern, camber and bottom surface conditioning (wax or no-wax).

---

[1]   An explanation of different ski performance characteristics is given in Table 20.2.

Table 20.2
SKI PERFORMANCE CHARACTERISTICS.

---

1. Straight-running Characteristics:

Ski Tracking: ability to maintain a straight course.

Ski Stability   - Vertical: ability to conform to bumps and dips in the vertical plane (smoothness).
- Lateral: ability to conform to bumps and dips in the lateral plane.
- Traversing: ability to remain in a traversing position without rotating up or down the slope.

2. Turning Characteristics:

Ski Turning   - Initiation: The ease with which a turn may be initiated.
- Execution (Steering Sensitivity): The ability of the ski to maintain a turn once initiated (overturning and edge catching are symptoms of turning execution difficulties).
- Completion: The ease with which a turn may be completed.
- Quickness: Ability to perform successive rapid directional changes (edge-setting reversals).

Ski Carving Ability: Ability to make a turn with minimal skidding or sideslipping on hard snows.

Ski Skidding Ability: Ability to hold an edge in a skidding turn without excessive chattering or side-slipping.

3. General Characteristics:

Edge Grip of the Ski: Ability to transmit braking and steering forces. Chattering and undesired side-slipping are normally symptoms of ineffectual edge grip.

Liveliness of Ski: Ability to react quickly to terrain changes and skier impulses. Ski deadness is the opposite of ski liveliness.

Quietness of Ski: Ability to dampen vibrations caused by bumps, dips, surface roughness features - related to ski stability and vibrational characteristics.

A study by Kopta (1974) of about 80 skiers at Snowbird and Alta in Utah, showed correlations of 75 percent between ski length and skier ability, 84 percent between ski length and skier weight, and 43 percent between ski length and skier velocity; there was also a correlation of 75 percent between skier ability and the centrifugal force overcome when turning.

Table 20.3 (after Glenne, 1978) attempts to match several important ski uses with ski properties for various snow conditions.

The materials used in the past often required that the ski have a thick shovel to prevent breakage or bending. The most important ski properties are:

Geometric - Length, width and thickness,
Camber and sidecamber,

Elastic - Flexural stiffness and distribution,
Torsional stiffness,

Table 20.3
**DEPENDENCE OF SKI PROPERTIES ON ACTIVITY AND
SNOW CONDITION.**

SKIING ACTIVITY		SNOW CONDITIONS	
		SOFT AND PACKED	HARD AND ICY
RECREATIONAL	LEARNER	Short Length Wide Ski Low Weight Gentle Flex	Short Length Wide Ski Medium Flex
	ADVANCED	Wide Ski Low Weight Gentle Flex	Medium Width Medium Flex High Damping
COMPETITIVE	FREESTYLE AND BALLET	Compact Length Low Weight Gentle Shovel Flex	Short Length Narrow Width High Sidecamber
	SLALOM AND GIANT SLALOM	Full-Length Medium Width Medium Torsion	Narrow Width Firm Forebody High Damping

Dynamic - Mass moments of inertia,
          (swingweight)
          Damping and ski-snow interactions.

Specific load-carrying materials may impart special characteristics to skis, especially for turning initiation and holding ability. Aluminum skis tend to be heavy, stable and good for carving turns while fiberglass skis are usually light, maneuverable and good for skidding turns. Hybrid skis, constructed of both materials, are an attempt to incorporate the advantages of each material into the performance characteristics.

Ski manufacturing essentially requires selection of materials and shape to produce a successful match of the geometric, elastic and dynamic ski properties. Normally, materials and thickness may be altered more easily than width. Therefore, even though the models from a given manufacturer often have different flex patterns and damping characteristics (as well as different cosmetics), they may have identical sidecuts.

Table 20.4 lists data on the geometric and elastic properties of several skis, as measured by ski engineers through ASTM (American Society for Testing and Materials) and ISO (International Standards Organization). Ski damping, although important, has been omitted because no simple measure of quantifying it has been found. The data show that modern skis are lighter, softer, narrower and have a higher sidecamber than the skis of the fifties and

Table 20.4
GEOMETRIC AND ELASTIC SKI PROPERTIES.

Type, Make & Model	Lengths Advert. cm	Contact cm	Weight kg	Min. Width mm	Sidecamber Angle deg.	Center Spring Constant N/cm	Spring Balance Constant	Torsional Stiffness Nm/deg.	Load Carrying Material
Oldtimers									
Kastle Slalom (1955)	210	181	2.2	69	0.34	52	0.85	2.2	Lam. Wood[b]
Head GS Comp. (1967)	210	186	2.6	72	0.34	58	1.03	3.3	Aluminum
Giant Slalom Skis									
Rossignol SM Comp.	207	184	2.1	68.5	0.46	43	1.01	2.8	Alum. & FG[c]
K2 255	210	189	2.3	69	0.43	49	1.00	3.4	Alum. & FG
Dynastar Acryglass	210	187	2.1	70	0.42	45	1.04	3.6	Aluminum
Slalom Skis									
Olin Mark VI	200	178	2.1	66	0.45	48	1.03	1.7	Fiberglass
Dynastar Omeglass	200	178	1.8	67	0.44	45	1.05	1.4	Fiberglass
Fischer C4 Comp.	200	177	2.1	66	0.43	47	1.08	1.5	Aluminum
Recreation Skis									
Dura-Fiber Enduro	196	174	1.8	68	0.43	38	1.14	1.6	Fiberglass
K2 244	195	175	1.8	66	0.45	33	1.05	1.5	Fiberglass
Olin Mark III S	200	173	2.0	67	0.50	37	1.09	1.0	Fiberglass
Freestyle Skis									
Hart Freestyle	180	157		65	0.50	34	1.15		Fiberglass
Kneissl Freestyle	185	161	1.8	70	0.48	36	1.21	1.1	Fiberglass

a Advertised
b Laminated wood
c Aluminum and fiberglass

sixties. The recreation and freestyle skis are relatively lighter and softer in flex and torsion. Shorter skis may be quite narrow or quite wide. The latter may be good for learning (GLM); however, compact skis are generally designed to act as if they were quite long.

Ski properties not only vary from make to make and model to model, but also from pair to pair and ski to ski. For example, the average flex difference between skis in a pair is usually 2-5%, while between pairs it can be somewhat larger.

MODES OF CONSTRUCTION

Apart from being classified according to their dominant material (wood, metal, fiberglass) skis may be classified with respect to their mode of construction. The two main modes are: (a) Sandwich Construction - a lamination of pre-made veneers, and (b) Box Construction - either wet-wrap fiberglass around core(s) or pre-impregnated fiberglass with spacer(s) (see Fig. 20.1). The sandwich construction is usually a dry lay-up (preformed parts) with a sheet of high-strength material near the top and bottom surfaces of a ski. Most metal skis are constructed in this manner. The unitized box construction, sometimes termed the torsion cell, has high-strength material close to all four surfaces of the ski; occasionally vertical stabilizers are embedded in the core.

Construction	Load Carrying Material	Core Material	Examples
Laminated Sandwich 	Aluminum	Wood	Fischer C4 Comp.
		Foam	Rossignol Roc
	Precured Fiberglass	Wood	Rossignol Strato
		Foam	Olin Mark V
Unitized Box 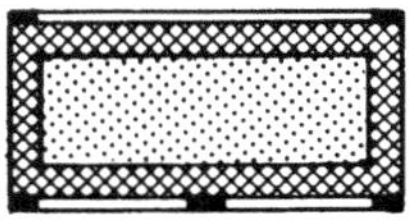	Wet-wrap Fiberglass	Wood	Dynamic VR-17
		Foam	K2 244
	Preimpregnated Fiberglass	Hollow Tubes	Dura-fiber
		Honeycomb	Hexcel

Fig. 20.1 Modes of ski construction.

Both modes of construction minimize the ski weight and the amount of high-strength materials needed close to the outside surfaces where the highest stresses occur. These modes are often modified or blended to produce skis having specific elastic or dynamic characteristics to minimize bonding or cost.

A ski of specific dimensions having a box construction is theoretically stiffer in torsion and sideflex than one of the same size having a sandwich construction if made of the same types and amounts of high-strength material. In reality the opposite is often true since the sandwich construction usually allows for better control of materials in ski manufacture and the use of stiffer materials.

The moment of inertia of the cross-sectional area, I, which describes the geometric bending resistance of a ski may be written as:

$$I_{sandwich} = (bd^3 - bd_1^3)/12, \qquad 20.1$$

$$I_{box} = (bd^3 - b_1d_1^3)/12, \qquad 20.2$$

where the symbols refer to the dimensions of the high-strength material;

i.e., b = maximum width (essentially the ski width),

d = maximum spacing (ski thickness minus thickness of the top and bottom layers of material,

b_1 = minimum width, and

d_1 = minimum spacing.

In theory, the larger the value of I, the smaller the deflection under a given load. In practice, however, Eqs. 20.1 and 20.2 rarely provide an accurate index of a ski's ability to resist deflection, since the steel edges, top edges, glue layers and cores may also carry loads. Hence for practical purposes, I, for the different constructions may be approximated by the expression,

$$I = k_1bd^m, \qquad 20.3$$

where k_1 and m are constants. Experiments by the author have shown that the exponent m has values from 2.7 to 2.9, and that k_1 depends on the construction mode. For a constant modulus of elasticity, the power law relationship between I and d implies that successive equal increments in ski thickness (spacing of high-strength material) will result in disproportionately larger increments in stiffness.

To obtain the necessary retention strength for the binding screws reinforcement is normally added to the ski in the binding area. Standards are also being promulgated for the resistance of different materials to pull-out of the binding screws.

Neoprene rubber is often used to modify the dynamic characteristics of more expensive skis by inserting a thin sheet of the rubber between the steel edge and the load-carrying material to isolate vibrations.

When designing a ski the load-carrying material and the construction mode are decided upon first. Then, the geometric dimensions of the ski are determined to give it the required elastic and dynamic characteristics. If the ski does not measure up to expectations, relatively simple alterations to the material or its placement are tried. If these modifications prove ineffective or unsuccessful the thickness of the ski or its mode of construction may be changed. In general, ski design is a trial and error process in which the use of analytical methods (for manufacturing) is just beginning.

GEOMETRIC FEATURES

A ski is a nonprismatic, heterogeneous, anisotropic prestressed beam. Fig. 20.2 shows its main geometric features; in addition to these certain other subtle features are significant, e.g., the location of the maximum shovel width relative to the shoulder contact — a ski's behaviour changes radically when it is put on edge. That is, its torsion and sideflex properties become important as well as its simple bending properties because when the ski is on edge its contact points with the snow also depend on the sidecamber as well as the bottom camber.

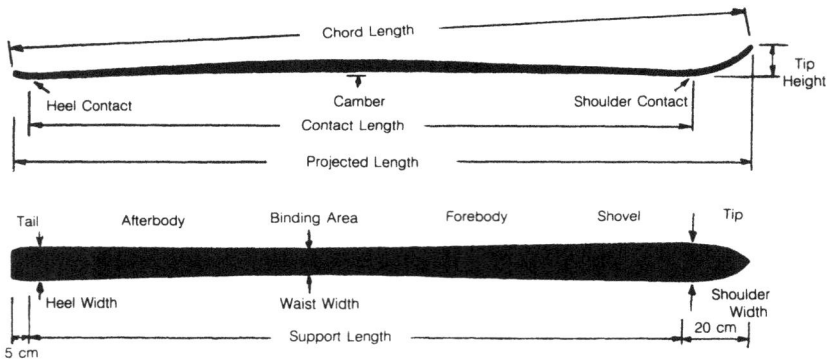

Fig. 20.2 Geometric characteristics of a ski.

The contact length is normally about 25 cm shorter than the chord length which is 2-3 cm shorter than the nominal (material) length — the length usually stamped on the ski (some manufacturers imprint the chord length). A ski should be long enough to give the desired stability and tracking yet short enough to be easily maneuverable. Obviously, therefore, the optimum ski length is a function of the type of skiing to be performed and the strength and skill of the skier. Recent improvements in ski-damping properties have facilitated the use of shorter alpine skis.

Ski length, slope angle and snow conditons determine a skier's turning radius which, in turn, determines the surface irregularities (mogul pattern) of a ski hill. Short radius turns aid in creating short and steep moguls.

On comparing the lengths of some famous old skis with those of modern skis, some interesting differences and similarities can be obtained (see Table 20.5). For example, although the chord length of the Sondre Norheim ski of 1868 is 48-cm longer than that of a modern slalom ski such as the Fischer C-4 Comp, the difference between their contact lengths is only 18 cm. The difference between their shovel lengths accounts for most of the remaining 20 cm. However, the waist widths of the two skis are identical.

The minimum ski width (important for stability and rapid turning) has been decreasing over the last two decades (Glenne, 1978). Racing cross-country skis today have a waist width of about 45 mm and touring skis have a waist

Table 20.5
SKI LENGTHS AND WIDTHS OF DIFFERENT VINTAGES.

Brand and Model	Vintage	Chord Length cm	Contact Length cm	Waist mm	Comments
Sondre Norheim	1868	245	195	68	from Telemark, Norway
California	1870	335	300	74	downhill
Bergendahl	1915	222	195	54	cross-country
Norge	1936	205	178	71	slalom
Marius Eriksen	1952	217	196	73	Stein's ski
Rossignol Strato	1968	213	191	70	giant slalom
Fischer C-4 Comp	1977	197	177	68	slalom
Kneissl Racing Team GT	1977	205	180	45	cross-country

width of 50-55 mm. Slalom and freestyle skis are normally narrower than giant slalom and recreational alpine skis. A typical modern alpine recreational ski may have a waist width of about 67 mm (see Table 20.4).

The sidecamber of skis, which affects the ease with which a turn can be initiated, has been increasing over the last two decades (Glenne, 1978). Several methods exist for measuring sidecamber, one is to form three ratios involving shoulder width, waist and heel width.

$$R_1 = \text{shoulder width/waist width,} \qquad\qquad 20.4$$

$$R_2 = \text{heel width/waist width, and} \qquad\qquad 20.5$$

$$R_3 = \text{shoulder width/heel width} = R_1/R_2. \qquad\qquad 20.6$$

Alpine skis typically have values of $R_1 = 1.30$, $R_2 = 1.14$ and $R_3 =$ of about 1.14 which means that the shoulder width is about 14% greater than the heel width. Another method for quantifying a ski's sidecamber is to measure the average angle a ski's sidewall makes with the ski's longitudinal centreline (see Fig. 20.2). The sidecamber angle may be calculated by the expression:

$$\text{Sidecamber angle} = \tan^{-1}\left[(\text{shoulder width}) + (\text{heel width}) - 2\,(\text{waist width})\right]/2\,(\text{contact length}). \qquad\qquad 20.7$$

Values of the sidecamber angle vary from zero for racing cross-country skis to about 0.50 degrees for alpine freestyle skis. The sidecamber angle facilitates comparison of the sidecamber for skis of various lengths. A large angle usually indicates a quick-turning but not very stable ski.

In practice, the amount of sidecamber needed in a ski is a function of several desired qualities — maneuverability, stability and holding ability — as well as several properties of the ski — bending stiffness, torsional stiffness, sideflex stiffness, and minimum width. Many modern cross-country racing skis which are designed to be used in a firm pre-made track have little or no sidecamber. Jumping skis, which are not designed for turning, also lack sidecamber. The graphs in Fig. 20.3 show the variations in width along the length of four popular slalom skis.

The camber of a ski is a prestress which aids in distributing a skier's weight properly onto the snow. The camber height and the ski stiffness determine the skier's weight distribution on the snow. Figure 20.4 shows the pressure distribution of two different alpine skis, one metal (Fischer President) the other fiberglass (Rossignol Strato), on a horizontal surface with a load of 356 N applied at the centre of the contact length. As shown, the pressure under either ski at a point corresponding to that at the front of a skier's foot is substantially lower than the pressure under the contact centre (an effect sometimes referred to as "double camber"); also the distribution for both types of skis is non-symmetrical about the contact centre with the majority of the load transferred to the afterbody.

A high camber tends to transfer a skier's weight to the shovel and heel of the ski thereby improving a ski's carving ability. A low camber distributes a skier's weight over the middle part of a ski and makes for easy swiveling.

The performance of a cross-country ski is quite sensitive to its camber and the change in camber along its length — a strong camber is essential to a good kick. The wax-wearing characteristics of a ski depend on both camber and stiffness. Cross-country skiing is basically a process, or movement, which involves maximizing the static friction coefficient while minimizing the sliding friction coefficient.

In designing and manufacturing a ski, changes in thickness can be made more easily than changes in the type of material along the length of the ski, so that the thickness is changed to attain a given stiffness distribution. The

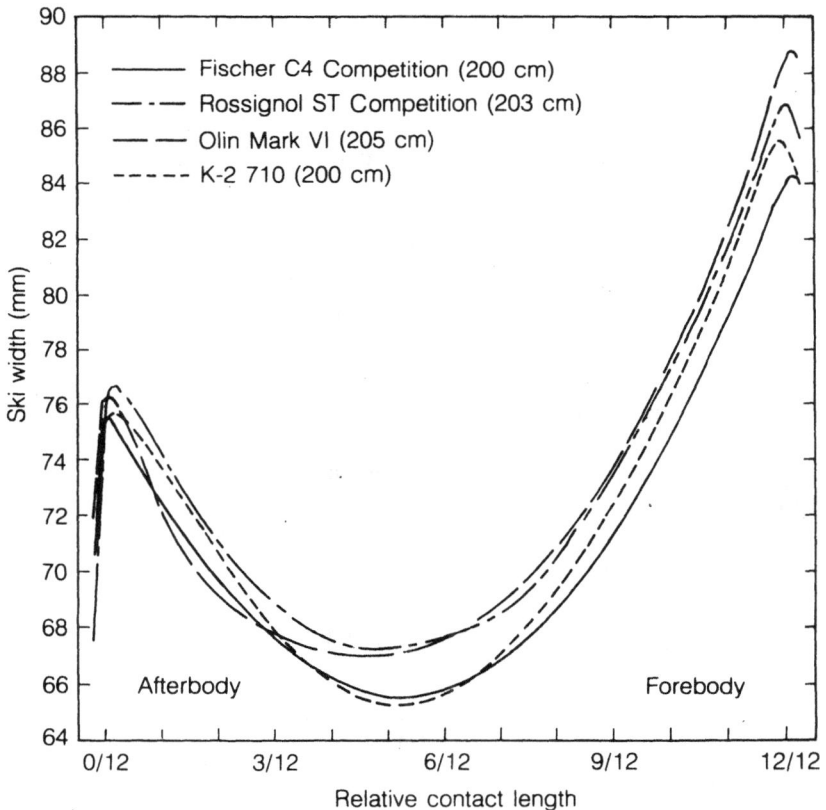

Fig. 20.3 Side geometry of slalom skis.

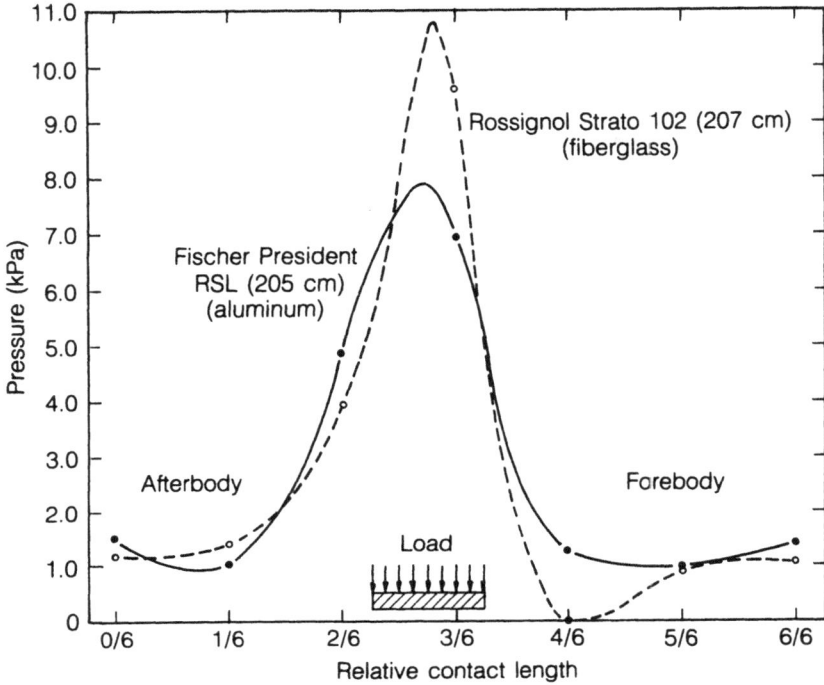

Fig. 20.4 Pressure distribution on a horizontal surface for a load of 356 N.

thickness distribution affects the load distribution of a surface. For ease of lateral movement, especially in cruddy snow, it is desirable to use a thin ski.

The shovel and tail turnup of skis have been increased in the last five years because of the presence of steeper moguls on most slopes and the increased popularity of ballet skiing. A constant shovel radius of 15-20 cm is sometimes used, or the shovel may be constructed with a geometry of an exponential curve to provide a more gradual transition.

ELASTIC PROPERTIES

A ski usually deforms in simple bending (flex), sideflex and torsion, so that strength and stiffness are its important properties along with its response to impact loading (creep), repetitive loading (fatigue) and environmental factors (temperature, humidity).

Skis may be bent, twisted, and vibrated in an assortment of ways in the laboratory to determine the significant elastic and dynamic characteristics

(Ohnishi, 1964; Piziali and Mote, 1970; Deak et al., 1974; Gardiner et al., 1974; Glenne et al., 1975). Subcommittee F8.14 of the American Society for Testing and Materials (1976), consisting mainly of engineers from the ski industry, is in the process of approving ten standards for ski testing (see Table 20.6).

High-strength materials used in skis are normally aluminum and a glass-epoxy composite (fiberglass). Aluminum is stiffer (higher modulus of elasticity) than fiberglass, but has a lower yield strength. Table 20.7, lists the elastic moduli of materials commonly used in ski construction. The tensile and compressive strengths of aluminum are approximately equal whereas fiberglass is stronger in tension than in compression. Most fiberglass skis therefore have more fibers in the compression layer (top layer) than in the tension layer (bottom layer) for effective utilization of materials. The neutral axis in these skis generally lies closer to the top surface than in skis where equal amounts of high-strength material are placed in the top and bottom layers.

Table 20.6
AMERICAN SOCIETY FOR TESTING AND MATERIALS (ASTM)
SUBCOMMITTEE F8.14.3: SKI DIMENSIONAL STANDARDS.

Doc. 01: Standard geometric terms and definitions for alpine skis.

Doc. 02: Standard specification for binding mounting area dimension on alpine skis.

Doc. 03: Standard specification for strength of binding mounting of alpine skis.

Doc. 04: Standard specification for test screw used in binding screw retention test.

Doc. 05: Standard specification for determination of the centre spring constant and spring constant balance of alpine skis.

Doc. 06: Standard specification for determination of the torsional characteristics of alpine skis.

Doc. 07: Standard specification for determination of vibration characteristics for forebody and shovel of alpine skis.

Doc. 08: Standard specification for determination of the effect of temperature on bottom camber of alpine skis.

Doc. 09: Standard specification for determination of deformation resistance of alpine skis.

Doc. 10: Standard specification for determination of fatigue characteristics of alpine skis.

Table 20.7
ELASTIC MODULUS (E) OF SKI
MATERIALS (Schultes, 1976).

Material	E(GPa)
Steel	220
Aluminium	70
Fiberglass	35
Hardwood (Ash)	13
Softwood (Okoume)	3
Thermoplastics (ABS)	2
PU Foam	1

A measure of the resistance of a ski to bending EI is the product of its geometry, as expressed by the moment of inertia I, and its material properties, as expressed by the elastic modulus E. The larger the value of EI the stiffer the ski. One test used to measure EI for a ski is to support it on rollers at the shoulder and heel contact positions and to measure the vertical deflections of the support (contact) centre with different loads applied at that point. The results from this test generally show that the load-deflection curve is almost linear over a large range of loads until large deflections are encountered. Figure 20.5 shows the maximum flexural stress plotted against the maximum strain for an aluminum ski (sandwich construction) and a fiberglass ski (box construction). The aluminum ski (Fischer C-4) is stiffer, whereas the fiberglass ski (Dura-Fiber XR-2) will take a larger strain without permanent deformation.

The ASTM has proposed that the centre spring constant be defined as the slope of the load-deflection curve obtained from a simply-supported bending test (over the load range 0 to 300 N). For modern alpine skis this constant is nearly invariant, independent of ski length.

An expression for the "respective stiffness" $\overline{EI}$ for a ski may be developed, based on the formula for the maximum deflection of a simply-supported beam having a concentrated load at its centre, to be:

$$\overline{EI} = C_m L^3 / 48, \qquad\qquad 20.8$$

in which C_m = centre spring constant (N/m), and
          L = support (contact) length (m).

Typical values of $\overline{EI}$ for 200-cm alpine skis fall between 400-500 N m², while values of the centre spring constants usually range from 30-45 N/cm (see Table 20.4).

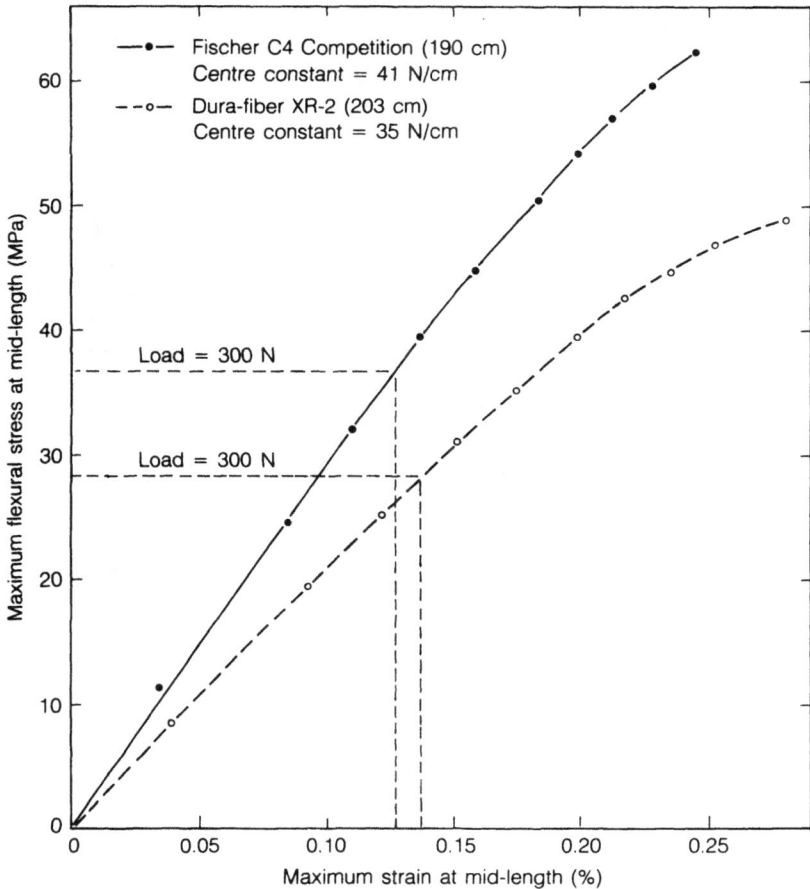

Fig. 20.5 Stress strain curves for an aluminum ski (Fischer C4) and fiberglass (Dura-Fiber XR-2) obtained from simply supported bending tests with concentrated load placed at the middle.

To obtain the actual stiffness distribution of a ski (Fig. 20.6) the deflection is measured at several points along the ski and the values are used in the elasticity equation to determine EI for different segments. It is evident from the figure that the maximum bending stiffness occurs over the segment of the ski that includes the contact center.

To perform a carving turn, the ski must be bent into an arc, whose geometry depends on the skier's weight, ski stiffness, ski length, turning radius, the skier's speed, and other secondary variables (skill, slope, etc.). An approximate relationship between the centre spring constant C_m and three variables affecting the "arc" has been derived:

$$C_m \; \alpha \; \frac{rW}{L^2} \cos \theta, \hspace{4cm} 20.9$$

in which   r = turning radius (m),
          W = skier's weight (N),
          L = ski's contact length (m), and
          θ = angle of ski slope.

The spring balance constant (bending) is obtained by dividing an afterbody spring constant by a forebody spring constant. Most modern skis have a spring balance constant larger than 1.0 (see Table 20.4) and for some of the freestyle skis it may be as high as 1.3. A spring balance constant greater than unity indicates the stiffness of the afterbody is equal to or greater than that of the forebody.

Ski properties not only vary with make and model, but also from pair to pair and ski to ski. Harder and Ayer (1975) tested 5 pairs of 10 models (50 skis) and found an average difference in flex of 2% between paired skis. Skis of the

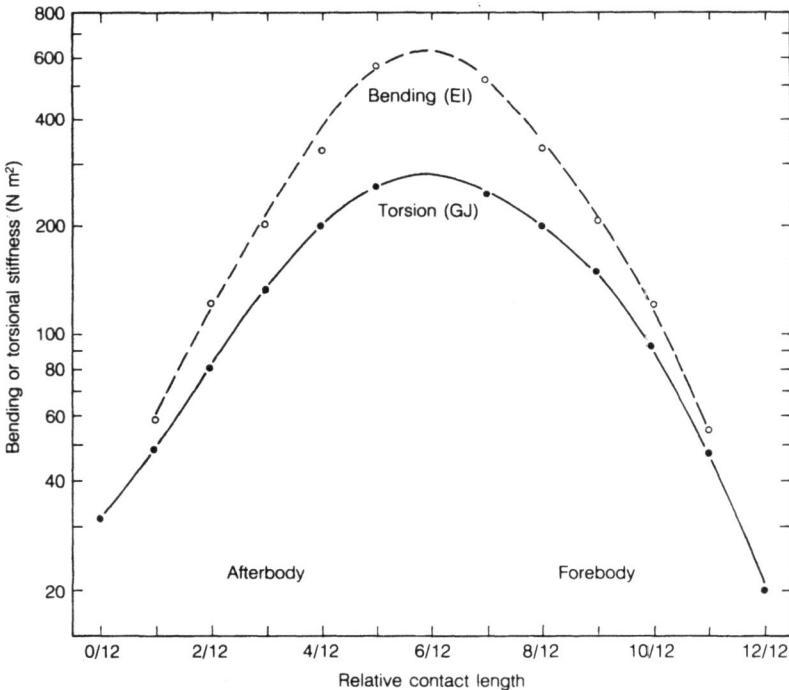

Fig. 20.6 Bending and torsional ski stiffness (ski length = 200 cm).

same make and model had coefficients of variation in weight and flex of 2.1 and 3.0% respectively.

Torsion is tested by twisting the shovel and heel after adjusting the ski to remove the camber. The most commonly used results from this test are the torsional spring constants for the forebody and afterbody. Some typical values are shown in Table 20.4. Metal skis usually are stiffer in torsion than fiberglass or wood skis. A ski, stiff in torsion, normally will have good heel and shovel control (carving ability) while a ski, soft in torsion, normally will have good control in the boot area (biting ability).

A ski's torsional stiffness GJ is the product of its polar moment of inertia J (geometrical properties) and shear modulus G (material properties), and may be calculated from the expression:

$$GJ = TL/2\psi, \qquad\qquad 20.10$$

where: T = moment applied (Nm),
         L = support (contact) length (m), and
         ψ = twist angle (radians).

The distribution in the torsional stiffness of a ski may be calculated from measurements of the twist angle at several points along its length. For typical values see Fig. 20.6 in which it is evident that the distributions of bending and torsional stiffness are similar.

When a skier performs a turn, each ski usually bends and twists simultaneously. A measure of the combined bending and torsional characteristics of a ski can be obtained by mounting it at different fixed angles to the horizontal and subjecting it to different vertical, upward-directed concentrated loads (see Fig. 20.7). The force-deflection curves of a Dynastar Omeglass (200 cm) ski obtained from several combined bending-torsion tests are shown in Fig. 20.8. The data show that the ski stiffness increases with edge angle; the stiffness of the ski (torsion and bending) when placed at an angle of 30° is 25% greater than that measured when placed flat (bending alone). Also, for a given load, the deflection decreases as the angle of inclination of the ski to the horizontal increases.

To obtain a measure of sideflex stiffness a ski is simply-supported on its sidewall and loaded at its support (contact) centre. A centre spring constant for sideflex analogous to that for simple bending may be calculated, yielding typical values of 40-100 kN/m. Metal skis usually are considerably stiffer in sideflex than fiberglass or wood skis.

Generally the stiffness properties of a ski increase with decreasing temperature; e.g., Lynard (1973) found that when the temperature of a Dura-Fiber, fiberglass ski was reduced from 0 to -16° C, its bending stiffness increased by 2.3% and its torsional stiffness by 4.3%. A small decrease in the weighted camber of the ski was also noticed.

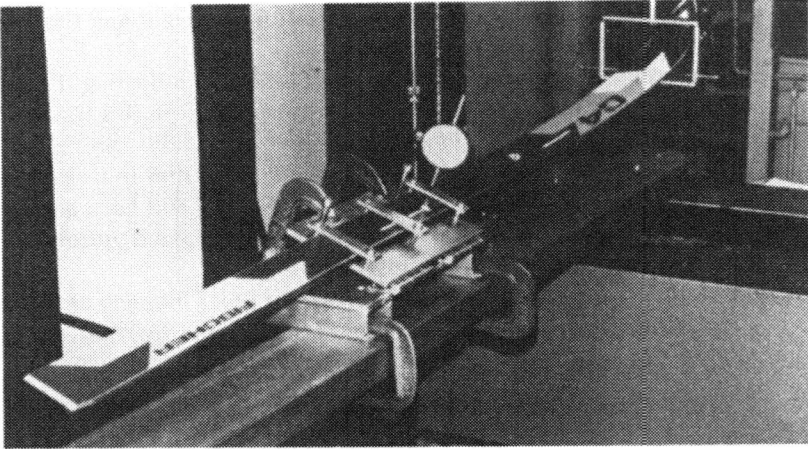

Fig. 20.7 Ski being tested in combined bending and torsion (University of Utah Ski Testing Program).

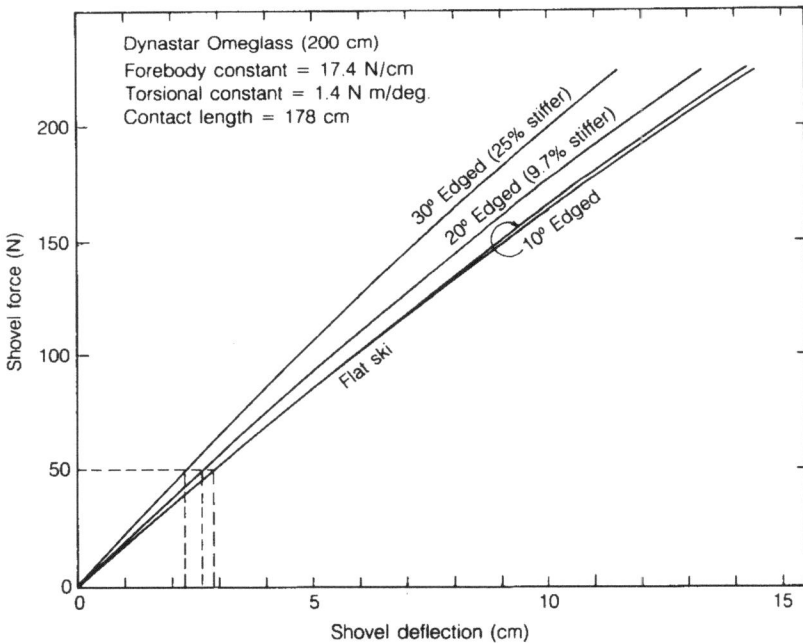

Dynastar Omeglass (200 cm)
Forebody constant = 17.4 N/cm
Torsional constant = 1.4 N m/deg.
Contact length = 178 cm

30° Edged (25% stiffer)
20° Edged (9.7% stiffer)
10° Edged
Flat ski

Shovel force (N)

Shovel deflection (cm)

Fig. 20.8 Load deflection curves for a 200-cm Dynastar Omeglass ski measured under a combined bending-torsion test. The ski was clamped at mid length and a concentrated load applied at the shovel (University of Utah).

The breaking strength of skis varies widely. A quasi-static laboratory strength test does not usually duplicate the impact loading conditions encountered in the field which frequently break skis; hence, the results from laboratory tests should be used only as relative or comparative measures of strength. Other factors, such as fatigue and stress-concentration also affect the breaking strength.

Ski stiffness in the binding area may be increased by the mounting of a ski binding and insertion of a ski boot in the binding. Typical heel-toe bindings (such as Salomon 555, Geze Jet) have been found to decrease ski deflections under the foot by 8-10% (Steward, 1976) compared to the deflections measured in concentrated load tests conducted on the same skis having no bindings.

Field measurements have been conducted with strain gauges attached to skis in an attempt to obtain better data on the actual strains and loads acting on a ski during skiing (Gardiner et al., 1974; Woehrle, 1975). The tests conducted by Gardiner et al. utilized eight strain gauges mounted in a rosette on the top surface of the ski's forebody to measure simple bending and torsional strains, and four strain gauges attached to the ski's sidewall to measure sideflex deformation. The ski used was a Dura-Fiber (205 cm) with a centre spring constant of 4.38 kN/m and a forebody torsion spring constant of 2.4 Nm/degree. Figure 20.9 shows smoothed curves of the variation in simple flex and bending with time, measured under different turning and snow conditions. The field measurements reported by Gardiner et al. are summarized in Table 20.8.

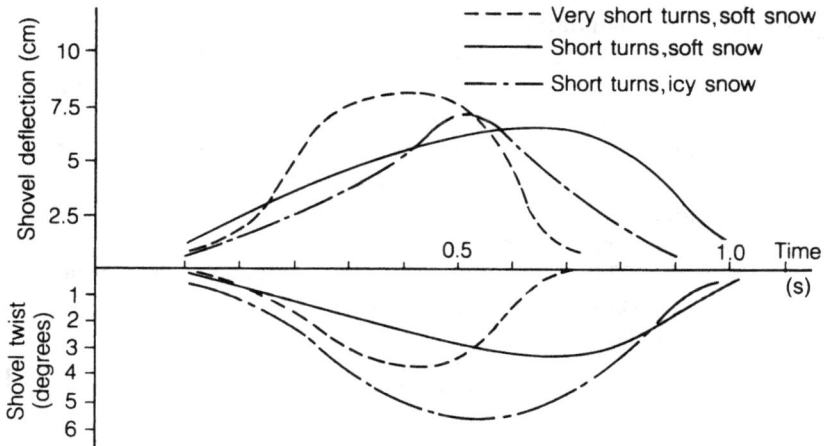

Fig. 20.9 Field measurements of the variation in shovel deflection and shovel twist with time of a Dura-Fiber ski under different turning and snow conditions (Gardiner et al., 1974).

Table 20.8

COMPARISON OF SKI AND SKIER PERFORMANCE
CHARACTERISTICS FOR TURNS CONDUCTED IN SOFT
AND HARD SNOW (Gardiner et al., 1974).

| Characteristic | Snow Condition | |
	Soft	Hard
Ski forebody bending and twisting	Occurred simultaneously for both long and short turns.	Occurred simultaneously for both long and short turns.
Skier weight bias	62% on outside ski; 38% on inside ski.	73% on outside ski; 27% on inside ski.
Proportion of strain energy to bending and torsion	98% to bending; 2% to torsion	~88% to bending; ~11% to torsion
Sideflex strain energy	Negligible.	Noticeable when chattering occurred.
Maximum shoulder bending and shoulder twist	Bending ~7.4 cm; twist ~3.7°.	Bending ~7 cm; twist ~5.5°.

DYNAMIC PROPERTIES

A ski travelling over snow is subjected to random forcing and damping functions which cause a complex interaction between snow and ski. The dynamic properties of a ski are especially important on hard snows and/or at high speeds when high damping is required.

Initially, studies of the dynamic properties have been directed toward investigating modes of resonance and free damping of skis in the laboratory. However, laboratory results of natural frequency and standard free vibration damping (logarithmic decay) do not correlate well with data gathered on snow. Hence, efforts have been directed toward using frequency response data and frequency spectra of skis, determined under field conditions, as measures of their dynamic properties.

Gardiner et al. (1974) showed that the natural frequency of free vibration of the forebody of a ski f_n is a function of its stiffness and weight and may be approximated by the equation:

$$f_n = k_2 \sqrt{8gC_B/w} \qquad\qquad 20.11$$

in which: k_2 = a constant, equal to about 0.45,
  g = acceleration of gravity (m/s^2),
  C_B = forebody spring constant (N/m), and
  w = weight of the ski (N).

Generally the natural frequency of the ski's forebody ranges from 9 to 13 Hz and that of the afterbody from 20 to 25 Hz. In torsion the natural frequency of the forebody is about 100 Hz, while in sideflex the natural frequency of the entire ski is about 50 Hz. High speed films have shown that higher modes of resonance often interfere with the skier's ability to control his skis on hard snow and at high velocities. Skis may have as many as five modes of resonance below 100 Hz (Piziali and Mote, 1970).

Laboratory measurements of ski damping usually have quantified the logarithmic decrement damping or the damping half-life (time to decrease the vibrational amplitude to one-half of its initial amplitude). The logarithmic decrement (D_d) is defined as:

$$D_d = \ln (X_2/X_1), \qquad\qquad 20.12$$

in which X_1 and X_2 are two successive vibrational amplitudes. The relative energy absorbed per vibrational cycle is proportional to the logarithmic decrement (Ohnishi, 1964). Committee TG 83/SC4 of the International Standards Organization (ISO) has defined damping half-life as the time required for a ski's forebody to reduce its free vibration from an amplitude of 2 mm to that of 1 mm when the ski is clamped at its body centre; typical half-lives are 0.6-1.2s. If a skier travels at 50 km/h (~14 m/s) his skis will encounter many disturbances and experience several vibrations in periods considerably less than the damping half-life in bending.

Field measurements of ski vibrations have been carried out by Piziali and Mote (1972), Gardiner et al. (1974) and Woehrle (1975). Gardiner et al. found that the maximum vibrational energies in forebody bending occurred between 0.25 to 2.0 Hz on soft snow, and 3.0 Hz on icy snow. Also, the maximum energies of torsional vibrations were at 3 Hz, and of sideflex vibrations at 53 to 55 Hz. They also found that the ski exhibited free vibrations. These were mainly sideflex vibrations which were associated with chattering.

Piziali and Mote (1972), found that straight skiing (no turning) had significant energy responses at all characteristic frequencies used in the laboratory (10, 14, 34, 51 and 70 Hz). This was not true for turning which had significant energies concentrated at frequencies of 9, 21 and 47 Hz and in a dominant band between 16 to 24 Hz.

FUTURE DEVELOPMENTS

At the present time attempts are being made in Germany and the United States to develop a better labelling system for skis to indicate length, performance category (competition, high-performance, recreation, learning) and suitability based on skier weight. Such a system should aid the consumer greatly in matching a ski to his or her use and ability.

The trends in ski development are towards a ski which is lighter and narrower having a high sidecamber, and which is also softer overall, particularly in the shovel area. Ski lengths seem to have stabilized at an average length 10-15 cm shorter than a decade ago.

Future advances in ski design will come in the ski/binding interface as well as in ski damping through improved understanding of the snow/ski interaction. Perhaps a ski with an integrated ski and binding unit (no binding mounting) which assurs a proper ski flex pattern will be developed. More skis for special purposes (i.e. ballet, mountaineering, powder skiing, etc.) are bound to appear.

The cross-country and touring ski will see rapid development with lighter weight, better bottoms and harder edges resulting from using modern materials. Significant improvements are also being made in no-wax cross-country skis.

LITERATURE CITED

American Society for Testing and Materials. 1976. *Proposals for ski testing.* Sub-comm. F8.14. 1916 Race St., Philadelphia, Penn.

Deak, A., J. Jorgensen and J. Vagners. 1975. *The engineering characteristics of snow skis.* J. Eng. Ind., Amer. Soc. Mech. Eng., pp. 131-137.

Gardiner, R., B. Glenne and W.E. Mason. 1974. *Dynamic modeling for ski design.* UTEC Rep. 74-165, Dept. Civil Eng., Univ. Utah, Salt Lake City.

Glenne, B. 1957. *Ski engineering.* Ski Mag., Vol. 1, Oct., New York, N.Y.

Glenne, B. 1978. *Ski performance and ski properties.* Student J., Am. Inst. Aeronaut. Astronaut., Spring Issue, New York, pp. 10-13.

Glenne, B., W. Mason and R. Gardiner. 1975. *Evolution of snow ski design.* Student J., Am. Inst. for Aeronaut. Astronaut., Summer Issue, New York, N.Y.

Harder, D.C. and D.C. Ayer. 1975. *A brief investigation of quality uniformity of skis.* Dept. Civil Eng., Univ. Utah, Salt Lake City, (Unpubl.).

Kopta, D. 1974. *The centripetal force produced by skiers.* Dept. Civil Eng., Univ. Utah, Salt Lake City, (Unpubl.).

Lund, M. 1975. *Eight classics.* Ski Magazine, Jan., New York, N.Y.

Lynard, W.G. 1973. *Flexural and dynamic properties of skis as a function of temperature.* Dept. Civil Eng., Univ. Utah, Salt Lake City, (Unpubl.).

Ohnishi, T. 1964. *Physical properties of skis.* (In German), Jpn. J. App. Phys., Vol. 3, No. 4, pp. 67-87.

Piziali, R.L. and C.D. Mote Jr. 1970. *Laboratory characteristics of snow skis.* Dept. Mech. Eng., Univ. Calif., Berkeley, (Unpubl.).

Piziali, R.L. and C.D. Mote Jr. 1972. *The snow ski as a dynamic system.* Pap. No. 72-WA/Aut-1, Am. Soc. Mech. Eng., Automatic Control Div.

Schultes, H. 1962. Europa-Sport/Sport-Bedarf, Jan. and Feb. issues.

Schultes, H. 1976. *Principles of modern alpine ski design.* Olin Ski Co., Middletown, Conn.

Steward, T. 1976. *Bindings and how they affect a ski.* Civil Constr. Tech., Okanagan College, Kelowna, B.C., (Unpubl.).

Vaage, J. 1967. *Kort historikk om Norsk skiloping.* Norsk Ski Instruksjon, Oslo, Norway.

Woehrle, M. 1975. *Ski engineering.* Seminar, Univ. Utah, Salt Lake City, (Unpubl.).

INDEX

www.ingramcontent.com/pod-product-compliance
Lightning Source LLC
Chambersburg PA
CBHW060415220326
41598CB00021BA/2183